AF324687

RECENT PROGRESS IN ATMOSPHERIC SCIENCES

Applications to the Asia-Pacific Region

RECENT PROGRESS IN ATMOSPHERIC SCIENCES

Applications to the Asia-Pacific Region

Editors

K. N. Liou

University of California, Los Angeles, USA

M.-D. Chou

National Taiwan University, Taiwan

Executive Editor

H.-H. Hsu

National Taiwan University, Taiwan

NEW JERSEY · LONDON · SINGAPORE · BEIJING · SHANGHAI · HONG KONG · TAIPEI · CHENNAI

Published by

World Scientific Publishing Co. Pte. Ltd.

5 Toh Tuck Link, Singapore 596224

USA office: 27 Warren Street, Suite 401-402, Hackensack, NJ 07601

UK office: 57 Shelton Street, Covent Garden, London WC2H 9HE

Library of Congress Cataloging-in-Publication Data
Recent progress in atmospheric sciences: applications to the Asia-Pacific region /
edited by K.N. Liou and M.-D. Chou.
    p. cm.
  Includes bibliographical references and index.
  ISBN 978-981-281-890-4 (alk. paper)
  1. Atmospheric sciences. I. Liou, Kuo-Nan. II. Chou, Ming-Dah
  QC861.3.R43 2008
  551.5--dc22

                        2008039270

British Library Cataloguing-in-Publication Data
A catalogue record for this book is available from the British Library.

Printed in Singapore by Mainland Press Pte Ltd

Preface

Recent Progress in Atmospheric Sciences: *Applications to the Asia-Pacific Region* contains 22 peer-reviewed articles, which cover a spectrum of contemporary subjects relevant to atmospheric sciences, with specific applications to the Asia-Pacific regions. The majority of these papers consist of a review of a scientific subfield in atmospheric sciences, while some contain original contributions. All of the accepted papers were subject to scientific reviews and revisions.

The book is divided into two traditional fields in atmospheric sciences: Atmospheric dynamics and meteorology, and atmospheric physics and chemistry. The authors of these papers are distinguished alumni of the Department of Atmospheric Sciences at the National Taiwan University, residing in the U.S.A and Taiwan. This book is dedicated to the 50th anniversary of the Department of Atmospheric Sciences that occurred in 2004, and to the 80th anniversary of the National Taiwan University, which took place in 2008.

Papers in atmospheric dynamics and metrology cover a range of subjects: EI Niño/Southern Oscillation; ocean-atmosphere-land feedbacks; intraseasonal oscillation and typhoon relationship; convection-radiation-mixing processes in the tropics and climatic impact; analysis of Meiyu frontal systems; modeling and observations of tropical cyclones/typhoons; storm track dynamics; understanding the quasi-equilibria state; assimilation of ocean surface winds for weather forecasting; and mesoscale modeling and applications. In atmospheric physics and chemistry, subjects include: interactions between aerosols and clouds, heat budgets in the context of air/sea interactions, atmospheric radiative transfer, satellite remote sensing of clouds, satellite remote sensing of typhoons and oceans, detection and tracking of Asian dust outbreaks, Doppler radar observations of vortex structures, an understanding of ice cloud microphysics in clouds, a review of zone trends in mega-cities, and analysis of COSMIC data for typhoon studies and weather predictions.

We are immensely grateful to the following colleagues, who graciously agreed to review the submitted manuscripts for this book volume and provided valuable and constructive comments and suggestions which led to substantial improvement in its presentation: Sim Aberson, Julio Bacmeister, Edmund Chang, Jen-Cheng Chang, Julius Chang, Simon Chang, Shu-Hua Chen, Yi-Leng Chen, Tai-Chi Chen, Chia Chou, Jim Coakley, Leo Donner, Robert Fovell, Qiang Fu, Ching-Yuang Huang, David Kratz, Tianming Li, Zhanqing Li, Xinzhong Liang, Yu-Chieng Liou, Tim Liu, Yangang Liu, Michael Mishchenko, Hisashi Nakamura, Vaughan Phillips, Rachael Pinker, Wayne Schubert, Chung-Hsiung Sui, Ming-Jeng Yang, Song Yang, Pao-Kuan Wang, Wei-Chyung Wang, Yuquing Wang, Zifa Wang, Chun-Chich Wu, Xiaoqing Wu, and Cheng-Ku Yu.

We would like to thank Tara Fickle for her dedicated assistance with various phases of this book project, including the initial editing of the accepted manuscripts, correspondence with the 22 lead authors, and correspondence with the publisher. Ms. Helen Jung is thanked for the design of the book cover. Finally, we thank Kim Tan of World Scientific Publishing for her interest and timely assistance in bringing this book project to completion.

K. N. Liou, *Los Angeles, USA*
M.-D. Chou and H.-H. Hsu, *Taiwan*

Contents

Part 2. Atmospheric Physics and Chemistry

Understanding the El Niño–Southern Oscillation and Its Interactions with the Indian Ocean and Monsoon

Jin-Yi Yu

Department of Earth System Science, University of California, Irvine, California, USA
jyyu@uci.edu

The Pacific and Indian Oceans are closely linked to each other through atmospheric circulation and oceanic throughflow. Climate variations in one ocean basin often interact with those in the other basin. This includes phenomena such as the El Niño–Southern Oscillation (ENSO), biennial monsoon variability, and the Indian Ocean zonal/dipole mode. Increasing evidence suggests that these interbasin interactions and feedbacks are crucial in determining the period, evolution, and pattern of ENSO and its decadal variability. This article reviews recent efforts in using a series of basin-coupling CGCM (coupled atmosphere–ocean general circulation model) experiments to understand the physical processes through which ENSO interacts with the Indian Ocean and monsoon, the impacts of interbasin interactions on the characteristics of ENSO, the relative roles of the Pacific and Indian Oceans in monsoon variability, and ENSO's role in the Indian Ocean zonal/dipole mode.

1. Introduction

The Pacific Ocean exhibits prominent sea surface temperature (SST) variations on time scales that range from interannual to interdecadal. The interannual fluctuations are primarily associated with the El Niño–Southern Oscillation (ENSO), which results from the interactions between the tropical Pacific Ocean and the overlying atmosphere (Bjerknes, 1969). Warm (El Niño) and cold (La Niña) ENSO events occur quasi-periodically and are generally associated with significant anomalies in global and regional climate patterns. The fundamental physical processes that give rise to ENSO are believed to reside within the tropical Pacific. ENSO research over the past few decades has focused mainly on the tropical Pacific Ocean, and has resulted in a significant understanding of this climate phenomenon. Coupled atmosphere–ocean models that include only the tropical Pacific Ocean have obtained encouraging success in ENSO simulations and prediction (e.g. Cane and Zebiak, 1985; Cane *et al.*, 1986; Schopf and Suarez, 1988; Battisti and Hirst, 1989; Delecluse *et al.*, 1998; Latif *et al.*, 2001; Yu and Mechoso, 2001).

Nevertheless, variations originating from the Indian Ocean and monsoons, such as weak and strong summer monsoons, interannual variability in the Indian Ocean SST, and volume fluctuations in the oceanic throughflow, are also potentially significant in the interaction of ENSO dynamics. The rising zone of the transverse and lateral circulation components of the Indian monsoon coincides with the ascending branch of the Walker circulation (Webster *et al.*, 1992). Variations in the strength of the monsoon can affect Pacific trade winds and, consequently, the period and magnitude of ENSO (Barnett, 1984; Wainer and Webster, 1996). In the ocean, ENSO manifests itself as a zonal displacement of warm water between the western and eastern tropical Pacific. The warm water pool in the western Pacific is connected to the eastern Indian Ocean through the Indonesian throughflow. This throughflow is widely known, on average, to carry warm and fresh Pacific waters through the Indonesian archipelago into the Indian Ocean. This leakage of upper ocean waters from the western Pacific to the Indian Ocean acts as a major heat sink for the Pacific Ocean and a heat source for the Indian Ocean (Godfrey, 1996). There were studies suggesting that the throughflow could affect the mean climate state of the tropical Pacific, such as its mean thermocline depth and zonal temperature gradients (Hirst and Godfrey, 1993; Murtugudde *et al.*, 1998; Lee *et al.*, 2002). Since ENSO's period and growth rate depend on the mean climate state over the Pacific (Fedorov and Philander, 2000), the throughflow provides another possible mechanism for the Indian Ocean to affect ENSO activity.

On the other hand, it is known that ENSO events have profound influences on the Indian monsoon (see Shukla and Paolino, 1983; Klein *et al.*, 1999). The relationship between ENSO and the Indian monsoon has long been an area of extensive research. The cause-and-effect relationship between ENSO and one of the strongest interannual signals of the monsoon variability, the tropospheric biennial oscillation (TBO), is still not yet fully understood. The TBO is referred to as the tendency for strong and weak monsoons to flip-flop back and forth from year to year (Mooley and Parthasarathy, 1984; Yasunari, 1990; Clarke *et al.*, 1998; Webster *et al.*, 1998; Meehl and Arblaster, 2002). It has been associated with SST anomalies in both the tropical Pacific and the Indian ocean (Rasmusson and Carpenter, 1982; Meehl, 1987; Kiladis and van Loon, 1988; Ropelewski *et al.*, 1992; Lau and Yang, 1996; Meehl and Arblaster, 2002). The associated SST anomalies in the Pacific Ocean are characterized by an ENSO-type pattern with a biennial ($\sim$2 years) periodicity. This association leads one to conclude that the TBO and the biennial ENSO component may be related. Some studies suggested that the TBO is forced by the biennial component of ENSO (e.g. Fasullo, 2004), but others argued that it has its own dynamics that are independent of ENSO (e.g. Li *et al.*, 2006).

The relationship between ENSO and the monsoon variations is further complicated by the existence of significant interannual variability in the Indian Ocean SST. The variability may be intrinsic to the coupled Indian Ocean–monsoon system, or remotely forced by ENSO. An increasing number of recent studies suggest that the Indian Ocean may play an active role in influencing ENSO characteristics (e.g. Yu *et al.*, 2000; Yu *et al.*, 2003; Yu, 2005; Wu and Kirtman, 2004; Kug and Kang, 2005; Kug *et al.*, 2006; Terray and Dominiak, 2005). It was suggested that the ENSO-forced basin-scale warming/cooling in the Indian Ocean might affect atmospheric circulation in the western Pacific to fasten the turnabout of the ENSO cycle and result in biennial ENSO (Kug and Kang, 2005). It was also postulated that the southeastern Indian Ocean SST anomalies act as persistent remote forcing to promote wind anomalies in the western equatorial Pacific and modulate the regional Hadley circulation in the south Pacific, both of which then affect the evolution of ENSO (Terray and Dominiak, 2005). These recent studies imply that the Indian Ocean may be a necessary part of the ENSO dynamics.

A thorough study of the interactions between ENSO and the Indian Ocean and monsoon is crucial for better understandings and successful forecasts of ENSO and monsoon activity. The complex coupling nature of these ENSO–Indian Ocean–monsoon interactions makes it difficult to determine their cause-and-effect relationships using observational analyses alone. Numerical model experiments that can isolate the coupling processes within and external to the Pacific Ocean are useful means for this purpose. In the past few years, a basin-coupling modeling strategy has been developed (Yu *et al.*, 2002) to perform these types of experiments with coupled atmosphere–ocean general circulation models (CGCM's). This article summarizes the findings obtained from these basin-coupling CGCM experiments.

The article is organized as follows. Section 2 describes the University of California, Los Angeles (UCLA) CGCM used for the experiments and the setup of the basin-coupling experiments. The ENSO cycle simulated by the model and its teleconnection patterns in the Indo-Pacific Oceans are presented in Sec. 3. The influences of the Indian Ocean on the ENSO cycle are then discussed in Sec. 4. ENSO's interactions with the biennial monsoon variability (TBO) are examined in Sec. 5. Section 6 reports the findings on the importance of ENSO influence on Indian Ocean SST variability. Section 7 concludes with a review of new understandings of the interactions between ENSO and the Indian Ocean and monsoon.

2. Basin-Coupling Modeling Strategy

The CGCM used in this study consists of the UCLA global atmospheric GCM (AGCM) (Suarez *et al.*, 1983; Mechoso *et al.*, 2000) and the Geophysical Fluid Dynamics Laboratory (GFDL) Modular Ocean Model (MOM) (Bryan, 1969; Cox, 1984; Pacanowski *et al.*, 1991). The AGCM includes the schemes of Deardorff (1972) for the calculation of surface wind stress and surface fluxes of sensible and latent heat, Katayama (1972) for short-wave radiation, Harshvardhan *et al.* (1987) for long-wave radiation, Arakawa and Schubert (1974) for parametrization of cumulus convection, and Kim and Arakawa (1995) for parametrization of gravity wave drag. The model has 15 layers in the vertical (with the top at 1 mb) and

a horizontal resolution of 5° longitude by 4° latitude. The MOM includes the scheme of Mellor and Yamada (1982) for parametrization of subgrid-scale vertical mixing by turbulence processes. The surface wind stress and heat flux are calculated hourly by the AGCM, and the daily averages passed to the OGCM. The SST is calculated hourly by the OGCM, and its value at the time of coupling is passed to the AGCM. The ocean model has 27 layers in the vertical with 10 m resolution in the upper 100 m. The ocean has a constant depth of about 4150 m. The longitudinal resolution is 1°; the latitudinal resolution varies gradually from 1/3° between 10°S and 10°N to almost 3° at 50°N. No flux corrections are applied to the information exchanged by model components.

A series of basin-coupling experiments were performed with the UCLA CGCM. In each of the experiments, the atmosphere–ocean couplings were restricted to a certain portion of the Indo-Pacific Ocean by including only that portion in the ocean model component of the CGCM. Three CGCM experiments were conducted in this study: the Indo-Pacific Run, the Pacific Run, and the Indian Ocean Run. In the Pacific Run, the CGCM includes only the tropical Pacific Ocean in the domain of its ocean model component. The Indian Ocean Run includes only the tropical Indian Ocean in the ocean model domain. The Indo-Pacific Run includes both the tropical Indian and Pacific Oceans in the ocean model domain. Outside the oceanic model domains, SST's and sea-ice distributions for the AGCM are prescribed based on a monthly varying climatology. Figure 1 shows the ocean model domain and the sea/land masks used in each of these experiments and the long-term (~50 years) means of their simulated SST's. All three CGCM runs produce a reasonable SST climatology in the Pacific and Indian Oceans, with a warm pool (SST greater than 28°C) that covers both the tropical western Pacific and eastern Indian Oceans.

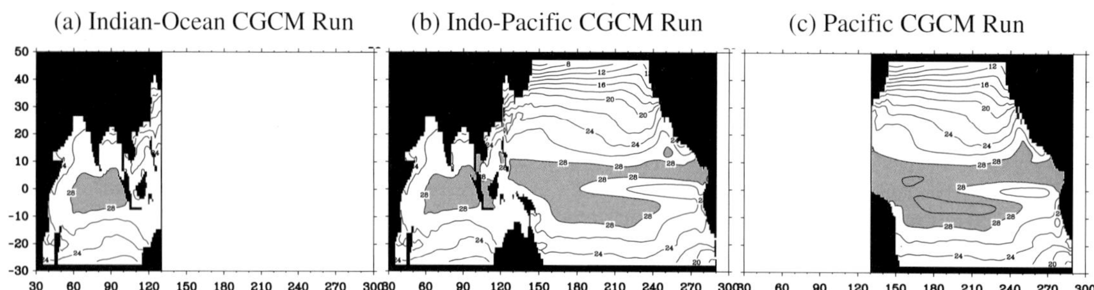

Figure 1. Ocean model domains used in the (a) Indian Ocean Run, (b) Indo-Pacific Run, and (c) Pacific Run. Also shown are the long-term mean SST's produced by these runs. Contour intervals are 2°C. Values greater than 28°C are shaded.

3. The Simulated ENSO Cycle

The UCLA CGCM has been shown to produce a realistic seasonal cycle in the tropical eastern Pacific (Yu *et al.*, 1999) and is considered one of the few coupled models capable of producing realistic ENSO simulations in an intercomparison project of ENSO simulation (Latif *et al.*, 2001). The ENSO cycle simulated by the Pacific Run was thoroughly examined by Yu and Mechoso (2001). It was found that ENSO-like SST variability is produced in the model approximately every 4 years, with a maximum amplitude of about 2°C. Large westerly wind stress anomalies are simulated to the west

of the maximum SST anomalies in all major warm events. A multichannel singular spectrum analysis (M-SSA) (Keppenne and Ghil, 1992) was used to extract the simulated ENSO cycle from the simulation. The M-SSA method has been shown to be capable of extracting near-periodicity, and their associated spatiotemporal structures, from short and noisy time series (Robertson *et al.*, 1995). Figure 2 shows the structure and evolution of the simulated ENSO cycle as revealed by the leading M-SSA mode along the equator. The ENSO cycle is found to be characterized by predominantly standing oscillations of SST in the eastern Pacific, almost simultaneous zonal wind stress anomalies

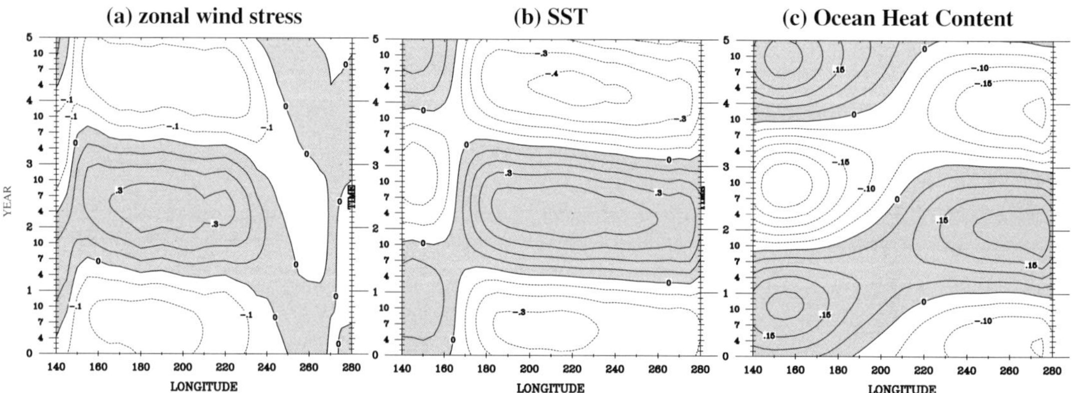

Figure 2. Structures of the simulated ENSO cycle along the equatorial Pacific extracted from the Pacific Run by the M-SSA method. Panel (a) shows the structure of zonal wind stress anomalies, (b) the structure of SST anomalies, and (c) the structure of ocean heat content anomalies. The coordinate is the 61-month (5-year) lag used in the M-SSA. Contour intervals are 0.1 dyn/cm^2 for (a), 0.10°C for (b), and 0.05°C for (c). Values shown in (a) are scaled by 10. Positive values are shaded. (From Yu and Mechoso, 2001.)

(a) zonal-mean component   (b) zonally asymmetric component

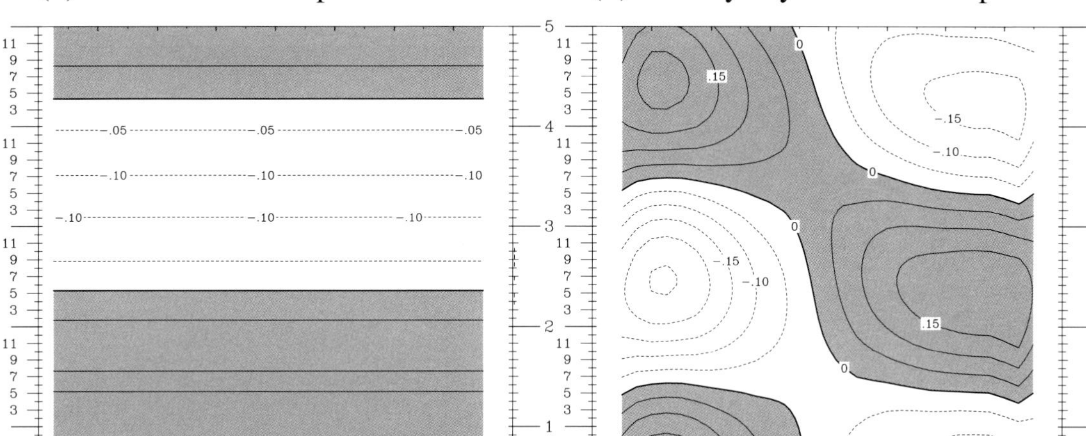

Figure 3. The (a) zonal-mean and (b) zonally asymmetric components of the ocean heat content structure shown in Fig. 2(c). The vertical coordinate spans the 61-month window used in the M-SSA. Contour intervals are 0.05°C. Positive values are shaded. (From Yu and Mechoso, 2001.)

to the west of the SST anomalies, and preceding ocean heat content anomalies near the eastern edge of the basin. These features are similar to those observed during ENSO events.

The ENSO dynamics in the UCLA CGCM were further examined by focusing on the relationship between SST and ocean heat content (i.e. the memory of ENSO). The ocean heat content is defined as the ocean temperature averaged in the upper 300 m. By separating the ocean heat content anomaly of the M-SSA mode into its zonal-mean and zonally asymmetric components (Fig. 3), it is found that the evolution of the zonal-mean component at the equator is 90° out of phase with that of the zonally asymmetric component, as well as that of the SST anomaly [compare Fig. 3(a) to Fig. 2(b)]. The onset of the warm ENSO phase occurs at the time when the mean ocean heat content anomaly grows (i.e. recharge) to a maximum value. The ocean heat content anomalies are then removed (i.e. discharge) as the warm phase develops toward

its mature stage. This phase lag indicates that the variation in the zonal-mean ocean heat content provides the oscillation memory for the ENSO cycle. The ENSO dynamics in the UCLA CGCM are consistent with the recharge oscillator theory (Wyrtki, 1975; Cane and Zebiak, 1987; Zebiak, 1989; Jin, 1997; Li, 1997). This theory is conceptually similar to the delayed oscillator theory in suggesting the importance of subsurface ocean adjustment processes in producing the needed delay for ENSO oscillation. The recharge oscillator, however, emphasizes the importance of the buildup (i.e. charge) and release (i.e. discharge) of zonal-mean ocean heat content in the equatorial band for the phase reversal of the ENSO cycle.

4. Teleconnection of ENSO

By applying the M-SSA to the SST variability produced by the Indo-Pacific Run, it is found

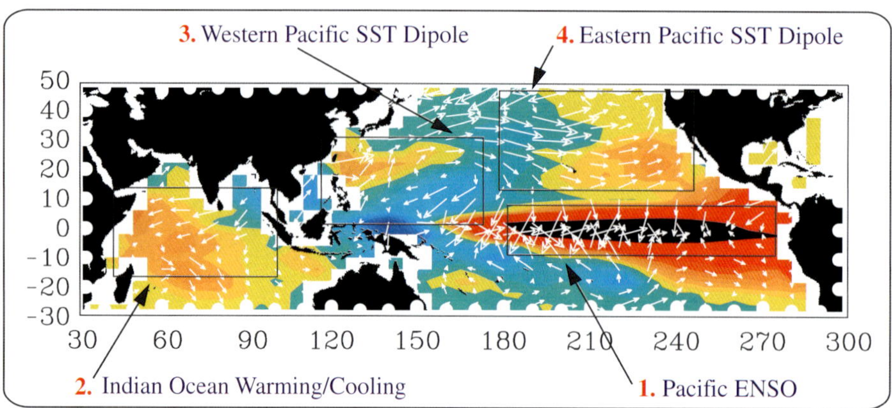

Figure 4. Structures of SST and surface wind stress anomalies of the leading M-SSA mode obtained from the Indo-Pacific Run. Contours represent SST anomalies. Surface wind stress anomalies are indicated by vectors. Contour intervals are 0.1°C. The structures shown are taken from the mature phase of ENSO in the leading M-SSA mode.

that the simulated ENSO cycle is accompanied by significant SST anomalies in many parts of the Indo-Pacific Ocean. Figure 4 shows the SST anomaly pattern of the leading M-SSA mode (i.e. the ENSO mode) extracted from this run. The major anomaly features in this figure include: (1) the Pacific ENSO, (2) the basinwide warming/cooling in the Indian Ocean, (3) an anomaly dipole in the northwestern Pacific (10°N–30°N and 120°E–160°E), and (4) an anomaly dipole in the northeastern Pacific (10°N–30°N and 170°E–120°W). The basinwide warming (cooling) during an El Niño (La Niña) event is a well-known remote response of the Indian Ocean to ENSO (e.g. Yu an Rienecker, 1999, 2000; Murtugudde *et al.*, 2000) via the "atmospheric bridge" mechanism (Lau and Nath, 1996) or the ENSO-induced tropospheric temperature and moisture variations (Chiang and Sobel, 2002; Neelin *et al.*, 2003). The SST dipole in the northwestern Pacific is accompanied by an anomalous anticyclonic (cyclonic) surface circulation during El Niño (La Niña) and is very similar to the Pacific–East Asia teleconnection pattern discussed by Wang *et al.* (2000) and Lau and Nath (2006). They suggested that this teleconnection pattern was

initially forced by ENSO, and was later maintained by a positive thermodynamic feedback between the circulation anomaly and the ocean mixed layer in the northwestern Pacific.

The northeastern SST anomaly dipole simulated by the CGCM is similar to the zonal SST dipole observed during the 1997–98 ENSO event (Liu *et al.*, 1998), which consisted of centers of anomalous warming along the coast of California and anomalous cooling further to the west in the central Pacific. The question arises as to whether or not the development of this subtropical SST anomaly feature is associated with that of ENSO. Furthermore, if such an association exists, what are the mechanisms that link the two phenomena? Yu *et al.* (2000) analyzed the Pacific Run to address these two questions. To concentrate on the relationship between ENSO and this SST dipole, an empirical orthogonal function (EOF) analysis was applied to the model SST anomalies in the Pacific domain between 30°S and 50°N. The leading EOF mode (not shown) is characterized by the ENSO in the tropics and the zonal SST dipole in the subtropics, similar to those shown in the Pacific portion of Fig. 4.

To examine their relationship, the lagged correlation coefficients were calculated between an index representing the strength of the northeastern Pacific SST dipole and an index representing the intensity of ENSO. The ENSO index is the averaged SST anomalies over the central equatorial Pacific (160°–130°W and 40°S–40°N), where the simulated SST anomalies are largest in Fig. 4. The dipole index is defined as the difference between the SST anomalies averaged over the eastern center (150°W–130°W and 24°N–36°N) of the dipole, and those averaged over the western center (180°E–160°W and 24°N–36°N). It was found that the maximum correlation coefficient between these two indices is 0.65 when ENSO leads the dipole by one month. These analyses suggested that the subtropical SST anomaly dipole is forced by ENSO.

To understand the generation mechanism of the SST dipole, Fig. 5 illustrates the relationships between the principal component of the leading EOF mode of SST and the anomalies in sea-level pressure, surface wind stress, and surface heat flux. For the sake of discussion, Fig. 5 is interpreted for the case of a warm ENSO event. Figure 5(a) shows that during El Niño events, the Aleutian Low is enhanced and results in an anomalous cyclonic circulation over the subtropical Pacific. Along the southwesterly branch of the cyclonic circulation, the surface heat flux is reduced [Fig. 5(b)]. This branch brings warm and moist air from the tropics to the subtropics, which is consistent with a reduction in sensible and latent heat flux off the North America coast. Similarly, the northwesterly branch of the cyclonic circulation brings dry and cold air from higher latitudes, which results in increased surface heat flux from the central Pacific. The regions of reduced and enhanced surface heat flux roughly coincide with the warm and cold centers of the subtropical SST anomaly dipole [Fig. 5(c)]. Figure 5, therefore, suggests that the SST anomaly dipole is driven primarily by anomalous surface heat

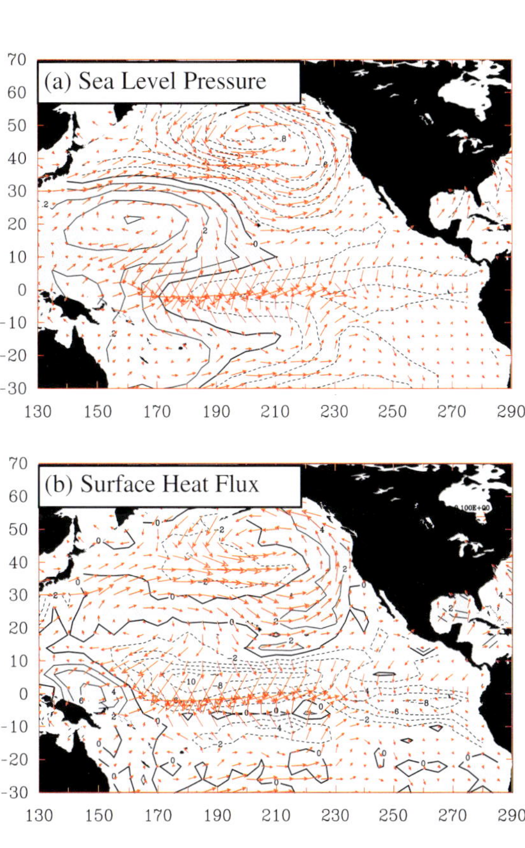

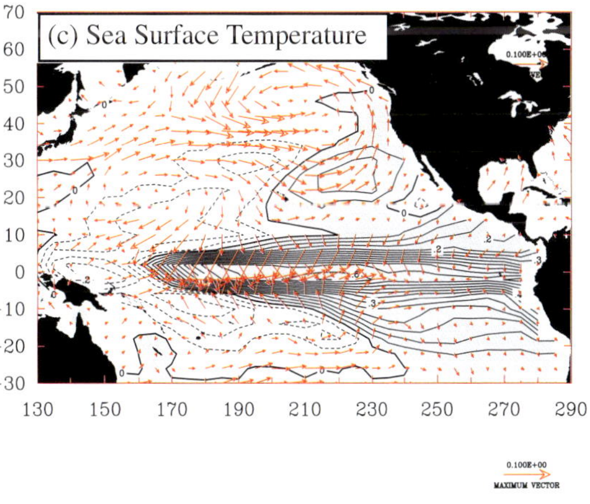

Figure 5. Linear regression between the principal component of the leading EOF mode and the anomalies in (a) sea level pressure, (b) surface heat flux, and (c) SST. Vectors represent regression coefficients between the principal component and zonal and meridional wind stress anomalies and are in units of dyn/cm². Contours intervals are 0.1 mb for (a), 0.2 W/m² for (b), and 0.05°K for (c). Positive values are shaded. Positive (negative) values in (b) indicate less (more) surface heat flux out of the ocean. (From Yu *et al.*, 2000.)

fluxes associated with the altered atmospheric circulation during ENSO events.

5. Impacts of Indian Ocean on the ENSO Cycle

As mentioned earlier, variations in the Indian Ocean may be capable of influencing ENSO activity through either the atmospheric circulation or the oceanic throughflow. To examine this possibility, ENSO simulations were contrasted between the Indo-Pacific Run and the Pacific Run. The former run includes the effects of Indian Ocean SST variability on ENSO, while the latter excludes that effect. Findings obtained from this comparative study were reported by Yu *et al.* (2002) and Yu (2005). It was found that the magnitude and frequency of the ENSO cycle are affected by the addition of an interactive Indian Ocean in the CGCM. Figure 6 compares the time series of the NINO3 index (SST anomalies in 90°–150°W; 5°S–5°N) calculated from the two runs. The first noticeable difference in this figure is that the Indo-Pacific Run produces stronger ENSO amplitudes than the Pacific Run. The standard deviation of the NINO3 SST anomalies is increased from 0.50°C for the Pacific Run to 0.78°C for the Indo-Pacific Run. The latter is closer to the observed value. The second noticeable difference is that the Indo-Pacific Run has a greater decadal variation in ENSO intensity than the Pacific Run. The former run can be broadly separated into two "strong variability decades" (years 10–21 and 38–52) with large warm and cold events and a "weak variability decade" (years 22–37) with weak warm and cold events. No such clear decadal differences are present in the Pacific Run. It appears that by including the Indian Ocean, the CGCM produces more realistic ENSO amplitude and stronger variability on decadal time scales.

The changes in the ENSO cycle when the Indian Ocean coupling is included may be expected, as the CGCM can now resolve

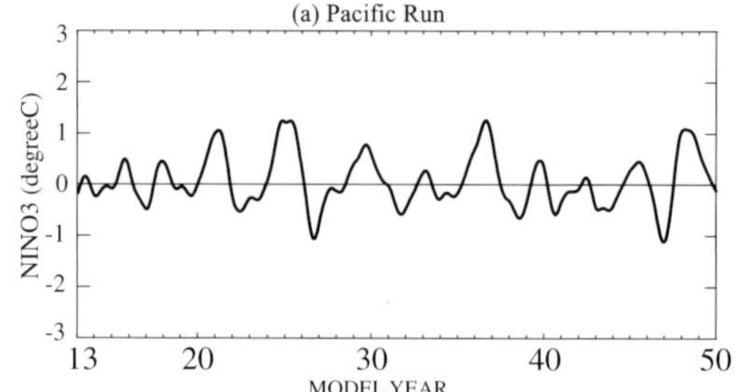

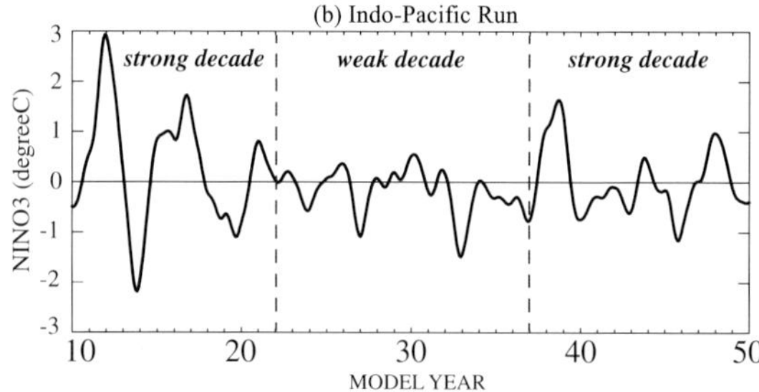

Figure 6. NINO3 SST anomalies calculated from (a) the Pacific Run and (b) the Indo-Pacific Run. These monthly values are low-pass filtered to remove variations shorter than 12 months. (From Yu *et al.*, 2002.)

interannual variability in the Indian Ocean and its interactions with the ENSO cycle. The Indian Ocean SST variability resulting from these interactions, such as the basinwide warming/cooling and the Indian Ocean zonal/dipole mode, may feed back to affect the ENSO evolutions. In addition, since the interactive part of the tropical warm pool covers both the western Pacific Ocean and the eastern Indian Ocean in the Indo-Pacific Run, the eastern Indian Ocean part of the warm pool can respond interactively to Pacific ENSO events. This can amplify the overall feedbacks from the atmosphere during ENSO events. It was also suggested by Kug *et al.* (2006) that an interactive Indian Ocean can affect surface winds in the western Pacific/maritime continent, which can further affect the ENSO evolution.

Yu (2005) noticed that the phase-locking of ENSO to the annual cycle is enhanced and becomes more realistic in the Indo-Pacific Run compared to the Indian Ocean Run (not shown).

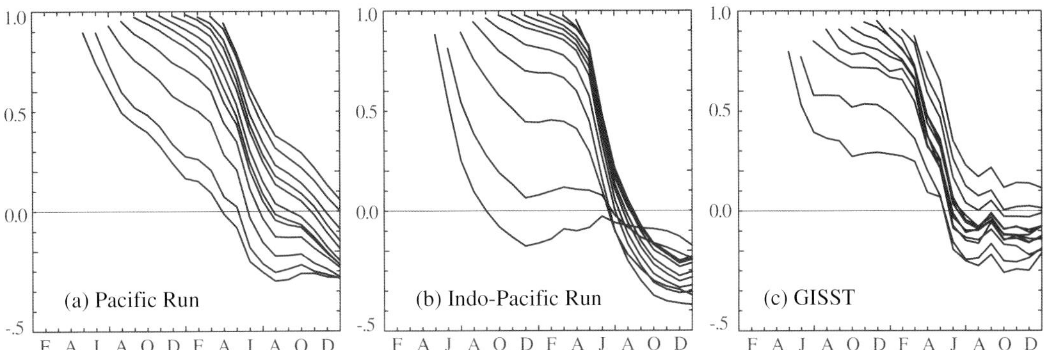

Figure 7. Lagged autocorrelation coefficients of NINO3 SST anomalies calculated from the (a) Pacific Run, (b) Indo-Pacific Run, and (c) GISST data. The curves are shifted to line up their one-month lag with the calendar month on the abscissa (From Yu, 2005.)

As a result of the phase-locking, the spring persistence barrier becomes more obvious in the Indo-Pacific than in the Pacific Run. The spring barrier is a well-known feature of the observed ENSO cycle. Lagged autocorrelation analyses with various ENSO indices, such as the NINO3 SST anomalies, Southern Oscillation pressure differences, and central Pacific rainfall anomalies, show sharp declines in the correlation coefficients in boreal spring (Troup, 1965; Wright, 1979; Webster and Yang, 1992; Torrence and Webster, 1998; Clarke and Gorder, 1999). Figure 7 shows the lagged correlation coefficients of the NINO3 calculated from the Pacific and Indo-Pacific runs and the observations. For the Pacific Run, Fig. 7(a) shows a more gradual decrease in the correlations and a weaker dependence of the decline on calendar months. This experiment produces a weaker spring persistence barrier than the observed [Fig. 7(c)]. Correlation coefficients in Fig. 7(c) are calculated using observed SST's from 1901 to 2000, based on the Global Sea-Ice and Sea Surface Temperature Data Set (GISST) (Rayner *et al.*, 1996). In the Indo-Pacific Run [Fig. 7(b)], the spring barrier is stronger and more realistic, with a rapid decline in the correlations occurring in March–May for most of the 12 curves.

It is well recognized that the ENSO has a low-frequency (3–7 years) and a biennial (~2 years) component (Rasmusson and Carpenter, 1982; Rasmusson *et al.*, 1990; Barnett, 1991; Gu and Philander, 1995; Jiang *et al.*, 1995; Wang and Wang, 1996). Yu (2005) found that the biennial ENSO component is very weak in the Pacific Run, but is significantly enhanced in the Indo-Pacific Run. This is clearly shown in the power spectra of NINO3 index of Fig. 8. By analyzing the persistence barrier in the decades of strong and weak biennial and low-frequency ENSO in the Indo-Pacific Run, Yu (2005) found that the overall amplitude of ENSO is not a primary factor in determining the strength of the persistence barrier. It is the amplitude of the biennial component of ENSO that affects the barrier the most. The persistence barrier is consistently strong (weak) when biennial ENSO variability is large (small). No such clear relationship is found between the strength of the barrier and the amplitude of the low-frequency ENSO component.

Results obtained from these two basin-coupling CGCM experiments (i.e. the Pacific Run and the Indo-Pacific Run) support the hypotheses that the spring persistence barrier is a result of the phase locking of ENSO (Torrence and Webster, 1998; Clarke and Gorder, 1999) and that the biennial ENSO component is crucial to the phase locking (Clarke and Gorder, 1999). Yu (2005) further suggests

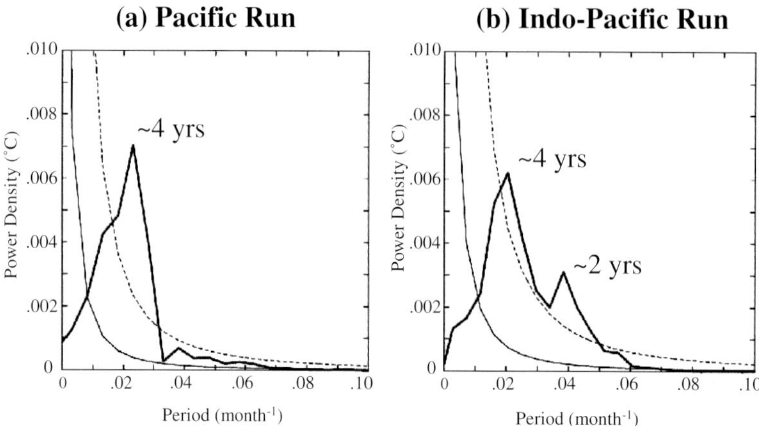

Figure 8. Power spectra of NINO3 SST anomalies calculated from the (a) Pacific Run and (b) Indo-Pacific Run. Dashed lines indicate the 95% significance level. Thin-solid lines are red-noise spectra. (From Yu, 2005.)

that the Indian Ocean coupling plays a key role in producing the biennial component of ENSO. The mechanisms for this are not yet understood, but are being studied. It is believed that the TBO in the Indian and Australian monsoons may be involved in the enhancement of the biennial ESNO component.

6. ENSO's Interactions with the Tropospheric Biennial Oscillation (TBO)

The TBO is a major climate variation feature of the Indian–Australian monsoon system. Years with above-normal summer rainfall tend to be followed by ones with below-normal rainfall and vice versa. The dynamics behind this phenomenon has not yet been fully understood. In work that suggests that the TBO has its own dynamics, the interactions between the monsoon and the tropical Indian and/or Pacific Oceans are emphasized to play a central role in the TBO (e.g. Nicholls, 1978; Meehl, 1987, 1993; Clarke et al., 1998; Chang and Li, 2000; Webster et al., 2002; Yu et al., 2003; Li et al., 2006). However, different theories emphasized different parts of the Indo-Pacific Oceans for the importance. Webster et al. (2002) argued that the TBO is resulted from the monsoon–ocean interaction

in the Indian Ocean. The wind-driven Ekman transport provides the needed phase reversal mechanism for the biennial oscillation. Meehl (1993) believed that the TBO involves the interactions between the monsoon and the Indian Ocean and both the western and the eastern Pacific Ocean. In contrast to this view, the TBO theory of Chang and Li (2000) assigned a passive role to the eastern Pacific Ocean. Instead, they emphasized monsoon–ocean interactions in the Indian and the western Pacific Ocean for the TBO.

The basin-coupling CGCM experiments are capable of isolating the monsoon–ocean interactions in the Pacific or the Indian Ocean and are, therefore, a useful tool for examining these TBO theories. Yu et al. (2003) contrasted the Indian monsoon variability produced in all three basin-coupling CGCM experiments (i.e. Pacific, Indo-Pacific, and Indian-Ocean runs) and noticed interesting differences among them. Figure 9 shows the power spectra of the Indian summer monsoon rainfall index (IMRI; rainfall averaged over an area between 10°N between 30°N, and between 65°E and 100°E) calculated from the experiments. The figure shows that there is virtually no biennial monsoon variation in the simulation including only the Pacific Ocean coupling (i.e. the Pacific Run).

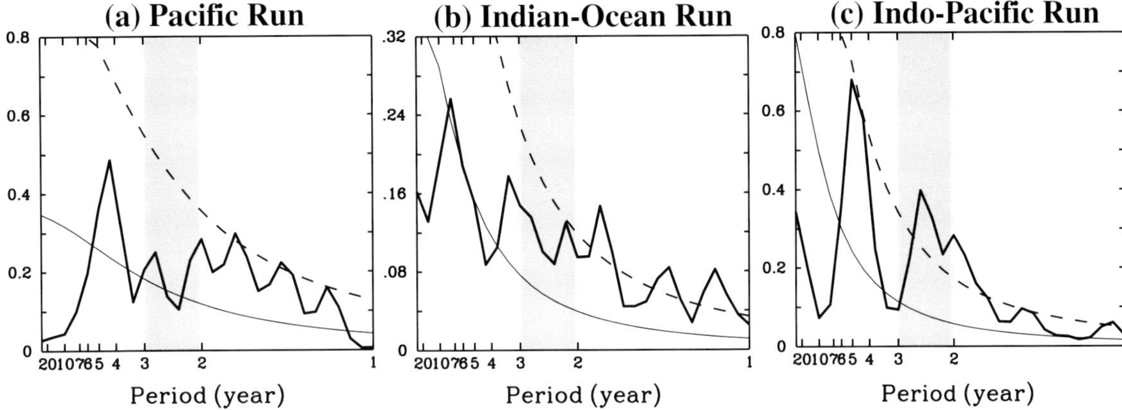

Figure 9. Power spectra of the Indian summer monsoon rainfall index calculated from the (a) Pacific Run, (b) Indian-Ocean Run, and (c) Indo-Pacific Run. The 95% significance levels are indicated by the dashed curves. The shaded area is the period for the quasi-biennial oscillation.

With only the Indian Ocean coupling, the biennial peak is enhanced but is not strong enough to pass the statistical significance level (i.e. the Indian Ocean Run). A statistically significant biennial peak shows up only in the CGCM simulation that includes both the Pacific and Indian Ocean couplings (Indo-Pacific Run). These results suggest that the monsoon–ocean interaction in the Indian Ocean is able to produce weak biennial monsoon variability, and the biennial variability is further enhanced when the interactions between the Indian and the Pacific Ocean are included. The interactions and feedbacks between TBO and the biennial ENSO component are probably responsible for this enhancement.

An important aspect of the TBO is that the biennial tendency in the Indian summer monsoon is related to the biennial tendency in the Australian summer monsoon (Meehl, 1987, 1993). A strong (weak) Indian summer monsoon is often followed by a strong (weak) Australian summer monsoon. The anomalies then reverse sign as they return to the northern hemisphere and lead to a weak (strong) Indian monsoon during the northern summer of the following year. The in-phase transition from Indian summer monsoon to Australian summer monsoon and the out-of- phase transition from Australian summer monsoon to Indian summer monsoon of the next year are two key features of the TBO.

Yu *et al.* (2003) examined the role of the Indian and Pacific Oceans in these two monsoon transitions of the TBO. They noticed that the in-phase monsoon transition was produced more often in the CGCM experiments that included the Pacific Ocean coupling (the Pacific and Indo-Pacific Runs), while the out-of-phase transition was produced more often in the experiments that included the Indian Ocean coupling (the Indian Ocean and Indo-Pacific Runs). These results are demonstrated in Fig. 10, which displays the lagged correlation coefficients between the simulated monthly IMRI and Australian monsoon rainfall index (AMRI; rainfall averaged over an area between 100°E–150°E and 20S–5°N) anomalies. The figure shows that both the observations and the Indo-Pacific CGCM Run produce two large correlation coefficients: a positive coefficient with the IMRI leads the AMRI by about two seasons, and a negative coefficient with the AMRI leads the IMRI by about two seasons. The large positive correlation represents the in-phase transition from Indian to Australian summer monsoons.

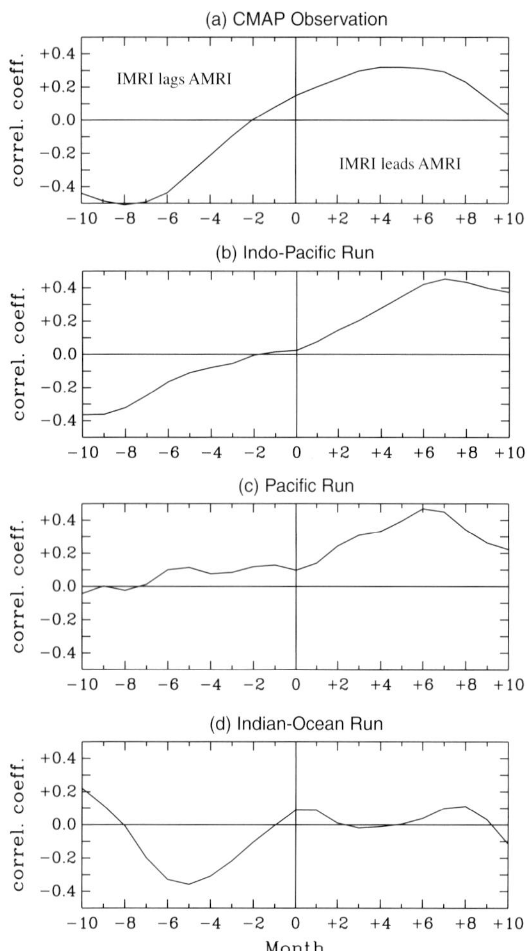

Figure 10. Time-lagged correlation coefficients between the monthly IMRI and AMRI calculated from the (a) CMAP observation, (b) Indo-Pacific Run, (c) Pacific Run, and (d) Indian Ocean Run. (From Yu *et al.*, 2003.)

The negative one represents the out-of-phase transition from Australian to Asian summer monsoons. The CGCM produces only the large positive correlations (i.e. the in-phase transition) in the Pacific Run, and produces only the large negative correlation (i.e. the out-of-phase transition) in the Indian Ocean Run. These results suggest that the Pacific Ocean coupling is crucial to the in-phase transition from the Indian summer monsoon to the Australian summer monsoon. The Indian Ocean coupling is crucial to the out-of-phase transition

from the Australian summer monsoon back to the Indian summer monsoon.

By analyzing SST evolutions during the monsoon transitions, Yu *et al.* (2003) noticed that the Indian and Pacific Oceans showed different and interesting relationships with the in-phase and out-of-phase monsoon transitions. Figure 11 is constructed to summarize these different SST evolutions and their relationships with the monsoon transitions. In this figure, the evolutions of SST anomalies in the central Indian Ocean (20°S–20°N; 40°E–80°E) and in the central Pacific Ocean (10°S–10°N; 150°E–170°W) are composited for the in-phase and out-of-phase monsoon transitions. The corresponding IMRI and AMRI values composited during those seasons are also shown in the figure. Figure 11(a) shows that after a strong (weak) Indian summer monsoon occurs, SST anomalies change sign in the Indian Ocean. SST anomalies are small throughout this in-phase monsoon transition. During the same period, SST anomalies in the Pacific Ocean are large and contribute to the in-phase monsoon transition. Figure 11(b) shows that, after changing sign in the Indian summer monsoon season, Indian Ocean SST anomalies continue to grow and reach large amplitudes during the out-of-phase monsoon transition. During this transition period, the Pacific SST anomalies change sign. Therefore, SST anomalies in the Pacific are small. The large Indian Ocean SST anomalies contribute to the out-of-phase monsoon transition from a strong (weak) Australian summer monsoon to a weak (strong) Indian summer monsoon.

The different evolutions of the Indian and Pacific Ocean SST anomalies are the key that reveals the different roles of these two oceans in the transition phases of the TBO. Figure 12 illustrates how the different monsoon–ocean interactions lead to the TBO. For the sake of discussion, the process begins with warm Indian Ocean SST anomalies and cold (La Niña-type) Pacific SST anomalies, an SST

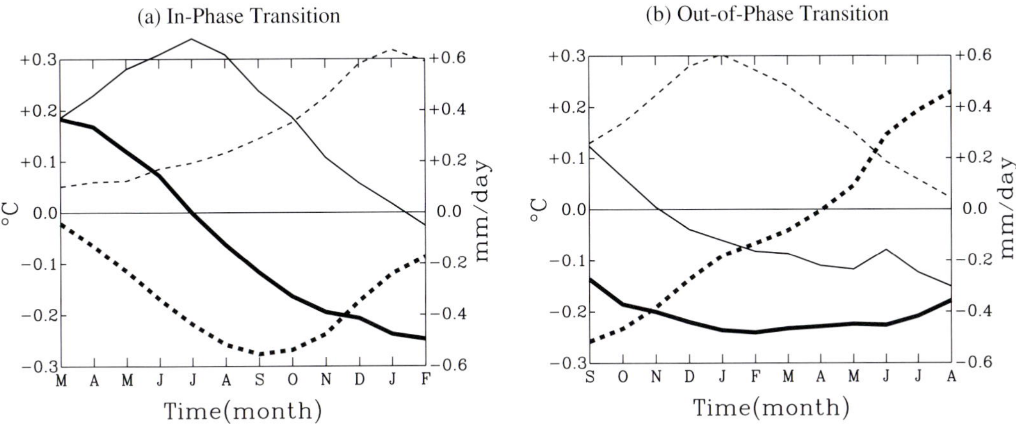

Figure 11. Temporal evolutions of the IMRI (thin-solid), AMRI (thin-dashed), central Pacific SST (Pac-SST; thick-dashed), and central Indian Ocean SST (Indo-SST; thick-solid) during the (a) in-phase monsoon transition and (b) out-of-phase monsoon transition. All values are composited from the Indo-Pacific Run.

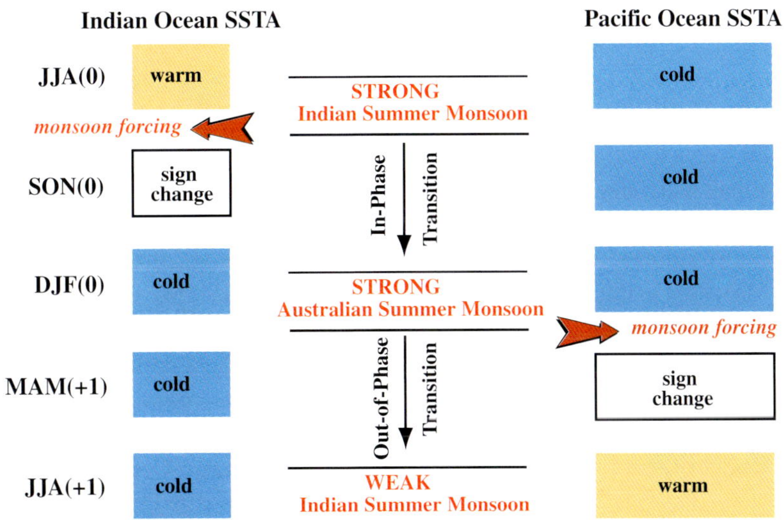

Figure 12. Schematic illustrating the relationship between the Indian–Australian summer monsoons and the Indo-Pacific Ocean SST anomalies (SSTA) during the TBO.

anomaly pattern typically associated with a strong Indian monsoon in June–July–August (JJA, Year 0) (Lau and Yang, 1996). The strong monsoon winds force the Indian Ocean SST anomalies to reverse sign. The Indian Ocean SST anomalies in September–October–November (SON, Year 0) are, therefore, in transition and have small amplitudes. During the same period, the large cold Pacific SST

anomalies persist and dominate the in-phase transition to a strong Australian summer monsoon in December–January–February (DJF, Year 0). This explains why the Pacific Ocean coupling is more important to the in-phase transition from the Indian summer monsoon to the Australian summer monsoon. The strong Australian summer monsoon winds then force the Pacific SST anomalies to change sign and to have small amplitudes in March–April–May (MAM, next year; +1). During this period, cold SST anomalies have been established in the Indian Ocean and have grown to large amplitudes. These cold Indian Ocean SST anomalies then dominate the out-of-phase transition and lead to a weak Asian summer monsoon in JJA of the next year. This explains why the Indian Ocean coupling is needed for the CGCM to produce the out-of-phase monsoon transition. The specific air–sea coupling processes that are involved in these monsoon–ocean interaction should be similar to the wind-evaporation/entrainment and cloud-radiating feedback processes discussed by Wang *et al.* (2003). They showed that these processes allow the remote forcing of ENSO in the Indo-Pacific warm pool to be amplified and maintained from the developing summer to the decaying summer of an ENSO event, to contribute to the for-mation of the TBO.

Yu *et al.* (2003) concluded that the Indian summer monsoon has a stronger impact on the Indian Ocean than on the Pacific Ocean, and that the Australian summer monsoon has a stronger impact on the Pacific Ocean than on the Indian Ocean. These seasonally dependent monsoon influences allow the Pacific and Indian Oceans to have different feedbacks during the in-phase and out-of-phase monsoon transitions, and thus lead to the TBO.

7. ENSO's Interactions with Indian Ocean SST Variability

The recent interest in the observed east–west contrast pattern in Indian Ocean SST anomalies has prompted the suggestion that the Indian Ocean has its own unstable coupled atmosphere–ocean mode similar to ENSO (e.g. Saji *et al.*, 1999; Webster *et al.*, 1999). This interannual SST variability is often referred to as the Indian Ocean zonal mode (IOZM) or Indian Ocean dipole. The IOZM is characte-rized by opposite polarities of SST anomalies between the western and eastern parts of the equatorial Indian Ocean, and is always accom-panied with zonal wind anomalies in the central Indian Ocean. The strong wind–SST coupling associated with the IOZM has been used to argue for the similarity of the phenomenon to the delayed oscillator of ENSO (Webster *et al.*, 1999). The fact that the temporal correlation between the observed IOZM and ENSO events is not strong, and that several significant IOZM events have occurred in the absence of large ENSO events, have led to the suggestion that the IOZM is independent of ENSO (Saji *et al.*, 1999). On the other hand, there are sugges-tions that the IOZM is not an independent phenomenon, but is forced by ENSO through changes in surface heat flux or Indian Ocean circulation (e.g. Klein *et al.*, 1999; Chambers *et al.*, 1999; Murtugudde and Busalacchi, 1999; Venzke *et al.*, 2000; Schiller *et al.*, 2000; Huang and Kinter, 2002; Xie *et al.*, 2002). It has also been suggested that the IOZM is a weak natural coupled mode of the Indian Ocean that can be amplified by ENSO during a particular season (e.g. Gualdi *et al.*, 2003; Annamalai *et al.*, 2003). The IOZM is also suggested to be a natural part of the Indian summer monsoon and the TBO (e.g. Meehl and Arblaster, 2002; Loschnigg *et al.*, 2003, Li *et al.*, 2006). The IOZM is argued to arise from the ocean–atmosphere interactions within the Indian Ocean with links to the Pacific involved with the TBO.

Yu and Lau (2004) examined the intrinsic and forced SST variability in the Indian Ocean by contrasting the Indian ocean SST variability between the Indo-Pacific and Indian Ocean Runs. The former run includes ENSO influences, while the latter one excludes the influences. The M-SSA was applied to the interannual anomalies of Indian Ocean SST to extract leading oscillatory modes from the simulations. One major advantage of the M-SSA is that it easily identifies oscillatory behavior, even if it is not purely sinusoidal (Robertson *et al.*, 1995). In the M-SSA, an oscillatory mode appears as a pair of M-SSA modes that have similar eigenvalues, similar sinusoidal principal components in quadrature with each other, and similar eigenvector structures.

In the Indo-Pacific Run, an oscillatory mode of the Indian Ocean SST variability was found with the M-SSA (not shown). The mode comprises two patterns that can be identified with the IOZM and a basinwide warming/cooling mode respectively. To link the oscillatory mode to the interannual SST variability in the Pacific Ocean, we calculated the time-lag correlation coefficients between the principal component of the leading M-SSA mode and SST anomalies in the entire Indo-Pacific Ocean. Figure 13 shows that the correlation is characterized by an IOZM pattern in the Indian Ocean and an ENSO pattern in the Pacific. The time sequence simulated in the Indo-Pacific Run appears close to the sequence observed during the 1997–98 ENSO event, although discrepancies exist. One discrepancy is that, in the Indo-Pacific Run, ENSO peaks earlier than does the IOZM. The

ENSO pattern peaks about 3.5 months before the IOZM peaks. This discrepancy may be due to the fact that the model ENSO tends to peak in early fall, rather than in winter as observed (Yu and Mechoso, 2001). The close connection between the IOZM and ENSO in the correlation analyses suggests that the IOZM-like oscillatory mode in the Indo-Pacific Run is related to ENSO. The correlation between the time series of the NINO3 index and the IOZM index from the Indo-Pacific Run is also examined. Following Saji *et al.* (1999), the IOZM index is defined as the SST anomaly difference between the western Indian Ocean (50°E–70°E and 10°S–10°N) and the eastern Indian Ocean (90°E–110°E and 10°S–0°). It is found that major IOZM and ENSO events coincide with each other during the simulation (not shown), although the simultaneous correlation coefficient between them is only 0.5.

No oscillatory mode can be found in the Indian Ocean when the ENSO influence is excluded in the Indian Ocean Run. However, IOZM-like features can still be found in the leading variability modes of the Indian Ocean SST. Examinations of these IOZM-like features in the Indian Ocean Run reveal similar ocean–atmosphere coupling associated with enhanced and weakened Indian summer monsoon circulations, as in the Indo-Pacific Run. Our modeling results indicate that IOZM-like features can occur even in the absence of large ENSO events in the Pacific. However, the oscillatory feature of the IOZM is forced by ENSO. ENSO acts as a strong pacemaker to the IOZM. The IOZM may be considered a coupled mode that is weakly damped and cannot be self-sustained when lacking external forcing, such as ENSO and the monsoon (Li *et al.*, 2003).

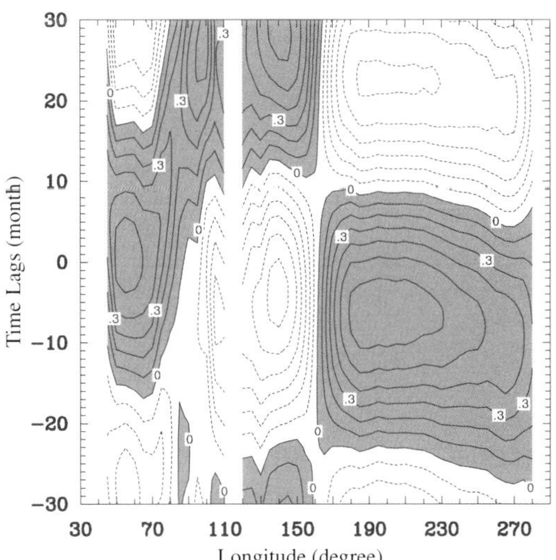

Figure 13. Time-lag correlation coefficients between the principal component of the leading M-SSA mode of the interannual Indian Ocean SST variability and the low-pass filtered SST anomalies in the Indo-Pacific Run. Values are averaged between 4°S and 4°N. Contour intervals are 0.1. Positive values are shaded. (From Yu and Lau, 2004.)

8. Conclusions

Climate changes and variations have strong impacts on human society, and are of common concern to people across national and regional boundaries. In the Indo-Pacific sector, ENSO and the monsoon are two of the most important climate features. These two features occur on opposite sides of the Indo-Pacific basin but have strong interactions with each other. A better understanding of their complex interactions and feedbacks is crucial for successful forecasts of ENSO and monsoon variability. This article reports a unique modeling effort toward that goal. By turning on and off the atmosphere–ocean coupling in various regions of the Indo-Pacific Ocean, the interactions between ENSO and the Indian Ocean and monsoon can be isolated in different CGCM experiments to study the ENSO–monsoon interaction and the ENSO–Indian Ocean interactions individually.

The results obtained from this series of basin-coupling CGCM experiments suggest that the Indian Ocean–monsoon system plays an active role in affecting the amplitude, frequency, and evolution of ENSO and in modulating their decadal variations. The Indian Ocean–monsoon system should be considered a crucial part of the ENSO dynamics. It is known that the Indian Ocean has been experiencing a gradual but significant warming trend in the past few decades (Nitta and Yamada, 1989; Terray, 1994; Wang, 1995). This trend may change the importance of the Indian Ocean to ENSO, and may be an additional reason for the decadal ENSO variability. In particular, the modeling results reported here indicate that the Indian Ocean–monsoon system is crucial to the selection of the dominant frequency of ENSO. An active Indian Ocean may favor a shorter period of ENSO, i.e. the biennial ENSO component. For the low-frequency ENSO component, influence from the Indian Ocean may be less important. The interactions between the Pacific and the Indian Ocean may be different between the decades that have strong biennial ENSO and the decades that have strong low-frequency ENSO. Much more can be learned about the decadal variability of ENSO by looking into the ENSO–Indian Ocean–monsoon

interactions from the perspective of biennial and low-frequency components of ENSO. The basin-coupling modeling strategy pioneered in these CGCM experiments will be a useful and effective tool for the investigations.

Acknowledgements

Support from NOAA (NA03OAR4310061), NSF (ATM-0638432), and NASA (NAG5-13248 and NNX06AF49H) is acknowledged. Model integrations and analyses were performed at the San Diego Supercomputer Center (SDSC) and the University of California Irvine's Earth System Modeling Facility, which is funded by the National Science Foundation (ATM-0321380).

[Received 15 December 2005; Revised 9 May 2007; Accepted 4 June 2007.]

References

Annamalai, H. R., M. J. Potemra, S. P. Xie, P. Liu, and B. Wang, 2003: Couple dynamics over the Indian Ocean: spring initialization of the zonal mode. *Deep-Sea Research II*, **50**, 2305–2330.

Arakawa, A., and W. H. Schubert, 1974: Interaction of a cumulus cloud ensemble with the large-scale environment, Part I. *J. Atmos. Sci.*, **31**, 674–701.

Barnet, T. P., 1984: Interaction of the monsoon and Pacific trade wind system at interannual time scales. *Mon. Wea. Rev.*, **112**, 2380–2387.

Barnet, T. P., 1991: The interaction of multiple time scales in the tropical climate system. *J. Climate*, **4**, 269–285.

Battisti, D. S., and A. C. Hirst, 1989: Interannual variability in the tropical atmosphere–ocean system: Influence of the basic state, ocean geometry, and nonlinearity. *J. Atmos. Sci.*, **46**, 1687–1712.

Bjerknes, J., 1969: Atmospheric teleconnections from the equatorial Pacific. *Mon. Wea. Rew.*, **97**, 163–172.

Bryan, K., 1969: A numerical method for the study of the circulation of the world ocean. *J. Comp. Phys.*, **4**, 347–376.

Cane, M. A., and S. E. Zebiak, 1985: A theory for El Niño and Southern Oscillation. *Science*, **288**, 1085–1087.

Cane, M. A., S. E. Zebiak, and S. C. Dolan, 1986: Experimental forecasts of EL Niño. *Nature*, **321**, 827–832.

Cane, M. A., and S. E. Zebiak, 1987: Model El Niño–Southern Oscillation. *Mon. Wea. Rev.*, **10**, 2262–2278.

Chambers, D. P., B. D. Tapley, and R. H. Stewart, 1999: Anomalous warming in the Indian Ocean coincident with El Niño. *J. Geophys. Res.*, **104**, 3035–3047.

Chang, C.-P., and T. Li, 2000: A theory for the tropical tropospheric biennial oscillation. *J. Atmos. Sci.*, **57**, 2209–2224.

Chiang, J. C. H., and A. H. Sobel, 2002: Tropical tropospheric temperature variations caused by ENSO and their influence on the remote tropical climate. *J. Climate*, **15**, 2616–2631.

Clarke, A. J., X. Liu, and S. V. Gorder, 1998: Dynamics of the biennial oscillation in the equatorial Indian and far western Pacific Oceans. *J. Climate*, **11**, 987–1001.

Clarke, A. J., and S. Van Gorder, 1999: The connection between the boreal spring Southern Oscillation persistence barrier and biennial variability. *J. Climate*, **12**, 610–620.

Cox, M. D., 1984: A primitive equation three-dimensional model of the ocean. GFDL Ocean Group Tech. Rep. No. 1.

Deardorff, J. W., 1972: Parameterization of the planetary boundary layer for use in general circulation models. *Mon. Wea. Rev.*, **100**, 93–106.

Delecluse, P., M. K. Davey, Y. Kitamura, S. G. H. Philander, M. Suarez, and L. Bengtsson, 1998: Coupled general circulation modeling of the tropical Pacific, *J. Geophys. Res.*, **103**, 14357–14374, 10.1029/97JC02546.

Fasullo, J., 2004: Biennial characteristics of Indian monsoon rainfall. *J. Climate*, **17**, 2972–2982.

Fedorov, A. V. and S. G. H. Philander, 2000: Is El Niño changing? *Science*, **288**, 1997–2002.

Godfrey, J. S., 1996: The effects of the Indonesian throughflow on ocean circulation and heat exchange with the atmosphere: A review. *J. Geophys. Res.*, **101**, 12217–12237.

Gu, D., and S. G. H. Philander, 1995: Secular changes of annual and interannual variability in the tropics during the past centry. *J. Climate*, **8**, 864–876.

Gualdi, S., E. Guilyardi, A. Navarra, S. Masina, and P. Delecluse, 2003: The interannual variability in the tropical Indian Ocean. *Clim. Dyn.*, **20**, 567–582.

Harshvardhan, D. A. Randall, and T. G. Corsetti, 1987: A fast radiation parameterization for general circulation models. *J. Geophys. Res.*, **92**, 1009–1016.

Hirst, A. C., and J. S. Godfrey, 1993: The role of Indonesian throughflow in a global ocean GCM. *J. Phys. Oceanog.*, **23**, 1057–1086.

Huang, B., and J. L. Kinter, 2002: The interannual variability in the tropical Indian Ocean and its relations to El Niño–Southern Oscillation. *J. Geophys. Res.*, **107**, 3199 (doi:10.1029/2001JC001278).

Jiang, N., J. D. Neelin, and M. Ghil, 1995: Quasi-quadrennial and quasi-biennial variability in the equatorial Pacific. *Clim. Dyn.*, **12**, 101–112.

Jin, F.-F., 1997: An equatorial recharge paradigm for ENSO, I. Conceptual model. *J. Atmos. Sci.*, **54**, 811–829.

Katayama, A., 1972: A simplified scheme for computing radiative transfer in the troposphere. Numerical Simulation of Weather and Climate Tech. Rep. 6, 77 pp. (Available from Department of Atmospheric Sciences, University of California, Los Angeles, CA 90024, USA.)

Keppenne, C. L., and M. Ghil, 1992: Adaptive spectral analysis and prediction of the southern oscillation index. *J. Geophys. Res.*, **97**, 20449–20454.

Kiladis, G. N., and H. van Loon, 1988: The Southern Oscillation. Part VII: Meteorological anomalies over the Indian and Pacific sectors associated with the extremes of the oscillation. *Mon. Wea. Rev.*, **116**, 120–136.

Kim, Y.-J. and A. Arakawa, 1995: Improvement of orographic gravity-wave parameterization using a mesoscale gravity-wave model. *J. Atmos. Sci.*, **52**, 1875–1902.

Klein, S. A., B. J. Soden, and N.-C. Lau, 1999: Remote sea surface temperature variations during ENSO: Evidence for a tropical atmospheric bridge. *J. Climate*, **12**, 917–932.

Kug, J.-S., and I.-S. Kang, 2005: Interactive feedback between ENSO and the Indian Ocean., *J. Climate*; Accepted.

Kug, J.-S., T. Li, S.-I. An, I.-S. Kang, J.-J. Luo, S. Masson, and T. Yamagata, 2006: Role of the ENSO–Indian Ocean Coupling on ENSO variability in a coupled GCM. *Geophys. Res. Lett.*, **33**, doi:10.1029/2005GL024916.

Latif, M., and Coauthors, 2001: The El Niño simulation intercomparison project. *Clim. Dyn.*, **18**, 255–276.

Lau, N.-C., and M. J. Nath, 1996: The role of the "atmospheric bridge" in linking tropical Pacific ENSO events to extratropical SST anomalies, *J. Climate*, **9**, 2036–2057.

Lau, N.-C., and M. J. Nath, 2006: ENSO modulation of the interannual and intraseasonal variability of the East Asian monsoon: A model study. *J. Climate*, **19**, 4508–4530.

Lau, K. M., and S. Yang, 1996: The Asian monsoon and predictability of the tropical ocean–atmosphere system. *Quart. J. Roy. Meteor. Soc.*, **122**, 945–957.

Lee, T., I. Fukumori, D. Menemenlis, Z. Xing, and L. L. Fu, 2002: Effects of the Indonesian throughflow on the Pacific and Indian Oceans. *J. Phys. Oceangr.*, **32**, 1404–1429.

Li, T., 1997: Phase transition of the El Niño–Southern Oscillation: A stationary SST mode. *J. Atmos. Sci.*, **54**, 2872–2887.

Li, T., B. Wang, C.-P. Chang, and Y. Zhang, 2003: A theory for the Indian Ocean dipole-zonal mode. *J. Atmos. Sci.*, **60**, 2119–2135.

Li, T., P. Liu, X. Fu, B. Wang, and G. A. Meehl, 2006: Tempo-spatial structures and mechanisms of the tropospheric biennial oscillation in the Indo-Pacific warm ocean regions. *J. Climate*, **19**, 3070–3087.

Liu, W. T., W. Tang, and H. Hu, 1998: Spaceborne sensors observe El Niño's effects on ocean and atmosphere in North Pacific. *EOS*, **79**, No. 21.

Loschnigg, J., G. A. Meehl, P. J. Webster, J. M. Arblaster, and G. P. Compo, 2003: The Asian monsoon, the tropospheric biennial oscillation and the Indian Ocean dipole in the NCAR CSM. *J. Climate*, **16**, 2138–2158.

Mechoso, C. R., J.-Y. Yu, and A. Arakawa, 2000: A coupled GCM pilgrimage: From climate catastrophe to ENSO simulations. Book chapter in *General Circulation Model Development: Past, Present, and Future.*, D. A. Randall (eds.), Academic, 807 pp.

Meehl, G. A., 1987: The annual cycle and interannual variability in the tropical Indian and Pacific Ocean regions. *Mon. Wea. Rev.*, **115**, 27–50.

Meehl, G. A., 1993: A coupled air–sea biennial mechanism in the tropical Indian and Pacific regions: role of oceans. *J. Climate*, **6**, 31–41.

Meehl, G. A., and J. M. Arblaster, 2002: The tropospheric biennial oscillation and Asian–Australian monsoon rainfall. *J. Climate*, **15**, 722–744.

Mellor, G. L., and T. Yamada, 1982: Development of a turbulence closure model for geophysical fluid problems. *Rev. Geophys. Space Phys.*, **20**, 851–875.

Mooley, D. A., and B. Parthasarathy, 1984: Fluctuations in all-India summer monsoon rainfall during 1871–1978. *Climate Change*, **6**, 287–301.

Murtugudde, R., A. Busalacchi, and J. Beauchamp, 1998: Seasonal-to-interannual effects of the Indonesian throughflow on the Indo-Pacific Basin. *J. Geophys. Res.*, **103**, 21245–21441.

Murtugudde, R., and A. J. Busalacchi, 1999: Interannual variability of the dynamics and thermodynamics of the tropical Indian Ocean. *J. Climate*, **12**, 2300–2326.

Murtugudde, R., J. P. McCreary, and A. J. Busalacchi, 2000: Oceanic processes associated with anomalous events in the Indian Ocean with relevance to 1997–1998. *J. Geophys. Res.*, **105**, 3295–3306.

Neelin, J. D., C. Chou, and H. Su, 2003: Tropical drought regions in global warming and El Niño teleconnections. *Geophys. Res. Lett.*, **30**, 2275, doi:10.1029/2003GL018625.

Nicholls, N., 1978: Air–sea interaction and the quasi-biennial oscillation. *Mon. Wea. Rev.*, **106**, 1505–1508.

Nitta, T., and S. Yamada, 1989: Recent warming of tropical sea surface temperature and its relationship to the northern hemisphere circulation. *J. Meteor. Soc. Japan*, **67**, 375–383.

Pacanowski, R. C., K. W. Dixon, and A. Rosati, 1991: The GFDL Modular Ocean Model user guide. Tech. Rep. 2, NOAA/GFDL, Princeton, NJ, 75 pp. (Available from GFDL/NOAA, Princeton University, Princeton, NJ 08540, USA.)

Rasmusson, E. M., and T. H. Carpenter, 1982: Variations in tropical sea surface temperature and surface wind fields associated with the Southern Oscillation/El Niño. *Mon. Wea. Rev.*, **110**, 354–384.

Rasmusson, E. M., X. Wang, and C. F. Ropelewski, 1990: The biennial component of ENSO variability. *J. Mar. Syst.*, **1**, 71–96.

Rayner, N. A., E. B. Horton, D. E. Parker, C. K. Folland, and R. B. Hackett, 1996: Version 2.2 of the Global Sea-Ice and Sea Surface Temperature Data Set, 1903–1994 (Clim. Res. Tech. Notes. 74, UK Meteorological Office, Bracknell).

Robertson, A. W., C.-C. Ma, C. R. Mechoso, and M. Ghil, 1995: Simulation of the tropical Pacific climate with a coupled ocean–atmosphere general circulation model. Part I: The seasonal cycle. *J. Climate*, **8**, 1178–1198.

Ropelewski, C. F., M. S. Halpert, and X. Wang, 1992: Observed tropospheric biennial variability and its relationship to the Southern Oscillation. *J. Climate*, **5**, 594–614.

Saji N. H., B. N. Goswami, P. N. Vinayachandran and T. Yamagata, 1999: A dipole mode in the tropical Indian Ocean, *Nature*, **401**, 360–363.

Schiller, A., J. S. Godfrey, P. C. McIntosh, G. Meyers, and R. Fielder, 2000. Interannual dynamics and thermodynamics of the Indo–Pacific Oceans. *J. Phys. Oceanog.*, **30**, 987–1012.

Schopf, P. S., and M. J. Suarez, 1988: Vacillations in a coupled ocean–atmosphere model. *J. Atmos. Sci.*, **45**, 549–566.

Shukla, J., and D. A. Paolino, 1983: The Southern Oscillation and long-range forecasting of the summer monsoon rainfall over India. *Mon. Wea. Rev.*, **111**, 1830–1837.

Suarez, M. J., A. Arakawa and D. A. Randall, 1983: The parameterization of the planetary boundary layer in the UCLA general circulation model: formulation and results. *Mon. Wea. Rev.*, **111**, 2224–2243.

Terray, P., 1994: An evaluation of climatological data in the Indian Ocean area. *J. Meteor. Soc. Japan*, **72**, 359–386.

Terray, P., and S. Dominiak, 2005: Indian Ocean sea surface temperature and El Niño–Southern Oscillation: A new perspective. *J. Climate*, **18**, 1351–1368.

Torrence, C., and P. J. Webster, 1998: The annual cycle of persistence in the El Niño/Southern Oscillation. *Quart. J. Roy. Meteor. Soc.*, **124**, 1985–2004.

Troup, A. J., 1965: The southern oscillation. *Quart. J. Roy. Meteor. Soc.*, **91**, 490–506.

Venzke, S., M. Latif, and A. Vilwock, 2000: The coupled GCM ECHO-2, II, Indian Ocean response. *J. Climate*, **13**, 1371–1383.

Wainer, I., and P. J. Webster, 1996: Monsoon–ENSO interaction using a simple coupled ocean–atmosphere model. *J. Geophys. Res.*, **101**, 25599–25614.

Wang, B., 1995: Interdecadal changes in El Niño onset in the last four decades. *J. Climate*, **8**, 267–285.

Wang, B., and Y. Wang, 1996: Temporal structure of the southern oscillation as revealed by waveform and wavelet analysis. *J. Climate*, **9**, 1586–1598.

Wang, B., R. Wu, and X. Fu, 2000: Pacific–East Asian teleconnection: How does ENSO affect East Asian climate? *J. Climate*, **13**, 1517–1536.

Wang, B., R. Wu, and T. Li, 2003: Atmosphere-warm ocean interaction and its impact on Asian–Australian monsoon variation. *J. Climate*, **16**, 1195–1211.

Webster, P. J., and S. Yang, 1992: Monsoon and ENSO: selectively interactive systems. *Quart. J. Roy. Meteor. Soc.*, **118**, 877–925.

Webster, P. J., S. Yang, I. Wainer, and S. Dixit, 1992: Processes involved in monsoon variability. In *Physical Processes in Atmospheric Models*, D. R. Sikka and S. S. Singh (eds.), Wiley Eastern, New Delhi, pp. 492–500.

Webster, P. J., V. O. Magana, T. N. Palmer, J. Shukla, R. A. Tomas, M. Yanai, and T. Yasunari, 1998: Monsoons: Processes, predictability, and the prospects for prediction. *J. Geophys. Res.*, **103**, 14451–14510.

Webster, P. J., A. Moore, J. Loschnigg and M. Leban: 1999: Coupled ocean–atmosphere dynamics in the Indian Ocean during 1997–98, *Nature*, **40**, 356–360.

Webster, P. J., C. Clark, G. Cherikova, J. Fasullo, W. Han, J. Loschnigg, and K. Sahami, 2002: The monsoon as a self-regulating coupled ocean–atmosphere system. Meteorology at the Millennium, International Geophysical Series, Vol. 83, Academic, pp. 198–219.

Wright, P. B., 1979: Persistence of rainfall anomalies in the central Pacific. *Nature*, **277**, 371–374.

Wu, R., and B. P. Kirtman, 2004: Impacts of the Indian Ocean on the Indian summer monsoon–ENSO relationship. *J. Climate*, **17**, 3037–3054.

Wyrtki, K., 1975: El Niño — the dynamic response of the equatorial Pacific Ocean to atmospheric forcing. *J. Phys. Oceanog.*, **5**, 572–584.

Xie, S. P., H. Annamalai, F. A. Schott, J. P. McCreary, 2002: Structure and mechanisms of south Indian Ocean climate variability. *J. Climate*, **15**, 867–878.

Yasunari, T., 1990: Impact of Indian monsoon on the coupled atmosphere–ocean system in the tropical Pacific. *Meteor. Atmos. Phys.*, **44**, 29–41.

Yu, J.-Y., and C. R. Mechoso, 1999: Links between annual variations of Peruvian stratocumulus clouds and of SSTs in the eastern equatorial Pacific. *J. Climate*, **12**, 3305–3318.

Yu, J.-Y., W. T. Liu, and C. R. Mechoso, 2000: The SST anomaly dipole in the northern subtropical Pacific and its relationship with ENSO. *Geophys. Res. Lett.*, **27**, 1931–1934.

Yu, J.-Y., and C. R. Mechoso, 2001: A coupled atmosphere–ocean GCM study of the ENSO cycle. *J. Climate*, **14**, 2329–2350.

Yu, J.-Y., C. R. Mechoso, J. C. McWilliams, and A. Arakawa, 2002: Impacts of the Indian Ocean on the ENSO cycle. *Geophys. Res. Lett.*, **29**, 46.1–46.4.

Yu, J.-Y., S.-P. Weng, and J. D. Farrara, 2003: Ocean roles in the TBO transitions of the Indian–Australian monsoon system. *J. Climate*, **16**, 3072–3080.

Yu, J.-Y., and K. M. Lau, 2004: Contrasting Indian Ocean SST variability with and without ENSO influence: A coupled atmosphere–ocean GCM study. *Meteor. Atmos. Phys.*, DOI: 10.1007/00703-004-0094-7.

Yu, J.-Y., 2005: Enhancement of ENSO's persistence barrier by biennial variability in a coupled atmosphere–ocean general circulation model. *Geophys. Res. Lett.*, **32**, L13707, doi:10.1029/ 2005GL023406.

Yu, L. S., and M. M. Rienecker, 1999: Mechanisms for the Indian Ocean warming during the 1997–98 El Niño. *Geophys. Res. Lett.*, **26**, 735–738.

Yu L. S, and M. M. Rienecker, 2000: Indian Ocean warming of 1997–1998. *J. Geophys. Res.*, **105**, 16923–16939.

Zebiak, S. E., 1989: Ocean heat content variability and El Niño cycles. *J. Phys. Oceanog.*, **19**, 475–486.

Ocean–Atmosphere–Land Feedbacks on the Western North Pacific–East Asian Summer Climate

Chia Chou

*Research Center for Environmental Changes, Academia Sinica and
Department of Atmospheric Sciences, National Taiwan University, Taipei, Taiwan
chiachou@rcec.sinica.edu.tw*

Jien-Yi Tu

*Department of Atmospheric Sciences, Chinese Culture University,
Yang-Ming Shan, Taipei, Taiwan*

In this review article, we summarize mechanisms limiting the poleward extent of the summer monsoon rain zones, particularly for the Asian summer monsoon. They include local processes associated with net heat flux into the atmosphere and soil moisture, ventilation by cross-continental flow, and the interactive Hodwell–Hoskins (IRH) mechanism, defined as the interaction between monsoon convective heating and baroclinic Rossby wave dynamics. The last two mechanisms, ventilation and the IRH mechanism, also induce an east–west asymmetry of the summer monsoon rain zones. Processes that change land–ocean heating contrast, and differences in net heat flux into the atmosphere rather than in surface temperature, are also discussed. Convection associated with the Asian summer monsoon is initiated by net heat flux into the atmosphere and modified by soil moisture via an evaporation process. In Asia, ventilation by moisture advection is particularly important, and the IRH mechanism tends to favor interior arid regions and east coast precipitation. Land surface conditions, such as surface albedo and topography, and ocean heat transport tend to modify land–ocean heating contrast, in terms of a tropospheric temperature gradient, and then change the Asian summer monsoon circulation and its associated rain zone. The stronger meridional gradient of tropospheric temperature tends to enhance the summer monsoon rainfall and extend the rain zone farther northward. Local SST, such as the warm SST anomalies in the western North Pacific during El Niño, is also important in the summer monsoon rainfall and its position.

1. Introduction

Land–ocean heating contrast is known to be fundamental to summer monsoon circulations (Webster, 1987; Young, 1987). However, examining the relation between solar heating of the continent and the associated monsoon rainfall, the rain zone does not extend as far poleward as the maximum heating would seem to indicate. This suggests that other mechanisms besides land–ocean heating contrast

determine the poleward extent of summer monsoons. Many studies (Lofgren, 1995; Meehl, 1994; Nicholson, 2000; Xue and Shukla, 1993; Xue *et al.*, 2004; Yang and Lau, 1998) have discussed the importance of land processes, such as surface albedo, soil moisture and vegetation, in affecting the magnitude and position of the monsoon. Topography, such as the Tibetan Plateau, also affects the monsoon circulation (Flohn, 1957; He *et al.*, 1987; Murakami, 1987; Meehl, 1992; Wu and Zhang, 1998; Yanai and

Li, 1994; Ye, 1981). Other effects, such as changes of the westerly jet stream in the upper troposphere, persist from the previous winter and spring into summer, and then affect the Asian summer monsoon (Chang *et al.*, 2001; Yang *et al.*, 2004). In this study, we examine dynamical mechanisms that mediate land–ocean contrasts and affect the poleward extent of summer monsoons. Figure 1 shows precipitation climatology in the boreal summer and winter. The associated summer monsoon rain zones, such as the Asia summer monsoon, tend to

have northeast–southwest tilting in the northern hemisphere, and northwest–southeast tilting in the southern hemisphere. Thus, not only can the summer monsoon rain zones not move as far poleward as the solar heating indicates, but also the rain zones show an east–west asymmetry with less poleward extent over the western part of the continents.

Among the summer monsoon systems in the world, the Asian summer monsoon is the most intense and also the most complicated because of the existence of the Himalayan–Tibetan

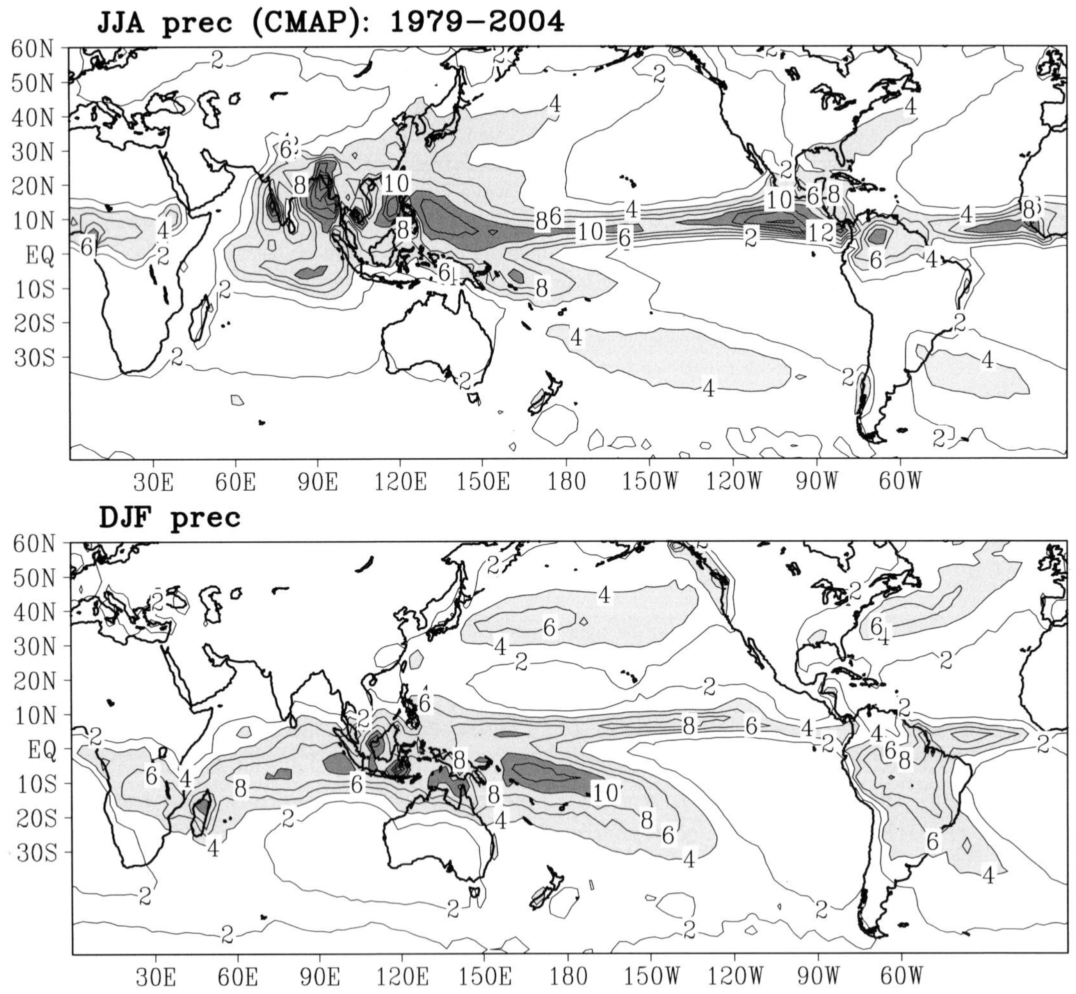

Figure 1. Precipitation climatology (1979–2004) in (a) the boreal summer (JJA) and (b) the boreal winter (DJF) from the Climate Precipitation Center (CPC) Merged Analysis of Precipitation (CMAP; Xie and Arkin, 1997).

Plateau. The Asian summer monsoon has several subregional systems, including the Indian summer monsoon, the western North Pacific summer monsoon, and the East Asian summer monsoon (Goswami *et al.*, 1999; Lau *et al.*, 2000; Wang and Fan, 1999; Wang and LinHo, 2002). Each subregional monsoon system has its own evolution. Furthermore, the Asian summer monsoon rain zone seems to extend even farther southwestward into Africa. Some studies indicate a connection between the African summer rainfall and the Indian summer rainfall (Liu and Yanai, 2001; Camberlin, 1997). Thus, the Asian summer monsoon system should be treated as one very broad system over the Eurasian–African continent and tropical oceans (Webster and Yang 1992; Yang *et al.*, 1992). On one hand, the Asian summer monsoon as a whole is associated with land–ocean contrasts between the Eurasian–African continent and neighboring oceans, such as the Pacific Ocean and the Indian Ocean. On the other hand, local forcings, such as local sea surface temperature (SST), may affect the Asian summer monsoon and create different subregional summer monsoon systems.

The Asian summer monsoon rain zone marches with the season which is associated with solar heating. The Asian summer monsoon also shows interannual variation in the magnitude and position of the monsoon rainfall with and without El Niño/Southern Oscillation (ENSO) influences. Figure 2 shows precipitation anomalies in the ENSO growing (year 0) and decaying (year 1) years and the non-ENSO year (Chou *et al.*, 2003; CTY hereafter). Under ENSO influences, a strong (weak) western North Pacific (WNP) summer monsoon occurs over the western North Pacific and the Philippine Sea in the El Niño (La Niña) growing year and the La Niña (El Niño) decaying year [Figs. 2(b) and 2(c)]. To its north, this strong WNP summer monsoon often accompanies a decrease of the northeast–southwest tilting summer monsoon rainfall. In other words,

the northeast–southwest tilting Asian summer monsoon rainfall decreases (increases) in the El Niño (La Niña) growing year [Fig. 2(b)] and the La Niña (El Niño) decaying year [Fig. 2(c)]. This interannual variation of the Asian summer monsoon indicates a dipole pattern of the precipitation anomalies along the coast of East Asia, and may affect the position of the northern edge of the Asian summer monsoon rain zone. Without ENSO influences, the Asian summer monsoon also shows a similar interannual variability [Fig. 2(a)]. This implies that ENSO affects some processes that can also affect the interannual variation of the Asian summer monsoon. Besides the impacts of ENSO on the Asian summer monsoon strength and location, ENSO also affects the WNP summer monsoon onset via the remote and local SST anomalies (Wu and Wang, 2000).

To examine dynamical mechanisms limiting the poleward extent of summer monsoons, an atmospheric model with intermediate complexity is coupled with a simple land-surface model and a mixed-layer ocean with prescribed ocean heat transport (Neelin and Zeng, 2000; Zeng *et al.*, 2000). A brief description of the model can be found in Sec. 2. Moisture and moist static energy budgets are used to discuss the summer monsoon mechanisms, and the derivations are shown in Sec. 3. We first use an idealized monsoon system to discuss those mechanisms in a general sense (Chou *et al.*, 2001; CNS hereafter), and then compare the impacts of the various mechanisms on the Asian summer monsoon, a realistic summer monsoon system (Chou and Neelin, 2003; CN03 hereafter; Chou, 2003, CHOU hereafter). Conclusions follow.

2. Quasi-equilibrium Tropical Circulation Model

To examine the ocean–atmosphere–land feedbacks on monsoons, a coupled ocean–atmosphere–land model of intermediate complexity

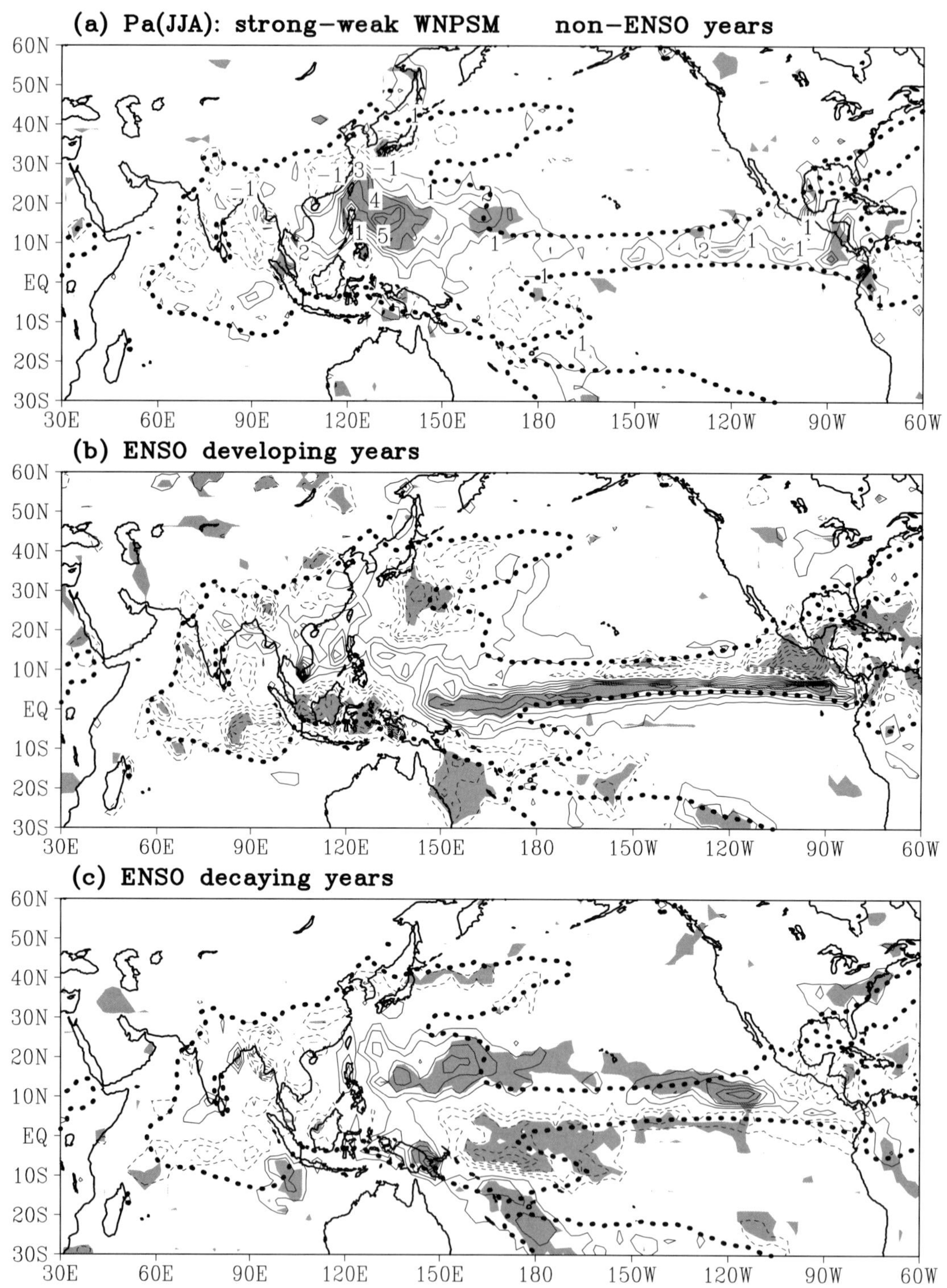

Figure 2. Composite difference of the CMAP summer rainfall anomalies between the strong and weak WNP summer monsoons for (a) the non-ENSO years, (b) the ENSO developing years and (c) the ENSO decaying years (from CTY). The contour interval is $1\,\mathrm{mm\,day^{-1}}$. Areas with the significance level at 5% by the two-sample t test are shaded. The thick dashed line shows the JJA precipitation climatology at $5\,\mathrm{mm\,day^{-1}}$.

(Neelin and Zeng, 2000; Zeng *et al.*, 2000) is used (CNS). Based on the analytical solutions derived from the Betts–Miller moist convective adjustment scheme (Betts and Miller, 1993), typical vertical structures of temperature, moisture and winds for deep convection are used as leading basis functions for a Galerkin expansion (Neelin and Yu, 1994; Yu and Neelin, 1994). The resulting primitive equation model makes use of constraints on the flow by quasi-equilibrium thermodynamic closures and is referred to as QTCM1 (quasi-equilibrium tropical circulation model with a single vertical structure of temperature and moisture for deep convection). Because the basis functions are based on vertical structures associated with convective regions, these regions are expected to be well-represented and similar to a general circulation model (GCM) with the Betts–Miller moist convective adjustment scheme. Far from convective regions, QTCM1 is a highly truncated Galerkin representation equivalent to a two-layer model. With its intermediate complexity, between a full GCM and simpler models, this model is easier to analyze, is numerically faster than a GCM, and has more physical processes than simpler models.

To accompany the representation of dynamics in this model, a cloud-radiation scheme (Chou and Neelin, 1996; Zeng *et al.*, 2000), simplified from the full GCM radiation schemes (Harshvardhan *et al.*, 1987; Fu and Liou, 1993), is included. The longwave radiation parametrization is derived from Harshvardhan *et al.* (1987) with the help of a Green's function method to calculate upward and downward longwave radiative fluxes. The shortwave radiation scheme is a simple linear calculation of shortwave fluxes at the surface and atmospheric absorption based on Fu and Liou (1993). The deep and cirrocumulus/cirrostratus cloud fraction is estimated with an empirical parametrization (Chou and Neelin, 1999). This physical parametrization package can give near GCM-like accuracy in the determination of radiative flux exchange at the surface under suitable circumstances. Caveats for the application here include lack of prognostic low and middle clouds. A simple atmospheric boundary layer that assumes a steady state and a vertically homogeneous mixed layer with fixed height (Stevens *et al.*, 2002) is implemented.

Similar in structure to the physical parametrization package, a land-surface model (Zeng *et al.*, 2000) with intermediate complexity includes the essentials of complex land-surface models, such as the biophysical control on evapotranspiration and surface hydrology, but retains computational and diagnostic simplicity. This model is designed to capture land-surface effects for climate simulation at time scales longer than a day. It uses a single land-surface layer for both the energy and water budget. Since the heat capacity of land is small, the net surface heat flux is essentially close to zero at a time scale longer than a day. Surface heat flux is balanced by solar radiation, longwave radiation, evaporation and sensible heat flux. The prognostic equation for soil moisture induces precipitation, evaporation, surface runoff and ground runoff. Surface runoff increases as the soil moisture reaches a saturation value that depends on the surface type.

To study the interaction between the atmosphere and land, fixed SST is commonly assumed in climate models in order to simplify the simulation. To avoid any artificial effects induced by the contrast of the fixed SST boundary condition and the interactive land surface, an ocean surface layer with active thermodynamics is included in the simulations presented here. Instead of coupling a complicated ocean general circulation model, a slab mixed-layer ocean model with a fixed mixed-layer depth of 50 m is used. By specifying Q flux, which crudely simulates divergence of ocean transport (Hansen *et al.*, 1988, 1997), SST can be determined by the energy balance between surface radiative flux, latent heat flux, sensible heat flux and Q flux. The Q flux can be obtained from observations or ocean model results (Doney *et al.*, 1998; Keith, 1995; Miller *et al.*, 1983; Russell *et al.*, 1985). In general, the Q flux varies from ocean to ocean as well as from season to season. QTCM version 2.3 is used here, with the solar radiation scheme slightly modified.

3. Moist Static Energy Analysis

3.1. *Moist static energy budget*

The vertically integrated thermodynamic and moisture equations can be written as

$$\partial_t \langle T \rangle + \langle v \cdot \nabla T \rangle + \langle \omega \partial_p s \rangle = \langle Q_c \rangle + H + S^{\text{net}} + R^{\text{net}}, \tag{1}$$

$$\partial_t \langle q \rangle + \langle v \cdot \nabla q \rangle + \langle \omega \partial_p q \rangle = \langle Q_q \rangle + E, \tag{2}$$

where v is horizontal velocity, ω is pressure velocity, H is sensible heat flux, and E is evaporation. Temperature T and specific humidity q are in J kg^{-1}, absorbing heat capacity at constant pressure (C_p) and latent heat per unit mass (L). The dry static energy s is defined as $s = T + \Phi$, with Φ the geopotential. The vertical integral $\langle X \rangle$ through the whole troposphere p_T is defined as $1/g \int X \, dp$. The longwave radiation R^{net} and the shortwave radiation S^{net} are the net radiative energy into the atmospheric column. When vertically integrated, the convective heating Q_c will cancel with the moisture sink Q_q, since horizontal moisture transport by convective motion is negligible. Thus, precipitation can be estimated through $P = -\langle Q_q \rangle = \langle Q_c \rangle$.

Because of the cancellation between $\langle Q_c \rangle$ and $\langle Q_q \rangle$, adding the thermodynamic equation (1) and the moisture equation (2) can derive the moist static energy (MSE) equation:

$$\partial_t \langle T + q \rangle + \langle v \cdot \nabla (T + q) \rangle + \langle \omega \partial_p h \rangle = F^{\text{net}}, \tag{3}$$

where the MSE is $h = s + q$.

The net heat flux into the atmospheric column F^{net}, positive when heating the atmosphere, can be derived from

$$F^{\text{net}} = S_t^{\downarrow} - S_t^{\uparrow} - R_t^{\uparrow}$$
$$- S_s^{\downarrow} + S_s^{\uparrow} - R_s^{\downarrow} + R_s^{\uparrow} + E + H \tag{4}$$
$$= F_t^{\text{net}} - F_s^{\text{net}}, \tag{5}$$

where subscripts s and t on the solar ($S^{\downarrow}$ and $S^{\uparrow}$) and longwave ($R^{\uparrow}$ and $R^{\downarrow}$) radiative terms denote the surface and model top, and $R_t^{\downarrow} \approx 0$ has been used. The net heat flux at the top of the atmosphere (TOA) is

$$F_t^{\text{net}} = S_t^{\downarrow} - S_t^{\uparrow} - R_t^{\uparrow} \tag{6}$$

and the net heat flux at the surface is

$$F_s^{\text{net}} = S_s^{\downarrow} - S_s^{\uparrow} + R_s^{\downarrow} - R_s^{\uparrow} - E - H. \tag{7}$$

Positive F_t^{net} and F_s^{net} indicate downward heat fluxes. The net longwave and shortwave radiative fluxes in (1) can then be estimated through $S^{\text{net}} = S_t^{\downarrow} - S_t^{\uparrow} - S_s^{\downarrow} + S_s^{\uparrow}$ and $R^{\text{net}} = -R_t^{\uparrow} - R_s^{\downarrow} + R_s^{\uparrow}$.

3.2. *Horizontal advection associated with v_ψ and v_χ*

In the experiments for examining the effect of horizontal transports, the terms $v \cdot \nabla T$ and $v \cdot \nabla q$ are usually set to zero (CNS; Chou and Neelin, 2001; CN01 hereafter). However, suppressing these terms creates errors in the conservation of energy and moisture. To refine the experiments, we divide the horizontal winds into nondivergent part v_ψ and irrotational part v_χ. By definition,

$$\nabla \cdot v_\psi = 0, \tag{8}$$

so

$$v_\psi \cdot \nabla q = \nabla \cdot (v_\psi q), \tag{9}$$

$$v_\psi \cdot \nabla T = \nabla \cdot (v_\psi T). \tag{10}$$

When taking an area average over the domain with $v_\psi \cdot \hat{n} \approx 0$ through the entire boundary,

$$\int_{\text{Domain}} v_\psi \cdot \nabla X \, dA = 0,$$

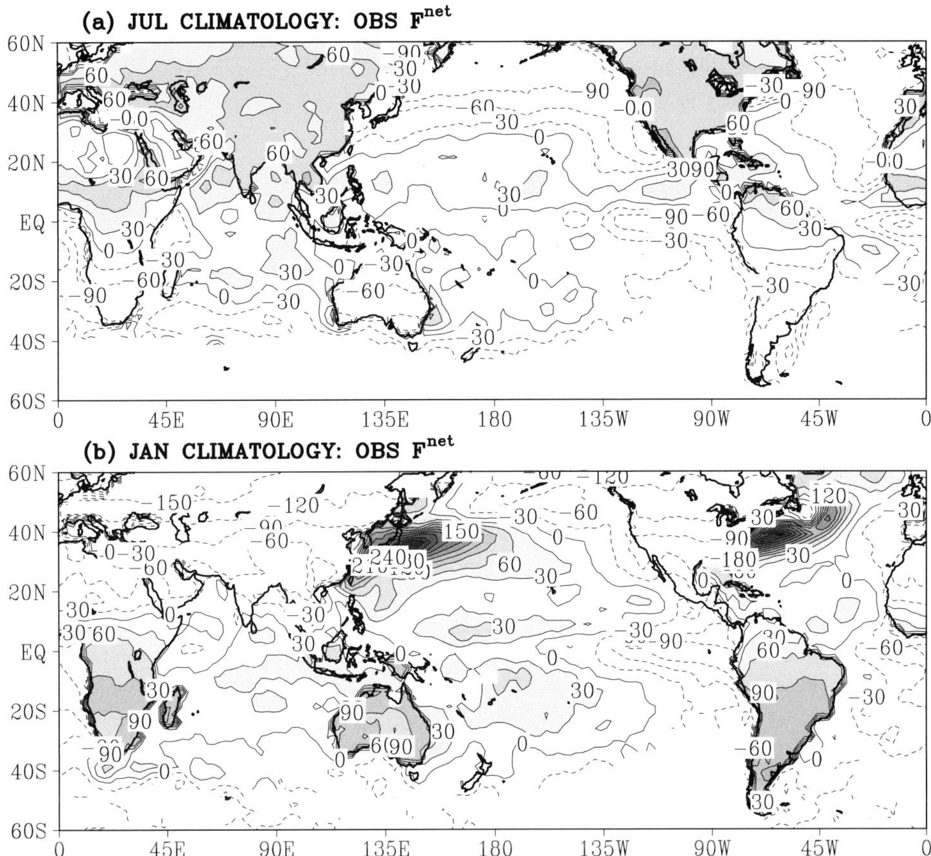

Figure 3. Net heat flux of energy into the atmospheric column, F^{net} (net surface and top-of-atmosphere shortwave and longwave radiation, sensible heat and latent heat fluxes), estimated from observations for (a) July (from CN03) and (b) January (from CN01).

where $\hat{n}$ is the normal direction of the boundary and X is T or q. Thus, suppressing the advection terms associated with v_ψ, e.g. $v_\psi \cdot \nabla q = 0$, will not affect domain-integrated conservation in the experiments for examining the effect of the horizontal transport. Usually, v_ψ is much larger than v_χ (six times larger in CN03), so the effect of suppressing the terms $v_\psi \cdot \nabla T$ and $v_\psi \cdot \nabla q$ is still substantial.

4. Mechanisms for the Poleward Extent of Summer Monsoons

Land–ocean heating contrast has been considered the heart of the summer monsoon circulation. Because of ocean heat storage and transport, the heating of the atmosphere initiated by solar radiation is much stronger over continental regions than over ocean regions. Figure 3 shows the net energy into the atmospheric column F^{net}. Strong heating (positive F^{net}) is found all over the summer continents, while weak heating or even cooling is over the ocean regions. Thus, strong land–ocean heating contrast is over the summer hemisphere. From the moisture budget (2) and MSE budget (3), the summer monsoon rainfall and the associated convection are affected not only by local processes via evaporation and F^{net}, but also by the moisture and MSE transports via cross-continent flow and a feedback from the summer monsoon circulation. The transport

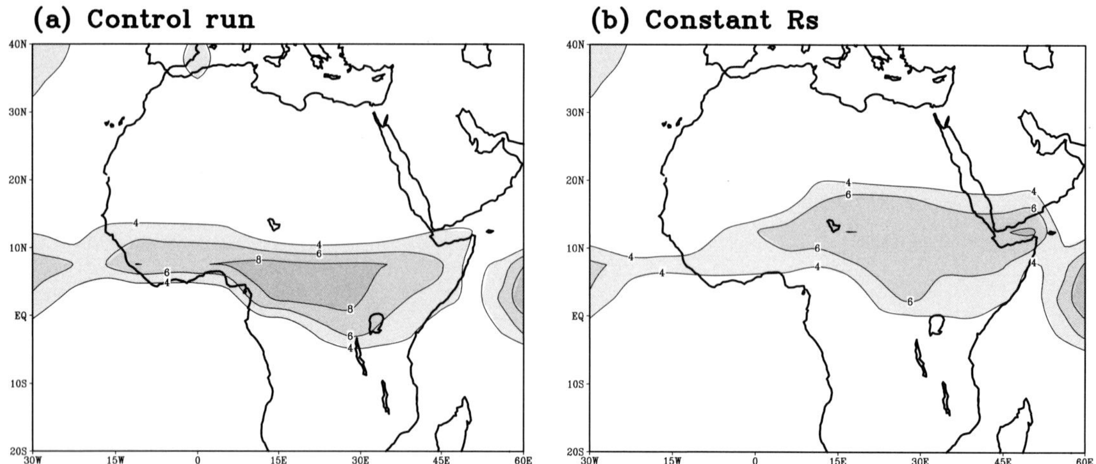

Figure 4. July precipitation for experiments over the African region: (a) control run and (b) constant surface albedo over Africa (from CN03).

effect associated with cross-continental flow is termed ventilation, and the interaction of the summer monsoon heating with its circulation is termed the interactive Rodwell–Hoskins (IRH) mechanism (CNS). To study mechanisms for the poleward extent of the summer monsoon rain zones, we examined those processes that affect the monsoon rainfall: the local processes, ventilation and the IRH mechanism. The magnitude of land–ocean heating contrast affects the monsoon circulation and then the monsoon rainfall, so the processes that change the land–ocean contrast are also examined.

4.1. *Local processes*

The summer monsoon rainfall is initiated by seasonal marching of solar heating via F^{net}. Positive F^{net} is found over summer continents except for Saharan Africa, where the surface albedo is high (Fig. 3). Based on (3), positive F^{net} heats the atmosphere and induces convection when the horizontal transport is small. Over regions with small F^{net}, such as the Sahara, the associated rainfall should be very low, even with enough soil moisture supply. When reducing the surface albedo over the Sahara in a pair of experiments for examining

the effect of F^{net}, the summer monsoon rainfall over northern Africa increases, and the monsoon rain zone extends farther northward due to stronger F^{net} (Fig. 4). The net surface heat flux F_s^{net} over land is near zero for an average longer than one day. When the low surface albedo reflects less shortwave radiation to the space, F^{net} increases. The variation of F^{net} over the summer continents also changes the land–ocean heating contrast that modifies the summer monsoon circulation and its feedback on the monsoon rainfall. For most regions, this feedback is stronger than the direct F^{net} effect, so the direct F^{net} effect is difficult to separate from other effects, except for regions with high surface albedo. This feedback associated with the monsoon circulation will be discussed in the following section.

Local soil moisture is another effect that affects the summer monsoon rainfall. Considering a steady-state moisture budget derived from (2), evaporation also contributes to the summer monsoon rainfall besides the moisture transport by vertical motion (ω term) and horizontal advection ($v \cdot \nabla$ term). Over continents, soil moisture is an important factor in the evaporation process, and it can be affected by the surface type and vegetation. A schematic

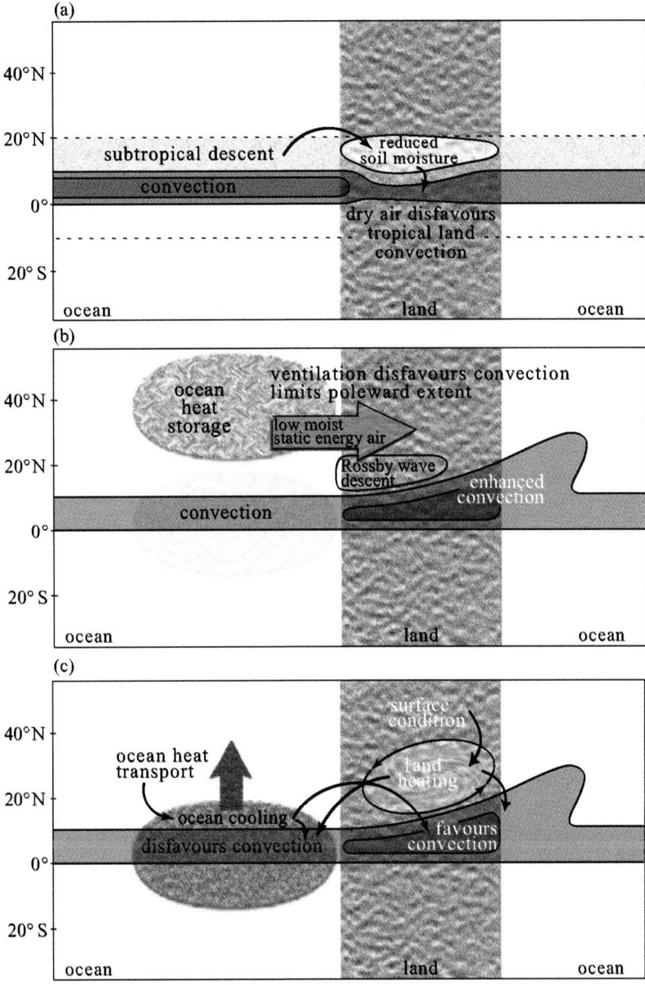

Figure 5. Schematic diagrams describing effects on continental convection zones (for northern hemisphere summer) of (a) soil moisture feedbacks, (b) ventilation and the interactive Rodwell–Hoskins mechanism, and (c) land–ocean contrast associated with ocean heat transport and land surface condition (adapted from CNS).

diagram is used to summarize the soil moisture effect [Fig. 5(a)]. Most summer monsoon rain zones over continents are in the descending branch of the Hadley cell, so soil moisture is relatively low prior to the summer season. This low soil moisture tends to decrease evaporation over land. With less moisture supply, the convective available potential energy (CAPE) reduces, so convection becomes less frequent. In other words, if soil moisture can be supplied indefinitely, such as in the saturated experiments discussed in CNS, the summer monsoon rain zones should spread across much wider areas. Soil moisture can also change the partitioning of evaporation and sensible heat fluxes, and then affects surface temperature. This change in the partitioning of surface heat fluxes does not affect F^{net} since the net surface heat flux F_s^{net} over continents is near zero on a time scale longer than one day. Thus, the monsoon circulation is modified very little by soil moisture, and so the feedback of soil moisture to the monsoon rainfall via the monsoon circulation is much weaker than the feedback of F^{net} discussed above.

4.2. Ventilation

The summer monsoon rainfall shown in Fig. 1 does not extend as far poleward as the corresponding F^{net} indicates in Fig. 3. Even with saturated soil moisture, the monsoon rain zone still cannot extend as far poleward as F^{net} indicates. This implies that mechanisms other than direct solar heating are working to limit the poleward extent of the summer monsoon rain zones. According to (3), the horizontal advection of MSE, $-v \cdot \nabla(T + q)$, is a possible cause. At midlatitudes, the dominant winds in the column average are westerly, and these westerly winds are largely set by global-scale dynamics. In summer, these cross-continental westerlies bring cold and low moist static-energy air from ocean regions due to ocean heat storage in the western part of the continent. The transport of the cold and low moist static-energy air limits the poleward extent of the summer monsoon rain zones. This import of low moist static-energy air into the continental regions from ocean regions is defined as ventilation (CNS). In the study by CNS, strong negative $-v \cdot \nabla(T + q)$ dominates the summer continent. This indicates that the ventilation mechanism removes the heating of F^{net} efficiently and limits the poleward extent of the summer monsoon rain zone. The advection of low moist static-energy air suppresses convection mostly over the western part of the continent, so the ventilation mechanism creates an east–west asymmetry in the summer monsoon precipitation [Fig. 5(b)].

4.3. Interactive Rodwell–Hoskins mechanism

Another mechanism controlling the distribution of the summer monsoon rainfall is the interactive Rodwell–Hoskins (IRH) mechanism. Rodwell and Hoskins (1996, 2001) proposed a monsoon-desert-like pattern to explain the association of the Asian summer monsoon and the Sahara desert. Monsoon heating induces a Rossy wave to the west associated with subsidence, which disfavors convection in this region. Thus, this process not only limits the poleward extent of the monsoon rain zone but also creates an east–west asymmetry of the monsoon rain zone. The term "interactive" emphasizes the interaction of the monsoon heating and the monsoon circulation via $-v \cdot \nabla(T + q)$, which further enhances the associated subsidence (CNS). The summer monsoon circulation is a cyclonic circulation with a continental scale associated with the convective heating of the summer monsoon. The eastern branch of the summer monsoon circulation transports moist and high MSE air from the south to the eastern part of the continent, and the convection over this region is enhanced. Meanwhile, the western branch of the circulation transports dry and low MSE air from the north to the western part of the continents, which suppresses convection over this region. The enhanced monsoon rainfall over the eastern part of the continent further strengthens the Rossby wave subsidence to the western part the continent, so the convection over the western part of the continent is suppressed more. This interaction with the monsoon circulation further enhances the east–west asymmetry [Fig. 5(b)]. This circulation-induced feedback to the asymmetry has also been discussed by Xie and Saiki (1999).

4.4. Land-ocean heating contrast

As discussed above, the feedback of the summer monsoon circulation plays an important role in the poleward extent of the summer monsoon rain zones. The strength of the monsoon circulation is determined by the horizontal pressure gradient between the continent and the neighboring oceans. This pressure gradient is strongly linked to the tropospheric temperature gradient induced by the land–ocean heating contrast, so the tropospheric temperature gradient is often used to represent the strength of the summer monsoon circulation (Li and Yanai, 1996). Thus, processes that control F^{net} over land and ocean, such as land surface conditions and ocean heat transport, are important in determining how strong the land–ocean heating contrast is. Land surface conditions, such as surface albedo associated with surface type and vegetation, can affect atmospheric temperature over land via F^{net} and change the pressure gradient between continental regions and ocean regions. For instance, small land surface albedo reflects less solar radiation into the space, and then F^{net} increases. The increased F^{net} not only enhances convection through the local processes discussed in Subsec. 4.1, but also increases the land–ocean heating contrast and the corresponding pressure and temperature gradients. Thus, the large-scale convergence over the continent is enhanced and the low-level cyclonic summer monsoon circulation is strengthened, and so the summer monsoon rainfall increases and the rain zones also extend farther poleward through the IRH mechanism [Fig. 5(c)]. The east–west asymmetry also becomes more significant.

Ocean heat transport, on the other hand, can also modify the land–ocean heating contrast via F_s^{net}. It is controlled by ocean circulation. Usually, tropical oceans are a heat sink to the atmosphere because of ocean heat transport from the tropics to higher latitudes and ocean heat storage, so tropical oceans need energy from the atmosphere to balance the loss, i.e. $F_s^{\mathrm{net}} > 0$. Due to $F_s^{\mathrm{net}} \approx 0$ over land, F^{net} over continental regions is usually positive and is larger than over ocean regions according to (5), and so a land–ocean heating contrast is created. The more efficient ocean heat transport at lower latitudes tends to induce the stronger cooling effect over ocean regions. The land–ocean heating contrast is then enhanced [Fig. 5(c)]. The enhanced land–ocean contrast favors convection over continental regions through the large-scale convergence. Thus, climate variations that can/change/F_s^{net} over the neighboring oceans, such as in ENSO events, affect the poleward extent of the summer monsoon rain zones via the feedback of the monsoon

circulation. This change of ocean heat transport can also directly modify local convection via the change of SST. Warmer SST implies less ocean heat transport out of the region. In other words, warm SST, such as in the western Pacific, tends to enhance convection via F^{net} and heats the troposphere above, and so the local precipitation is enhanced.

5. Examination on the Asian Summer Monsoon System

Based on the mechanisms discussed in the previous section, we next examine the impacts of those mechanisms on the Asian summer monsoon, which is the biggest and strongest summer monsoon in the world. The Asian continent is the largest continent in the world and contains the highest mountains, namely the Tibetan Plateau. To the south of the Asian continent, the Indian Ocean covers the tropical region. This land–sea configuration is different from those in the idealized monsoon study (CNS) that was used to discuss these summer monsoon mechanisms in the previous section. Thus, the impacts of those mechanisms on the summer monsoon discussed in the idealized case may be modified when applied to the Asian summer monsoon case. Before using QTCM for studying a more realistic summer monsoon, we note that due to the simplicity of QTCM, the summer monsoon simulated by QTCM might not be as good as those in GCMs, especially at local scales. For instance, QTCM does not have distinct maximum precipitation over the Bay of Bengal. However, the large-scale aspects of the rainfall pattern, including the northward extent along the east coast, tend to be simulated, so QTCM is good enough for studying the mechanisms discussed in the previous section for the Asian summer monsoon.

5.1. *Local processes*

Without the feedback of large-scale circulation, two local effects that can affect the summer monsoon rainfall are F^{net} and soil moisture via evaporation. Unlike the example of the African case discussed in Subsec. 4.1, the effect of F^{net} cannot be easily separated from the feedback of the monsoon circulation. In the African case, the observed F^{net} is close to zero because of high surface albedo. Modifying F^{net} can create an impact that is stronger than the feedback via the monsoon circulation, so the effect of F^{net} can be examined. However, F^{net} over the summer Asian continent is much larger than that over Africa, so as the associated land–ocean contrast. Thus, modifying F^{net} in Asia produces a dominant effect from the feedback of the Asian summer monsoon circulation, and the direct F^{net} effect via convection in the MSE equation is weaker. Therefore, we discuss only the local effect of soil moisture here, not the F^{net} effect.

The existence of the Tibetan Plateau is important for producing a realistic Asian summer monsoon (e.g. Flohn, 1968; Hahn and Manabe, 1975; He *et al.*, 1987; Murakami, 1987; Yanai *et al.*, 1992; Ye, 1981). However, experiments without the topographic effect can still give us some idea of the mechanisms for the Asian summer monsoon. Thus, QTCM without topography is used to examine the soil moisture effect, and the result of the control run is shown in Fig. 6(a). Comparison with the observed Asian summer monsoon (Fig. 1), excluding the Tibetan Plateau, may produce some caveats, such as no distinct feature of maximum precipitation over the Bay of Bengal and the west of the Indian peninsula and a little higher precipitation over the northeast coast of Asia. When the soil moisture over the Asian continent is saturated, the corresponding precipitation not only increases but also extends farther northward. Only the very center of the Asian continent is still dry. In the saturated soil moisture experiment, the evaporation over the Asian continent is enhanced due to larger moisture supply from soil and the local precipitation is then increased. The stronger local evaporation also cools the

surface temperature over the Asian continent. Note that the colder surface temperature shows little influence on changing the monsoon circulation because of the near-zero net surface heat flux F_s^{net}. Meanwhile, the tropospheric temperature over South Asia becomes warmer due to the convective heating of the enhanced summer monsoon rainfall. This warmer tropospheric temperature creates a low pressure anomaly, and the associated anomalous cyclonic circulation over South Asia. This anomalous circulation transports moisture from the south (not shown) and further enhances the rainfall over the southern and southeastern parts of Asia. Both local soil moisture and the feedback of the convection-induced anomalous circulation affect the Asian summer monsoon rainfall and its northward extension.

5.2. *Ventilation and interactive Rodwell–Hoskins mechanisms*

To examine the ventilation mechanism, the horizontal advection of moisture associated with v_ψ is suppressed in the QTCM simulations. We note that such experiments suppress not only the effect of ventilation but also the partial effect of the IRH mechanism that includes the effect of the moisture and MSE transports by the monsoon circulation. The summer monsoon rainfall is increased substantially, and the rain zone extends much farther northward [Fig. 6(c)]. However, the rainfall over the northeast coast of the Asian continent disappears when $v_\psi \cdot \nabla (T+q)$ is suppressed. The ventilation mechanism does suppress the Asian summer monsoon rainfall in most Asian monsoon regions, except for East Asia including Korea and Japan. Over this region, the moisture transport is associated with the monsoon circulation in the IRH mechanism, so the horizontal moisture transport is a moisture source for the monsoon rainfall, not a moisture sink. Thus, the precipitation over East Asia is weakened when suppressing $v_\psi \cdot \nabla (T + q)$.

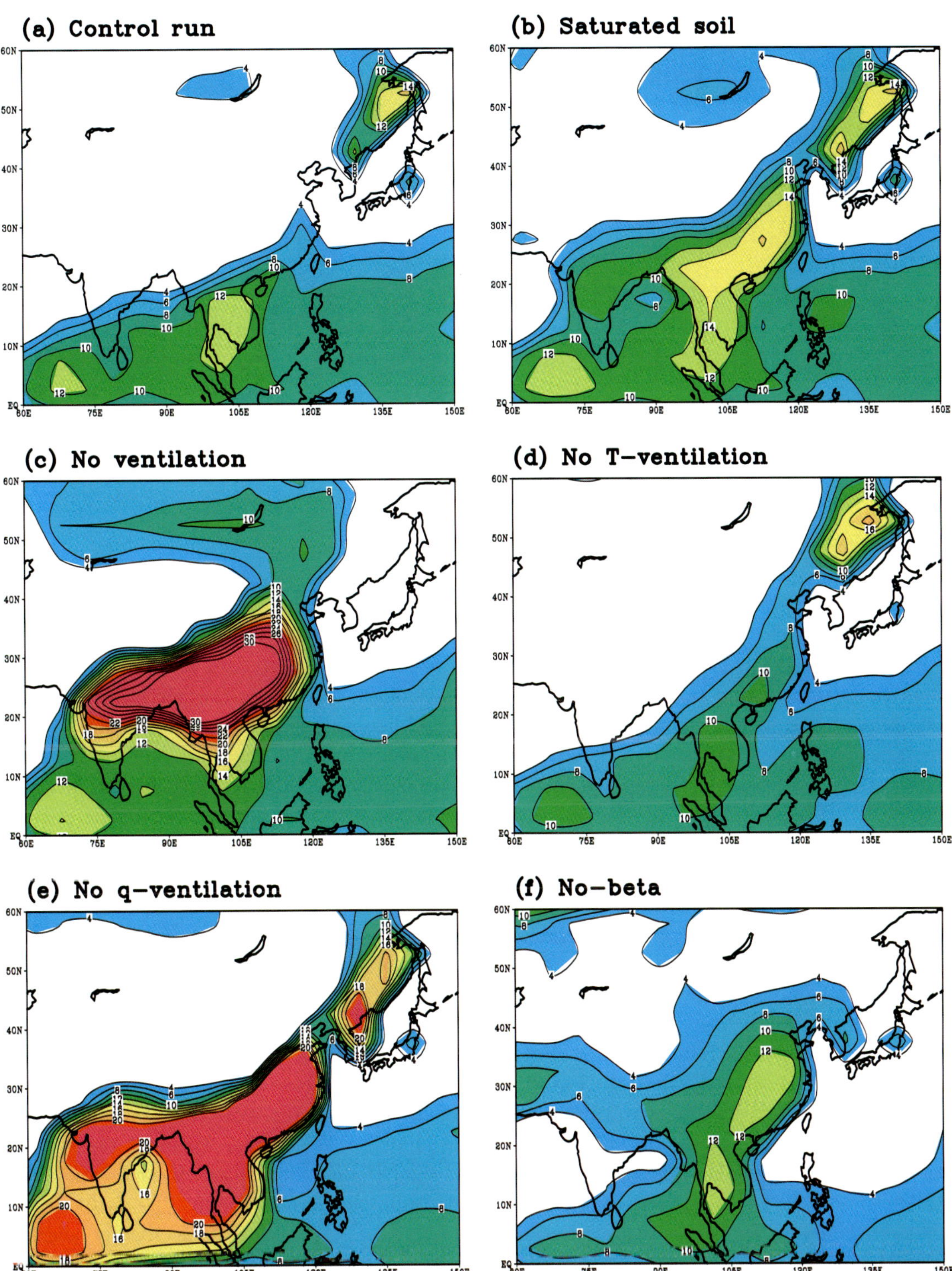

Figure 6. July precipitation for experiments over the Asian region: (a) control run, (b) saturated soil moisture, (c) no ventilation, (d) no T ventilation, (e) no q ventilation, (f) no β, (g) no ventilation and no β, and (h) no ventilation and partial β (from CN03).

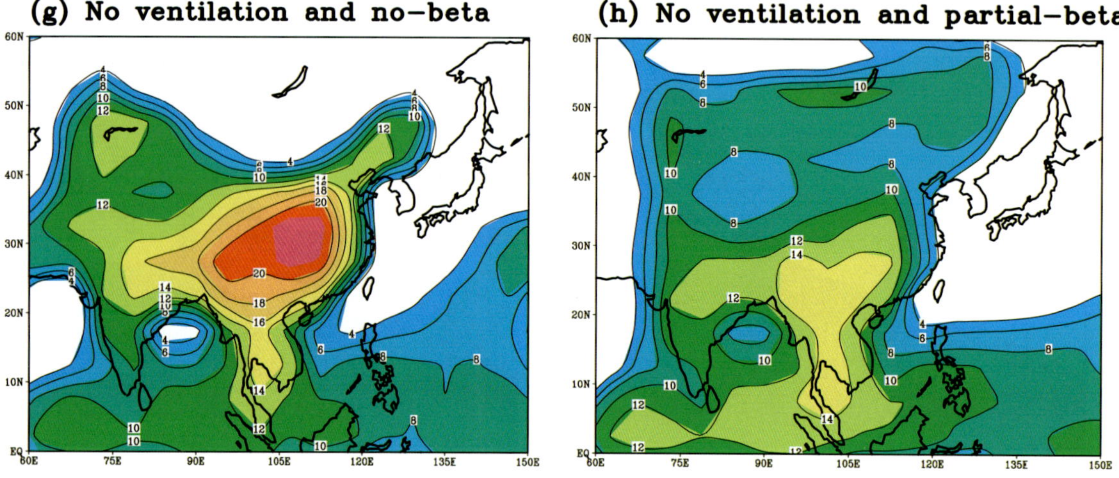

Figure 6. (*Continued*)

The term $v_\psi \cdot \nabla(T + q)$ includes two components: the moisture transport $v_\psi \cdot \nabla q$ and the temperature transport $v_\psi \cdot \nabla T$. To examine the effects of these two components respectively, $v_\psi \cdot \nabla T$ and $v_\psi \cdot \nabla q$ are suppressed separately in Figs. 6(d) and 6(e). When $v_\psi \cdot \nabla T$ is suppressed, the Asian summer monsoon rain zone does extend a little northward, but concentrates over the southeastern part of Asia [Fig. 6(d)]. The amplitude of the monsoon rainfall does not show any sign of enhancement. When $v_\psi \cdot \nabla q$ is suppressed, on the other hand, the change of the Asian summer monsoon is substantial [Fig. 6(e)]. The monsoon rainfall increases significantly, and the rain zone extends northward not only over the southeastern part of Asia but also South Asia. The pattern of the Asian summer monsoon rainfall in Fig. 6(e) is very similar to the pattern shown in Fig. 6(c) when both moisture and temperature transports are suppressed, except for the rainfall over East Asia. This implies that the ventilation mechanism in the Asian summer monsoon is contributed mainly by the moisture transport rather than the temperature transport.

To examine the IRH mechanism, the β effect is suppressed by dropping the terms that contain $f - f_0$ in the momentum equation, where f_0

is a reference value of Coriolis parameter f. Figure 6(f) shows that the Asian summer monsoon rain zone extends much farther northward, particularly over the western part of the Asian continent. The Rossby-wave-induced subsidence does affect the northward extent of the Asian summer monsoon rain zone. However, the amplitude of the monsoon rainfall does not change too much. Over higher latitudes, between 40°N and 50°N in particular, the precipitation is still relatively low, which may be due to the ventilation by cross-continental westerly flow. When both of the ventilation and IRH mechanisms are suppressed, the rainfall over the target region becomes very similar to what F^{net} indicates [Fig. 6(g)]. In other words, F^{net} initiates the summer monsoon rainfall, but the ventilation and IRH mechanisms determine the northern boundary of the Asian summer monsoon rain zone and induce the east–west asymmetry. We further examine the effect of a local Hadley circulation. In this experiment, we suppress only the term $(f - f_0)k \times \bar{v}$, not $(f - f_0)k \times v'$, where $\bar{v}$ is zonally averaged across the target region and v' is the departure from this zonal average. Figure 6(h) shows a similar pattern of the summer monsoon rainfall to Fig. 6(g). With or without the local Hadley circulation, the summer monsoon rainfall does not change too much. This implies that descent in a local Hadley circulation does not limit the northward extent of the summer monsoon rain zone.

5.3. *Land–ocean heating contrast*

5.3.1. *Land surface condition*

Two variables that affect land–ocean heating contrast are examined here: surface albedo and the Tibetan Plateau. Surface albedo can be affected by the surface type, such as snow cover. Many studies (e.g. Bamzai and Shukla, 1999; Dickson, 1984; Douville and Royer, 1996; Hahn and Shukla, 1976; Kripalani *et al.*, 1996; Liu and Yanai, 2002; Meehl, 1994; Yanai and Li, 1994) discussed an inverse relationship between the Eurasian snow cover and the succeeding Asian summer monsoon. Strong snowfall in the previous winter and spring cools the Asian continent and delays the Asian summer monsoon onset in the coming season. Besides snow cover, vegetation is also related to surface albedo. For instance, desert regions have higher surface albedo, while grass has lower surface albedo.

Figure 7(a) shows the rainfall changes when the surface albedo is reduced over Eurasia: the less the surface albedo, the stronger the monsoon rainfall. The cyclonic summer monsoon circulation is enhanced when the surface albedo is small, while the Pacific subtropical high extends more westward. The monsoon rain zone is then pushed northward, especially over the eastern part of the Asian continent, and so the tilting angle of the monsoon rain zone becomes larger. Meanwhile, the upper tropospheric temperature (200–500 hPa) over the Asian continent becomes warmer, and so the meridional temperature gradient is enhanced [Fig. 7(b)]. The maximum warm tropospheric temperature anomalies are over the east coast of the Asian continent, with a little eastward shift relative to the area with the changed surface albedo, which is over the entire Eurasian

continent. This is due to the temperature advection by the cross-continental flow, which cools the troposphere over the western part of the continent and heats the troposphere over the eastern part of the continent [Fig. 7(c)]. This increased meridional temperature gradient is associated with the strengthened Asian summer monsoon rainfall and circulation (CHOU; Li and Yanai, 1996; Webster *et al.*, 1998). The positive moisture transport is also consistent with the increased rainfall over the eastern part of the continent. This indicates that the enhanced monsoon rainfall is associated with the feedback of the monsoon circulation via the horizontal moisture transport. Besides positive precipitation anomalies, negative precipitation anomalies are also found over tropical Africa and the western North Pacific including the South China Sea, just south of the enhanced monsoon rainfall. The negative rainfall anomalies over Africa are associated with the IRH mechanism: a Rossby-wave-induced subsidence (Rodwell and Hoskins, 1996, 2001) and the feedback of the anomalous moisture transport [Fig. 7(d)]. The negative rainfall anomalies over the western North Pacific, on the other hand, are induced by dry advection associated with the strengthened Pacific subtropical high.

Next we examine the effect of the Tibetan Plateau. Here we focus only on the thermal effect of the Tibetan Plateau, an elevated heating source, and exclude the mechanical effect. Thus, some caveats, such as the turning flow around the Plateau and the associated advection, may not be correctly simulated. However, we should still be able to examine the land–ocean heating contrast induced by the Tibetan Plateau in such experiments. When the mountains are uplifted, the monsoon rainfall is enhanced and the rain zone extends farther north [Fig. 8(a)]. The upper tropospheric temperature increases over the Asian continent with a more concentrated warming pattern than that in the surface albedo experiments. A maximum warming is found over the eastern part of the

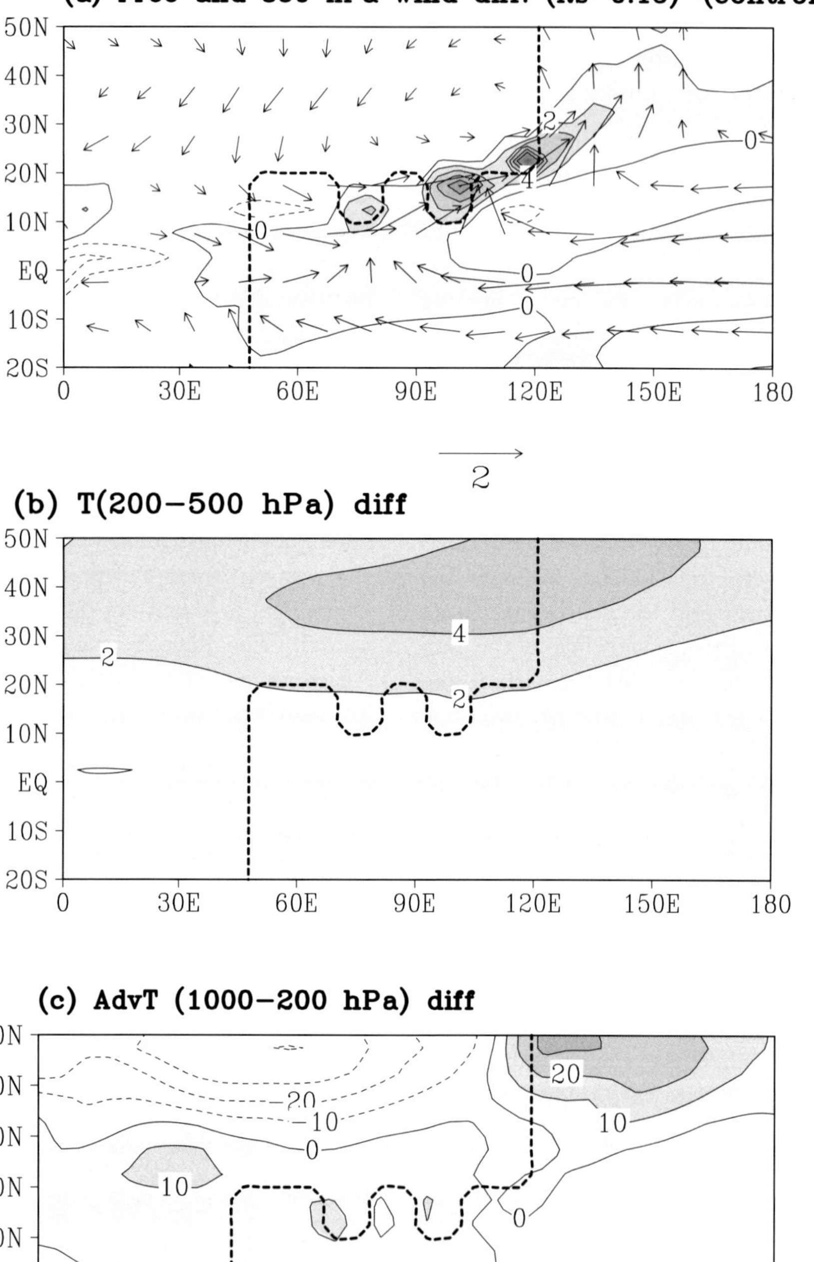

Figure 7. Differences between surface albedo $R_s = 0.15$ and $R_s = 0.25$ (control run) for (a) July precipitation and winds at 850 hPa (from CN03), (b) the upper tropospheric temperature (from CN03), (c) horizontal advection of temperature $-\langle v \cdot \nabla T \rangle$ and (d) horizontal advection of moisture $-\langle v \cdot \nabla q \rangle$.

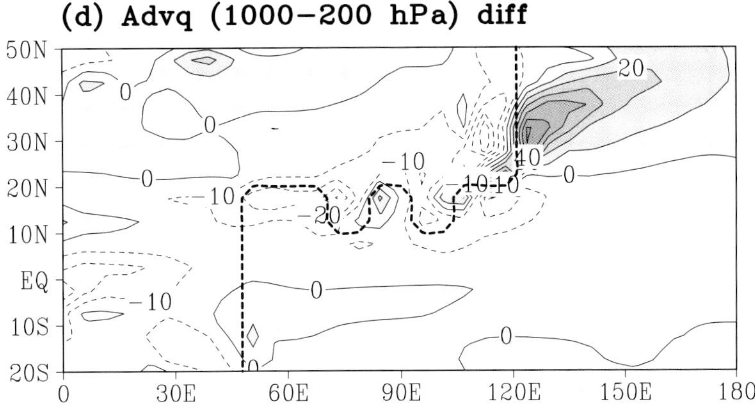

Figure 7. (*Continued*)

Tibetan Plateau. This warming associated with the uplift of the Tibetan Plateau enhances the cyclonic monsoon circulation over the Asian continent and induces the westward extent of the Pacific subtropical high. This low-level wind anomaly pattern is similar to the surface albedo experiments in Fig. 7(a). However, the anomalous cyclonic monsoon circulation [Fig. 8(a)] is much stronger and more concentrated than that in Fig. 7(a). The anomalous monsoon circulation did not reach as far west as in the surface albedo experiments. This is due to the more concentrated warm tropospheric temperature anomalies. The anomalous monsoon circulation enhances the precipitation over the eastern part of the Asian continent, and reduces the precipitation over the western part of the Asian continent via the IRH mechanism [Figs. 8(c) and 8(d)]. The ventilation effect also suppresses convection over the western part of the Asian continent [Figs. 8(c) and 8(d)]. The westward extent of the subtropical high transports dry air from the Pacific to reduce the precipitation over the southeastern part of the Asian continent and the neighboring oceans, including the South China Sea and the western North Pacific [Fig. 8(d)]. Comparing both the advection terms of temperature and moisture, $-\langle v \cdot \nabla q \rangle$ is more important than $-\langle v \cdot \nabla T \rangle$. The term $-\langle v \cdot \nabla q \rangle$ is also responsible for inducing

the dipole pattern of the rainfall anomalies over the east coast of the Asia continent (Fig. 8), since their patterns are very similar over this region.

5.3.2. *Ocean heat transport*

To examine the effects of ocean heat transport, zonally symmetric Q flux anomalies are prescribed with a maximum anomaly at the equator. These Q flux anomalies induce roughly symmetric SST anomalies with maximum negative SST anomalies at the equator [Fig. 9(a)]. The cold SST anomalies create colder tropospheric temperature with maximum anomalies at the equator [Fig. 9(b)]. These cold tropospheric temperature anomalies increase the meridional gradient of the tropospheric temperature globally, so the responses to the enhanced meridional temperature gradient cover entire continental areas including Eurasia and Africa. Figure 9(c) shows that positive precipitation anomalies extend from East Asia southwestward to Africa. This long positive precipitation anomaly band indicates a northward extent of the monsoon rain zones from Asian to Africa. The corresponding anomalous cyclonic circulation is not only over Eurasia and Africa but also extends farther westward into the Atlantic. The easterly trade winds at lower

Chia Chou and Jien-Yi Tu

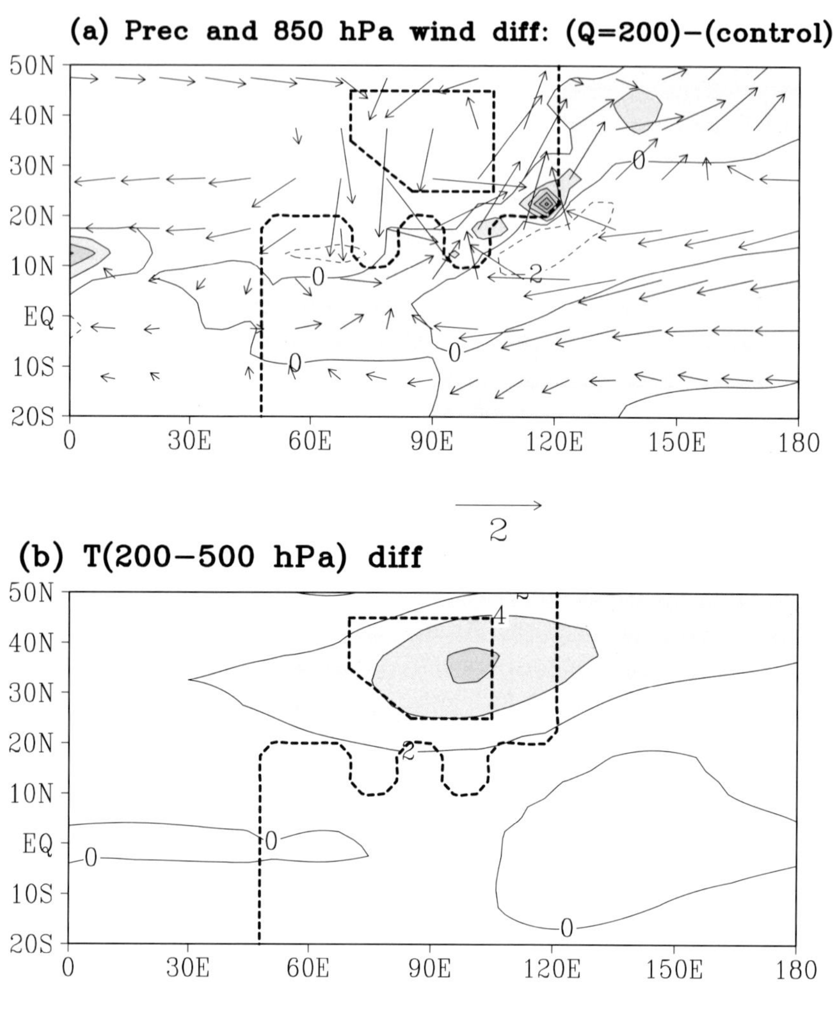

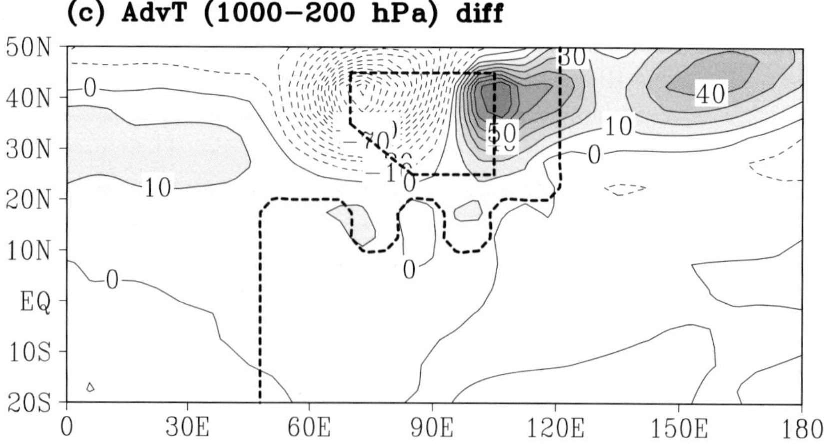

Figure 8. As in Fig. 7, but for differences between the experiments with and without the Tibetan Plateau. The dashed line on the Asian continent indicates the range of the Tibetan Plateau.

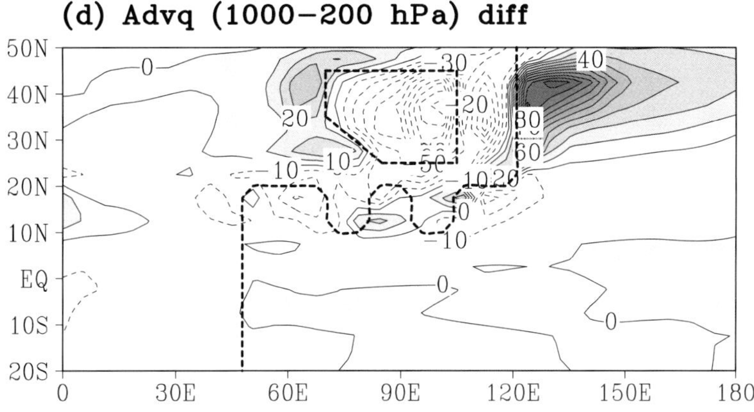

Figure 8. (*Continued*)

latitudes are also enhanced, and the anomalies extend westward to those in the Atlantic. These easterlies are associated with the North Pacific subtropical high and transport dry air to suppress convection over the south coast of Asia and the Indian Ocean. Overall, the changes in precipitation and winds are similar to those in the previous experiments for examining surface albedo and topography effects (Figs. 7 and 8), but with a much broader scale. Further examining the horizontal advection terms, the moisture transport $-\langle v \cdot \nabla q \rangle$ is also the dominant factor for the precipitation anomalies [Figs. 9(d) and 9(e)].

In a more realistic case, such as ENSO, the effect of ocean heat transport is also examined. By specifying anomalous Q flux, typical El Niño–like SST anomalies are simulated [Fig. 10(a)] and these SST anomalies create longitudinal gradients of SST in the tropics. The associated tropospheric temperature anomalies also show a longitudinal gradient anomaly over the north of 20°N [Fig. 10(b)]. Because of wave dynamics in the tropics (Chiang and Sobel, 2002; Su and Neelin, 2002), widespread warm tropospheric temperature anomalies (Wallace *et al.*, 1998) reduce the meridional tropospheric temperature gradient over the Asian continent. The anomalous meridional and longitudinal tropospheric temperature gradients induce

a weak anomalous anticyclonic circulation over the Asian continent and reduce the strength of the Asian summer monsoon [Fig. 10(c)] when one is judging from a monsoon index that is often used to present the Asian summer monsoon as a whole (Webster and Yang, 1992). However, this anomalous anticyclonic circulation exists only over the north of 20°N. Over the south of 20°N, another anticyclonic circulation anomaly is found over South Asia and the Indian Ocean. This anomalous anticyclone is associated with a Rossby wave response to the weaker convection induced by the cold SST anomalies over the western Pacific and the Indian Ocean. The cold SST anomalies reduce the precipitation over the Indian Ocean. The decreased monsoon heating induces an anticyclonic circulation anomaly associated with the Rossby wave to its west. The anticyclonic circulation anomaly over South Asia and the Indian Ocean pushes the monsoon rain zone northward. Thus, the monsoon rain zone extends northward when the meridional tropospheric temperature gradient and the summer monsoon circulation are weakened. This implies that local SST is important to the subregional monsoon system when the tropospheric temperature gradient does not change too much. Further examining the horizontal advection terms, $-\langle v \cdot \nabla q \rangle$ is still the dominant effect on the positive precipitation

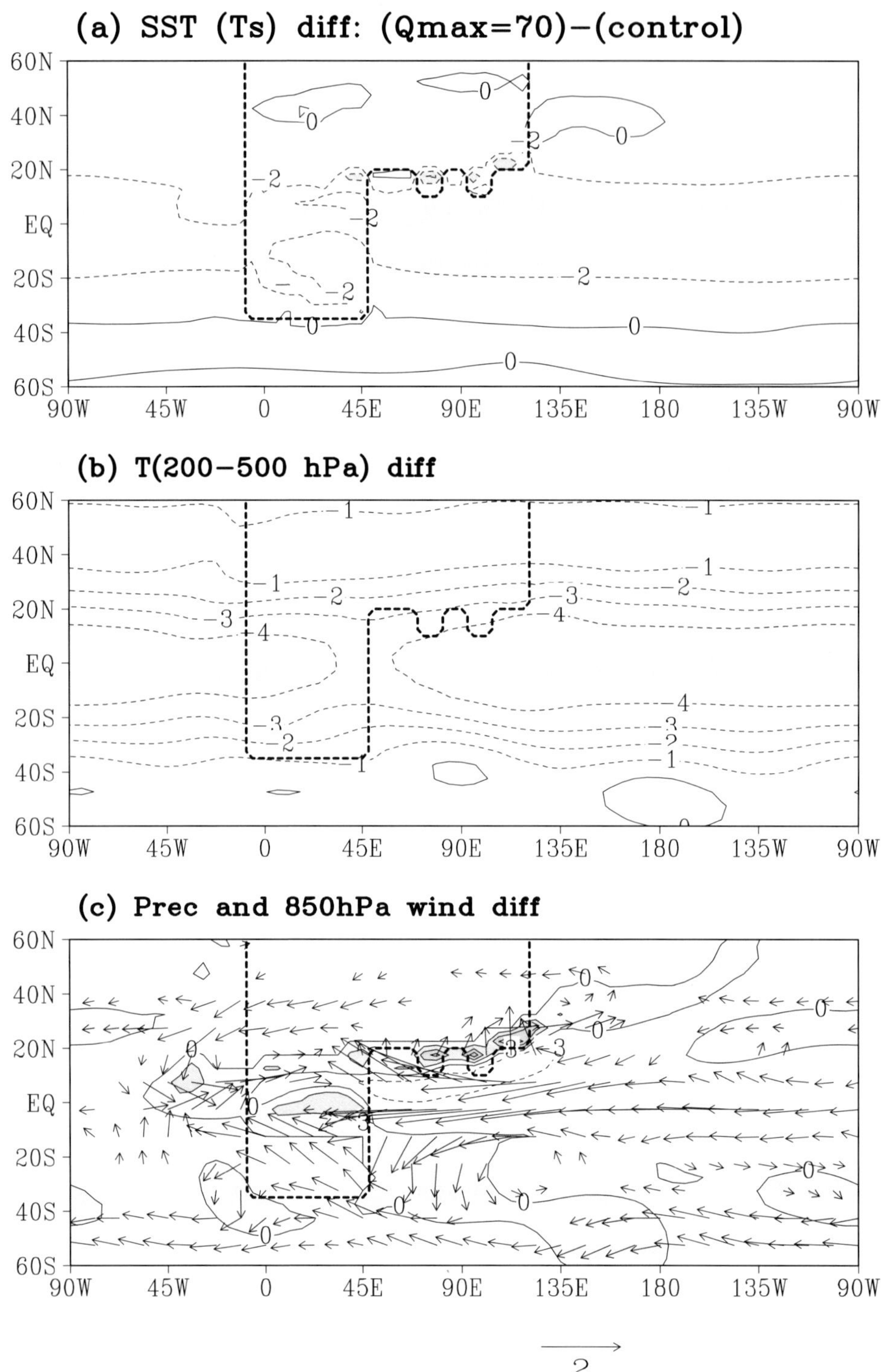

Figure 9. Differences between maximum Q flux = $70\,\mathrm{Wm}^{-2}$ and maximum Q flux = $50\,\mathrm{Wm}^{-2}$ (control run) for (a) July SST anomalies (from CHOU), (b) the upper tropospheric temperature (from CHOU), (c) July precipitation and winds at 850 hPa (from CHOU), (d) horizontal advection of temperature, $-\langle v \cdot \nabla T \rangle$ and (e) horizontal advection of moisture $-\langle v \cdot \nabla q \rangle$.

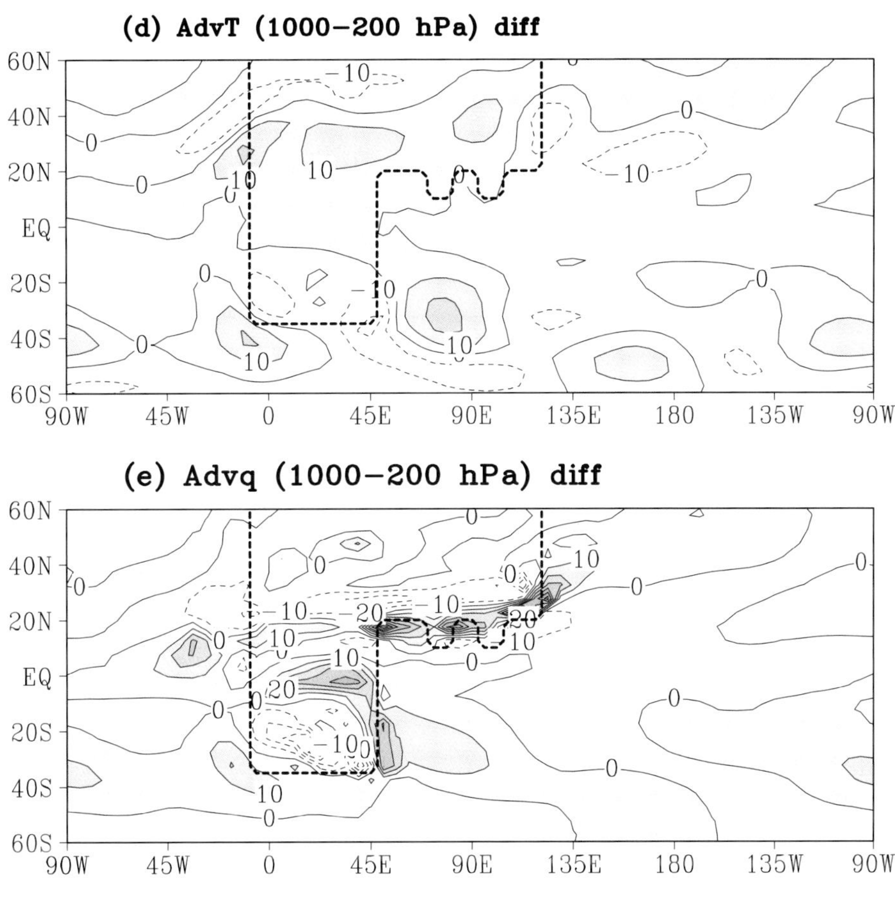

Figure 9. (*Continued*)

anomalies. Note that the El Niño like case discussed here does not correspond to either the ENSO growing or decaying phase, because the experiments here are all at equilibrium.

6. Conclusion

The monsoon has great differences in detail in different regions, but at its heart is the seasonal movement of land convection zones. The convection is directly affected by evaporation in the moisture equation (2). More importantly, it is also associated with a dynamical feedback via the net energy into the atmosphere F^{net} in the MSE equation (3). However, the summer monsoon includes not only the convection but also the associated monsoon circulation which

is related to land–ocean heating contrast. We have argued here that the differences in the net energy into the atmospheric column F^{net} are the leading cause of the land–ocean contrast, not surface temperature. Land–ocean heating contrast creates pressure gradients between continents and the neighboring oceans and induces the cyclonic summer monsoon circulation over the continental regions. The pressure gradient can be represented by the tropospheric temperature gradient. This monsoon circulation, along with mean flow, the cross-continental westerly winds in the subtropics and at midlatitudes, modifies the summer monsoon rainfall pattern via the horizontal advection of temperature and moisture $-\langle v\cdot\nabla T\rangle$ and $-\langle v\cdot\nabla q\rangle$. The advection associated with the monsoon circulation is

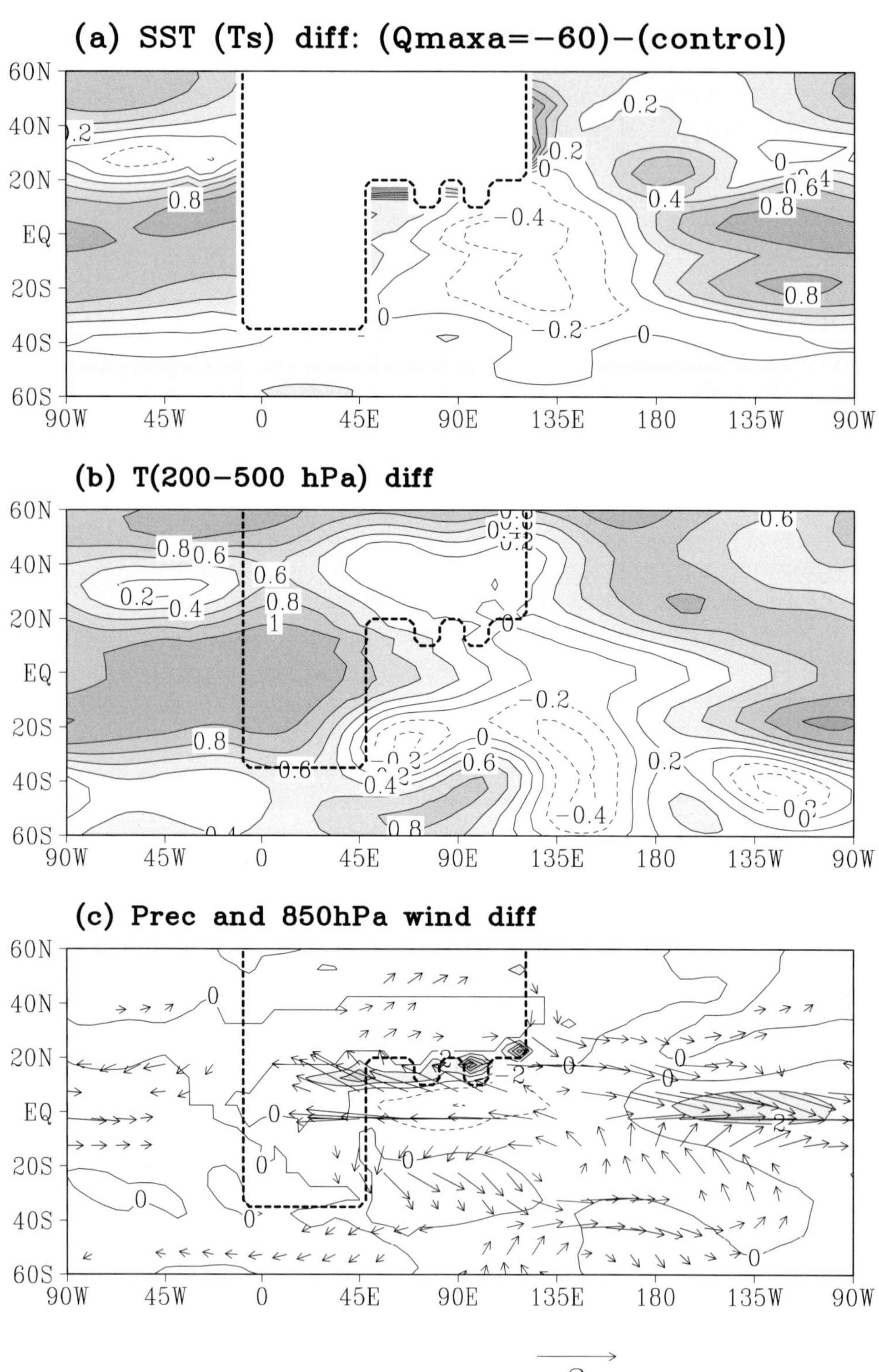

Figure 10. As in Fig. 9, but for differences between anomalous maximum Q flux $= -60\,\mathrm{W\,m}^{-2}$ and the control run.

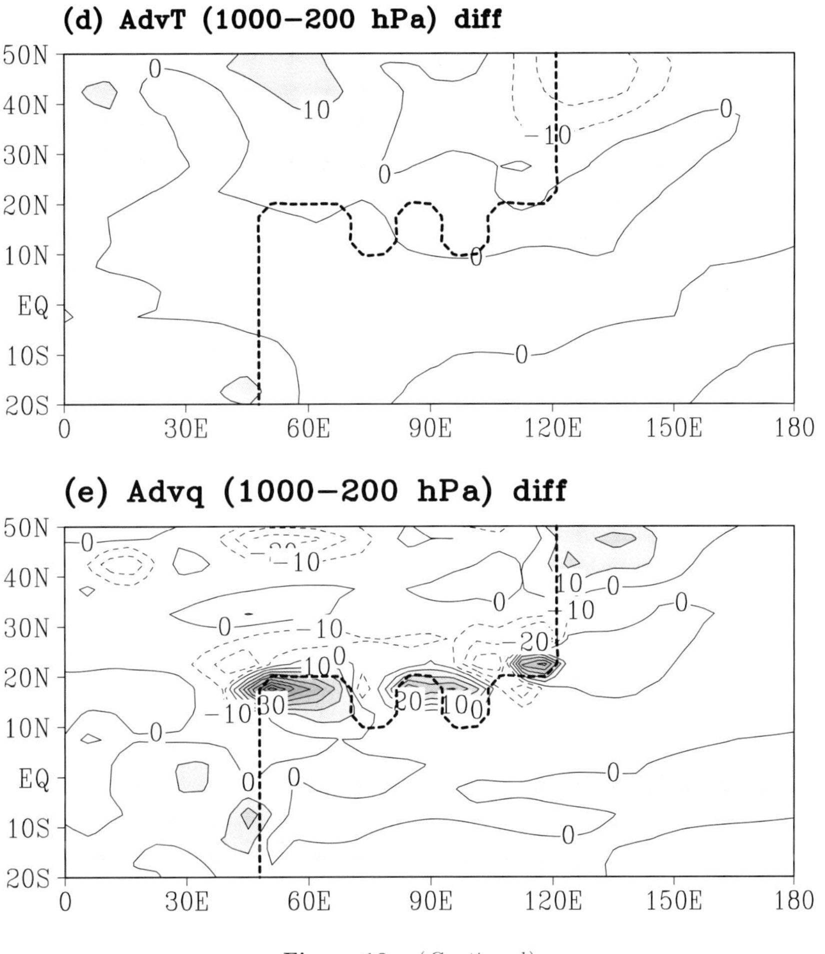

Figure 10.   (*Continued*)

termed the interactive Rodwell–Hoskin (IRH) mechanism, while the advection associated with the cross-continental winds is termed ventilation. Thus, the local effects of evaporation and F^{net} initiate the convection and the effects of ventilation, and the IRH mechanism modifies the summer monsoon rainfall pattern.

In this study, we first analyzed the local effects, including evaporation and the dynamical feedback of F^{net}. The evaporation over land is associated with soil moisture. Soil moisture does affect the summer monsoon rainfall and extend the rain zone farther northward, but its impact is not as dominant as the IRH mechanism and ventilation. We then examined the ventilation

and the IRH mechanism. Both mechanisms not only determine the poleward extent of the summer monsoon rain zone but also induce the east–west asymmetry of the rain zone. In ventilation, the cross-continental flow transports low moist static-energy air from ocean regions to the western part of continent and disfavors convection over the region. In the IRH mechanism, the Rossby wave subsidence induced by the monsoon convective heating and the interaction with the monsoon circulation suppress convection over the western part of the continent and enhance convection over the eastern part of the continent. In the IRH mechanism, the horizontal transport of temperature and moisture is a feedback via the monsoon circulation, whose strength is determined by land–ocean heating contrast. In other words, land–ocean heating contrast also affects the poleward extent of the summer monsoon rain zone through the IRH mechanism. Both land and ocean conditions affect land–ocean heating contrast. The land condition can be influenced by topography and surface type associated with soil moisture and surface albedo. The ocean condition is affected by ocean heat transport and ocean heat storage.

We further examined impacts of those mechanisms on the Asian summer monsoon. The increase of soil moisture enhances the Asian summer monsoon rainfall and extends its rain zone northward. However, ventilation and the IRH mechanism have more substantial impacts on the Asian summer monsoon rainfall. For these mechanisms, the effect associated with the moisture transport is particularly important to the Asian summer monsoon. When suppressing ventilation and the IRH mechanism, the pattern of the Asian summer monsoon becomes similar to the F^{net} pattern even with unsaturated soil moisture. In land–ocean heating contrast, the elevated heating source of the Tibetan Plateau enhances the meridional gradient of tropospheric temperature and strengthens the Asian summer monsoon circulation. Land surface albedo associated with snow cover and ocean heat transport also modify land–ocean heating contrast and change the Asian summer monsoon circulation and the northward extent of the monsoon rain zone. Local SST can sometimes be important to subregional monsoon systems when the variation of land–ocean heating contrast is weak, and thus it can also affect the northward extent of the Asian summer monsoon rain zone.

Acknowledgements

This work was supported under National Science Council grant 93-2111-M-001-001. It has been a pleasure to work with Prof. J. D. Neelin, Dr. H. Su, Prof. J.-Y. Yu and others.

[Received 29 December 2005; Revised 5 September 2007; Accepted 17 September 2007.]

References

Bamzai, A. S., and J. Shukla, 1999: Relation between Eurasian snow cover, snow depth, and the Indian summer monsoon: an observational study. *J. Climate*, **12**, 3117–3132.

Betts, A. K., and M. J. Miller, 1993: The Betts–Miller scheme. In *The Representation of Cumulus Convection in Numerical Models of the Atmosphere*, Meteor. Monogr., No. 46 (Amer. Meteor. Soc.), pp. 107–121.

Camberlin, P., 1997: Rainfall anomalies in the source region of the Nile and their connection with the Indian summer monsoon. *J. Climate*, **10**, 1380–1392.

Chang, C.-P., P. Harr, and J. Ju, 2001: Possible roles of Atlantic circulations on the weakening Indian monsoon rainfall–ENSO relationship. *J. Climate*, **14**, 2376–2380.

Chiang, J. C. H., and A. H. Sobel, 2002: Tropical tropospheric temperature variations caused by ENSO and their influence on the remote tropical climate. *J. Climate*, **15**, 2616–2631.

Chou, C., and J. D. Neelin, 1996: Linearization of a longwave radiation scheme for intermediate tropical atmospheric models. *J. Geophys. Res.*, **101**, 15129–15145.

Chou, C., and J. D. Neelin, 1999: Cirrus detrainment-temperature feedback. *Geophys. Res. Lett.*, **26**, 1295–1298.

Chou, C., and J. D. Neelin, 2001: Mechanisms limiting the southward extent of the South American summer monsoon. *Geophys. Res. Lett.*, **28**, 2433–2436.

Chou, C., J. D. Neelin, and H. Su., 2001: Ocean–atmosphere–land feedbacks in an idealized monsoon. *Quart. J. Roy. Meteor. Soc.*, **127**, 1869–1891.

Chou, C., 2003: Land–sea heating contrast in an idealized Asian summer monsoon. *Climate Dyn.*, **21**, 11–25.

Chou, C., and, J. D. Neelin, 2003: Mechanisms limiting the northward extent of the northern summer monsoons over North America, Asia, and Africa. *J. Climate*, **16**, 406–425.

Chou, C., J.-T. Tu, and J.-Y. Yu, 2003: Interannual variability of the western North Pacific summer monsoon: differences between ENSO and non-ENSO years. *J. Climate*, **16**, 2275–2287.

Dickson, R. R., 1984: Eurasian snow cover versus Indian monsoon rainfall — an extension of the Hahn–Shukla results. *J. Climate Appl. Meteorol.*, **23**, 171–173.

Doney, S. C., W. G. Large, and F. O. Bryan, 1998: Surface ocean fluxes and water-mass transformation rates in the coupled NCAR climate system model. *J. Climate*, **11**, 1420–1441.

Douville, H., and J. F. Royer, 1996: Sensitivity of the Asian summer monsoon to an anomalous Eurasian snow cover within the Météo-France GCM. *Climate Dyn.*, **12**, 449–466.

Flohn, H., 1957: Large-scale aspects of the "summer monsoon" in south and east Asia. *J. Meteoro. Soc. Japan*, **35**, 180–186.

Flohn, H., 1968: Contributions to a meteorology of the Tibetan Highlands. Atmospheric Science Paper No. 130 (Colorado State University, Fort Collins), 120 pp.

Fu, Q., and K. N. Liou, 1993: Parameterization of the radiative properties of cirrus clouds. *J. Atmos. Sci.*, **50**, 2008–2025.

Goswami, B. N., B. Krishnamurthy, and H. Annamalai, 1999: A broad-scale circulation index for interannual variability of the Indian summer monsoon. *Quart. J. Roy. Meteorol. Soc.*, **125**, 611–633.

Hahn, D. G., and S. Manabe, 1975: The role of mountains in the south Asian monsoon circulation. *J. Atmos. Sci.*, **32**, 1515–1541.

Hahn, D. G., and J. Shukla, 1976: An apparent relationship between Eurasian snow cover and Indian monsoon rainfall. *J. Atmos. Sci.*, **33**, 2461–2462.

Hansen, J., I. Fung, A. Lacis, D. Rind, S. Lenedeff, R. Ruedy, and G. Russell, 1988: Global climate changes as forecast by Goddard Institute for Space Studies three-dimensional model. *J. Geophys. Res.*, **93D**, 9341–9364.

Hansen, J., M. Sato, and R. Ruedy, 1997: Radiative forcing and climate response. *J. Geophys. Res.*, **102**, 6831–6864.

Harshvardhan, R. Davies, D. A. Randall, and T. G. Corsetti, 1987: A fast radiation parameterization for general circulation models. *J. Geophys. Res.*, **92**, 1009–1016.

He, H., J. W. McGinnis, Z. Song, and M. Yanai, 1987: Onset of the Asian monsoon in 1979 and the effect of the Tibetan Plateau. *Mon. Wea. Rev.*, **115**, 1966–1995.

Keith, D. W., 1995: Meridional energy transport: uncertainty in zonal means. *Tellus*, **47A**, 30–44.

Kripalani, R. H., S. V. Singh, A. D. Vernekar, and V. Thapliyal, 1996: Empirical study on Nimbus-7 snow mass and Indian summer monsoon rainfall. *Int. J. Climatol.*, **16**, 23–34.

Lau, K.-M., K.-M. Kim, and S. Yang, 2000: Dynamical and boundary forcing characteristics of regional components of the Asian summer monsoon. *J. Climate*, **13**, 2461–2482.

Li, C., and M. Yanai, 1996: The onset and interannual variability of the Asian summer monsoon in relation to land–sea thermal contrast. *J. Climate*, **9**, 358–375.

Liu, X., and M. Yanai, 2001: Relationship between the Indian monsoon rainfall and the tropospheric temperature over the Eurasian continent. *Quart. J. Roy. Meteor. Soc.*, **127**, 909–937.

Liu, X., and M. Yanai, 2002: Influence of Eurasian spring snow cover on Asian summer rainfall. *Int. J. Climatol.*, **22**, 1075–1089.

Lofgren, B. M., 1995: Surface albedo-climate feedback simulated using two-way coupling. *J. Climate*, **8**, 2543–2562.

Meehl, G. A., 1992: Effect of tropical topography on global climate. *Annu. Rev. Earth Planet. Sci.*, **20**, 85–112.

Meehl, G. A., 1994: Influence of the land surface in the Asian summer monsoon: external conditions versus internal feedbacks. *J. Climate*, **7**, 1033–1049.

Miller, J. R., G. L. Russell, and L.-C. Tsang, 1983: Annual oceanic heat transports computed from an atmospheric model. *Dyn. Atmos. Oceans*, **7**, 95–109.

Murakami, T., 1987: Orography and monsoons. In *Monsoons*, eds. J. S. Fein and P. L. Stephens, (John Wiley, New York), pp. 331–364.

Neelin, J. D., and J.-Y. Yu, 1994: Modes of tropical variability under convective adjustment and the Madden–Julian Oscillation. Part I: Analytical theory. *J. Atmos. Sci.*, **51**, 1876–1894.

Neelin, J. D., and N. Zeng, 2000: A quasi-equilibrium tropical circulation model — formulation. *J. Atmos. Sci.*, **57**, 1741–1766.

Nicholson, S., 2000: Land surface processes and Sahel climate. *Rev. Geophys.*, **38**, 117–139.

Rodwell, M. J., and B. J. Hoskins, 1996: Monsoons and the dynamics of deserts. *Quart. J. Meteor. Soc.*, **122**, 185–1404.

Rodwell, M. J., and B. J. Hoskins, 2001: Subtropical anticyclones and summer monsoons. *J. Climate*, **14**, 3192–3211.

Russell, G. L., J. R. Miller, and L.-C. Tsang, 1985: Seasonal oceanic heat transports computed from an atmospheric model. *Dyn. Atmos. Oceans*, **9**, 253–271.

Stevens, B., J. Duan, J. C. McWilliams, M. Munnich, and J. D. Neelin, 2002: Entrainment, Rayleigh friction, and boundary layer winds over the tropical Pacific. *J. Climate*, **15**, 30–44.

Su, H., and J. D. Neelin, 2002: Teleconnection mechanisms for tropical Pacific descent anomalies during El Ñino. *J. Atmos. Sci.*, **59**, 2694–2712.

Wang, B., and Z. Fan, 1999: Choice of south Asian summer monsoon indices. *Bull. Amer. Meteor. Soc.*, **80**, 629–638.

Wang, B., and LinHo, 2002: Rainy season of the Asian-Pacific summer monsoon. *J. Climate*, **15**, 386–398.

Webster, P. J., 1987: The elementary monsoon. In *Monsoons*, eds. J. S. Fein and P. L. Stephens, (John Wiley, New York), pp. 3–32.

Webster, P. J., and S. Yang, 1992: Monsoon and ENSO: selectively interactive systems. *Quart. J. Roy. Meteor. Soc.*, **118**, 877–926.

Webster, P. J., V. O. Magañna, T. N. Palmer, J. Shukla, R. A. Tomas, M. Yanai, and T. Yasunari, 1998: Monsoons: processes, predictability, and the prospects for prediction. *J. Geophys. Res.*, **103**, 14451–14510.

Wu, G., and Y. Zhang, 1998: Tibetan Plateau forcing and the timing of the monsoon onset over South Asia and the South China Sea. *Mon. Wea. Rev.*, **126**, 913–927.

Wu, R., and B. Wang, 2000: Interannual variability of summer monsoon onset over the western North Pacific and the underlying processes. *J. Climate*, **13**, 2483–2501.

Xie, P., and P. A. Arkin, 1997: Global precipitation: a 17-year monthly analysis based on gauge observations, satellite estimates, and numerical outputs. *Bull. Amer. Meteor. Soc.*, **78**, 2539–2558.

Xie, S.-P., and N. Saiki, 1999: Abrupt onset and slow seasonal evolution of summer monsoon in an idealized GCM simulation. *J. Meteoro. Soc. Japan*, **77**, 949–968.

Xue, Y., and J. Shukla, 1993: The influence of land surface properties on Sahel climate. Part I: Desertification. *J. Climate*, **5**, 2232–2245.

Xue, Y., M.-H. Juang, W. Li, S. Prince, R. DeFries, Y. Jiao, and R. Vasic, 2004: Role of land surface processes in monsoon development: East Asia and West Africa. *J. Geophy. Res.*, **109**, 3105, doi:10.1029/2003JD003556.

Yanai, M., C. Li, and Z. Song, 1992: Seasonal heating of the Tibetan Plateau and its effects on the evolution of the Asian summer monsoon. *J. Meteor. Soc. Japan*, **70**, 319–351.

Yanai, M., and C. Li, 1994: Mechanism of heating and the boundary layer over the Tibetan Plateau. *Mon. Wea. Rev.*, **122**, 305–323.

Yang, S., P. J. Webster, and M. Dong, 1992: Longitudinal heating gradient: another possible factor influencing the intensity of the Asian summer monsoon circulation. *Adv. Atmos. Sci.*, **9**, 397–410.

Yang, S., and K.-M. Lau, 1998: Influences of sea surface temperature and ground wetness on Asian summer monsoon. *J. Climate*, **11**, 3230–3246.

Yang, S., K.-M. Lau, S.-H. Yoo, J. L. Kinter, K. Miyakoda, and C.-H. Ho, 2004: Upstream subtropical signals preceding the Asian summer monsoon circulation. *J. Climate*, **17**, 4213–4229.

Ye, D., 1981: Some characteristics of the summer circulation over the Qinghai–Xizang (Tibet) Plateau and its neighborhood. *Bull. Amer. Meteor. Soc.*, **62**, 14–19.

Young, J. A., 1987: Physics of monsoons: the current view. In *Monsoons*, eds. J. S. Fein and P. L. Stephens (John Wiley, New York), pp. 211–243.

Yu, J.-Y., and J. D. Neelin, 1994: Modes of tropical variability under convective adjustment and the Madden–Julian Oscillation. Part II: Numerical results. *J. Atmos. Sci.*, **51**, 1895–1914.

Yu, J.-Y., C. Chou, and J. D. Neelin, 1998: Estimating the gross moist stability of the tropical atmosphere. *J. Atmos. Sci.*, **55**, 1354–1372.

Zeng, N., J. D. Neelin, and C. Chou, 2000: A quasi-equilibrium tropical circulation model — implementation and simulation. *J. Atmos. Sci.*, **57**, 1767–1796.

Coupling of the Intraseasonal Oscillation with the Tropical Cyclone in the Western North Pacific during the 2004 Typhoon Season

Huang-Hsiung Hsu, An-Kai Lo, Ching-Hui Hung, Wen-Shung Kau and Chun-Chieh Wu

Department of Atmospheric Sciences, National Taiwan University, Taipei, Taiwan
hsu@atmos1.as.ntu.edu.tw

Yun-Lan Chen

Weather Forecast Center, Central Weather Bureau, Taipei, Taiwan

A strong in-phase relationship between the intraseasonal oscillation (ISO) and the tropical cyclone (TC) was observed in the tropical western North Pacific from June through October 2004. The ISO, which is characterized by the fluctuations in the East Asian monsoon trough and the Pacific subtropical anticyclone, modulated the TC activity and led to the spatial and temporal clustering of TCs during its cyclonic phase. This clustering of strong TC vortices contributed significant positive vorticity during the cyclonic phase of the ISO and therefore enlarged the intraseasonal variance of 850 hPa vorticity. This result indicates that a significant percentage (larger than 50%) of observed intraseasonal variance along the clustered TC tracks in the tropical western North Pacific came from TCs. Numerical simulation confirmed that the presence and enhancement of TCs in the models enlarged the simulated intraseasonal variance. This implies that the contribution of TCs has to be taken into account to correctly estimate and interpret the intraseasonal variability in the tropical western North Pacific.

1. Introduction

The tropical western North Pacific is an active region for the tropical cyclone (TC) and the intraseasonal oscillation (ISO) in the northern summer (e.g. Lau and Chan, 1986; Wang and Rui, 1990; Elsberry, 2004; Hsu, 2005). Many studies have revealed the modulation effect of the ISO on the TC activity in this region (e.g. Nakazawa, 1986; Heta, 1990; Liebmann, *et al.*, 1994; Maloney and Dickinson, 2003). According to these studies, TCs tend to cluster in the westerly and positive vorticity phase of the ISO in the lower troposphere. Under the circumstances, one would wonder whether the clustering of strong TC vortices may in turn increase the overall amplitude of positive vorticity in this phase, and therefore increase the intraseasonal variance (ISV) of vorticity. If this effect is in action, the ISV may not result entirely from the ISO itself. Instead, part of the variance may come from the clustered TC activity. This possibility has been explored and confirmed by Hsu *et al.* (2008), using the European Centre for Medium-range Weather Forecast reanalysis (ERA40, 1958–2002). This new finding suggests that the TC contribution must be considered to obtain a better understanding of the intraseasonal variability in the tropical western North Pacific.

The typhoon season (defined as June–October, JJASO) of 2004 was a unique season in the western North Pacific, in terms of the strong TC and ISO activity, and the in-phase relationship between the two. The most significant phenomenon was the record-breaking number (10) of typhoon landfalls in Japan. Another interesting feature was the temporal clustering, on the intraseasonal time scale, of the tropical cyclone genesis during the summer and early autumn. It was this strong in-phase relationship that led to the record-breaking typhoon landfalls in Japan, because the clustering effect of the ISO on TC resulted in the tendency for the typhoons to move along similar tracks (Nakazawa, 2006). In view of this close relationship between the TC and the ISO, this unique season enables an excellent case study of the coupling of the ISO and the TC. It is the goal of this study to explore the relationship between the TC and the ISO in this particular season, and to evaluate the contribution of the TC to the ISV.

In this study, an unconventional approach applied by Hsu *et al.* (2008) was employed in removing TCs from the global analysis. The potential contribution of TCs was estimated, which was defined as the variance difference between the original and TC-removed vorticity fields at 850 hPa. Numerical experiments based on a regional model and a general circulation model (GCM), with/without TCs and with enhanced TCs, were also carried out. Since similar results were obtained in GCM and regional model simulations, only the GCM results will be shown here for the sake of brevity. The approaches adopted in this study were designed to shed light on the TC contribution to the ISV along the TC tracks. Our results indicate that the differences in variance (i.e. the effect of TCs on the ISV) are large enough to be of concern on the intraseasonal time scale.

The arrangement of this article is as follows. Section 2 describes the data and methodology. ISO modulation on TCs is presented in Sec. 3, and the TCs' potential contribution to the ISV is reported in Sec. 4. Simulation results and conclusions are presented in Secs. 5 and 6, respectively.

2. Data and Methodology

The wind and mean sea level pressure (MSLP) data used in this study were retrieved from National Centers for Environmental Prediction (NCEP) Reanalysis I on a 2.5°-by-2.5° grid, while the TC statistics and best-track data were obtained from the Japanese Meteorological Agency (JMA) and the Joint Typhoon Warning Center (JTWC), respectively. A 32–76-day Butterworth band-pass filter (Kaylor, 1977) was applied to NCEP Reanalysis I to extract the intraseasonal fluctuations from the daily mean vorticity. The reason for the choice of the 32–76-day band will be presented in a later section. To reduce the end effect of the Butterworth filter, only data from 16 June to 15 October were used to calculate the ISV.

In reality, due to the possible nonlinear TC–climate interaction, it is impossible to exactly quantify the TC contribution. This essentially holds good for all studies involving multiple temporal and spatial scales. Despite this concern, temporal and spatial filtering has often been used to decompose a total field into perturbation and mean flow, or even into several subfields of distinct temporal or spatial scales. This linear-thinking approach has proved useful in providing important insights into climate processes, such as the wave–mean flow interaction. In order to evaluate the possible TC contribution to the ISV in the western North Pacific during the 2004 typhoon season, a spatial filtering approach was taken to remove the TC vortices from the 850 hPa vorticity. This filtering procedure is similar to the decomposition of the total field into perturbation and background flow, but is performed in a more sophisticated manner. The potential contribution of TCs to the ISV was estimated by calculating the difference between the ISV of the original and the TC-removed vorticity at 850 hPa.

Removal of TCs from the analysis data has been a common practice in typhoon and hurricane simulation and forecasting. The procedure, which has been used in the Geophysical Fluid Dynamics Laboratory (GFDL) hurricane prediction system (Kurihara *et al.*, 1993, 1995) and in typhoon simulations (Wu *et al.*, 2002) for enhancing the representation of the environmental field in the initial condition, has proved effective in improving the overall TC track forecast. Following the procedure proposed by Kurihara *et al.* (1993, 1995), the four-time daily 850 hPa winds associated with each TC, based on the JTWC best track, were subtracted from the 850 hPa wind field during the typhoon season.

The basic procedure, demonstrated in Fig. 1, is briefly described as follows. [See Kurihara *et al.* (1993, 1995) for details.] The zonal and meridional winds were individually separated into basic and disturbance fields using a smoothing operator. The winds associated with a TC were isolated in the filter domain, which

defines the extent of a TC in the global analysis, and subtracted from the disturbance field to create a non-TC component. In the procedure, 1200 km is specified as the radius of the filter domain. This does not mean that everything in the domain is identified as the TC component. It is simply the longest distance for the procedure to automatically search the effective radius of a TC in 24 directions (for every 15°) surrounding the TC center. When the radius of a TC, e.g. 500 km, is identified in a particular direction, the searching stops, and only the TC winds within the radius are subtracted. The reason for choosing a larger domain is to avoid missing the TC circulation in a TC (or typhoon) that has a large radius, and which would lead to underestimation of the TC circulation. After removing the TC component from the disturbance field, the non-TC component was added to the basic field to form the environmental flow, which is the TC-removed wind field in this study. The environmental flow outside the filter domain is identical to the original global analysis. The original and TC-removed vorticity fields were calculated based on the corresponding wind fields.

Hsu *et al.* (2008) demonstrated that the TC-removing procedure is able to correctly separate the TC and environmental components. One example is shown in Fig. 2 three typhoons appeared simultaneously in the western North Pacific on 28 August 2004. They appear as three isolated vortices (i.e. the TC component) in Fig. 2(b), while the environmental flow (background flow plus non-TC component) is clearly characterized by the zonally elongated monsoon trough [Fig. 2(c)] across the South China Sea and the Philippine Sea. This result, as in many other cases, indicates that the TC removal procedure removes mainly the TC vortex and accurately retains the large-scale circulation, along with the climate variability contributed by the large-scale fluctuations.

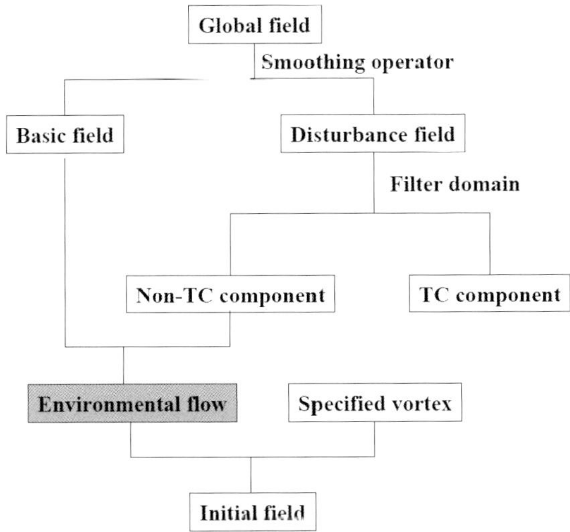

Figure 1.   Flow chart of the TC removal procedure.

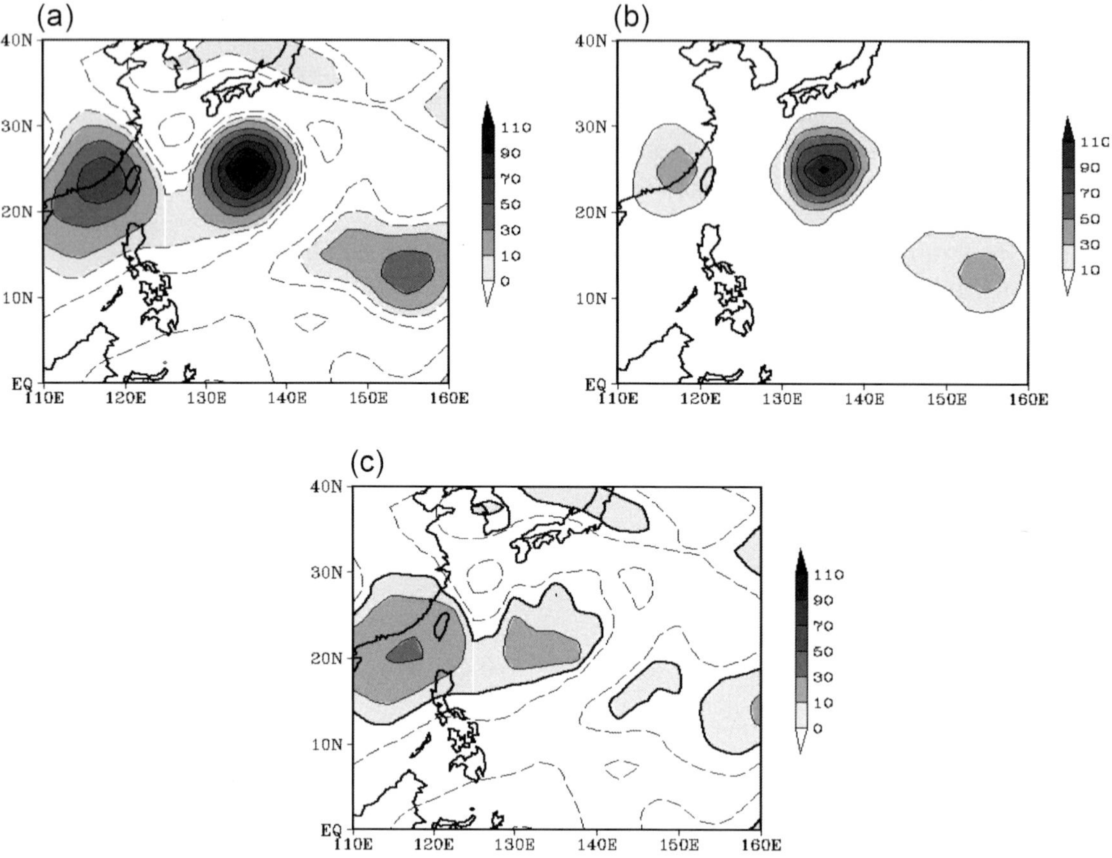

Figure 2. 850 hPa vorticity fields on 26 August 2004, when three typhoons were observed in the tropical western North Pacific: (a) total field, (b) TC and (c) environmental component. The unit for the contour is $10^{-6}\,\mathrm{s}^{-1}$.

One potential problem in removing TCs from the global analysis, such as NCEP Reanalysis I, is the possible mismatch between the JTWC TC track, which defines the center of a TC, and the position of the TC-corresponding vortex in NCEP Reanalysis I. This is partially attributed to the coarse resolution of the data set commonly used in climate study. The JTWC best track and NCEP Reanalysis I were examined, and a general consistency, although not an exact match, was found. The main goal of this study is to statistically assess the gross contribution of TCs to the climate variability in a large domain covering the entire tropical western North Pacific. A mismatch of a few degrees in latitude and longitude would not seriously affect the overall results. As will be seen later, the removed vorticity is collocated nicely with the TC tracks. The possible mismatch does not seem to cause problems in this study. TCs may be presented differently in different global analyses or in different resolutions. Hsu *et al.* (2008), who compared the results derived from the European Centre for Medium-range Weather Forecast reanalysis and the NCEP reanalysis, and also between different spatial resolutions, demonstrated the general consistency between the analyses and the effectiveness of the TC-removing procedure adopted in this study. Their results suggest that the overall results presented here are not affected by different analyses and spatial resolutions, although certain quantitative differences may be found.

Another concern is the underrepresentation of the wind speed and vorticity of TCs in the global analysis. This is an existing problem, which cannot be solved in this study. The study's goal is to demonstrate how TCs contribute significantly to the ISV, based on presently available global analysis. The results should be viewed as the TC effect represented in currently available global analysis, which has been widely used to estimate the ISV. If the exact location and strength of TCs are represented in the global analysis, the actual contribution of TCs will likely be larger than what was found in this study.

3. ISO and TC

Based on the JMA statistics, there were 29 named TCs (including tropical storms and typhoons) in the western North Pacific during the 2004 typhoon season, slightly more than the climatological mean of 26.7. Most of the TCs in this period tended to appear in clusters quasi-periodically, as noted in previous studies (e.g. Gray, 1979; Nakazawa, 2006). This clustering phenomenon is closely associated with the fluctuations of the monsoon trough and subtropical anticyclone (e.g. McBride, 1995; Harr and Elsberry, 1998; Elsberry, 2004). The tendency of TC occurrence in or near the East Asian (EA) monsoon trough was particularly evident in 2004. Climatologically, the movement of TCs can be roughly classified into two types: straight-moving track and recurving track. The former type of TC moves across the Philippine Sea in the northwest direction toward southern China and the Indochina peninsula, while the latter type recurves northeastward following a certain period of northwestward movement over the Philippine Sea. During the June–October period, the ratio between the straight-moving and the recurving track is about 1:1 for those TCs formed south of 20°N (Harr and Elsberry, 1998). However, most of the TCs in JJASO 2004 moved northwestward over the Philippine Sea, recurving northeastward when approaching the subtropics. This clustering phenomenon can be seen clearly in Fig. 3, which presents the

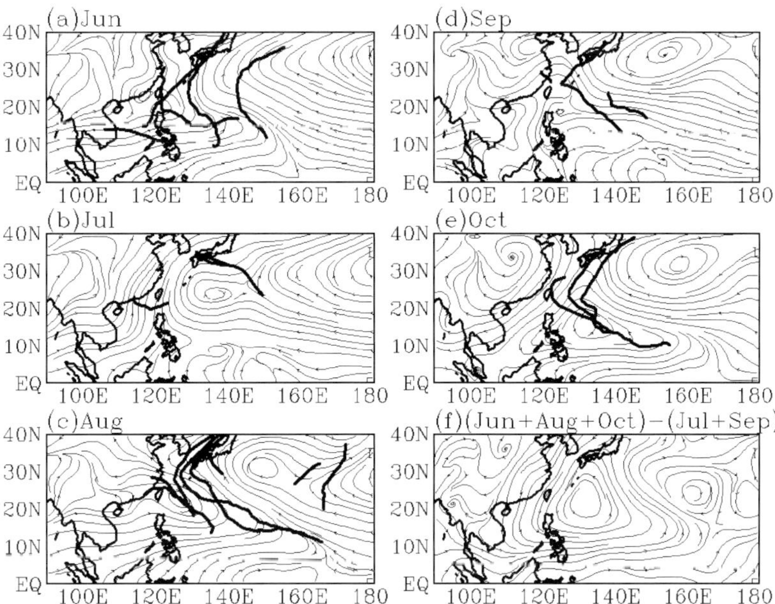

Figure 3. Monthly mean 850 hPa streamline and named TC tracks from (a) June to (e) October 2004, and (f) the difference between the June/August/October and July/September mean circulation at 850 hPa.

monthly-mean 850 hPa streamlines and the TC tracks for each month from June to October. Note that the recurving occurred most evidently in June, August, and October, when TCs were more active.

The EA monsoon trough extended southeastward from the Indochina peninsula to the Philippine Sea, and contracted westward intermittently from June to October. There were more-than-average numbers of tropical storms in June (5), August (8), and October (4), compared to the 1971–2000 climatological mean numbers of 1.7, 5.5, and 2.8, when the EA monsoon trough was active and extended further southeastward than normal. Fewer tropical storms occurred in July (2) and September (3), compared to the climatological mean numbers of 4.1 and 5.1, respectively, when the EA monsoon trough was weak and contracted westward. It is known that TCs tend to occur in or near the EA monsoon trough (e.g. Elsberry, 2004; Chen *et al.*, 2004). This relationship seems particularly evident in JJASO 2004. Most of the TCs also seemed inclined to move in a clockwise direction along the southern and western peripheries of the Pacific anticyclone. This movement pattern is similar to the recurve-south track identified by Harr and Elsberry (1998). It appears that the fluctuations of the EA monsoon trough and the Pacific anticyclone strongly modulated the TC activity in this particular season.

The intermittent occurrence of the eastward extension and westward contraction of the EA monsoon trough in JJASO 2004 was associated with strong ISO activity. To identify this intraseasonal signal, an index was designed to represent the fluctuation in the EA monsoon trough. The index is defined as the MSLP averaged over the region (120°E–150°E, 10°N–20°N), where the extension and contraction of the EA monsoon trough is most evident (Fig. 3). Wavelet analysis (Torrence and Compo, 1998) on this daily index from April to December 2004 was performed to identify the dominant

periodicity. As shown in Fig. 4(a), large fluctuations are evident in the 32–76-day and 11–27-day bands throughout the period. The accumulated variance explained by the 32–76-day perturbations accounted for 54.5% of the total variance in JJASO 2004, while the 11–27-day period accounted for 29.3%. The 32–76-day ISO was apparently the major fluctuation to affect the EA monsoon trough. In view of its dominance in variance, the following discussion will focus on the 32–76-day ISO. To further reveal the uniqueness of the this ISO in 2004, wavelet analysis was performed on the EA monsoon trough index during the June–October season from 1951 to 2004 annually. The result shown in Fig. 4(b) reveals that the 32–76-day variance in 2004 was the largest from 1951 to 2004, indicating the strongest intraseasonal fluctuation of the EA monsoon trough in this 54-year period. This variance (almost $3\,hPa^2$) is much larger than the second-largest variance (about $2\,hPa^2$) occurring in the 1979 summer, which was known as a summer of strong ISO activity (Lorenc, 1984). The JJASO of 2004 was obviously a unique season for the 32–76-day ISO. The reason for this large ISV is not clear and will be explored in other studies.

Spatial distribution of the 32–76-day ISV of MSLP in JJASO 2004, shown in Fig. 5(a), exhibits two centers of maximum variance: one over the Philippine Sea, and in the extratropical North Pacific. A comparison between Fig. 3 and Fig. 5(a) indicates that the large variances over the Philippine Sea and the extratropical North Pacific were associated with the movement and fluctuation of the EA monsoon trough and the Pacific anticyclone during the season, respectively. The shading shown in Fig. 5(a) indicates the percentile of the 32–76-day variance in JJASO 2004 at every point, compared to all JJASO variances in the 54-year period. The area exceeding the 90th percentile covers most of East Asia and the western North Pacific, while small variance appears in the Indochina peninsula, the South China Sea, and the equatorial

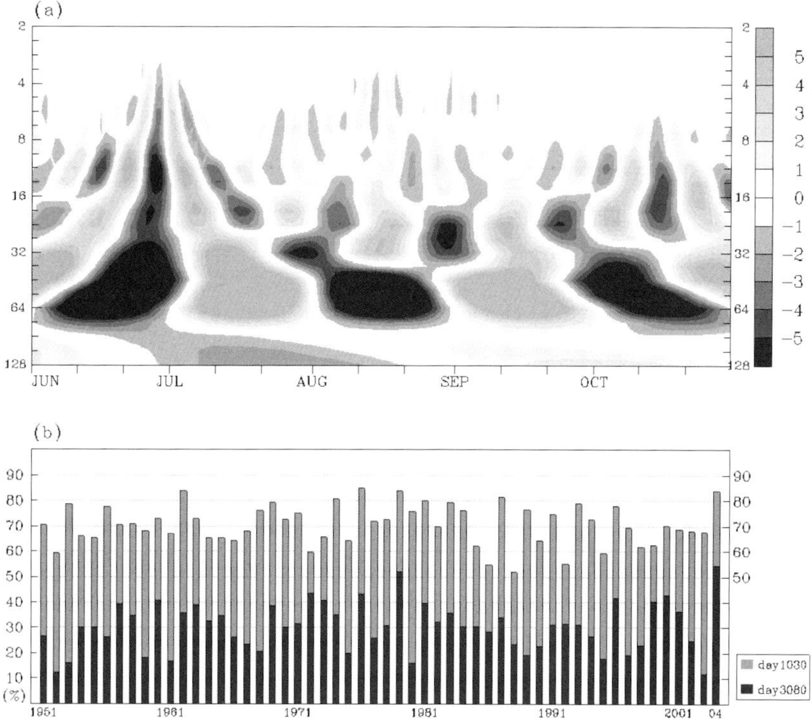

Figure 4. (a) Wavelet coefficients for the EA monsoon trough index (see text), (b) the percentages of the index variance explained by the 11–27-day (blue bar) and 32–76-day (red bar) bands for the 54 years from 1951 to 2004. The vertical axis in (a) and (b) denotes the period in days and percentage, respectively, while the horizontal axis in (a) and (b) denotes dates from June to October 2004 and years from 1951 to 2004, respectively.

western Pacific. The area chosen to construct the index is well inside the 95% region. The index designed to reflect the large fluctuation in the monsoon trough appears adequately chosen. The distribution shown in Fig. 5(a) indicates that the unusually active ISO in JJASO 2004 occurred not only in the EA monsoon trough but also in other EA summer monsoon regions (e.g. eastern China, Taiwan, Japan, and Korea).

The spatial distribution of the 32–76-day OLR variance is shown in Fig. 5(b) to reveal the ISV in convection. A large variance area exceeding the 90th percentile is observed in the tropical western North Pacific, located to the southeast of its MSLP counterpart. A large variance and high percentile area is also found near Japan, reflecting the large number of typhoons affecting Japan and possibly the

clustering effect of the ISO on the typhoon activity. These results indicate that the anomalously active intraseasonal fluctuations existed in both the circulation and convection fields during JJASO 2004.

The strong ISV of the EA monsoon trough and the Pacific anticyclone apparently resulted in the spatial and temporal clustering of TCs. This close relationship is shown in Fig. 6, which presents the composite of MSLP and the TC tracks during three cyclonic phase months (June, August, and October — upper panel) and two anticyclonic phase months (July and September — lower panel) appearing intermittently as shown in Fig. 3, respectively. The majority of TCs during JJASO occurred in the cyclonic phase, as indicated by the negative MSLP anomaly, and tended to take a recurring track, while only a few appeared in the anticyclonic

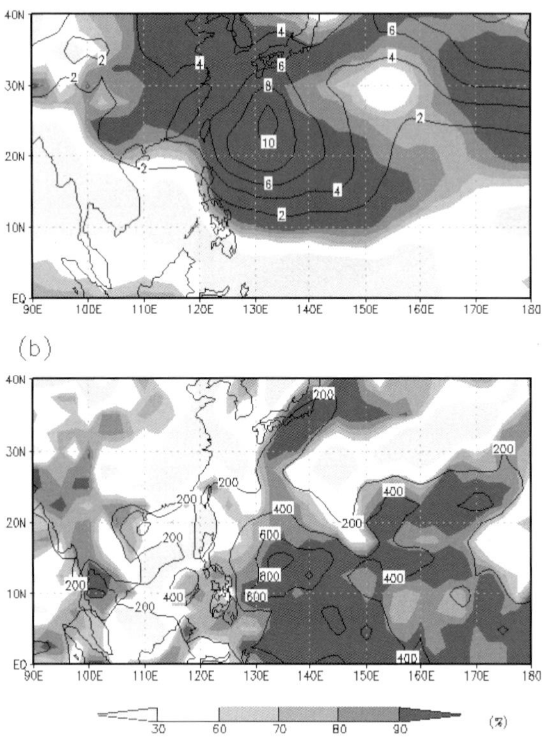

Figure 5. Spatial distribution of the 32–76-day filtered (a) MSLP and (b) OLR variance in JJASO 2004. The contour intervals are $2\,\mathrm{hPa}^2$ and $200\,(\mathrm{W/m}^2)^2$ for MSLP and OLR, respectively. Shading indicates the percentile of the 2004 variance in the 54-year period from 1951 to 2004.

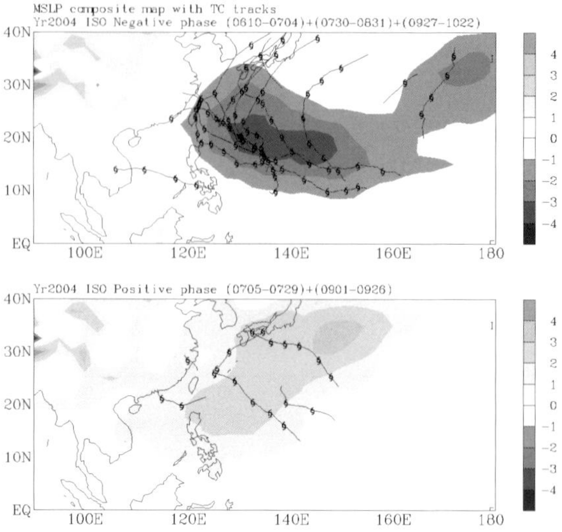

Figure 6. Composites of MSLP for (upper) three cyclonic phases and (lower) two anticyclonic phases. The TC tracks during these two phases are also marked. Periods chosen for composites are shown at the top of each figure.

phase in an environment of the positive MSLP anomaly.

The close ISO–TC relationship is further demonstrated in Fig. 7, which presents the 32–76-day fluctuation in the MSLP averaged between 120°E and 150°E, and the latitudinal positions of TSs that happened to be situated in this longitudinal band. A sequence of the 32–76-day ISO propagated northward from 10°N to 30°N regularly from June to October. Evidently, many more TSs appeared in the negative phase than in the positive phase of the ISO and moved northward in an anomalous low-pressure environment. As found in many previous studies, the genesis and track of the TCs in the tropical western North Pacific during JJASO 2004 were strongly modulated by the intermittent extension and retraction of the monsoon trough, which was in turn affected by the unusually strong ISO.

4. TC Contribution to the ISV

The potential contribution of TCs to the ISV was estimated by calculating the difference between the ISV was the original and the TC-removed vorticity at 850 hPa. As demonstrated in Fig. 2 and Hsu *et al.* (2008), the Kurihara scheme cleanly separates TCs from the background large-scale circulation. The variance contributed by TCs and large-scale circulation are also well separated. The original 32–76-day variance of 850 hPa vorticity presented in Fig. 8(a) reveals two major ISV regions along the TC tracks: one elongated region in the Philippine Sea and the other south of Japan. The former corresponds to the northwestward TC tracks, while the latter corresponds to the recurring TC tracks toward Japan. After the removal of TCs, the variance in these two regions reduced dramatically, as shown in Fig. 8(b), while the variance outside these two regions

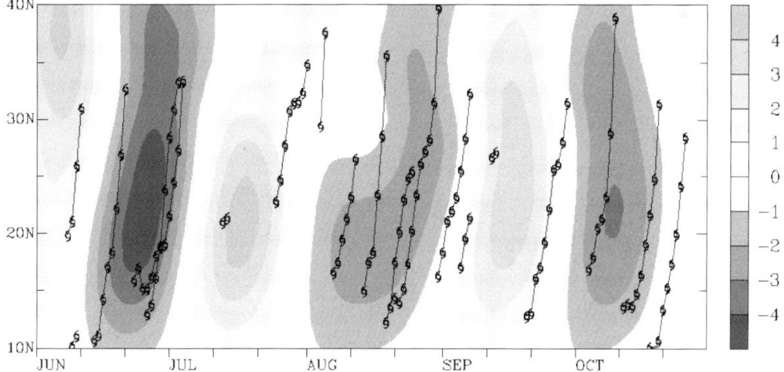

Figure 7. Hovmüller diagrams for the 32–76-day filtered MSLP averaged between 120°E and 150°E. The contour interval is 1 hPa. The latitudinal positions of the TCs that happened to be in 120°E–150°E are marked.

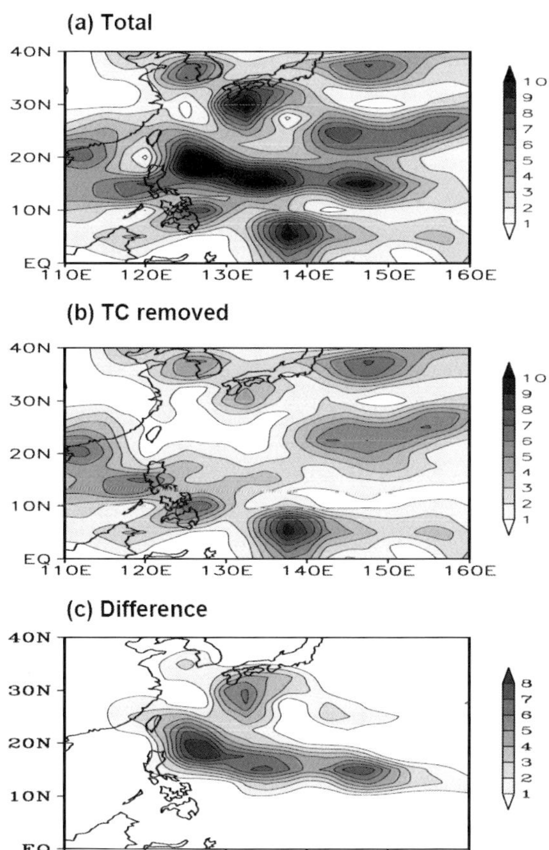

Figure 8. Intraseasonal (32–76-day) variance of the (a) original and (b) TC-removed 850 hPa vorticity. The difference between (a) and (b) is shown in (c). Only values greater than $1 \times 10^{-11}\,\mathrm{s}^{-2}$ are plotted.

remained almost unchanged. The variance difference shown in Fig. 8(c) indicates that the amount of the variance contributed by the clustered TCs can be as large as 50–80%.

To illustrate how the TCs contribute to the ISV in the statistical sense, the time series of the original and TC-removed 850 hPa vorticity was averaged over 125°E–140°E, 15°N–22.5°N, where the maximum variance is observed, and the difference between the two is shown in Fig. 9(a). Positive vorticity peaks occurred in groups in June, August, and October, indicating the clustering of TCs, while weak negative vorticity was observed during the TC-inactive period (July and September). Removal of TCs apparently results in a significantly reduced amplitude of the positive vorticity peaks, but it has no effect on negative vorticity for an obvious reason. This leads to the overall enlargement of positive vorticity during the TC-active periods, which recurred on the intraseasonal time scale, and therefore the increased ISV for the whole JJASO season. For example, the ISV in the above region drops from $5.9 \times 10^{-11}\,\mathrm{s}^{-1}$ to $2.2 \times 10^{-11}\,\mathrm{s}^{-1}$ after the TC removal. This significant reduction can also be seen in the spectra of the area-averaged vorticity presented in Fig. 9(b). While the original and the TC-removed spectral density exhibit qualitatively

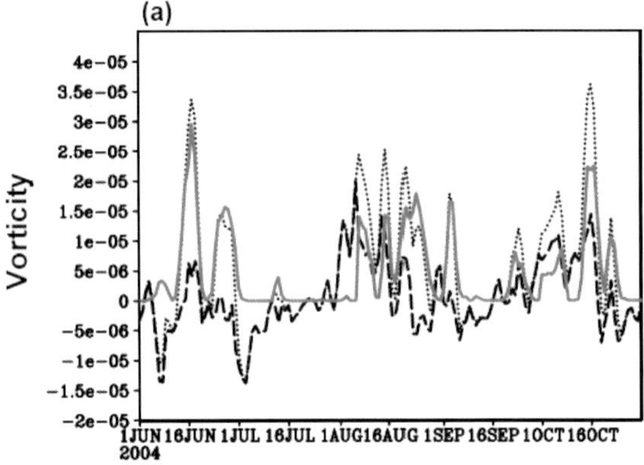

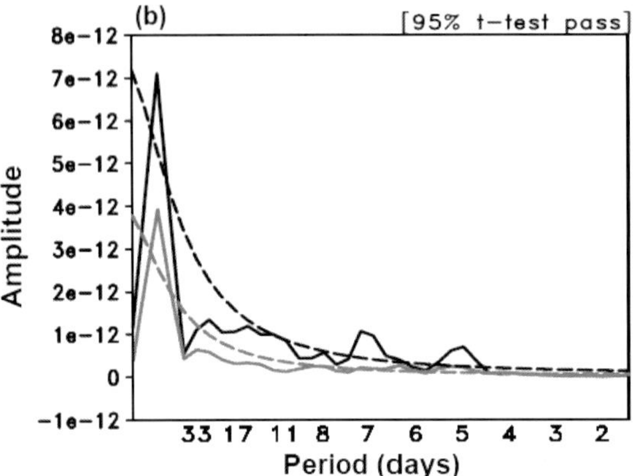

Figure 9. (a) Time series of the area-averaged 850 hPa vorticity over 125°E–140°E, 15°N–22.5°N from June to October 2004. The dotted and dashed lines represent the original and TC-removed vorticity fields, respectively, while the thick solid line represents the difference. (b) Spectra density of the original (dark solid line) and the TC-removed (light solid line) time series shown in (a). Dashed lines represent the 95% confidence limits. Fluctuations with periods longer than 120 days were removed before calculating the spectral density.

similar distribution in frequency, the intraseasonal peak (statistically significant at the 0.05 level) drops by about 50% in the TC-removed case. In addition, removal of TCs also results in a reduction of the seasonal mean vorticity. For example, the seasonal mean vorticity in the region chosen for Fig. 4(a) is $2.1 \times 10^{-5}\,\mathrm{s}^{-1}$ in the original vorticity but reduces to almost zero in the TC-removed vorticity. These results indicate that the presence of TCs in the tropical western North Pacific not only enlarges the ISV

but also increases the seasonal mean vorticity along the TC tracks. In a recent study, Hsu *et al.* (2008) demonstrated that the presence of TCs enhances not only the ISV but also the interannual variance.

In view of the large reduction in the ISV after the removal of TCs, one would wonder whether the propagation of the ISO is affected. Figure 10 presents the latitude-time Hovmüller diagram of the 850 hPa vorticity averaged between 120°E and 150°E, where the ISO exhibits obvious northward propagation. The original ISO exhibited two cycles of oscillation, with the largest amplitudes near 5°N and 15°N, and a node near 10°N [Fig. 10(a)]. In between the occurrence of maximum amplitudes, northward propagation from 5°N to 15°N was evident. After the removal of TCs, the amplitude of the vorticity weakens significantly between 10°N and 20°N, where TC tracks were located, while the pattern and amplitude at other latitudes are nearly unchanged [Fig. 10(b)]. The northward propagation from 5°N to 15°N, which is the major characteristic of the ISO in this region during the boreal summer, is still evident.

Since the TCs' effect on the ISO mainly occurred in the 10°N–20°N latitudinal band, the longitude-time Hovmüller diagram of the 850 hPa vorticity averaged between 10°N and 20°N was examined. In Fig. 10(c), two cycles of the ISO with the maximum amplitude at various longitudes are evident. There are signs of westward propagation in the early half of summer, especially over 120°E–140°E, and eastward propagation in the latter half of summer. After the removal of TCs, the amplitude of the ISO reduces significantly, while the pattern remains similar [Fig. 10(d)], although there are indications of weakened eastward propagation between 140°E and 155°E in the latter half of summer. Overall, the most significant effect of TCs on the ISO is the large reduction in the amplitude. In comparison, the TCs' effect on the phase and propagation of the ISO is minimal.

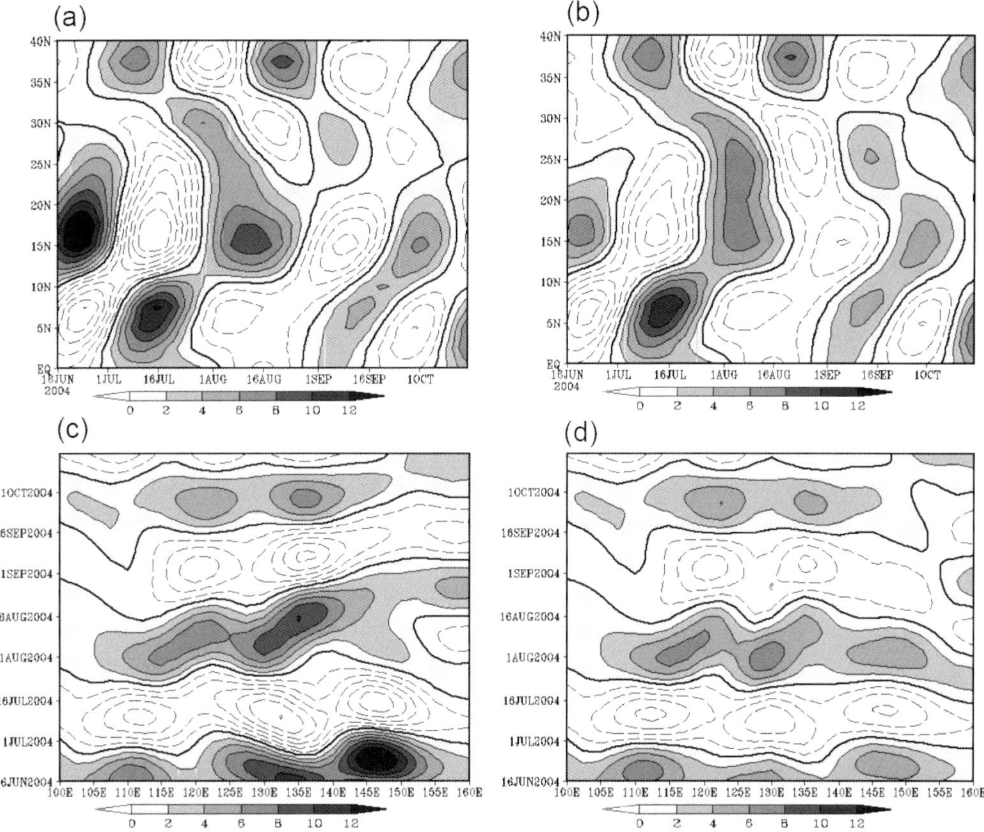

Figure 10. Hovmüller diagrams for the (upper left) original and (upper right) the TC-removed 32–76-day filtered 850 hPa vorticity averaged over 120°E–150°E, and for the (lower left) original and the (lower right) TC-removed 32–76-day filtered 850 hPa vorticity averaged over 10°N–20°N. The contour intervals are $3 \times 10^{-6}\,\mathrm{s}^{-1}$ and $2 \times 10^{-6}\,\mathrm{s}^{-1}$ for the upper panel and the lower panel, respectively.

5. Numerical Experiment

The empirical results presented above reveal the significant effect on the amplitude and variance of the ISO. This section presents numerical simulations performed to see whether similar effects could be realized in the numerical models. Two models were used in this study. The first was the Purdue regional model (PRM), which had been used for the study of various mesoscale phenomena (e.g. Sun *et al.*, 1991; Hsu and Sun, 1994; Sun and Chern, 1993) and climate simulation (e.g. Bosilovich and Sun, 1998, 1999; Hsu *et al.*, 2004; Sun *et al.*, 2004). The second one was the National Taiwan University's general circulation model (NTUGCM; Hsu *et al.*, 2001).

The resolution for the PRM was 60 km in the horizontal and 28 levels in the vertical. The model domain is 90°E–160°E and 0–40°N, with a 15-point buffer zone on the lateral boundary. The ECMWF advanced analysis at 1.125° resolution was used as the initial condition and the lateral boundary condition. The latter was updated every 6 hours. The resolution for the NTUGCM was T42 in the horizontal and 13 levels in the vertical. The ECMWF basic analysis at 2.5° resolution was used as the initial condition.

Both models show a certain degree of ability to simulate TCs in the first few days. One example for the NTUGCM is shown in Fig. 11 for the forecast with initial condition at 00Z

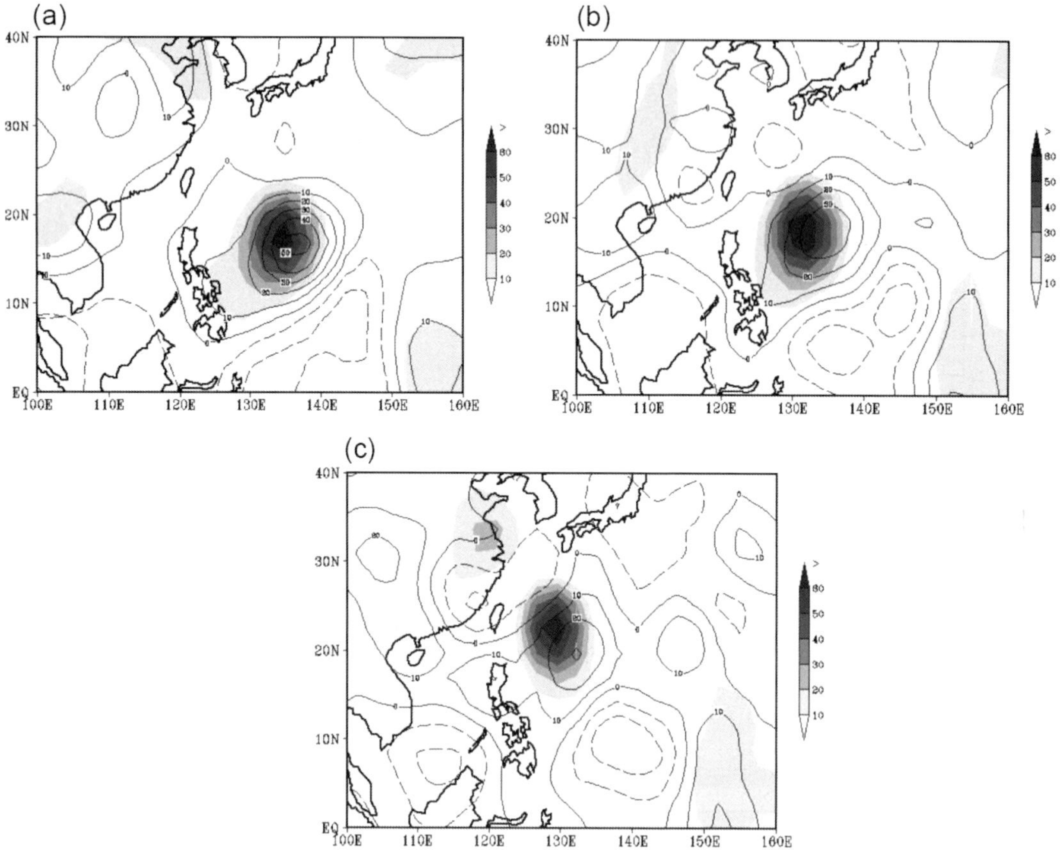

Figure 11. (a) 24-hour, (b) 48-hour, and (c) 72-hour hindcast of 850 hPa vorticity (contour) by the NTUGCM, initialized from 00Z 16 June 2004. The observed vorticity is plotted in shading.

16 June. The 24-hour hindcast [Fig. 11(a)] was able to simulate the location of TCs, although with a much weaker amplitude. The TC vortex looks larger and smoother because of the low resolution. In the 48-hour and 72-hour hindcasts [Figs. 11(b) and 11(c)], the simulated TC-like vortex is still evident, but lags behind the observed TC vortex. The 48-hour and 72-hour hindcasts generally show large track bias and smaller amplitudes.

It appears that both models are able to produce satisfactory simulation in the 24-hour hindcast. To evaluate the TCs' effect on the simulated ISV, different numerical experiments were performed. A series of 24-hour hindcasts were performed daily, using both models, from 1 June to 31 October 2004. They yielded a

dataset of the control experiment in JJASO, which was used for diagnostics like the real data. Three series of the NTUGCM experiment were performed. The first is a series of 24-hour hindcasts starting with the observed initial conditions. The second is the same as the first, except that TCs have been removed from the initial condition. The third is a 24-hour hindcast experiment, starting with the observed initial condition, in which the vorticity in the lower troposphere was artificially enhanced where the TCs are located. The vorticity was enhanced at the grid points, where the surface pressure was lower than 980 hPa and the 900 hPa/850 hPa/700 hPa mean vorticity was larger than $0.00005\,\mathrm{s}^{-1}$. The vorticity at 900 hPa/850 hPa/700 hPa was multiplied by

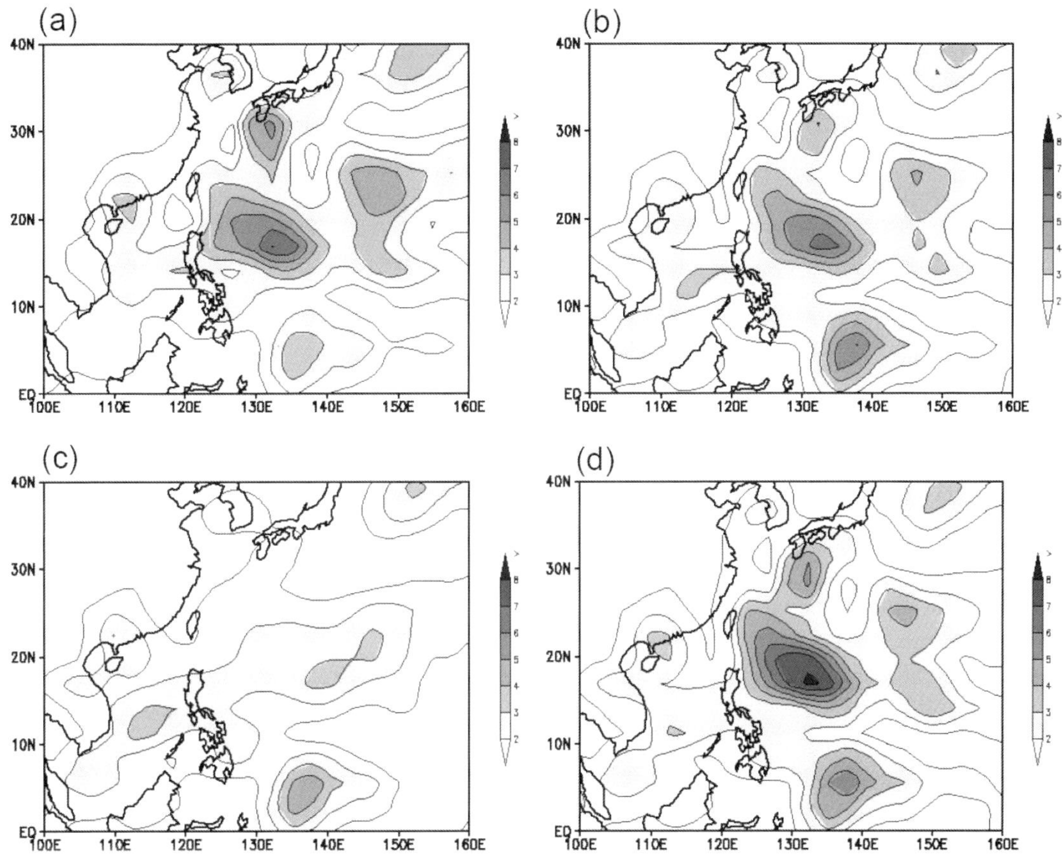

Figure 12. Variance of the 32–76-day filtered 850 hPa vorticity for (a) the observed and the (b) control, (c) TC-removed, and (d) vorticity-enhanced simulations using the NTUGCM. The contour interval is $0.5 \times 10^{-10}\,\mathrm{s}^{-2}$.

1.15/1.01875/1.00555, respectively. These three experiment series were performed daily from 1 June to 31 October, and the 32–76-day band-pass filter was applied to the three simulation datasets. Both the simulated and the observed ISV of the 850 hPa vorticity are shown in Fig. 12. The observed ISV, which is plotted at the NTUGCM spatial resolution, looks much smoother because of the low spatial resolution [2.5° by 2.5°; Fig. 12(a)]. The control experiment [Fig. 12(b)] was able to reproduce a realistic spatial distribution of the ISV with a slightly weaker magnitude. In contrast, the TC-removed experiments [Fig. 12(c)] failed completely to simulate the maximum variance along the TC tracks. The vortex-enhanced experiments [Fig. 12(d)], on the contrary, not only

simulated well the ISV distribution but also enhanced the magnitude to the observed level. A comparison between the results obtained from the three experiment series indicates that the presence and enhancement of TC-like disturbance, although fluctuating at a much higher frequency, enhance the ISV.

A series of heating-enhanced hindcast experiments using the PRM were performed, by adding a prescribed heating profile to the simulated TC-like vortices. The heating profile mimics the Q1 profile (not shown), which was calculated from the reanalysis at those grid points where category-4 typhoons passed. The profile was characterized by the maximum heating at 600 hPa, and the decreasing linearly upward and downward. However, the heating below 800 hPa was kept constant, to obtain better simulation. During the simulation, the heating was multiplied by 2 and held constant at the center of TC-like vortices, and exponentially decreased outward for 10 grid points. Since this was an idealized experiment, the prescribed heating is the same for all cases at all times, disregarding the different sizes and strengths of the vortices. The purpose was to artificially and significantly enhance the strength of the TC-like vortices in the model and, through a comparison with the controlled experiments, to evaluate whether the TCs can enhance the ISV in the simulation.

The 32–76-day band-pass filter was applied to the 850 hPa vorticity of the control and heating-enhanced hindcast experiments to extract the intraseasonal fluctuation. The control experiment was able to reproduce the overall distribution of the ISV, but the values are only about 1/3 of the observed variance. This is because the simulated TC-like vortices are weaker than the observed, and tend to weaken quickly. The simulated variance in the heating-enhanced experiments is raised to the equivalent level of the observed magnitude, while maintaining a realistic spatial distribution. This contrast indicates that the enhanced TC-like disturbance in the regional model also contributed to enhancing the ISV. The results of numerical experiments confirm the hypothesis, which was proposed based on the empirical results, that the presence of TCs in clusters can enhance the ISV.

6. Conclusions and Discussion

This study has revealed the strong ISO–TC in-phase relationship in the tropical western North Pacific during JJASO 2004. The repeated appearance of the ISO during JJASO resulted in the fluctuation of the EA monsoon trough and the Pacific anticyclone, which in turn modulated the TC activity and led to the spatial and temporal clustering of TCs. While TCs occurred in groups during the cyclonic phase of the ISO, the clustering of these strong TC vortices significantly increased the overall amplitude of positive vorticity during the cyclonic phase of the ISO. On the contrary, the overall amplitude of negative vorticity during the anticyclonic phase of the ISO remained unaffected, because of poorly organized TC activity in the anticyclonic phase of the ISO. The ISV was therefore enlarged with the occurrence of TCs. This result reveals that a significant percentage (larger than 50%) of the observed ISV in the tropical western North Pacific was contributed by TCs along the clustered TC tracks. This large contribution, which has not been well recognized, was clearly present in the 2004 typhoon season and is likely to have occurred in other years as well. Hsu *et al.* (2008) reported similar results, based on a study using multiyear datasets. A series of 24-hour hindcast numerical experiments with various designs were performed to evaluate the effect of TC-like vortices on the simulated ISV. Results of both the regional model and the GCM indicated that the experiments with larger TC-like vortices produced a stronger ISV and better simulations in terms of both spatial distribution and magnitude. This finding is consistent with the conclusion inferred from the empirical results.

The extraordinarily strong ISV and its coupling with TCs have been clearly demonstrated. This unusual phenomenon apparently led to the record-breaking number of typhoon landfalls in Japan. However, what mechanism led to these unusual features (e.g. the largest ISV in 54 years, the strong coupling between ISO and TC, the location of strong ISV) is still unknown. One may suspect that the warm SST might have an effect on the ISO and TC since it was an El Niño summer. On the other hand, the El Niño was a minor one and the SST anomaly was not particularly strong. It appears that there should be other unknown mechanisms that were responsible for the unusual phenomena in the western North Pacific during the summer of 2004.

An unconventional analysis procedure was carried out in this study to contrast the variance with and without TCs. The estimated variance difference is not likely to be the real contribution from TCs, but can probably be viewed as a quick estimation of the TCs' contribution to the ISV. The results shown above indicate that the contribution of TCs has to be taken into account to correctly estimate and interpret the ISV, especially during those years when TCs are strongly clustered by the ISO. Traditional wisdom usually assumes that the fluctuations on a shorter-time scale can be removed by time averaging or low-pass filtering. This may be true when the fluctuations in the positive and the negative phase are symmetric in both amplitude and recurrent frequency. Such a practice may be problematic for the intraseasonal variability analysis in the tropical western North Pacific, where TCs are the strongest cyclonic vortices and may not be canceled out by much weaker anticyclonic vortices. It is likely that the occurrence of TCs leaves footprints in the intraseasonal variability through the clustering effect.

The procedure adopted in this study has been used by the TC simulation and prediction community to remove the inadequate representation of TCs in the global analysis and plant an idealized vortex to represent TCs in the model. On the other hand, it is also true that TC circulation, although underestimated and sometimes unrealistic, does exist in the global analysis. It is this TC-like component that this study would like to evaluate its possible contribution to the large-scale climate variability, such as those on the intraseasonal time scale. Although the TC circulation is likely underestimated in the global analysis, the results presented above indicate that this inadequately represented TC component already contributes a significant amount of intraseasonal variability. If the TCs are accurately resolved in the global analysis, its contribution will likely be even more significant. This information will be valuable for the intraseasonal variability study, and perhaps even for the climate variability study on other time scales, as shown by Hsu *et al.* (2008).

In the tropical western North Pacific, the tropical cyclone often has a life span longer than 10 days. As the moving tropical cyclones have a strong and long-lasting energy source, large-scale circulations are likely to be induced, followed by energy emanation to remote regions. This feedback may affect the environmental flow, such as the subtropical anticyclone and monsoon trough, and leave notable footprints in the intraseasonal variability in the regions away from TC tracks. This potential effect cannot be estimated by the methodology adopted in the present study and will be explored in future works. Another problem is the coarse resolution of the datasets used in this study. Although the TCs are reasonably represented in the global analysis in the qualitative sense, the strength of the TCs is underestimated and the spatial structure is not as sharp as in the real world. As a result, the actual contribution of TCs to the ISV cannot be accurately estimated. Moreover, the contribution reported in this study is probably underestimated. Despite not being able to exactly quantify the effect, this research explains in a qualitative sense why the contribution from TCs cannot be overlooked.

Most of the general circulation models used to simulate past climate suffers from the poor simulation of the climate variability in the tropical western North Pacific during the boreal summer (Wang *et al.*, 2004). The results reported here imply that the inability to resolve and simulate TCs may be one of the key weaknesses of the GCM leading to poor simulation. Using the coarse-resolution GCM may lead to inaccurate climate simulation and prediction on the intraseasonal, interannual, and perhaps even climate change time scales. High-resolution models that are able to reasonably simulate the ensemble effect of TCs, at least in the statistical sense, seem necessary for resolving the multiscale interaction and producing better simulations of the ISV in TC-prone regions, such as the tropical western North Pacific. Although this study does not reveal any mechanism relating to the TC–ISO interaction (if there is one), it seems to imply the following. For numerical models that cannot explicitly simulate TCs in the tropical western North Pacific, the simulated ISV is probably underestimated. Whether an improved TC simulation would improve the ISO simulation in models is an interesting issue for further study. Another implication is that the clustering effect and the TC contribution to the ISV imply a possible two-way interaction between the TC and the ISO in the tropical western North Pacific. Well-designed numerical experiments and theory development are needed to quantify and understand this process.

Our finding also raises a question about the definition of the ISV in the TC-prone regions. Traditionally, the ISV is interpreted as the variability of intraseasonally filtered perturbations. Our results, however, point out that these intraseasonal perturbations may contain the contribution from TCs, which fluctuate in much shorter periods and on a smaller spatial scale. In a region such as the tropical western North Pacific, where the mesosynoptic-scale and large-scale systems are closely intertwined, and the multiscale interaction is likely one of the key processes affecting the climate variability, the contribution from severe weather systems like TCs has to be taken into account in order to understand the mechanisms leading to the intraseasonal (and climate) variability. The present results suggest that the TC and the ISO be viewed as an integrated system to improve our understanding of the intraseasonal variability in the tropical western North Pacific.

Acknowledgements

This study was jointly supported by the Central Weather Bureau and the Nation Science Council, Taiwan, under Grant MOTC-CWB-94-6M-03 and NSC 93-2111-M-002-004-AP4, respectively.

[Received 20 September 2007; Revised 24 January 2008; Accepted 25 January 2008.]

References

Bosilovich, M. G., and W.-Y. Sun, 1998: Monthly simulation of surface layer fluxes and soil properties during FIFE. *J. Atmos. Sci.*, **55**, 1170–1184.

Bosilovich, M. G., and W.-Y. Sun, 1999: Numerical simulation of the 1993 Midwestern flood: land–atmosphere interactions. *J. Climate*, **12**, 1490–1505.

Chen, T. C., S. Y. Wang, M.-C. Yen, and W. A. Gallus Jr., 2004: Role of the monsoon gyre in the interannual variation of tropical cyclone formation over the western North Pacific. *Weather and Forecasting*, **19**, 776–785.

Elsberry, R. L., 2004: Monsoon-related tropical cyclones in Eats Asia. In *East Asian Monsoon* (ed. C.-P. Chang), *World Scientific Series on Meteorology of East Asia*, Vol. 2, World Scientific, Singapore, pp. 463–498.

Gray, W. M., 1979: Hurricanes: their formation, structure and likely role in the tropical circulation. In *Meteorology over the Tropical Oceans*, D. B. Show (ed.), Royal Meteorological Society, pp. 155–218.

Harr, P. A., and R. L. Elsberry, 1998: Large-scale circulation variability over the tropical western North Pacific. Part I: Spatial patterns and tropical cyclone characteristics. *Mon. Wea. Rev.*, **123**, 1225–1246.

Heta, Y., 1990: An analysis of tropical wind fields in relation to typhoon formation over the Western Pacific. *J. Meteor. Soc. Japan*, **68**, 65–76.

Hsu, H.-H., Y.-L. Chen, and W. S. Kau, 2001: Effects of ocean–atmosphere interaction on the winter temperature in Taiwan and East Asia. *Climate Dynamics*, **17**, 305–316.

Hsu, H.-H., Y.-C. Yu, W.-S. Kau, W.-R. Hsu, W.-Y. Sun, and C.-H. Tsou, 2004: Simulation of the 1998 East Asian summer monsoon using Purdue regional model. *J. Meteor. Soc. Japan*, **82**, 1715–1733.

Hsu, H.-H., 2005: East Asian monsoon. In *Intraseasonal Variability in the Atmosphere–Ocean Climate System*, W. K. M. Lau and D. E. Waliser (eds.), Praxis, Springer, Berlin, Heidelberg, pp. 63–94.

Hsu, H.-H., C.-H. Hung, A.-K. Lo, C.-W. Hung, and C.-C. Wu, 2008: Influence of tropical cyclone on the estimation of climate variability in the tropical Western North Pacific. *J. Climate* (in press).

Hsu, W.-R., and W.-Y. Sun, 1994: A numerical study of a low-level jet and its accompanying secondary circulation in a Mei-Yu system. *Mon. Wea. Rev.*, **122**, 324–340.

Kaylor, R. E., 1977: Filtering and decimation of digital time series. Tech. Note BN 850, Institute of Physical Science Technology, University of Maryland, College Park, 42 pp.

Kurihara, Y., M. A. Bender, and R. J. Ross, 1993: An initialization scheme of hurricane models by vortex specification. *Mon. Wea. Rev.*, **121**, 2030–2045.

Kurihara, Y., M. A. Bender, R. E. Tuleya, and R. J. Ross, 1995: Improvements in the GFDL hurricane prediction system. *Mon. Wea. Rev.*, **123**, 2791–2801.

Lau, K.-M., and P. H. Chan, 1986: Aspects of the 40–50-day oscillation during the northern summer as inferred from outgoing longwave radiation. *Mon. Wea. Rev.*, **114**, 1354–1367.

Liebmann, B., H. H. Hendon, and J. D. Glick, 1994: The relationship between tropical cyclones of the Western Pacific and Indian Ocean and the Madden–Julian oscillation. *J. Meteor. Soc. Japan*, **72**, 401–412.

Lorenc, A. C., 1984: The evolution of planetary scale 200-mb divergences during the FGGE year. *Quart. J. R. Meteor. Soc.*, **110**, 427–441.

Maloney, E. D., and M. J. Dickinson, 2003: The intraseasonal oscillation and the energetics of summertime tropical Western North Pacific synoptic-scale disturbances. *J. Atmos. Sci.*, **60**, 2153–2168.

McBride, J. L., 1995: Tropical cyclone formation. In Chap. 3, *Global Perspectives on Tropical Cyclones*. Tech. Doc. WMO/TD No. 693, World Meteorological Organization, Geneva, Switzerland, pp. 63–105.

Nakazawa, T., 1986: Intraseasonal variations of OLR in the tropics during the FGGE year. *J. Meteor. Soc. Japan*, **64**, 17–34.

Nakazawa, T., 2006: Madden–Julian oscillation activity and typhoon landfall on Japan in 2004. *SOLA*, **2**, 136–139, doi:10.2151/sola.2006-035.

Sun, W.-Y., J.-D. Chern, C.-C. Wu, and W. R. Hsu, 1991: Numerical simulation of mesoscale circulation in Taiwan and surrounding area. *Mon. Wea. Rev.*, **119**, 2558–2573.

Sun, W.-Y., and J.-D. Chern, 1993: Diurnal variation of lee vortices in Taiwan and surrounding area. *J. Atmos. Sci.*, **51**, 191–209.

Sun, W. Y., J. D. Chern, and M. Bosilovich, 2004: Numerical study of the 1988 drought in the United States. *J. Meteorol. Soc. Japan*, **82**, 1667–1678.

Torrence, C., and G. P. Compo, 1998: A practical guide to wavelet analysis. *Bull. Amer. Meteor. Soc.*, **79**, 61–78.

Wang, B., and H. Rui, 1990: Synoptic climatology of transient tropical intraseasonal convection anomalies: 1975–1985. *Meteor. Atmos. Phys.*, **44**, 43–61.

Wu, C.-C., T.-H. Yen, Y.-H. Kuo, and W. Wang, 2002: Rainfall simulation associated with Typhoon Herb (1996) near Taiwan. Part I: The topographic effect. *Wea. Forecasting*, **17**, 1001–1015.

Convective–Radiative-Mixing Processes in the
Tropical Ocean–Atmosphere

Chung-Hsiung Sui

Institute of Hydrological and Oceanic Sciences and Department of Atmospheric Sciences,
National Central University, Taoyuan, Taiwan
sui@ncu.edu.tw

Xiaofan Li

Joint Center for Satellite Data Assimilation, NESDIS, NOAA,
Camp Springs, Maryland, USA

William K.-M. Lau and Wei-Kuo Tao

Laboratory for Atmospheres, Goddard Space Flight Center, Greenbelt, Maryland, USA

Ming-Dah Chou

Department of Atmospheric Sciences, National Taiwan University, Taipei, Taiwan

Ming-Jen Yang

Institute of Hydrological and Oceanic Sciences and Department of Atmospheric Sciences,
National Central University, Taoyuan, Taiwan

Understanding convective–radiative-mixing processes is crucial in making better predictions about tropical climate. The cloud-resolving model and the mixed-layer model, combined with observations, are powerful tools for studying these physical processes interacting with climate. In this article, the authors' research work of the past 15 years on tropical climate processes is reviewed. The topics reviewed include climate equilibrium study, tropical convective responses to radiative and microphysical processes, the diurnal cycle, cloud clustering and associated cloud-microphysical processes, precipitation efficiency, air–sea exchanges and ocean-mixing processes at diurnal-to-intraseasonal scales, and coupled boundary layer and forced oceanic responses. Representation of these processes in climate models and future perspectives are also discussed.

1. Introduction

Clouds play an important role in regulating weather and climate in the tropical ocean–atmosphere system through convective–radiative-mixing processes. The incoming solar radiative flux and the internal climate oscillations provide the environment with a large amount of unstable energy, which initiates the formation of cloud clusters. The development of convection, in turn, significantly modifies the environment by redistributing momentum, temperature, moisture, and salinity vertically through radiative, microphysical, and dynamic mixing processes. These processes are fundamentally important in maintaining tropical oscillations and the climate state.

To improve our knowledge of the convective–radiative-mixing processes, appropriate models like cloud-resolving models are powerful tools. Soong and Ogura (1980a) performed a pioneering study using their two-dimensional slab-symmetric numerical cloud model to examine the statistical properties of cumulus clouds that respond to the given large-scale forcing, which is mainly a vertical velocity. Their nonhydrostatic model with an anelastic approximation includes prognostic equations for momentum, temperature, specific humidity, and cloud species. This model was greatly improved later at the NASA Goddard Space Flight Center (GSFC), and named the Goddard cumulus ensemble (GCE) model (Tao and Simpson, 1993; Tao, 2003; Tao et al., 2003). The GCE model includes detailed solar and infrared parametrization schemes (Chou et al., 1991, 1997; Chou and Suarez, 1994a), cloud-microphysical parametrization schemes for cloud water, raindrops, cloud ice, snow, and graupel (Rutledge and Hobbs, 1983, 1984; Lin et al., 1983; Tao et al., 1989; Krueger et al., 1995), and subgrid-scale turbulence closure (Klemp and Wilhelmson, 1978).

The GCE model was originally designed to study the cumulus response to large-scale forcing (a prospective from a grid box of general circulation models), which can be imposed in the model in two ways: one with a vertical velocity and the other with heat and moisture source/sink. The model has been applied to case-oriented short-term simulations such as deep tropical cumulus clouds (Soong and Tao, 1980b), tropical squall line (Tao and Simpson, 1989), cloud interaction and merging (Tao and Simpson, 1984), and three-dimensional tropical clouds (Tao and Soong, 1986; Tao and Simpson, 1989).

In addition to the one-way response to large-scale forcing, convective–radiative processes may interact with climate change to form a feedback loop, like the feedback mechanisms of water vapor-cloud radiative forcing and surface evaporative cooling to climate warming (e.g. Newell, 1979; Fu et al., 1992; Hartmann and Michelsen, 1993; Lau et al., 1994a; Prabhakara et al., 1993; Ramanathan and Collins, 1991; Lindzen, 2001). Some of these climate feedback mechanisms may be investigated using a cloud resolving model. Lau et al. (1993, 1994a) and Sui et al. (1994) integrated the two-dimensional GCE model imposed with the large-scale vertical velocity to reach quasi-equilibrium states, applying the cloud-resolving model as a new tool for studying the effects of the convective–radiative interaction on tropical climate.

Based on extensive knowledge acquired through previous studies, longer simulations were performed to study the cumulus ensemble response of the model to the vertical velocity derived from observational data such as Marshall Island data (Yanai et al., 1973; Sui et al., 1994). Other research groups employed different cloud-resolving models, and successfully simulated the deep convective response to large-scale forcing observed in the Global Atmosphere Research Programme Atlantic Tropical Experiment (GATE) (e.g. Xu and Randall, 1996; Grabowski et al., 1996) and the TOGA COARE (e.g. Wu et al., 1998; Li et al, 1999; Johnson et al., 2002).

In addition to the studies just mentioned, the GCE model has been further extended to study various convective–radiative processes related to cumulus ensemble responses to large-scale forcing (Li et al., 1999, 2002b, 2005), diurnal cycle (Sui et al., 1998a), microphysical processes and cloud clusters (Peng et al., 2001; Li et al., 2002c; Sui and Li, 2005), and precipitation efficiency (Li et al., 2002a; Sui et al., 2005, 2007b). Similarly, an ocean mixed layer model has been used to study air–sea exchange and ocean mixing at diurnal to intraseasonal scales (Sui *et al.*, 1997b; Lau and Sui, 1997; Li *et al.*, 1998; Sui *et al.*, 1998b).

This article will highlight major scientific findings made by the authors during the past 15 years at NASA/GSFC, USA, and the National Central University, Taiwan. In the next section, the roles of convective–radiative processes in climate equilibrium states, cumulus ensemble responses, the diurnal cycle, microphysical processes and the development of cloud clusters, and precipitation efficiency will be reviewed. Air–sea exchange and ocean mixing at diurnal-to-intraseasonal scales will be discussed in Sec. 3. The coupled boundary layer at the atmosphere–ocean interface and forced oceanic responses will be addressed in Sec. 4. Representation of convective–radiative processes in climate models will be discussed in Sec. 5. A summary and a discussion are given in Sec. 6.

2. Convective–Radiative Processes

2.1. *Cloud-resolving modeling for climate equilibrium and feedback study*

Lau *et al.* (1993) and Sui *et al.* (1994) studied tropical water and energy cycles, and their roles in the tropical convective systems, by integrating the two-dimensional cloud-resolving model to the climate equilibrium states. The model is imposed with a time-invariant, horizontally uniform, large-scale vertical velocity and a fixed SST at 28°C, in which the simulated atmosphere is conditionally unstable below the freezing level and close to neutral above the freezing level. After the adjustment, in about 20 days, the simulations reach the quasi-equilibrium states. In the convective–radiative equilibrium conditions, two-thirds and one-third of surface rainfall

come from convective and stratiform clouds, respectively. The vertically integrated moisture budget shows that three-fourths of the total moisture supply is from the moisture advection associated with the imposed large-scale vertical velocity, whereas one-third of the total moisture supply is from the surface evaporation flux. The total moisture supply is completely converted into surface rainfall. The heat budget displays that the cooling from radiation, and temperature advection associated with the imposed large-scale vertical velocity, are mainly balanced by the latent heat release associated with the precipitation processes.

Ramanathan and Collins (1991) conducted an observational analysis of measurements from NASA's Earth Radiation Budget Experiment (ERBE) during the 1987 El Niño and proposed a cirrus cloud thermostat effect. They proclaimed that cloud-radiative cooling by cirrus counteracts the super-greenhouse warming, and limits SST over the western Pacific warm pool to a rather uniform distribution between 29°C and 30°C. Lau *et al.* (1994a) further used the cloud-resolving model to assess the cirrus-cloud thermostat effect for tropical SST by analyzing the net radiation flux at the top of the atmosphere and the net heat exchanges at the ocean–atmosphere interface. The model is integrated with the SST's of 28°C and 30°C, and with and without the large-scale forcing, respectively. The net radiation flux at the top of the atmosphere comprises the net absorbed solar radiation averaged over clear sky regions, the longwave radiation emitted by the ocean surface, atmospheric greenhouse effect, and longwave and shortwave cloud forcings. The net heat flux at the ocean–atmosphere interface consists of solar and longwave radiation, sensible and latent heat fluxes. The comparison of the experiments with the same large-scale forcing but different SST's shows that the largest changes in the components contributing to the net radiation flux at the top of atmosphere are due to the emission of the surface longwave radiation, and greenhouse warming by the increase in water vapor, which to a large degree offset each other. The magnitude of the emission of the surface longwave radiation is smaller than the greenhouse warming, suggesting no apparent "super-greenhouse" effect. The changes in longwave and shortwave cloud forcings are small, and are insensitive to the changes in the SST's, because the change in the SST induces the change in low and mid-tropospheric clouds, but does not have an impact on upper tropospheric clouds. The change in the net heat flux at the ocean–atmosphere interface is due mainly to the change in the surface latent heat flux. The increase in the SST induces surface cooling by increasing surface evaporation. The increase in the SST produces a 13% increase in surface precipitation. The comparison of the experiments with the same SST's but different large-scale forcings (with and without the forcing) shows that the largest changes in the budget at the top of the atmosphere occur in the shortwave and longwave cloud forcings, which in large part cancel each other out. The experiment without the forcing undergoes a large reduction of the greenhouse effect by decreasing the moisture. At the ocean–atmosphere interface, the largest change appears in the surface radiative flux as a result of the largest difference of clouds between the experiments with and without large-scale forcing. More discussions on the cloud-resolving modeling assessment of the cirrus cloud thermostat hypothesis can be found in Ramanathan *et al.* (1994) and Lau *et al.* (1994b).

2.2. *Cumulus ensemble responses to radiative and microphysical processes*

Li *et al.* (1999) conducted two experiments to study cloud–radiation interaction. The cloud single scattering albedo and asymmetry factor varied with clouds and environmental thermodynamic conditions in one experiment, whereas they were fixed at 0.99 and 0.843, respectively, in the other experiment. A comparison of solar radiation calculations between the two experiments showed that the experiment with the varying single scattering albedo and asymmetry factor had stronger solar radiation absorption by ice clouds in the upper troposphere than did the experiment with the constant single scattering albedo and asymmetry factor. The difference in temperatures between the two experiments further showed that the temperature around 200 mb was 2°C warmer in the experiment with the variable single scattering albedo and asymmetry factor than in the experiment with constant values.

A statistical analysis of the clouds and surface rain rates revealed that stratiform (convective) clouds contributed to 33 (67) % of the total rain in the experiment with the variable cloud optical properties and 40 (60) % in the experiment with the constant cloud optical properties. The fractional cover by stratiform clouds increased from 64% in the experiment with the variations to 70% in the experiment with the constants. These sensitivity tests show the cloud–radiation interaction processes for stabilizing the atmosphere, in which the change in the vertical heating gradient by solar radiation due to variations of cloud optical properties stabilizes the middle and upper troposphere and contributes to the reduction of stratiform clouds, which further stabilizes the cloud system by reducing infrared cloud top cooling and cloud base warming.

Li *et al.* (2005) carried out two experiments to investigate the role of precipitation–radiation interaction in thermodynamics. One experiment includes the precipitation–radiation interaction, while the other excludes it. The experiment excluding the interaction produces 1–2°C colder and 1–1.5 g kg^{-1} drier than the experiment including the interaction. The comparison of the heat budget between the two experiments shows that the experiment excluding the interaction exhibits a more stable upper troposphere (above 500 mb) and a more unstable lower troposphere (below 500 mb) compared to the experiment including the interaction. The more stable upper troposphere suppresses the development of ice clouds that are responsible for the cooling bias, whereas more radiative cooling accounts directly for a cooling bias in the mid- and lower troposphere in the experiment excluding the interaction. The analysis of moisture budgets shows that the suppression of rain evaporation as a result of a less stable mid- and lower troposphere induces a drying bias when the experiment excludes the precipitation–radiation processes.

Cloud-microphysical processes determine conversion between environmental moisture and cloud hydrometeors. The microphysical parametrization of cloud ice and snow proposed by Hsie *et al.* (1980) was originally used in the cloud-resolving model, which produced a relatively small amount of cloud ice and snow compared to the observations. Hsie *et al.* (1980) modified the work of Orville and Kopp (1977) based on the equation of the rate of growth of ice crystals by deposition proposed by Koenig (1971), and formulated the depositional growth of cloud ice in mass and size to become snow by the mixing ratio divided by a time scale that is needed for an ice crystal to grow from a radius of $40\,\mu\mathrm{m}$ to $50\,\mu\mathrm{m}$. Based on the aircraft observations, Krueger *et al.* (1995) suggested that the time scale in the depositional growth of snow from cloud ice should be for a crystal to grow from a radius of $40\,\mu\mathrm{m}$ to $100\,\mu\mathrm{m}$, which increases the amount of cloud ice and snow significantly, as indicated by Li *et al.* (1999). More cloud ice leads to more infrared cooling at the cloud top, and less heating below the cloud top. Li *et al.* (1999) conducted additional experiments in which the cloud–radiation interaction is excluded, and found that the exclusion of cloud–radiation interaction and the reduction of ice clouds have a similar thermal effect,

whereas the two experiments have different impacts on moisture. The simulation excluding cloud–radiation interaction causes drying by enhancing condensation, whereas the simulation with reduced ice clouds by the microphysics scheme induces moistening by suppressing condensation.

Li *et al.* (2005) further examined the role of the depositional growth of snow from cloud ice by conducting the comparison study with two experiments: one with the snow depositional growth, and the other without. The results show that the experiment without the snow depositional growth produces a much larger amount of cloud ice than the experiment with the snow depositional growth. The budget of cloud ice further reveals that the vapor deposition rate is balanced by the conversion from cloud ice to snow and the depositional growth of snow from cloud ice. When the growth of snow from cloud ice is absent, the cloud ice could be accumulated and its amount becomes anomalously large. The analysis of the heat budget indicates that the anomalous amount of cloud ice reflects a large amount of solar radiation, and the upper tropospheric atmosphere becomes anomalously cold whereas it traps a large amount of infrared radiation and the lower tropospheric atmosphere becomes anomalously warm. Thus, the depositional growth of snow from cloud ice is an important sink for cloud ice.

Tropical convection (surface rainfall) occurs as a result of instability (convective available potential energy, CAPE) in the environment. Since the environmental time scales (a few days and longer) are much longer than the convective time scales (a few hours or less), the "quasi-equilibrium" between the rate of production of available potential energy by the large-scale processes and the rate of consumption of the available potential energy by the convection is the basic assumption which Arakawa and Schubert (1974) used to develop their cumulus parametrization scheme. A decrease in the CAPE often coincides with the development of convection so that the CAPE and rain rate are negatively correlated (e.g. Thompson *et al.*, 1979; Cheng and Yanai, 1989; Wang and Randall, 1994; Xu and Randall, 1998). The phase relation between the CAPE and rainfall is due to the coupling between the environmental dynamic and thermodynamic fields (Cheng and Yanai, 1989). The phases of the CAPE and rainfall could be different, because it takes time for clouds to develop. This phase difference can be included with relaxation of the quasi-equilibrium assumption in cumulus parametrization (e.g. Betts and Miller, 1986; Randall and Pan, 1993). The minimum CAPE typically occurs a few hours after the maximum rainfall, as indicated by the observational analysis. Xu and Randall (1998) interpreted the maximum phase lag as the adjustment time scale from disequilibrium to equilibrium states in the presence of time-varying large-scale forcing. Since the CAPE is calculated in a Lagrangian framework and the relevant equations cannot be derived, the physical processes responsible for the phase difference between the CAPE and the surface rain rate cannot be examined. However, an alternative for studying the phase difference has been developed in an Eulerian framework by Li *et al.* (2002b), in which potential and kinetic energy in an Eulerian framework represent the CAPE and surface rain rate in a Lagrangian framework, respectively.

Lorenz (1955) introduced the concept of available potential energy for a dry atmosphere that represents the portion of the potential energy that can be transferred into kinetic energy. He defined the available potential energy for a dry atmosphere as the difference between the actual total enthalpy and the minimum total enthalpy that could be achieved by rearranging the mass under adiabatic flow. The dry enthalpy per unit mass is defined as the product of the temperature and the specific heat at constant pressure. In the absence of energy sources and sinks, the total kinetic energy and total enthalpy are conserved during adiabatic expansion. In a moist atmosphere, latent heat energy should be included in the energy conservation. The latent heat energy per unit mass is defined as the product of the specific humidity and the latent heat of vaporization at $0°C$. In the absence of energy sources and sinks, the total kinetic energy, enthalpy and latent heat energy are conserved during dry and subsequent saturated adiabatic expansion. Therefore, the moist available potential energy is defined as the difference between the actual moist potential energy (sum of the enthalpy and latent heat energy) and the minimum moist potential energy that could be achieved by rearranging the mass under moist adiabatic processes. Li *et al.* (2002b) derived a set of equations for conversions between the moist available potential energy and kinetic energy in an Eulerian framework. Their equations were demonstrated to be the same as those derived by Lorenz (1955) in the absence of moisture.

Lag correlation analysis by Li *et al.* (2002b) showed that the maximum perturbation kinetic energy associated with the simulated convective systems and its maximum growth rate lags and leads the maximum imposed large-scale upward motion by about 1–2 hours respectively, indicating that the convection is phase-locked with the imposed large-scale forcing. Their imposed large-scale vertical velocity had time scales longer than the diurnal cycle, whereas the simulated convective systems had an average lifetime of about 9 hours. The imposed large-scale upward motion decreases the horizontal-mean moist available potential energy by the associated vertical advective cooling, providing a favorable environment for the development of convection.

They further showed that the maximum latent heating and vertical heat transport by perturbation circulations cause maximum growth of perturbation kinetic energy to lead maximum loss of perturbation available potential energy by about 3 hours. The maximum vertical advective cooling, the horizontal-mean cloud-related heating, and perturbation radiative processes cause maximum loss of perturbation moist available potential energy to lead maximum loss of the horizontal-mean moist available potential energy by about 1 hour. Consequently, the maximum gain of perturbation kinetic energy leads the maximum loss of horizontal-mean moist available potential energy by about 4–5 hours (about half of the lifetime of the simulated convection).

2.3. *Diurnal variation of tropical oceanic convection*

The diurnal variation of tropical oceanic convection is one of most important phenomena in tropical variability, and plays a crucial role in regulating tropical hydrological and energy cycles. The dominant diurnal signal is the nocturnal peak in precipitation that occurs in the early morning (see the review in Sui *et al.*, 1997a). Kraus (1963) emphasized the role of radiative forcing in the diurnal variation, and suggested that solar heating and IR cooling tend to suppress convection during daytime and enhance convection during nighttime respectively. Gray and Jacobson (1977) suggested that the diurnal variation of convection is a result of a synoptic-scale dynamic response to cloud radiative forcing (the radiational differences between cloudy regions and clear-sky regions). The cloud radiative forcing causes upward motion and convection during nighttime through the low-level convergence.

Sui *et al.* (1997a) conducted an analysis using the observational data from TOGA COARE. The data are first categorized into the disturbed and undisturbed periods by calculating the standard deviation of brightness temperature measured by the Geostationary Meteorological Satellite (GMS), operated by the Japanese Meteorological Agency. Over the disturbed periods, the total surface rain rate as well as convective and stratiform rain rates reach the maxima at 0300 local standard time (LST). Fractional cover for stratiform clouds has a

maximum at 0300 LST whereas fractional cover for convective clouds does not show a significant diurnal variation. Diurnal variation of the rain rate histogram shows that the evolution of nocturnal rainfall has a growing phase from 2200 to 0300 LST, when a wide range of convection (the rain rate is larger than $0.5\,\mathrm{mm\,h^{-1}}$) becomes enhanced with most occurrences within 0.5–$5\,\mathrm{mm\,h^{-1}}$. The nocturnal rainfall is associated with anomalous ascending motion in the layer between 500 and 200 mb at 0400 LST. Over the undisturbed periods, the surface rain rate is very small, but shows a maximum from 1200 to 1800 LST. Diurnal variation of the rain rate histogram shows that the evolution of afternoon rainfall has a growing phase from 1200 to 1800 LST, while most occurrences of rain rates are within 0.5 to $5\,\mathrm{mm\,h^{-1}}$. The afternoon rainfall peak is associated with the maximum SST after the solar radiation flux reaches the maximum. A schematic summary of the nocturnal maximum convection in the disturbed period and afternoon clouds and showers in the undisturbed period is shown in Fig. 1. Based on the observational analysis, the nocturnal rainfall peak is suggested to be related to the destabilization by radiative cooling during nighttime and the falling temperature that makes more precipitable water available for the surface precipitation.

Xu and Randall (1995) performed a study using a cumulus ensemble model (CEM), and suggested that nocturnal convection results from a direct radiation–convection interaction in which solar absorption by clouds stabilizes the atmosphere (Randall *et al.*, 1991). Tao *et al.* (1996) performed a study of cloud–radiation mechanisms in the tropics and mid-latitudes using the Goddard cumulus ensemble (GCE) model. They emphasized the increase of surface precipitation by IR cooling as a result of increased relative humidity.

Sui *et al.* (1998a) conducted the cloud-resolving simulations to test their nocturnal rainfall mechanism. An experiment with the

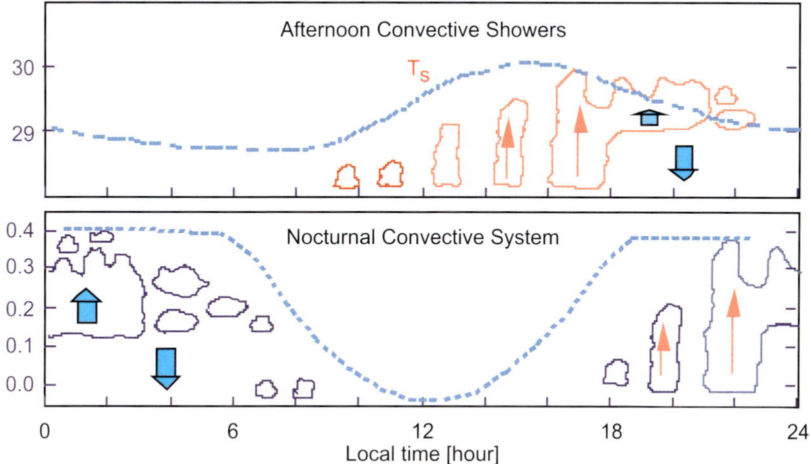

Figure 1. Schematic diagram of diurnal variations of convection during the disturbed (upper panel) and undisturbed (lower panel) periods. The dashed curve in the lower panel indicates the time rate of change of the saturation columnar water vapor amount, $-\partial W^*/\partial t$, corresponding to the diurnal cycle of temperature distribution. This quantity represents a direct effect of the radiative cooling/heating cycle on available precipitable water (APW), or a change of APW in the first step. The convective response to the direct forcing can induce further changes in temperature and moisture that lead to a corresponding change of APW in the second step. Since observed and simulated diurnal variations of convection are evidently in phase with the idealized cycle, the curve is regarded as a good theoretical limit for diurnal rainfall. The dashed curve in the lower panel indicates the diurnal cycle of sea surface temperature.

imposed large-scale ascending motion and a time-invariant SST generated a positive rainfall anomaly in the night and a negative rainfall anomaly in the day. The simulated maximum rain rate occurs around 0200 LST. Two additional experiments are carried out: one experiment with a zero imposed vertical velocity and a time-invariant SST, and the other with the cloud–radiation interaction suppressed. All three experiments show a dominant nocturnal rainfall maximum despite the experiments with very different external forcings and interaction processes. The results imply that cloud–radiation interaction does not play a crucial role in the formation of the nocturnal rainfall peak. The common feature in all of these experiments is the falling temperature induced by the nocturnal radiative cooling in the absence of the solar radiative heating. Thus, these numerical experiments support the suggestion of Sui *et al.* (1997a) that the nocturnal rainfall peak is related to more (less) available precipitable water in the night (day) as a result of the diur-

nal cooling/heating cycle (Fig. 1, lower panel). Sui *et al.* also conducted the experiment with zero imposed vertical velocity and a zonally uniform, diurnally varied SST and found that the simulated diurnal variations still have a nocturnal rainfall maximum but with a weaker magnitude and a secondary rainfall peak in the afternoon. This indicates that the maximum SST in the afternoon induces the unstable atmosphere that eventually leads to the rainfall peak (Fig. 1, upper panel).

2.4. *Cloud clustering and microphysical processes*

Lau *et al.* (1991) analyzed infrared radiance measurements at the cloud top from the Japanese GMS to study the structure and propagation of tropical cloud clusters over the equatorial western Pacific. The observed cloud clusters display a hierarchy of collective motions at time scales of 1 day, 2–3 days, and 10–15 days, respectively. The 1–15-day time scale is closely related to the intraseasonal oscillation, and their super cloud clusters propagate eastward along the equator from the Indian Ocean to the western Pacific all around the global tropics. The cloud clusters embedded in the super cloud clusters have the 2–3-day time scale, and propagate in the opposite direction to the super cloud clusters. The diurnal time scale is most significant in the cloud clusters, in which the signals are more pronounced over the continent than over the open ocean.

Sui and Lau (1992) analyzed the First GARP Global Experiment IIIb circulation data along with the Japanese GMS-1 IR data to study the multiscale variability in the atmosphere over the tropical western Pacific during the 1979 Northern Hemisphere. Two intraseasonal oscillations propagate eastward from the Indian Ocean to the western Pacific. Over the western Pacific warm pool, the intraseasonal oscillations develop into quasi-stationary systems with the enhanced rotational circulations. The intraseasonal oscillations interact with regional and synoptic-scale systems such as monsoon circulations. The intraseasonal oscillations also excite the 2–4-day disturbances. Sui and Lau also found that the diurnal signal becomes strong when the intraseasonal oscillation loses the intensity whereas the opposite is true.

To investigate the relevant cloud clustering processes (formation and evolution of cloud ensembles), Peng *et al.* (2001) performed a numerical experiment using a two-dimensional cloud ensemble model covering a basin-scale domain (15 360 km) with prescribed warm SST surrounded by cold SST, to mimic the equatorial western Pacific. The model used an open lateral boundary. Under the condition of no prescribed basic zonal flow and no initial perturbation, deep convective clouds develop in hierarchical clustered patterns, which are limited to the area of warm SST above 28°C. The most fundamental cloud cluster in the model has a horizontal scale of a few hundred kilometers, in which new cumulus clouds are generated at the leading

edge of a propagating surface cold pool, the "gust front." It may last for days and propagate for a long distance if the background flow is broad and persistent, as is the case in the low-level convergence zone of the SST-induced background flow.

The largest hierarchical propagating cloud systems in the model have horizontal scales of up to 3000 km and consist of up to four cloud clusters that are generally of the gust-front type. The constituent cloud clusters are generated intermittently and have life spans of 12–36 h. The internal heating of the constituent clusters collectively induces an overall troposphere-deep gravity wave (Mapes and Houze, 1993; Mapes, 1993). The overall wave travels in the direction of the tropospheric deep shear at a speed determined by the thermodynamic asymmetry in the wave created by the transition from warm and moist incoming air in the front to drier and cooler air in the rear.

The development of new cumulus clusters in the gust-front region of the hierarchical system is due to the combined effect of the overall wave and the gravity waves excited by the constituent clusters on the lower-tropospheric stability. When there are no interruptions from outside the cloud system, new cloud clusters develop intermittently from shallow disturbances hundreds of kilometers ahead of the existing deep convection. The resulting hierarchical cloud pattern resembles the observed equatorial super cloud cluster (SCC) in the time–longitude diagram. However, the life spans of the constituent clusters of the simulated system are shorter than that in the observed SCC.

The dynamic processes for cloud clustering are intimately coupled to microphysical processes. This coupling may be revealed by water- and ice-cloud contents and their corresponding microphysics. This is investigated in several companion papers. Li *et al.* (2002c) simulated the cloud clusters using the two-dimensional cloud-resolving model with the imposed forcing from the TOGA COARE data during the disturbed period. The cloud clusters move westward, while the embedded individual clouds propagate eastward. Along with the westward propagation and during the development of tropical convection, the area-mean vertical velocity profiles exhibit the major ascending motion below 500 mb in the western half of the cloud, and the maximum ascending motion between 300 and 500 mb in the eastern half, indicating that the western half of the cloud undergoes the deep convective development whereas the anvil cloud grows in the eastern half. The surface rainfall is much larger in the western half than in the eastern half. The amount of water hydrometeors is much larger than that of ice hydrometeors in the western half, whereas ice and water hydrometeors have similar amounts in the eastern half. The analysis of the rainwater budget reveals that the collection of cloud water by raindrops is a major process for the surface rainfall, and thus the water hydrometeor processes are dominant in the deep convective clouds in the western half, whereas both the collection of cloud water by raindrops and the melting of precipitation ice into raindrops are responsible for the surface rainfall in the anvil clouds in the eastern half.

The simulations show that the performance of cloud-microphysical parametrization schemes has the direct, crucial impacts on the simulations of cloud clusters in the genesis, evolution, propagation, and amplitudes. However, the computation of the full set of cloud-microphysical equations is time-consuming. Li *et al.* (2002c) found from their analysis of cloud-microphysical budgets that in the deep tropical convective regime, the magnitudes of 12 terms out of the total of 29 cloud-microphysical processes are negligibly small. Thus, they proposed a simplified set of cloud-microphysical equations, which saves 30–40% of CPU time. The neglected terms in the simplified set include the accretion of cloud ice and snow by raindrops, the evaporation of melting snow, the accretion of cloud water and raindrops by snow, the accretion of raindrops

and the homogeneous freezing of cloud water by cloud ice, the accretion and freezing of raindrops by graupel, the growth of cloud water by the melting of cloud ice, and the growth of cloud ice and snow by the deposition of cloud water. An experiment with the simplified set of cloud-microphysical equations was conducted and compared to an experiment with the original set of cloud-microphysical equations. The two experiments show similar time evolution and magnitudes of temperature and moisture profiles, surface rain rates including stratiform percentage and fractional coverage of convective, raining and nonraining stratiform clouds. This suggests that the original set of cloud-microphysical equations could be replaced by the simplified set in simulations of tropical oceanic convection.

Sui and Li (2005) analyzed the same TOGA COARE experiment performed by Li *et al.* (2002c) to show that interaction between ice and water clouds is crucial in determining the life span of deep convective and stratiform clouds and the evolution of cloud clusters. They defined a cloud ratio that is the ratio of the vertically integrated content of ice clouds (ice water path, IWP) to the liquid water path (LWP) to study the ice–water-hydrometeor interaction processes and their impacts in the development of convective and stratiform clouds. Clouds become more stratiform when the tendency of the cloud ratio is positive whereas they become more convective when the tendency of the cloud ratio is negative. The advantage of the definition of the cloud ratio is to mathematically derive a tendency equation of the cloud ratio based on the prognostic cloud equations in the cloud-resolving framework. The derived cloud ratio budget indicates that the tendency of the cloud ratio is determined by the vapor condensation and deposition (cloud sources), rainfall and evaporation (cloud sinks), and conversion between ice and water hydrometeors including melting of precipitation ice to raindrops and accretion of cloud water by precipitation ice. The analysis reveals that the tendency of the cloud ratio is mainly determined by the vapor condensation and deposition during the genesis and decay of the tropical convection when the system is relatively weak, whereas the tendency is controlled by the convection process during the development of tropical convection, when the system is relatively strong.

Sui *et al.* (2007a) proposed using threshold values of the cloud ratio to separate the convective component of the precipitation from the remainder. The cloud variables (IWP, LWP, and their ratio) are physically linked to the cloud microphysics, as demonstrated by an analysis of simulated cloud microphysics budgets in the same two-dimensional cloud-resolving model experiment subject to the imposed forcing from the TOGA COARE. Their analysis suggests that rainfall can be designated convective when the corresponding value of the cloud ratio is smaller than 0.2, or the value of IWP is larger than 2.55 mm. The remaining grids are classified as mixed and stratiform when the corresponding range of the cloud ratio is 0.2–1.0, and greater than 1, respectively. The new partition method is evaluated by the vertical velocity (w) data. The frequency distribution of w shows that w in the convective region has a wide distribution, with maximum values exceeding $15 \, \mathrm{m \, s^{-1}}$. In the designated stratiform region, the distribution is narrow, with absolute values of w confined within $5 \, \mathrm{m \, s^{-1}}$. The statistics of w and the budgets of cloud microphysics are consistent with corresponding physical characteristics of the convective and stratiform regions of precipitation. The w distribution in the mixed region exhibits features more convective than stratiform, indicating a transition stage of convective development. But the consideration of features like fractional cloud covers, rain rates, vertical velocity profiles, and the corresponding wave response leads us to regard the mixed and stratiform regions as the nonconvective region.

2.5. *Precipitation efficiency*

Precipitation efficiency is an important physical parameter for measuring the interaction between convection and its environment (Doswell *et al.*, 1996; Ferrier *et al.*, 1996; Tao *et al.*, 2004). Its definition may vary. For large-scale applications involving cumulus parametrization (e.g. Kuo, 1965, 1974), the precipitation efficiency is defined as the ratio of the surface rain rate to the sum of the surface evaporation and vertically integrated moisture convergence, and is referred to as large-scale precipitation efficiency (LSPE). LSPE is similar to the precipitation efficiency defined by Braham (1952). For cloud-resolving models with cloud-microphysical parametrization schemes (e.g. Li *et al.*, 1999), the precipitation efficiency can be defined as the ratio of the surface rain rate to the sum of the vertically integrated condensation and deposition rates. This is referred to as cloud microphysics precipitation efficiency (CMPE). CMPE is similar to the precipitation efficiency defined by Weisman and Klemp (1982) and Lipps and Hemler (1986).

Li *et al.* (2002a) analyzed the domain-mean CMPE and found that the LSPE could be more than 100%, whereas the CMPE is less than 100%. The statistical analysis shows that the ratio of the CMPE to the LSPE is 1.2. The precipitation efficiency may depend on the environmental conditions and the strength of convection. Ferrier *et al.* (1996) showed that wind shear and updraft structure play an important role in determining the precipitation efficiency. Li *et al.* (2002a) showed that the CMPE increases with increasing mass-weighted mean temperature and surface rain rate. This suggests that precipitation processes are more efficient for the heavy rain regime in a warm environment.

Since the LSPE and the CMPE are expected to be the same based on physical considerations, the difference in Li *et al.* (2002a) is attributed in Sui *et al.* (2005) to the horizontal hydrometeor advection that is excluded in the domain-mean CMPE due to the cyclic lateral boundary condition. Sui *et al.* (2005) analyzed the grid data from the two-dimensional cloud-resolving simulations with the imposed TOGA COARE forcing and the three-dimensional MM5 cloud-resolving simulation of Typhoon Nari (Yang and Huang, 2004). The analysis of two-dimensional grid data through the root-mean-square differences and linear correlation coefficients shows that the sum of vapor condensation and deposition rates is approximately balanced by the sum of surface evaporation and vertically integrated moisture convergence. This relation leads to the statistical equivalence between the CMPE and the LSPE.

Analysis of the two-dimensional simulation further shows that the additional hydrometeor converging into the atmospheric column would make the precipitation efficiency larger. When the hydrometeor convergence becomes the dominant term in the cloud budget, the CMPE can be larger than 100%, as found in light-rain conditions ($<5\,\mathrm{mm\,h^{-1}}$). On the other hand, a loss of clouds due to hydrometeors diverging out to the neighboring columns would make the CMPE smaller. This occurs mostly in heavy-rain conditions ($>5\,\mathrm{mm\,h^{-1}}$). The three-dimensional simulation of Typhoon Nari (2001) with more intense precipitation (compared to the TOGA COARE tropical convection) generally supports the two-dimension results.

Sui *et al.* (2007b) revisited the issue using the same two-dimensional cloud-resolving model simulation. They proposed more complete definitions of precipitation efficiency based on either the moisture budget (LSPE2) or the hydrometeor budget (CMPE2), which include all sources related to surface rainfall processes. They reached the following conclusions: (1) LSPE2 and CMPE2 range from 0 to 100%; (2) LSPE2 and CMPE2 are highly correlated; (3) LSPE2 and CMPE2 are insensitive to the spatial scales of averaged data, and moderately sensitive to the time period of averaged

data; (4) the simplified precipitation efficiencies of LSPE1 and CMPE1 appear to be good enough measures of precipitation efficiency in the heavy-rain conditions. CMPE2 (or CMPE1) is a physically more straightforward definition of precipitation efficiency than LSPE2 (or LSPE1). But the former can only be estimated in models with explicit parametrization of cloud microphysics which is model dependent, while the latter may be estimated based on currently available assimilation data of satellite and sounding measurements.

3. Air–Sea Exchange and Ocean Mixing at the Diurnal and Intraseasonal Scales

Lau and Sui (1997) and Sui *et al.* (1997b) analyzed the COARE data and found that the dry and wet phases in the atmospheric precipitable water are associated with the passage of the intraseasonal oscillations during the TOGA COARE. The passage of the intraseasonal oscillations also affects the surface radiative and heat fluxes that impact the ocean mixing processes and thus the upper-ocean stratification. The 2–3-day disturbances and diurnal cycles are phenomena with the dominant time scales during the wet phase, whereas the diurnal cycles in the SST are regularly forced by the diurnal solar radiative flux in the day and outgoing radiative and heat fluxes in the night during the dry phase (Fig. 2). This implies that the vertical solar absorption profile in the upper ocean plays an important role in determining the mixed-layer depth and the amount of solar absorption in the mixed layer, which eventually control the evolution of ocean mixed-layer temperature.

Sui *et al.* (1997b) employed a mixed-layer model to study the role of the vertical solar absorption profile in the diurnal ocean temperature simulations and their impacts on the intraseasonal variability. Owing to the asymmetric diurnal variation for shoaling and deepening

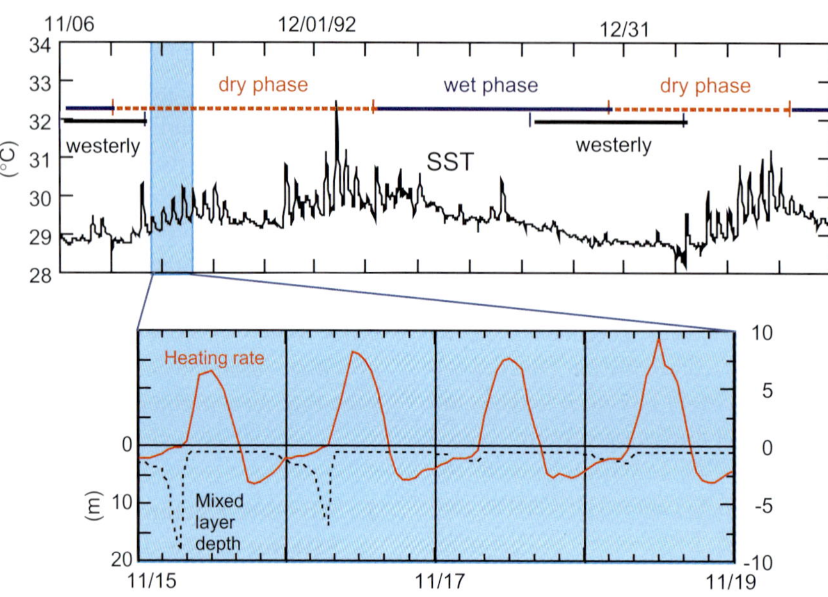

Figure 2. Time series of observed SST during the TOGA COARE (upper panel) and the simulated heating rate (solid) and mixed-layer depth (dashed) during selected COARE periods (lower panel).

of the mixed layer, the cumulative effect of diurnal mixing cycles is essential to maintaining a stable upper-ocean thermal stratification and to simulating a realistic evolution of mixed-layer and temperature at the intraseasonal time scale. Further sensitivity tests of mixed layer to diurnal cycles indicate that the inclusion of diurnal convective–radiative processes in the atmosphere–ocean system in the coupled models affects the capability of simulating intraseasonal variability.

The salinity contributes more to the density stratification than does the temperature, which is responsible for upper-ocean stability. Li *et al.* (1998) further included the salinity in the ocean mixed-layer modeling to examine the impacts of the precipitation and associated upper-ocean salinity stratification on the ocean mixed layer. The inclusion of salinity and precipitation-induced fresh water flux in the simulation shows that much deeper mixing occurs when rainfall appears during the night since the fresh water flux induces a much shallower mixed layer with a large deepening rate, which is

consistent with the observations. The inclusion of the salinity stratification could cause warmer water entrained into the ocean mixed layer since the salinity stratification maintains the upper-ocean stability, whereas the exclusion of salinity in the simulation only shows entrainment of cold water into the ocean mixed layer since the thermal stratification accounts for the upper-ocean stability. Because the Kraus–Tuner mixing parametrization scheme (Niiler and Kraus, 1977) requires both thermal and saline stratifications to determine the mixed-layer depth, Li *et al.* carried out decoupled salinity experiments to examine the effect of thermal stratification on the saline structure. The experiments reveal that when the fresh water input is large, the salinity variations simulated with and without the thermal stratification can be significantly different. The difference in the salinity could be 0.2 PSU. The simulations indicate that the inclusion of precipitation-induced fresh water flux and salinity stratification improves the simulation of thermal evolution in the ocean mixed layer.

Sui *et al.* (1998b) used a mixed-layer model along with the estimate of surface forcing obtained from the TOGA COARE to estimate the amount of heat absorbed in the observed mixed layer to maintain the observed amplitudes of SST which is largely dependent on the depth-dependent solar heating. The simulated amplitude of the diurnal cycle of mixed-layer temperature ($< 1°C$) is significantly smaller than the observed amplitude ($1–3°C$), implying that the mixed layer in the simulation does not absorb enough solar heat. They found that more than 39% of the net surface solar irradiance is absorbed within the first 0.45 meters in order to maintain the observed SST, which is higher than previous estimates. The vertical profile of solar absorption is then modified, and the simulation with the modified solar profile yields more realistic amplitudes of the SST at both the diurnal and intraseasonal time scales.

4. Coupled Boundary Layer and Forced Oceanic Responses

Sui *et al.* (1991) developed an equilibrium model to study the coupled ocean–atmosphere boundary layer in the tropics, which consists of a one-dimensional thermodynamic atmospheric model for a partially mixed, partially cloudy convective boundary layer (CBL), Betts and Ridgway, 1988; 1989 and an oceanic mixed-layer (OML) model. Two experiments were performed with sea surface temperature (SST) specified. They solve the equilibrium state of the coupled system as a function of SST for a given surface wind, and as a function of surface wind for a given SST. The increases in SST lead to the increase in the depth of the convective boundary layer owing to the increase in the water vapor. The moistened and deepened CBL leads to a reduced net surface heat flux which is balanced by weakened upwelling and causes a deepened ocean mixed layer. The increase in surface wind also causes the increase in the depth of the ocean mixed layer and the

decrease in the upwelling below the ocean mixed layer. But this is due to the generation of turbulence kinetic energy and the decrease in the net downward heat flux. The latter is due to the nonlinear change of air–sea humidity difference with increasing surface wind, such that a deepening CBL reduces the downward solar radiation and increases the downward longwave radiation at the surface. In another two experiments, the coupled model was solved iteratively as a function of surface wind for a fixed upwelling, and for a fixed mixed-layer depth (h). SST falls with increasing wind in both experiments, but the fall is gradual in the fixed upwelling condition because the depth is allowed to deepen and the cooling is spread over a larger mass of water, while the fall is steeper in the other experiment because h is fixed. The decrease of evaporation with increasing wind in fixed h condition leads to a very dry and shallow CBL. More experiments with surface wind and SST (upwelling) prescribed as a function of longitude similar to the observed values across the Pacific give realistic gradients of mixed-layer depth and upwelling (SST). The work quantifies the sensitivity of the equilibrium state of the coupled system to the coupling of the boundary layers, and provides a framework for understanding physical processes in the CBL and OML in coupled models.

Li *et al.* (2000) developed a coupled ocean–cloud-resolving atmosphere model to study effects of precipitation on ocean mixed-layer temperature and salinity. When the effects of fresh water flux and salinity were included in the coupled model, differences in the horizontal-mean mixed-layer temperature and salinity between 1D and 2D experiments were about $0.4°C$ and 0.3 PSU, respectively. The mean salinity difference was larger than the mean temperature difference in terms of their contributions to the mean density difference. In the 2D experiment, the surface heat flux showed a significant diurnal signal with the dominance of downward solar radiation during daytime and upward flux (longwave, sensible and latent heat fluxes) during nighttime at each grid, although the amplitude was affected by precipitation. Thus, there was a strong thermal correlation between grids. Narrow cloudy areas were surrounded by broad, cloud-free areas. Horizontal-mean precipitation could occur, whereas the precipitation may not occur in most of the integration period. Thus, there is a very low correlation between horizontal-mean and grid values of the fresh water fluxes. Since the rain rates have significant spatial variations, the fresh water flux has much larger spatial fluctuations than the saline entrainment. Therefore, the fresh water flux determines large spatial salinity fluctuations, which contributes to a large mean salinity difference between the 1D ocean model experiment and the 2D ocean model experiment.

Sui *et al.* (2003) investigated the impacts of high-frequency surface forcing on the upper ocean over the equatorial Pacific by conducting a nonlinear reduced-gravity isopycnal ocean circulation simulations with the daily and monthly mean forcings, respectively, and found that the daily-forcing experiment produces a colder SST than does the monthly-forcing experiment, and the difference in the SST between the two experiments is generated in the first year integration. The negative difference in the SST between the daily-forcing and monthly-forcing experiments in the western Pacific is primarily caused by enhanced latent heat loss due to the transient winds. Over the eastern Pacific, the zonal thermal advection accounts for the difference, while other terms are large but cancel each other out. Relative to the monthly forcing, the effect of daily forcing is found to (1) enhance vertical mixing and reduce vertical shear in the upper ocean; (2) reduce net heat into the ocean through two contrasting processes — increased surface latent heat loss induced by transient winds, and a colder SST (due to stronger mixing) reducing surface heat loss; (3) weaken meridional thermal advection through more active instability waves; (4) change mixed-layer depth so that the temperature in the simulation

with the daily forcing is warmer around the thermocline.

5. Representation of Convective–Radiative Processes in Climate Models

The research results discussed in this review represent a one-way interaction approach for process models to study the physical processes interacting with climate oscillations. The acquired knowledge forms a basis for improving the representation of these physical processes in climate models, and also motivates the next-step approach to examining the radiative–convective (two-way) interaction with climate dynamics in large-scale models. Indeed, since the original attempt by Sundqvist (1978) to include prognostic cloud water content for parametrizing nonconvective condensation, there has been increasing research attempts to incorporate detailed microphysical processes in weather and climate models. The trend is to treat cloud water and/or ice content as a prognostic variable which is governed by microphysical processes (e.g. Tiedtke, 1993; Del Genio *et al.*, 1996; Fowler *et al.*, 1996; Sud and Walker, 1999, 2003; Zhao and Carr, 1997). This approach allows the storage and full life cycle of cloud water, and a better cloud–radiation linkage through interactive cloud optical properties. While this approach provides a more physically based framework for representing the physical processes, it also introduces a number of microphysical parameters absent from the simpler approaches. Thus, a great deal of effort is required to implement a prognostic cloud parametrization scheme in climate models, and to validate the scheme against various observational data sets.

Another approach to incorporating detailed convective processes in weather and climate models is through directly resolving convective dynamics in the GCM, such as "super-parametrization" (Grabowski and Smolarkiewicz, 1999; Grabowski, 2001;

Khairoutdinov and Randall, 2001). This approach is to implement a cloud-resolving model (CRM) inside each grid box of a global model. The CRM (so far two-dimensional) does not fill the global model's grid box. Instead, it occupies only part of the grid box. The advection terms computed at each grid of the global model is imposed in the CRM as heat and moisture source/sink. In return, the CRM computes cloud ensemble statistics. In this way, the super-parametrization provides a framework for coupling convective–radiative processes with large-scale dynamics all at the physical time and space scales of the convective process. But this approach also introduces many problems different from those of simpler cumulus parametrization. See Randall *et al.* (2003) for details.

At the GSFC, continual efforts have been made to improve the representation of convective–radiative processes on a multi–model framework. One approach is to run the finite-volume general circulation model (fvGCM; Lin, 2004) at 1/8 of a degree (about 12 km) to resolve convective vortices in the global model context. The model is capable of simulating realistic tropical cyclones in the weather prediction model (Altas *et al.*, 2005). An attempt at super-parametrization in the fvGCM is also being examined.

Another important effort at the GSFC is to utilize satellite measurements to advance cloud-climate feedback study in climate models. Lau and Wu (2003) performed an analysis of satellite data from the Tropical Rainfall Measuring Mission (TRMM; Simpson *et al.*, 1988). They found that warm rain accounts for 31% of the total rain amount and 72% of the total rain area in the tropics, and that there is a substantial increase in the precipitation efficiency of light warm rain as the sea surface temperature increases, but the precipitation efficiency of heavy rain associated with deep convection is independent of the sea surface temperature. This implies that in a warmer climate, there may be more warm rain, at the expense of less cloud

water available for middle and high level clouds. The study points out a possible need to pay attention to resolving the melting/freezing zone in convection to simulate and better understand the role of cumulus congestus in tropical convection, and its sensitivity to SST, and global warming. Lau *et al.* (2005) performed a sensitivity test of GCM dynamics to the microphysics parameter of autoconversion. The result shows that a faster autoconversion rate leads to enhanced deep convection, more warm rain but less cloud over oceanic regions, and an increased convective-to-stratiform rain ratio over the entire tropics. The resultant vertical differential heating destabilizes the tropical atmosphere, producing a positive feedback, resulting in more rain and an enhanced atmospheric water cycle over the tropics. The feedback is maintained via secondary circulations between convective tower and anvil regions (cold rain), and adjacent middle-to-low cloud (warm rain) regions. The lower cell is capped by horizontal divergence and maximum cloud detrainment near the freezing/melting (0°C) level, with rising motion (relative to the vertical mean) in the warm rain region connected to sinking motion in the cold rain region. The upper cell is found above the 0°C level, with induced subsidence in the warm rain and dry regions, coupled to forced ascent in the deep convection region. The above result reveals that warm rain plays an important role in regulating the time scales of convective cycles, and in altering the tropical large-scale circulation through radiative dynamic interactions. Reduced cloud–radiation feedback by a faster autoconversion rate results in intermittent but more energetic eastward-propagating Madden and Julian oscillations (MJO's). Conversely, a slower autconversion rate, with increased cloud radiation produces MJO's with more realistic westward-propagating transients embedded in eastward-propagating supercloud clusters.

Super-parametrization is an intermediate approach to representing convective–radiative processes in a global model, between a prognostic cloud scheme and a global cloud-resolving model. If computational resources allow, it is most straightforward to develop an ultrahigh-resolution nonhydrostatic climate model that can resolve clouds with explicit microphysics. Global nonhydrostatic models are being developed at many institutions. In particular, a global nonhydrostatic grid model with icosahedral structure is being developed in Japan (Satoh *et al.*, 2005). This model is intended for high-resolution climate simulation, so the numerical scheme is designed for conserving mass and energy. Global simulations with cloud-resolving physical processes (cloud microphysics, radiation, and boundary layer processes) have been performed on an aqua planet setup with grid intervals of 7 km and 3.5 km. The model simulates reasonable features in the tropics, like the diurnal cycle of precipitation, hierarchical structure of clouds, and intraseasonal oscillations (Tomita *et al.*, 2005). The model's response to SST warming has also been investigated by Miura *et al.* (2005).

6. Relevance to Climate Variability and Future Perspectives

In this article, the authors' research results in the past 15 years in the tropics are highlighted. Reviewed are convective–radiative processes including climate equilibrium study, tropical convective responses to radiative and microphysical processes, the diurnal cycle, cloud clustering and associated cloud-microphysical processes, precipitation efficiency, air–sea exchanges and ocean mixing processes at diurnal-to-intraseasonal scales, and coupled boundary layer and forced oceanic responses. These physical processes coupled with the tropical wave dynamics have been recognized as key mechanisms in maintaining climate variability in the tropical ocean atmosphere, as exemplified in the following scientific issues.

One of the important issues is the response of water vapor and clouds to anthropogenic changes of greenhouse gases and aerosols. While many observational analyses and GCM simulations support a positive water vapor feedback on a global scale (e.g. Zhang *et al.*, 1996; Soden, 1997; Inamdar and Ramanathan, 1998), some studies suggest that this water vapor feedback may be overestimated or even negative (Lindzen, 1990; Lindzen *et al.*, 2001). Being inseparable from water vapor feedback, cloud-radiative forcing is another important climate feedback process. However, the effect of cloud feedback on climate change is equally if not more controversial than water vapor feedback. Studies of cirrus (or high) clouds and associated radiative effect on tropical climate do not even agree on the sign of the cloud feedbacks (e.g. Prabhakara *et al.*, 1993; Ramanathan and Collins, 1991; Kiehl, 1994). A synthesis of the above studies reveals that water vapor and cloud feedbacks depend on the relative areas of cloudy/moist regions versus clear/dry regions, as well as on cloud properties (type, height, optical thickness) and water vapor distribution within the clear and cloudy regimes.

Another issue is the impact on the simulations of global mean climate and climate variability through the improved representation of cloud-related processes in GCM's using the CRM simulations. Wu and Moncrieff (2001) used the CRM to identify the biases in the radiation and cloud scheme used in the GCM. Liang and Wu (2005) used the CRM to evaluate the treatment of subgrid cloud distribution in the radiation scheme; Wu and Liang (2005) used the CRM results to improve the climate simulations with the inclusion of effects of subgrid cloud–radiation interaction. With the inclusion of the convective momentum transport (CMT) scheme derived from the CRM simulations, the tropical convection and the Hadley circulation are better-represented in GCM's (Zhang and Wu, 2003; Wu *et al.*, 2003).

The other issue is the multiscale air–sea interaction processes associated with the

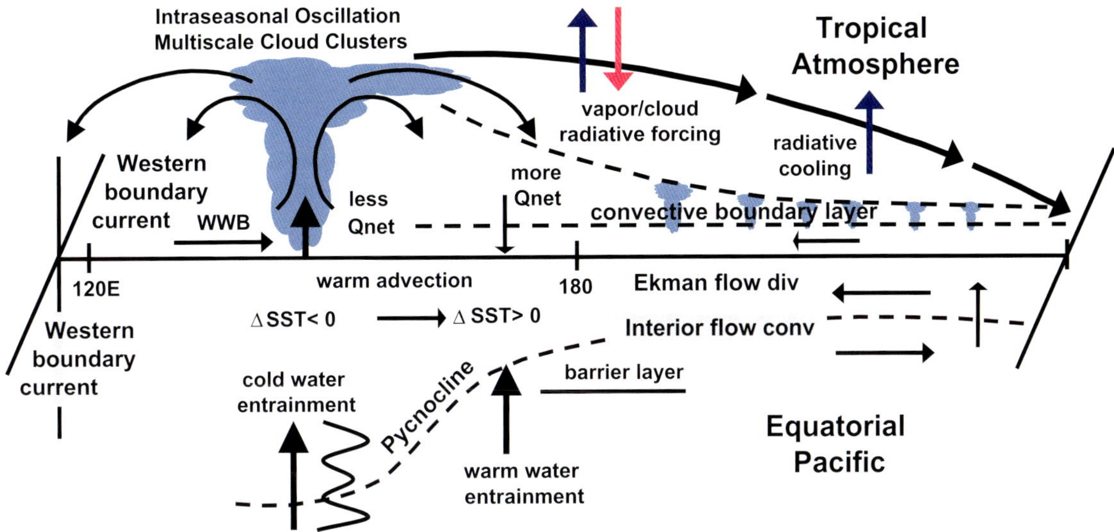

Figure 3. Summary diagram for the convective–radiative-mixing processes in the tropical ocean–atmosphere.

intraseasonal oscillations in the warm pool region illustrated in Fig. 3. Because the precipitation rate is greater than the evaporation rate over the equatorial western Pacific warm pool, whereas it is smaller over the eastern Pacific, a barrier layer appears year round within the warm pool (e.g. Lukas and Linstrom, 1991; Ando and McPhaden, 1997; Vialard and Delecluse, 1998ab). The barrier layer insulates the penetrated solar energy in the barrier layer from the mixed layer. As an intraseasonal oscillation propagates eastward to the warm pool, the strong westerly wind burst causes strong vertical mixing in the upper ocean that destroys the barrier layer and entrains cold water beneath into the mixed layer. The rapid decay of the intraseasonal oscillation near the dateline means that there is no impact on the barrier layer there and vertical entrainment would entrain the warm water beneath into the mixed layer. These entrainment processes together with the surface air–sea interaction processes illustrated in Fig. 3 may result in a cool SST anomaly over the warm pool and a warm SST anomaly near the dateline. The SST anomalies in turn induce an eastward extension of atmospheric intraseasonal oscillations, and reduce easterly trade winds by weakening the overall zonal SST gradient. The air–sea interaction might strengthen and prolong the intraseasonal oscillation and play an important role in the development stage of ENSO, which could be phase-locked to the annual cycle.

The above discussion proposes some key climate processes to be addressed by more advanced models with more physical representations of convective, radiative, and mixing processes. In addition, it is important to utilize new observations like satellite measurements to advance cloud-climate feedback study in climate models (e.g. Lau et al., 2005). There are also some critical issues related to convective–radiative-microphysical processes not discussed in this review article, and they deserve further investigation. For example, turbulence mixing in the boundary layer is still crudely parametrized in CRMs and needs to be improved. There are different time scales in the cloud-microphysical processes (condensation versus coalescence) and the probabilistic nature of microphysical processes that should be included in the microphysics parametrization (e.g. Chen and Liu, 2004). The cloud–aerosol interaction, which is not considered in most CRM's, needs to be studied comprehensively in the future.

Acknowledgements

We thank the two anonymous reviewers for their constructive comments, and are grateful for the continuous support of our research from the National Research Council in Taiwan, the National Aeronautics and Space Administration, the National Science Foundation, and the National Ocean and Atmosphere Administration in the US in the past 15 years. The GCE modeling work is mainly supported by the NASA Headquarters Physical Climate Program and the NASA TRMM. In particular, Dr. R. Kakar at NASA headquarters has continuously supported the model improvements and applications.

[Received 24 April 2006; Revised 16 May 2007; Accepted 27 July 2007.]

References

Ando, K., and M. J. McPhaden, 1997: Variability of surface layer hydrography in the tropical Pacific Ocean. *J. Geophys. Res.*, **102**, 23063–23078.

Arakawa, A., and W. H. Schubert, 1974: Interaction of a cumulus cloud ensemble with the large-scale environment. Part I. *J. Atmos. Sci.*, **31**, 674–701.

Atlas, R., O. Reale, B.-W. Shen, S.-J. Lin, J.-D. Chern, W. Putman, T. Lee, K.-S. Yeh, M. Bosilovich, and J. Radakovich, 2005: Hurricane forecasting with the high-resolution NASA finite volume general circulation model. *Geophys. Res. Lett.*, **32**, L03807, doi:10.1029/ 2004GL021513.

Betts, A. K., and M. J. Miller, 1986: A new convective adjustment scheme. Part II: Single column tests using GATE wave, BOMEX, ATEX and arctic airmass data sets. *Quart. J. Roy. Meteor. Soc.*, **112**, 692–709.

Betts, A. K., and W. Ridgway, 1988: Coupling of the radiative, convective and surface fluxes over the equatorial Pacific. *J. Atmos. Sci.*, **45**, 522–536.

Betts, A. K., and W. Ridgway, 1989: Climatic equilibrium of the atmospheric convective boundary layer over a tropical ocean. *J. Atmos. Sci.*, **46**, 2621–2641.

Braham, R. R., Jr., 1952: The water and energy budgets of the thunderstorm and their relation to thunderstorm development. *J. Meteor.*, **9**, 227–242.

Chen, J.-P., and S.-T. Liu, 2004: Physically based two-moment bulkwater parametrization for warm-cloud microphysics. *Quart. J. Roy. Meteor. Soc.*, **130**, 51–78.

Cheng, M.-D., and M. Yanai, 1989: Effects of downdrafts and mesoscale convective organization on the heat and moisture budgets of tropical cloud cluster. Part III: Effects of mesoscale convective organization. *J. Atmos. Sci.*, **56**, 3028–3042.

Chou, M.-D., D. P. Kratz, and W. Ridgway, 1991: IR radiation parametrization in numerical climate studies. *J. Climate*, **4**, 424–437.

Chou, M.-D., and M. J. Suarez, 1994: An efficient thermal infrared radiation parametrization for use in General Circulation Model. *NASA Technical Memorandum 104606*, Vol. 3, 85 pp.

Chou, M.-D., M. J. Suarez, C.-H. Ho, M. M.-H. Yan, and K.-T. Lee, 1997: Parameterizations for cloud overlapping and shortwave single-scattering properties for use in General Circulation and Cloud Ensemble Models. *J. Climate*, **11**, 202–214.

Del Genio, A. D., N.-S. Yao, W. Kovari, and K.-W. Lo, 1996: A prognostic cloud water parametrization for general circulation models. *J. Climate*, **9**, 270–304.

Doswell, C. A., III, H. E. Brooks, and R. A. Maddox, 1996: Flash flood forecasting: an ingredients-based methodology. *Wea. Forecasting*, **11**, 560–581.

Ferrier, B. S., J. Simpson, and W.-K. Tao, 1996: Factors responsible for different precipitation efficiencies between midlatitude and tropical squall simulations. *Mon. Wea. Rev.*, **124**, 2100–2125.

Fowler, L. D., D. A. Randall, and S. A. Rutledge, 1996: Liquid and ice cloud microphysics in the CSU general circulation model. Part I: Model description and simulated microphysical processes. *J. Climate*, **9**, 489–529.

Fu, R., A. D. Del Genio, W. B. Rossow, and W. T. Liu, 1992: Cirrus cloud thermostat for tropical sea surface temperatures tested using satellite data. *Nature*, **358**, 394–397.

Grabowski, W. W., X. Wu, and M. W. Moncrieff, 1996: Cloud-resolving model of tropical cloud systems during Phase III of GATE. Part I: Two-dimensional experiments. *J. Atmos. Sci.*, **53**, 3684–3709.

Grabowski, W. W., 2001: Coupling cloud processes with convection parameterizaiton (CRCP). *J. Atmos. Sci.*, **58**, 978–997.

Grabowski, W. W., and P. K. Smolarkiewicz, 1999: CRCP: A cloud-resolving convection parametrization for modeling the tropical convective atmosphere. *Physica D*, **133**, 171–178.

Gray, W. M., and R. W. Jacobson, 1977: Diurnal variation of deep cumulus convection. *Mon. Wea. Rev.*, **105**, 1171–1188.

Hartmann, D. L., and M. L. Michelsen, 1993: Large-scale effects on the regulation of tropical sea surface temperature. *J. Climate*, **6**, 2049–2062.

Hsie, E. Y., R. D. Farley, and H. D. Orville, 1980: Numerical simulation of ice phase convective cloud seeding. *J. Appl. Meteor.*, **19**, 950–977.

Inamdar, A. K., and V. Ramanathan, 1994: Physics of greenhouse effect and convection in warm oceans. *J. Climate*, **7**, 715–731.

Johnson, D., W.-K. Tao, J. Simpson, and C.-H. Sui, 2002: A study of the response of deep tropical clouds to mesoscale processes: Part I: Modeling strategies and simulations of TOGA-COARE convective systems. *J. Atmos. Sci.*, **59**, 3492–3518.

Khairoutdinov, M. F., and D. A. Randall, 2001: A cloud-resolving model as a cloud parametrization in the NCAR Community Climate System Model: Preliminary results. *Geophys. Res. Lett.*, **28**, 3617–3620.

Kiehl, J. T., 1994: On the observed near cancellation between longwave and shortwave cloud forcing in tropical regions. *J. Climate*, **7**, 559–565.

Klemp, J. B., and R. B. Wilhelmson, 1978: The simulation of three-dimensional convective storm dynamics. *J. Atmos. Sci.*, **35**, 1070–1093.

Koenig, L. R., 1971: Numerical modeling of ice deposition. *J. Atmos. Sci.*, **28**, 226–237.

Kraus, E. B., 1963: The diurnal precipitation change over the sea. *J. Atmos. Sci.*, **20**, 546–551.

Krueger, S. K., Q. Fu, K. N. Liou, and H.-N. S. Chin, 1995: Improvement of an ice-phase microphysics parametrization for use in numerical simulations of tropical convection. *J. Appl. Meteor.*, **34**, 281–287.

Kuo, H. L., 1965: On formation and intensification of tropical cyclones through latent heat release by cumulus convection. *J. Atmos. Sci.*, **22**, 40–63.

Kuo, H. L., 1974: Further studies of the parametrization of the influence of cumulus convection on large-scale flow. *J. Atmos. Sci.*, **31**, 1232–1240.

Lau, K.-M., T. Nakazawa, and C. H. Sui, 1991: Observations of cloud cluster hierarchy over the tropical western Pacific. *J. Geophys. Res.* **96**, 3197–3208.

Lau, K.-M., C.-H. Sui, and W.-K. Tao, 1993: A preliminary study of the tropical water cycle using the Goddard Cumulus Ensemble model. *Bull. Amer. Meteor. Soc.* **74**, 1313–1321.

Lau, K.-M., C.-H. Sui, M.-D. Chou, and W.-K. Tao, 1994a: An inquiry into the cirrus-cloud thermostat effect for tropical sea surface temperature. *Geophys. Res. Lett.*, **21**, 1157–1160.

Lau, K.-M., C.-H. Sui, M.-D. Chou, and W.-K. Tao, 1994b: Reply to the comment on the paper "An inquiry into the cirrus-cloud thermostat effect for tropical sea surface temperature," by V. Ramanathan, W. D. Collins, and B. Subasilar. *Geophys. Res. Lett.*, **21**, 1187–1188.

Lau, K.-M., and C.-H Sui, 1997: Mechanisms of short-term sea surface temperature regulation: observations during TOGA COARE. *J. Climate*, **10**, 465–472.

Lau, K. M., and H. T. Wu, 2003: Warm rain over tropical oceans and climate implications. *Geophys. Res. Lett.*, **30**, doi:10.1029/ 2003GL018567.

Lau, K. M., H. T. Wu, Y. C. Sud, and G. K. Walker, 2005: Effects of cloud microphysics on tropical atmospheric hydrologic processes and intraseasonal variability in the GEOS GCM, *J. Climate*, **18**, 4731–4751.

Liang, X.-Z., and X. Wu, 2005: Evaluation of a GCM subgrid cloud–radiation interaction parametrization using cloud-resolving model simulations. *Geophys. Res. Lett.*, **32**, L06801, doi:10.1029/2004GL022301.

Li, X., C.-H. Sui, D. Adamec, and K.-M. Lau, 1998: Impacts of precipitation in the upper ocean in the western Pacific warm pool during TOGA COARE. *J. Geophys. Res.*, **103**, C3, 5347–5359.

Li, X., C.-H. Sui, K.-M. Lau, and M.-D. Chou, 1999: Large-scale forcing and cloud–radiation interaction in the tropical deep convective regime. *J. Atmos. Sci.*, **56**, 3028–3042.

Li, X., C.-H. Sui, K.-M. Lau, and D. Adamec, 2000: Effects of precipitation on ocean mixed-layer temperature and salinity as simulated in a 2-D coupled ocean–cloud-resolving atmosphere model. *J. Meteor. Soc. Japan*, **78**, 647–659.

Li, X., C.-H. Sui, and K.-M. Lau, 2002a: Precipitation efficiency in the tropical deep convective regime: a 2-D cloud-resolving modeling study. *J. Meteor. Soc. Japan*, **80**, 205–212.

Li, X., C.-H. Sui, and K.-M. Lau, 2002b: Interactions between tropical convection and its environment: an energetics analysis of a 2-D cloud resolving simulation. *J. Atmos. Sci.*, **59**, 1712–1722.

Li, X., C.-H. Sui, and K.-M. Lau, 2002c: Dominant cloud-microphysical processes in a tropical oceanic convective system: a 2-D cloud resolving modeling study. *Mon. Wea. Rev.*, **130**, 2481–2491.

Li, X., C.-H. Sui, K.-M. Lau, and W.-K. Tao, 2005: Tropical convective responses to microphysical and radiative processes: a sensitivity study with a 2D cloud-resolving modeling study. *Meteor. Atmos. Phys.*, **90**, 245–259.

Lin, S.-J., 2004: A "vertically Lagrangian" finite-volume dynamical core for global models. *Mon. Wea. Rev.*, **132**, 2293–2307.

Lin Y. L., R. D. Farley, and H. D. Orville, 1983: Bulk parametrization of the snow field in a cloud model. *J. Climate Appl. Meteor.*, **22**, 1065–1092.

Lipps, F. B., and R. S. Hemler, 1986: Numerical simulation of deep tropical convection associated with large-scale convergence. *J. Atmos. Sci.*, **43**, 1796–1816.

Lindzen, R. S., 1990: Some coolness concerning global warming. *Bull. Amer. Meteor. Soc.*, **71**, 288–299.

Lindzen, R. S., M. D. Chou, and A. Hou, 2001: Does the earth have an adaptive infrared Iris? *Bull. Am. Meteor. Soc.*, **82**, 417–432.

Lorenz, E. N., 1955: Available potential energy and the maintenance of the general circulation. *Tellus*, **7**, 157–167.

Lukas, R., and E. Lindstrom, 1991: The mixed layer of the western equatorial Pacific Ocean. *J. Geophys. Res.*, **96**, 3343–3457.

Mapes, B. E., 1993: Gregarious tropical convection. *J. Atmos. Sci.*, **50**, 2026–2037.

Mapes, B. E., and R. A. Houze, Jr., 1993: Cloud clusters and superclusters over the oceanic warm pool. *Mon. Wea. Rev.*, **121**, 1398–1415.

Miura, H., H. Tomita, T. Nasuno1, S. Iga1, M. Satoh, and T. Matsuno, 2005: A climate sensitivity test using a global Cloud Resolving Model under an aqua planet condition. *Geophys. Res. Lett.*, **32**, L19719, dot:10.1029/2005GL023672.

Newell, R. E., 1979: Climate and the ocean. *Amer. Sci.*, **67**, 405–416.

Niiler, P. P., and E. B. Kraus, 1977: One-dimensional models. In *Modeling and Prediction of the Upper Layers of the Ocean*, E. B. Kraus (ed.), Pergamon, New York, pp. 143–172.

Orville, H. D., and F. J. Kopp, 1977: Numerical simulation of the life history of a hailstorm. *J. Atmos. Sci.*, **34**, 1596–1618.

Peng, L., C.-H. Sui, K.-M. Lau, and W.-K. Tao, 2001: Genesis and evolution of hierarchical cloud clusters in a two-dimensional cumulus-resolving model. *J. Atmos. Sci.*, **58**, 877–895.

Prabhakara, C., D. P. Kratz, J.-M. Yoo, G. Dalu, and A. Vernekar, 1993: Optically thin cirrus clouds: radiative impact on the warm pool. *J. Quant. Spectrosc. Radiat. Transfer.*, **49**, 467–483.

Ramanathan, V., and W. Collins, 1991: Thermodynamic regulation of ocean warming by cirrus clouds deduced from observations of the 1987 El Niño. *Nature*, **351**, 27–32.

Ramanathan, V., W. D. Collins, and B. Subasilar, 1994: Comments on the paper "An inquiry into the cirrus-cloud thermostat effect for tropical sea surface temperature." *Geophys. Res. Lett.*, **21**, 1185–1186.

Randall, D. A., and D.-M. Pan, 1993: Implementation of the Arakawa–Schubert cumulus parametrization with a prognostic closure. In *The Representation of Cumulus Convection in Numerical Models of the Atmosphere*, K. A. Emanuel and D. J. Raymond, (eds.) *Meteor. Monogr.*, No. 46, pp. 37–144.

Randall, D. A., Harshvardhan and D. A. Dazlich, 1991: Diurnal variability of the hydrologic cycle in a general circulation model. *J. Atmos. Sci.*, **48**, 40–62.

Randall, D. A., M. Khairoutdinov, A. Arakawa, and W. Grabowski, 2003: Breaking the cloud parametrization deadlock.*Bull. Amer. Meteor. Soc.*, 1547–1564.

Rutledge, S. A., and R. V. Hobbs, 1983: The mesoscale and microscale structure and organization of clouds and precipitation in midlatitude cyclones. Part VIII: A model for the "seeder–feeder" process in warm-frontal rainbands. *J. Atmos. Sci.*, **40**, 1185–1206.

Rutledge, S. A., and R. V. Hobbs, 1984: The mesoscale and microscale structure and organization of clouds and precipitation in midlatitude cyclones. Part XII: A diagnostic modeling study of precipitation development in narrow cold-frontal rainbands. *J. Atmos. Sci.*, **41**, 2949–2972.

Satoh, M., H. Tomita, H. Miura, S. Iga, and T. Nasuno, 2005: Development of a global cloud resolving model — a multi-scale structure of tropical convections. *J. Earth Simulator*, **3**, 1–9.

Simpson, J., R. Adler, and G. North, 1988: A proposed Tropical Rainfall Measuring Mission

(TRMM) satellite. *Bull. Amer. Meteor. Soc.*, **69**, 278–295.

Soden, B. J., 1997: Variations in the tropical greenhouse effect during El Niño. *J. Climate*, **10**, 1050–1055.

Soong, S. T., and Y. Ogura, 1980a: Response of tradewind cumuli to large-scale processes. *J. Atmos. Sci.*, **37**, 2035–2050.

Soong, S. T., and W. K. Tao, 1980b: Response of deep tropical cumulus clouds to mesoscale processes. *J. Atmos. Sci.*, **37**, 2016–2034.

Sud, Y. C., and G. K. Walker, 1999: Microphysics of clouds with the relaxed Arakawa–Schubert cumulus scheme (McRAS); Part I: Design and evaluation with GATE Phase III data. *J. Atmos. Sci.*, **56**, 3196–3220.

Sud, Y. C., and G. K. Walker, 2003: Influence of ice-phase physics of hydrometeors on moist convection. *Geophys. Res. Lett.* **30**, 14, 1758–1761.

Sui, C.-H., K.-M. Lau, and A. K. Betts, 1991: An equilibrium Model for the coupled ocean–atmosphere boundary layer in the tropics. *J. Geophys. Res.*, **96**, 3151–3163.

Sui, C.-H., and K.-M. Lau, 1992: Multi-scale phenomena in the tropical atmosphere over the western Pacific. *Mon Wea. Rev.*, **120**, 407–430.

Sui, C.-H., K.-M. Lau, W.-K. Tao, and J. Simpson, 1994: The tropical water and energy cycles in a cumulus ensemble model. Part I: Equilibrium climate. *J. Atmos. Sci.*, **51**, 711–728.

Sui, C.-H., K.-M. Lau, Y. Takayabu, and D. Short, 1997a: Diurnal variations in tropical oceanic cumulus ensemble during TOGA COARE. *J. Atmos. Sci.*, **54**, 639–655.

Sui, C.-H., X. Li, K.-M. Lau, and D. Adamec, 1997b: Multi-scale air–sea interactions during TOGA COARE. *Mon. Wea. Rev.*, **125**, 448–462.

Sui, C.-H., X. Li, and K.-M. Lau, 1998a: Radiative–convective processes in simulated diurnal variations of tropical oceanic convection. *J. Atmos. Sci.*, **55**, 2345–2359.

Sui, C.-H., X. Li, and K.-M. Lau, 1998b: Selective absorption of solar radiation and upper ocean temperature in the equatorial western Pacific. *J. Geophys. Res.*, **103**, C5, 10313–10321.

Sui, C.-H., X. Li, M. M. Rienecker, K.-M. Lau, I. Laszlo, and R. T. Pinker, 2003: The impacts of daily surface forcing in the upper ocean over tropical Pacific: a numerical study. *J. Climate*, **16**, 756–766.

Sui, C.-H., and Li, X., 2005: A tendency of cloud ratio associated with the development of tropical water and ice clouds. *Terr. Atmos. Oceanic Sci.*, **16**, 419–434.

Sui, C.-H., X. Li, M.-J. Yang, and H.-L. Huang, 2005: Estimation of oceanic precipitation efficiency in cloud models. *J. Atmos. Sci.*, **62**, 4358–4370.

Sui, C.-H., C.-T. Tsay, and X. Li, 2007a: Convective–stratiform rainfall separation by cloud content. *J. Geophys. Res.* **112**, D14213, doi:10.1029/2006JD008082.

Sui, C.-H., X. Li, and M.-J. Yang, 2007b: On the definition of precipitation efficiency. *J. Atmos. Sci.*, **64**, 4506–4513.

Sundqvist, H., 1988: Parameterization of condensation and associated clouds in models from weather prediction and general circulation simulation. In *Physical Based Modeling and Simulation of Climate and Climate Change*, M. E. Schlesinger (ed.), Reidel, pp. 433–461.

Sundqvist, H., 1978: A parametrization of non-convective condensation including prediction of cloud water content. *Quart. J. Roy. Meteor. Soc.*, **104**, 677–690.

Tao, W.-K, 2003: Goddard Cumulus Ensemble (GCE) model: application for understanding precipitation processes. In *Cloud Systems, Hurricanes, and the Tropical Rainfall Measuring Mission (TRMM): A Tribute to Dr Joanne Simpson, Meteor. Monogr.,* No. 51, *Amer. Meteor. Soc.*, pp. 107–138.

Tao, W.-K., and J. Simpson, 1984: Cloud interactions and merging: numerical simulations. *J. Atmos Sci.*, **41**, 2901–2917.

Tao, W.-K., and S.-T. Soong, 1986: The study of the response of deep tropical clouds to mesoscale processes: three-dimensional numerical experiments. *J. Atmos. Sci.*, **43**, 2653–2676.

Tao, W.-K., and J. Simpson, 1989: A further study of cumulus interactions and merging: three-dimensional simulations with trajectory analyses. *J. Atmos. Sci.*, **46**, 2974–3004.

Tao, W.-K., and J. Simpson, 1993: The Goddard cumulus ensemble model. Part I: Model description. *Terr., Atmos. Oceanic Sci.*, **4**, 35–72.

Tao, W.-K., J. Simpson, and M. McCumber, 1989: An ice-water saturation adjustment. *Mon. Wea. Rev.*, **117**, 231–235.

Tao, W. K., S. Lang, J. Simpson, C.-H Sui, B. S. Ferrier, and M.-D. Chou, 1996: Mechanisms of cloud–radiation interaction in the tropics and midlatitude. *J. Atmos. Sci.*, **53**, 2624–2651.

Tao, W.-K., J. Simpson, D. Baker, S. Braun, M.-D. Chou, B. Ferrier, D. Johnson, A. Khain, B. Lynn, S. Lang, C.-L. Shie, C.-H. Sui, Y. Wang, and P. Wetzel, 2003: Microphysics, radiation, and surface processes in a non-hydrostatic model. *Meteor. Atmos. Phys.*, **82**, 97–137.

Tao, W.-K., D. Johnson, C.-L. Shie, and J. Simpson, 2004: The atmospheric energy budget and large-scale precipitation efficiency of convective systems during TOGA COARE, GATE, SCSMEX, and ARM: cloud-resolving model simulations. *J. Atmos. Sci.*, **61**, 2405–2423.

Tiedtke, M., 1993: Representation of clouds in large scale models. *Mon. Wea. Rev.*, **121**, 3040–3061.

Tomita, H., H. Miura, S. Iga, T. Nasuno, and M. Satoh, 2005: A global cloud-resolving simulation: preliminary results from an aqua planet experiment, *Geophys. Res. Lett.*, **32**, L08805, doi:10.1029/2005GL022459.

Thompson, R. M., Jr., S. W. Payne, E. E. Recker, and R. J. Reed, 1979: Structure and properties of synoptic-scale wave disturbances in the intertropical convergence zone of the eastern Atlantic. *J. Atmos. Sci.*, **36**, 53–72.

Vialard, J., and P. Delecluse, 1998a: An OGCM study for the TOGA decade. Part I: Role of salinity in the physics of the western Pacific fresh pool. *J. Phys. Oceanog.*, **28**, 1071–1088.

Vialard, J., and P. Delecluse, 1998b: An OGCM study for the TOGA decade. Part II: Barrier layer formation and variability. *J. Phys. Oceanog.*, **28**, 1089–1106.

Wang, J., and D. A. Randall, 1994: The moist available energy of a conditionally unstable atmosphere. Part II: Further analysis of GATE data. *J. Atmos. Sci.*, **51**, 703–710.

Weisman, M. L., and J. B. Klemp, 1982: The dependence of numerically simulated convective storms on vertical wind shear and buoyancy. *Mon. Wea. Rev.*, **110**, 504–520.

Wu, X., and X.-Z. Liang, 2005: Effect of sub-grid cloud–radiation interaction on climate simulations. *Geophys. Res. Lett.*, **32**, L06801, doi:10.1029/2004GL022301.

Wu, X., and M. W. Moncrieff, 2001: Long-term behavior of cloud systems in TOGA COARE and their interactions with radiative and surface processes. Part III: Effects on the energy budget and SST. *J. Atmos. Sci.*, **58**, 1155–1168.

Wu, X., X.-Z. Liang, G.-J., Zhang 2003: Seasonal migration of ITCZ precipitation across the equator: Why can't GCMs simulate it? *Geophys. Res. Lett.*, **30**, 1824, doi:10.1029/2003GL017198.

Wu, X., W. W. Grabowski, and M. W. Moncrieff, 1998: Long-term evolution of cloud systems in TOGA COARE and their interactions with radiative and surface processes. Part I: Two-dimensional cloud-resolving model. *J. Atmos. Sci.*, **55**, 2693–2714.

Xu, K.-M., and D. A. Randall, 1995: Impact of interactive radiative transfer on the macroscopic behavior of cumulus ensembles. Part II: Mechanisms for cloud–radiation interactions. *J. Atmos. Sci.*, **52**, 800–817.

Xu, K.-M., and D. A. Randall, 1996: Explicit simulation of cumulus ensembles with the GATE Phase III data: comparison with observations. *J. Atmos. Sci.*, **53**, 3710–3736.

Xu, K.-M., and D. A. Randall, 1998: Influence of large-scale advective cooling and moistening effects on the quasi-equilibrium behavior of explicitly simulated cumulus ensembles. *J. Atmos. Sci.*, **55**, 896–909.

Yanai, M., S. Esbensen, and J.-H. Chu, 1973: Determination of bulk properties of tropical cloud clusters from large-scale heat and moisture budgets. *J. Atmos. Sci.*, **30**, 611–627.

Yang, M.-J., and H.-L. Huang, 2004: Precipitation processes associated with the landfalling Typhoon Nari (2001). In *Preprints, The 26th Conference on Hurricanes and Tropical Meteorology*, Miami, 3–7 May 2004, *Amer. Meteor. Soc.*, pp. 134–135.

Zhang, G.-J., and X. Wu, 2003: Convective momentum transport and perturbation pressure field from a cloud-resolving model simulation. *J. Atmos. Sci.*, **60**, 1120–1139.

Zhang, M. H., R. D. Cess, and S. C. Xie, 1996: Relationship between cloud radiative forcing and sea surface temperatures over the entire tropical oceans. *J. Climate*, **9**, 1374–1384.

Zhao, Q., and F. H. Carr, 1997: A prognostic cloud scheme for operational NWP models. *Mon. Wea. Rev.*, **125**, 1931–1953.

Understanding Atmospheric Catastrophes

Winston C. Chao

Laboratory for Atmospheres, Goddard Space Flight Center, Greenbelt, Maryland, USA
Winston.c.chao@nasa.gov

The atmosphere, like other parts of nature, is full of phenomena that involve rapid transitions from one (quasi-)equilibrium state to another, i.e. catastrophes. These (quasi-)equilibria are the multiple solutions of the same dynamical system. Unlocking the mystery behind a catastrophe reveals not only the physical mechanism responsible for the transition, but also how the (quasi-)equilibria before and after the transition are maintained. Each catastrophe is different, but they do have some common traits. Understanding these common traits is the first step in studying these catastrophes. In this article, three examples are reviewed to show how atmospheric catastrophes can be studied.

1. Introduction

Changes in the physical world are of two basic types: gradual and sudden. Gradual changes are the norm; sudden changes are less frequent, larger in magnitude, and easily recognizable. A physical system may experience gradual change over a long period of time, and then "all of a sudden," a rapid change may occur. For example, at a typical tropical island station in the Caribbean, the surface air temperature gradually rises after sunrise and continues to rise till late afternoon, but then suddenly drops owing to a thunderstorm. Often a sudden change occurs without any sudden rapid external stimulus, such that a slow gradual change in a system gives way to a sudden change. Water suddenly starting to boil in a heated pot is an example. At other times a sudden change is triggered by external events, but the amount of energy provided by the trigger is usually much less than that associated with the rapid change in the physical system initiated by the trigger. An example is the triggering of cumulus convection by air flowing over hills.

When the dependence of the equilibrium state of a dynamical system on an external parameter is nonlinear, multiple equilibrium states may occur. [The equilibrium state may be stationary (fixed point), time-periodic (limit cycle), or time-aperiodic (quasi-equilibrium, including strange attractor)]. Figure 1 illustrates such a conceptual picture. When the externally adjustable parameter ε is $\varepsilon_1 < \varepsilon < \varepsilon_2$, two (quasi-)equilibria exist. If initially the system is on the lower branch when ε is raised passing ε_2, the system jumps from the lower branch to the upper branch. The speed of the transition and the degree of overshooting before the system settles down on the new solution branch vary from system to system. Regardless, a transition such as this from one (quasi-)equilibrium solution to another when an external parameter passes a critical value, is called a catastrophe. Alternative to an external parameter passing a critical value, a catastrophe can be triggered by an external disturbance providing enough energy for the system to overcome the energy hump, which is of course larger when the external parameter is further away from the

                                        Winston C. Chao

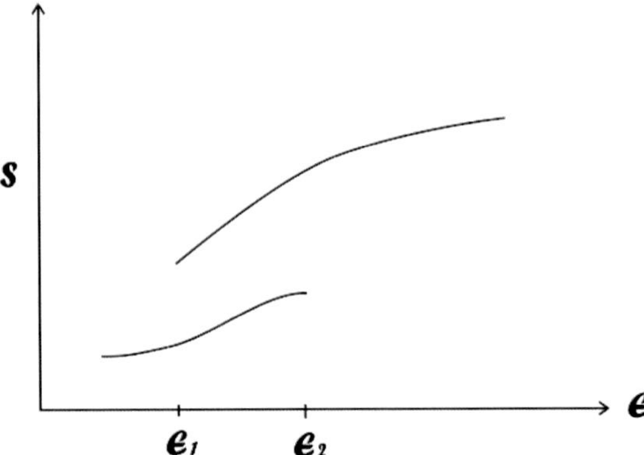

Figure 1. Schematic diagram showing the state of a system as a function of an external parameter ε. The system has two states between ε_1 and ε_2.

critical value of ε_2. During the jump, the system may overshoot the new equilibrium and bounce back and oscillate about the new equilibrium before settling down. The overshooting may or may not occur, depending on the transient response of the system and the frictional force it experiences. A simple example to further illustrate a catastrophe is as follows (Fig. 2). A particle is held in a potential well as the structure of the well changes, the well eventually disappears, and the particle falls into a neighboring well. This fall takes place suddenly; it is called a spontaneous catastrophe. The fall can also take place without the well which the particle resides in disappearing, if the particle is pushed over the top of the well (energy barrier). This is called a triggered catastrophe.

The concept of a catastrophe (Thom, 1972; Poston and Stewart, 1978; Saunders, 1980; Iooss and Joseph, 1980) exists in many disciplines, and has been discovered by many researchers independently. Thus, it is not surprising that catastrophes have been given many other names.

Figure 2. Schematic diagram showing the state of a particle in a potential well. As an external parameter changes, the shape of the potential well changes such that the particle changes its position suddenly.

In many disciplines a catastrophe is called a "subcritical instability" (e.g. Drazin, 1992). The name "hysteresis" is also used when the emphasis is on the occurrence of multiple (quasi-) equilibria instead of the suddenness of the transitions. In engineering, it is called a "structural instability." In the physical sciences, it is often called a "critical phenomenon," owing to the fact that a parameter must exceed a critical value for the catastrophe to occur; or it may be called a "nonlinear instability," owing to the fact that, unlike in a linear instability, the growth rate is not constant. The term "nonlinear instability" includes subcritical and supercritical instabilities (Drazin, 1992). The forcing acting on a system to pull it away from an equilibrium that has just disappeared (in the case of a spontaneous catastrophe) increases as the system moves further away from the equilibrium. This is followed by a reduction of the forcing as the system approaches the other equilibrium. Thus the jump is under positive and then negative acceleration after the system overshoots the new equilibrium. Such accelerated growth lasts, of course, only for a finite time period and will eventually have to stop owing to the finite energy supply of the basic state; that is, when the system reaches the other quasi-equilibrium. The name "explosive instability" (Sturrock, 1966) is sometimes used, because the forcing pushing the system away from the initial (quasi-)equilibrium increases with time initially. Since it can be triggered by a finite-amplitude initial disturbance (a trigger), it is also called "finite-amplitude instability." In using the term "finite-amplitude instability," one should note that triggering is merely an alternative way to initiate the instability and is not necessary when an external parameter is changed to exceed a critical value. Nonetheless, in many cases triggering is the more likely way an instability gets started, owing to the frequent occurrence of perturbations in the atmosphere. Before a triggered catastrophe occurs, the system is stable, and this stability is called "metastability."

The term "catastrophe," for a while, was considered an oversold concept and it incurred some connotation of a fraud when it was overly publicized in the 1970's. It has since regained its respectability after the initial commotion died down. It is now accepted as a standard mathematical term (Arnold, 1981) without the implication of something necessarily negative happening. We use it instead of terms equivalent to it — "explosive instability," "subcritical instability," "finite-amplitude instability," "critical phenomenon," and "structural instability" — for the sake of brevity.

In attempting to identify if a particular phenomenon is a catastrophe, one strives to ascertain if the common traits of catastrophes can be found. These common traits are bimodality (or multiple equilibria with at least two stable equilibria and an unstable one in between), suddenness of occurrence and spontaneous or triggered onset. Each catastrophe has its own cause for multiple equilibria, feedback processes in the transition, and triggering mechanisms (if they exist). The majority of the catastrophes can be easily identified through observation. One looks first for suddenness of the change. Suddenness means that the duration of the change is short, relative to the less eventful periods before and after the event. Bimodality — the next thing one looks for — means that the states before and after the change are clearly different, i.e. the change is sizable relative to the changes that take place before and after the transition. A phenomenon termed "onset" or "genesis" is very likely a catastrophe. Identification of the triggers of a catastrophe is not always very easy, since triggers do not always exist. Moreover, events concurrent with a catastrophe may be a result of that catastrophe rather than the triggers of it. Thus identifying triggers when they are not obvious can only be reliably accomplished through theoretical and modeling work, as in

the case of a complete understanding of each catastrophe.

As far as understanding a catastrophe is concerned, the terms in the governing equation of a system that exhibits catastrophic behavior can be grouped (at least conceptually) into two competing sets (or two forcings), A and B. Thus,

$$\frac{\partial S}{\partial t} = A - B \qquad (1)$$

In a multi-dimensional system, S is a judicially chosen single variable that represents the *gross* state of the system. Figure 3 is an example of Eq. (1). Initially, the state is at the stable equilibrium represented by point S_1 in Fig. 3.[a] Point S_2 is an unstable (quasi-)equilibrium state and point S_3 is another stable equilibrium state. As one or more parameters change such that curve A moves right ward while keeping its slope and/or the peak of curve B diminishes to such a degree that the point S_1 (quasi-)equilibrium can no longer be sustained, the state moves rapidly to (quasi-)equilibrium point S_3. The movement

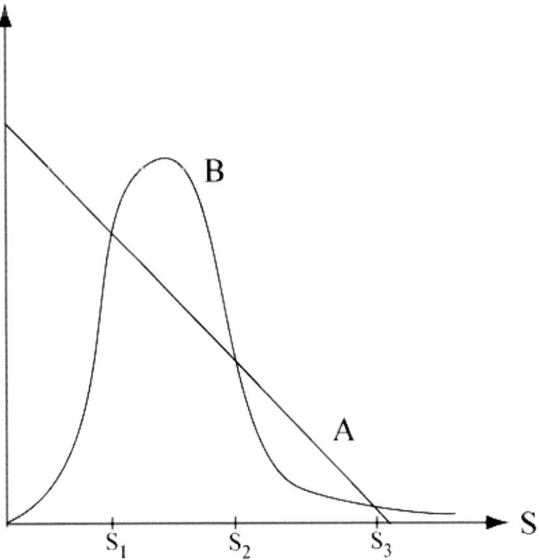

Figure 3. A and B as functions of S. The equilibrium states S_1 and S_3 are stable, but not the one at S_2.

---

[a]Since we have used S to denote a gross state, points 1 and 3 may be quasi-equilibria, rather than equilibria, in a multidimensional system.

is propelled by the difference between A and B in a "free fall" (more on this shortly). Figure 3 was used by Held (1983) to explain the topographically induced Rossby wave instability of Charney and DeVore (1979), which is a catastrophe. It is obvious that the existence of multiple quasi-equilibria alone is not sufficient for a catastrophe to occur. The relative movement of A and B induced by changes in external parameters or a trigger is necessary for a catastrophe to occur.

Figure 3 is a useful conceptual figure; however, it needs some clarification. According to Eq. (1), when the state reaches S_3, $\partial S/\partial t = 0$ and the state should stop changing and no overshooting should occur. But overshooting is often observed. The correct description is that A and B are functions of not only S but also $\partial S/\partial t$; what is depicted in Fig. 3 is A and B when $\partial S/\partial t = 0$, i.e. steady state forcings. Held (1983) pointed out that the curve B he used represents the drag exerted on the zonal flow by the *steady* forced waves generated by topography in the presence of dissipation. Thus the "free fall" is driven by a forcing greater than what appears in Fig. 3 as A-B.

Moreover, within the domain of the S-variable where one of the forcings varies highly nonlinearly, the other forcings usually do not vary highly nonlinearly and can thus be represented by a linear or a quasi-linear line; however, there can be exceptions.

Other examples of catastrophes are abundant. A simple example is the buckling found in flipping a wall switch or an earthquake. Another example is an explosion of any type. Other atmospheric examples are also abundant. In this article, three atmospheric catastrophes — tropical cyclogenesis, stratospheric warming, and monsoon onset — chosen based on the author's research interest are reviewed in some detail, to illustrate how one can study atmospheric catastrophes. Our understanding of these phenomena is still not complete; we will try to point out areas that require further study. Many more atmospheric catastrophes are enumerated below with varying degrees of commentary. In studying these catastrophes, one strives to identify the forces or forcings within that explain the associated multiple equilibria to arrive at a schematic diagram like Fig. 3. Also, one must support diagrams such as this by theoretical arguments and/or numerical experiments.

2. Tropical Cyclogenesis

Tropical cyclogenesis is the transition from a cloud cluster to a tropical cyclone (TC). Prior to the transition, the cloud cluster can last for days with little change in its characteristics. Thus, one can claim that the cloud cluster is in a quasi-equilibrium state. After the transition, the tropical cyclone can also last for days without changing its basic characteristics. Hence, one can identify the tropical cyclone as a quasi-equilibrium state as well.

Tropical cyclogenesis typically takes only about two or three days. Relative to the duration of either the cloud cluster or the TC, this transition period is very short. Tropical cyclogenesis occurs when the cloud cluster (quasi-equilibrium) state can no longer be sustained (a spontaneous catastrophe) or when a trigger acts on a cloud cluster (a triggered catastrophe). A spontaneous tropical cyclogenesis must be associated with a certain condition being met; in a triggered tropical cyclogenesis this condition is very nearly being met. Although the precise details of this condition are still unknown, it is generally associated with what is already known: SST higher than $26.5°C$, low background vertical wind shear, and a sufficiently high Coriolis parameter (Chap. 15 of Palmen and Newton, 1969). When the condition is met, all cloud clusters can turn into TC's. It is well known that a series of TC's often occur concurrently. Through these identified characteristics — two quasi-equilibrium states and the rapid transition between the two, etc., — tropical cyclogenesis clearly can be identified as a catastrophe.

It is heuristic to construct a schematic diagram similar to Fig. 3 for understanding the catastrophic nature of tropical cyclogenesis. One can use the 3D mass-weighted average temperature in the core region (e.g. within a 30 km radius) of a disturbance minus that of the environment — i.e. the degree of warming of the core region — as the gross state variable S. Curve A represents the diabatic heating — which is mostly cumulus heating — in the core region. Curve B is the adiabatic cooling due to upward motion in the same region. Assuming that the thermal balance is mainly between A and B, A-B is zero at both quasi-equilibria: the cloud cluster and the TC. Figure 4(a) shows A-B at the moment the cloud cluster loses its equilibrium status. A-B is zero at both the cloud cluster quasi-equilibrium and the TC quasi-equilibria. As shown in Fig. 4(a) at the moment the cloud cluster loses it equilibrium status, A-B does not cross the abscissa at the cloud cluster quasi-equilibrium S_1 but only touches it tangentially. A-B crosses zero at the TC quasi-equilibrium S_3, which is a stable quasi-equilibrium, since a perturbation away from this quasi-equilibrium will be reduced by A-B to zero. The positive value of A-B on the left side of the TC quasi-equilibrium S_3 ensures that the state starting from the cloud cluster status will move to the TC status.

Curve B can be roughly represented by a straight line through the origin. This is because a disturbance with a higher core temperature has a stronger meridional circulation, which implies stronger adiabatic cooling in the core region. As a result of these considerations, A and B at the moment of the cloud cluster's losing its quasi-equilibrium status can be represented as in Fig. 4(b). Before that moment, the picture is as depicted in Fig. 4(c). In this figure, the cloud cluster corresponds to S_1 and the TC S_3. S_2 is an unstable quasi-equilibrium. As the boundary conditions change (e.g. the

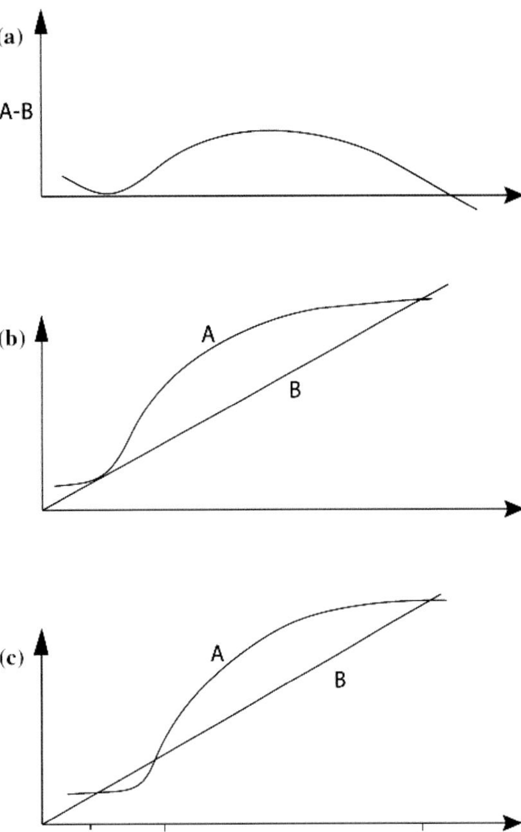

Figure 4. Schematic diagram showing the strength of the forcings on a tropical disturbance. The abscissa is the state of the system, S_1 denotes the cloud cluster state and S_3 the TC. Curve A is the adiabatic heating and line B is the cooing due to adiabatic upward motion in the core region. See text for more details.

SST increases), curve A moves upward and S_1 and S_2 merge and then both disappear, i.e. the cloud cluster can no longer be maintained. The system rapidly moves toward the TC state in a "free fall" pulled by a forcing greater than A-B. There can be overshooting.

The remaining task is to explain the shape of curve A. As S increases, curve A increases, because a higher S causes a stronger meridional circulation, which brings in more moisture and causes more evaporation through the associated stronger surface winds, thus resulting in more convective heating. The rate of increase of A is initially modest, but becomes greater as a result of the increase in wind speed in the boundary layer and the corresponding increase in evaporation at the surface as the vortex spins up. A increases at a reduced rate as the state gets close to S_3, as a result of the higher temperature in the core region, which reduces vertical instability and hinders further increase in convective activity. Moreover, with increased convergence at low levels and increased divergence at high levels, inertial stability provides further inhibition of meridional circulation. Also, as the boundary-layer air approaches saturation, the evaporation rate cannot keep increasing, thus limiting the further intensification of convective activity.

The foregoing interpretation of tropical cyclogenesis as a catastrophe was presented by Chao *et al.* (2003) and needs to be supported with modeling effort.

3. Stratospheric Sudden Warming

In the polar region of the stratosphere, the temperature in winter can rise by more than $40°K$ in as little as a week followed by a substantial drop, which is just as fast. The corresponding zonal mean zonal wind, consistent with the thermal wind relationship, can turn easterly in a major warming event. This phenomenon is called "stratospheric sudden warming" (SSW). It may occur more than once during a winter; the last episode, called the "final warming," is often the strongest and precedes the spring flow pattern.

SSW is a catastrophe, due to the interaction between the zonal flow and planetary waves. Chao (1985) proposed a mechanism explaining SSW, as depicted in Fig. 3. This figure depicts the "forcings," or attractions, acting on the zonally averaged zonal wind $[u]$ (the abscissa, i.e., $S = [u]$) in middle and high latitudes in the stratosphere. The effect of a stationary baroclinic Rossby wave of a zonal wave number on the zonal mean flow is represented by curve B; a positive value denotes westward acceleration. This acceleration occurs when the Rossby waves (i.e. the eddies) carry an upward westward momentum flux that is then deposited in the stratosphere owing to radiative damping in a quasi-equilibrium situation. However, the transient effect of wave–zonal wind interaction (Sec. 12.4 of Holton, 2004)[b] is prominent during the jump. curve B has a peak at the zonal mean zonal wind speed corresponding to the resonance speed, i.e. $c = u - \beta/k^2 = 0$ for the stationary Rossby wave, where c is the phase speed of the Rossby wave, u the zonally averaged zonal wind, k the zonal wavenumber, $\beta = \partial f/\partial y$; when the phase speed of the Rossby wave matches the speed of the mountain, which is zero. At resonance, the momentum flux due to planetary-scale topography transported upward by the Rossby wave reaches its maximum.

The forcing on $[u]$ due to the meridional gradient of the radiative–convective equilibrium temperature is depicted by line A. Line A pulls $[u]$ toward the state of radiative equilibrium (a positive A represents eastward acceleration), which is represented by the point where line A intersects the x axis, S_3. In midwinter, this point has a very high zonal mean zonal wind, and it is the state that exists when there is no topographical forcing (i.e. when curve B is zero) and the high zonal mean zonal wind reflects, through the thermal wind relationship, the large meridional gradient of the radiative equilibrium temperature. In midwinter the state is at S_3. As winter progresses toward spring, the zonal mean zonal wind corresponding to the radiative equilibrium meridional temperature gradient becomes smaller, and thus line A moves toward the left in Fig. 3 while retaining its slope. Eventually S_2 and S_3 merge and both then disappear. At this time, the difference between line A and curve B reflects a westward acceleration of the zonally averaged zonal wind. This difference becomes large as $[u]$ diminishes and approaches the resonance speed. In the meantime, the amplitude of the Rossby wave increases owing to the

---

[b]Related to A and B being functions of $\partial S/\partial t$, as mentioned in the Introduction.

resonance. Thus, $[u]$ races toward the other equilibrium state, S_1. Chao's (1985) model experiments showed S_1 to be a quasi-equilibrium state rather than an equilibrium state; in fact, an oscillatory state in the simplified model used. Therefore, Fig. 3 can only serve as a conceptual aid, not an actual depiction of the catastrophe. Owing to the tremendous difference between curve B and line A at resonance, overshooting can occur; thus $[u]$ can pass S_1 and, in some extreme cases, become negative. That is why the westerly wind $[u]$ weakens quickly and may even turn easterly; correspondingly, the temperature in the polar region rises rapidly. However, the warmer polar temperature exists only momentarily, since the state of the system soon bounces back from the overshooting. Such bouncing back can be as fast as the overshooting itself; hence, sudden warming is followed immediately by sudden cooling.

The description above is based on a model that has only one zonal wave number (see Chao, 1985, for details). When all zonal wave numbers are included, how the description should be modified still remains to be worked out. Moreover, what is described above explains the final warming well; warmings earlier in the season remain to be explained.

4. Abrupt Latitudinal Movement of the ITCZ (Monsoon Onset)

The intertropical convergence zone (ITCZ) is the location where surface winds converge and rise into the upward branch of the Hadley/Walker circulation. The latent heat released in the ITCZ drives the Hadley/Walker circulation. In other words, the ITCZ is a part of the Hadley/Walker circulation. The latitudinal location and the intensity of the ITCZ impact the tropical surface wind distribution, which, in turn, impacts the air–sea interaction. The latter is known to play a crucial role in El Niño. The convection within the ITCZ in the Indian Ocean

and western Pacific has an intraseasonal oscillation, which is called the Madden–Julian oscillation (MJO). The MJO is strongest when the ITCZ is over the equator. Thus the annual cycle of the ITCZ latitudinal movement determines the seasonal variation of the MJO intensity.

Because of its importance the latitudinal location of the ITCZ has attracted the attention of many prominent researchers, such as Charney (1971). Charney's theory of the latitudinal location of the ITCZ was built on his CISK theory (conditional instability of the second kind; Charney and Eliassen, 1964; and Ooyama 1964), which he used to explain tropical cyclogenesis. Chaney's ITCZ theory is a zonal mean version of his CISK idea. According to CISK, synoptic-scale convection relies on frictionally induced boundary-layer convergence (i.e. the Ekman pumping), and thus favors higher latitudes, i.e. the poles. Charney also invoked the idea that the moisture supply is higher over the equatorial tropics than over the polar regions. The compromise of these two factors gives the ITCZ a latitudinal location close to, but not at, the equator.

The Charney ITCZ theory was quite influential. Many people adopted or embellished it (see the appendix of Chao and Chen, 2004). However, Charney's ITCZ theory and its offshoots turned out to be in conflict with the aqua-planet (AP) experiments of Sumi (1991), Kirtman and Schneider (1996), Chao (2000) and Chao and Chen (2001a, 2004). These researchers used AP models with uniform sea surface temperature (SST) and solar angle — which eliminated the second factor in Charney's theory — and obtained the ITCZ over or near the equator (at about $14°$ N or S) instead of over the polar region, as the first factor of Charney's ITCZ theory would predict.[c] What is relevant to our topic is that the region adjacent to the equator (e.g. from $4°$ to $8°$) is not, or rarely, accessible by the AP ITCZ, which gives a clear indication of multiple equilibria.

---

[c]Charney realized the problem with the CISK theory in the last years of his life (Arakawa, personal communication).

The following is the author's explanation (from Chao and Chen, 2004) for the AP model results. An AP model with uniform SST and solar angle would have uniform time-averaged precipitation if the earth's rotation rate, Ω, were set to zero, since in the absence of Ω, one location on the globe is indistinguishable from another. When the earth's true rotation rate is used, convection finds preferred latitudes. To explain these preferred latitudes, one should look at the cause of convection, which is that vertical instability in the presence of the Coriolis parameter turns negative; in other words, the squared frequency of the inertial gravity wave, $\omega^2 = f^2 + \alpha^2 N^2 + |F|$, turns negative, and convection occurs [Eq. (8.4.23) of Gill, 1982]. Here, N^2 is the vertical stability, $|F|$ is the stabilization due to friction, and α is the ratio of the vertical scale to the horizontal scale of the wave or convective cell. The fact that f^2 is added to a term proportional to N^2 in the definition of ω^2 indicates that f^2 has an equivalent effect on convection to N^2; see Chao and Chen (2001a) for an explanation of this equivalence. When α is large, as in the case of individual clouds with small horizontal scales, f^2 can be ignored. However, when we consider synoptic-scale convective systems in the ITCZ, f^2 is not negligible. The equivalence of f to N makes the equator an attractor for the ITCZ. Thus when N^2 is globally uniform, ω^2 is a minimum at the equator. This implies that convection, or the ITCZ, should occur at the equator — this being a first effect of f on the ITCZ. In an AP model convection must occur somewhere, given the destabilizing effects of radiative cooling and the surface sensible and latent heat fluxes. Convection, or more generally the ITCZ, occurs at the latitude where the atmosphere is most unstable, i.e. $\partial\omega^2/\partial\phi = 0$, where ϕ is the latitude. This means that the ITCZ occurs at the latitude where $-\partial f^2/\partial\phi$ is balanced by $\partial(\alpha^2 N^2)/\partial\phi$, if $|F|$ is ignored. N^2 is reduced by the boundary evaporation, which is enhanced by the tangential wind component in the boundary layer, which, in turn, is induced by the Coriolis

parameter — which is a second effect of f. Thus the poles are additional attractors of the ITCZ.

$\partial(\alpha^2 N^2)/\partial\phi$ is *the latitudinal gradient of the f-modified surface wind-evaporation feedback mechanism (the role of α^2 is yet to be explored)*. Unfortunately, it remains a very difficult challenge, if not an impossibility, to obtain an analytical expression for $\partial(\alpha^2 N^2)/\partial\phi$. For the stability of the individual clouds, N^2 is equal to $g\partial\ln\theta/\partial z$, where θ is the potential temperature (or equivalent potential temperature for a saturated atmosphere); but for synoptic-scale cloud systems (or as in GCMs, where individual clouds are not simulated and cumulus parametrization is used to represent an ensemble of clouds), N^2 should be the vertical stability for cumulus ensembles rather than for individual clouds. Thus, N^2 has no known simple tractable analytical expression. Therefore, one has to seek other means to push the investigation forward, as we will soon discuss.

The gradient $\partial f^2/\partial\phi$ can be identified as the forcing due to the ITCZ attractor at the equator, and $\partial(\alpha^2 N^2)/\partial\phi$ as the forcing due to the other attractors at the poles. The former gradient is equal to $8\Omega^2 \sin\phi \cos\phi$, and is represented by curve A in Fig. 5(a); the latter gradient is represented by curve B in Fig. 5(a). In this figure the abscissa S, is the latitude of the ITCZ. Since curve A has a known analytic form and does not depend on the model design, it is easily understood. Curve B, however, has no known analytic form; it has been constructed through numerical experimental results and theoretical arguments. Curve B depends on the way N^2 is affected by the model design and, in particular, by the design of the model physics. Since curve B represents the attraction on convective systems due to the attractor at the poles through the second effect of f, it is zero at the poles — the center of the attractor. Also, curve B depends on the gradient of f — remember that it is the gradient of $\alpha^2 N^2$, and $\alpha^2 N^2$ is affected by f. In other words, curve B depends on β. This gives Curve B a maximum at the equator. However, since the convective system is fairly sizeable,

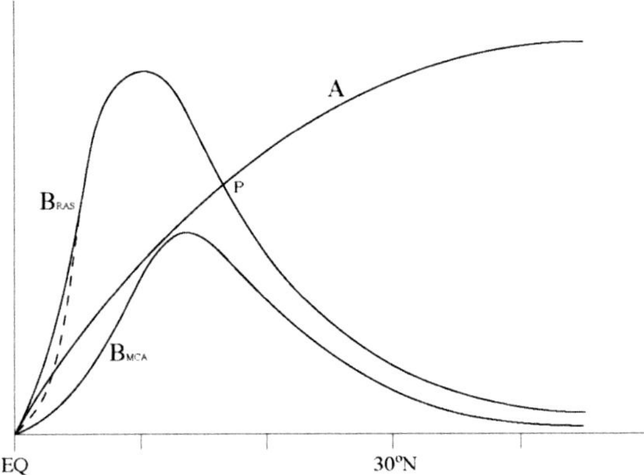

Figure 5(a). Schematic diagram showing the strength of the two types of attraction acting on the ITCZ. The difference between the solid and dashed curve B_{RAS} lies in their slopes at the equator.

when its center is close to the equator a part of the system is in the other hemisphere; therefore, it also experiences the attraction by the other pole. Thus, curve B has to be zero at the equator, where the attractions due to both poles cancel. This is shown in Fig. 5(b), with the dashed curve being the attraction due to the second effect of f if the size of the convective system is not accounted for. The solid curve, curve B — being the running average of the

dashed curve with an averaging window the size of a tropical synoptic system — represents the net attraction experienced by the convective system.

The dependence of curve B on the cumulus scheme is illustrated in Fig. 5(a). B_{RAS} and B_{MCA}, represent curve B when the relaxed Arakawa and Schubert scheme (RAS, Moorthi and Suarez, 1992) and Manabe's moist convective adjustment (MCA) scheme are used, respectively. As for the relative height of B_{RAS} and B_{MCA}, let us first recall that curve B represents the latitudinal gradient of f-modified surface heat fluxes. When f is larger, the surface winds that converge toward the center of a synoptic-scale convective system (which is a constituent of the ITCZ) develop a greater tangential wind component, which adds to the wind speed and thus enhances the surface heat fluxes. Convective cells simulated with MCA are usually smaller than those simulated with RAS and have faster surface wind speed toward the center of a convective system which does not allow enough time for the surface wind to be fully modified by the Coriolis parameter to allow the tangential wind component to fully develop. Therefore, curve B under MCA is smaller than

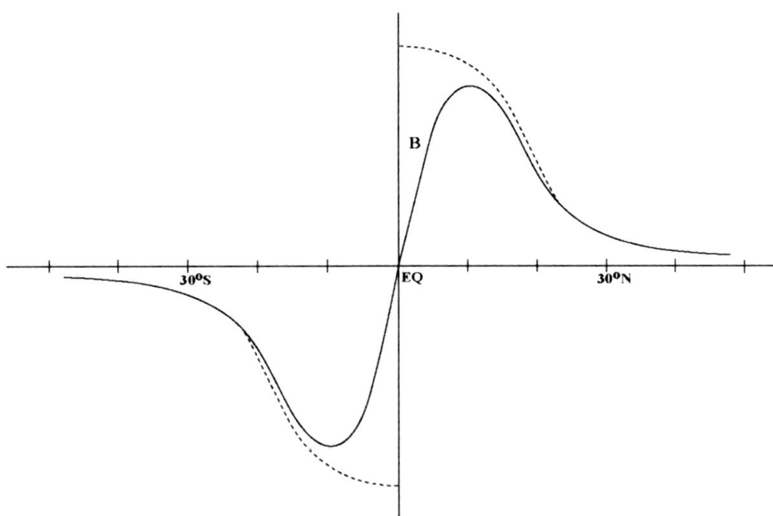

Figure 5(b). Schematic diagram showing curve B (solid line) as the latitudinal running mean of the magnitude of the attraction toward the poles.

that under RAS. See Chao and Chen (2004) for a more detailed discussion. If a condition is imposed on the RAS cumulus parametrization scheme such that the boundary layer relative humidity must exceed a critical value in order for the scheme to operate, then this critical value is increased from 90% to 95%, and curve B will change from B_{RAS} to B_{MCA}. At 90%, the criterion does not impose much restriction on RAS, but at 95% the restriction is strong enough to make RAS behave more like MCA. MCA requires both neighboring levels of the model to be saturated for it to operate, and thus is quite restrictive. In Fig. 5(a), as curve B rises from B_{MCA} to B_{RAS}, the location of the intercept between curve A and curve B — i.e. the location of the ITCZ — remains at the equator until the slope of curve B at the equator exceeds that of curve A at the equator. Then the equator is no longer a stable equilibrium location, and the ITCZ jumps to the latitude of the other intercept, P. This jump is very fast, because the ITCZ is pulled in a "free fall" by the large difference between the two curves. On the other hand, when curve B decreases from B_{RAS} to B_{MCA}, point P — or the ITCZ — moves

gradually toward the equator until the peak of Curve B gets below curve A. At this moment P disappears, and the state jumps toward the equator. But, in this case, the difference between the two curves is much smaller than in the case of the ITCZ jumping away from the equator, which involves a rising curve B. Thus, the move of the ITCZ back to the equator is not as abrupt.

The above deductions are borne out in experiments shown in Fig. 6(a), where the boundary layer relative humidity criterion is held at 90% for the first 200 days and is changed linearly in time to 95% over the next 100 days and then kept unchanged for the reminder of the experiment. As expected, at the beginning of the experiment shown in Fig. 6(a) the ITCZ is away from the equator and then moves to the equator gradually as the RH criterion is increased. The fact that the simulated ITCZ is asymmetric with respect to the equator in the first phase of the experiment remains to be explained. Figure 6(b) shows the results of an identical experiment except that the values of 90% and 95% are switched. In this experiment the move away from the equator is abrupt, characteristic of a catastrophe. Also, in another experiment, the

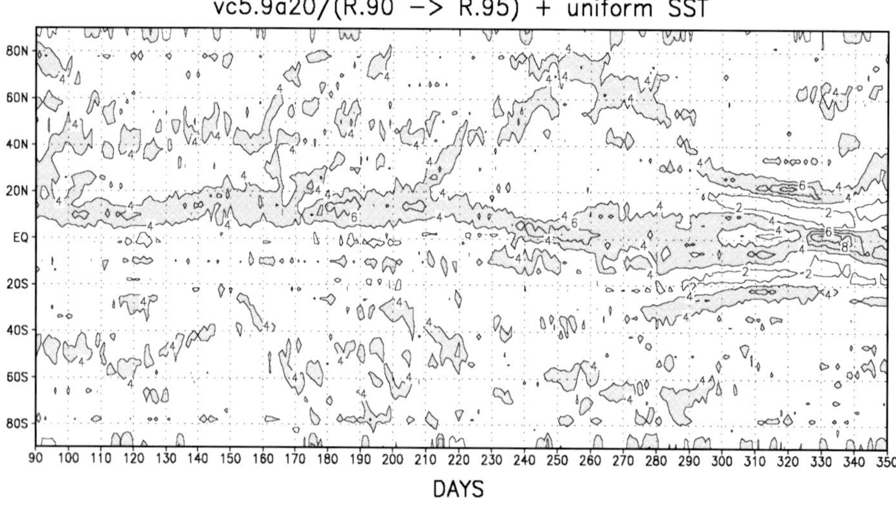

Figure 6(a). Zonally averaged precipitation rate (mm/day) in an AP experiment with globally and temporally uniform SST (29°C) and solar angle using RAS. The critical boundary layer humidity is set at 90% in the first 200 days and is changed to 95% linearly in the next 100 days.

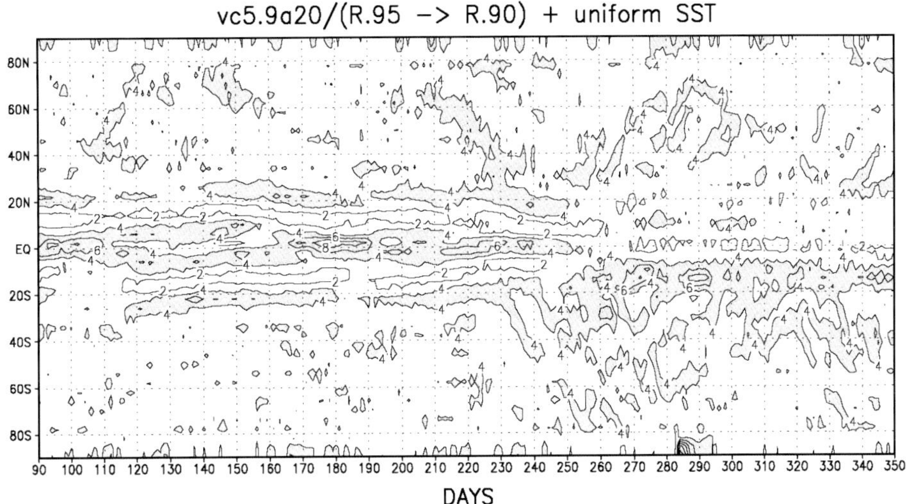

Figure 6(b). Same as Fig. 5(a) except that the values of 90% and 95% are switched.

wind speed used in computing the surface sensible and latent fluxes is fixed at 5 m/s everywhere. Because curve B has been eliminated, the ITCZ is located at the equator. A further experiment shows that radiation can have impact on the ITCZ location through its effect on N^2 and in turn on curve B. Different radiation packages lead to different amounts of forcing for the cumulus convection, resulting in differing convective systems of different intensity. This further means different magnitudes of surface winds and thus different heights for curve B.

The above interpretation of the abrupt ITCZ latitudinal movement may be expanded to explain monsoon onset. Chao and Chen (2001b) argued that a monsoon is no more than an ITCZ sufficiently removed from the equator during its annual cycle of latitudinal movement. This movement is not completely smooth; there can be a sudden large shift away from the equator, which can be interpreted as a monsoon onset (Chao, 2000; Chao and Chen, 2001a). This concept also led to the argument that a monsoon is an ITCZ after it suddenly shifts away from the equator and it does not have to rely on land–sea thermal contrast for its existence (Chao and Chen, 2001b.)

The location of the ITCZ is obviously also influenced by land–sea distribution of the globe and by ocean–atmosphere interaction, which makes the subject more complex and interesting than it would be otherwise. The reader is referred to Philander *et al.* (1996) and Xie and Saito (2001) about these additional factors.

5. Other Examples of Atmospheric Catastrophes

In this section we will mention more atmospheric catastrophes, with varying degrees of commentary. These catastrophes either are already well known as catastrophes or can be easily recognized as such. The underlying dynamics of some of them are not very clear; therefore, many of them are good research topics.

Polar ice cap instability in energy-balance climate models

This best-known example of a catastrophe in climatology is found in energy-balance climate models. Budyko (1969) and Sellers (1969) demonstrated, in a one-dimensional model, a catastrophic transition from a globe that has ice in the polar regions only to an ice-covered globe by lowering the solar constant by a few percent. A description of this catastrophe can be found in textbooks on climatology, such as Hartmann (1994).

Abrupt climate changes

Ocean–atmosphere interaction gives rise to additional catastrophes. The onset of an ice age, a long-term drought, and abrupt changes in ocean general circulation are some examples.

Barotropic and baroclinic instabilities

The textbook explanation for these instabilities presents only a linear growth rate depiction based on linear analyses, which is of limited use. The linear analysis approach starts from a basic (quasi-)equilibrium state, such as S_1, S_2 and S_3 in Fig. 3, adds perturbation and determines if the perturbation can grow and, if so, its growth rate. If a state such as S_2 in Fig. 3 is unstable, the perturbation will then grow at a constant growth rate, and can be of either sign. The obvious limitations of linear analysis are that the unstable equilibrium is not physically realizable and that the analysis is valid only for a very short period of time, before nonlinear effects become sizeable. Therefore, to get a more complete picture of these instabilities, one should turn to the concept of a catastrophe. In studying these instabilities as catastrophes, one needs to ask what the equilibria are before and after these instabilities and what the forcings sustaining the equilibria are, in order to arrive at a complete picture of the whole life span of the phenomenon. In a catastrophe, the growth rate is initially zero, when the system loses its equilibrium, and then increases and finally decreases over the life span of the catastrophe. The growth can be in one direction only. The trigger in a triggered catastrophe has to be in the same direction as the growth of the catastrophe.

Middle-latitude explosive cyclogenesis

This is a good research topic. Linear baroclinic instability studies (Sec. 6.e of Hoskins *et al.*, 1985) are not adequate to explain cyclogenesis, owing to its nonlinear nature and the convective heating involved.

Jumping of the Mei-yu front

The Mei-yu front in China jumps in its movement northward. This is a good research topic.

Boundary layer instability

There can be catastrophes in the boundary layer (e.g. Randall and Suarez, 1983).

Blocking onset

The topographically induced Rossby wave instability has been used to explain blocking onset (Charney and DeVore, 1979).

Onset of cumulus convection

This is the most obvious, but not necessarily the most understood, atmospheric catastrophe.

Low-order models

Lorenz's celebrated equations, exhibiting sudden changes in the flow characteristics of the system he studied, are very well researched (e.g. Sparrow, 1986; Lorenz, 1993).

ENSO

ENSO has been studied intensely. The suggestion that ENSO can be triggered by the MJO (Yu and Rienecker, 1998) implies the possibility that the onset of ENSO is a catastrophe.

Tornado-genesis

This is an excellent research topic, since so little is known about its dynamics (Church, 1993; Mak, 2001).

Annulus experiments

The various flow regimes in annulus experiments show a considerable amount of overlap in the regime diagram (Lorenz, 1967; Holton, 2004) and the transition among them is often quite abrupt.

Transition to turbulence

Routes to turbulence have many catastrophes.

Martian polar warming

The stratospheric sudden warming mechanism reviewed in Sec. 3 has been used to explain Martian polar warming (Barnes and Hollingsworth, 1987).

Onset of Martian dust storms

Dust storms lasting months can engulf nearly the whole Martian atmosphere within weeks of their onset. Their growth is explosive and certainly fits the description of a catastrophe. There have been attempts to explain this growth as a catastrophe, but no consensus has been reached (Fernandez, 1997).

6. Remarks

The above examples demonstrate that the concept of a catastrophe is fundamental to atmospheric dynamics. While some of the specifics of these examples may later turn out to be incomplete or even incorrect, the general concept of a catastrophe will still remain useful. When we study (quasi-)equilibria such as Rossby waves, cloud clusters, and gravity waves, we need to ask not only how they are maintained, but also whether and how they can become unstable, and how they transition to other (quasi-)equilibria.

Why are there so many catastrophes in the atmosphere? The answer has two components. First, as seen from the examples in the preceding section, multiple (quasi-)equilibria exist because one or both of the forcings that maintain a quasi-equilibrium are highly nonlinear functions of a gross measure of the physical system.

The atmosphere is full of factors contributing to nonlinearity. Among them are the earth's rotation, resonance, the structure of the external heating, air–sea interaction, etc. Second, there is an abundance of mechanisms that can act as triggers in moving a system from one (quasi-) equilibrium to another. Among these mechanisms are the annual cycle, the diurnal cycle (in spontaneous catastrophes), and various perturbations (in triggered catastrophes.) (The annual cycle and the diurnal cycle can be viewed as triggers in a larger sense.)

The reasons for nonlinearity are different for different atmospheric catastrophes. Finding them is the core challenge of studying atmospheric catastrophes. The examples given in this review may not be sufficient to serve as a complete guide to studying other atmospheric catastrophes, but they do point in the general direction.

Finally, how can we turn the knowledge gained from studying atmospheric catastrophes into practical use, such as improving the performance of forecast models? One possibility is, rather than assessing the performance of global models by comparing the seasonal mean state of the model with observations — the most common practice — one should compare the catastrophes simulated by the models with those observed — such as stratospheric sudden warming and monsoon onset. Most models have difficulties in simulating these events with fidelity. Through studying the models' failures, one can strive for improvements.

[Received 27 October 2006; Revised 29 March 2007; Accepted 27 May 2007.]

References

Arnold, V. I., 1984: *Catastrophe Theory* (translated by R. K. Thomas), Springer-Verlag, 79 pp.

Barnes, J. R., and J. L. Hollingsworth, 1987: Dynamic modeling of a planetary wave mechanism for a Martian polar warming. *ICARUS*, **71**, 313–334.

Brown, R. A., 1970: A secondary flow model for the planetary boundary layer. *J. Atmos. Sci.*, **27**, 742–757.

Budydo, M. I., 1969: The effect of solar radiation variation on the climate of the earth. *Tellus*, **21**, 613–629.

Chao, W. C., 1985: Stratospheric sudden warmings as catastrophes. *J. Atmos. Sci.*, **42**, 1631–1646.

Chao, W. C., 2000: Multiple quasi-equilibria of the ITCZ and the origin of monsoon onset. *J. Atmos. Sci.*, **57**, 641–651.

Chao, W. C., and B. Chen, 2001a: Multiple quasi-equilibria of the ITCZ and the origin of monsoon onset. Part II. Rotational ITCZ attractors. *J. Atmos. Sci.*, **58**, 2820–2831.

Chao, W. C., and B. Chen, 2001b: The origin of monsoons. *J. Atmos. Sci.*, **58**, 3497–3507.

Chao, W. C., B. Chen, and W.-K. Tao, 2003: Tropical cyclogenesis as a catastrophe. *AMS Conf. Atmos. Ocean Fluid Dynamics*, San Antonio, preprint vol., paper 11.

Chao, W. C., and B. Chen, 2004: Single and double ITCZ in an aqua-planet model with constant sea surface temperature and solar angle. *Clim. Dyn.*, **22**, 447–459.

Chao, W. C., and L. Deng, 1998: Tropical intraseasonal oscillation, super cloud clusters and cumulus convection schemes. Part II: 3D aqua-planet simulations. *J. Atmos. Sci.*, **55**, 690–709.

Charney, J. G., 1971: Tropical cyclogenesis and the formation of the ITCZ. In *Mathematical Problems of Geophysical Fluid Dynamics.*, W. H. Reid, (ed.), *Lectures in Applied Mathematics.* Vol. 13, *Amer. Math. Soc.*, 355–368.

Charney, J. G., and J. G. DeVore, 1979: Multiple flow equilibria in the atmosphere and blocking. *J. Atmos. Sci.*, **36**, 1205–1216.

Charney, J. G., and A. Eliassen, 1964: On the growth of the hurricane depression. *J. Atmos. Sci.*, **21**, 68–75.

Church, C., 1993: Ed., *The Tornado: Its Structure, Dynamics Prediction and Hazards.* Amer. Geophys. Union, 637 pp.

Drazin, P. G., 1992: *Nonlinear System*, Cambridge University Press, 317 pp.

Fernandez, W., 1997: Martian dust storms: A review. *Earth, Moon and Planets*, **77** 19–46.

Hartmann, D. L, 1994: *Global Physical Climatology*, Academic, 409 pp.

Held, I. M., 1983: Stationary and quasi-stationary eddies in the extratropical troposphere: Theory. Chap. 6 of *Large-Scale Dynamical Processes*,

B. J. Hoskins and R. P. Pearce (eds.), Academic, 397 pp.

Holton, J. R., 2004: *An Introduction to Dynamic Meteorology*, 4th edn., Academic, 531 pp.

Hoskins, B. J., M. E. McIntyre, and A. W. Robertson, 1985: On the use and significance of isentropic potential vorticity maps. *Quart. J. Roy. Meteor. Soc.*, **111**, 877–946.

Iooss, G., and D. D. Joseph, 1980: *Elementary Stability and Bifurcation Theory*, Springer-Verlag, 286 pp.

Kirtman, B. P., and E. K. Schneider, 2000: A spontaneously generated tropical atmospheric general circulation. *J. Atmos. Sci.*, **57**, 2080–2093.

Lorenz, E. N., 1967: *The Nature and Theory of the General Circulation of the Atmosphere*, WMO, Geneva.

Lorenz, E., 1993: *The Essence of Chaos*, University of Washington Press.

Mak, M., 2001: Non-hydrostatic barotropic instability: Applicability to nonsupercell tornado genesis. *J. Atmos. Sci.,* **58**, 1965–1977.

Ooyama, K., 1964: A dynamical model for the study of tropical cyclone development. *Geofis. Int.*, **4**, 187–198.

Palmen, E., and C. W. Newton, 1969: *Atmospheric Circulation Systems*, Academic, 603 pp.

Philander, S. G. P., D. Gu, D. G. Halpern, G. Lambert, N. C. Lau, T. Li, and R. C. Pacanowski, 1996: Why the ITCZ is mostly north of the equator. *J. Climate*, **9**, 2958–2972.

Poston, T., and I. N. Stewart, 1978: *Catastrophe Theory and Its Applications*, Pitman, London, 491 pp.

Randall, D. A., and M. J. Suarez, 1984: On the dynamics of stratocumulus formation and dissipation. *J. Atmos. Sci.*, **41**, 3052–3057.

Rotunno, R., and K. A. Emanuel, 1987: An air–sea interaction theory for tropical cyclones. Part II: Evolutionary study using a nonhydrostatic axi-symmetric numerical model. *J. Atmos. Sci.*, **44**, 542–561.

Sellers, W. D., 1969: A global climate model based on the energy balance of the earth–atmosphere system. *J. Appl. Meteor.*, **8**, 392–400.

Sparrow, C., 1982: The *Lorenz Equations: Bifurcations, Chaos, and Strange Attractors. Applied Math. Sci.*, Vol. **41**, Springer–Verlag, pp. 269.

Sturrock, P. A., 1966: Explosive and non-explosive onsets of instability. *Phys. Rev. Lett.* **16**, 270–273.

Sumi, A., 1992: Pattern formation of convective activity over the aqua-planet with globally uniform sea surface temperature. *J. Meteor. Soc. Japan,* **70**, 855–876.

Thom, R., 1972: *Structural Stability and Morphogenesis* (translated by D. H. Fowler, 1975), Benjamin, Reading, Massachusetts.

Xie, S. P., and K. Saito, 2001: Formation and variability of a northerly ITCZ in a hybrid coupled AGCM: continental forcing and oceanic–atmospheric feedback. *J. Climate,* **14**, 1262–1276.

Yu, L. A, and M. M. Rienecker, 1998: Evidence of an extratropical atmospheric influence during the onset of the 1997–98 El Niño. *Geoph. Res. Lett.,* **25** (18), 3537–3540.

Relative Intensity of the Pacific and Atlantic Storm Tracks in a Maximally Simplified Model Setting

Mankin Mak

Department of Atmospheric Sciences,
University of Illinois at Urbana-Champaign, Urbana, Illinois, USA
mak@atmos.uiuc.edu

The climatological winter Atlantic storm track is distinctly more intense (by about 10%) than the Pacific storm track even though the Atlantic jet is about 30% weaker than the Pacific jet. It is hypothesized that this counterintuitive feature is partly attributable to having statistically stronger seeding disturbances upstream of the Atlantic jet. The difference in the seeding disturbances may stem from the geographical distribution of continents and oceans in the northern hemisphere, especially when differential friction over land versus water surfaces is taken into consideration. This hypothesis is shown to be valid even in the context of a barotropic model. The proxy forcing in the model is introduced in the form of relaxation of the instantaneous flow toward a reference flow, which has two localized jets broadly resembling the winter Pacific and Atlantic jets. A linear modal instability analysis of these jets is first presented. The nonlinear model Atlantic storm track is found to be more intense than its counterpart over the Pacific. The difference in the relative intensity of the two model storm tracks becomes more pronounced when the drag coefficient over the land sectors is several times larger than that over the ocean sectors. These results may be taken as evidence in support of the hypothesis.

1. Introduction

The Pacific storm track and the Atlantic storm track are two major features of the general circulation of the northern hemisphere in winter. They stem from repeated passage of synoptic disturbances over those two oceanic sectors. It has been documented with 40 years of NCEP reanalysis data that the Atlantic storm track is distinctly more intense than the Pacific storm track in 30 out of 40 winters. On the average, the former is about 10% more intense, although the Atlantic jet is about 30% weaker than the Pacific jet (see Figs. 1 and 3 in Mak and Deng, 2007).

This counterintuitive characteristic may be accounted for if the seeding disturbances entering the Atlantic jet are statistically stronger than those entering the Pacific jet. A case can be made that the difference in the seeding disturbances is a natural consequence of the geographical distribution of continents and oceans in the northern hemisphere. Euro-Asia is a huge landmass, with the Pacific being on the downstream side, while the Atlantic is on the downstream of the much smaller North America. Disturbances typically suffer greater dissipation over land than water because of the differences in their surface roughness. It follows that eddies (seeding disturbances) reaching the Pacific jet after having traversed through Euro-Asia would typically be weaker than those reaching the Atlantic jet after having traversed through North America, so much so that eddies over the North Atlantic may intensify to a greater equilibrated intensity, even though the Pacific jet is substantially stronger. This is the hypothesis for the phenomenon under consideration. The difference in the horizontal structure of the two jets may be an additional factor.

The literature on storm tracks has been extensively reviewed by Chang *et al.* (2002). The constituent disturbances of the storm tracks are largely baroclinic. One would naturally use a baroclinic model to investigate storm track dynamics (e.g. Lee and Mak, 1996; Whitaker and Sardeshmukh, 1998; Deng and Mak, 2005). Orlanski (2005) and Zurita-Gotor and Chang (2005) have examined the role of seeding disturbances in the context of a single storm track. The latter authors conjecture that greater damping over Asia could result in a weaker Pacific storm track. The hypothesis for the relative intensity of the two winter storm tracks has been specifically shown to be valid by Mak and Deng (2007) in the context of a two-layer baroclinic model.

The purpose of this study is to further demonstrate that the hypothesis is valid even in a barotropic model setting, which is a maximally simplified model. Large-scale barotropic and baroclinic dynamics have much in common at the most fundamental level. Barotropic instability and baroclinic instability can both be succinctly interpreted in terms of mutual reinforcement of the constituent elements in an unstable disturbance (the so-called wave resonance mechanism, e.g. Mak, 2002). Indeed, some intrinsic aspects of storm track dynamics can be delineated with a barotropic model (Swanson *et al.*, 1997; Lee, 2000).

The design of a particular forced dissipative barotropic model with two jets is presented in Sec. 2. Subsection 3.1 reports the inviscid modal instability properties of such a flow, and Subsec. 3.2 reports the instability properties under the influence of differential friction. The properties of eddy feedback and nonlinear model

storm tracks are presented in Sec. 4 with uniform friction (Subsecs. 4.1 and 4.2) and with differential friction (Subsec. 4.3). The article ends with concluding remarks in Sec. 5.

2. Design of a Barotropic Storm Track Model

The circumpolar midlatitude region is modeled as a reentrant channel on a beta-plane between two rigid lateral boundaries $\left(-\frac{X}{2} \leq x \leq \frac{X}{2}\right.$ and $\left.-\frac{Y}{2} \leq y \leq \frac{Y}{2}\right)$. The domain is 30,000 km long and 8,000 km wide. A steady external forcing is introduced to support a reference flow, which has a zonally nonuniform velocity field ($u_{\rm ref}$, $v_{\rm ref}$) with corresponding stream function $\psi_{\rm ref}$ and vorticity $\zeta_{\rm ref}$, viz.

$$J(\psi_{\rm ref}, \zeta_{\rm ref} + f) + a\zeta_{\rm ref} = \text{Forcing},$$

where a is a friction coefficient, $f = f_{\rm o} + \beta y$ and $J(A, B) = A_x B_y - A_y B_x$ is the Jacobian. We particularly prescribe a reference flow with two localized jets, which plays the role of a proxy forcing.

A localized westerly jet has cyclonic vorticity in its northern flank and anticyclonic vorticity in its southern flank. We therefore introduce a reference flow that has the following idealized distribution of vorticity:

$$\zeta_{\rm ref} = \sum_{j=1}^{2} A^{(j)} y \exp(-y^2) F_j(x), \qquad (1)$$

where x and y are nondimensional coordinates with $F_j(x) = 1$ in two sectors, $x_W^{(j)} \leq x \leq x_E^{(j)}$, and $F_j(x) = 0$ elsewhere. In the special case of $F_j(x) = 1$ for the whole domain, the reference flow would be a zonally uniform Gausian jet centered at $y = 0$, namely $u_{\rm ref} = B \exp(-y^2)$, with B being its strength. The relative strengths and geometrical characteristics of the winter mean Pacific and Atlantic jets at an upper tropospheric level are used as a guide for prescribing $A^{(j)}$, $x_E^{(j)}$ and $x_W^{(j)}$. These two jets are referred to by superscripts $j = 1$ and $j = 2$ respectively.

Hence, (1) is the vorticity field of two approximately zonally oriented Gaussian jets separated by a distance $(x_W^{(2)} - x_E^{(1)})$. Furthermore, since the vorticity is only a function of the spatial derivatives of the velocity components, a reference flow would not be completely prescribed until a domain average zonal wind component, u_{oo}, is also specified. Then the stream function field of the reference state would be $\psi_{\text{ref}} = -u_{oo}y + \nabla^{-2}\zeta_{\text{ref}}$, which can be readily determined by inversion subject to well-known boundary conditions. There are quite a few parameters that specify the relative strength, length, shape, orientation and locations of the two idealized localized jets. It would be sufficient to focus on one particular set of parameter values that supports a reference flow that has the primary features of the observed time mean flow in the upper troposphere.

Distance, velocity and time are measured in units of $L = 10^6\,\text{m}$, $U = 60\,\text{ms}^{-1}$ and $LU^{-1} = 0.17 \times 10^5\,\text{s}$ respectively. Hence, the nondimensional domain is $-15 \leq x \leq 15$ and $-4 \leq y \leq 4$. To mimic the relative strengths and lengths of the Atlantic and Pacific jets, we use $\frac{A^{(2)}}{A^{(1)}} = 0.7$ and $\frac{x_E^{(2)} - x_W^{(2)}}{x_E^{(1)} - x_W^{(1)}} = 0.6$. We specifically use $A^{(1)} = 2.55$, $x_W^{(1)} = -10$, $x_E^{(1)} = -4$, $x_W^{(2)} = 4.9$ and $x_E^{(2)} = 8.5$. The nondimensional value of the variation of the Coriolis parameter with latitude in the extratropics is $\beta = 0.229$. A value of $u_{oo} = 0.2$ is used as a representative additional domain average zonal velocity component corresponding to 12 m/s.

The nonlinear evolution of the flow in this model is governed by

$$\zeta_t + J(\psi_{\text{ref}}, \zeta) + J(\psi, \zeta_{\text{ref}} + f)$$
$$+ J(\psi, \zeta) = -a\zeta - r\nabla^4\zeta \qquad (2)$$

where $\psi_{\text{total}} = \psi_{\text{ref}} + \psi$, $\zeta_{\text{total}} = \zeta_{\text{ref}} + \zeta$, with an auxiliary relation

$$\zeta = \nabla^2\psi. \qquad (3)$$

A fourth-order hyperdiffusion with a coefficient $r = 2.5 \times 10^{-4}$ is incorporated in the model for computational purposes. The frictional effect is represented by a linear drag. It may be viewed as a linearized form of a bulk aerodynamic formula for representing a surface stress. A reasonable value of the drag coefficient is $a = 0.03$, corresponding to a damping time of about 5 days (Whitaker and Sardeshmurkh, 1998). When we take into account the influence of continents and oceans, we use $a = 0.015$ in the oceanic sectors and $a = 0.045$ in the land sectors.

3. Linear Instability

It is instructive to first examine the dynamical properties of weak perturbations in this system. They are governed by the linearized version of (6.2), viz.

$$\zeta_t + J(\psi_{\text{ref}}, \zeta) + J(\psi, \zeta_{\text{ref}} + f) = -a\zeta. \qquad (4)$$

The absolute vorticity and stream function fields of a pertinent basic state with $u_{oo} = 0.2$ are shown in Fig. 1. The maximum nondimensional velocity is 1.05. The model Pacific jet has a maximum speed of $\sim$63 m/s and the model Atlantic jet has a maximum speed of $\sim$37 m/s. The two localized jets are made up of a zonal mean shear flow and a spectrum of planetary scale waves. They are ultimately sustained by radiative forcing in the presence of zonal inhomogeneous surface conditions on earth. The proxy forcing is meant to mimic the net effect of all such factors.

The first task is to solve for the normal modes in the form of $\psi = \xi(x, y)\exp(t\sigma)$. The eigenvalue σ and the eigenfunction ξ can only be solved numerically. The domain is depicted with $I = 140$ grid points in the x direction and $J = 49$ grid points in the y direction. The corresponding dimensional resolution is $\delta x = 220\,\text{km}$ and $\delta y = 160\,\text{km}$. Such resolution has been verified to be adequate in the sense that the solution changes little when the domain is only depicted with $I = 100$ and $J = 29$. The finite difference version of (4) written in a matrix form is

$$A\vec{\psi} = \sigma \quad B\vec{\psi}, \qquad (5)$$

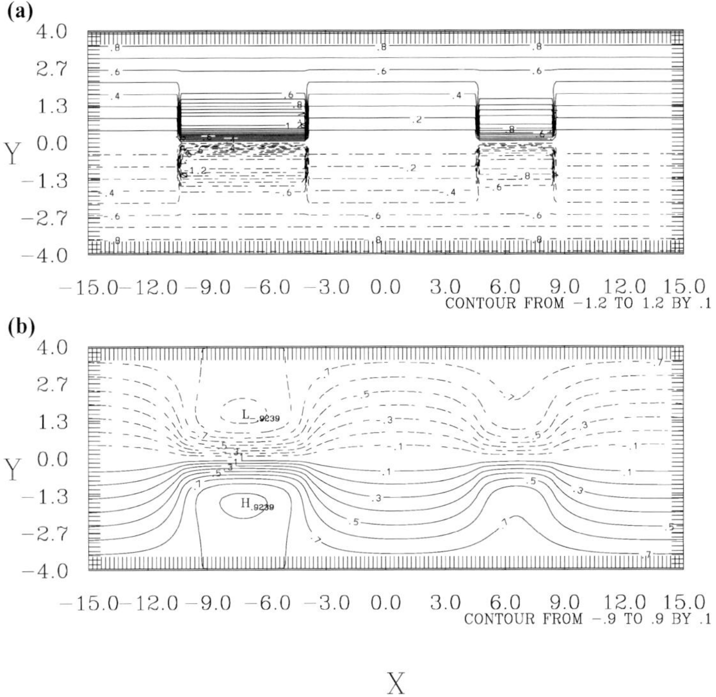

Figure 1. Distribution of (a) the nondimensional absolute vorticity and (b) the nondimensional stream function of the reference state. The absolute vorticity value does not include the value of the Coriolis parameter in the middle of the domain.

where A and B are $M \times M$ matrices with $M = IJ$. $\vec{\psi}$ is an M vector containing the unknown ξ at all grid points. In passing, it is noteworthy that for $M = 6860$, solving (5) by the matrix inversion method is a computationally demanding task. Such computation would not be feasible if we were to do so with a counterpart two-layer model.

The purpose of presenting a linear modal instability analysis is to delineate some fundamental and relevant properties of a basic two-localized-jet flow system. These properties help us infer the plausible locations of the storm tracks in a corresponding nonlinear system. They also help us develop a "feel" for what the individual disturbances might intrinsically look like. But they do not directly enable us to deduce any quantitative properties of the storm tracks in an equilibrated state, partly because the process of nonmodal instability

rather than modal instability seems to be more relevant to storm track dynamics and, more important, storm tracks are a product of nonlinear dynamics.

3.1. *Inviscid modal instability properties, $a = 0$*

All eigenvalues for each parameter setting can be plotted as an ensemble of points on a (σ_r, σ_i) plane. Associated with a complex eigenvalue, $\sigma = \sigma_r + i\sigma_i$, we would have an eigenvector with complex value, $\vec{\psi} = \vec{\psi}_r + i\vec{\psi}_i$. The normal mode can be evaluated as

$$\psi = (\psi_r \cos \sigma_i t - \psi_i \sin \sigma_i t) \exp(\sigma_r t), \qquad (6)$$

where $\psi_r(x,y)$ and $\psi_i(x,y)$ are the counterpart scalar fields of $\vec{\psi}_r$ and $\vec{\psi}_i$. It embodies the information about the structure of the unstable mode under consideration. We will normalize ψ_r and

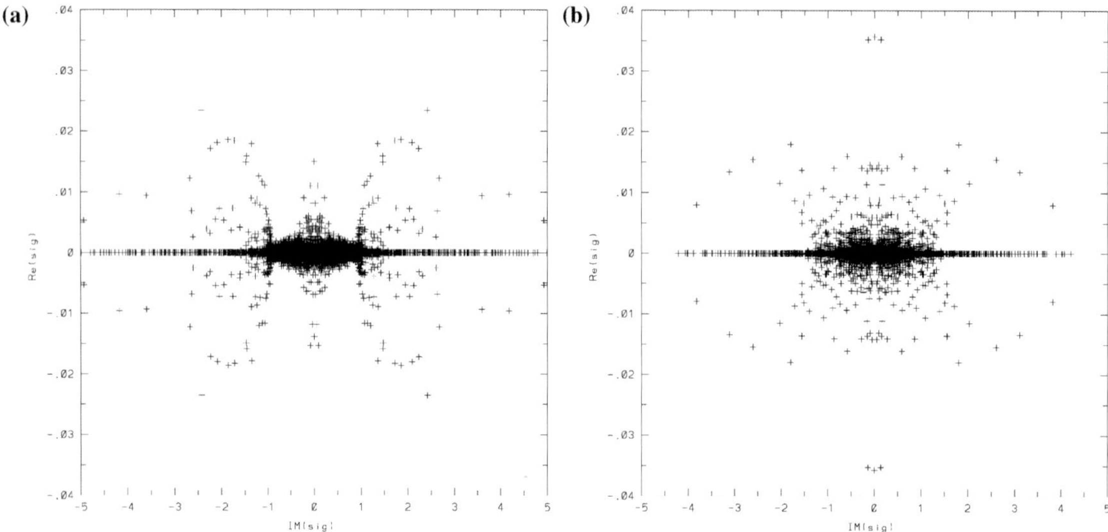

Figure 2. Distribution of eigenvalues of the inviscid normal modes plotted as points on a ($\sigma_r \equiv \mathrm{Re}\{\sigma\}$, $\sigma_i \equiv \mathrm{Im}\{\sigma\}$) plane, for (a) $u_{oo} = 0.2$ corresponding to $14\,\mathrm{m/s}$ and (b) $u_{oo} = 0.0$. The unit of time is $LU^{-1} = 0.17 \times 10^5$ s.

ψ_i of each normal mode to have unit domain-integrated kinetic energy.

The eigenvalues of the normal modes for the case of $u_{oo} = 0.2$ without the influence of friction are shown in Fig. 2(a). These eigenvalues consist of two sets symmetrically distributed with respect to the σ_r axis. This symmetry stems from the fact that we may use either positive zonal wave numbers or negative zonal wave numbers in a spectral representation of the flow. It has been verified that each pair of corresponding modes in these two sets are structurally indistinguishable. It would therefore suffice to focus on the modes with $\sigma_i > 0$.

One subset of modes has distinctly higher frequency. The most unstable mode belongs to this subset and has a frequency $\sigma_i \sim 2.4$, which corresponds to a period of $\sim$0.6 day. The structure of these high frequency modes reveals that they are almost exclusively associated with just one of the two jets. One example of such modes is shown in Fig. 3(a), which has $(\sigma_r, \sigma_i) = (0.0235, 2.42)$ and arises solely from the model Pacific jet. This mode has quite a short length scale and straddles the jet. It grows

by extracting energy mostly from the meridional shear of the jet. We therefore refer to the high frequency modes as *single-jet modes*.

The other subset of normal modes in Fig. 2(a) has nondimensional frequencies ranging from 0.0 to $\sim$ 0.4. The corresponding range of periods is quite large, from virtually being infinitely long to $\frac{2\pi}{\sigma_i} \frac{L}{U} \approx 2$ days. These modes consist of two groups. One group has lower frequency and larger growth rates. For example, one of these modes has a growth rate of $\sigma_r \approx 0.011$ and a period of $\sigma_i = 0.08$. The corresponding e-folding time is $\sim$11 days and the corresponding period is 12 days. Their structures suggest that these lower frequency modes tap energy from both jets. An example of them is shown in Fig. 3(b). The most intense parts of this mode are located at the exit region of the two jets where the stretching deformation is strong. Its length scale is longer than the synoptic scale. Such a mode is a counterpart of the normal modes found by Simmons *et al.* (1983) for an observed 300 mb January mean flow in their study of low frequency variability. We will therefore refer to those modes in this system as *low-frequency modes*.

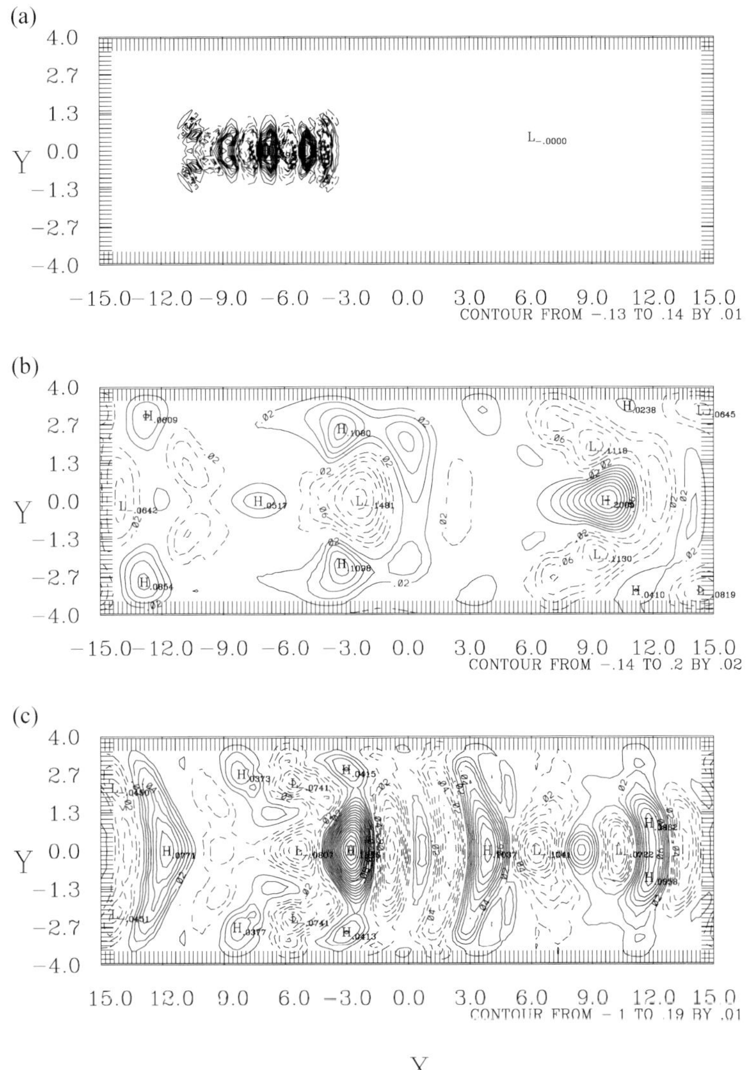

Figure 3. Structure of (a) a single-jet mode with $(\sigma_r, \sigma_i) = (0.0235, 2.42)$, (b) a low-frequency mode $(\sigma_r, \sigma_i) = (0.011, 0.08)$ and (c) a synoptic-frequency mode with $(\sigma_r, \sigma_i) = (0.009, 0.22)$ for the control case of $u_{oo} = 0.2$ without friction.

One example of the remaining group of modes in Fig. 2(a) is characterized by $(\sigma_r, \sigma_i) = (0.009, 0.22)$ with a period of ~ 5 days. The structure of this mode is shown in Fig. 3(c). It has synoptic temporal and spatial scales extending from one model oceanic region to the other. As such, we may think of them as the counterpart of storm track disturbances. We refer to them as *synoptic-frequency modes*. This

particular mode has stronger intensity downstream of the model Pacific jet than downstream of the model Atlantic jet. But there is no basis to infer from this result that there will necessarily be a stronger model Pacific storm track since linear modal instability may not be pertinent.

A domain average zonal wind component, u_{oo}, has a trivial effect on the instability of a zonally uniform shear flow because it merely advects each normal mode by that speed. However, the effect of u_{oo} on the instability of a zonally varying shear flow is less obvious, and can be significant. We can ascertain the impact of this factor by comparing the result for $u_{oo} = 0.2$ with that for $u_{oo} = 0$ [Fig. 2(a) vs Fig. 2(b)]. Since the single-jet modes are largely under the local influence of one particular jet, the impact of u_{oo} on their growth rates is relatively small. The impact of u_{oo} on the low-frequency modes and synoptic-frequency modes is, however, much greater, as they are subject to the influence of both jets. For example, the most unstable mode for the case of $u_{oo} = 0.0$ is a low-frequency mode with a growth rate of 0.036, whereas the corresponding low-frequency mode for the case of $u_{oo} = 0.2$ has a growth rate of only 0.016. The presence of u_{oo} reduces the growth rate of this mode by more than 50%. The stabilizing effect of $u_{oo} > 0$ may be interpreted as a reduction of the residence time of disturbances in a two-jet flow (Pierrehumbert, 1984; Mak and Cai, 1989).

3.2. *Modal instability with the influence of friction, $a \neq 0$*

The effect of linear drag would be trivially simple if the friction coefficient was spatially uniform. There would simply be a reduction of the growth rate by an amount equal to the frictional coefficient a. In that case, the distribution of eigenvalues on a (σ_r, σ_i) plane would be identical to Fig. 2(a), except that the (σ_r) axis is to be relabeled as $(\sigma_r + a)$. There would be no change in the structure of the modes. According to Fig. 2(a), there would be no unstable mode if the uniform friction coefficient is $a = 0.03 \geq (\sigma_r)_{\max}^{\text{inviscid}} = 0.023$.

To determine the impacts of spatially non-uniform friction on the instability, we need to make additional computations. The frictional coefficient for generic land surfaces may be several times larger than that for water surfaces at moderate wind speeds (Arya, 1988). Thus, we present the result for $a_{\text{ocean}} = 0.015$ and

$a_{\text{land}} = 0.045$. The corresponding damping time scales are about 10 days and 3 days respectively. It is seen that the distribution of eigenvalues on the (σ_r, σ_i) plane is substantially different when there is differential friction (Fig. 4 vs Fig. 2(a)). Since we have $a_{\text{ocean}} = 0.015 \ll a_{\text{land}} = 0.045$, it is not surprising to find only one remaining weakly unstable mode which has a small growth rate of about $\sigma_r \approx 0.003$. That weakly unstable mode is a single-jet mode solely associated with the model Pacific jet [similar to Fig. 3(a)]. Figure 5 is an example of the structure of one synoptic-frequency mode under the influence of differential friction. It has a synoptic length scale. It should be emphasized that this is a weakly *stable* mode with $\sigma_r = -0.0095$ and has a frequency of $\sigma_i = 0.22$. Its appearance is broadly similar to that of Fig. 3(c). One might then ask: "Would there still be a storm track(s) in this model under a parameter condition where there is no relevant unstable normal mode?" It is possible, if nonmodal growth of transient disturbances is sufficiently strong. The answer to this question can only be ascertained by performing nonlinear experiments with this model.

4. Nonlinear Model Storm Tracks

A number of numerical integrations of (2) are performed to investigate the relative intensity of the nonlinear model storm tracks under various parameter conditions. We use a time step of $\Delta t = 0.05$ in units of LU^{-1}, corresponding to $\Delta t = 0.8 \times 10^3 s \approx 14\,\text{min}$. Each integration starts with an unbiased initial disturbance consisting of ten waves,

$$\psi(x, y, 0) = \sum_{m=1}^{10} \psi_{oo} \sin\left(\frac{xm2\pi}{X}\right) \cos\left(\frac{y\pi}{Y}\right),$$
$$(7)$$

with a finite magnitude $\psi_{oo} = 0.05$. Each integration lasts for 1800 nondimensional time units (equivalent to $\sim$354 days).

Let us first consider a control run in which we use $u_{oo} = 0.2$ with a uniform friction coefficient for the whole domain, $a = 0.03$. The

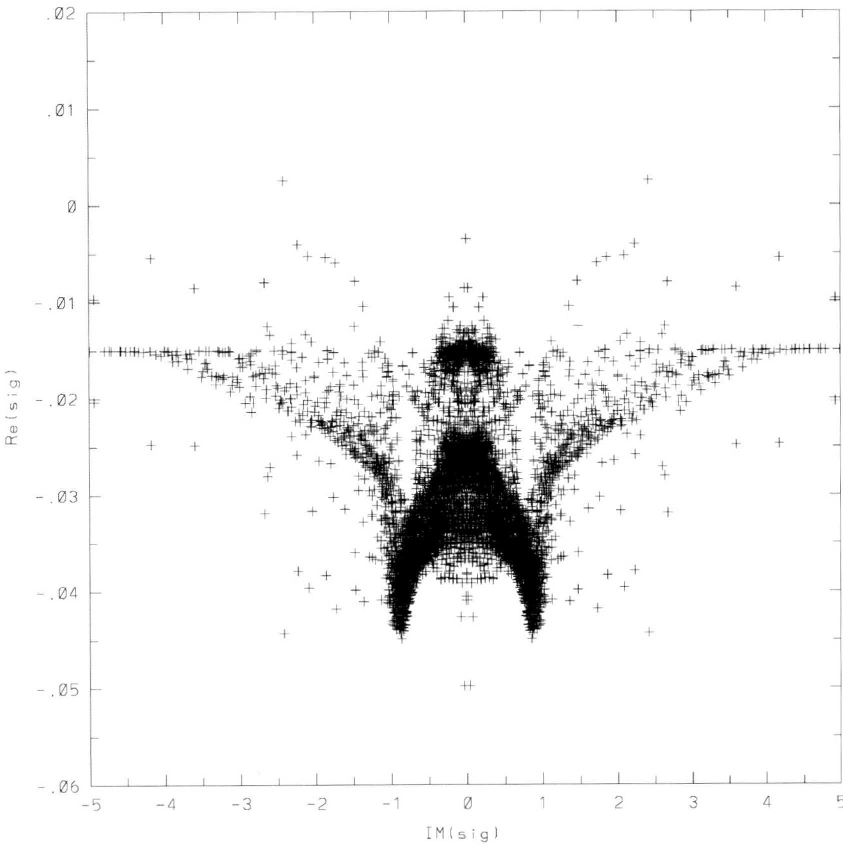

Figure 4.   Distribution of eigenvalues of normal modes under the influence of differential friction ($a_{\text{ocean}} = 0.015$, $a_{\text{land}} = 0.045$). The unit of time is $LU^{-1} = 0.17 \times 10^5$ s.

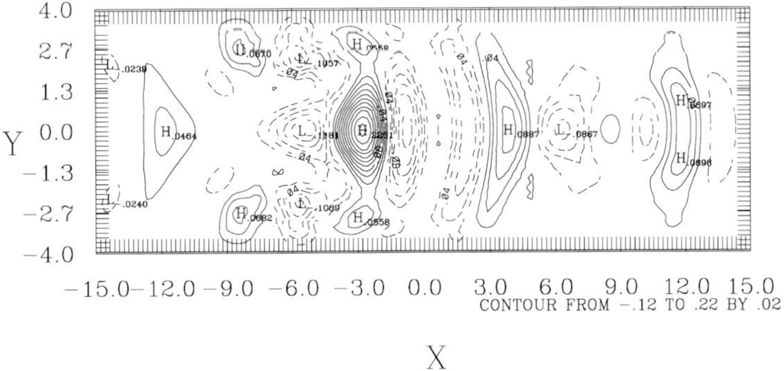

Figure 5.   Structure of a normal mode with $\sigma_r = -0.0095$, $\sigma_i = 0.22$ under the influence of differential friction.

corresponding damping time scale is about 5 days. An overall measure of the intensity of an instantaneous disturbance field is the spatial variance of the departure stream function from that of the reference state. Figure 6 shows that this system has evolved to an equilibrated state for a sufficiently long time interval that robust statistical properties can be deduced

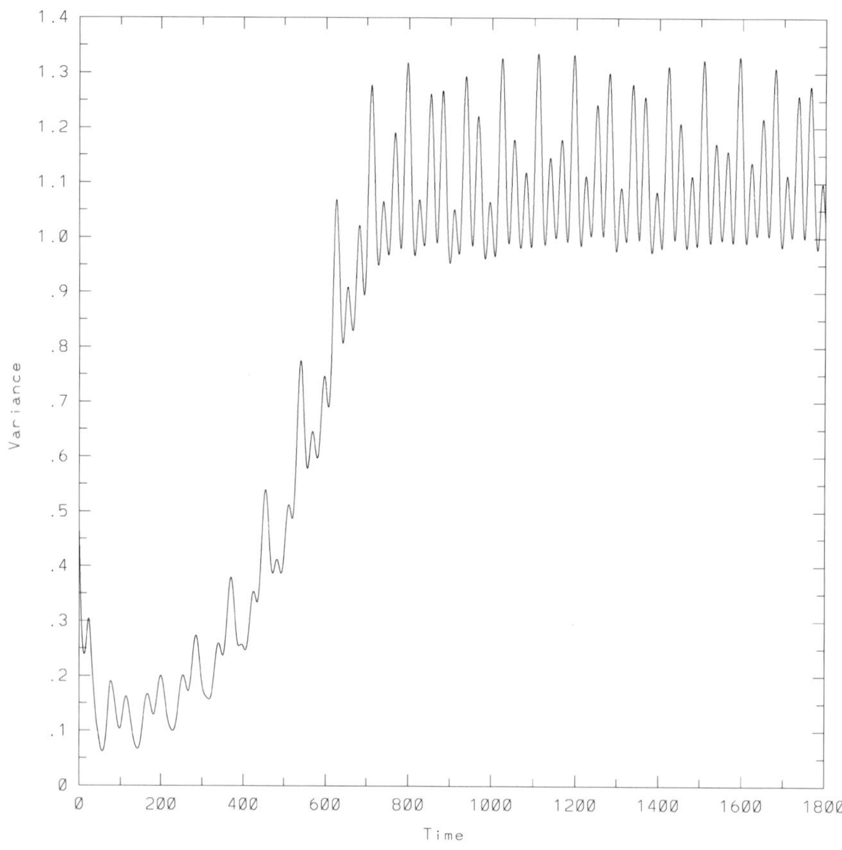

Figure 6. Evolution of the variance of the stream function for the case of $a = 0.03$. The unit of time is $LU^{-1} = 0.17 \times 10^5$ s.

from the model output of the last 1000 time units. Even though none of the normal modes are unstable because $a = 0.03 > (\sigma_r)_{\max}^{\text{inviscid}} = 0.023$, an initial disturbance field is nevertheless able to intensify via transient growth (alternatively referred to as nonmodal instability) to a sufficiently significant magnitude that finite disturbances are repeatedly regenerated by continually extracting energy from the external forcing. The equilibrated state is established after about 100 days. The equilibrated flow is also maintained by continual self-sustained transient growth. This is compatible with the explicit assumption invoked in a linear stochastic model of storm tracks (e.g. Whitaker and Sardeshmukh, 1998). The output for the last 250 days is used to compute the statistics of the flow.

We will discuss the characteristics of the related model storm tracks in Subsec. 4.2.

4.1. *Eddy feedback*

Let us first examine the time mean equilibrated departure stream function from that of the reference state, $\bar{\psi} = (\bar{\psi}_{\text{total}} - \psi_{\text{ref}})$ (Fig. 7). The velocity vector field of the reference state is also plotted in this figure for comparison. This departure field, $\bar{\psi}$, is induced by the mean feedback effect of the eddies. A dipole is induced in the region immediately downstream of each jet. It consists of an anticyclonic gyre to its north and a cyclonic gyre to its south. The dipole over the Atlantic sector is slightly stronger than the one over the Pacific sector (maximum

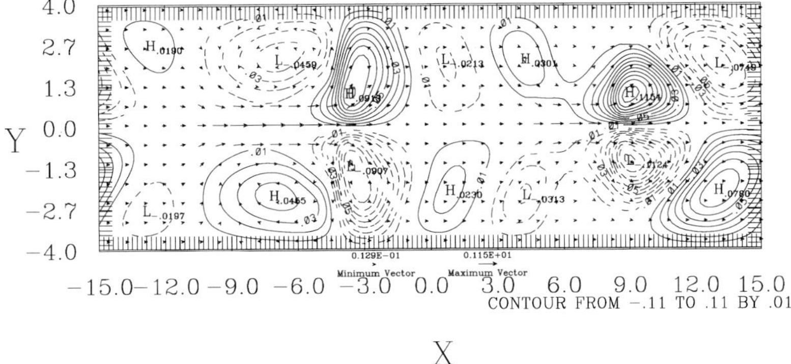

Figure 7. Velocity field of the reference state (vectors) and the time mean equilibrated departure stream function from that of the reference flow (contours) for the case of $u_{oo} = 0.2$ and a drag coefficient $a = 0.03$.

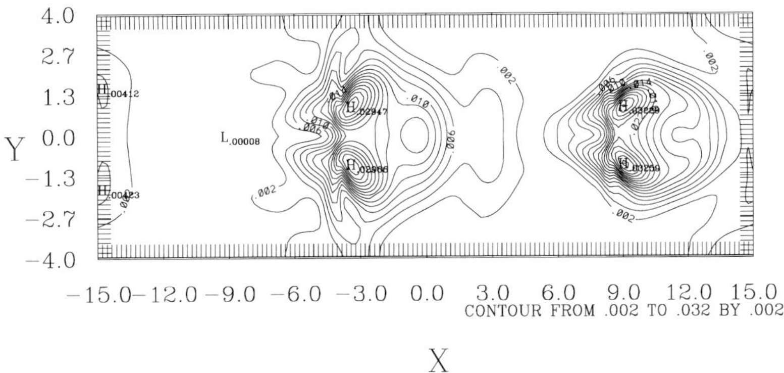

Figure 8. Distribution of nondimensional average kinetic energy of the fluctuations for the case of $a = 0.03$.

value 0.11 vs 0.09). The exit region of each modified background jet, on the average, is therefore more diffluent than the corresponding jet in the reference state. There would be stronger local stretching deformation in the exit region of each jet, which contributes to a stabilizing influence on the eddies. The meridional shear on both sides of each jet is also reduced. Such feedback effect is comparable to the counterpart found in GCM simulations (Black and Dole, 2000). The modification is stronger for the Atlantic jet. The strength of modification itself is relatively mild. This is *a posteriori* verification that the form of forcing used in the model will yield relevant time mean jets.

4.2. *Modal storm tracks with uniform friction*

Now let us examine the model storm tracks. As a measure of the intensity of variability, we plot in Fig. 8 the distribution of the average kinetic energy of the equilibrated disturbance field in the control run using $a = 0.03$. There are two well-defined model storm tracks. The high degree of meridional symmetry of this statistical property about the central latitude of the domain is an indication that the integration is sufficiently long. The maximum intensities of the model Pacific and Atlantic storm tracks are 0.029 vs 0.032 respectively. A value of

Mankin Mak

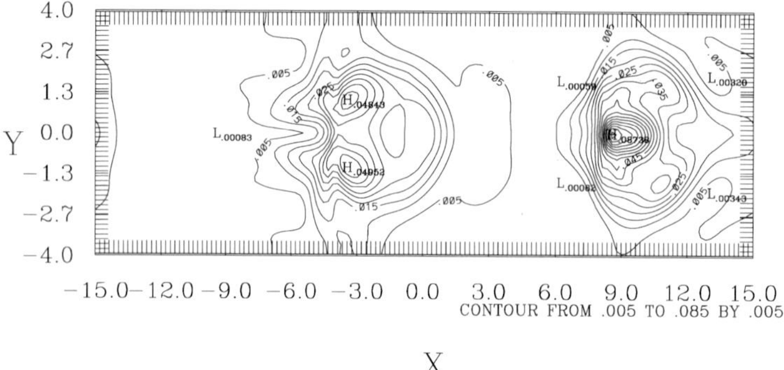

Figure 9. Distribution of nondimensional average kinetic energy of the fluctuations in the experiment using $a = 0.015$ in the oceanic sectors and $a = 0.045$ in the land sectors.

0.03 corresponds to a fluctuating wind speed of about 11 m/s. The storm track downstream of the stronger jet (model Pacific jet) is therefore weaker than the storm track downstream of the weaker jet (model Atlantic jet) by about 10%. This may be taken as evidence in support of the hypothesis concerning the role of seeding disturbances in determining the relative intensity of the two nonlinear storm tracks.

4.3. *Modal storm tracks with differential friction*

In the second set of experiments, we consider the impact of differential friction. Very little is really known empirically about the dependence of the frictional coefficient on the surface roughness. Therefore, we simply use two different plausible values of frictional coefficients for the land and ocean sectors in this model. Different combinations of them have been tried. The results obtained with the use of $a = 0.045$ for the land sectors in the model (North America, Euro-Asia) and $a = 0.015$ for the ocean sectors (Pacific and Atlantic) are representative of this set of experiments. While the drag coefficient over land is three times larger than that over the ocean in this experiment, the domain average value of the drag coefficient in the two experiments is about the same as in the control run. The time mean equilibrated departure stream

function field also has dipole structure downstream of each jet similar to that for the case of $a = 0.03$ everywhere. One difference is that the time mean departure flow in the Atlantic sector is twice as strong as that in the Pacific sector (0.21 vs 0.09; plot not shown for brevity). There are also two equilibrated storm tracks in this case, arising from continual self-sustained transient growth of disturbances. Figure 9 shows the distribution of average kinetic energy of the fluctuating flow component in this experiment. The model storm track downstream of the model Atlantic jet is found to be about 38% more intense than that downstream of the model Pacific jet. Their maximum values are 0.08 vs 0.05 respectively. The key point is that differential friction over land versus over oceans can significantly accentuate the relative intensity of the two storm tracks in favor of the Atlantic storm track for the reason elaborated in the Introduction. This result is a further demonstration that seeding disturbances of the model Atlantic jet are statistically stronger than those of the model Pacific jet, leading to the relative intensity of the two storm tracks.

5. Concluding Remarks

It has been found that the set of inviscid normal modes of a two-localized-jet basic flow that broadly resembles the Pacific and Atlantic jets consists of three groups: single-jet modes, low-frequency modes and synoptic-frequency modes. The synoptic-frequency modes are most relevant to the disturbances in the model storm tracks jointly influenced by the two jets. The differential friction associated with the land and ocean sectors in the model is shown to have significant impact on the eigenvalues as well as the structure of the normal modes. It is noteworthy that some synoptic-frequency modes do not have a stronger localized magnitude downstream of the Atlantic jet. The relative intensity of the Pacific and Atlantic storm tracks therefore cannot be interpreted from the perspective of modal linear dynamics.

Nonlinear simulation of this forced flow produces a slightly more intense model Atlantic storm track with uniform friction. Seeding disturbances arise from wave–wave interactions among the major constituent wave components in a general flow field. Differential friction enhances the difference in the seeding disturbances, leading to an even greater difference in the relative intensity of the model storm tracks. We may conclude from this analysis that, given two pertinent localized jets as a proxy forcing, their locations and the related differential friction coefficients are the most important factors.

There are obvious limitations to the model results. For example, the model height variability has two regions of maximum values straddling each jet, with pronounced symmetric tilts against the shear of the background flow. Such features are intrinsic characteristics of barotropic unstable disturbances. This symmetry also reflects the constraint due to the use of β-plane approximation. In a spherical domain, wave disturbances would be dispersed preferentially toward the low latitudes. This study illustrates that the validity of the hypothesis for the relative intensity between the Pacific and Atlantic storm tracks can be demonstrated even in a barotropic model setting. While this is a significant contributing

factor, it is unlikely to be the only important factor.

Acknowledgements

The National Science Foundation under Award ATM-0301120 provides support for this research. This study was presented at the 15th Conference on Atmospheric and Oceanic Dynamics of the American Meteorological Society at Cambridge, MA, USA, on 15 June 2005 and at IUGG in Beijing in August 2005. The helpful comments received from two reviewers are greatly appreciated.

[Received 14 February 2006; Revised 4 May 2007; Accepted 30 May 2007.]

References

Arya, S. P., 1988: *Introduction to Micrometeorology.* Academic, New York, 307 pp.

Black, R. X., and R. Dole, 2000: Storm tracks and barotropic deformation in climate models. *J. Climate*, **13**, 2712–2728.

Chang, E. K. M., S. Lee, and K. Swanson, 2002: Storm track dynamics. *J. Climate*, **15**, 2163–2183.

Deng, Y., and M. Mak, 2005: An idealized model study relevant to the dynamics of the midwinter minimum of the Pacific storm track. *J. Atmos. Sci.*, **62**, 1209–1225.

Lee, S., 2000: Barotropic effects on atmospheric storm tracks. *J. Atmos. Sci.*, **57**, 1420–1435.

Lee, W.-J., and M. Mak. 1996: The role of orography in the dynamics of storm tracks. *J. Atmos. Sci.*, **53**, 1737–1750.

Mak, M., and M. Cai, 1989: Local barotropic instability. *J. Atmos. Sci.*, **46**, 3289–3311.

Mak, M., 2002: Wave-packet resonance: Instability of a localized barotropic jet. *J. Atmos. Sci.*, **59**, 823–836.

Mak, M., and Y. Deng, 2007: Diagnostic and dynamical analyses of two outstanding aspects of storm tracks. *Dyn. Atmos. Oceans*, **43**, 80–99.

Orlanski, I. 2005: A new look at the Pacific storm track variability: sensitivity to tropical SSTs and to upstream seeding. *J. Atmos. Sci.*, **62**, 1367–1390.

Pierrehumbert, R. T., 1984: Local and global baroclinic instability of zonally varying flow. *J. Atmos. Sci.*, **41**, 2141–2162.

Simmons, A. J., J. M. Wallace, and G. W. Branstator, 1983: Barotropic wave propagation and instability, and atmospheric teleconnection patterns. *J. Atmos. Sci.*, **40**, 1363–1392.

Swanson, K., P. J. Kushner, and I. M. Held, 1997: Dynamics of barotropic storm tracks. *J. Atmos. Sci.*, **54**, 791–810.

Whitaker, J. S., and P. Sardeshmukh, 1998: A linear theory of extratropical synoptic eddy statistics. *J. Atmos. Sci.*, **55**, 237–258.

Zurita-Gotor, P., and E. K. M. Chang, 2005: The impact of zonal propagation and seeding on the eddy–mean flow equilibrium of a zonally varying two-layer model. *J. Atmos. Sci.*, **62**, 2261–2273.

The Role of Cumulus Heating in the Development and Evolution of Meiyu Frontal Systems

George Tai-Jen Chen

Department of Atmospheric Sciences, National Taiwan University, Taipei, Taiwan

george@webmail.as.ntu.edu.tw

Meiyu is a unique feature of East Asia, and the Meiyu frontal system is the key synoptic feature which causes the maximum seasonal rainfall. Numerous studies have been focused on various aspects of the Meiyu frontal system. The main purpose of this article is to present an overview on the recent research on the Meiyu frontal system, which includes the structure and dynamics of the related phenomena, such as Meiyu frontogenesis, frontal movement, frontal disturbances, and low-level jets (LLJ's) during the Meiyu season. Particularly, the recent studies of the role of convective latent heating in frontogenesis, cyclogenesis, and LLJ formation using the piecewise PV inversion technique will be emphasized.

1. Introduction

The Meiyu (Baiu) rainy season starts concurrently with the East Asian summer monsoon onset in the South China Sea, which tends to occur before the South Asian monsoon development over the Bay of Bengal during late May and over the western coast of the Indian subcontinent in early June (Ding, 1992; Tao and Chen, 1987). It is a unique climatological feature in East Asia, and occurs in the period of mid-May to mid-June over South China and Taiwan, late May to late June in Japan (Baiu), and mid-June to mid-July over the Yangtze River Valley. The seasonal rainfall maximum during the Meiyu period is caused by the organized distribution of upward motion and the moisture flux convergence along the Meiyu frontal zone (G. Chen, 1977a, 1979; G. Chen and Tsay, 1978). The occurrence and variation of this climatological phenomenon has significant impacts on regional agriculture, water resources, and human activities.

Satellite pictures usually reveal a long stratiform cloud band along the Meiyu front with vigorous convection embedded within the band (G. Chen, 1978a,b). Figure 1 presents a case of a Meiyu frontal cloud band with stratiform and convective clouds on 9 June 2006 as revealed by JMA MTSAT satellite infrared imagery. The organized mesoscale convective systems (MCS's) developed along and to the south of the Meiyu front over the southern China coast, the northern South China Sea, Taiwan and the vicinity. The MCS's produced more than 300 mm daily rainfall over central and southern Taiwan on the same day in this case. Synoptically, the rainfall in Meiyu season is associated with the repeated occurrence of a front which develops in the deformation wind field between a migratory high to the north, and the subtropical Pacific high to the south (G. Chen 1977b, 1983). The Meiyu front often moves southward or southeastward slowly in the early stage of its lifetime, and appears as a quasi stationary front in the late stage, with an average lifetime of 8 days (G. Chen and Chi, 1980). The characteristics of different types of Meiyu fronts,

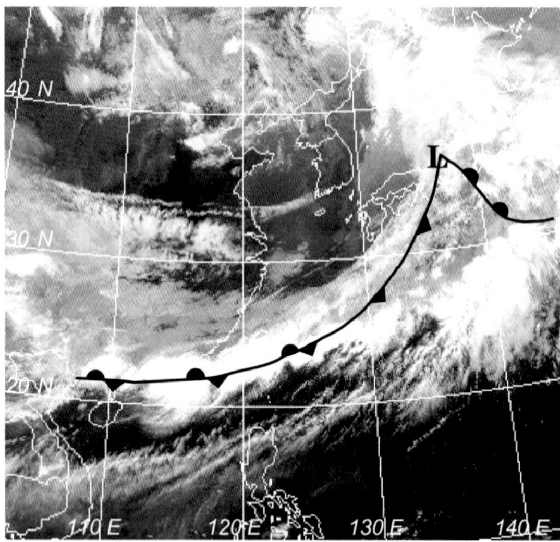

Figure 1. JMA MTSAT-1R infrared satellite image at 0000 UTC 9 June 2006. The surface Meiyu front is indicated.

as compared to the polar front occurring in the midlatitudes, can be found in a review paper by G. Chen (2004).

One of the very important features accompanying a Meiyu front was the existence of a low-level jet (LLJ) to the south or southeast of the 850/700 hPa trough (or shear line) (G. Chen, 1977a, 1978a; Ding, 1992; Tao and Chen, 1987). A close relationship between extremely heavy rainfall and an LLJ in the 850–700 hPa layer during the Meiyu season was found in many observational case studies over various geographic locations (Akiyama, 1973; G. Chen, 1979, 1983; G. Chen and Chi, 1978; G. Chen and Yu, 1988; G. Chen et al., 2005; Matsumoto, 1972; Ninomiya and Akiyama, 1974; Tao and Chen, 1987; Tsay and Chain, 1987). Over China, the heavy rainfall during the Meiyu period is mainly generated by the meso-α- and meso-β- scale disturbances, which are embedded within and propagated along the Meiyu cloud and rain band or frontal zone with a horizontal length scale of several thousand kilometers (Ding, 1992). One example of these disturbances is the intermediate-scale cyclone which forms along

the Meiyu front with a horizontal scale of 1000–3000 km.

In view of the important roles of the Meiyu front, LLJ, and the frontal disturbances in producing the seasonal maximum rainfall in the Meiyu period, this article intends to present an overview of the recent studies of the synoptic and dynamical aspects of these features. Particularly, studies of the role of latent heat release from cumulus convection in Meiyu frontogenesis, frontal movement, development of frontal disturbances, and LLJ formation from a potential vorticity (PV) perspective will be emphasized. Earlier research on these features can be found in the review papers by G. Chen (1992, 2004), Ding (1992), Ding and Chan (2005), Tao and Chen (1987), and others.

2. Meiyu Frontogenesis

Frequency distribution of Meiyu frontogenesis during the Meiyu season of South China and Taiwan as presented in Fig. 2 revealed that the Meiyu front tends to form over the subtropical latitudes to the south of 35°N (G. Chen and Chi, 1980). The front that forms in the midlatitudes to the north of 35°N is the polar front, which does not penetrate into the subtropical latitudes. Different frontal characteristics were found before and after the seasonal transition in mid-June for Meiyu fronts over southern China. Before the transition, it was common for Meiyu fronts to process appreciable baroclinicity and penetrate into the subtropics (e.g. Chen et al., 1989; Trier et al., 1990; Chen and Hui, 1990; G. Chen, Wang, and Wang, 2007). After the transition, Meiyu fronts in southern China usually had strong moisture, but only weak temperature gradients (e.g. G. Chen and Chang, 1980; G. Chen et al., 2003; G. Chen et al., 2006).

Theoretical studies by Cho and G. Chen (1994, 1995) suggested that the frontogenetic process is initiated and maintained by the conditional instability of the second kind (CISK) mechanism (G. Chen and Chang, 1980) through

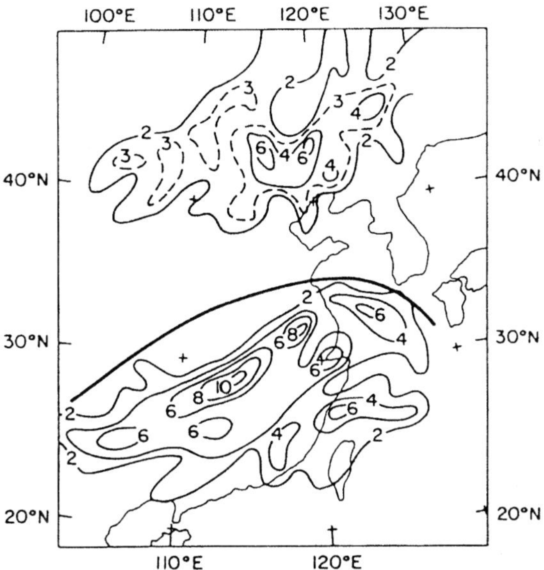

Figure 2. Frequency of frontogenesis at 1° lat × 1° long grid intervals during the Meiyu season of 15 May–15 June 1968–77 (1970 and 1975 excluded) over South China and Taiwan. The heavy solid line marks the boundary between the formation of the polar front and the Meiyu front (from G. Chen and Chi, 1980).

the interaction between the PV anomaly and the convection induced by Ekman layer pumping, as the Meiyu front over South China is characterized by a positive low-level PV anomaly with a weak baroclinicity. The scale contraction produced by the convergence flow associated with convection provides the basic frontogenetic forcing. This is quite different from the classical frontal theory, in that frontogenesis is primarily due to the deformation field and non-linear positive feedback mechanism from the induced thermally direct secondary circulation to enhance the temperature gradient and scale contraction. The crucial role of the convective latent heating in the Meiyu frontogenesis was also demonstrated in a modeling case study by Chen *et al.* (1998) and recently in a diagnostic case study using the piecewise PV inversion technique by G. Chen *et al.* (2003).

Figure 3 presents the total PV at 850 hPa from 1200 UTC 12 June to 0000 UTC 13 June 1990 of the case studied by G. Chen *et al.* (2003).

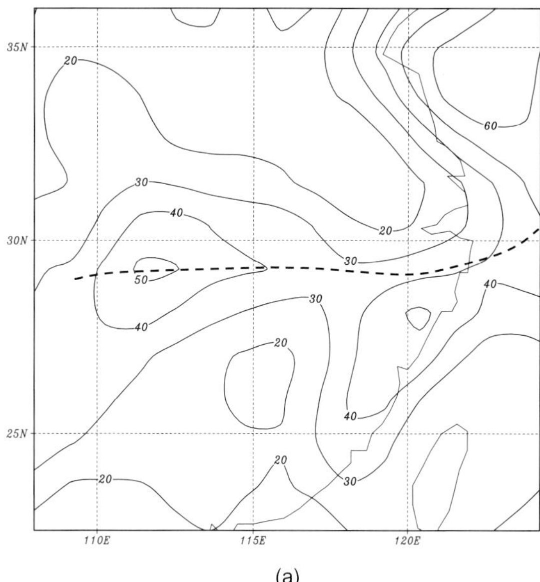

(a)

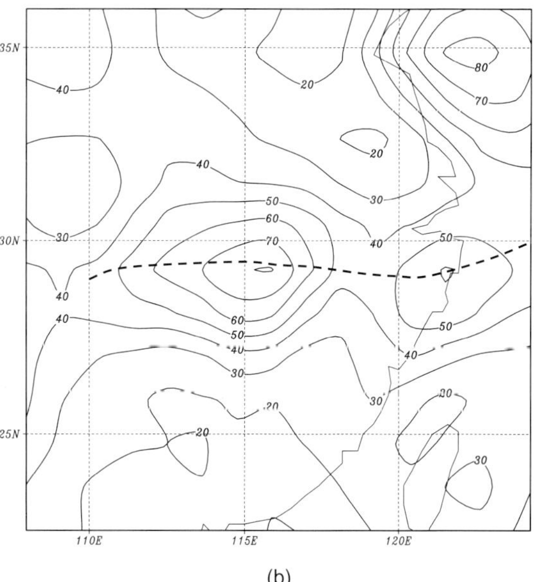

(b)

Figure 3. Total 850-hPa potential vorticity (10^{-2} PVU) at (a) 1200 UTC 12 June, and (b) 0000 UTC 13 June 1990. Contour intervals are 10×10^{-2} PVU ($1\ \mathrm{PVU} = 10^{-6}\,\mathrm{K\,m^2\,kg^{-1}\,s^{-1}}$). The dashed line indicates the location of the 850-hPa Meiyu front from synoptic analysis (from G. Chen *et al.*, 2003).

The Meiyu front was accompanied by a clear PV maximum with a weak baroclinicity, and remained stationary near 29.5°N to the west of about 117°E. Meiyu frontogenesis was indicated by the intensification of PV along the front. In that study, G. Chen *et al.* (2003) partitioned PV perturbations (q') into those from different physical processes, largely following the piecewise PV inversion technique of Davis and Emanuel (1991) and Davis (1992). The schematic diagram is presented in Fig. 4 for partitioning the PV anomaly (perturbation) q'. The potential temperature perturbation (θ') at 925 hPa (interpolated from values at 1000 and 850 hPa) was defined as the component at the lower boundary (denoted by 1b). The upper-level (ul) component included q' and θ' from 400 to 150 hPa. In the lower to middle troposphere between 850 and 500 hPa, q' was further partitioned into perturbations related to latent heat release (saturated; denoted by ms) and those not related (unsaturated, mu). They further partitioned the ms perturbations into two parts; one from deep convection (denoted as msd) and the other from shallower stratiform clouds (mss).

Partitioning results among individual processes [Fig. 5(a)] of G. Chen *et al.* (2003)

indicate that the contribution from PV perturbations related to midlevel latent heat release (ms) was nearly identical to the total contribution. Therefore, latent heating was suggested to be the major contributing process in frontogenesis and explained almost all the frontal intensity in this case, while the remaining processes had a combined contribution of nearly zero. Over the 18 h period, both upper-level (ul)

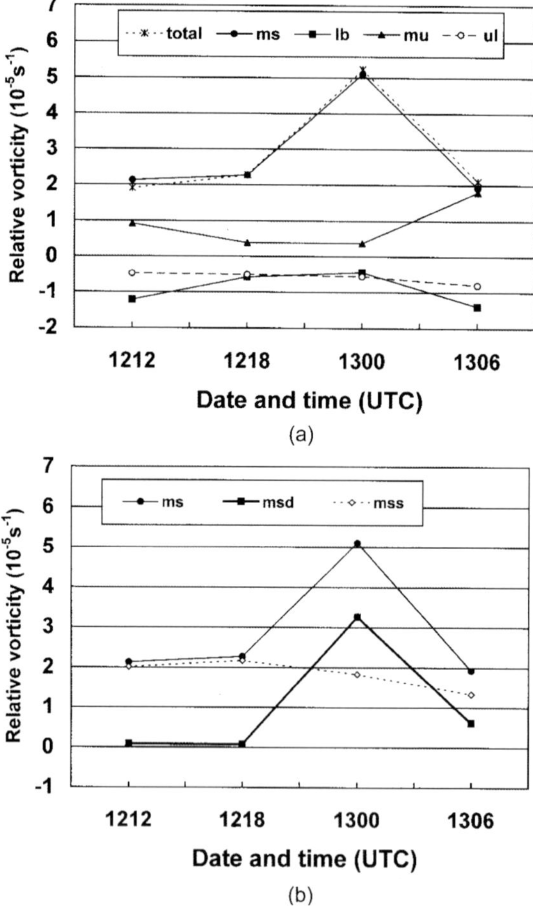

Figure 5. Time series of averaged relative vorticity (ζ) over the frontal region (29.25°–30.375° N, 109.125°–117° E) at 850 hPa. (a) Partitioning of contributions from different processes (lb, ul, ms, and mu) toward total ζ and (b) partitioning of contribution from ms between deep convection (msd) and stratiform clouds (mss). The abscissa indicates date and time (UTC), and the ordinate indicates vorticity ($10^{-5}\,s^{-1}$) (from G. Chen *et al.*, 2003).

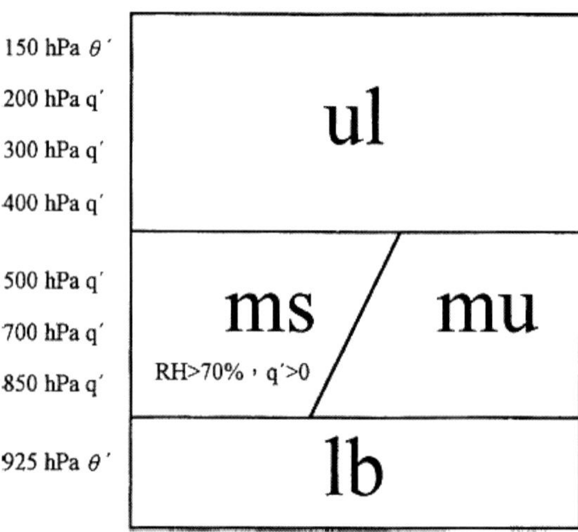

Figure 4. Scheme for partitioning the potential vorticity anomaly q' (or potential temperature perturbation θ') into perturbations associated with different processes using piecewise inversion. Data levels and types are indicated along the vertical axis. See text for details. (From G. Chen *et al.*, 2003).

and lower boundary (1b) perturbations had weak negative contributions toward frontal vorticity, and their effects were roughly canceled by the positive contribution from midlevel perturbations not related to latent heating (mu). The major contributor to 850-hPa frontal vorticity, ms, was further partitioned into perturbations associated with deep convections (msd) and those associated with shallower stratiform clouds [mss, Fig. 5(b)]. During the 18 h period analyzed, it is clear that the significant intensification of the Meiyu front at 0000 UTC 13 June was mainly attributable to the heating associated with deep convection. Apparently, the convective latent heating played a major role in Meiyu frontogenesis in this case. Similar diagnosis was carried out by G. Chen *et al.* (2006) on a retreating Meiyu front with a weak baroclinicity over the Taiwan area. Figure 6 presents the temporal variations of averaged relative vorticity (ζ) over the region of maximum frontal ζ and the partition of different components from the piecewise PV inversion. Again, the latent heating effect associated with the organized MCS appeared to be the primary mechanism for strengthening the Meiyu front.

Meiyu frontogenesis was also studied recently by G. Chen, Wang, and Wang (2007) for a Meiyu front case with a relatively strong baroclinicity using the 2 D frontogenetical function of Ninomiya (1984). The frontogenetical function F can be written as follows:

$$F \equiv \frac{d}{dt}|\nabla_H \theta| = FG1 + FG2 + FG3 + FG4, \quad (1)$$

where the four forcing terms on the right hand side, respectively, are

$$FG1 = \frac{1}{|\nabla_H \theta|}\left[\nabla_H \theta \cdot \nabla_H\left(\frac{d\theta}{dt}\right)\right], \quad (2)$$

$$FG2 = -\frac{1}{2|\nabla_H \theta|}\left[\left(\frac{\partial\theta}{\partial x}\right)^2 + \left(\frac{\partial\theta}{\partial y}\right)^2\right]$$

$$\times \left(\frac{\partial u}{\partial x} + \frac{\partial v}{\partial y}\right), \quad (3)$$

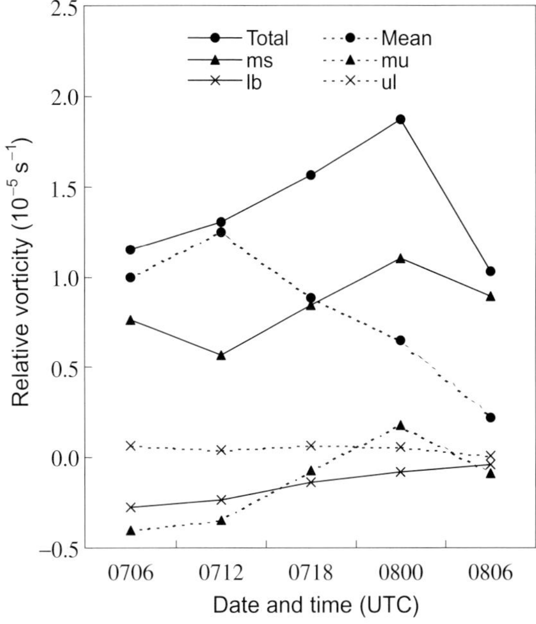

Figure 6. Temporal variation of relative vorticity (ζ; ordinate; $10^{-5}\,\mathrm{s}^{-1}$) averaged over the area centered at the maximum relative vorticity on the Meiyu front at 850 hPa from 0600 UTC 7 to 0600 UTC 8 June 1998 (abscissa). "Total" and "Mean" represent vorticity of balanced total wind and time mean wind (15 May–15 June 1998), respectively. See text for the definition of other terms (from G. Chen *et al.*, 2006).

$$FG3 = -\frac{1}{|\nabla_H \theta|}\left\{\frac{1}{2}\left[\left(\frac{\partial\theta}{\partial x}\right)^2 - \left(\frac{\partial\theta}{\partial y}\right)^2\right]\right.$$

$$\left.\times\left(\frac{\partial u}{\partial x} - \frac{\partial v}{\partial y}\right) + \frac{\partial\theta}{\partial x}\frac{\partial\theta}{\partial y}\left(\frac{\partial v}{\partial x} + \frac{\partial u}{\partial y}\right)\right\}, \quad (4)$$

$$FG4 = -\frac{1}{|\nabla_H \theta|}\left[\frac{\partial\theta}{\partial p}\left(\frac{\partial\theta}{\partial x}\frac{\partial\omega}{\partial x} + \frac{\partial\theta}{\partial y}\frac{\partial\omega}{\partial y}\right)\right]. \quad (5)$$

The terms $FG1$, $FG2$, $FG3$, and $FG4$ represent effects from diabatic processes, horizontal convergence, deformation, and tilting, respectively, while $FG3$ includes both stretching ($\partial u/\partial x - \partial v/\partial y$) and shearing ($\partial v/\partial x + \partial u/\partial y$) deformation.

The analyses at the 925 hPa surface is presented in Fig. 7. The thermal gradient of the Meiyu front increased from 8 June to reach a maximum at 1200 UTC 10 June 2000, then

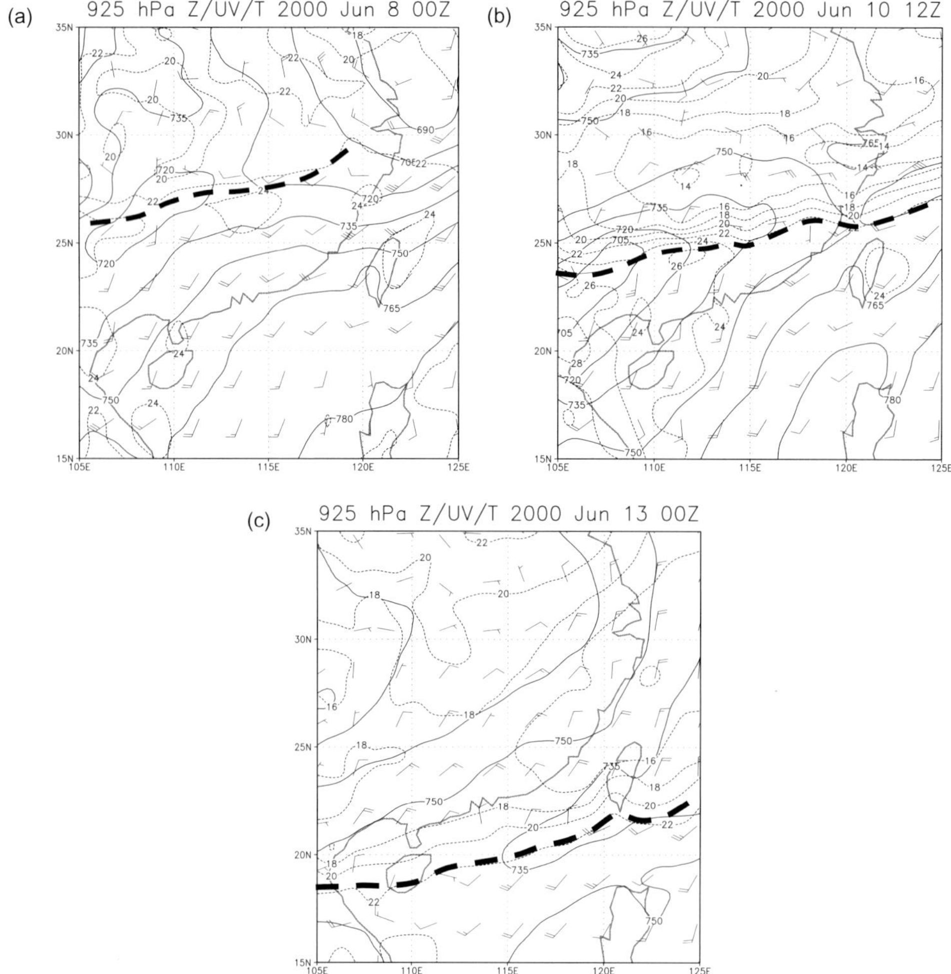

Figure 7. 925-hPa ECMWF analyses of geopoential height (gpm; solid), temperature (°C; dashed), and horizontal winds (m s^{-1}) over the domain of 15°–35°N, 105°–125°E at (a) 0000 UTC 8 June, (b) 1200 UTC 10 June, and (c) 0000 UTC 13 June, 2000. Contour intervals are 15 gpm for geopotential height and 2°C for temperature, respectively. Thick dashed lines indicate the position of the 925-hPa Meiyu front based on temperature gradient and winds (from G. Chen, Wang, and Wang, 2007).

remained quite strong until after 12 June. The along-front averages of the frontogenetical function and contributing terms over 108°–120°E from 5.625° south to 7.875° north of the 925 hPa front are presented in Fig. 8. During the initial stage [Fig. 8(a)], all three terms, including diabatic processes, horizontal convergence, and deformation, were in phase with the frontal zone, with the front mainly maintained through diabatic effects. During the intensification stage [Fig. 8(b)], the frontal θ gradient strengthened and maximized about 120 km north of the front. Diabatic effects become strongly frontolytic, while convergence frontogenesis nearly collocated with the frontal zone. The deformation, while also frontogenetic with roughly the same contribution as convergence, had a tendency to peak slightly to the south of the maximum θ gradient. During the decaying stage [Fig. 8(c)], the front entered the South China Sea. The

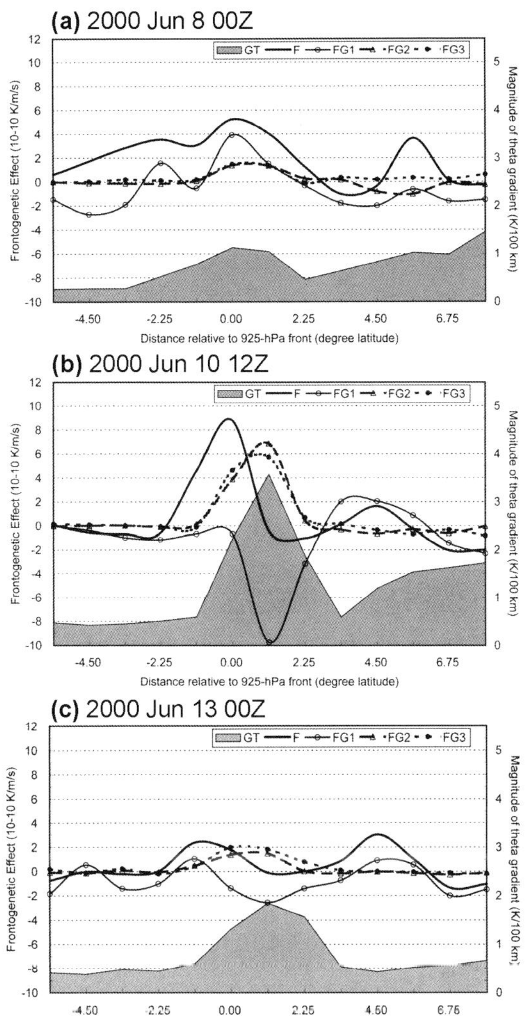

Figure 8. Averaged values of frontogenetical function $(d|\nabla_H\theta|/dt, F)$, its contributing terms $FG1$ (diabatic processes), $FG2$ (horizontal convergence), and $FG3$ (deformation) (all in 10^{-10} K m^{-1}s^{-1}, scale on left side), and the magnitude of the horizontal potential temperature gradient ($|\nabla_H\theta|$, GT, shaded, scale on right side) at 925 hPa from $-5.625°$ (south) to $7.875°$ (north) relative to the 925-hPa front (at $0°$) at (a) 0000 UTC 8 June, (b) 1200 UTC 10 June, and (c) 0000 UTC 13 June, 2000. The average is performed over $108°$–$120°$E, and curves for frontogenetical terms (F, $FG1$, $FG2$, and $FG3$) are smoothed (from G. Chen, Wang, and Wang, 2007).

frontogenetical effects of convergence and deformation were reduced, the frontal θ gradient weakened mainly owing to the continuous sensible heat flux over warmer water into the postfrontal

cold air. Overall, the distribution of the frontogenetical function in this case indicated that the Meiyu frontogenesis and the maintenance of the front were attributed to both deformation and convergence.

In summary, the latent heat release and the associated CISK mechanism were suggested to play major roles in Meiyu frontogenesis for one type of Meiyu front, which is characterized by a weak baroclinicity and a strong PV anomaly. For the Meiyu front with a relatively strong baroclinicity, on the other hand, it was suggested that the diabatic effects become strongly frontolytic and the horizontal convergence as well as deformation wind fields play major roles in Meiyu frontogenesis similar to what would be expected from the classical frontal theory. Indeed, further studies will be needed to clarify the role of latent heat release in frontogenesis for the Meiyu front with a strong baroclinicity.

3. Movement of the Meiyu Front

As pointed out in the introduction, Meiyu fronts often move southward or southeastward slowly and then become stationary. Climatologically, about 20% of the Meiyu fronts retreat northward after they become stationary in the vicinity of Taiwan. G. Chen, Wang, and Wang (2007) studied the southward movement of the Meiyu front case as presented in Fig. 7. They analyzed the distribution of frontal strength ($|\nabla_H\theta|$, GT), frontogenetical function $(d|\nabla_H\theta|/dt$, F), local tendency of $|\nabla_H\theta|$, $(\partial|\nabla_H\theta|/\partial t$, LT), and horizontal advection of $|\nabla_H\theta|$ $(-\vec{V}\cdot\nabla_H|\nabla_H\theta|$, ADV). The local tendency of the frontal strength $\partial|\nabla_H\theta|/\partial t$ can be written as follows:

$$\frac{\partial}{\partial t}|\nabla_H\theta| = F - \vec{V}\cdot\nabla_H|\nabla_H\theta|. \qquad (6)$$

Results are presented in Fig. 9. Note that the earth-relative frontal motion exists when there is a phase difference between LT and the frontal

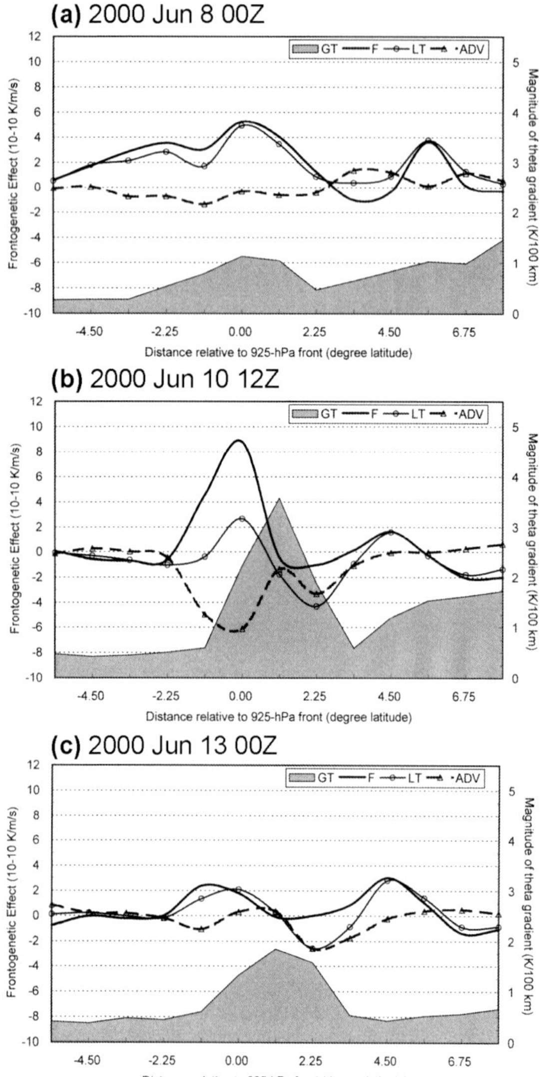

Figure 9. Same as Fig. 8 except for averaged values of frontogenetical function $(d|\nabla_H\theta|/dt,\ F)$, local tendency $(\partial|\nabla_H\theta|/\partial t,\ LT)$, and horizontal advection $(-\vec{V}\cdot\nabla_H|\nabla_H\theta|,\ ADV)$ of the magnitude of the horizontal potential temperature graident (all in $10^{-10}\ \mathrm{K\,m^{-1}\,s^{-1}}$, scale on left side), and the magnitude of the horizontal potential temperature gradient $(|\nabla_H\theta|,\ GT,\ \text{shaded},$ scale on right side). Curves for F, LT, and ADV are smoothed (from G. Chen, Wang, and Wang, 2007).

zone, and is related to the propagation represented by F and the transport by advection. On 8 June [Fig. 9(a)], the total F contributed toward a positive LT that was roughly in phase

with the frontal gradient, resulting in intensification of the front, as discussed in Sec. 2. When the front grew stronger [Fig. 9(b)], the effect of ADV was negative near the front, especially immediately ahead of and behind the zone of the maximum θ gradient. The total F peaked ahead of the frontal zone, and resulted in a positive local tendency also slightly ahead of the front, and thus the forward propagation of the front was contributing toward its total southward movement. On 13 June [Fig. 9(c)], the advection became positive ahead of and negative behind the front, indicating that the postfrontal cold air advection was dominant at this time. The local tendency was roughly 90° ahead of the front, in agreement with the rapid southward movement during the period. Apparently, for the Meiyu frontal movement, the total frontogenetical function (F) that peaked ahead of the frontal zone contributed toward the southward propagation of the Meiyu front, in addition to the advection by the postfrontal cold air.

G. Chen *et al.* (2006) studied a case of a retreating Meiyu front which caused heavy rainfall over the Taiwan area. Synoptic analyses at 850 hPa (Fig. 10) indicate the northward movement of the Meiyu front, as defined by the zone of maximum relative vorticity. The nonlinearly balanced fields of geopotential height and wind associated with PV perturbation of midlevel latent heat release (ms) are presented in Fig. 11. Based on the overall analyses using the piecewise PV inversion technique, it was suggested that the convective latent heating was the primarily cause of the formation and intensification of the LLJ and the subsequent northward retreat of the Meiyu front. They also studied the vorticity budget for this retreating Meiyu front case. Figure 12 presents the time variations of vorticity budget terms across the front at 850 hPa. The stationary nature of the front before 1200 UTC 7 June is clearly illustrated by the near-zero local tendency along the front, while its rapid northward movement afterward is indicated by the large positive tendency north

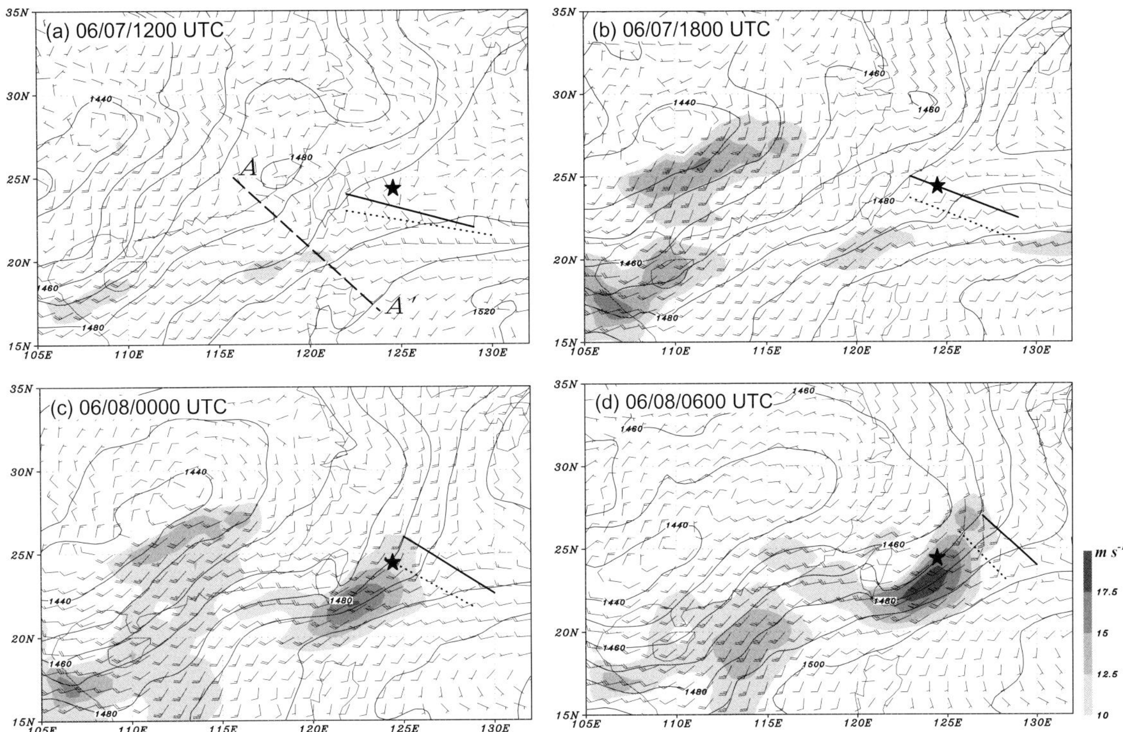

Figure 10. Synoptic maps at 850 hPa analyzed using ECMWF gridded data every 6 h from (a) 1200 UTC 7 to (d) 0600 UTC 8 June 1998. Geopotential heights (gpm, contour) are analyzed at 10 gpm intervals. Full and half barbs represent 5 and 2.5 m s^{-1}, respectively, and areas with wind speed $\geq$10 m s^{-1} are shaded (gray scale shown at lower right corner). Heavy solid (dotted) lines indicate a Meiyu front as defined by the maximum θ_e gradient (relative vorticity), and asterisks ($\star$) mark the location of Ishigakijima station (from G. Chen *et al.*, 2006).

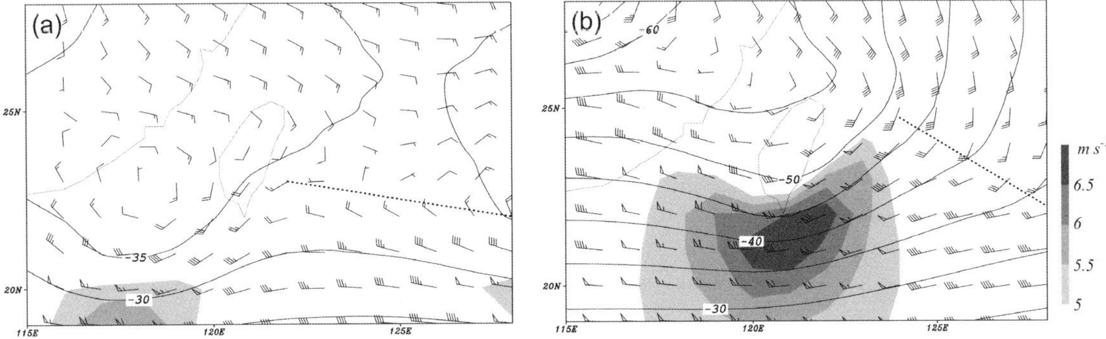

Figure 11. Nonlinearly balanced fields of geopotential height (gpm) and wind (m s^{-1}) associated with the ms component at 850 hPa for (a) 1200 UTC 7 June and (b) 0000 UTC 8 June, 1998. Geopotential heights are analyzed at 5 gpm intervals. The pennant, full barb, and half barb represent 5, 1, and 0.5 m s^{-1}, respectively, and areas with winds $\geq$5 m s^{-1} are gray-shaded (scale shown on right hand side). The dotted lines indicate frontal position (from G. Chen *et al.*, 2006).

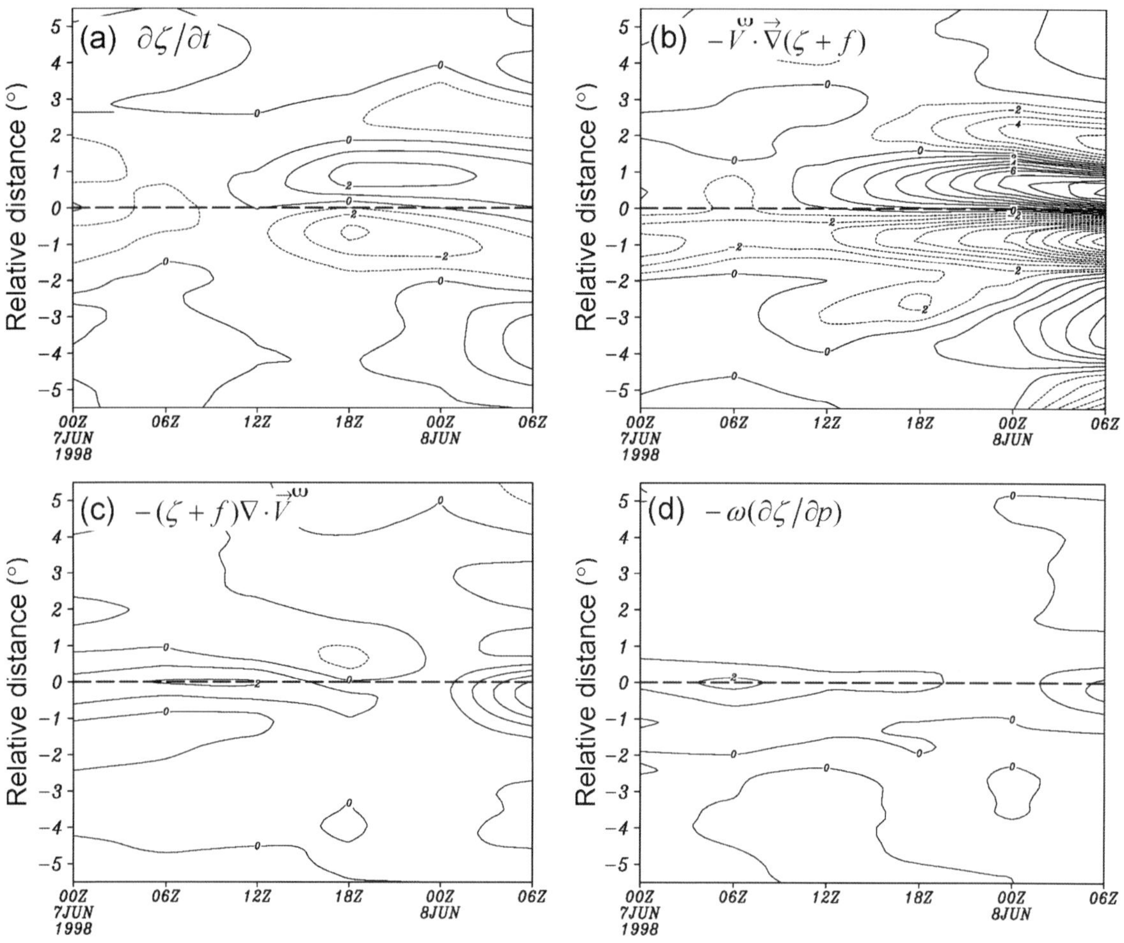

Figure 12. Time variations of vorticity budget terms (a) local rate of change ($\partial\zeta/\partial t$), (b) horizontal advection $[-\vec{V}^\omega \cdot \nabla(\zeta + f)]$, (c) divergence $[-(\zeta + f)\nabla \cdot \vec{V}^\omega]$, and (d) vertical advection $[-\omega(\partial\zeta/\partial p)$, all in $10^{-5}\,\text{s}^{-1}\,(6\,\text{h})^{-1}]$ across the front at 850 hPa during 0000 UTC 7-0600 UTC 8 June 1998 (abscissa). Values were averaged along the front from $5.5°$ south (negative, ordinate) to 5.5 north (positive). Isolines are analyzed at $1 \times 10^{-5}\,\text{s}^{-1}\,(6\,\text{h})^{-1}$, and heavy dashed lines indicate frontal position (from G. Chen *et al.*, 2006).

of the front. Based on the overall diagnoses, they suggested the vital role of the vorticity advection process associated with LLJ in causing the northward retreat of the Meiyu front in this case.

4. Low-Level Jet (LLJ) Formation

The frequency distribution of the 850 hPa migratory LLJ's which affect northern Taiwan in the Meiyu season is presented in Fig. 13 (G. Chen *et al.*, 2005). The LLJ's tended to form over southern China between $20°$ and 30°N [Fig. 13(a)], with the highest frequency (near 23°N, 109°E) just to the east of the Tibetan Plateau. The main axis of the jets was oriented from southwest to northeast, also roughly parallel to the terrain. As the LLJ's migrated eastward at t_{-24} and t_{-12}, their overall shape became much more elongated, with the area of high frequency extending farther downstream [Figs. 13(b) and 13(c)]. In addition, the orientation of the maximum axis also turned slightly from west-southwest to east-northeast. From t_{-12} to t_0, the LLJ's moved

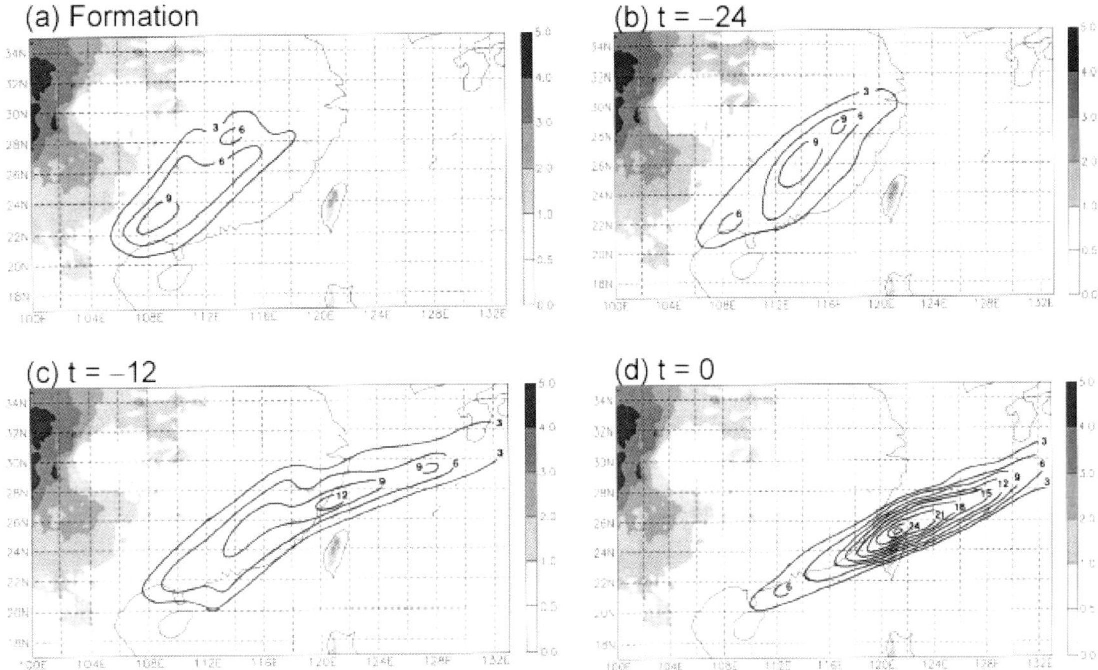

Figure 13. Geographical distribution of frequencies of wind speed reaching $10\,\mathrm{m\,s^{-1}}$ at 850 hPa associated with a total of 30 migratory LLJ cases in May–June, 1985–1994, at (a) the time of LLJ formation, (b) 24 h before ($t = -24$), (c) 12 h before ($t = -12$), and (d) at the synoptic time ($t = 0$) when the LLJ moved over northern Taiwan (inside the area of $24°$–$26°$N, $120.5°$–$122.5°$E). Regions with a terrain higher than 0.5 km are shaded (from G. Chen *et al.*, 2005).

southward significantly, and the frequency converged toward the main axis passing through northern Taiwan [Fig. 13(d)].

Observational study by G. Chen and Yu (1988) suggested that an LLJ might form to the south of the heavy rainfall area. They proposed that the reversed secondary circulation to the south of the Meiyu front as observed in the composite by G. Chen and Chi (1978), presumably driven by convective latent heating, was a possible formation mechanism for an LLJ. This mechanism was also suggested by both theoretical study (Chen, 1982) and numerical experiments (Chen *et al.*, 1998; Chen *et al.*, 2000; Chou *et al.*, 1990). Results of Chen's theoretical study (1982) suggested that the existence of an unstable inertiogravity wave in the ascending motion region caused a thermally direct circulation beneath the upper-level jet. Latent heating in the area of ascending motion

accelerated the thermally direct circulation to the north and induced a reversed circulation to the south through the geostrophic adjustment process. It was concluded that the LLJ formed through the Coriolis acceleration of northward flow in the lower branch of this reversed circulation. Numerical study by Chou *et al.* (1990) using a two-dimensional frontogenesis model simulated the formation of an LLJ to the south of the area of strong convection and suggested the importance of cumulus convection, especially a slantwise structure, in developing the reversed circulation and the LLJ.

A weak thermally indirect circulation was observed to the south of the Meiyu front for TAMEX IOP 5 case, as presented in Fig. 14(a) (Chen *et al.*, 1994). In that case, the LLJ was observed to the south of the Meiyu front within the lower return branch of the secondary circulation and the thermally direct circulation across the front was observed similar to that of G. Chen and Chi (1978) and G. Chen and Chang (1980). Numerical study by Chen *et al.* (1997) of this LLJ case revealed that the LLJ developed through the Coriolis force acting on the cross contour ageostrophic winds in response to the increased pressure gradients related to the development of the cyclone, and was enhanced by latent heating. A recent case study using momentum budget computations of numerical

simulation by Zhang *et al.* (2003) also suggested that the pressure gradient force and the horizontal advection are the main contributors to the development of mesoscale LLJ's. Numerical simulation study of a Meiyu frontal case by Chen *et al.* (1998) suggested that the existence of an upper-level jet will induce the development of the thermally direct circulation, as illustrated in Fig. 14(b). It was also suggested that the southward branch of the thermally

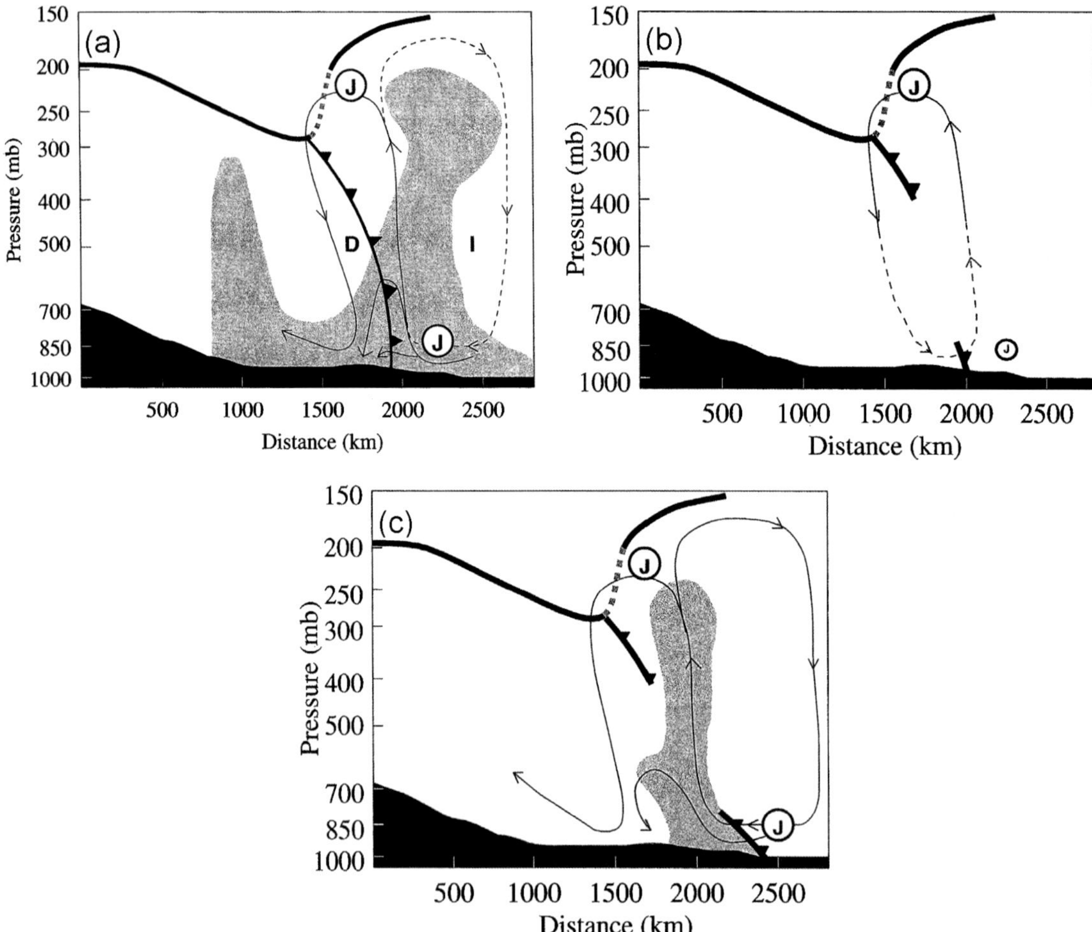

Figure 14. Schematic diagram delineating the secondary circulations across the Meiyu front (a) from Chen *et al.* (1994), (b) the upper-level jet, and (c) the vertical and slantwise convection in the development of the LLJ from Chen *et al.* (1998). In (a), thin solid and dashed lines depict the strong direct (D) and the weak indirect (I) circulations, respectively. The heavy solid line with triangles shows the frontal position. The J indicates the upper-level and low-level jet positions, and the boldness of the J represents the jet strength in (b) and (c). Thick heavy lines show the positions of the tropopause boundary. The dashed line depicts the weaker circulation and the thin solid line represents the stronger circulation. Regions with relative humidity greater than 70% are shaded.

indirect circulation, with its rising motion in the frontal region and equatorward sinking motion, is driven by the frontal vertical deep convection, as presented in Fig. 14(c). The return flow of this circulation at low level can produce an LLJ through geostrophic adjustment. Observational case studies (Lin and Chiou, 1985; G. Chen *et al.*, 1986; Lin, 1988; Lin and Tsai, 1989; Pu and G. Chen, 1988; Qian *et al.*, 2004) also showed that an LLJ tended to form or to intensify to the south of an MCS in South China. A numerical study of a TAMEX case by Hsu and Sun (1994), on the other hand, suggested that cumulus heating may not have played a critical role in the LLJ formation and the latent heat release by stratiform clouds contributed to the development of the LLJ.

In a recent case study of a retreating Meiyu front (Fig. 10), G. Chen *et al.* (2006) observed that diabatic latent heating from the MCS, large enough in scale, generated positive PV and height fall at low levels. The enhanced height gradient induced northwestward-directed ageostrophic wind toward the area of the MCS (Fig. 15) and the LLJ formed to the southeast of the MCS through Coriolis torque. The piecewise PV inversion contributions from different components to the LLJ at 850 hPa in this case are presented in Fig. 16. Apparently, the formation and intensification of the LLJ in this case could be largely attributed to the latent heating effects of the organized MCS, superimposed upon a background southwesterly monsoonal flow of about $6\,\mathrm{m\,s^{-1}}$. Similar results were obtained for a Meiyu system over South China by G. Chen, Wang, and Chang (2007) using piecewise PV inversion and ageostrophic wind analysis techniques.

In summary, results of observational, theoretical, and numerical studies all suggested that a thermally indirect circulation with rising motion in the frontal region and equatorward sinking motion is driven by the latent heat release associated with frontal deep convection. The LLJ could then form to the south of the convective heavy rainfall area through the Coriolis acceleration of the lower branch of an induced thermally indirect secondary circulation.

5. Development of the Frontal Cyclone

A theoretical study by Du and Cho (1996) proposed that the growth rate of the most unstable wave along the Meiyu front depends on the intensity of cumulus heating. Their results suggested that when the cumulus heating parameter is below a critical value, the wavelength is about 8–15 times the cross-front width scale of the PV anomaly, and the structure of the wave is of the barotropic type but modified by convection. When the heating parameter is above the critical value, the disturbance draws its energy almost entirely from heating and the structure of the wave resembles a system driven purely by cumulus heating. The wavelength of the most unstable wave is about 1700–2100 km and bears little relationship to the width of the background PV anomaly.

Observationally, one kind of low-level vortex during the Meiyu season is the intermediate-scale cyclone (1000–3000 km), which forms along the Meiyu front, and the latent heat release was found to play a major role in its development (Chang and Chen, 2000; Zhao *et al.*, 1982). It was also found that the vertical coupling processes compounded the diabatic heating effect, and could lead to the strong intensification of the Meiyu frontal cyclone (Chang *et al.*, 1998). In a recent study, Takahashi (2003) also found that the coupling of the upper-level migratory trough and the lower-level shear line can be important for the evolution of the Meiyu frontal disturbances. Lee *et al.* (2006) examined 20 tropical cyclone formations in the South China Sea in May and June of 1972–2002 and found that 11 of them were associated with the weak baroclinic environment of a Meiyu front, while the remaining 9 were nonfrontal. They also

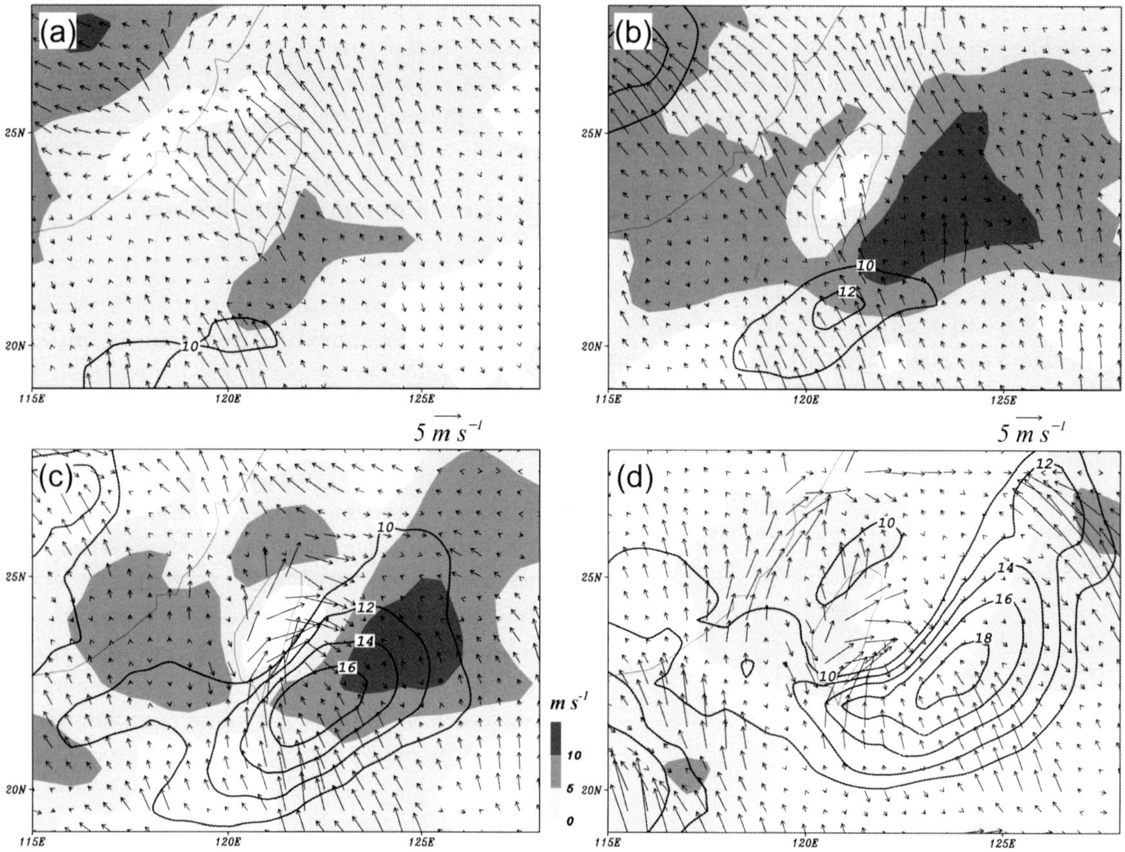

Figure 15. Ageostrophic wind component (arrow, $m\,s^{-1}$) perpendicular to geopotential height contours and local tendency of the wind component along the geopotential height contours (shaded, $m\,s^{-1}$ per 12 h, gray scale shown at bottom) at 850 hPa at 6 h intervals from (a) 1200 UTC 7 to (d) 0600 UTC 8 June, 1998. The 850-hPa LLJ is depicted by isotachs (thick solid) analyzed at intervals of $2\,m\,s^{-1}$, starting from $10\,m\,s^{-1}$ (from G. Chen et al., 2006).

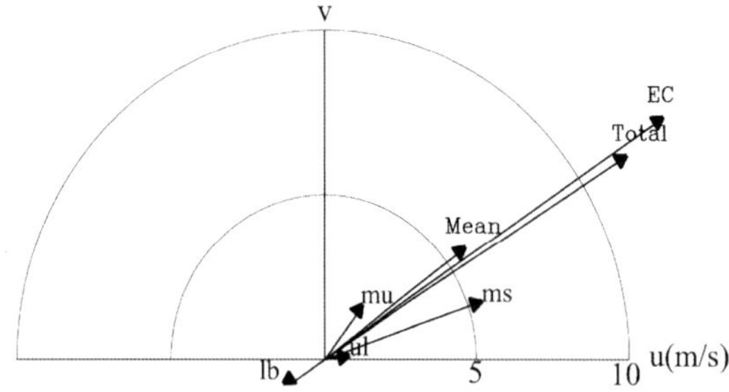

Figure 16. Wind vectors averaged over a hexagonal domain centered along the axis of the LLJ from different components at 0000 UTC 8 June 1998. "EC" depicts the observed wind of ECMWF gridded analysis, and "Total" and "Mean" represent balanced total wind and time mean wind (15 May–15 June 1998), respectively. See text for the definition of other terms (from G. Chen et al., 2006).

found that the strengthening of northeasterlies to the north of the Meiyu front was important for increased cyclonic vorticity of the frontal cyclone, and finally caused the detachment of the cyclone from the Meiyu front to become a tropical storm.

A recent diagnostic case study by G. Chen, Wang, and Chang (2007) suggested that the growth of frontal disturbances was a result of the nonlinear mechanism similar to the CISK, in which the frontal PV centers and cumulus convection reinforce each other through a positive feedback process. Figure 17 presents 850 hPa analyses of that case. At 1200 UTC 6 June [Fig. 17(a)], the Meiyu front had developed

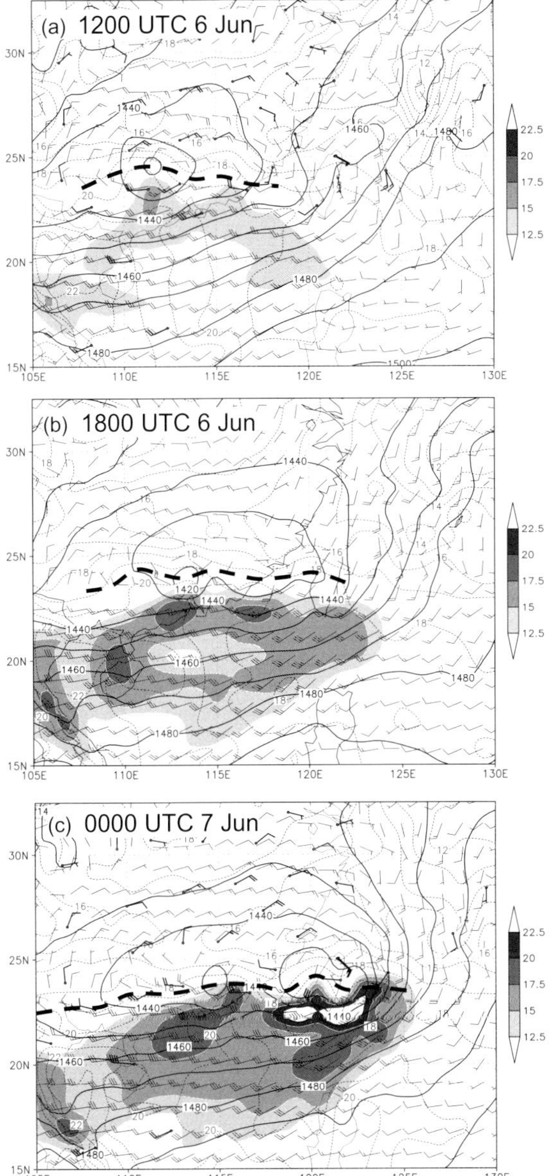

Figure 17. 850-hPa ECMWF analysis of geopotential height (solid, gpm), temperature (dashed, °C), and horizontal wind (m s^{-1}) at (a) 1200 UTC, (b) 1800 UTC 6 June, and (c) 0000 UTC 7 June, 2003. Geopotential heights and temperatures are analyzed at intervals of 10 gpm and 1°C, respectively. For winds, full (half) barbs are 5 (2.5) m s^{-1}, and wind speed $\geq$ 12.5 m s^{-1} is shaded. Thick wind flags in (a) and (c) are sounding data, and thick dashed lines indicate the position of the Meiyu front at 850 hPa (from G. Chen, Wang, and Chang, 2007).

along 24°N with the deepening of a frontal cyclone and the eastward extension of its trough. The development of the frontal cyclone was also evident at 1800 UTC 6 June [Fig. 17(b)]. At 0000 UTC 7 June, three individual vortices could be seen near 115°, 119°, and 122°E as the Meiyu front reached 125°E while moving slowly southward [Fig. 17(c)]. Figure 18 presents the nonlinear balanced 850 hPa height, wind, and ζ fields from the component of latent heat release "ms." From 1800 UTC 6 June to 0000 UTC

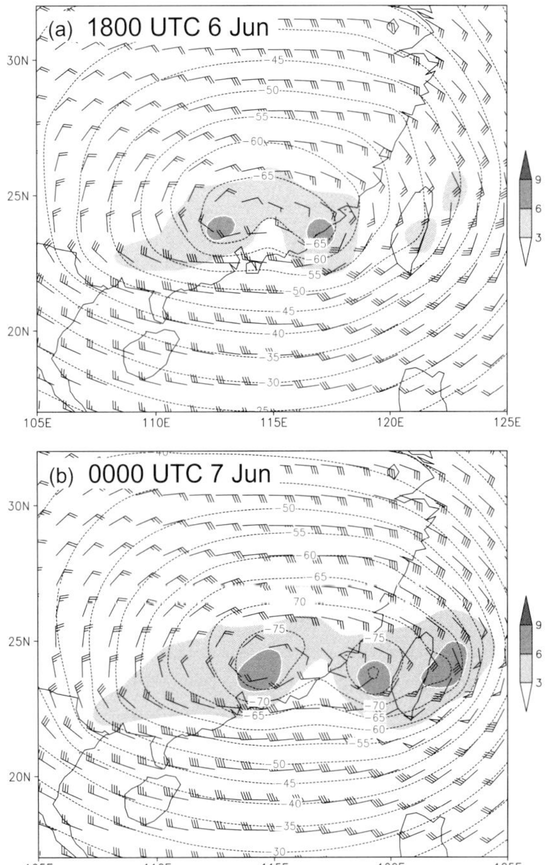

Figure 18. 850-hPa nonlinear balanced geopotential height (gpm, contour), horizontal wind (m s^{-1}), and $\zeta(10^{-5}$ s^{-1}, shading) inverted from the low- to mid-level q' related to latent heat release (ms) at 6 h intervals at (a) 1800 UTC 6 June and (b) 0000 UTC 7 June, 2003. Contour intervals are 5 gpm and dashed lines are used for negative values, while full (half) barbs in wind flags correspond to 2 (1) m s^{-1}, respectively (from G. Chen, Wang, and Chang, 2007).

7 June, the distribution of ζ from "ms" was very similar to that derived from the observed wind field (not shown). The frontal cyclonic circulation grew stronger at 0000 UTC 7 June [Fig. 17(b)], with a pattern and an evolution very similar to those appearing in Fig. 17.

In summary, it is evident that the latent heat release from the MCS's played a vital role in the strengthening and maintenance of the frontal disturbances in this case. However, whether the Meiyu frontal cyclogenesis is driven by a CISK process or not is still an open question. Further research efforts are needed to shed some light on the mechanism of the Meiyu cyclogenesis.

6. Concluding Remarks

The Meiyu frontal system is the key synoptic feature responsible for the seasonal rainfall maximum occurring during mid-May to mid-June over South China and Taiwan. The important components of the Meiyu frontal system include the Meiyu front, LLJ, and frontal cyclone. Recently, numerous studies have been focused on various aspects of these components, particularly on the role of cumulus heating in the development and evolution of the Meiyu frontal system from a PV perspective. What we have learned about the Meiyu frontal system recently can be summarized as follows.

Diagnoses for weak baroclinic Meiyu front cases using the piecewise PV inversion technique all revealed that the cumulus heating played a major role in frontogenesis. Whereas the diagnosis of the frontogenetical function for the relatively strong baroclinic Meiyu front case indicated that the frontogenesis and the maintenance of the front were due to both horizontal convergence and deformation similar to what would be expected from the classical frontal theory. Based on the piecewise PV inversion technique and vorticity budget analyses, it was also suggested that the cumulus heating was the cause of the formation and intensification of the LLJ, and the subsequent northward retreat of the Meiyu front was caused by the vorticity advection processes associated with the LLJ. Diagnosis of the frontogenetical function indicated that the Meiyu frontal movement was caused by the southward frontal propagation due to frontogenetical processes in addition to the advection of the postfrontal cold air.

Case studies of the formation of LLJ's using the piecewise PV inversion technique all suggested that the formation and intensification of LLJ's could be attributed to the cumulus heating effects of the organized MCS over a heavy rainfall area through the Coriolis acceleration of the lower branch of an induced secondary circulation superimposed upon the background southwesterly monsoonal flows. Using a similar technique, it was also found that the growth of the frontal cyclone was a result of a nonlinear mechanism similar to the CISK, in which the frontal PV centers and cumulus convection reinforce each other through the positive feedback process.

In this paper, we have attempted to give an overview of the recent research results on the role of cumulus heating in the development and evolution of the Meiyu frontal system, particularly from a PV perspective. It is clear that our knowledge is far from complete. More research efforts will be needed to conduct observational, theoretical, and modeling studies in order to improve our understanding of the Meiyu frontal system including the Meiyu front, LLJ, and frontal cyclone. Particularly, the role of latent heat release in the Meiyu frontogenesis for the strong baroclinic Meiyu fronts deserves further studies in both theoretical and numerical simulation aspects. Also, more studies are needed to shed some more light on the role of the nonlinear positive feedback mechanism between the frontal PV center and the cumulus convection in the cyclogenesis process along the Meiyu front.

Acknowledgements

Thanks are due to Miss Man-Chun Chiu and Mr. Chih-Sheng Chang for preparing this manuscript. This article is partially supported by NSC 96-2111-M-002-011 and NSC 96-2111-M-002-010-MY3.

[Received 12 January 2007; Revised 31 May 2007; Accepted 5 June 2007.]

References

Akiyama, T., 1973: Ageostrophic low-level jet stream in the Baiu season associated with heavy rainfalls over the sea area. *J. Meteor. Soc. Japan,* **51**, 205–208.

Chang, C. P., S. C. Hou, H. C. Kuo, and G. T. J. Chen, 1998: The development of an intense East Asian summer monsoon disturbance with strong vertical coupling. *Mon. Wea. Rev.,* **126**, 2692–2712.

——, L. Yi, and G. T. J. Chen, 2000: A numerical simulation of vortex development during the 1992 East Asian summer monsoon onset using the Navy's regional model. *Mon. Wea. Rev.,* **128**, 1604–1631.

Chen, C., W. K. Tao, P. L. Lin, G. S. Lai, S. F. Tseng, and T. C. Chen Wang, 1998: The intensification of the low-level jet during the development of mesoscale convective systems on a Mei-Yu front. *Mon. Wea. Rev.,* **126**, 349–371.

Chen, G. T. J., 1977a: An analysis of moisture structure and rainfall for a Meiyu regime in Taiwan. *Proc. Natl. Sci. Counc.,* **1**, **11**, 1–21.

——, 1977b: A synoptic case study on mean structure of Mei-Yu in Taiwan. *Atmos. Sci.,* **4**, 38–47.

——, 1978a: The structure of a subtropical Mei-Yu system in Southeast Asia. *Sci. Rep., Dept. Atmos. Sci., Natl. Taiwan Univ.,* **2**, 9–23. (Available from the Dept. of Atmos. Sci., Natl. Taiwan Univ., Taipei, Taiwan, ROC.)

——, 1978b: On the meso-scale systems for the Mei-Yu regime in Taiwan. *Proc. Conf. Severe Weather in Taiwan Area,* NSC and Academia Sinica, pp. 150–157 (in Chinese, with English abstract). (Available from the Dept. of Atmos. Sci., Natl. Taiwan Univ., Taipei, Taiwan, ROC.)

——, 1979: On the moisture budget of a Mei-Yu system in southeastern Asia. *Proc. Natl. Sci. Counc.,* **3**, **1**, 24–32.

——, 1983: Observational aspects of the Mei-Yu phenomena in subtropical China. *J. Meteor. Soc. Japan,* **61**, 306–312.

——, 1992: Mesoscale features observed in the Taiwan Mei-Yu season. *J. Meteor. Soc. Japan,* **70**, 497–516.

——, 2004: Research on the phenomena of Meiyu during the past quarter century: an overview. In *World Scientific Series for Meteorology of East Asia Vol. 2: East Asian Monsoon,* C. P. Chang, (ed.), World Scientific, pp. 357–403.

——, and C. P. Chang, 1980: The structure and vorticity budget of an early summer monsoon trough (Mei-Yu) over southeastern China and Japan. *Mon. Wea. Rev.,* **108**, 942–953.

——, and S. S. Chi, 1978: On the meso-scale structure of Mei-Yu front in Taiwan. *Atmos. Sci.,* **5**, **1**, 35–47 (in Chinese, with English abstract).

——, and S. S. Chi, 1980: On the frequency and speed of Mei-Yu front over southern China and the adjacent areas. *Papers Meteor. Res.,* **3**, **1&2**, 31–42.

——, and C. Y. Tsay, 1978: A synoptic case study of Mei-Yu near Taiwan. *Papers Meteor. Res.,* **1**, 25–36.

——, C. C. Wang, and S. W. Chang, 2007: A diagnostic case study of Meiyu frontogenesis and development of wave-like frontal disturbances in the subtropical environment. *Mon. Wea. Rev.* (in press).

——, ——, and D. T. W. Lin, 2005: Characteristics of low-level jets over northern Taiwan in Mei-Yu season and their relationship to heavy rain events. *Mon. Wea. Rev.,* **133**, 20–43.

——, ——, and L. F. Lin, 2006: A diagnostic study of a retreating Mei-Yu front and the accompanying low-level jet formation and intensification. *Mon. Wea. Rev.,* **134**, 874–896.

——, ——, and S. C. S. Liu, 2003: Potential vorticity diagnostics of a Mei-yu front case. *Mon. Wea. Rev.,* **131**, 2680–2696.

——, ——, and A. H. Wang, 2007: A case study of subtropical frontogenesis during a blocking event. *Mon. Wea. Rev.,* **135**, 2588–2609.

——, C. W. Wu, and S. S. Chi, 1986: Climatological aspects of the mesoscale convective systems over subtropical China and the western North Pacific during Mei-Yu season of 1981–1983. *Atmos. Sci.,* **13**, 33–45 (in Chinese, with English abstract).

——, and C. C. Yu, 1988: Study of low-level jet and extremely heavy rainfall over northern Taiwan in the Mei-Yu season. *Mon. Wea. Rev.,* **116**, 884–891.

Chen, Q., 1982: The instability of the gravity-inertia wave and its relation to low-level jet and heavy rainfall. *J. Meteor. Soc. Japan,* **60**, 1041–1057.

Chen, S. J., Y. H. Kuo, W. Wang, Z. Y. Tao, and B. Cui, 1998: A modeling case study of heavy rainstorms along the Mei-Yu front. *Mon. Wea. Rev.*, **126**, 2330–2351.

——, W. Wang, K. H. Lau, Q. H. Zhang, and Y. S. Chung, 2000: Mesoscale convective systems along the Meiyu front in a numerical model. *Meteor. Atmos. Phys.*, **75**, 149–160.

Chen, Y. L., X. A. Chen, S. Chen, and Y. H. Kuo, 1997: A numerical study of the low-level jet during TAMEX IOP 5. *Mon. Wea. Rev.*, **125**, 2583–2604.

——, ——, and Y. X. Zhang, 1994: A diagnostic study of a low-level jet during TAMEX IOP 5. *Mon. Wea. Rev.*, **122**, 2257–2284.

——, and N. B. F. Hui, 1990: Analysis of a shallow front during the Taiwan Area Mesoscale Experiment. *Mon. Wea. Rev.*, **118**, 2649–2667.

——, Y. X. Zhang, and N. B. F. Hui, 1989: Analysis of a surface front during the early summer rainy season over Taiwan. *Mon. Wea. Rev.*, **117**, 909–931.

Cho, H. R., and G. T. J. Chen, 1994: Convection and Mei-Yu front. *Terrestrial Atmos. Oceanic Sci.*, **5**, 121–136.

——, and ——, 1995: Mei-Yu frontogenesis. *J. Atmos. Sci.*, **52**, 2109–2120.

Chou, L. C., C. P. Chang, and R. T. Williams, 1990: A numerical simulation of the Mei-Yu front and the associated low-level jet. *Mon. Wea. Rev.*, **118**, 1408–1428.

Davis, C. A., 1992: A potential-vorticity diagnosis of the importance of initial structure and condensational heating in observed extratropical cyclogenesis. *Mon. Wea. Rev.*, **120**, 2409–2427.

——, and K. A. Emanuel, 1991: Potential vorticity diagnostics of cyclogenesis. *Mon. Wea. Rev.*, **119**, 1929–1953.

Ding, Y. H., 1992: Summer monsoon rainfalls in China. *J. Meteor. Soc. Japan*, **70**, 373–396.

——, and J. C. L. Chan, 2005: The East Asian summer monsoon: an overview. *Meteor. Atmos. Phys.*, **89**, 117–142.

Du, J., and H. R. Cho, 1996: Potential vorticity anomaly and mesoscale convective systems on the Baiu (Mei-Yu) front. *J. Meteor. Soc. Japan*, **74**, 891–908.

Hsu, W. R., and W. Y. Sun, 1994: A numerical study of a low-level jet and its accompanying secondary circulation in a Mei-Yu system. *Mon. Wea. Rev.*, **122**, 324–340.

Lee, C. S., Y. L. Lin, and K. K. W. Cheung, 2006: Tropical cyclone formations in the South China Sea associated with the Mei-Yu front. *Mon. Wea. Rev.*, **134**, 2670–2687.

Lin, S. C., 1988: The life cycle and structure of a mesoscale convective system occurring in the southern China area during Mei-Yu season. *Papers Meteor. Res.*, **11**, 1–26.

——, and T. K. Chiou, 1985: Objective scale separation technique and its application on the mesoscale convective system diagnosis. *Papers Meteor. Res.*, **8**, **2**, 69–94.

——, and C. M. Tsai, 1989: Kinetic energy budgets of a mesoscale convective system occurring in the Mei-Yu season. *Atmos. Sci.*, **17**, 187–209 (in Chinese, with English abstract).

Matsumoto, S., 1972: Unbalanced low-level jet and solenoidal circulation associated with heavy rainfalls. *J. Meteor. Soc. Japan*, **50**, 194–203.

Ninomiya, K., 1984: Characteristics of Baiu front as a predominant subtropical front in the summer northern hemisphere. *J. Meteor. Soc. Japan*, **62**, 880–894.

——, and T. Akiyama, 1974: Band structure of mesoscale echo cluster associated with low-level jet stream. *J. Meteor. Soc. Japan*, **52**, 300–313.

Pu, C. P., and G. T. J. Chen, 1988: A preliminary analysis of the low-level jet and mesoscale convective systems over subtropical China during Mei-Yu season: the case of June 1–3, 1983. *Meteor. Bull.*, **34**, 285–297 (in Chinese, with English abstract).

Qian, J. H., W. K. Tao, and K. M. Lau 2004: Mechanisms for torrential rain associated with the Mei-Yu development during SCSMEX 1998. *Mon. Wea. Rev.*, **132**, 3–27.

Takahashi, H. 2003: Observational study on the initial formation process of the Mei-Yu frontal disturbance in the eastern foot of the Tibetan Plateau in middle-late June 1992. *J. Meteor. Soc. Japan*, **81**, 1303–1327.

Tao, S. Y., and L. X. Chen, 1987: A review of recent research on the East Asian summer monsoon in China. In *Monsoon Meteorology*, C. P. Chang and T. N. Krishnamurti, (eds.), Oxford University Press, pp. 60–92.

Trier, S. B., D. B. Parsons, and T. J. Matejka, 1990: Observations of a subtropical cold front in a region of complex terrain. *Mon. Wea. Rev.*, **118**, 2449–2470.

Tsay, C. Y., and B. F. Chain, 1987: A composite study of low level jet and its relationship with heavy rainfall in Taiwan area during Mei-Yu season. *Atmos. Sci.*, **15**, **1**, 1–16 (in Chinese, with English abstract).

Zhang, Q. H., K. H. Lau, Y. H. Kuo, and S. J. Chen, 2003: A numerical study of a

mesoscale convective system over the Taiwan Strait. *Mon. Wea. Rev.*, **131**, 1150–1170.

Zhao, S. X., X. P. Zhou, K. S. Zhang, and S. H. Liu, 1982: Numerical simulation experiment of the formation and maintenance of meso-scale low. *Scientia Atmospherica Sinica*, **6**, 109–119 (in Chinese, with English abstract).

Advance in the Dynamics and Targeted Observations
of Tropical Cyclone Movement

Chun-Chieh Wu

Department of Atmospheric Sciences, National Taiwan University, Taipei, Taiwan
cwu@typhoon.as.ntu.edu.tw

The advance in the dynamics and targeted observations of hurricane movement is reviewed in this article. In celebration of the 50th anniversary of the Department of Atmospheric Sciences, National Taiwan University, special emphasis is put on the author's major scientific contributions to the following issues: the baroclinic effect on tropical cyclone motion, the potential vorticity diagnosis of the tropical cyclone motion, and the targeted observations from DOTSTAR (Dropwindsonde Observations for Typhoon Surveillance near the Taiwan Region) in understanding and improving the tropical cyclone track predictability.

1. Introduction

The dynamics of tropical cyclone (TC) motion are rather complex. As pointed out by Holland (1984), a complete description would require at least detailed knowledge of the interactions between the cyclone circulation, the environmental wind field, the underlying surface, and the distribution of moist convection. It has generally been proposed that TC motion is governed by the tropospheric average steering flow and a drift caused by the presence of a background potential vorticity gradient. However, there are many other factors that can affect the cyclone's motion. Thanks to scientific advances in the past few decades, the understanding of the dynamics of TC motion is rather comprehensive, along with the improvement of observations (Aberson, 2003; Wu *et al.*, 2005) and numerical models (Kurihara *et al.*, 1998); the forecasting of TC motion has also been in steady progress.

Some reviews on this issue have been well conducted, by Wang *et al.* (1998) and Chan (2005). To celebrate the 50th anniversary of the Department of Atmospheric Sciences, National Taiwan University, this article reviews the author's major contributions to the current understanding of the dynamics and targeted observations of TC movement. The following topics are covered: the baroclinic effect on TC motion (Wu and Emanuel, 1993, 1994), the potential vorticity perspective of the TC motion (Wu and Emanuel, 1995a,b; Wu and Kurihara, 1996; Wu *et al.*, 2003, 2004), and the targeted observations in improving TC track prediction (Wu *et al.*, 2005, 2006, 2007a,b). In Sec. 2, the baroclinic effect on TC motion is shown. The potential vorticity perspective of the TC motion and the targeted observations from DOTSTAR (Dropwindsonde Observations for Typhoon Surveillance near the Taiwan Region) in improving the forecasting of TC motion are discussed in Secs. 3 and 4, respectively. Finally, a summary appears in Sec. 5.

2. Baroclinic Effect on TC Motion (Wu and Emanuel, 1993, 1994)

2.1. *Background*

Studies of TC motion have focused mainly on steering by the mean flow and the effect of background potential vorticity gradients, i.e. the evolution of barotropic vortices in a barotropic flow. These effects, taken together, suggest that TCs should follow the mean large scale (steering) flow (George and Gray, 1976), but with a westward and poleward relative drift (Fiorino and Elsberry, 1989). Other, minor factors may also affect TC motion, such as the asymmetric convection (Willoughby, 1988) and the vortex interaction (Holland and Dietachmayer, 1993).

However, observations have shown that real TCs are strongly baroclinic, with broad anticyclones aloft, and the distribution of the large-scale potential vorticity gradient in the tropical atmosphere is very nonuniform. As indicated by Wu and Emanuel (1993), these properties may substantially influence the movement of storms. The upper anticyclone, though weak in terms of wind velocity relative to the lower-layer cyclonic circulation, can be very extensive. Slight displacements of the upper region of anticyclonic flow from the low-level cyclone can conceivably lead to large mutual interaction, and the potential vorticity gradient may act on these two flows in very different ways.

Based on the concepts of vortex interaction, Wu and Emanuel (1993) proposed that a baroclinic TC, which is structured like a vertically distributed pair of vortices of opposite signs, would experience a mutual propagation if the vortex dipole is tilted (Fig. 1). In other words, the background vertical wind shear can tilt the vortex pair by blowing the upper potential vorticity anomaly downshear. As members of the vortex pair are displaced, they begin to interact with each other and thus move at right angles to the axis connecting them. On this basis, it is hypothesized that Northern Hemispheric (Southern Hemispheric) TCs should drift with respect to the mean winds in a direction to the left (right) of the background vertical shear vector.

2.2. *Methodology*

The intention is to isolate the effect of background vertical shear. The hurricane is represented in a two-layer quasi-geostrophic model as a point source of mass and zero potential vorticity air in the upper layer, collocated with a point cyclone in the lower layer. The model is integrated by the method of contour dynamics and contour surgery. The contour dynamics is a

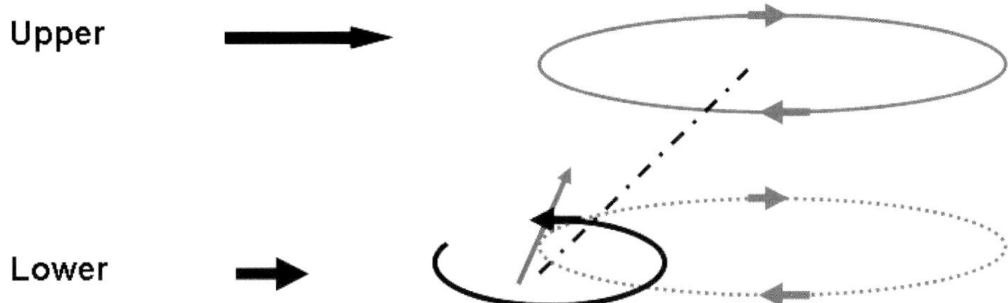

Figure 1.   The vertical shear (black arrows) of the environmental flow results differential advection of the potential vorticity associated with a storm. Black and gray ellipses indicate the lower- and upper-level storm circulations, respectively. The upper layer is tilted due to the presence of vertical shear, and its projection (dashed gray ellipse) results in the storm drift to the left (Northern Hemisphere) of the shear vector (gray arrow).

Lagrangian computational method used to integrate flows associated with patches of piecewise constant potential vorticity. This method leads to a closed dynamical system within which the evolution of the flow can be uniquely determined by the contours bounding the patches. The method of contour surgery improves the resolution of the contour by adding nodes (called node adjustment) in regions of high curvature or small node velocity. It is also more efficient and prevents unlimited enstrophy cascades to a small scale by removing contour features (called contour adjustment) thinner than some prescribed tolerance. A detailed description of the methods of contour dynamics and contour surgery can be found in Dritschel (1989).

In this work, the simplest analog of a mature TC is considered to be a diabatically, frictionally maintained point vortex of constant strength in the lower layer, and a patch of uniform, zero potential vorticity air in the upper layer, surrounded by an infinite region of constant potential vorticity (it is assumed that the mean meridional gradient of potential vorticity associated with the coriolis parameter can be canceled out by introducing upper and lower boundaries with gentle meridional slopes in the two-layer model). The diabatic sink of potential vorticity in the upper layer is represented as the expansion of the upper potential vorticity anomaly, owing to a radial outward potential (irrotational) flow emanating from a point mass source collocated with the lower vortex. According to the principle of mass continuity, the potential flow can be calculated from the lower boundary frictionally driven mass influx.

The case of a vanishing ambient potential vorticity gradient is explored as well. Therefore, the upper vortex patch is advected by the rotational flows (associated with both the upper-layer contour itself and the lower-layer vortex), the divergent flow (associated with the mass source), and the mean shear flow. The work of Wu and Emanuel (1993, 1994) is meant to describe the first-order effects of vertical shear, given the approximation inherent in the model.

2.3. *Results*

For the case with no vertical shear, it is found that, as expected, the upper patch expands with time and remains circularly symmetric with no lower vortex movement. The wind distributions in each layer show that the upper flow is anticyclonic and outward, similar to a real hurricane outflow, and the lower layer flow is symmetric around the vortex center, so that no vortex drift is induced.

For the case with shear, the vortex patch expands and is advected downshear. Also, roll-up of the vortex patch occurs on the downshear side, essentially due to barotropic instability. For cases with larger shear, the patch is advected rapidly downshear and becomes zonally elongated. The low potential vorticity anomaly behaves like a passive plume.

The trajectories of the lower vortex in three experiments with westerly shear of various magnitudes are shown in Fig. 2. In all cases,

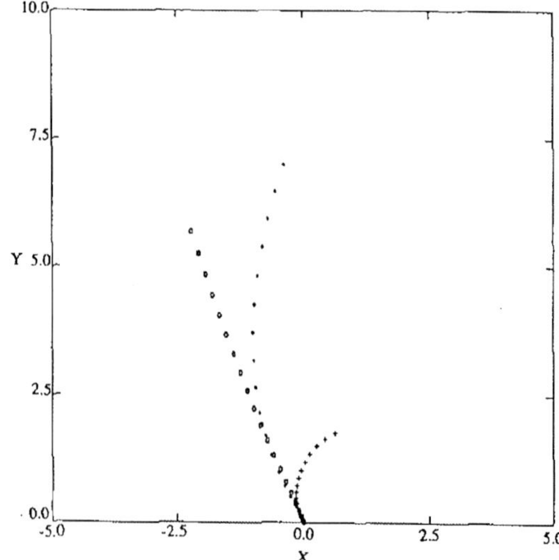

Figure 2. Trajectories (units of 500 km) of the lower-layer vortex encountering weaker (shown as +); fair (shown as *); and stronger (shown as ○) vertical wind shear. (From Wu and Emanuel, 1993).

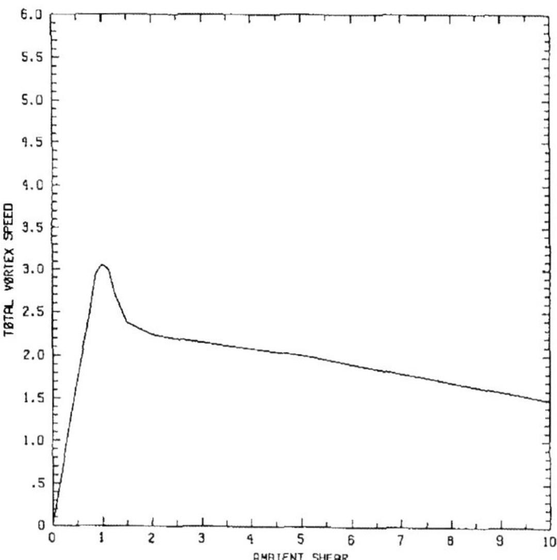

Figure 3. The relation between the maximum induced vortex speed and the magnitude of the vertical shears. (From Wu and Emanuel, 1993.)

distinct northward vortex drifts associated with different magnitudes of the mean westerly shears are found, as expected. Also, the drift in the zonal direction is a function of the background shear, i.e. more eastward drift is associated with weaker shear. Figure 3 shows the maximum total drift speed as a function of the ambient shear. The vortex drift increases initially as the shear increases, and there exists an optimal shear that maximizes the vortex drift. Above that optimal shear, the drift speed decreases with increasing shear, and approaches a constant, since the impact to the lower-layer vortex drift by the upper-layer plumed advected far downstream would be very limited.

2.4. *Summary*

Observations show that TCs have broad anticyclones aloft, and that the distribution of potential vorticity gradients in the tropical atmosphere is highly inhomogeneous. There are indications that the potential vorticity gradients in the subtropical troposphere are very weak (see Fig. 2 in Wu and Emanuel, 1993), perhaps

having been rendered so by the action of synoptic scale disturbances.

It is found that the direct effect of ambient vertical shear is to displace the upper-level plume of anticyclonic relative potential vorticity downshear from the lower layer cyclonic point potential vortex, thus inducing a mutual interaction between the circulations associated with each other. This results in a drift of the point vortex broadly to the left of the vertical shear vector (in the Northern Hemisphere).

3. Potential Vorticity Perspective of TC Motion (Wu and Emanuel, 1995a,b; Wu and Kurihara, 1996; Wu *et al.*, 2003, 2004)

The above papers are some pioneering works in adopting the potential vorticity (PV) diagnostics to evaluate the control by the large-scale environment of hurricane movement and, more importantly, to assess the storm's influence on its own track. By using the PV framework, the exploration of the dynamics of hurricanes is feasible with the validity of balanced dynamics in the tropics. In this section, the use of PV diagnostics is to understand the hurricane steering flow, and to demonstrate the interaction between the cyclone and its environment, including the investigation of the binary TC interaction and the major factors affecting the analyses and forecast bias of the TC track forecasts.

3.1. *Background*

PV methods have proven useful in understanding synoptic- and large-scale midlatitude dynamics (Hoskins *et al.*, 1985), and are widely applied to tropical systems (e.g. Molinari, 1993; Montgomery and Farrell, 1993). A hurricane is a localized yet robust vortex, and can possibly change the surrounding environmental flow field substantially by its strong circulation, which in turn affects the evolution of the track, intensity, and structure of the hurricane. This interaction between the TC and its environment is nonlinear. Previous studies have investigated how the large-scale flow fields affect the track, intensity (e.g. Molinari *et al.*, 1995), and structure of the hurricane, and how the storm circulation changes the environmental flow field (e.g. Ross and Kurihara, 1995). However, these studies consider only a one-way interaction between the storm and its environment. In this work, the two-way hurricane–environment interaction is studied, i.e. how the change of the environment due to the storm feeds back to affect the storm's track.

The concept of binary interaction has been well described by tank experiments (Fujiwhara, 1921, 1923, 1931), observations (Brand, 1970; Dong and Neumann, 1983; Lander and Holland, 1993; Carr *et al.*, 1997; Carr and Elsberry, 1998), and modeling studies (Chang, 1983, 1984; DeMaria and Chan, 1984; Ritchie and Holland, 1993; Holland and Dietachmayer, 1993; Wang and Holland, 1995). All of these studies indicate that two TCs can interact and subsequently develop mutual orbiting and possible approaching, merging, and escaping processes (Lander and Holland, 1993).

The observational evidences Brand (1970) has provided suggest that there is a correlation between the separation distance and the angular rotation rate of two TCs. Brand also showed that the effect of such binary interaction depends not only on differences in storm size and intensity, but also on variations of the currents in which the tropical storm systems are imbedded.

Work from Carr *et al.* (1997) and Carr and Elsberry (1998) has proposed detailed conceptual models to categorize the binary interaction processes, namely: (1) the direct TC interaction with one-way influence, mutual interaction or merger of the two TCs; (2) the semidirect TC interaction involving another TC and an adjacent subtropical anticyclone; and (3) the indirect TC interaction involving the anticyclone between the two TCs. In spite of a success rate of 80% in distinguishing the modes of the binary interactions from analyses of eight-year samples of western North Pacific TCs, the ways to quantify the binary interaction of TCs are still arguable.

This section has attempted to demonstrate the quantitative use of PV diagnostics in understanding the hurricane steering flow, in addition to the interaction between the cyclone and its environment, along with the binary TC interaction and the evaluation of the forecast bias.

3.2. *Methodology*

(1) *Defining the hurricane advection flow*
This is an original work to define the hurricane advection (steering) flow based on the PV diagnostics. One conventional problem in estimating the steering flow based on the azimuthal average is that the resulting wind is highly sensitive to the exact choice of the hurricane center (if the storm center is misrepresented, the averaged flow would be contaminated by the asymmetric high wind near and off the storm center). To avoid such a problem, the hurricane advection flow is defined as the balanced flow (at the storm center) associated with the entire PV field in the troposphere, except for the PV anomaly of the hurricane itself.

(2) *PV inversion*
The PV inversion is that given a distribution of PV, a prescribed balances condition, and boundary conditions, the balanced mass and wind fields can be recovered. Formulated on the $\pi\,[\pi = C_p(p/p_0)^\kappa]$ coordinate and spherical coordinates, the two equations to be solved are

$$q = \frac{gk\pi}{p}\left[(f + \nabla^2\Psi)\frac{\partial^2\Phi}{\partial\pi^2} - \frac{1}{a^2\cos^2\phi}\frac{\partial^2\Psi}{\partial\lambda\partial\pi}\right.$$

$$\left. \times \frac{\partial^2\Phi}{\partial\lambda\partial\pi} - \frac{1}{a^2}\frac{\partial^2\Psi}{\partial\phi\partial\pi}\frac{\partial^2\Phi}{\partial\phi\partial\pi}\right], \qquad (1)$$

$$\nabla^2 \Phi = \nabla \cdot (f \nabla \Psi) + \frac{2}{a^4 \cos^2 \phi}$$
$$\times \frac{\partial(\partial \Psi/\partial \lambda, \partial \Psi/\partial \phi)}{\partial(\lambda, \phi)}, \qquad (2)$$

where q represents PV, Φ represents geopotential height, and Ψ represents stream function, whereas a is the Earth's radius, f is the Coriolis parameter, $\kappa = R_d/C_p$, ϕ is longitude, and λ is latitude. Given the distribution of q, the lateral boundary of Φ and Ψ, along with θ on the upper and lower boundaries, the distribution of Φ and Ψ can be solved. Therefore, the nondivergent wind and potential temperature can also be obtained through the following two relations:

$$\vec{V} = \hat{k} \times \nabla \Psi, \qquad (3)$$

$$\theta = -\frac{\partial \Phi}{\partial \pi}. \qquad (4)$$

In the PV inversion method, another robust strength is the piecewise PV inversion, i.e. when the flow field is divided into the mean and perturbation components, the above equation can be rederived (Davis, 1992) to obtain the balanced fields associated with each individual PV perturbation. By taking the perturbation field as $\Psi' = \Psi - \bar{\Psi}$, $\Phi' = \Phi - \hat{\Phi}$, and $q' = q - \hat{q}$, the piecewise PV inversion is to calculate the balanced flow and mass fields associated with each PV perturbation. Such a method describes how the different PV features, the environment perturbations, affect the TC's track (Wu and Emanuel, 1995a,b; Shapiro, 1996; Shapiro and Franklin, 1999; Wu and Kurihara, 1996; Wu et al., 2003, 2004).

(3) Defining AT: the normalized steering effect associated with each PV perturbation in the along-track direction
In order to evaluate the steering flow due to various PV perturbations, the time series of the deep-layer-mean (DLM) flow associated with the total PV perturbation $[\tilde{\mathbf{V}}_{\text{SDLM}}(q')]$ and each PV perturbation $[\tilde{\mathbf{V}}_{\text{SDLM}}(q'_a)]$ is defined. Note that $q' = q'_a + q'_{na}$ while the subscript "a" represents some specific perturbation of

interest and "na" stands for the remaining perturbation. The DLM (925–300-hPa-averaged) steering wind is defined as

$$\vec{\mathbf{V}}_{\text{SDLM}} = \frac{\int_{925\,\text{hPa}}^{300\,\text{hPa}} \vec{\mathbf{V}}_S(p)\,dp}{\int_{925\,\text{hPa}}^{300\,\text{hPa}} dp}, \qquad (5)$$

where

$$\vec{\mathbf{V}}_S(p) = \frac{\int_0^3 \int_0^{2\pi} \vec{\mathbf{V}}\, r\,dr\,d\theta}{\int_0^3 \int_0^{2\pi} r\,dr\,d\theta}. \qquad (6)$$

To realize the influence of the steering flow associated with each perturbation, the ratio of the steering flow associated with each perturbation to that associated with all perturbations has been calculated. Following Wu *et al.* (2003), this quantity is defined as

$$AT(q'_a) = \frac{|\vec{\mathbf{V}}_{\text{SDLM}}(q'_a) \cdot \vec{\mathbf{V}}_{\text{SDLM}}(q')|}{|\vec{\mathbf{V}}_{\text{SDLM}}(q')|^2}, \qquad (7)$$

where $\tilde{\mathbf{V}}_{\text{SDLM}}(q'_a)$ and $\tilde{\mathbf{V}}_{\text{SDLM}}(q')$ are as defined in Eq. (5). Note that by definition, $AT(q') = AT(q'_a) + AT(q'_{na})$.

(4) Defining the new centroid for plotting the centroid-relative track
Conventionally, the centroid-relative tracks based on geography centers are often plotted to illustrate the binary interaction. This picture is often misleading, since the actual binary interaction is generally a function of the strengths of both vortices, and would not rotate with respect to the geographic center (Lander and Holland, 1993; Carr et al., 1997; Carr and Elsberry, 1998; Wu et al., 2003). In this article, the definition of the centroid from Wu et al. (2003) is discussed, i.e. the centroid between storms A and B is defined as the location weighted according to the strength of the steering flow induced by the PV perturbation associated with each storm:

$$\vec{\mathbf{r}}_c = \frac{|\vec{\mathbf{V}}_{\text{SDLM}}(q'_A)|}{|\vec{\mathbf{V}}_{\text{SDLM}}(q'_A)| + |\vec{\mathbf{V}}_{\text{SDLM}}(q'_B)|}\, \vec{\mathbf{r}}_A$$

$$+ \frac{|\vec{\mathbf{V}}_{\text{SDLM}}(q'_B)|}{|\vec{\mathbf{V}}_{\text{SDLM}}(q'_A)| + |\vec{\mathbf{V}}_{\text{SDLM}}(q'_B)|}\, \vec{\mathbf{r}}_B, \qquad (8)$$

where q'_A (q'_B) represents the perturbation PV of storm A (B), while $\vec{\mathbf{r}}_c$, $\vec{\mathbf{r}}_A$, and $\vec{\mathbf{r}}_B$ stand for the position vectors of the centroid, storm A, and storm B, respectively.

3.3. Results

(1) *PV diagnostics of the motion of Hurricane Bob (1991), Tropical Storm Ana (1991), and Hurricane Andrew (1992) (Wu and Emanuel, 1995a,b)*

In Wu and Emanuel (1993, 1994), the results of the advection flow of Bob show that Bob's movement is due not only to the climatological (July–September) mean balanced flow, but also to significant contributions from the balanced flow associated with the upper-tropospheric PV perturbations (U) [Fig. 4(a)]. This implies that Bob's motion is strongly influenced by the mid-latitude systems and the disturbances in the upper troposphere. However, in the case of Ana, the advection of this tropical storm is mainly associated with the mean flow and the PV perturbation in the lower troposphere. Unlike the case of Bob, due to the cancellation effects between the PV anomalies (i.e. the balanced flows associated with different pieces of the upper PV anomalies tend to aim toward different directions), the upper PV perturbation

does not have a large effect on Ana's motion [Fig. 4(b)]. And for Hurricane Andrew, the climatological mean, upper and lower PV perturbations have about the same magnitude of contribution to the advection flow [Fig. 4(c)].

Compared to the annular mean winds, the analysis by Wu and Emanuel (1995a,b) provides a more dynamically consistent method of determining the advection flow through the hurricane center. In addition, the PV framework is conceptually more concise (Hoskins *et al.*, 1985), and allows one to study the essential dynamical mechanism responsible for hurricane motion.

(2) *A new look at the binary interaction: potential vorticity diagnosis of the unusual southward motion of Tropical Storm Bopha (2000) and its interaction with Supertyphoon Saomai (2000) (Wu et al., 2003)*

Tropical Storm Bopha (2000) showed a very unusual southward course parallel to the east coast of Taiwan, mainly steered by the circulation associated with Supertyphoon Saomai (2000) to Bopha's east.

To quantitatively measure the influence of the steering flow associated with other PV features in this binary interaction, the method shown in Subsec. 3.2(2) is adopted. Results show that the total PV perturbation provides

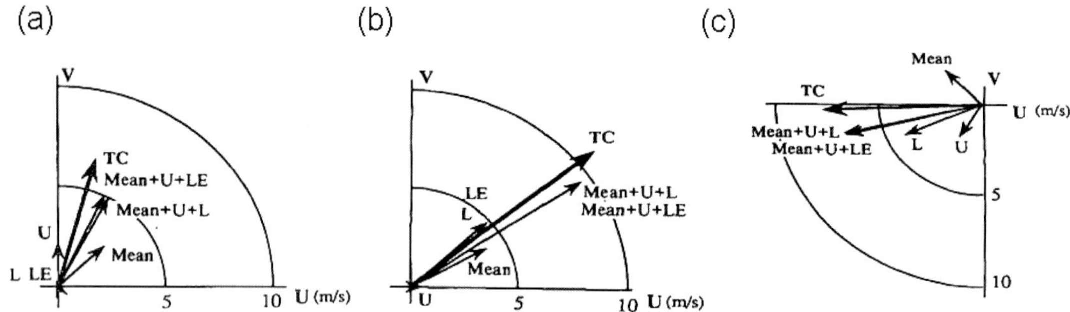

Figure 4. Interpolation of the balanced wind fields to the 850–500 mb pressure-averaged balanced vortex center. Mean, U, L, and LE represent the 850–500 mb pressure-averaged balanced flows associated with the mean potential vorticity, and potential vorticity perturbations of U, L, and LE, respectively. Mean + U + LE represents the total hurricane advection flow. TC indicates (a) Hurricane Bob's motion at 1200 UTC, 18 August 1991; (b) Tropical Storm Ana's motion at 1200 UTC, 3 July 1991; (c) Hurricane Andrew's motion at 1200 UTC, 23 August 1992. (From Wu and Emanuel, 1995a,b.)

a good approximation of the actual motion of both storms, Bopha and Saomai. Besides, Saomai plays the dominant role in advecting Bopha southward while the influence of Bopha on Saomai is rather limited (Fig. 5).

To better indicate the binary interaction processes, the new method for ploting the centroid-relatice track is devised. As shown in Fig. 6, Bopha and Saomai mutually interacted by rotating cyclonically around each other at a distance of about 1200 km, with the centroid much closer to Saomai. The unusual southward drift of Bopha appears to agree with the proposed mechanism as the direct binary interaction with one-way influence (Carr *et al.*, 1997). Clearly, the above PV analysis with the AT values can nicely quantify the binary interaction processes, and the newly defined centroid-relative track can describe the appropriate interaction of the TCs.

(3) PV diagnosis of the key factors affecting the motion of Typhoon Sinlaku (2002) (Wu et al., 2004)

Typhoons over the western North Pacific often move westward or northwestward due to the dominating steering flow associated with the Pacific Subtropical High (SH). However, forecasts of typhoons are sometimes difficult during the late season as typhoons approach about 130°E, where some storms may slow down, stall, or even recurve due to the weakening of the SH, as well as the strengthening of the Continental High (CH) over China and the presence of the deep midlatitude baroclinic wave or trough (TR). The stalling scenario appeared in the case of Sinlaku, where forecasts from some major operational centers failed to predict its slowdown and mistakenly predicted its southward dip before it approached the offshore northeastern Taiwan.

The PV diagnosis indicates that the initial deceleration of Sinlaku was mainly associated with the retreat of the SH under the influence of the TR. The upper-level cold-core low (CCL) played only a minor role in impeding Sinlaku from moving northward, while the CH over mainland China strongly steered Sinlaku westward. On account of the steering effects from the above four systems (SH, TR, CCL, and CH), which tend to cancel one another, the subtle interaction therein makes it difficult to make a precise track forecast (Fig. 7). This proposes another challenge to TC track forecasts in this region.

3.4. *Summary*

The validity of balance dynamics in the tropics allows us to explore the dynamics of hurricanes using the PV framework. Subsection 3.3 demonstrated the use of PV diagnostics in understanding the hurricane steering flow, and the interaction between the cyclone and its environment. The results are consistent with the previous finding that the hurricane advection flow, defined by inverting the entire PV distribution which excludes the storm's own positive anomaly, is a good approximation to real cyclone movement, even though the original data cannot capture the actual hurricane strength.

Moreover, the binary interaction between two typhoons is well demonstrated by the PV diagnosis. The quantitative description of the Fujiwhara effect is not easy for real-case TCs, because it tends to be masked by other environmental steering flows. A new centroid-relative track is proposed, with the position weighting based on the induced steering flow due to the PV anomaly of another storm. Such analysis is used to understand more complicated vortex merging and interacting processes between and among multiple TCs from either observational data or specifically designed numerical experiments. Finally, the PV inversion concept has been successfully applied to real-time analysis and prediction systems, and to quantitatively evaluate the factors affecting the storm motions.

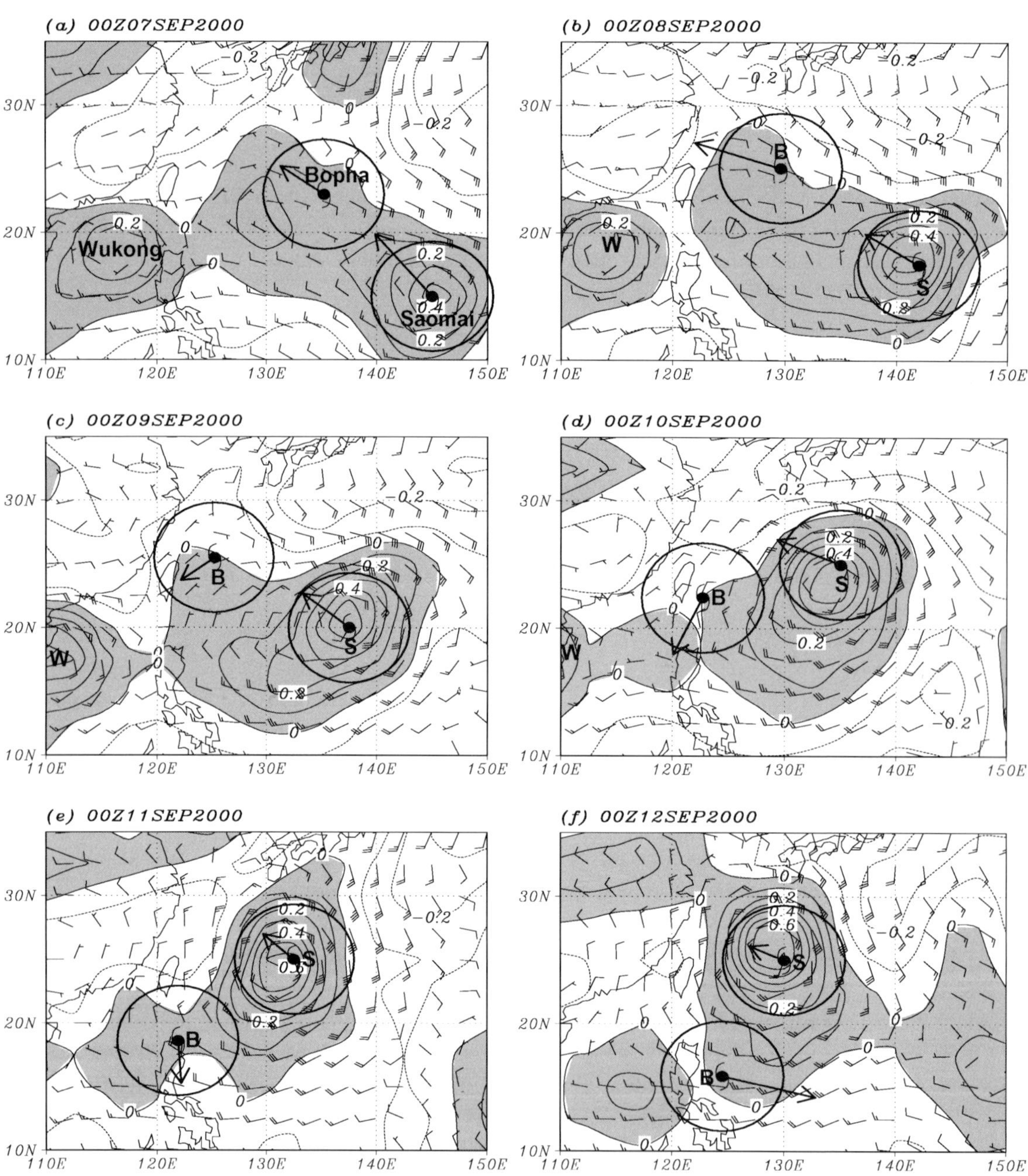

Figure 5. Total potential vorticity perturbation (contour interval of 0.1 PVU; the positive PV perturbation is shaded) at 500 hPa, and the 925–400 hPa deep-layer wind while Bopha is the mean part (one full wind barb = $5 \, \mathrm{ms}^{-1}$) at 0000 UTC: (a) 7 September, (b) 8 September, (c) 9 September, (d) 10 September, (e) 11 September, and (f) 15 September 2000. The instantaneous movement of Bopha, as well as Saomai, is indicated by the bold arrow, whose length represents the actual translation velocity, and the sold circle shows the scale of $5 \, \mathrm{ms}^{-1}$. B, S, and W indicate Bohpa, Saomai, and Wukong, respectively. (From Wu *et al.*, 2003.)

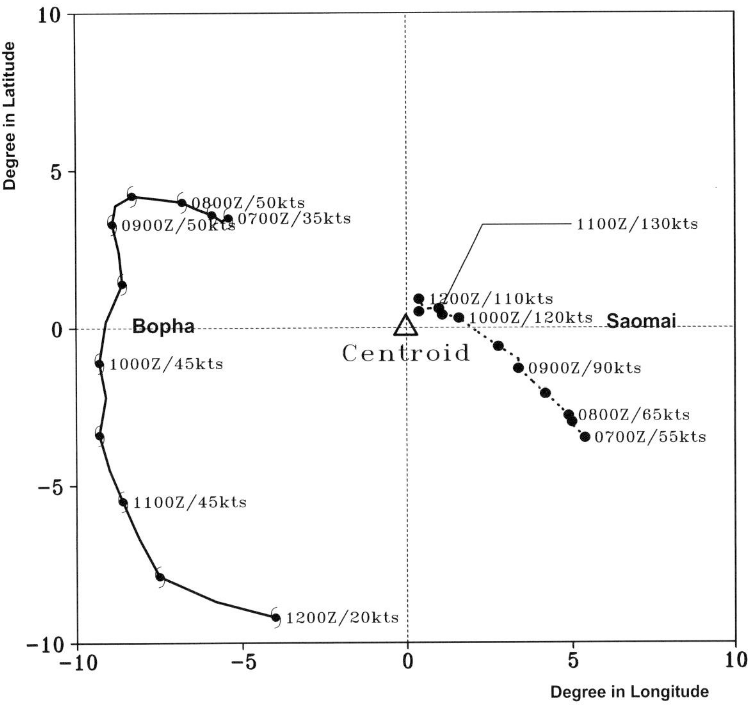

Figure 6. Centroid-relative tracks of Bopha (TC symbol) and Saomai (solid dot) for every 12 h from 0000 UTC, 7 September, to 0000 UTC, 12 September 2000. (From Wu *et al.*, 2003.)

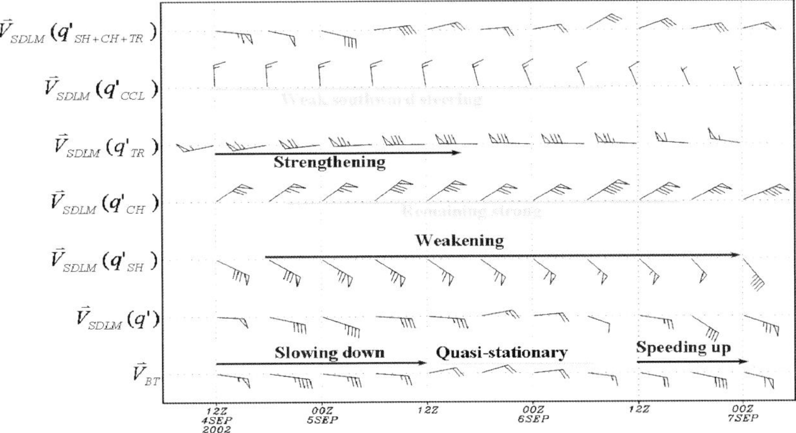

Figure 7. The time series of the movement of Sinlaku ($\vec{V}_{BT}$) and the steering flow (averaged in the inner three latitude degrees between 975 and 300 hPa) associated with the total PV perturbation ($\vec{V}_{SDLM}(q')$], and the PV perturbations of SH [$\vec{V}_{SDLM}(q'_{SH})$], CH [$\vec{V}_{SDLM}(q'_{CH})$], TR [$\vec{V}_{SDLM}(q'_{TR})$], and CCL [$\vec{V}_{SDLM}(q'_{CCL})$], individually. One full wind barb (a flag) represents 1 (5) ms^{-1}. (From Wu *et al.*, 2004.)

4. Targeted Observations and Data Assimilation for TC Motion (Wu *et al.* 2005, 2006, 2007a,b)

4.1. *Background*

Over the past 30 years, steady progress in the track forecasts of TCs has been made through the improvement of the numerical models, the data assimilation system, and the new data available to the forecast system (Wu, 2006). In addition to the large amount of satellite data, the special dropwindsonde data deployed from the surveillance aircraft have provided significant added value in improving the track forecasts.

In order to optimize the limited aircraft resources, targeted observations in the critical areas which have the maximum influence on numerical weather forecasts of TCs are of great importance. Therefore, targeted observing strategies for aircraft missions must be further developed. And it is a prerequisite for the device of observing strategy to identify the sensitive areas that have the greatest influence in improving the numerical forecast, or minimizing the track forecast error.

To make use of the available data or the potentially new data, it is important to evaluate the potential impact and to test the sensitivity of the simulation and prediction of TCs to different parameters. This understanding can be of great use in designing a cost-effective strategy for targeted observations of TCs (Morss *et al.*, 2001; Majumdar *et al.* 2002a,b; Aberson, 2003; Wu *et al.*, 2005).

In this section, three issues related to the author's works are addressed: (1) the impact of the DOTSTAR data; (2) results from a set of Observation System Simulation Experiments (OSSEs); (3) an innovative development of the new targeted observation strategy, the Adjoint-Derived Sensitivity Steering Vector (ADSSV).

4.2. *Impact of dropwindsonde data on TC track forecasts from DOTSTAR*

Since 2003, the research program of DOTSTAR (Wu *et al.*, 2005, 2007b) has continuously conducted dropwindsonde observations of typhoons in the western North Pacific (Fig. 8). Three operational global and two regional models were used to evaluate the impact of the dropwindsonde on TC track forecasting (Wu *et al.*, 2007b). Based on the results of 10 missions conducted in 2004 (Wu *et al.*, 2007b), the use of the dropwindsonde data from DOTSTAR has improved the 72 h ensemble forecast of three global models, i.e. the Global Forecasting System (GFS) of National Centers for Environmental Prediction (NCEP), the Navy Operational Global Atmospheric Prediction System (NOGAPS) of the Fleet Numerical Meteorology and Oceanography Center (FNMOC), and the Japanese Meteorological Agency (JMA), Global Spectral Model (GSM), by 22% (Fig. 9).

Wu *et al.* (2007b) showed that the average improvement of the dropwindsonde data made by DOTSTAR to the 72 h typhoon track prediction in the Geophysical Fluid Dynamics Laboratory (GFDL) hurricane models is an insignificant 3%. It is very likely that the signal of the dropwindsonde data is swamped by the bogusing procedure used during the initialization of the GFDL hurricane model. Chou and Wu (2007) showed a better way of appropriately combining the dropwindsonde data with the bogused vortex in the mesoscale model in order to further boost the effectiveness of dropwindsonde data with the implanted storm vortex.

4.3. *OSSE study (Wu et al., 2006)*

As the conventional observations usually have far less degrees of freedom than the models, the four-dimensional variational data assimilation

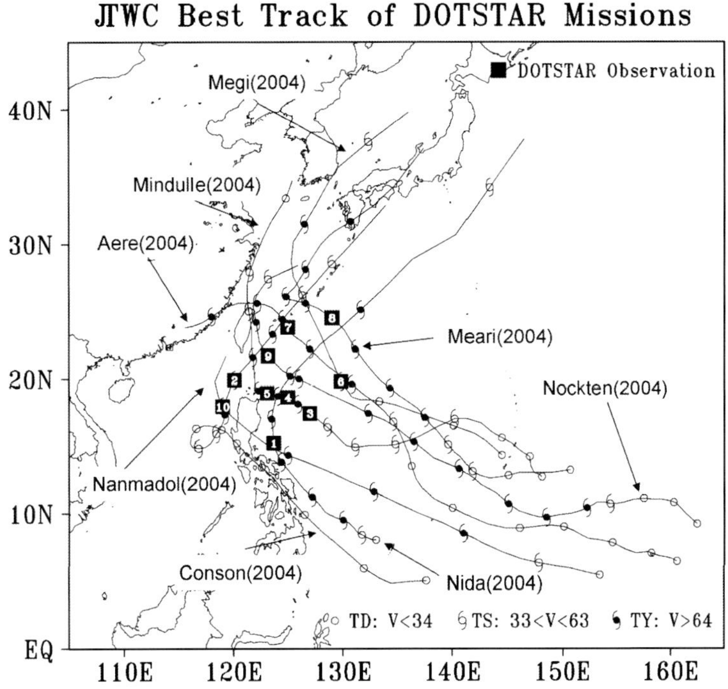

Figure 8. Best tracks (in typhoon symbols for every 24 h) of the eight typhoons with ten DOTSTAR observation missions in 2004. The squares indicate the storm locations where the DOTSTAR missions were taken. The numbers on the squares represent the sequence of the missions. (From Wu *et al.*, 2007b.)

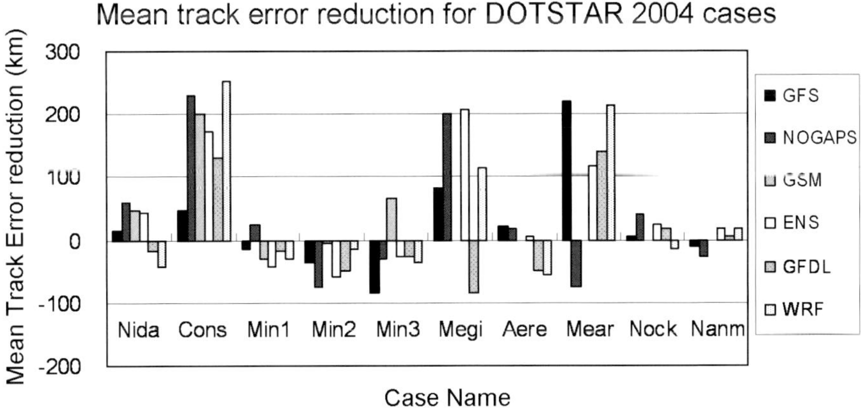

Figure 9. 6–72 h mean track error reduction (in km) after the assimilation of the dropwindsonde data into each of the ten models. The storm's name is abbreviated to the first four letters, while Min1, Min2, and Min3 stand for the first, second, and third cases in Mindulle. (From Wu *et al.*, 2007b.)

(4DVAR) has become one of the most advanced approaches to combining the observations with the model in such a way that the initial conditions are consistent with the model dynamics and physics (Guo *et al.*, 2000). Based on 4DVAR, a bogus data assimilation method has been developed by Zou and Xiao (2000) to improve the initial conditions for TC simulation.

A set of OSSEs have been performed to identify the critical parameters and the improved procedures for the initialization and prediction of TCs. A control experiment is carried out to create the imaginary "nature" data for Typhoon Zane (1996), using the fifth-generation Pennsylvania State University–National Center for Atmospheric Research Mesoscale Model (MM5). Then the initial data from the control experiment are degraded to produce the new initial condition and simulation, which mimics typical global analysis that resolves the Zane circulation. By assimilating some variables from the initial data of the control experiment into the degraded initial condition based on 4DVAR, the insight into the key parameters for improving the initial condition and prediction of TCs is attained (Wu *et al.*, 2006).

It is shown that the wind field is critical for maintaining a correct initial vortex structure of TCs. The model's memory of the pressure field is relatively short. Therefore, when only the surface pressure field is assimilated, due to the imbalance between the pressure and wind fields, the pressure field adjusts to the wind field and the minimal central sea-level pressure of the storm rises quickly.

It has been well demonstrated that taking the movement of the TC vortex into consideration during the data assimilation window can improve the track prediction, particularly in the early integration period. When the vortex movement tendency is taken into account during the bogus data assimilation period, it can partially correct the steering effect in the early prediction and the simulation period (Fig. 10). This concept provides a new and possible approach to the improvement of TC track prediction.

4.4. *Targeted observations for TCs (Wu et al., 2007a)*

(1) *Adjoint-Derived Sensitivity Steering Vector (ADSSV)*
By appropriately defining the response functions to represent the steering flow at the verifying

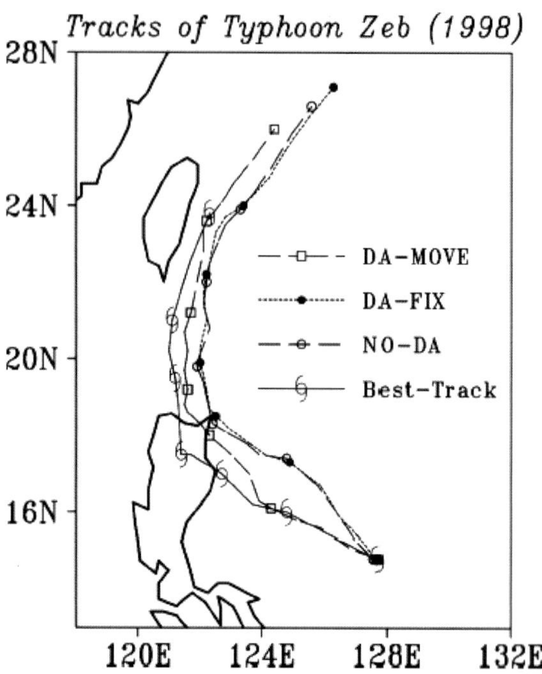

Figure 10. The 72 h JTWC best track (indicated with the typhoon symbol) and the simulated storm tracks from the experiments NO-DA, DA-FIX, and DA-MOVE for Typhoon Zeb (1998), shown for 12 h intervals from 0000 UTC, 13 October, to 0000 UTC, 16 October 1998. NO-DA: A standard MM5 simulation with an initial bogused vortex following Wu *et al.* (2002) without data assimilation. DA-FIX: An experiment assimilating the above bogused vortex (fixed in location) based on a 30-min-window 4DVAR data assimilation. DA-MOVE: An experiment in which the vortex is assimilated with the same initial data except that it moved in 3-h-window assimilation. (From Wu *et al.*, 2006.)

time, a simple innovative vector, Adjoint-Derived Sensitivity Steering Vector (ADSSV), has been designed to clearly demonstrate the sensitive locations and the critical direction of the typhoon steering flow at the observing time.

Because the goal is to identify the sensitive areas at the observing time that will affect the steering flow of the typhoon at the verifying time, the response function is defined as the deep-layer-mean wind within the verifying area. A 600-km-by-600-km-square area centered on the MM5-simulated storm location is used to calculate the background steering flow as defined by Chan and Gray (1982), and two response functions are defined: R_1, the 850–300 hPa deep-layer area average (Wu *et al.*, 2003) of the zonal component (u), along with R_2, the average of the meridional component (v) of the wind vector, i.e.

$$R_1 \equiv \frac{\int_{850\,\text{hPa}}^{300\,\text{hPa}} \int_A u \, dxdydp}{\int_{850\,\text{hPa}}^{300\,\text{hPa}} \int_A dxdydp}, \qquad (9)$$

$$R_2 \equiv \frac{\int_{850\,\text{hPa}}^{300\,\text{hPa}} \int_A v \, dxdydp}{\int_{850\,\text{hPa}}^{300\,\text{hPa}} \int_A dxdydp}. \qquad (10)$$

By averaging, the axisymmetric component of the strong cyclonic flow around the storm center is removed, and thus the vector of (R_1, R_2) represents the background steering flow across the storm center at the verifying time. To interpret the physical meaning of the sensitivity, a unique new parameter, ADSSV, is designed to relate the sensitive areas at the observing time to the steering flow at the verifying time. The ADSSV with respect to the vorticity field (ς) is

$$\text{ADSSV} \equiv \left(\frac{\partial R_1}{\partial \varsigma}, \frac{\partial R_2}{\partial \varsigma} \right), \qquad (11)$$

where the magnitude of the ADSSV at a given point indicates the extent of the sensitivity, and the direction of the ADSSV represents the change in the response of the steering flow due to a vorticity perturbation placed at that point. For example, an increase in the vorticity at the observing time would be associated with an increase in the eastward steering of the storm at the verifying time, given that the ADSSV at one particular grid point aims toward the east at the forecast time.

The ADSSV, based on the MM5 forecast (Fig. 11), extends about 300–600 km from the north to the east of Typhoon Meari (2004). The directions of the ADSSVs indicate greater sensitivity in affecting the meridional component of the steering flow.

(2) *Recent techniques for targeted observations of TCs*
To optimize the aircraft surveillance observations using dropwindsondes, targeted observing strategies have been developed and examined. The primary consideration in devising such strategies is to identify the sensitive areas in which the assimilation of targeted observations is expected to have the greatest influence in improving the numerical forecast, or minimizing the forecast error. Since 2003, four objective methods have been tested for operational surveillance missions in the environment of Atlantic hurricanes conducted by National Oceanic and Atmospheric Administration (NOAA) (Aberson, 2003) and DOTSTAR (Wu *et al.*, 2005). These products are derived from four distinct techniques: the ensemble deep-layer mean (DLM) wind variance (Aberson, 2003), the ensemble transform Kalman filter (ETKF; Majumdar *et al.*, 2002), the total-energy singular vector (TESV) technique (Peng and Reynolds, 2006), and the Adjoint-Derived Sensitivity Steering Vector (ADSSV) (Wu *et al.*, 2007a). These techniques have been applied in a limited capacity to identify locations for aircraft-borne dropwinsondes to be collected in the environment of the TCs. For the surveillance missions in Atlantic hurricanes conducted by NOAA Hurricane Research Division (HRD; Aberson, 2003) and DOTSTAR (Wu *et al.*, 2005), besides the ADSSV method shown above, three other sensitivity techniques have been used to determine the observation strategies:

(i) *Deep-layer mean (DLM) wind variance*
Based on the DLM (850–200-hPa-averaged) steering flows from the NCEP Global Ensemble Forecasting System (EFS; Aberson, 2003), areas with the largest (DLM) wind ensemble spread represent the sensitive regions at the observing time. The DLM wind ensemble spread is chosen because TCs are generally steered by the environmental DLM flow, and the dropwindsondes from the NOAA Gulfstream IV sample this flow.

(ii) *Ensemble transform Kalman filter (ETKF)*
The ETKF (Bishop *et al.*, 2001) technique predicts the reduction in forecast error variance for a variety of feasible flight plans for deployment

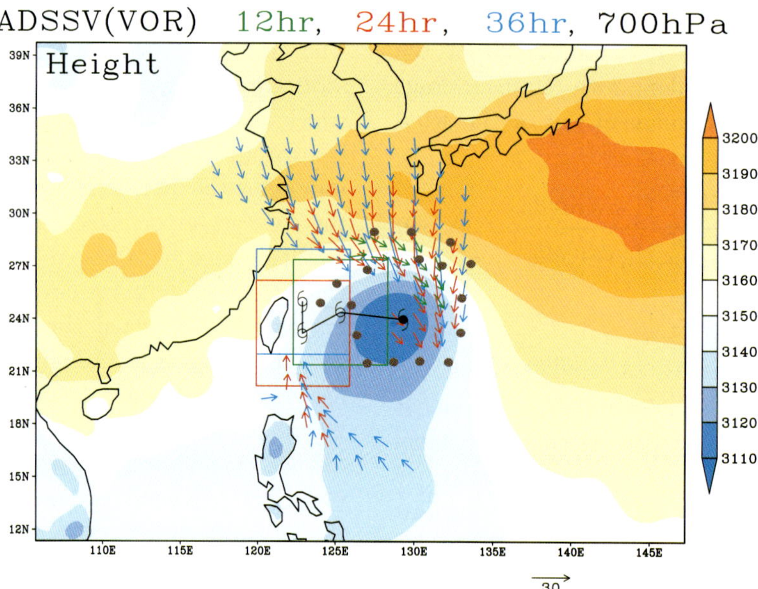

Figure 11. The ADSSV with respect to the vorticity field at 700 hPa with 12 h (in green), 24 h (in red), and 36 h (in blue) as the verifying time, superposed with the geopotential height field (magnitude scaled by the color bar to the right; the unit is m) at 700 hPa and the deployed locations of the dropsondes in DOTSTAR (brown dots). The scale of the ADSSV is indicated as the arrow to the lower right (unit: m). The 36 h model-predicted track of Meari is indicated with the typhoon symbol in red for every 12 h. The three square boxes represent the verifying areas at three different verifying times. (From Wu *et al.*, 2007a.)

of targeted observations based on the 40-member NCEP EFS (Majumdar *et al.*, 2006). That is, the ETKF uses the differences among ensemble members to estimate regions for observational missions. It takes the approach of DLM wind variance further. While DLM wind variance indicates areas of forecast uncertainty at the observation time, it does not correlate initial condition uncertainty with the errors in the forecasts. The ETKF explicitly correlates errors at the observation time with errors of the forecasts and identifies ensemble variance that impacts the forecasts in the verifying area at the verifying time.

(iii) Singular vector (SV) technique
The SV technique maximizes the growth of the total energy or kinetic energy norm (e.g. Palmer *et al.*, 1998; Peng and Reynolds, 2006) using the adjoint and forward-tangent models of the NOGAPS; Rosmond, 1997; Gelaro *et al.*, 2002), along with the ensemble prediction system

(EPS) of the JMA and the SV products from European Center for Medium-Range Weather Forecasts (ECMWF). Peng and Reynolds (2006) have demonstrated the capability of the SV technique in identifying the sensitive regions suitable for targeted observations of TCs.

The above techniques have been applied in a limited capacity to identify locations for aircraft-borne dropwinsondes to be collected in the environment of the TCs. To gain more physical insights into these targeted techniques, studies to compare and evaluate the techniques have been conducted by Majumdar *et al.* (2006), Etherton *et al.* (2006), and Reynolds *et al.* (2007).

4.5. *Summary*

DOTSTAR, a TC surveillance program using dropwindsondes, has been successfully launched since 2003. To capture the sensitive areas which may influence the TC track, a newly designed vector, ADSSV, has been proposed (Wu *et al.*, 2007a). Aside from being used to conduct research on the impact of targeted observations, DOTSTAR's tropospheric soundings around the TC environment may also prove to be a unique dataset for the validation and calibration of remotely sensed data for TCs in the Northwest Pacific region.

Five models (four operational models and one research model) were used to evaluate the impact of dropwindsonde data on TC track forecasts during 2004. All models except the GFDL hurricane model show positive impacts from the dropwindsonde data on TC track forecasts. In the first 72 h, the mean track error reductions in the three operational global models, NCEP GFS, NOGAPS and JMA GSM, are 14%, 14%, and 19%, respectively, and the mean track error reduction of the ensemble of the three global models is 22%.

Along with the development of the ADSSV in DOTSTAR, an important issue on the targeted observations based on various techniques has been raised and should be further studied.

4.6. *Further thoughts (Wu, 2006)*

Experiments, such as the assimilation process of Numerical Weather Prediction (NWP) models conducted in recent years, have demonstrated their value in significantly reducing track forecast errors, suggesting that targeting observations are in need of more efforts. As reported at the 6th International Workshop on Tropical Cycones (Wu, 2006), some issues worth further exploration are:

(1) Research on targeted data should be extended to other observing systems and data (e.g. satellite-derived soundings). Application of new concepts in predictability and data assimilation should be tested.

(2) More studies of varying definitions, interpretations, and significance of sensitive regions (e.g. different methods, metrics) could be made.

(3) More work on metrics to assess the impact of targeting — or, more generally, on any changes in the observation network — should be done.

(4) Emphasis on the potential value of OSEs and OSSEs (e.g. Wu *et al.*, 2006) in assessing potential observing system impacts prior to actual field programs is needed.

(5) Stronger effort is needed to develop alternative observing platforms (other than the dropwindsondes) for targeting, especially adaptively selecting satellite observations by revising the data-thinning algorithms currently used.

(6) Improvement and continuous refinement of targeted observing strategies are required.

(7) The focus not only be on synoptic surveillance missions but also on inner core missions, especially in cyclone basins that have not been investigated yet.

(8) In addition to the targeted observations for TC motion, the targeting for TC intensity, structure, and rainfall prediction would be another challenging issue for further examination.

5. Concluding Remarks

Typhoons are the most catastrophic weather phenomenon in Taiwan, and ironically, the rainfall that typhoons bring is also a crucial water resource. Over the past 30 years, steady progress the motion of TCs has been well made through the improvement of the numerical models, the data assimilation and bogusing systems (Kurihara *et al.*, 1995; Xiao *et al.*, 2000; Zou and Xiao, 2000; Pu and Braun, 2001; Park and Zou, 2004; Wu *et al.*, 2006), the targeted observations (Aberson, 2003; Wu, 2006; Wu *et al.*, 2007a), and the new data available to the forecasting systems (Velden, 1997; Soden *et al.*, 2001; Zou *et al.*, 2001; Pu *et al.*, 2002; Zhu *et al.*, 2002; Chen *et al.*, 2004; Wu *et al.*, 2005, 2007b).

The author's perspective on the research works of TC motion has been presented in this article, and the major scientific contributions on TC motion issues from the author are summarized below:

(1) *The baroclinic effect on TC motion (Wu and Emanuel 1993, 1994)*
The dynamic properties of potential vorticity have been extensively utilized in observational work in meteorology. Potential vorticity concepts are first applied to the understanding of hurricane movement and hurricane structure, to a certain extent. Observations have shown that real TCs are strongly baroclinic, with broad anticyclones aloft. The interaction of the baroclinic hurricane vortex with background vertical shear may lead to storm drift, relative to the background mean flow, to the left (Northern Hemisphere) of the shear vector, and to a strong deformation of the outflow potential vorticity, resulting in jetlike outflow structure.

(2) *The potential vorticity perspective of the TC motion (Wu and Emanuel, 1995a,b; Wu and Kurihara, 1996; Wu et al., 2003, 2004)*
The potential vorticity diagnostics have been first designed to understand the controlling factors affecting the motion of typhoons. A newly proposed centroid-relative track, with the position weighting based on the steering flow induced by the PV anomaly associated with the other storm, has been plotted to highlight the binary interaction processes (Wu *et al.*, 2003). More detailed work has been conducted to evaluate and to quantify the physical factors that lead to the uncertainty of the typhoon movement, such as Sinlaku (2002) (Wu *et al.*, 2004). Further work is proposed to get into the physics of the statistical behavior of typhoon tracks in the whole of the western North Pacific region.

(3) *The targeted observations in improving TC motion (Wu **et al.**, 2005; Wu 2006; Wu et al., 2006, 2007a,b)*
As described by Wu and Kuo (1999, BAMS), the improvement of the understanding of typhoon dynamics and typhoon forecasting in the Taiwan area hinges very much on the ability to incorporate the available data into high-resolution numerical models through advanced data assimilation techniques. The considerable efforts with data assimilation research are shown in Wu *et al.* (2006), which demonstrates the wind field to be the most important key variable affecting the initialization and simulation of typhoons. In addition, to obtain sufficient typhoon data and to improve TC track forecasts, a pioneering research and surveillance program for TCs in the western North Pacific, DOTSTAR, has been conducted (Wu *et al.*, 2005). The flight routes enable observations in the most sensitive region around TCs, which is an area shown by several targeted observation methods, including the ADSSV, an innovative method developed by Wu *et al.* (2007a). It is believed that this adjoint sensitivity can be used to identify important areas and dynamical features affecting the TC track, and is helpful in constructing the strategy for adaptive observations. The ADSSV has been used among other methods for the targeted observations of typhoons in the western North Pacific (DOTSTAR) and hurricanes in the Atlantic (HRD's G-IV surverillance program).

Following the recommendation by Wu (2006), an intercomparison project has been initiated to evaluate the similarities and differences of all the different targeted techniques available. More analyses are ongoing to identify the similarities and differences of all these methods and to interpret their meaning dynamically. Results from this work would not only provide better insights into the physics of the targeted techniques, but also offer very useful information to assist future targeted observations. All this work is scheduled to be presented in a special selection on "Targeted Observation, Data Assimilation, and Tropical Cyclone Predictability" in the Monthly Weather Review of the American Meteorological Society.

Overall, DOTSTAR has had significant impact on the typhoon research and operation community in the international arena, and is also one of the key components of the newly developed project of T-PARC (THORPEX — Pacific Regional Campaign). Meanwhile, the Japanese team has also decided to conduct experiments (Typhoon Hunting 2008 — TH08) (T. Nakazawa, personal communication 2007) on targeted observations of TCs in 2008. It has been planned to fly two jets from both DOTSTAR and TH08 to target the same typhoon near the east of Taiwan and the south of Okinawa under T-PARC in 2008, so as to obtain the ideal and optimum dataset around the typhoon environment. With the potential international joint efforts of DOTSTAR and TH08 associated with T-PARC in 2008, the outlook for a further advance in the targeted observations and the predictability of the TC track is promising.

Acknowledgements

The author would like to thank all the collaborators for their contributions in the papers reviewed in this article. The helpful remarks of Yuqing Wang and another, anonymous reviewer are also appreciated. Special thanks also go to Prof. Po-Hsiung Lin and the COOK team on the DOTSTAR project, and the colleagues and students in my research group, "Typhoon Dynamics Research Center," at the Department of Atmospheric Sciences, National Taiwan University. The recent work is supported by the National Science Council of Taiwan through the grants NSC92-2119-M-002-009-AP1 and NSC93-2119-M-002-013-AP1, the Office of Naval Research grant N00014-05-1-0672, and MOTC-CWB-95-6M-03.

[Received 10 April 2007; Revised 13 September 2007; Accepted 17 September 2007.]

References

Aberson, S. D., 2003: Targeted observations to improve operational tropical cyclone track forecast guidance. *Mon. Wea. Rev.*, **131**, 1613–1628.

Bishop, C. H., and Z. Toth, 1999: Ensemble transformation and adaptive observations. *J. Atmos. Sci.*, **56**, 1748–1765.

Brand, S., 1970: Interaction of binary tropical cyclones of the western North Pacific Ocean. *J. Appl. Meteor.*, **9**, 433–441.

Carr, L. E., III, M. A. Boother, and R. L. Elsberry, 1997: Observational evidence for alternate modes of track-altering binary tropical cyclone scenarios. *Mon. Wea. Rev.*, **125**, 2094–2111.

Carr, L. E., III, and R. L. Elsberry, 1998: Objective diagnosis of binary tropical cyclone interactions for the Western North Pacific basin. *Mon. Wea. Rev.*, **126**, 1734–1740.

Chan, J. C.-L., and W. M. Gray, 1982: Tropical cyclone movement and surrounding flow relationship. *Mon. Wea. Rev.*, **110**, 1354–1376.

Chan, J. C.-L., 2005: The physics of tropical cyclone motion. *Annu. Rev. Fluid. Mech.*, **37**, 99–128.

Chen S.-H., F.C. Vandenberghe, G. W. Petty, and J. F. Bresch, 2004: Application of SSM/I satellite data to a hurricane simulation. *Quart. J. Roy. Meteor. Soc.*, **130**, 801–825.

Chang, S. W. J., 1983: A numerical study of the interaction between two tropical cyclones. *Mon. Wea. Rev.*, **112**, 1806–1817.

Chang, S. W. J., 1984: Reply. *Mon. Wea. Rev.*, **112**, 1646–1647.

Chou, K.-H., and C.-C. Wu, 2007: Development of the typhoon initialization in a mesoscale model — combination of the bogused vortex and the dropwindsonde data in DOTSTAR. *Mon. Wea. Rev.* (in press).

Davis, C. A., 1992: Piecewise potential vorticity inversion. *J. Atmos. Sci.*, **49**, 1397–1411.

DeMaria, M., and C. L. Chan, 1984: Comments on "A numerical study of the interactions between two tropical cyclones." *Mon. Wea. Rev.*, **112**, 1643–1645.

Dong, K., and C. J. Neumann, 1983: On the relative motion of binary tropical cyclones. *Mon. Wea. Rev.*, **111**, 945–953.

Dritschel, D. G., 1989: Contour dynamics and contour surgery: numerical algorithms for extended, high resolution modelling of vortex dynamics in two-dimensional inviscid compressible flows. *Computer Phys. Rep.*, **10**, 77–146.

Etherton, B., C.-C. Wu, S. J. Majumdar, and S. D. Aberson, 2006: A comparison of the targeting techniques for 2005 Atlantic tropical cyclones. Preprints, 27th Conf. on Hurricanes and Tropical Meteorology (Monterey, CA, USA, Amer. Meteor. Soc.).

Fiorino, M., and R. L. Elsberry, 1989: Some aspects of vortex structure related to tropical cyclone motion. *J. Atmos. Sci.*, **46**, 975–990.

Fujiwara, S., 1921: The mutual tendency towards symmetry of motion and its application as a principle in meteorology. *Quart. J. Roy. Meteor. Soc.*, **47**, 287–293.

Fujiwara, S., 1923: On the growth and decay of vortical systems. *Quart. J. Roy. Meteor. Soc.*, **49**, 75–104.

Fujiwara, S., 1931: Short note on the behavior of two vortices. *Proc. Phys. Math. Soc. Japan, Ser. 3*, **13**, 106–110.

Gelaro, R., T. E. Rosmond, and R. Daley, 2002: Singular vector calculations with an analysis error variance metric. *Mon. Wea. Rev.*, **130**, 1166–1186.

George, J. E., and W. M. Gray, 1976: Tropical cyclone motion and surrounding parameter relationships. *J. Appl. Meteor.*, **15**, 1252–1264.

Guo, Y.-R., Y.-H. Kuo, J. Dudhia, D. Parsons, and C. Rocken, 2000: Four-dimensional variational data assimilation of heterogeneous mesoscale observations for a strong convective case. *Mon. Wea. Rev.*, **128**, 619–643.

Holland, G. J., 1984: Tropical cyclone motion: a comparison of theory and observation. *J. Atmos. Sci.*, **41**, 68–75.

Holland, G. J., and G. S. Dietachmayer, 1993: On the interaction of tropical-cyclone-scale vortices. III: Continuous barotropic vortices. *Quart. J. Roy. Meteor. Soc.*, **119**, 1381–1398.

Hoskins, B. J., M. E. McIntyre, and A. W. Robertson, 1985: On the use and significance of isentropic potential-vorticity maps. *Quart. J. Roy. Meteor. Soc.*, **111**, 877–946.

Kurihara, Y., M. A. Bender, R. E. Tuleya, and R. J. Ross, 1995: Improvements in the GFDL hurricane prediction system. *Mon. Wea. Rev.*, **123**, 2791–2801.

Kurihara, Y., C. L. Kerr, R. E. Tuleya, and M. A. Bender, 1998: The GFDL hurricane prediction system and its performance in the 1995 hurricane season. *Mon. Wea. Rev.*, **126**, 1306–1322.

Lander, M., and G. J. Holland, 1993: On the interaction of tropical-cyclone-scale vortices. I: Observations. *Quart. J. Roy. Meteor. Soc.*, **119**, 1347–1361.

Majumdar, S. J., C. H. Bishop, R. Buizza, and R. Gelaro, 2002a: A comparison of ensemble-transform Kalman-filter targeting guidance with ECMWF and NRL total-energy singular-vector guidance. *Quart. J. Roy. Meteorol. Soc.*, **128**, 2527–2549.

Majumdar, S. J., C. H. Bishop, B. J. Etherton, and Z. Toth, 2002b: Adaptive sampling with the ensemble transform Kalman filter. Part II: Field program implementation. *Mon. Wea. Rev.*, **130**, 1356–1369.

Majumdar, S. J., S. D. Aberson, C. H. Bishop, R. Buizza, M. S. Peng, and C. A. Reynolds, 2006: A comparison of adaptive observing guidance for Atlantic tropical cyclones. *Mon. Wea. Rev.* (in press).

Molinari, J., 1993: Environmental controls on eye wall cycles and intensity change in Hurricane Allen (1980), in *Proc. ICSU/WMO Int. Symp. on Tropical Cyclone Disasters* (Beijing, China, World Meteorological Society) (Peking University Press), pp. 328–337.

Molinari, J., S. Skubis, and D. Vollaro, 1995: External influences on hurricane intensity. Part III: Potential vorticity structure. *J. Atmos. Sci.*, **52**, 3593–3606.

Montgomery, M. T., and B. F. Farrell, 1993: Tropical cyclone formation. *J. Atmos. Sci.*, **50**, 285–310.

Morss, R. E., K. A. Emanuel, and C. Snyder, 2001: Idealized adaptive observation strategy for improving numerical weather prediction. *J. Atmos. Sci.*, **58**, 210–232.

Palmer, T. N., R. Gelaro, J. Barkmeijer, and R. Buizza, 1998: Singular vectors, metrics, and adaptive observations. *J. Atmos. Sci.*, **55**, 633–653.

Park, K., and X. Zou, 2004: Toward developing an objective 4DVAR BDA scheme for hurricane initialization based on TPC observed parameters. *Mon. Wea. Rev.*, **132**, 2054–2069.

Peng, M. S., and C. A. Reynolds, 2006: Sensitivity of tropical cyclone forecasts. *J. Atmos. Sci.*, **63**, 2508–2528.

Pu, Z.-X., and S. A. Braun, 2001: Evaluation of bogus vortex techniques with four-dimensional variational data assimilation. *Mon. Wea. Rev.*, **129**, 2023–2039.

Pu, Z.-X., W.-K. Tao, S. Braun, J. Simpson, Y. Jia, H. Halverson, W. Olson, and A. Hou, 2002: The impact of TRMM data on mesoscale numerical simulation of Supertyphoon Paka. *Mon. Wea. Rev.*, **130**, 2448–2458.

Reynolds, C. A., M. S. Peng, S. J. Majumdar, S. D. Aberson, C. H. Bishop, and R. Buizza, 2007: Interpretation of adaptive observing guidance for Atlantic tropical cyclones. Submitted to *Mon. Wea. Rev.*

Ritchie, E. A., and G. J. Holland, 1993: On the interaction of tropical-cyclone-scale vortices. II: Discrete vortex patches. *Quart. J. Roy. Meteor. Soc.*, **119**, 1363–1379.

Rosmond, T. E., 1997: A technical description of the NRL adjoint model system. NRL/MR/7532/97/7230, Naval Research Laboratory, Monterey, CA, USA, 93943, 62 pp.

Ross, R. J., and Y. Kurihara, 1995: A numerical study of influences of Hurricane Gloria (1985) on the environment. *Mon. Wea. Rev.*, **123**, 332–346.

Shapiro, L. J., 1996: The motion of Hurricane Gloria: a potential vorticity diagnosis. *Mon. Wea. Rev.*, **124**, 1497–2508.

Shapiro, L. J., and J. L. Franklin, 1999: Potential vorticity asymmetries and tropical cyclone motion. *Mon. Wea. Rev.*, **127**, 124–131.

Soden, B. J., C. S. Velden, and R. E. Tuleya, 2001: The impact of satellite winds on experimental GFDL hurricane model forecasts. *Mon. Wea. Rev.*, **129**, 835–852.

Velden, C. S., C. M. Hayden, S. J. Nieman, W. P. Menzel, S. Wanzong, and J. S. Goerss, 1997: Upper-tropospheric winds derived from geostationary satellite water vapor observations. *Bull. Amer. Meteor. Soc.*, **78**, 173–195.

Wang Y., and G. J. Holland, 1995: On the interaction of tropical-cyclone-scale vortices. IV: Baroclinic vortices. *Quart. J. Roy. Meteor. Soc.*, **121**, 95–126.

Wang, B., R. Elsberry, and Y. Wang, 1998: Dynamics in tropical cyclone motion: a review. *Chinese J. Atmos. Sci.*, **22**, 416–434.

Wu, C.-C., and K. A. Emanuel, 1993: Interaction of a baroclinic vortex with background shear: application to hurricane movement. *J. Atmos. Sci.*, **50**, 62–76.

Wu, C.-C., and K. A. Emanuel, and 1994: On hurricane outflow structure. *J. Atmos. Sci.*, **51**, 1995–2003.

Wu, C.-C., and K. A. Emanuel, 1995a: Potential vorticity diagnostics of hurricane movement. Part I: A case study of Hurricane Bob (1991). *Mon. Wea. Rev.*, **123**, 69–92.

Wu, C.-C., and K. A. Emanuel, 1995b: Potential vorticity diagnostics of hurricane movement. Part II: Tropical Storm Ana (1991) and Hurricane Andrew (1992). *Mon. Wea. Rev.*, **123**, 93–109.

Wu, C.-C., and Y. Kurihara, 1996: A numerical study of the feedback mechanism of hurricane–environment interaction on hurricane movement from the potential vorticity perspective. *J. Atmos. Sci.*, **53**, 2264–2282.

Wu, C.-C., and Y.-H. Kuo, 1999: Typhoons affecting Taiwan: current understanding and future challenges. *Bull. Amer. Met. Soc.*, **80**, 67–80.

Wu, C.-C., T.-S. Huang, W.-P. Huang, and K.-H. Chou, 2003: A new look at the binary interaction: potential vorticity diagnosis of the unusual southward movement of Typhoon Bopha (2000) and its interaction with Typhoon Saomai (2000). *Mon. Wea. Rev.*, **131**, 1289–1300.

Wu, C.-C., T.-S. Huang, and K.-H. Chou, 2004: Potential vorticity diagnosis of the key factors affecting the motion of Typhoon Sinlaku (2002). *Mon. Wea. Rev.*, **132**, 2084–2093.

Wu, C.-C., P.-H. Lin, S. D. Aberson, T.-C. Yeh, W.-P. Huang, J.-S. Hong, G..-C. Lu, K.-C. Hsu, I.-I. Lin, K.-H. Chou, P.-L. Lin, and C.-H. Liu, 2005: Dropwindsonde observations for typhoon surveillance near the Taiwan Region (DOTSTAR): an overview. *Bull. Amer. Meteor. Soc.*, **86**, 787–790.

Wu, C.-C., 2006: Targeted observation and data assimilation for tropical cyclone track prediction. In *Proc. 6th Int. Workshop on Tropical Cyclones* (WMO/CAS//WWW, San Jose, Costa Rica, 21–28, Nov.), pp. 409–423.

Wu, C.-C., K.-H. Chou, Y. Wang, and Y.-H. Kuo, 2006: Tropical cyclone initialization and prediction based on four-dimension variational data assimilation. *J. Atmos. Sci.*, **63**, 2383–2395.

Wu, C.-C., J.-H. Chen, P.-H. Lin, and K.-H. Chou, 2007a: Targeted observations of tropical cyclone movement based on the Adjoint-Derived Sensitivity Steering Vector. *J. Atmos. Sci.*, **64**, 2611–2626.

Wu, C.-C., K.-H. Chou, P.-H. Lin, S. D. Aberson, M. S. Peng, and T. Nakazawa, 2007b: The impact of dropwindsonde data on typhoon track forecasts in DOTSTAR. *Wea. Forecasting*, **22**, 1157–1176.

Xiao, Q., X. Zou, and B. Wang. 2000: Initialization and simulation of a landfalling hurricane using a variational bogus data assimilation scheme. *Mon. Wea. Rev.*, **128**, 2252–2269.

Zhu, T., D.-L. Zhang, and F. Weng, 2002: Impact of the advanced microwave sounding unit measurements on hurricane prediction. *Mon. Wea. Rev.,* **130**, 2416–2432.

Zou, X., and Q. Xiao, 2000: Studies on the initialization and simulation of a mature hurricane using a variational bogus data assimilation scheme. *J. Atmos. Sci.,* **57**, 836–860.

Zou, X., Q. Xiao, A. E. Lipton, and G. D. Modica, 2001: A numerical study of the effect of GOES sounder cloud-cleared brightness temperatures on the prediction of Hurricane Felix. *J. Appl. Meteor.,* **40**, 34–55.

Vortex Interactions and Typhoon Concentric Eyewall Formation

H.-C. Kuo

Department of Atmospheric Sciences, National Taiwan University, Taipei, Taiwan
kuo@as.ntu.edu.tw

L.-Y. Lin

National Science and Technology Center for Disaster Reduction, Taipei, Taiwan

C.-L. Tsai

Department of Atmospheric Sciences, National Taiwan University, Taipei, Taiwan

Y.-L. Chen

Central Weather Bureau, Taipei, Taiwan

An important issue in the formation of concentric eyewalls in a typhoon is the development of a symmetric structure from asymmetric convection. As an idealization of the interaction of a tropical cyclone core with nearby weaker vorticity of various spatial scales, we consider nondivergent barotropic model integrations to illustrate that concentric vorticity structures result from the interaction between a small and strong inner vortex (the tropical cyclone core) and neighboring weak vortices (the vorticity induced by the moist convection outside the central vortex of a tropical cyclone). In particular, the core vortex induces a differential rotation across the large and weak vortex, to strain out the latter into a vorticity band surrounding the former without a merging of the two. The straining out of a large, weak vortex into a concentric vorticity band can also result in the contraction of the outer tangential wind maximum. The dynamics highlight the essential role of the vorticity strength of the inner core vortex in maintaining itself, and in stretching, symmetrizing and stabilizing the outer vorticity field.

Our binary vortex experiments from the Rankine vortex suggest that the formation of a concentric vorticity structure requires: (1) a very strong core vortex with a vorticity at least six times stronger than the neighboring vortices, (2) a neighboring vorticity area that is larger than the core vortex, and (3) a separation distance between the neighboring vorticity field and the core vortex that is within three to four times the core vortex radius. On the other hand, when the companion vortex is four times larger than the core vortex in radius, a core vortex with a vorticity skirt produces concentric structures when the separation distance is five times greater than the smaller vortex. At this separation distance, the Rankine vortex produces elastic interaction. Thus, a skirted core vortex of sufficient strength can form a concentric vorticity structure at a larger radius than what is allowed by an unskirted core vortex. This may explain the wide range of radii for concentric eyewalls in observations.

1. Introduction

Aircraft observations of Hurricane Gilbert (1988) (e.g. Willoughby *et al.*, 1982; Black and Willoughby, 1992, hereafter BW92) show that intense tropical cyclones often exhibit concentric eyewall patterns in their radar reflectivity. Approximately 12 hours after reaching its minimum sea level pressure of 888 hPa, the lowest recorded so far in the Atlantic basin (Willoughby *et al.*, 1989), Hurricane Gilbert displayed concentric eyewalls. BW92 estimated the radius to be 8–20 km for the inner eyewall and 55–100 km for the outer eyewall. Between the two eyewalls, an echo-free gap (or moat) of about 35 km exists where the vorticity is low. In this pattern, deep convection within the inner, or primary, eyewall is surrounded by a nearly echo-free moat, which in turn is surrounded by a partial or complete ring of deep convection. Both convective regions typically contain well-defined local wind maxima and thus vorticity field. The primary wind maximum is associated with the inner core vortex, while the secondary wind maximum is usually associated with enhanced vorticity field embedded in the outer rainband (see e.g. the Hurricane Gilbert example given in Fig. 1 of Kossin *et al.*, 2000).

Figure 1 shows the radar observations of Typhoon Lekima of 2001 (it has been adapted from Kuo *et al.*, 2004; hereafter K04), and the passive microwave satellite data of Typhoon Imbudo (2003). The figure indicates that a huge area of convection outside the core vortex wraps around the inner eyewall to form the concentric eyewalls in a time scale of 12 hours for both cases. Figure 1 and that of BW92, along with many microwave images reported by Kuo and Schubert (2006; hereafter K06), suggest the formation of a concentric eyewall from the organization of asymmetric convection outside

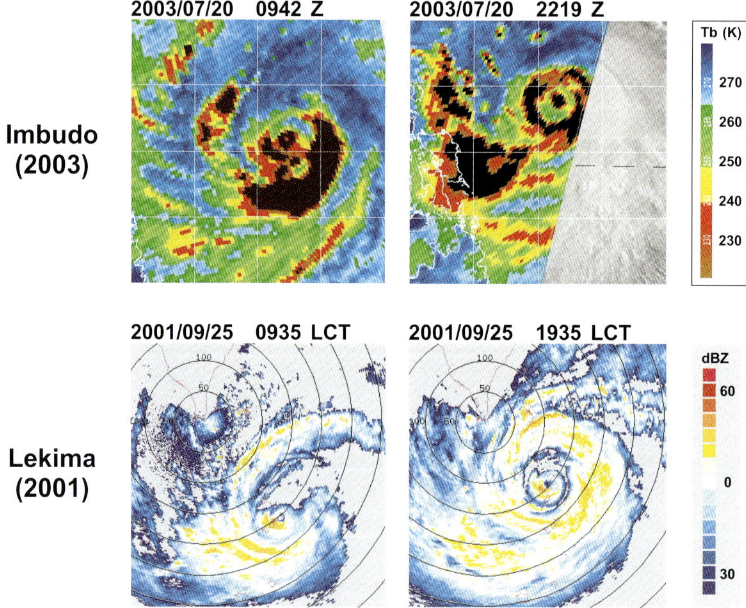

Figure 1.   The top images are from passive microwave data for western Pacific Typhoon Imbudo (2003) at 0942 Z July 20 and at 2219 Z July 20 (image courtesy of Naval Research Laboratory, Monterey, California). The bottom images are reflectivity at 0.5 elevation angle for Typhoon Lekima (2001) from the Central Weather Bureau WSR-88D (10 cm) radar at Kung-Ting at local time 0935 September 25 and 1935 September 25. The radial increment of the circles centered at the radar station is 50 km. The time interval of concentric eyewalls formation is approximately 12 hours in both cases.

the primary eyewall into a symmetric band that encircled the eyewalls. In a concentric eyewall, both the deep convection and the potential vorticity are enhanced. Thus, in terms of potential vorticity dynamics, the formation of a concentric eyewall would appear to have two important aspects: (i) the organization of an existing asymmetric potential vorticity distribution into a symmetric potential vorticity distribution with a moat; (ii) the diabatic enhancement of potential vorticity during the organizational process. Although both aspects are probably important in most examples, it is useful to isolate the first aspect through the study of highly idealized dynamical models. Namely, one way to produce the concentric eyewall structure is through a binary vortex interaction such as those reported in K04 and K06.

This article reviews some important points from K04 and K06 in the dynamics of tropical cyclone concentric eyewall formation. We argue that the organization of the asymmetric convection into a symmetric concentric eyewall can be accomplished through pure advective processes of vorticity dynamics. We address the dynamics of why there is a wide range of radii for concentric eyewalls, and the roles of the moat and the turbulent vorticity scale in the formation of concentric eyewall structure. Section 2 describes the model and the solution method. The physical parameters are discussed in Sec. 3. The numerical results of binary vortex interactions and the results of concentric eyewall formation solely from a turbulent background are presented in Sec. 4. Section 4 also contains a discussion on the two-dimensional turbulence characteristic of the model as well as the regime phase space diagrams. Concluding remarks are made in Sec. 5.

2. Nondivergent Barotropic Model

The basic dynamics considered is two-dimensional nondivergent barotropic with ordinary diffusion, i.e. $D\zeta/Dt = \nu\nabla^2\zeta$, where $D/Dt = \partial/\partial t + u(\partial/\partial x) + v(\partial/\partial y)$. Expressing the velocity components in terms of the stream function by $u = -\partial\psi/\partial y$ and $v = \partial\psi/\partial x$, we can write the nondivergent barotropic model as

$$\frac{\partial\zeta}{\partial x} + \frac{\partial(\psi,\zeta)}{\partial(x,y)} = \nu\nabla^2\zeta, \tag{1}$$

where

$$\zeta = \nabla^2\psi, \tag{2}$$

is the invertibility principle and $\partial(,)/\partial(x,y)$ is the Jacobian operator. The diffusion term on the right hand side of (1) controls the spectral blocking associated with the enstrophy cascade to higher wave numbers. As discussed in K04 and Prieto $et\ al.$ (2001), we have avoided the use of hyperviscosity (higher iterations of the Laplacian operator on the right hand side of the vorticity equation) because of the unrealistic oscillations it can cause in the vorticity field. We perform calculations on a doubly periodic square domain. The Fourier pseudospectral method is employed for the discretization, with 512×512 equally spaced collocation points on a 300 km $\times$ 300 km domain for the vortex interaction experiments. The code was run with a dealiased calculation of quadratic nonlinear terms with 170×170 Fourier modes. The fourth-order Runge–Kutta method with a 3s time step is used for the time differencing. The diffusion coefficient, unless otherwise specified, was chosen to be $\nu = 6.5$ m^{2}s^{-1}. For the 300 km $\times$ 300 km domain, this value of ν gives an e^{-1} damping time of 3.37 h for all modes having total wave number 170, and a damping time of 13.5 h for modes having total wave number 85. Some of the experiments were repeated at increased resolution and/or with a larger domain size. From these experiments, we conclude that the results shown here are insensitive to the domain size and to the resolution employed. The use of such a simple model obviously precludes the simulation of the complete secondary eyewall cycle, but it allows for some simple numerical experiments concerning the initial symmetrization dynamics involved in secondary eyewall formation.

3. Physical Parameters

Our experiments can be viewed as an extension of those of Dritschel and Waugh (1992; hereafter DW92) and Dritschel (1995). DW92 described the general interaction of two barotropic vortices with equal vorticity but different sizes. They performed experiments on the f-plane by varying the ratio of the vortex radii and the distance between the edges of the vortices normalized by the radius of the larger vortex. The resulting structures can be classified into elastic interaction, merger, and straining-out regimes. The elastic interaction involves distortion to the vortices in a mutual cyclonic rotation. The merger regime involve part of the smaller vortex being removed, and some of it being incorporated into the larger vortex. In the complete straining-out regime, a thin region of filamented vorticity bands surrounding the central vortex with no incorporation into the central vortex. This appeared to resemble a concentric vorticity structure with a moat. However, the outer bands which result from the smaller vortex are much too thin to be identified with that observed in the outer eyewall of a tropical cyclone.

There are four parameters used in the DW92 study: the vorticity ζ, the vortex sizes R_1 and R_2, and the separation distance between the edges of the two vortices Δ. With the specification of R_1 and ζ, the end states of binary vortex interaction are summarized by the remaining degrees of freedom: R_2/R_1 and Δ/R_1. The interaction of a small and strong vortex with a large and weak vortex was not studied by DW92 as their vortices are of the same strength (i.e. the same ζ), and their larger vortex was always the "victor" and the smaller vortex was the one being partially or totally destroyed. Observations of Typhoons Lekima and Imbudo, as seen in Fig. 1, indicate that there may be a huge area of convection with weak cyclonic vorticity outside the core vortex that wraps around the inner eyewall, rather than the other way around. An extension of the complete straining-out regime

to include a finite-width outer band is needed to explain the interaction of a small and strong vortex (representing the tropical cyclone core) with a large and weak vortex (representing the vorticity induced by the moist convection outside the central vortex of a tropical cyclone).

We have introduced the sharp-edged vorticity patches ζ_1 and ζ_2 to replace the DW92 ζ parameter in K04. With the specification of the vorticity of the companion vortex ζ_2 and the size of the core vortex $R_1 = 10$ km, the number of parameters is reduced to three, which we take to be the vorticity strength ratio $\gamma = \zeta_1/\zeta_2$, the vortex radius ratio $r = R_1/R_2$, and the dimensionless gap Δ/R_1. Using aircraft flight-level data, Mallen *et al.* (2005) demonstrated that tropical cyclones are often characterized by a relatively slow decrease of tangential wind outside the radius of maximum wind, and hence by a corresponding cyclonic vorticity skirt. To include the skirt effect, we have set the radial profile of vorticity of constant vorticity ζ_1 in the inner region and of the skirted structure $\frac{1}{2}\zeta_1(1 - \alpha)(r/R_1)^{-\alpha-1}$ in the outer region. The constant R_1 is a measure of the core vortex size, and α is the nondimensional skirt parameter. In the vorticity skirt region, the azimuthal wind behaves as $r^{-\alpha}$. The $\alpha = 1$ profile is the Rankine vortex profile, which has zero vorticity gradient and rapid decrease of angular velocity with radius outside the core. Aircraft observations of the azimuthal winds in hurricanes (e.g. Shea and Gray, 1973; Mallen *et al.*, 2005) suggest that a reasonable range for the skirt parameter is $0.5 \leq \alpha \leq 1$. In contrast to the strong core vortex, the companion vortex is not skirted. The omission of a vorticity skirt on the companion vortex is based on our experience that the vorticity skirt on the "victorious" core vortex is more important. In addition, it is desirable to keep the number of parameters to a manageable level.

Vorticity skirts play a role in several aspects of tropical cyclone dynamics. For example, vortex mergers can occur owing to vortex propagation on the outer vorticity gradients associated

Table 1. Summary of the experimental parameters.

	Parameters	Conditions	Dimensionless parameters
(a)	ζ, R_1, R_2, Δ	$R_1 = 1$ $0.1 \leq R_2 \leq 1.0$ $(\zeta = 2\pi)$	$\dfrac{R_2}{R_1}, \dfrac{\Delta}{R_1}$
(b)	$\zeta_1, \zeta_2, R_1, R_2, \Delta$	$R_1 \leq R_2$ $\zeta_1 > \zeta_2$ $(R_1 = 10 \text{ km})$ $(\zeta_2 = 3 \times 10^{-3}\text{s}^{-1})$	$\dfrac{\zeta_1}{\zeta_2}, \dfrac{R_1}{R_2}, \dfrac{\Delta}{R_1}$
(c)	$\alpha, \zeta_1, \zeta_2, R_1, R_2, \Delta$	$0 < \alpha \leq 1$ $R_1 \leq R_2$ $\zeta_1 > \zeta_2$ $(R_1 = 10 \text{ km})$ $(\zeta_2 = 3 \times 10^{-3}s^{-1})$	$\alpha, \dfrac{\zeta_1}{\zeta_2}, \dfrac{R_1}{R_2}, \dfrac{\Delta}{R_1}$

with each vortex (DeMaria and Chan, 1984). In addition, the radial vorticity gradient associated with the vorticity skirt provides a means for the radial propagation of vortex Rossby waves (Montgomery and Kallenbach, 1997; Balmforth et al., 2001). Finally, skirted vortices can be more resilient to the destructive effects of large-scale vertical wind shear (Reasor et al., 2004). As far as the concentric eyewall fomation dynamics are concerned, the skirt on the core vortex provides a shielding effect which may lead to the formation of outer bands further away. In addition, the slower radial decrease of angular velocity associated with the vorticity skirt lengthens the filamentation time and slows moat formation.

In summary, we have used the skirt parameter α, the vorticity strength ratio $\gamma = \zeta_1/\zeta_2$, the vortex radius ratio $r = R_1/R_2$, and the dimensionless gap Δ/R_1 to classify our model end states. Table 1 gives a summary of the parameters, the specified conditions, and dimensionless parameters that were used for the end state classification for the DW92, K04, and K06 papers. Figure 2 shows the initial schematic configuration of two circular vortices in DW92, K04, and K06. With the introduction of the vorticity strength ratio and the skirt parameter α into the binary vortex interaction problem, we have

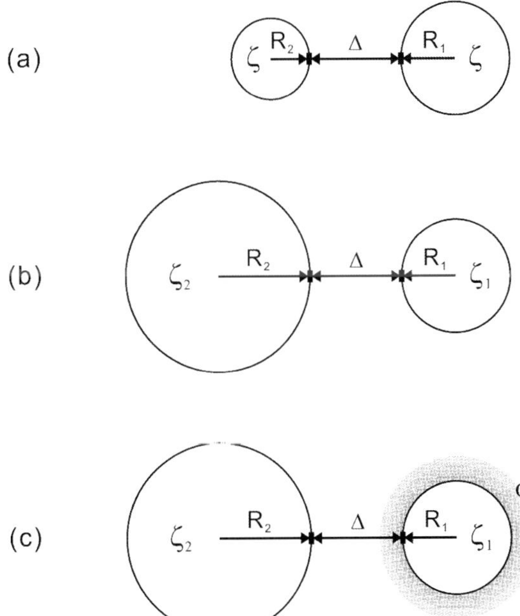

Figure 2. Initial configuration of two circular vortices with radii R_1 and R_2, vorticity ζ_1 and ζ_2, the skirt parameter α and the gap Δ. The $R_1 \geq R_2$ and $\zeta_1 = \zeta_2 = \zeta$ in DW 92, and $R_1 \leq R_2$ and $\zeta_1 > \zeta_2$ in K04 and K06.

added a new dimension to the Dritschel–Waugh vortex interaction scheme which provides a proper concentric vorticity structure of various sizes as well as the tripole vortex structure.

4. Numerical Results

4.1. *Binary vortex interaction*

Figure 3 shows the vorticity field in the binary vortex experiments with $r = 1/4$ for the (a) concentric case ($\gamma = 8$, $\Delta/R_1 = 0$, $\alpha = 1.0$), (b) merger case ($\gamma = 8$, $\Delta/R_1 = 0$, $\alpha = 0.5$), (c) elastic interaction ($\gamma = 6$, $\Delta/R_1 = 5$, $\alpha = 1.0$), and (d) concentric case ($\gamma = 6$, $\Delta/R_1 = 5$, $\alpha = 0.5$). The figure suggests the formation of concentric vorticity structures for both skirted $\alpha = 0.5$ and unskirted ($\alpha = 1.0$) vortices. The Rankine core vortex ($\alpha = 1.0$) produces the concentric structure with the zero initial separation distance. On the other hand, the skirted core vortex ($\alpha = 0.5$) produces a merger with the same zero initial separation distance. With a separation distance that is five times the core vortex radius, the Rankine vortex produces inelastic interaction, and the $\alpha = 0.5$ vortex produces concentric structure. Thus, Fig. 3 suggests that a skirted core vortex of sufficient strength can form a concentric vorticity structure at a larger radius than what is allowed by an unskirted core vortex. This may explain the wide range of radii for concentric eyewalls in observations.

Even though the change of sign of the vorticity gradient across the outer band satisfies the Rayleigh necessary condition for barotropic stability, the band is stabilized by the Fjørtoft

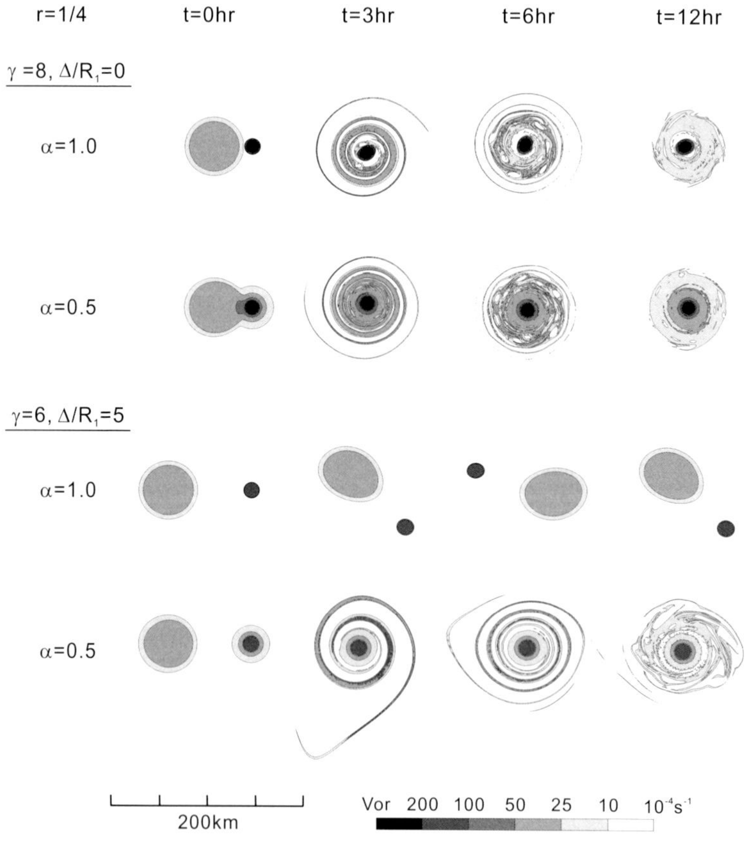

Figure 3. Vorticity fields in the binary vortex experiments with the $r = 1/4$ parameters for the (a) concentric case ($\gamma = 8$, $\Delta/R_1 = 0$, $\alpha = 1.0$), (b) merger case ($\gamma = 8$, $\Delta/R_1 = 0$, $\alpha = 0.5$), (c) elastic interaction ($\gamma = 6$, $\Delta/R_1 = 5$, $\alpha = 1.0$), and (d) concentric case ($\gamma = 6$, $\Delta/R_1 = 5$, $\alpha = 0.5$).

sufficient condition for stability. Namely, the strong inner vortex causes the wind to be stronger at the inner edge than at the outer edge, allowing the vorticity band and therefore the concentric structure to be sustained. A similar mechanism was discussed by Dritschel (1989) and Polvani and Plumb (1992), who showed how thin filaments can be stabilized by the flow field of the main vortex. They argued that the filament is linearly stable and appears circular in the presence of sufficiently strong "adverse shear." The adverse shear is an externally controlled parameter with the opposite sense to that produced by the filament's vorticity alone. In general, the inner vortex must also possess high vorticity not only to be maintained against any deformation field induced by the outer vortices, but also to maintain a smaller enstrophy cascade and to resist the merger process into a monopole. A detailed discussion can be found in K04.

Figure 4 shows the tangential wind speed from the aircraft observations of Hurricane Gilbert (BW92), and from a K04 model concentric eyewall simulation, along radial arms emanating from the core vortex center. These observations show the contraction of the outer tangential wind maximum from a distance of 90 km from the storm center to 60 km in approximately 12 hours. The core vortex intensity remained approximately the same during the contraction of the outer tangential wind maximum. Wind profiles from the model also clearly show the contraction of secondary maxima in the tangential wind field. It appears that the contraction of the secondary wind maxima in the model are in general agreement with observations of Hurricane Gilbert. The contraction mechanism for outer bands is often argued to be a balanced response to an axisymmetric ring of convective heating (Shapiro and Willoughby, 1982). The results shown in Fig. 4 suggest that the nonlinear advective dynamics involved in the straining-out of a large, weak vortex into a concentric vorticity band can also result in contraction of the secondary wind maximum. No moist convection is involved in this process. However, moist convection could substantially enhance the strength of such a band.

4.2. *Two-dimensional turbulence on the f-plane*

We now address the two-dimensional turbulence characteristics that are relevant to our tripole and concentric eyewall formation dynamics. We restrict our attention to f-plane dynamics, so the Rhines scale dynamics will not be discussed. Three important integral properties from our model equations (1) and (2) during the numerical simulations are the energy $E = \iint \frac{1}{2}\nabla\psi \cdot \nabla\psi\,dxdy$, the enstrophy $Z = \iint \frac{1}{2}\zeta^2\,dxdy$, and the palinstrophy $P = \iint \frac{1}{2}\nabla\zeta \cdot \nabla\zeta\,dxdy$. The palinstrophy is a measure of the overall vorticity gradient in the domain. If there is no flux across the boundary, as can be shown from (1) and (2), these three integral properties are related by

$$\frac{dE}{dt} = -2\nu Z, \qquad (3)$$

$$\frac{dZ}{dt} = -2\nu P, \qquad (4)$$

and the palinstrophy by

$$\frac{dP}{dt} = \iint \left(\frac{\partial(\psi,\zeta)}{\partial(x,y)}\right)\nabla^2\zeta\,dxdy$$

$$- \iint \nu(\nabla^2\zeta)^2\,dxdy. \qquad (5)$$

The last term in (5), which represents the effect of *mixing*, is always negative; mixing always reduces the overall vorticity gradient in the domain. The first term on the right hand side of (5), which represents *stirring*, is usually positive. The reason is that, in an average sense, positive local vorticity extremes (i.e. $\nabla^2\zeta < 0$) near the typhoon core are likely to be coupled with the positive vorticity advection (i.e. $\frac{\partial(\psi,\zeta)}{\partial(x,y)} < 0$). In other words, stirring directly increases the palinstrophy. The velocity field in our vortices, especially the wind field associated

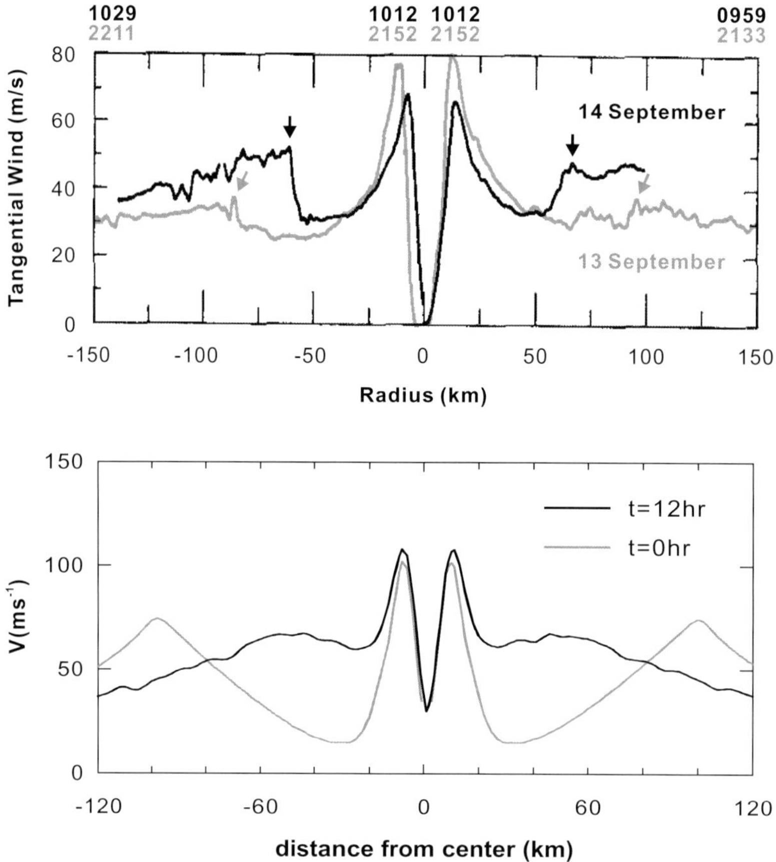

Figure 4. Tangential wind speed along a radial leg from the aircraft observations of Hurricane Gilbert (adapted from BW92) and tangential wind speed along radial arms emanating from the center of the strong core vortex for an experiment in K04.

with the core vortex, provides the stirring for the initial increase of P. The mixing will reduce the P in the later stage. Figure 5 shows the time dependence of kinetic energy, enstrophy, and palinstrophy for the binary vortex experiment with $\gamma = 5$, $r = 1/3$, and $\Delta/R_1 = 2.5$ parameters. Two values of diffusivity ν, 3.25 m² s⁻¹ and 6.5 m² s⁻¹, are employed in the experiments. The scale for the palinstrophy (kinetic energy and enstrophy) is on the left (right) side of the figure.

The near-conservation of the kinetic energy, the damping of the enstrophy field, and the initial increase (stirring) and the eventual decrease (mixing) of the palinstrophy field in the vortex interaction experiments, all possess the characteristics of two-dimensional turbulence. Batchelor (1969) argued that in the case of a nearly inviscid fluid (ν small enough), the vorticity contours can pack close together before diffusion or mixing is effective. The closely packed contours increase $|\nabla \zeta|$, and greatly enhance the palinstrophy. Even when ν is small, the $-2\nu P$ term on the right-hand side of (4) may not be small owing to the increase of palinstrophy. We then have a significant enstrophy cascade. With a significant enstrophy cascade (area thus a smaller enstrophy later), the right hand side of the kinetic energy equation (3) is small and the kinetic energy is nearly conserved. This is the phenomenon of *selective decay*, i.e. the

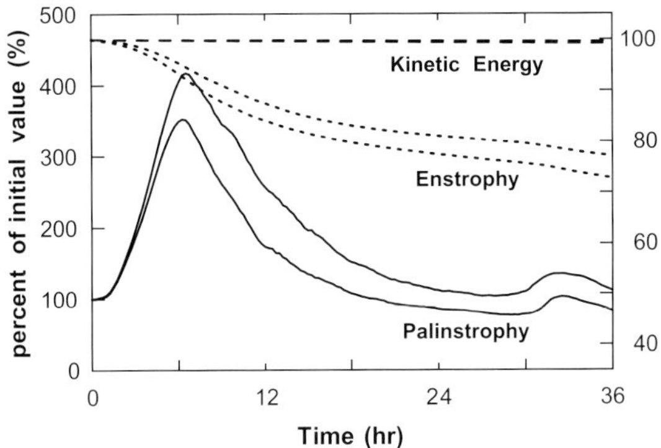

Figure 5. Time dependence of kinetic energy, enstrophy, and palinstrophy with respect to their initial values for experiments with the vorticity strength ratio ($\gamma = 5$), skirt parameter ($\alpha = 1$), vortex radius ratio $r = 1/3$ and the dimensionless gap $\Delta/R_1 = 2.5$. The diffusivity ν employed are 3.25 m^2 s^{-1} and 6.5 m^{2}s^{-1}. Note that the palinstrophy (kinetic energy and enstrophy) scale is in the left (right) ordinate. The curves with higher values are the 3.25 m^2 s^{-1} case.

enstrophy is selectively decayed over kinetic energy in two-dimensional turbulence (Cushman-Roisin, 1994). One of the important implications of selective decay is merger dynamics. That is, on the average, the vortices become larger, stronger, and fewer.

Merger dynamics can be understood with the following argument. In the presence of strong rotation, the wind field is nearly geostrophic, so that

$$u \sim \frac{g^*}{f_0} \frac{\Delta p}{l}, \tag{6}$$

the kinetic energy is

$$E \sim u^2 \sim \left(\frac{g^*}{f_0} \frac{\Delta p}{l} \right)^2, \tag{7}$$

and the enstrophy is

$$Z \sim \left(\frac{u}{l} \right)^2 \sim \left(\frac{g^*}{f_0} \frac{\Delta p}{l} \right)^2, \tag{8}$$

where Δp is the pressure perturbation, g^* is the reduced gravity, and l is the vortex scale. The near conservation of kinetic energy, according

to (7) with constant f_0 and g^* values (i.e. the heating/cooling does not alter the stratification appreciately), requires that $\Delta p/l$ remain approximately constant. This implies that a steady increase or decrease of the length scale l should be associated with a proportional increase or decrease in the eddy amplitude Δp. The cascade of enstrophy according to (8), along with the conservation of kinetic energy (7), imply a steady increase of the spatial scale l, with a proportional increase in Δp. There is thus a natural tendency toward larger structures in two-dimensional turbulence with successive eddy mergers. Figure 6 shows the evolution of turbulent vorticity fields with initial spatial scales of 5–10 km ($r^* = 1$) and 30–40 km ($r^* = 4$). It is a striking fact that for the random initial state of different scales, a two-dimensional fluid will rapidly organize itself into a system of coherent, interacting vortices swimming through a sea of passive filamentary structure produced from earlier vortex interactions (McWilliams, 1984). To conserve angular momentum and/or kinetic energy during the merger process, the inward merger of the vorticity field toward the core vortex must be accompanied also by some outward vorticity redistribution (Schubert *et al.*, 1999). The outward redistribution can be seen in the form of filaments that orbit the core vortex. The coherent vorticity structures, such as the concentric vortex in Fig. 3, can prolong the merger process into a monopole.

4.3. *Concentric eyewall formation solely from a turbulent background*

Given the fact that there is a natural tendency toward larger structures in two-dimensional turbulence with successive eddy mergers, we now address the question of whether the rapid development of a strong vortex within a turbulent background vorticity field can lead to concentric vorticity patterns similar to those observed during binary vortex interactions. Rankine-type

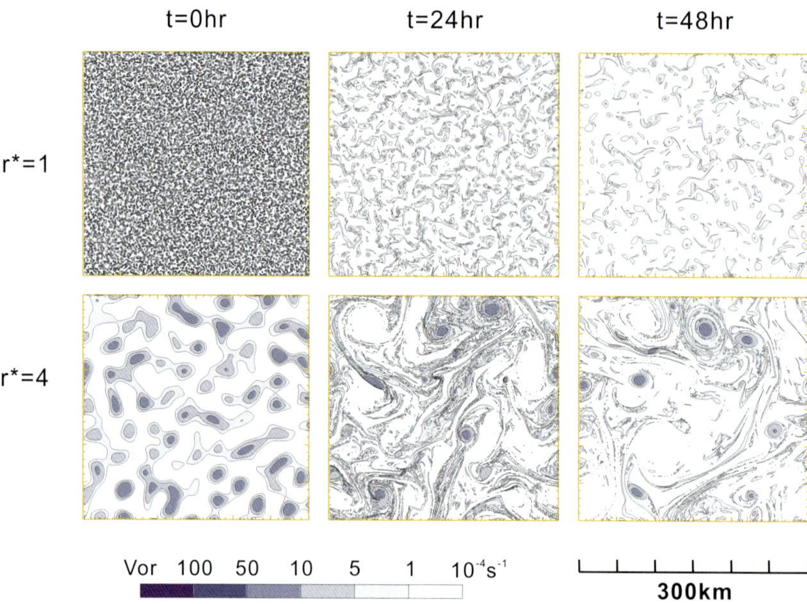

Figure 6. Vortex emergence and evolution for two different initial vorticity spatial scales, $r^* = 1$ (5–10 km scale) and $r^* = 4$ (30–40 km scale).

core vortices are used in the experiments. The parameter γ is now taken to be the ratio of the vorticity in the core vortex to the peak turbulent background vorticity, which is set to $3 \times 10^{-3} \mathrm{s}^{-1}$. This is the same amplitude used for the companion vortex in the binary vortex experiments. We have avoided the use of stronger turbulent background vorticity in order to retain the basic nature of the problem we are investigating.

In their analysis of this problem, Rozoff *et al.* (2006, hereafter R06) used a core vortex that approximates a Category 5 hurricane (75 ms^{-1} tangential winds near 25 km radius) embedded within turbulent background vorticity elements having horizontal scales between 20 km and 40 km and a peak amplitude of $1 \times 10^{-3} \mathrm{s}^{-1}$. The vorticity in the R06 core vortex is $6.5 \times 10^{-1} \mathrm{s}^{-1}$, so that the corresponding vorticity strength ratio is $\gamma = 6.5$, as in the present article. Their integrations (e.g. their Fig. 8) result in a concentric structure, but with a rather weak vorticity halo due to the vorticity mixing that occurs. We have used the formula of R06 in the specification of the characteristic spatial scale of the

turbulent background. In addition, we have extended the R06 experiments by including a vorticity clear zone (moat) near the core vortex. In real hurricanes the moat can be viewed as being the result of rapid filamentation (R06) and of strong subsidence induced by surrounding convection (e.g. Dodge *et al.*, 1999).

Figure 7 shows results from the turbulent background experiments for the (1) straining-out case ($r^* = 1$, $\gamma = 2$), (2) merger case ($r^* = 4$, $\gamma = 2$), (3) tripole case ($r^* = 4$, $\gamma = 2$), and (4) concentric case ($r^* = 4$, $\gamma = 6$). A vorticity clear zone (moat) of 10 km width is introduced near the core vortex for the last two experiments. Note that the outer vorticity bands in the concentric case in Fig. 7, unlike the results in R06 which are a magnitude weaker, are of similar strength compared to the binary vortex interaction. Figure 7 indicates that the core vortex is able to stir and to mix the turbulent background vorticity into a concentric eyewall end state. Of particular interest in Fig. 7 is the tripole case. A tripole is a linear arrangement of three regions of distributed

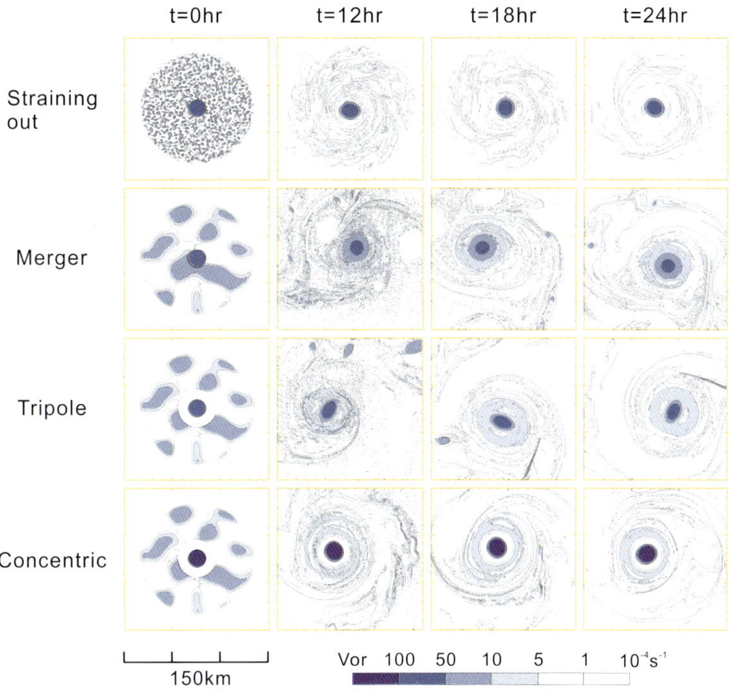

Figure 7. Turbulent background experiments for the (1) straining-out case ($r^* = 1$, $\gamma = 2$), (2) merger case ($r^* = 4$, $\gamma = 2$), (3) tripole case ($r^* = 4$, $\gamma = 2$), and (4) concentric case ($r^* = 4$, $\gamma = 6$). A vorticity clear zone (moat) of 10 km is introduced near the core vortex for the last two experiments.

vorticity of alternate signs, with the whole configuration steadily rotating in the same sense as the vorticity of the elliptically shaped central core (Carton *et al.*, 1989; Polvani and Carton, 1990; Carton and Legras, 1994; Kloosterziel and Carnevale, 1999). Examples of elliptical eyes that might be associated with tripolar vorticity structure were reported by Kuo *et al.* (1999) for the case of Typhoon Herb (1996) and by Reasor *et al.* (2000) for the case of Hurricane Olivia (1994).

4.4. *Phase diagrams*

DW92 and Dritschel (1995) described the general interaction of two barotropic vortices with equal vorticity but different sizes. The resulting structures can be classified into elastic interaction, merger, and straining-out regimes in terms of the radius ratio $r = R_1/R_2$ and the dimensionless gap Δ/R_1. Figure 8, which is reproduced from DW92, summarizes the regimes with the simulated vorticity field in the merger and straining-out regimes. The figure suggests, in general, that binary vortex interaction is prone to produce the merger regime (straining-out regime) if the two vortices are of similar (different) size. Larger separation distances will result in the incomplete merger and incomplete straining-out. The incomplete merger and straining-out regimes result in part of the smaller vortex being torn away incompletely and a tiny vortex being left behind. A separation parameter greater than 1.5 gives elastic interaction. The outer bands in the complete straining-out regime, which result from the smaller vortex, are much too thin to be identified as the outer band of concentric eyewalls. Moreover, there is no tripole regime produced in the end states.

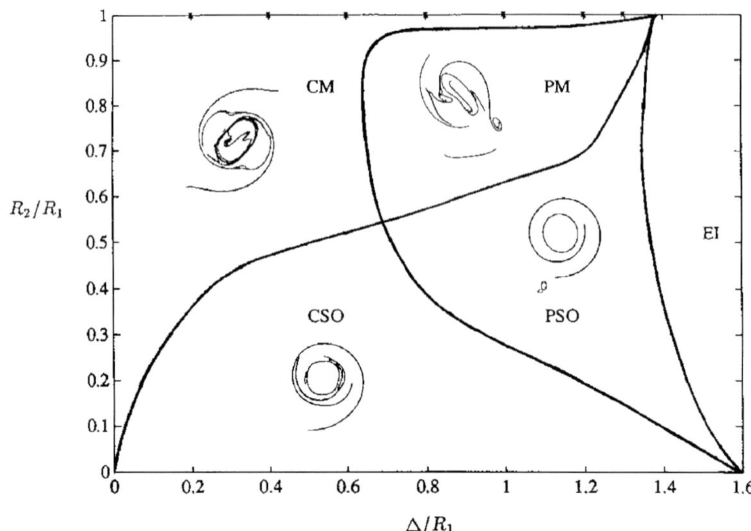

Figure 8. Interaction regimes for binary vortices calculated as a function of the dimensionless gap Δ/R_1 and the vortex radius ratio R_2/R_1. The structures are categorized into complete merger, partial merger, complete straining-out, partial straining out and elastic interaction regimes. (Adapted from DW92.)

Figure 9, adapted from K04, illustrates the Rankine vortex binary interaction regimes as a function of the dimensionless gap Δ/R_1, and the vorticity strength ratio $\gamma = \zeta_1/\zeta_2$ for the radius ratios $r = R_1/R_2 = 1/2, 1/3$, and $1/4$. The abscissa of Fig. 9 is the dimensionless gap Δ/R_1, which ranges from 0 to 4, and the ordinate is the vorticity strength ratio γ, which ranges from 1 to 10. The classifications are based on the scheme devised by DW92 and K04. The structures are categorized into the "concentric," "tripole," "merger," and "elastic interaction" regimes, with the concentric end states shaded. Figure 9 indicates that the tripole vortex structure often serves as the transition between the concentric and merger structures. The results suggest that the formation of a concentric vorticity structure requires: (i) a core vortex with vorticity at least six times stronger than the neighboring vorticity patch; (ii) a neighboring vorticity area that is considerably larger than the core vortex; (iii) a separation distance between the core vortex and the neighboring vorticity patch that is less than three to four times the core vortex radius.

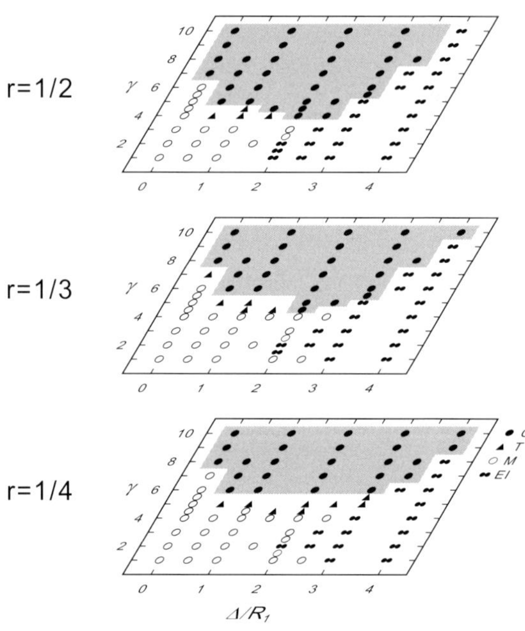

Figure 9. Summary of the Rankine vortex binary interaction regimes as a function of the dimensionless gap Δ/R_1, and the vorticity strength ratio $\gamma = \zeta_1/\zeta_2$ for the radius ratios $r = R_1/R_2 = 1/2, 1/3$, and $1/4$. As indicated by the code at the lower right, the structures are categorized as follows: elastic interaction (EI), merger (M), straining-out (S), tripole (T), concentric (C). The concentric regime is shadowed. (Adapted from K04.)

5. Concluding Remarks

There are observations and model simulations of binary tropical cyclone interactions that resemble the theoretical work of DW92 (see e.g. Larson, 1975; Lander and Holland. 1993; Kuo *et al.*, 2000; Khain *et al.*, 2000; Prieto *et al.*, 2003). We have enriched the DW92 dynamics with the tripole and concentric structures. We have used vorticity dynamics to explain the formation of the concentric eyewall structure. Implicitly assumed in our approach is that deep convection, such as that observed in Fig. 1, may be a signature of vorticity. It is reasonable to expect that the $(f + \zeta)\nabla \cdot u$ term in the vorticity equation allows vorticity to be generated by the lower tropospheric convergence associated with convection in the neighborhood of the vortex core. There is evidence that convection and vorticity are highly correlated. For example, Simpson *et al.* (1997) reported an elastic interaction (mutual rotation) of two 100 km scale convective systems and Hendricks *et al.* (2004) reported a "vortical hot tower" of deep convection that had a scale of several tens of kilometers. Zhang *et al.* (2005) also found that enhanced vorticity is associated with convective bands in MM5 simulations of the concentric eyewalls of Typhoon Winnie.

Since tropical cyclones often weaken after the formation of a concentric eyewall, intensity forecasting would obviously benefit from an improvement of concentric eyewall prediction. From the work presented here, it may be important to observe and understand the spatial and temporal characteristics of the vorticity field outside the cyclone core, as well as the detailed core structure, in order to better predict the formation of concentric eyewalls. The convection in the cyclone environment has been treated as an existing asymmetric vorticity in our model. The mesoscale vorticity-generating processes that determine the scale and strength of vorticity in the cyclone environment require further exploration. In this regard, there are several studies describing vorticity/convection initiation processes that can occur in concert with the organizational dynamics discussed here. For example, Montgomery and Kallenbach (1997) have identified the mechanism of radial propagation of linear Rossby waves in the presence of a critical radius outside the radius of maximum wind, which may be important for the formation of concentric eyewalls. Nong and Emanuel (2003) discussed the formation of concentric eyewalls in their axisymmetric model via an upper level external forcing-triggered finite amplitude WISHE instability.

In closing, we note that the appearance of tripoles in our study adds to a long list of papers in which these structures are documented. For example, tripoles emerge from unstable initial states in laboratory experiments with both rotating fluids (Kloosterziel and van Heijst, 1991; van Heijst *et al.*, 1991; Denoix *et al.*, 1994) and pure electron plasmas (Driscoll and Fine, 1990). They also are found in two-dimensional turbulence simulations as coherent structures (Legras *et al.*, 1988), as a result of collisions of two dipoles (Larichev and Reznik, 1983; Orlandi and van Heijst, 1992), as a result of finite amplitude quadrapolar (i.e., azimuthal wave number 2) distortions of a monopolar Gaussian vorticity distribution (Rossi *et al.*, 1997), as a result of the barotropic instability across the annular region separating a strong core vortex from a weaker vorticity ring (Kossin *et al.*, 2000), and as the end state of an initial vorticity distribution in which a low vorticity eye is uncentered within a region of high vorticity (Prieto *et al.*, 2001). Given their robustness and multitude of production methods, it is possible that tripoles may actually play a role in tropical cyclone dynamics. The tripoles may be an indication of incomplete mixing in the tropical cyclone core. This follows from the statistical mechanics arguments, such that tripoles are a restricted statistical equilibrium far from the end state of strong mixing (Robert and Rosier, 1997; Chavanis and Sommeria, 1998). The idea of incomplete mixing appears to be in agreement with our results that the tripoles serve as a demarcation for the merger and the concentric eyewall regimes.

Acknowledgements

We thank Wayne Schubert, Terry Williams, C.-P. Chang, Yu-Ming Tsai, and Li-Huan Hsu for their helpful comments and suggestions. This research was supported by the National Research Council of Taiwan through the grants NSC 96-2111-M-002-002, NSC94-2745-P-002-002, NSC95-2745-P-002-004 and MOTC-CWB-97-2M-01 to the National Taiwan University.

[Received 14 February 2007; Revised 22 May 2007; Accepted 22 May 2007.]

References

Balmforth, N. J., S. G. Llewellyn Smith, and W. R. Young, 2001: Disturbing vortices. *J. Fluid Mech.*, **426**, 95–133.

Batchelor, G. K., 1969: Computation of the energy spectrum in homogeneous two-dimensional turbulence. *Phys. Fluids Suppl. II*, **12**, 233–239.

Black, M. L., and H. E. Willoughby, 1992: The concentric eyewall cycle of Hurricane Gilbert. *Mon. Wea. Rev.*, **120**, 947c957.

Carton, X., G. R. Flierl, and L. M. Polvani, 1989: The generation of tripoles from unstable axisymmetric isolated vortex structure. *Europhys. Lett.*, **9**, 339–344.

Carton, X., and B. Legras, 1994: The life-cycle of tripoles in two-dimensional incompressible flows. *J. Fluid Mech.*, **267**, 53–82.

Chavanis, P. H., and J. Sommeria, 1998: Classification of robust isolated vortices in two-dimensional hydrodynamics. *J. Fluid Mech.*, **356**, 259–296.

Cushman-Roisin, B., 1994: *Introduction to Geophysical Fluid Dynamics*. Prentice Hall, Englewood Cliffs, New Jersey, 320 pp.

DeMaria, M., and J. C. L. Chan, 1984: Comments on "A numerical study of the interactions between two tropical cyclones." *Mon. Wea. Rev.*, **112**, 1643–1645.

Denoix, M.-A., J. Sommeria, and A. Thess, 1994: two-dimensional turbulence: The prediction of coherent structures by statistical mechanics. In *Progress in Turbulence Research*, H. Branover, and Y. Unger (eds.), AIAA, pp. 88–107.

Dodge, P., R. W. Burpee, and F. D. Marks Jr., 1999: The kinematic structure of a hurricane with sea level pressure less than 900 mb. *Mon. Wea. Rev.*, **127**, 987–1004.

Driscoll, C. F., and K. S. Fine, 1990: Experiments on vortex dynamics in pure electron plasmas. *Phys. Fluids*, **2**, 1359–1366.

Dritschel, D. G., 1989: On the stabilization of a two-dimensional vortex strip by adverse shear. *J. Fluid Mech.*, **206**, 193–221.

Dritschel, D. G., 1995: A general theory for two-dimensional vortex interactions. *J. Fluid Mech.*, **293**, 269–303.

Dritschel, D. G., and D. W. Waugh, 1992: Quantification of the inelastic interaction of unequal vortices in two-dimensional vortex dynamics. *Phys. Fluids*, **A4**, 1737–1744.

Hendricks, E. A., M. T. Montgomery, and C. A. Davis, 2004: The role of "vertical" hot towers in the formation of tropical cyclone Diana (1984). *J. Atmos. Sci.*, **61**, 1209–1232.

Khain, A., I. Ginis, A. Falkovich, and M. Frumin, 2000: Interaction of binary tropical cyclones in a coupled tropical cyclone–ocean model. *J. Geophys. Res.*, **105**, D17, 22337–22354.

Kloosterziel, R. C., and G. F. Carnevale, 1999: On the evolution and saturation of instabilities of two-dimensional isolated circular vortices. *J. Fluid Mech.*, **388**, 217–257.

Kloosterziel, R. C., and G. J. F. van Heijst, 1991: An experimental study of unstable barotropic vortices in a rotating fluid. *J. Fluid Mech.*, **223**, 1–24.

Kossin, J. P., W. H. Schubert, and M. T. Montgomery, 2000: Unstable interactions between a hurricane's primary eyewall and a secondary ring of enhanced vorticity. *J. Atmos. Sci.*, **57**, 3893–3917.

Kuo, H.-C., R. T. Williams, and J.-H. Chen, 1999: A possible mechanism for the eye rotation of Typhoon Herb. *J. Atmos. Sci.*, **56**, 1659–1673.

Kuo, H.-C., G. T.-J. Chen, and C.-H. Lin, 2000: Merger of tropical cyclones Zeb and Alex. *Mon. Wea. Rev.*, **128**, 2967–2975.

Kuo, H.-C., L.-Y. Lin, C.-P. Chang, and R. T. Williams, 2004: The formation of concentric vorticity structures in typhoons. *J. Atmos. Sci.*, **61**, 2722–2734.

Kuo, H.-C., and W. H. Schubert, 2006: Vortex interactions and the barotropic aspects of concentric eyewall formation. Preprints, 27th Conference on Hurricane and Tropical Meteorology, Monterey, CA, *Amer. Meteor. Soc.*, 6B.2.

Lander, M., and G. J. Holland, 1993: On the interaction of tropical-cyclone-scale vortices. I: Observations. *Quart. J. Roy. Meteor. Soc.*, **119**, 1347–1361.

Larichev, V. D., and G. M. Reznik, 1983: On collisions between two-dimensional solitary Rossby waves. *Oceanology*, **23**, 545–552.

Larson, R. N., 1975: Picture of the month – hurricane twins over the Eastern North Pacific Ocean. *Mon. Wea. Rev.*, **103**, 262–265.

Legras, B., P. Santangelo, and R. Benzi, 1988: High resolution numerical experiments for forced two-dimensional turbulence. *Europhys. Lett.*, **5**, 37–42.

Mallen, K. J., M. T. Montgomery, and B. Wang, 2005: Reexamining the near-core radial structure of the tropical cyclone primary circulation: Implications for vortex resiliency. *J. Atmos. Sci.*, **62**, 408–425.

McWilliams, J. C., 1984: The emergence of isolated coherent vortices in turbulent flow. *J. Fluid Mech.*, **146**, 21–43.

Montgomery, M. T., and R. J. Kallenbach, 1997: A theory for vortex Rossby waves and its application to spiral bands and intensity changes in hurricane. *Quart. J. Roy. Meteor. Soc.*, **123**, 435–465.

Nong, S., and K. A. Emanuel, 2003: A numerical study of the genesis of concentric eyewalls in hurricane. *Quart. J. Roy. Meteor. Soc.*, **129**, 3323–3338.

Orlandi, P., and G. J. F. van Heijst, 1992: Numerical simulations of tripolar vortices in 2d flow. *Fluid Dyn. Res.*, **9**, 179–206.

Polvani, L. M., and X. J. Carton, 1990: The tripole: a new coherent vortex structure of incompressible two-dimensional flows. *Geophys. Astrophys. Fluid Dyn.*, **51**, 87–102.

Polvani, L. M., and R. A. Plumb, 1992: Rossby wave breaking, microbreaking, filamentation, and secondary vortex formation: the dynamics of a perturbed vortex. *J. Atmos. Sci.*, **49**, 462–476.

Prieto, R., J. P. Kossin, and W. H. Schubert, 2001: Symmetrization of lopsided vorticity monopoles and offset hurricane eyes. *Quart. J. Roy. Meteor. Soc.*, **127**, 1–17.

Prieto, R., B. D. McNoldy, S. R. Fulton, and W. H. Schubert, 2003: A classification of binary tropical-cyclone-like vortex interactions. *Mon. Wea. Rev.*, **131**, 2656–2666.

Reasor, P. D., M. T. Montgomery, F. D. Marks, and J. F. Gamache, 2000: Low-wavenumber structure and evolution of the hurricane inner core observed by airborne dual-Doppler radar. *Mon. Wea. Rev.*, **128**, 1653–1680.

Reasor, P. D., M. T. Montgomery, and L. D. Grasso, 2004: A new look at the problem of tropical cyclones in vertical shear flow: vortex resiliency. *J. Atmos. Sci.*, **61**, 3–22.

Robert, R., and C. Rosier, 1997: The modeling of small scales in two-dimensional turbulent flows: a statistical mechanics approach. *J. Stat. Phys.*, **86**, 481–515.

Rossi, L. F., J. F. Lingevitch, and A. J. Bernoff, 1997: Quasi-steady monopole and tripole attractors for relaxing vortices. *Phys. Fluids*, **9**, 2329–2338.

Rozoff, C. M., W. H. Schubert, B. D. McNoldy, and J. P. Kossin, 2006: Rapid filamentation zones in intense tropical cyclones. *J. Atmos. Sci.*, **63**, 325–340.

Schubert, W. H., M. T. Montgomery, R. K. Taft, T. A. Guinn, S. R. Fulton, J. P. Kossin, and J. P. Edwards, 1999: Polygonal eyewalls, asymmetric eye contraction, and potential vorticity mixing in hurricanes. *J. Atmos. Sci.*, **56**, 1197–1223.

Shapiro, L. J., and H. E. Willoughby, 1982: The response of balanced hurricanes to local sources of heat and momentum. *J. Atmos. Sci.*, **39**, 378–394.

Shea, D. J., and W. M. Gray, 1973: The hurricane's inner core region. I. Symmetric and asymmetric structure. *J. Atmos. Sci.*, **30**, 1544–1564.

Simpson, J., E. A. Ritchie, G. J. Holland, J. Halverson and S. Stewart, 1997: Mesoscale interactions in tropical cyclone genesis. *Mon. Wea. Rev.*, **125**, 2643–2661.

van Heijst, G. J. R., R. C. Kloosterziel, and C. W. M. Williams, 1991: Laboratory experiments on the tripolar vortex in a rotating fluid. *J. Fluid Mech.*, **225**, 301–331.

Willoughby, H. E., J. M. Masters, and C. W. Landsea, 1989: A record minimum sea level pressure observed in Hurricane Gilbert. *Mon. Wea. Rev.*, **117**, 2824–2828.

Willoughbyd, H. E., J. A. Clos, and M. Shoreibah, 1982: Concentric eye walls, secondary wind maxima, and the evolution of the hurricane vortex. *J. Atmos. Sci.*, **39**, 395–411.

Zhang, Q.-H., S.-J. Chen, Y.-H. Kuo, K.-H. Lau, and R. A. Anthes, 2005: Numerical study of a typhoon with a large eye: Model simulation and verification. *Mon. Wea. Rev.*, **133**, 725–742.

Rare Typhoon Development near the Equator

C.-P. Chang

*Department of Atmospheric Sciences, National Taiwan University, Taipei, Taiwan and
Department of Meteorology, Naval Postgraduate School, Monterey, California, USA
cpchang@nps.edu*

Teo Suan Wong

Meteorological Services Division, National Environment Agency, Singapore

The formation of Typhoon Vamei on 27 December 2001 in the southern South China Sea was the first-observed tropical cyclogenesis within 1.5 degrees of the equator. This rare event was first detected by observations of typhoon strength winds from a US navy ship, and the existence of an eye structure was confirmed by satellite and radar imageries. This paper reviews these observations, and discusses the dynamic theory that may explain the process suggested by Chang *et al.* (2003) in which a strong cold surge event interacting with the Borneo vortex led to the equatorial development. As pointed out by Chang *et al.*, the most intriguing question is not how Vamei could form so close to the equator, but is why such a formation was not observed before then.

1. Introduction

One of the generally accepted conditions for tropical cyclone formation has been that the location is "away from the equator." This condition is based on the lack of Coriolis effect at the equator, and supported by observations over more than a century that show most tropical cyclogeneses to occur poleward of 5° latitude (Gray, 1968; McBride, 1995). The previous record was set by Typhoon Sarah in 1956 at 3.3°N (Fortner, 1958). Typhoon Vamei formed at 1.5°N at the southern tip of the South China Sea at 00 UTC 27 December 2001, a latitude that most textbooks (e.g. Anthes, 1982) ruled out for development. The cyclone was named by the Japan Meteorological Agency, which initially identified it as a tropical storm with estimated winds of $21\,\mathrm{m\,s^{-1}}$. It was upgraded

to a typhoon by the Joint Typhoon Warning Center (JTWC) in Hawaii. Figure 1 shows the best track and intensity of Vamei, published by the JTWC. The storm made landfall over southeast Johor at the southern tip of Peninsular Malaysia, about $50\,\mathrm{km}$ northeast of Singapore, at 0830 UTC 27 December 2001. Upon making landfall, it weakened rapidly to a tropical depression. It continued in its west–northwest track across southern Johor, the Malacca Straits, and made landfall again in Sumatra. Upon entering the Bay of Bengal, the storm regenerated and continued in its northwest track before dissipating in the central Bay of Bengal on 31 December 2001. During the short period of 12 h as a typhoon and another 12 h as a tropical storm, Vamei caused damage to two US Navy ships, including a carrier, and flooding

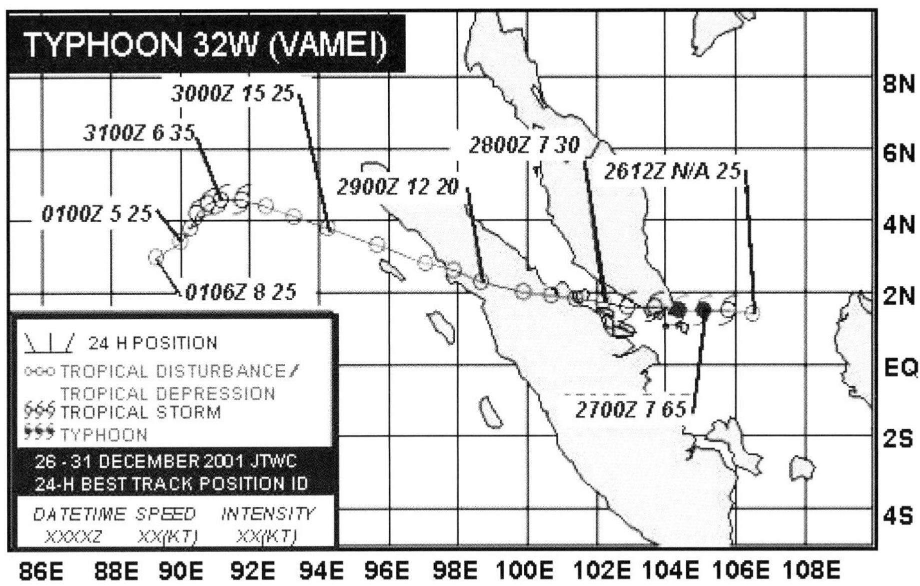

Figure 1. Best track and intensity of Vamei from 1200 UTC 12 December 2001 to 00 UTC 1 January 2002. (Diagram courtesy of JTWC.)

and mudslides in southern Peninsular Malaysia's Johor and Pahang states. More than 17,000 people were evacuated and 5 lives were lost.

The upgrade of Vamei to the typhoon category by the JTWC was based mainly on the shipboard observations from several US Navy ships within the small eyewall, with reports of sustained winds of $39\,\mathrm{m\,s}^{-1}$ and gusts of up to $54\,\mathrm{m\,s}^{-1}$. Because of its equatorial latitude, there was considerable interest among tropical cyclone forecasters regarding the typhoon's structure and the process of its development. This article will review the relevant data used to observe the development of Typhoon Vamei and discuss some theoretical considerations regarding its possible formation mechanism.

2. Background Flow and the Observed Development

Vamei developed in late December 2001, near the middle of the Asian winter monsoon season, which is characterized by strong baroclinicity in the middle latitudes and northeasterly winds at lower levels. Freshening of the northeasterly winds, or cold surges (Chan and Li, 2004; Chang et al., 2004, 2005), occur sporadically and spread equatorward. Although cold surge winds are typically dry, they are moistened by the over-water trajectory (Johnson and Houze, 1987) and have been associated with increased deep convection and enhanced upper-tropospheric outflow over the Maritime Continent, which is related to an enhanced East Asian local Hadley cell (Lau and Chang, 1987). The cold surge air can reach the equator in about two days (Chang et al., 1983). Conservation of potential vorticity causes the air to turn eastward after it crosses the equator. These Southern Hemisphere equatorial westerlies may enhance the Australian monsoon trough farther south, between 10°S and 20°S, where tropical cyclogenesis occurs frequently (e.g. Holland, 1984; McBride, 1995).

Synoptic-scale disturbances are also found to occur in the vicinity of the island of Borneo (Johnson and Houze, 1987; Chang et al., 2005).

Over this region, the low-level basic-state background vorticity is cyclonic, due to the mean northeasterly wind maximum over the South China Sea and the equatorial westerlies associated with the Asian winter monsoon. Therefore, perturbations in this basic state often amplify into synoptic-scale cyclonic circulations. These disturbances are often found southeast of the primary region of cold surge northeasterly winds. Often, the circulation is present as a quasi-stationary, low-level cyclonic circulation, which is a persistent feature of the boreal winter climatology (Johnson and Houze, 1987; Chang *et al.*, 2005). Although the circulation may not be completely closed on the east side over the island, it has been referred to as the Borneo vortex (Chang *et al.*, 2004, 2005). The mean location of the vortex along the northwest coast of Borneo may be seen in Fig. 2, which shows the 1999/2000–2001/2002.

December–February mean 850 hPa vorticity from the $1° \times 1°$ Navy Operational Global Atmospheric Prediction System (NOGAPS) analysis, overlaid with the surface vorticity derived from the QuikSCAT satellite scatterometer winds. The Borneo vortex is often associated with deep convection and intense latent heat release, and upper-level divergence is often present. However, because most of the time a significant part of the vortex circulation is over land (Fig. 3), even when a vortex drifts to northern Borneo between 5°N and 7°N, which are latitudes considered more favorable for tropical cyclone development, it is very difficult for the vortex to develop into a tropical cyclone (Chang *et al.*, 2003).

Chang *et al.* (2003) provided the following description of the synoptic events preceding the development of Vamei. Starting from 19 December 2001, a cold surge developed rapidly over the South China Sea while the center of the Borneo vortex was located near 3°N on the northwest coast (not shown). The 850 hPa NOGAPS wind analysis and vorticity in Fig. 4 depict the southwestward movement of

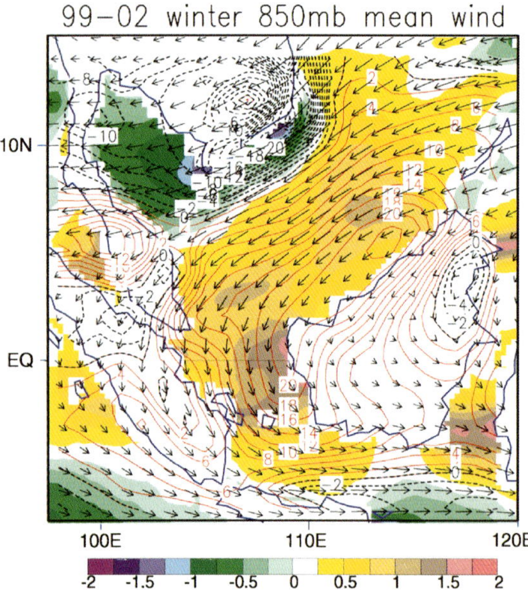

Figure 2. 1999/2000–2001/2002 boreal winter (DJF) mean of 850 hPa NOGAPS $1° \times 1°$ wind and vorticity (contours: solid — positive; dashed — negative; interval $2 \times 10^{-5}\,\mathrm{s^{-1}}$), and surface vorticity based on 25 km resolution QuikSCAT winds (yellow — positive; green — negative). [Diagram from Chang *et al.* (2003) by permission of American Geophysical Union.]

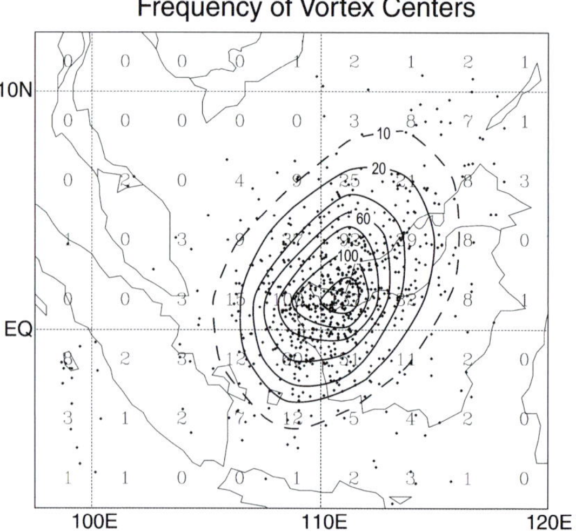

Figure 3. Analyzed Borneo vortex center locations based on streamlines of unfiltered 925 hPa winds. (NCEP/NCAR reanalysis winds at 925 hPa at $2.5° \times 2.5°$ grids, for 21 boreal winters (December 1980–February 2001.) [Diagram from Chang *et al.* (2005) by permission of American Meteorological Society.]

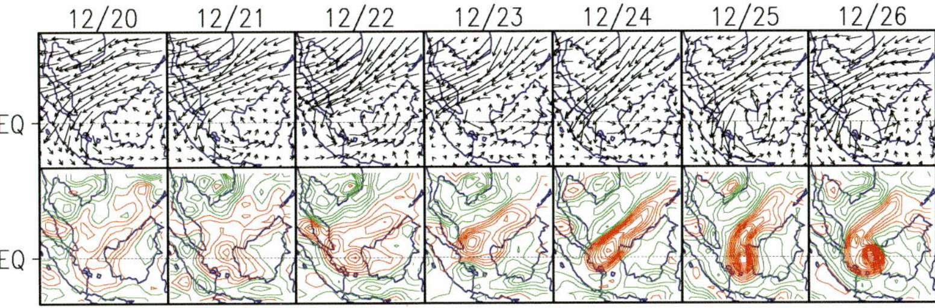

Figure 4. NOGAPS $1° \times 1°$ 850 hPa wind and vorticity (red — positive; green — negative) at 00 UTC 20–26 December 2001.

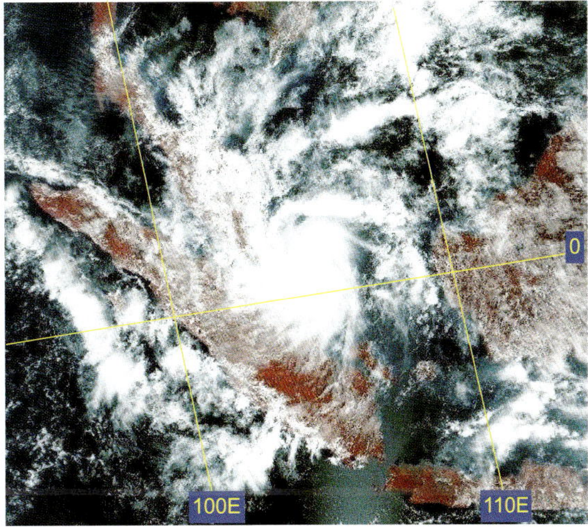

Figure 5. MODIS satellite image on 27 December 2001, showing Typhoon Vamei near Singapore. (Diagram courtesy of Professor Lim Hock, National University of Singapore.)

the vortex from along the Borneo coast toward the equator. By 21 December, the center of the vortex had moved off the coast over water, where the open sea region at the southern end of the South China Sea narrows to about 500 km, with Borneo to the east and the Malay Peninsula and Sumatra to the west. This over-water location continued for several days. While the vortex center remained in the narrow equatorial sea region, the strong northeasterly surge persisted, and was slightly deflected to the northwest of the vortex. This near "trapping" of the Borneo vortex by a sustained surge is unusual, because normally the vortex center would be pushed eastward by the strengthening surge that streaks southwestward in the middle of the South China Sea. Consequently, the cross-equatorial flow wrapped around the vortex and provided a background area of cyclonic relative vorticity with a magnitude of $>1 \times 10^{-5}\,\mathrm{s}^{-1}$, which is comparable to that of the Coriolis parameter $5°$ or more away from the equator.

Figure 5 shows the MODIS satellite image on 27 December 2001. Vamei's circulation center can be estimated to be just north of $1°$N, but an eye is not observable under the clouds. Even though the size of the typhoon is quite small, which is considered a special characteristic of low latitude TCs by some researchers

Figure 6.   Japanese Geostationary Meteorological Satellite image at 0232 UTC 27 December 2001. All national weather services in the region reported it as a tropical storm.

(e.g. DeMaria and Pickle, 1988), the spiral cloud bands emanating out from around the center clearly indicate that the storm circulation was on both sides of the equator. Figure 6 shows the Japanese Geostationary Meteorological Satellite image at 0232 UTC of the same day. Feeder bands from both sides of the equator spiral into the center of Vamei, where a small eye is visible. An eye was also observed in TRMM and SSM/I images within the preceding two hours (not shown). The diameter of the eye estimated from different sensors ranges from 28 km to 50 km. Vamei's small size as it formed at the southern end of the South China Sea made it difficult to observe its highest wind speed from ground-based observations or to estimate its intensity from satellite images. As a result, all national weather services in the region reported it as a tropical storm. Without the chance passage by the USS Carl Vinson carrier group through its

eyewall, the JTWC may not be able to operationally upgrade the intensity of the storm to that of a typhoon either.

Figure 7 shows Doppler weather radar images from Singapore's Changi Airport during the 12 hours prior to the arrival of Vamei. The rapid development of the eye of Vamei can be readily seen in 3-hour intervals. The eye was just starting to form with an irregular boundary when the storm moved into radar range at 1930 UTC 26 December 2001 (right panel). It became quite well organized 3 hours later, at 2230 UTC (middle panel), with a geometric center of the eye near 1.4°N. By 0130 UTC 27 December (left panel), the eye had become a nearly symmetric round feature.

Evidence of the strength of Vamei and its relationship with the cold surge can be revealed from QuickSCAT satellite scatterometer wind data. Figure 8 shows that the QuickSCAT wind

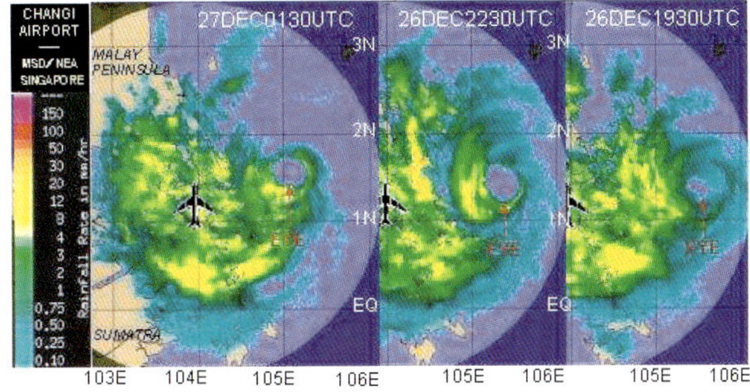

Figure 7. Changi Airport (Singapore) Doppler weather radar images (the color legend indicates estimated rain rates, in mm/h) in three-hour intervals, with the time sequence of the images organized from right to left: 1930 UTC 26 December (right panel), 2230 UTC (middle panel) and 0130 UTC 27 December (left panel).

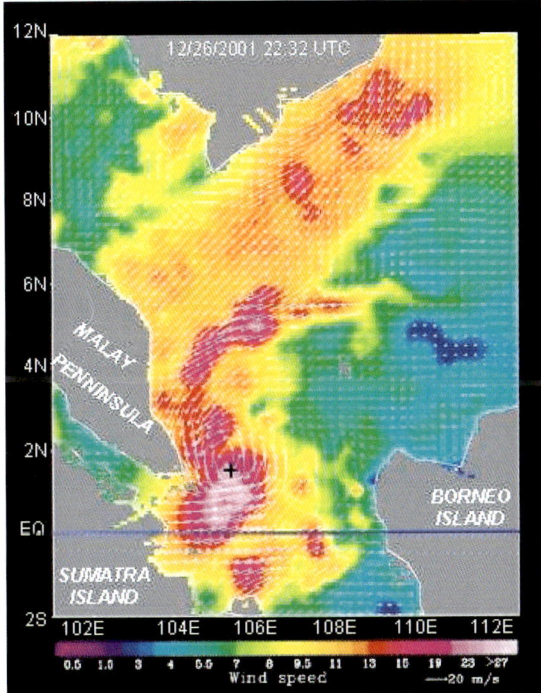

Figure 8. QuickSCAT satellite scatterometer wind direction and speed (color shading and arrow length) at 2232 UTC 26 December 2001, showing the typhoon strength of Vamei and the remnant of the continuing surge wind upstream in the northern South China Sea. See text for details. (Diagram courtesy of Jet Propulsion Laboratory/NASA.)

direction and speed at 2232 UTC 26 December 2001 captured both the signal of Vamei as it developed to typhoon strength, and the remnant of the continuing surge wind upstream in the northern South China Sea. At the southern perimeter, the wind speed at a 10 m height has already reached above $27\,\mathrm{m\,s^{-1}}$ over an area of about $1°$ latitude $\times$ $1°$ longitude. The northern spiral band extends to about $6°$N and is detached from the cold surge wind belt further north.

3. Roles of the Winter Monsoon and Possible Mechanisms

Based on the synoptic sequence of the low-level circulations, Chang *et al.* (2003) suggested that Vamei formed as a result of an interaction between two prominent features of the Asian winter monsoon: a weak Borneo vortex that drifted into, and remained at, the southern tip of the South China Sea; and a strong and persistent cold surge that created the large background cyclonic vorticity at the equator.

A similar equatorial generation process was proposed two decades ago in the cold surge theory of Lim and Chang (1981), who used the framework of the equatorial beta-plane equatorial wave theory. In their barotropic theory, geostrophic adjustment and potential vorticity conservation following a cross-equatorial surge spin up counterclockwise rotation to the east of the surge axis, where in the real world the Borneo vortex is located. A comparison of Lim and Chang's cold surge theory and the observed low-level flow during the development

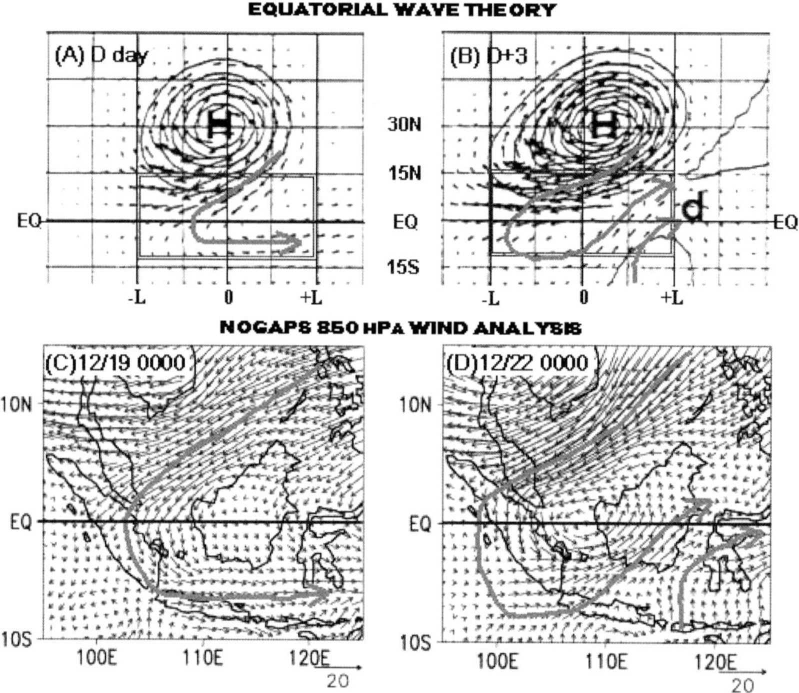

Figure 9. Comparison of Lim and Chang's (1981) barotropic equatorial beta-plane cold surge theory (panels A and B) and observed NOGAPS 850 hPa wind analysis (panels C and D), each for two time periods separated by three days. Because the narrow width of the South China Sea confined the width of the intense surge belt to about one half of that in the terrain-free equatorial beta-plane solution, a comparison of the theory and the actual development may be made by scaling the east–west dimension of the upper panels to one half of the original size (L = 15° longitude instead of 30°), or treating the highlighted rectangular area in panels A and B as being comparable to the domain of the NOGAPS plots. The location of the high center in panel B was also shifted eastward by 0.4 L to account for the reduced zonal scale, the typical eastward movement of the East Asian surface high center, and the geographical restriction of the surge belt by the South China Sea. See text for details. [Diagram adapted from Lim and Chang (1980) and Chang *et al.* (2003) by permission of American Meteorological Society and American Geophysical Union, respectively.]

of Typhoon Vamei is shown in Fig. 9, where the top panels (A and B) display the theoretical solutions three days apart in a pressure-induced surge, and the bottom panels (C and D) display the NOGAPS 850 hPa wind analysis for 19 December and 22 December 2007, respectively.

Panel A shows the theoretical solution of a case of an equatorward surge that is initiated by a high-pressure anomaly centered at 30°N, with no mean flow. The pattern resembles the typical cold surge event that follows the southeastward movement of an East Asian surface high center with the development of a northeast–southwest tilt. This tilt is due to the dispersive properties of equatorial beta-plane Rossby waves in which the lower meridional modes have larger amplitudes closer to the equator, and therefore propagate westward more quickly. As the northeasterly wind strengthens south of the high center, it streams southward, and after crossing the equator, it turns eastward between the equator and 15°S. Panel B shows the solution three days later, in which the northeast–southwest tilt becomes even more pronounced. To the southeast of the northeasterly surge streak, southwesterly cross-equatorial winds produce a wave (area d) as they swing back south to merge with the equatorial easterlies. The area between the surge streak and area d is a northeast–southwest-oriented counter-clockwise circulation belt over the equator. The flow pattern (west of the equatorial easterlies) is mainly the manifestation of a dispersive Rossby wave group. The lower panels show the NOGAPS 850 hPa wind analysis at the beginning of the actual cold surge (panel C; 0000 UTC 19 December), and three days later (panel D; 0000 UTC 22 December). Because of the narrow width of the South China Sea, the width of the intense surge belt is confined to about 750 km, which is approximately one half of that in the equatorial beta-plane solution in panel B which is not subject to any terrain restriction. Thus, a comparison of the theory and the actual development may be made by scaling the east–west dimension of the upper panels to one half of the original size (L = 15° longitude instead of 30°), or treating the highlighted rectangular area in panels A and B as being comparable to the domain of the NOGAPS plots. The location of the high center in panel B was also shifted eastward by 0.4 L to account for the slower propagation due to the reduced zonal scale and two factors in real cold air outbreak events: the eastward movement of the East Asian surface high center due to the westerly mean flow, and the fixed location of the surge belt that is restricted geographically by the South China Sea.

4. Concluding Remarks: Key Question Posed by Typhoon Vamei

Since the observation of Typhoon Vamei, a number of modeling studies have successfully simulated this case of equatorial formation (e.g. Chambers and Li, 2007; Juneng et al., 2007; Koh, 2006). This is not surprising, since the interaction between the winter monsoon circulation and the complex terrain and the moisture from the warm ocean surface provided the vorticity and latent heat sources for development. However, the strong cold surge and Borneo vortex that led to the development of Typhoon Vamei are both regularly observed, major systems of the Asian winter monsoon in the South China Sea, and abundant low-level warm and moist air is present every winter. So the most interesting question is more than just how or why Typhoon Vamei could form so close to the equator, but rather, why more typhoon formation was *not* observed in the equatorial South China Sea.

Chang *et al.* (2003) postulated that the answer lies in the narrow extent of the equatorial South China Sea. Prior to Typhoon Vamei's formation, a strong cold surge persisted for nearly one week over the narrowing South China Sea, providing a source for background cyclonic vorticity as the surge wind crossed the equator. The anomalous strength and persistence of this surge was related to the anomalously strong meridional gradient of sea-level pressure in the equatorial South China Sea during December 2001 (*Bureau of Meteorology Northern Territory Region*, 2002). The narrowing of the South China Sea at the equator plays two counteracting roles that combine to make the occurrence of the typhoon formation possible but rare. On the one hand, the channeling and strengthening of the cross-equatorial surge winds helps to produce the background cyclonic vorticity at the equator. On the other hand, the open water region of approximately 5° longitude is just sufficient to accommodate the diameter of a small tropical cyclone. However, it is too small for most synoptic-sized disturbances to remain over the water for more than a day or so. In the unusual case of Typhoon Vamei, the durations of the intense cold surge and the Borneo circulation remaining over water were both significantly longer than normal, which allowed the interaction to continue for nearly a week until the storm was formed.

In an analysis of the NCEP/NCAR reanalysis during the boreal winters of

1951/52–2001/02, Chang *et al.* (2003) found that a total of 61 strong surge events lasting one week or more in the southern South China Sea occurred. The total number of days under these persistent surges is 582. Assuming that the vortex needs at least a 3-day overlap with the surge to develop, the sustained cyclogenesis due to a strong background relative vorticity is estimated to be present at the equator on about 10% of the boreal winter days. If the minimum persistent surge duration required is reduced from 7 days to 5, the available time of the spinning top effect is increased to 14%. During the 51 boreal winters, the frequency of a pre-existing Borneo vortex staying over the equatorial water continuously for 4 days or more is 6, or a probability of 12% in a given year. Whether a pre-existing disturbance develops into a tropical cyclone depends on background vertical shears of wind and vorticity, upper level divergence, and a variety of environmental factors (Anthes, 1982; McBride, 1995). In the more favored tropical cyclone basins of the western Pacific and North Atlantic, the percentage of pre-existing synoptic disturbances developing into tropical cyclones during their respective tropical cyclone seasons ranges between 10% and 30%. Thus, of all the conditions that led to the formation of Vamei, Chang *et al.* (2003) estimated the probability of an equatorial development from similar conditions to be about once in a century or longer. This estimate appears consistent with the history of observations. However, it is not known whether other near-equatorial developments have occurred but were not observed during the presatellite era.

Acknowledgements

Interesting discussions were provided by H. Lim and S. L. Woon (Singapore); S. H. Ooi and Y. F. Hwang (Malaysia); C.-H. Liu and H.-C. Kuo (Taiwan); and R. Edson and M. Lander (Guam). This work was supported by ONR contract N0001402WR20302, NSF Grant ATM0101135 and the National Science Council/National Taiwan University.

[Received 15 August 2007; Revised 19 January 2008; Accepted 22 January 2008.]

References

Anthes, R. A., 1982: *Tropical Cyclones: Their Evolution, Structure and Effects*, American Meteorological Society Boston, 208 pp.

Chan, J., and C. Li, 2004: The East Asia winter monsoon. In *East Asian Monsoon. C.-P. Chang* (ed.), *World Scientific Series on Earth System Science in Asia*, Vol. 2, pp. 54–106.

Chambers, C. R. S., and T. Li, 2007: Simulation of a near-equatorial typhoon Vamei (2001). *Meteorol. Atmos. Phys.*, **97**, 67–80.

Chang, C.-P., J. E. Millard, and G. T. J. Chen, 1983: Gravitational character of cold surges during Winter MONEX. *Mon. Wea. Rev.*, **111**, 293–307.

Chang, C.-P., C.-H. Liu, and H.-C. Kuo, 2003: Typhoon Vamei: an equatorial tropical cyclone formation. *Geophys. Res. Lett.*, **30**, 1151–1154.

Chang, C.-P., P. A. Harr, J. McBride, and H. H. Hsu, 2004: Maritime Continent monsoon: annual cycle and boreal winter variability. *East Asian Monsoon*, C.-P. Chang (ed.), *World Scientific Series on Earth System Science in Asia*, Vol. 2, pp. 107–150.

Chang, C.-P., P. A. Harr, and H. J. Chen, 2005: Synoptic disturbances over the equatorial South China Sea and western Maritime Continent during boreal winter. *Mon. Wea. Rev.*, **133**, 489–503.

DeMaria, M., and J. D. Pickle, 1988: A simplified system of equations for simulations of tropical cyclones. *J. Atmos. Sci.* **45**, 1542–1554.

Fortner, L. E., 1958: Typhoon Sarah, 1956. *Bull. Amer. Meteor. Soc.*, **39**, 633–639.

Gray, W. M., 1968: Global view of tropical disturbances and storms. *Mon. Wea. Rev.*, **96**, 669–700.

Holland, G. J., 1984: On the climatology and structure of tropical cyclones in the Australian Southwest Pacific Region. *Aus. Meteor. Mag.*, **32**, 17–31.

Johnson, R. H., and R. A. Houze, Jr., 1987: Precipitating clouds systems of the Asian monsoon. In *Monsoon Meteorology*, C.-P. Chang and T. N.

Krishnamurti (eds.), Oxford University Press, pp. 298–353.

Juneng, L., F. T. Tangang, C. J. Reason, S. Moten, and W. A. W. Hassan, 2007: Simulation of tropical cyclone Vamei (2001) using the PSU/NCAR MM5 model. *Meteorol. Atmos. Phys.*, **97**, 273–290.

Koh, T.-Y., 2006: Numerical weather prediction research in Singapore and case study of tropical cyclone vamei. In *Winter MONEX: A Quarter Century and Beyond* (WMO Asian Monsoon Workshop, 4–7 April 2006, Kuala Lumpur, Malaysia).

Lau, W. K.-M., and C.-P. Chang, 1987: Planetary scale aspects of winter monsoon and teleconnections. In *Monsoon Meteorology*, C.-P. Chang and T. N. Krishnamurti (eds.), Oxford University Press, pp. 161–202.

Lim, H., and C.-P. Chang, 1981: A theory for mid-latitude forcing of tropical motions during winter monsoon. *J. Atmos. Sci.*, **38**, 2377–2392.

McBride, J. L., 1995: Tropical cyclone formation. In *Global Perspective on Tropical Cyclones*, R. L. Elsberry (ed.), Tech. Docu. 693, World Meteorological Organization, pp. 63–105.

Assimilation of Satellite Ocean Surface Winds at NCEP and Their Impact on Atmospheric Analyses and Weather Forecasting

Tsann-Wang Yu

NOAA Center for Atmospheric Sciences, Howard University, Washington, D.C., USA
tyu@Howard.edu

The improvement in the forecasting skills of short range numerical weather prediction is attributable to two major advances since the inception of operational numerical weather prediction in the late 1950's: the advent of satellite remote sensing since the 1970's, which has vastly increased the number of observations in the operational database, and the improvement in the treatment of model physics, numerics and spatial resolution of the global forecast models, and their associated atmospheric data analysis and data assimilation systems, which are capable of effectively using the vast amounts of satellite data observations for NWP operations. The evolution of the operational global data assimilation systems at NCEP is briefly described, followed by a review of various satellites during the last two decades that have been designed to provide global coverage of ocean surface winds. These include SEASAT, NSCAT, ERS-1/2, SSM/I and the most current operational satellite, QuikSCAT. On board these satellites, data characteristics of two microwave instruments for measuring the ocean surface winds from both an active scatterometer and a passive radiometer are discussed. The procedures for effectively using these satellite ocean surface wind data in the global data assimilation experiments, and that for assessing the impact of any particular data set, are described. Results of preimplementation impact investigations on the use of these satellite surface winds are discussed, and based on the results of these investigations, these data are implemented in NCEP's NWP operations. It is fair to state that from the gross statistics based on many cases of forecasts, the impact of satellite ocean surface winds on the short range NWP forecasts is mostly positive and significant, albeit small because the data are of a single level nature and are available only over the ocean surface. A case study is presented which shows that the use of satellite scatterometer ocean surface winds has a significantly large positive impact on the storm intensity and circulation over the southwestern Pacific Ocean.

1. Introduction

The forecasting skills of short range numerical weather prediction (NWP) have been steadily improving since the beginning of the NWP operations in the late 1950's. These improvements are undoubtedly attributable to improvements in model physics and numerics, increase in spatial resolutions, and advances in the fields of satellite remote sensing, atmospheric analysis and data assimilation. At operational NWP centers such as the National Center for Environmental Prediction (NCEP) and the European Center for Medium Range Weather Forecasts, (ECMWF), the numbers of conventional and satellite data used in a synoptic cycle analysis are typically more than 2 million, of which the majority (more than 90%) are from satellite observations. Figures 1 and 2 respectively show 500 mb geopotential height anomaly correlations for NCEP (Lord, 2004) and ECMWF

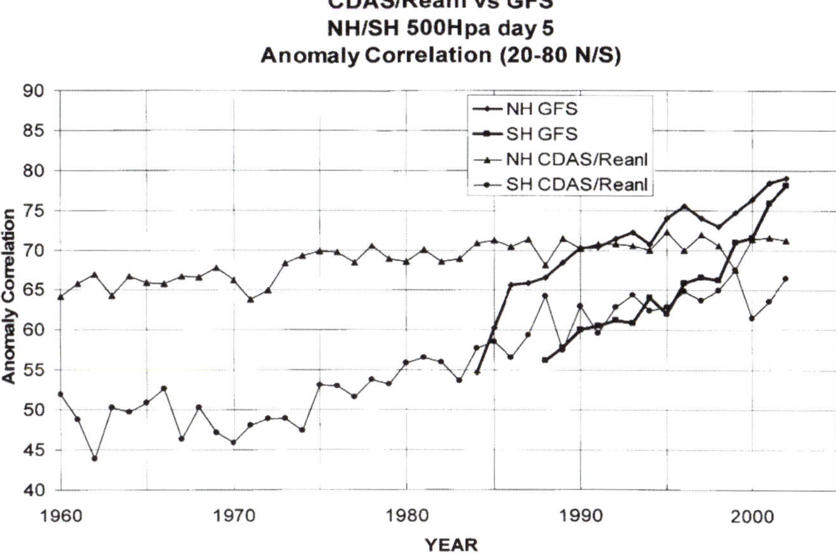

Figure 1. Evolution of anomaly correlations for 500 hPa geopotential height forecasts at NCEP for the northern and southern hemispheres.

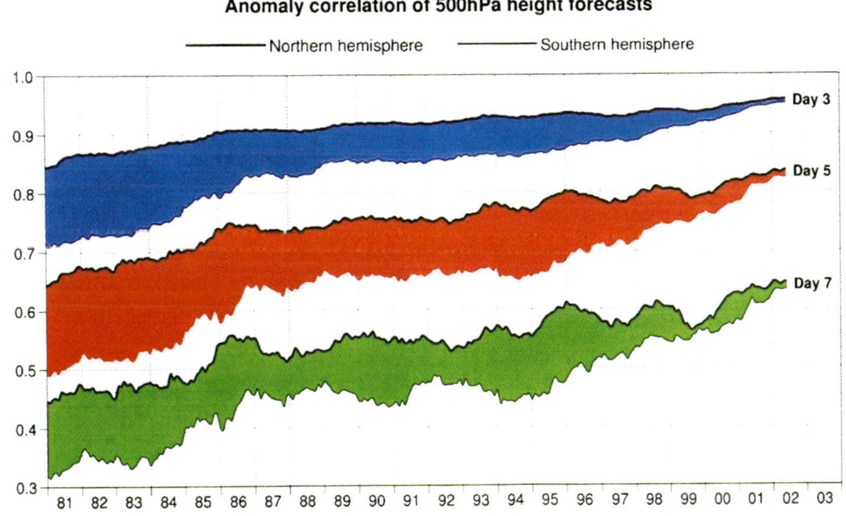

Figure 2. Evolution of anomaly correlations for 500 hPa geopotential height forecasts at ECMWF for the northern and southern hemispheres.

(Hollingsworth, 2004). Both figures demonstrate the current state of NWP forecasting skills, and convey the main conclusion that forecasting skills at Day 5 have steadily been improved for both the northern and southern hemispheres at these two major NWP operational centers over the last 40 years. Furthermore, they show that in the early days of NWP, owing to a lack of data in the southern hemisphere, the forecasting skill in the SH is far less than that skill in the NH. However, this gap in the forecasting skill has been considerably narrowed between the two hemispheres owing to an increase in satellite observations during the last few decades.

The advent of vast numbers of satellite data introduces a critical issue regarding the design of a viable atmospheric analysis and data assimilation scheme in order to most effectively use these satellite data for NWP. In the early days of NWP operations during the 1960's, the Cressman analysis scheme (Cressman, 1958) was used operationally at NMC (National Meteorological Center, the predecessor of NCEP). It was designed mainly for its efficiency. However, the main deficiency of the Cressman scheme is its arbitrariness of weights assigned to each observation as a function of distance between the observation and the grid point to be analyzed. With the rapid improvement in the computing speed and data storage during the last two decades, the current operational atmospheric analysis methods have evolved from the Cressman scheme during the 1960's and 1970's, and the Optimum Interpolation scheme (Gandin, 1963) during the 1980's and early 1990's, to the current variational analysis schemes (Sasaki, 1972; Derber and Parrish, 1993) from the late 1990's to the present. Most of these schemes are designed so that they can make use of various kinds of global observations and take into account unique characteristics for each type of satellite remotely sensed data.

Remote sensing satellite data can basically be vertical profile data, such as temperature and humidity soundings from TOVS (TIROS Operational Vertical Sounder); or surface level data, such as ocean surface winds, sea surface temperatures, albedo, and vegetation indices. Satellite data can also include many other types, such as those on vertically integrated total precipitable water, and cloud-tracked winds, etc. In general, profile data on temperatures and humidity are retrieved from the observed values of radiance from satellites. The radiance data, when used directly in the atmospheric variational analysis, are found to be most useful for NWP operations owing to their great coverage in three spatial dimensions. Single level surface wind data from various satellites during the last 20 years or so generally fall into two categories: one category contains ocean surface vector winds retrieved from active scatterometer measurements, and the other category contains wind speed information only observed from passive radiometer measurements. Owing to the wind directional ambiguity problems associated with the scatterometer wind data, and the incomplete wind information (only wind speed information, but no wind directions) associated with the passive radiometer wind data, methodologies designed to use these single level satellite surface wind data in the atmospheric analysis and data assimilation systems are different from those for the use of radiance data. A brief discussion on the evolution of the atmospheric analysis schemes at operational NWP centers such as NCEP and ECMWF is given in Sec. 2, with a particular emphasis on data assimilation methodologies designed for the use of the single level surface wind data from satellites.

The use of satellite ocean surface winds in the operational global data assimilation systems (GDAS's) at NCEP and ECMWF has been found to have a positive impact on short range numerical weather forecasts. The first satellite to measure surface wind vectors over the oceans was SEASAT, which was launched in 1978, and had on board both an active scatterometer and a passive multichannel scanning radiometer, among other instruments. This was a proof-of-concept oceanographic satellite, which for the first time demonstrated that global ocean surface wind fields could be measured by a remote sensing technology. Following SEASAT, many other satellites have been launched, such as GEOSAT, ERS-1/2, SSM/I, NSCAT, and QuikSCAT, which were designed for observing ocean surface winds. Satellite ocean surface winds from ERS-1/2 were used operationally at NCEP and in ECMWF NWP operations during the late 1980's, while QuikSCAT winds are currently being used operationally at both centers. During the last two decades or so, there have been a considerable number of global data assimilation experiments conducted to test the impact of these satellite surface wind data before their implementation in the NWP operations. Some of the recent preimplementation impact experiments conducted at NCEP, NASA, and ECMWF on the use of ERS-1/2 and QuikSCAT wind data in various GDAS's will be presented in Sec. 3. However, only gross statistics over a large number of analysis and forecasts of these preimplementation tests will be presented. For details of impact investigations into the use of satellite ocean surface winds, see Baker *et al.* (1984), Yu and McPherson (1984), Duffy *et al.* (1984), Yu (1987), Hoffman *et al.* (1990), Stoffelen and Cats (1991), Ingley and Bromley (1991), Hoffman (1993), Thepaut *et al.* (1993), Yu *et al.* (1994), Yu and Derber (1995), Yu *et al.* (1996), Stoffelen and Anderson (1997), Yu *et al.* (1997), Andrews and Bell (1998), Yu (2000), Atlas *et al.* (2001a), Atlas *et al.* (2001b), Yu (2003), and many others.

The main purpose of this article is to review recent progress in the use of satellite ocean surface wind data in atmospheric analysis and data assimilation, to discuss the impact and current status of operational applications of these wind data in NWP, and to demonstrate that in some synoptic cases where applications of satellite ocean surface winds can lead to very significant improvements in the ocean surface wind and sea level pressure analyses over the southwestern Pacific region. The outline of this article is as follows. In Sec. 2, the instrument error characteristics and data coverage of various satellites of the last two decades designed for measuring ocean surface winds are described. Then, a brief account of the evolution of the atmospheric analysis schemes at NCEP is given, with special emphasis on the

use of satellite ocean surface winds and their quality control procedures in the global data assimilation experiments. Section 3 discusses results of impact investigation related to various satellite ocean surface wind data, such as ERS-1/2, SSM/I, and QuikSCAT wind data, followed by the current status on the use of satellite ocean surface winds in NCEP and other operational centers' NWP operations. In Sec. 4 a synoptic case is discussed where use of satellite surface winds shows the most significant impact.

2. Assimilation of Satellite Ocean Surface Winds in Atmospheric Analyses

2.1. *Characteristics of satellite ocean surface winds*

As stated earlier, the two microwave instruments that can measure ocean surface winds are the scatterometer and the radiometer. The main principle of these two instruments lies in their ability to detect very small scales of motions of gravity and capillary waves over the ocean surface, from which surface winds can be deduced. Scatterometer measurements of winds contain both wind speed and direction, with wind directions consisting of as many as four values, of which only one is correct. This wind direction ambiguity problem is inherent in the scatterometer measurements for all the follow-up satellites, such as ERS-1/2 from the European Space Agency during the late 1980's, and most recently NASA's QuikSCAT during the late 1990's. On the other hand, radiometer measurements from SEASAT and various follow-up satellites such as GEOSAT during the 1980's, and SSM/I (Special Sensor Microwave Imager) of the Defense Meteorological Satellites Program (DMSP) during the 1990's and up to the present time, contain only wind speed, without wind direction information.

Satellite ocean surface wind specifications for various platforms such as radiometer

Table 1.   Satellite ocean surface wind specifications for various platforms.

Satellite	DMSP f11, f13, f14	ERS-1/2	QuikSCAT
Sensor	Passive microwave	Active microwave	Active microwave
Number of antennas	7 channels	3 antennas	4 antennas
Observed field of view	Scanning	1-sided	2-sided
Frequency/Bands	19 (H,V), 22V, 37(H,V), 85(H,V) GHZ	C-band radar	Ku-band radar
Swath	1492 km	500 km	1800 km
Number of cells	64	19	72
Footprint	25 km	50 km	25 km
Speed range	3–25 m/s	4–24 m/s	4–25 m/s
Speed accuracy	2 m/s, for greater than 20 m/s (10%)	2 m/s, for greater than 20 m/s (10%)	2 m/s, for greater than 20 m/s (10%)
Direction accuracy	No directions	$+/- 20°$	$+/-20°$

measurement from DMSP and scatterometer measurements from ERS-1/2 and QuikSCAT, are shown in Table 1. It is important to note that the accuracy of satellite ocean surface winds is about 2 m/s or 10% for wind speed, and about 20° for wind direction for both the scatterometer and radiometer measurements. These wind error statistics are borne out by validation studies which compare satellite-derived surface winds with *in situ* buoy reports (e.g. Gemmill *et al.*, 1999). From Table 1, one can also see that the resolution of satellite ocean surface winds ranges from 25 km to 50 km, and the coverage of the satellite swath varies, ranging from about 500 km for ERS-1/2, 600 km for NS CAT, to 900 km for QuikSCAT on both sides of the spacecraft. Figure 3 shows typical satellite swaths for these three satellites. As such, during a 6-hour window within a synoptic analysis cycle, the number of satellite ocean surface wind data can range from several thousands to hundreds of thousands of data points, depending on the resolution of the footprint design for any particular satellite. When compared with the conventional marine observations of ships and buoys, which are typically less than 1000 for a synoptic analysis cycle, these satellite ocean surface winds constitute the most important source of wind data to fill the void of the global oceans.

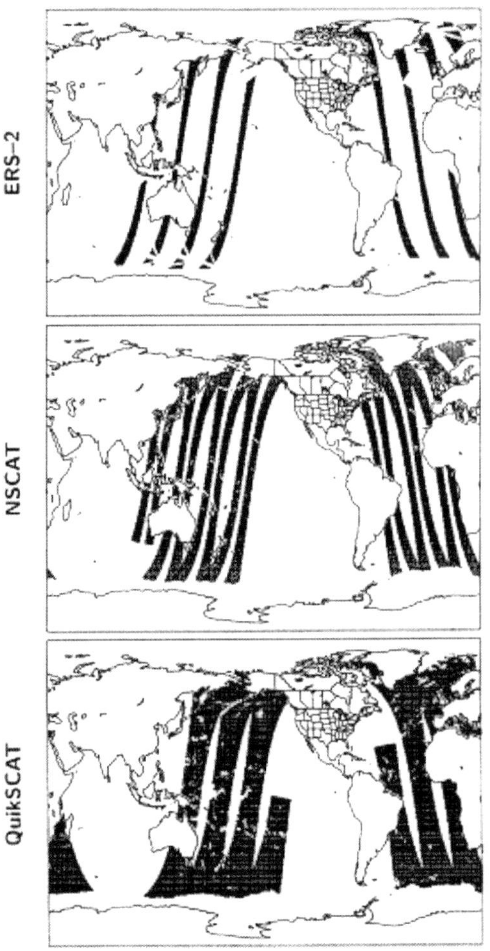

Figure 3.   Scatterometer data coverage in six hours during a synoptic analysis cycle for ERS-2 (top panel), NSCAT (middle panel), and QuikSCAT (bottom panel).

2.2. *Atmospheric analysis and data assimilation systems*

In the early days of NWP operations during the late 1950's and the 1960's, with limited computer speed and storage space, the Cressman analysis (Cressman, 1958) scheme was the operational scheme at NMC (National Meteorological Center, the predecessor of NCEP) and other NWP operational centers. The Cressman analysis scheme is a basically two-dimensional spatial interpolation scheme, and therefore it is a very efficient scheme. However, it suffers a deficiency in its arbitrariness of weights assigned to each observation as a function of the distance between observations and grid points to be analyzed. With the rapid advance of computer power and storage during the 1970's and 1980's, the optimum interpolation (OI) scheme (Gandin, 1965) replaced the Cressman scheme and became the main operational atmospheric analysis scheme at most of the operational NWP operational centers in the world. Then, during the early 1990's, owing to the vast improvement in computer power, it became feasible to apply the variational analysis principles of Sasaki (1970) to a viable atmospheric analysis scheme (Le DiMet and Telegrand, 1986), including some of the attributes of the OI scheme for treating the observations from various satellites. The main advantage of OI and variational analysis schemes lies in their ability to handle a variety of satellite data with various error attributes, including ocean surface wind data retrieved from scatterometers and radiometers.

The NCEP global operational forecasting model was changed from a grid point model to a spectral model in 1980, starting with R30 resolution and 12 vertical levels (Sela, 1980), hereafter referred to as (R30, L12). With the increase of computer power, the NCEP operational spectral model has improved both the horizontal and vertical resolutions throughout the last two decades, with the most current operational spectral model having a configuration of T354 and L64, respectively, for its

Table 2. Evolution of NCEP global models and atmospheric analysis schemes from the 1980's to the present time.

Horizontal Resolution	Vertical Resolution	Analysis Scheme	Date of Operations
R30	L12	OI	August 1980
T80	L18	OI	August 1987
T126	L18	3D-VAR	March 1991
T126	L28	3D-VAR	August 1993
T170	L42	3D-VAR	January 2000
T254	L64	3D-VAR	October 2002
T384	L64	3D-VAR	May 2005

horizontal and vertical resolution (see Table 2). Furthermore, as mentioned, by the time the ERS-1 scatterometer wind data were ready for a preimplementation test in the early 1990's, the atmospheric analysis scheme also changed from an OI scheme into a spectral statistical interpolation scheme (Derber and Parrish, 1991; Parrish and Derber, 1992), which is essentially a three-dimensional variational analysis scheme (3D-VAR). It is important to note that the beginning of the operational use of satellite ocean surface winds from ERS-1/2, and SSM/I wind speed data, occurred in the 1990's, and the current operational NASA QuikSCAT wind data were implemented in early 2002. Thus, 3D-VAR has played a very critical role in the use of these satellite surface wind data in the NCEP operational GDAS. A similar 3D-VAR scheme (Courtier *et al.*, 1998; Rabier *et al.*, 1998) was designed for using the ERS-1/2 wind data in ECMWF's GDAS during the late 1990's and early 2000's. However, it should be noted that currently a new 4D-VAR scheme (Rabier *et al.*, 1997) has been implemented for the use of QuikSCAT wind data in ECMWF's GDAS.

As stated earlier, satellite wind data from the scatterometer's measurements of ERS-1/2 and QuikSCAT contain both wind speeds and directions. However, since the wind vectors retrieved from the scatterometer measurements suffer the wind directional ambiguity problem, the quality of scatterometer wind data in the current operational application is to treat them as though they were of the same quality as ship wind data, and as such they can be readily applied to the 3D-VAR analysis of the NCEP GDAS. Fundamentally, scatterometer measurements are backscattered radiance from C-band or Ku-band radars. In principle, both the 3D-VAR and 4D-VAR variational assimilation systems at NCEP and ECMWF can make use of the brightness temperatures and backscattered radiance information in the atmospheric analysis. The current operational use of scatterometer wind data in 4D-VAR of the ECMWF GDAS is to use the backscattered radiance directly. This direct use of backscattered radiance from scatterometer measurements is similar to the use of TOVS radiance data, and in general requires greater computational resources. In a synoptic case study to be discussed in Sec. 5, Yu and Derber (1996) and Yu (1996) have demonstrated the use of backscattered measurements and the use of retrieved wind speed and wind direction information from ERS-1 in the NCEP 3D-VAR analysis. Results of these two studies show that the use of ERS-1 wind data did improve the central intensity of cyclonic pressure and location of the circulation center. However, a comparison of results between the assimilation of backscattered measurements and those from the assimilation of the retrieved winds shows that there is little difference in the analysis results for storm center pressure intensity and location in this case study. This is the reason that NCEP elected to use the retrieved wind vectors from scatterometer measurements from ERS-1/2 and QuikSCAT wind data in the current operational GDAS.

Satellite wind data from the radiometer measurements from SEASAT in 1978, and those from the follow on SSM/I of the DMSP satellite from the 1980's to the present time contain only wind speeds, and no wind directions. In an effort to make use of the SEASAT passive radiometer wind speed data in the OI analysis scheme, Yu (1987) has developed a technique which can deduce a unique wind direction for each of the radiometer-measured wind speed data, provided that a large scale sea level pressure analysis field is given. Such fields are typically available at an NWP operational center, such as NCEP and ECMWF, and they are in general of high quality, and the deduced wind directions are quite realistic and compare well with those from collocated buoy reports (Yu, 1987). Another way to use the radiometer-measured wind speed data is to assign wind direction to the speed data from the first-guess (background) wind field of a six-hour model forecast in the assimilation cycle. In fact, this is the method that has been in operational use for the SSM/I wind speed data at the current operational GDAS at NCEP (Yu *et al.*, 1997). It should also be noted that the brightness temperature data from the SSM/I radiometer measurements can in principle be used directly in the 3D-VAR analysis system using a forward model associated with the variational analysis methodology, without having to derive any ancillary wind direction information for the radiometer wind speed data.

3. Impact of Satellite Ocean Surface Wind Data

Observing system experiments (OSE's) have been a benchmark designed for investigating the impact of a satellite data set. Figure 4 is a schematic showing an example of an OSE designed to test QuikSCAT winds and infrared sea surface temperature data from the AVHRR and GOES satellites. An OSE typically invokes a parallel experiment in which a test run and a control run are conducted. The control run uses the conventional data, while the test run contains all conventional data and the additional satellite data set to be investigated in the data assimilation and forecast experiments. Before any new satellite data are implemented in the NCEP's NWP operations, two parallel experiments are conducted, one with the new satellite data, the other without, to assess the impact of the new data set. To be discussed in this

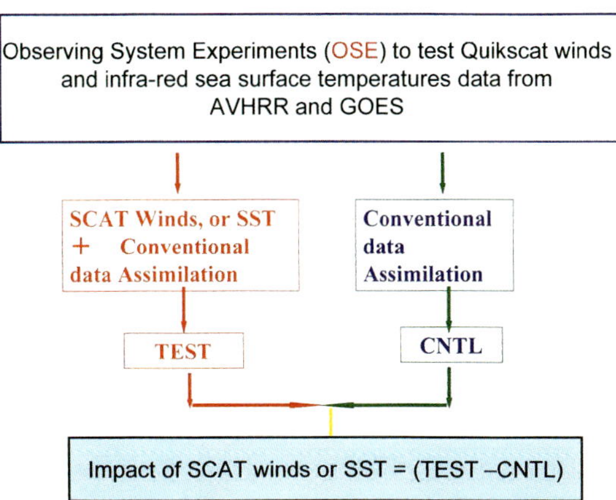

Figure 4. Schematic for the design of parallel global data assimilation experiments to test satellite data impact.

section are summaries of main results from scatterometer data impact experiments that have been conducted to investigate ERS-1 in 1995 (Yu *et al.*, 1996), and QuikSCAT winds in 2001 and 2003 (Yu, 2003), and those from SSM/I radiometer wind impact experiments in 1997 (Yu, 2001).

The forecast RMS vector wind errors shown in Table 3 are the model's 10 m wind forecasts when compared to midlatitude deep ocean buoys (between 25 North and 60 North) during two parallel data assimilation experiments conducted to investigate the impact of ERS-1 scatterometer wind data on NCEP's operational global data assimilation system. As discussed in

Table 3. Forecast RMS vector wind errors of model 10 m winds when compared to midlatitude deep ocean buoys (25 North–60 North) during data assimilation experiments (from November 21 to December 31, 1995) conducted to investigate impact of ERS-1 scatterometer wind data on NCEP's operational global data assimilation system.

Forecast Hours	Number of Buoys	Without ERS-1 winds (m/s)	With ERS-1 winds (m/s)
24	3713	5.09	5.02
48	3754	6.16	6.04
72	3750	7.42	7.38
96	3750	7.97	8.00
120	3705	8.75	8.70

Table 1, the data coverage for ERS-1 is about 500 km, with a footprint of 50 km. Before the ERS-1 wind data were implemented operationally, two data assimilation experiments — one with the use of ERS-1 scatterometer wind data (the test run), and one without (the control run) — were conducted for a period of about 40 days, starting on November 21, 1995, and ending on December 31, 1995. The results shown in Table 3 are based on 40 cases of forecasts. As one can see from the table, inclusion of ERS-1 scatterometer winds in the NCEP GDAS leads to a small positive impact on the forecast of 10 m winds during the five days of forecasts. Note that the improvement in the forecasts is very small and may not be significant from the overall statistics. However, significant improvements are most often seen in some special synoptic situation where the inclusion of scatterometer winds can make a large improvement in identifying the storm intensity and circulation center, as will be discussed in the next section. Based on results of these parallel experiments, the ERS-1 wind data were implemented operationally in NCEP's global data assimilation experiment in the spring of 1997.

QuikSCAT, a NASA oceanographic satellite, was launched in July 1999. The main mission of the satellite was to provide global coverage of ocean surface winds for NAVY and NOAA applications. The QuikSCAT scatterometer single cell measures a footprint of 25 km, and with a total of 72 cells it can cover a data swath of 1800 km, and can provide more than 100 000 data points in a 6-hour window during a synoptic analysis cycle (see Table 1). Early impact experiments with the full resolution (25 km) QuikSCAT wind data in the NCEP GDAS and NASA GDAS did not lead to any positive impact. One of the reasons for the lack of early impact investigations was the incompatibility of the fine data resolution (25 km) and the much coarser model and analysis resolution (of greater than 100 km) configured in the NCEP and NASA global data assimilation systems.

Follow-up experiments using coarser resolution QuikSCAT wind data by a super-averaging approach subsequently led to a positive impact on the global data assimilation systems of NCEP (Yu, 2003) and NASA (Atlas *et al.*, 2001).

The super-averaging approach, sometimes also called the supper-ob approach, is designed mainly to thin out large numbers of satellite data. In order to use thinned-out QuikSCAT data in NCEP GDAS experiments, the data are averaged over a $1° \times 1°$ longitude–latitude grid box, reducing the total data numbers to nearly one-fourth of the original, and resulting in a 100 km effective resolution for the data. Table 4 shows mean sea level pressure forecast errors with respect to midlatitude deep ocean buoys during data assimilation experiments conducted from October 2, 2001, to November 10, 2001, to investigate the impact of QuikCAT

winds (100 km resolution) on the NCEP operational global data assimilation system (T170, L42). The results, based on about 40 forecast cases, clearly show that use of QuikSCAT winds significantly improves the sea level pressure forecasts when compared to a buoy's observations. Similar conclusions are found when comparing the forecasts of model 10 m winds with *in situ* observations from midlatitude deep ocean buoys (Table 5) and tropical TOGA buoys (Table 6). Based on these statistics, QuikSCAT winds of 100 km resolution were implemented operationally in the NCEP's global data assimilation system on January 15, 2002. About one week later, ECMWF also implemented the QuikSCAT wind data of a coarser resolution in its global data assimilation, on January 22, 2002.

After the operational implementation of the coarse resolution (100 km) of QuikSCAT wind

Table 4. Mean sea level pressure forecast errors (mb) with respect to about 45 midlatitude deep ocean buoys during data assimilation experiments (from October 2 to November 10, 2001) conducted to investigate the impact of QuikSCAT scatterometer winds (100 km resolution) on the NCEP operational global data assimilation system (T170, L42).

Forecast Hours (hours)	Without Use of QuikSCAT	With Use of QuikSCAT	Improvement (%)	Improvement (hours)
06	1.15	1.13	1.7	—
24	1.91	1.59	16.8	7.6
48	2.43	2.19	9.9	11.1
72	3.07	2.87	6.5	8.5
96	3.71	3.63	2.2	3.0
120	4.18	4.18	0.0	0.0

Table 5. RMS vector wind forecast errors (m/s) at 10 m with respect to about 45 midlatitude deep ocean buoys during data assimilation experiments (from October 2, 2001 to November 10, 2001) conducted to investigate the impact of QuikSCAT scatterometer winds (100 km resolution) on the NCEP operational global data assimilation system (T170, L42).

Forecast Hours (hours)	Without Use of QuikSCAT	With Use of QuikSCAT	Improvement (%)	Improvement (hours)
06	3.10	3.08	1.0	—
24	3.94	3.63	7.9	6.6
48	4.44	4.29	3.4	7.2
72	5.46	5.08	7.0	8.9
96	6.16	5.99	2.8	5.8
120	6.96	6.94	0.3	0.6

Table 6. RMS vector wind forecast errors (m/s) at 10 m with respect to about 12 TOGA deep ocean buoys during data assimilation experiments (from October 2 to November 10, 2001) conducted to investigate the impact of QuikSCAT scatterometer winds (100 km resolution) on the NCEP operational global data assimilation system (T170, L42).

Forecast Hours (hours)	Without Use of QuikSCAT	With Use of QuikSCAT	Improvement (%)	Improvement (hours)
06	3.04	3.02	0.7	—
24	3.35	3.26	2.7	5.6
48	3.81	3.55	6.8	13.6
72	4.37	4.15	5.0	12.0
96	4.58	4.61	−0.7	−3.4
120	5.33	5.22	2.0	3.5

data in January 2002, NCEP implemented a higher resolution global forecast model of T254, L64 configuration in October 2002. This effectively increased the horizontal resolution to about 60 km from the previous T170, L42, which is 100 km in the horizontal resolution. It was then deemed necessary to investigate the impact of higher resolution (50 km) with the purpose of making effective use of the inherent mesoscale features associated with the QuikSCAT wind data. Two parallel data assimilation experiments were conducted — one with the finer resolution QuikSAT wind (50 km) data, and the other without the data — for a total period of about 60 days, starting January 8, 2004, and ending March 8, 2003. The major findings of these two parallel experiments based on the average of the 60 cases of forecasts reveal a significant positive impact on heights and winds at all levels for both the northern and southern hemispheres, especially for winds over the tropical oceans (Yu, 2003). Figure 5 shows anomaly correlations of zonal and meridional winds at 850 hPa over the tropics for forecasts with the use of QuikSCAT winds and those without the use of QuikSCAT winds. Note that the most significant improvement occurs in Day 4 and Day 5 forecasts and over the mesoscale features (of waves from 10 to 20).

Radiometer wind speed data from SSM/I measurements of the US Defense Meteorological Satellite Programs (DMSP) since the 1980's constitute a continuing source of ocean surface wind data. The SSM/I wind speed data from DMSP has been used in the NCEP operational global data assimilation system since March 1993. The justification for implementing the SSM/I wind speed data was based on the results from a number of impact studies showing that the assimilation of wind speed data was slightly beneficial to the NCEP numerical analysis and short-range forecasts (Yu and Deaven, 1991; Yu et al., 1992). However, these operational SSM/I wind speeds were derived using an algorithm developed by Goodberlet et al. (1989), which assumed a linear dependence of wind speed on brightness temperatures. This assumption is acceptable when the level of moisture, both water vapor and liquid water, in the atmosphere is very low. As soon as the level of moisture increases, the dependence of the wind speed on brightness temperatures becomes significantly nonlinear, and errors in wind speeds retrieved by the Goodberlet algorithm become very large. For this reason, during the early days of operations at NCEP, only SSM/I wind speed data over the clear-sky area were used, and a large number of data points over active weather regions had to be rejected. This was rather unfortunate, since it is these developing weather systems that are most interesting and important in weather forecasting.

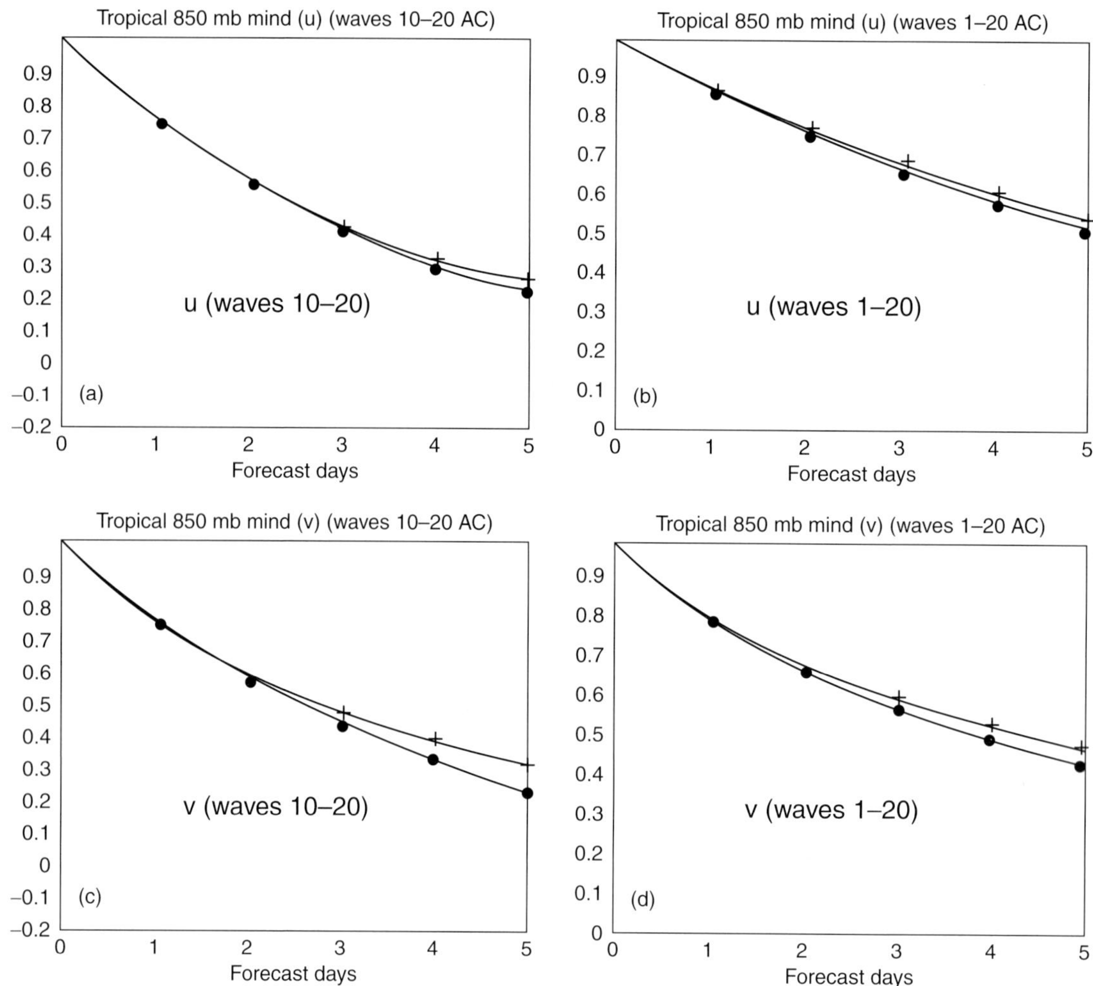

Figure 5. Anomaly correlations of zonal and meridional winds at 850 hPa over the tropics for forecasts with QuikSCAT winds (curves shown with a + sign), and those without QuikSCAT winds (curves shown with a • sign).

To remedy the deficiency associated with the Goodberlet algorithm mentioned above, Krasnopolsky (1995a, 1995b) developed nonlinear algorithms based on neural network architectures, which were shown to be capable of modeling the nonlinear dependence of wind speed on brightness temperatures. The performance of neural network algorithms in the retrieval of SSM/I wind speed data discussed in Krasnopolsky (1995a, 1995b) has been tested mainly by using a well-prepared matchup database, i.e. their wind data were tested only in an experimental retrieval procedure, but not in any operational forecast system. For this reason, Yu *et al.* (1997) further conducted a global data assimilation experiment to investigate the impact of SSM/I wind speeds derived from the neural network algorithm. The assimilation period for this experiment was about three weeks, from May 16 to June 4, 1996. Detailed results of this investigation can be found in Yu *et al.* (1997). Figure 6 shows anomaly correlations for 500 hPa and 1000 hPa geopotential heights calculated from the control run (using SSM/I wind speeds derived from the Goodberlet algorithm) and the parallel run (using SSM/I

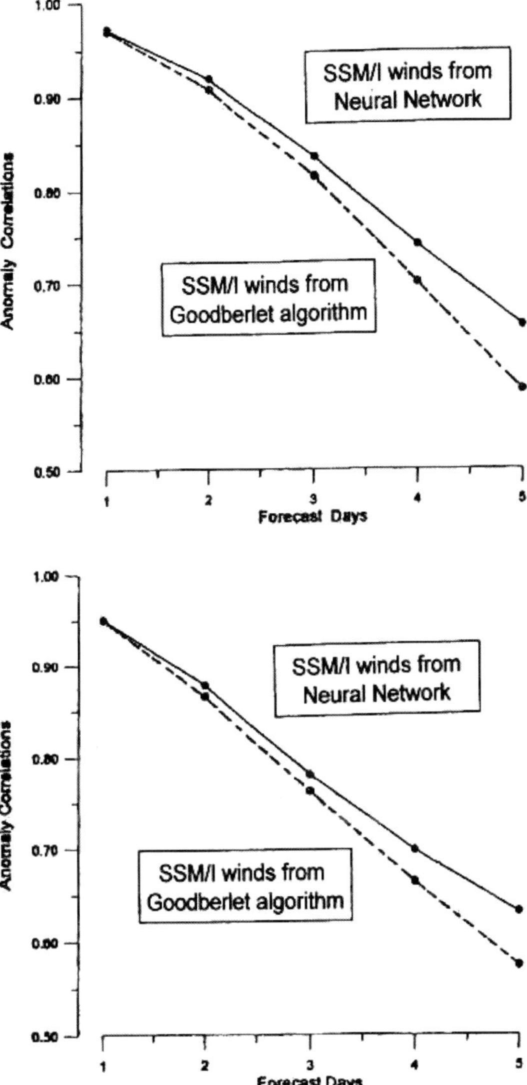

Figure 6. Anomaly correlations for 500 hPa (top panel) and 1000 hPa (bottom panel) geopotential heights calculated from the control run (with use of SSM/I winds from the Goodberlet algorithm — dashed line) and the parallel run (with use of SSM/I winds from the neural network algorithm — solid line) over the southern hemisphere.

wind speeds derived from a neural network algorithm) over the southern hemisphere. The results clearly show that use of SSM/I wind speeds derived from a neural network algorithm improves the height forecasts at the 1000 hPa and 500 hPa levels over those derived by the

Goodberlet algorithm. Based on these results, the neural-network-derived SSM/I wind speeds were implemented in the NCEP global data assimilation system in 1997. It should be noted that the current operational SSM/I wind speeds at NCEP are still derived by the same neural network algorithm tested by Yu *et al.* (1997).

4. Applications of Ocean Surface Wind Data in the Southwestern Pacific Region

The results from the previous section on data assimilation experiments suggest that routine assimilation and forecast experiments may not show a very significant impact of satellite ocean surface winds from the gross statistics of anomaly correlations and root-mean-squared errors over many cases of analysis and forecasts. Nonetheless, they are important statistics, from which new data sets such as satellite ocean surface winds of ERS-1/2, SSM/I, and QuikSCAT were implemented operationally at NCEP, as has been discussed in the previous section. However, the impact of any satellite ocean surface winds may be most significant in some selected synoptic situations where a satellite has provided data over the region that is not covered by the conventional observations. Atlas *et al.* (1999) have shown an example of a very significant improvement to the cyclonic circulation due to the influence of NSCAT wind data over the extratropical Pacific oceans of the southern hemisphere. In the same paper, they also show an example of a substantial improvement in the forecast of the Christmas Day storm crossing northern Europe by using QuikSCAT wind data in the analysis. Similar case studies abound where use of satellite ocean surface winds has led to significant improvements in the analysis and forecasts of wind and sea level pressure fields over the global oceans.

The synoptic case chosen for the following discussions happens to occur on 0000 UTC,

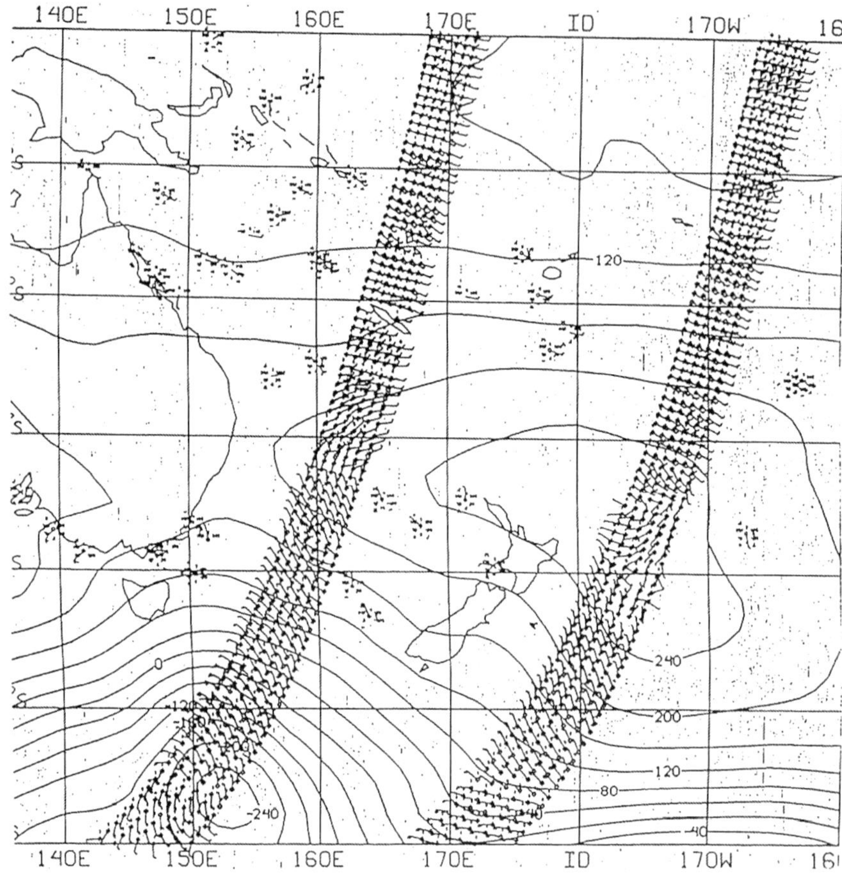

Figure 7. ERS-1 satellite data swaths for the synoptic case study on 0000 UTC, May 2, 1994.

May 2, 1994, over the southwestern Pacific region, and clearly demonstrates the important application of satellite ocean surface wind data (in this case, of ERS-1 scatterometer wind data) in identifying a closed cyclonic circulation over this region (Yu and Derber, 1995). Two analysis cases will be shown; one analysis case uses the ERS-1 scatterometer wind data (SCAT case), whereas the other does not use the wind data (control case) in the analyses. During the six-hour window centered at this analysis time, there were two swaths of ERS-1 scatterometer wind data passing through a well-developed cyclonic pressure circulation centered at a location between 150 and 155 east longitudes and between 55 and 60 south latitudes southeast of Tasmania in the southern

hemisphere (see Fig. 7). The low pressure center is also well identified in the NOAA-12 visible imagery (Fig. 8), which serves as a ground truth for the assessment of analysis results. For this synoptic case, the NCEP surface wind analysis failed to depict a closed circulation center when compared to the satellite imagery. It is therefore of particular interest to see if additional ERS-1 observations of ocean surface wind data will improve the low level wind analysis in better defining the center of the storm circulation.

The vector winds at the lowest model level (40 m above the ocean surface) from the analysis which includes ERS-1 scatterometer wind data are shown in Fig. 9 (SCAT case). They should be compared with the analysis which were generated without the use of the ERS-1

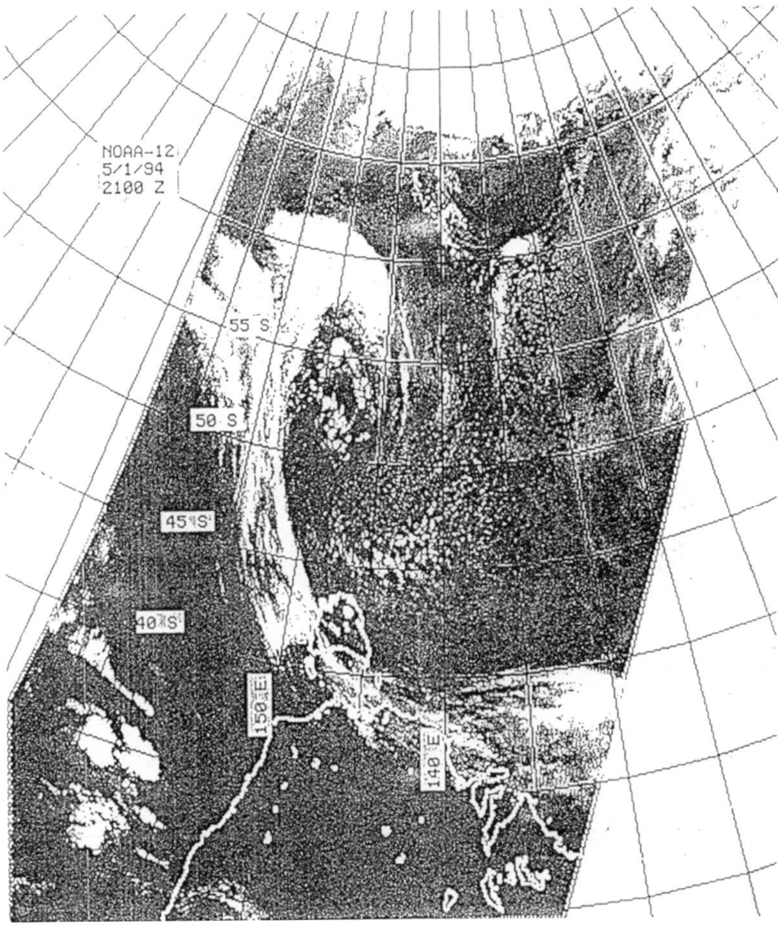

Figure 8. NOAA-12 satellite imagery over the southern Asia–Pacific ocean on 2100 UTC, May 2, 1994.

scatterometer wind data shown in Fig. 10 (control case). One can see from comparing Figs. 9 and 10 that the analysis with the inclusion of the ERS-1 wind data shows a better defined circulation for the storm center than the control case. The increase in the cyclonic circulation contributed by the addition of the ERS-1 wind data is clearly shown in vector wind differences between the two analysis (see Fig. 11). Close inspection of the two analysis and their differences reveals that there are areas of large vector wind (about 20 m/s) differences between the two analysis, and these differences occurred near the center of the storm over the passes of the two satellite swaths.

5. Summary and Conclusions

The forecasting skills of short range numerical weather predictions have been steadily improved since the beginning of NWP operations in the late 1950's. During the last two decades the forecasting skills have seen more drastic improvement, primarily owing to advances in the fields of satellite remote sensing and atmospheric analysis and data assimilation systems. This article first reviewed the evolution of atmospheric analysis schemes and global models at major NWP operational centers at NCEP and ECMWF during the last two decades. The procedures for the operational use of ocean surface

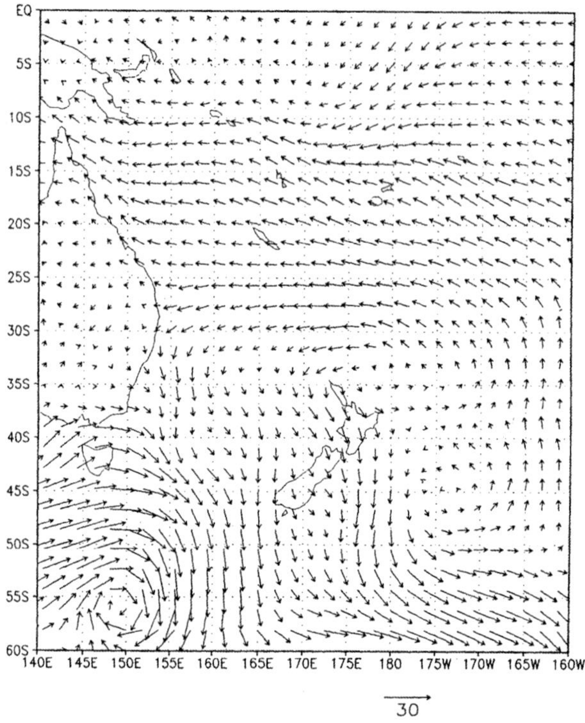

Figure 9. SCAT analysis of ocean surface winds (m/s) at 40 m over the southern Asia–Pacific ocean for 0000 UTC May 2, 1994.

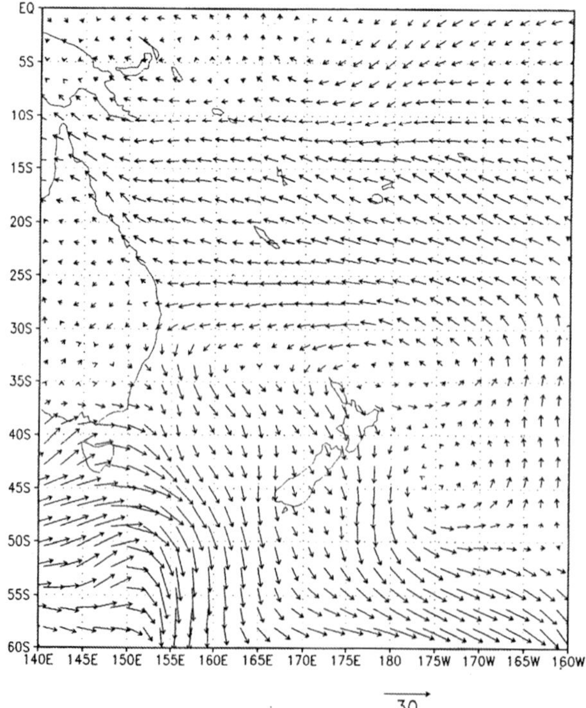

Figure 10. Control analysis of ocean surface winds (m/s) at 40 m over the southern Asia–Pacific ocean for 0000 UTC, May 2, 1994.

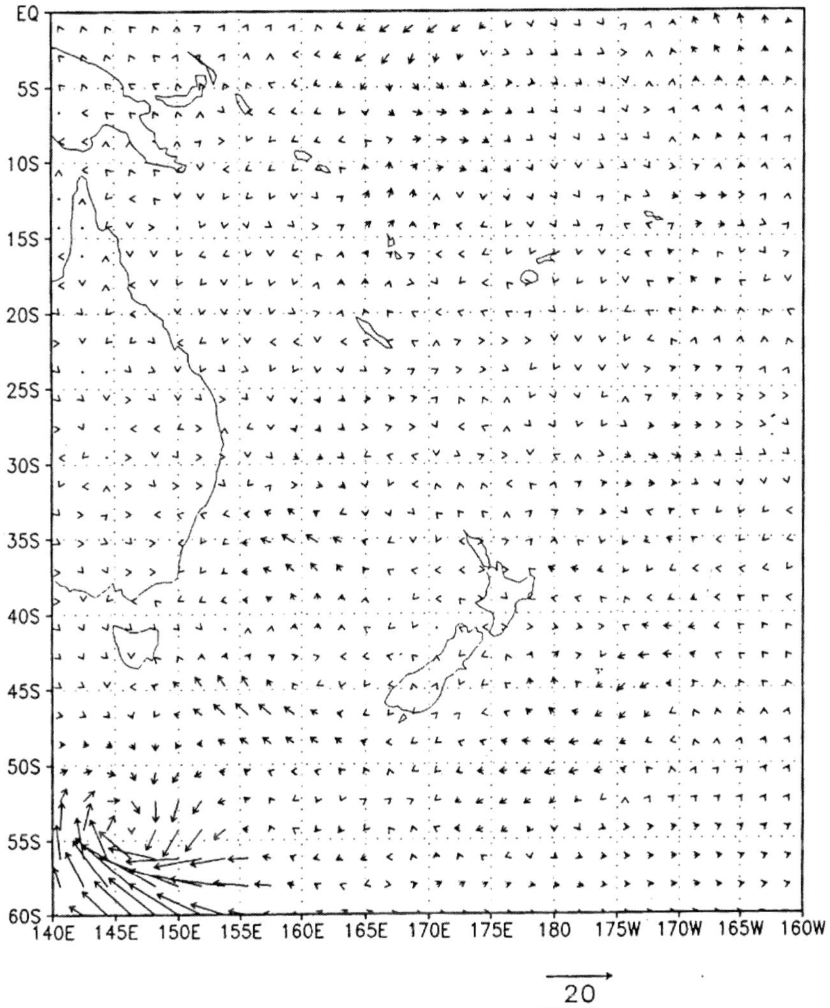

Figure 11. Vector wind differences (m/s) at 40 m between control and SCAT analysis over the southern Asia–Pacific ocean for 0000 UTC, May 2, 1994.

winds retrieved from the passive microwave radiometer of SSM/I on board the DMSP satellites, and from active microwave scatterometer measurements on board SEASAT, NSCAT, ERS-1/2 and QuikSCAT satellites were discussed.

Of special import are the summaries of pre-implementation test results on the use of these various satellite ocean surface wind data at NCEP global data assimilation systems, based on which these surface wind data have been operationally implemented at various timelines

during the last two decades. It is fair to state that from gross statistics based on many cases of forecasts, the impact of the satellite ocean surface winds on NWP short range forecasts is mostly positive, small but significant because the data are of a single level nature and only available at the ocean surface. Large impacts of the satellite ocean surface winds may only occur over some special synoptic situations and over the areas of severe weather storms when conventional data are lacking. This is well demonstrated in a special case investigation over the southwestern Pacific Ocean, where ERS-1 scatterometer winds are found to be most significant in identifying the storm center intensity and circulation.

Acknowledgements

The author would like express his sincere appreciation to Dr Stephen Lord of NCEP and Dr A. Hollingsworth of ECMWF for permission to use their centers' forecast performance statistics on anomaly correlations as shown in Figs. 1 and 2 of this article. He is also grateful to the two anonymous reviewers for their constructive comments and suggestions, which improved the final version of the manuscript.

[Received 15 January 2006; Revised 3 May 2007; Accepted 2 June 2007.]

References

Atlas, R., R. Hoffman, S. Leider, J. Sienkiewicz, T.-W. Yu, S. Bloom, E. Brin, J. Ardizzone, J. Terry, D. Bungato, and J. Jusem, 2001: The effects of marine winds from scatterometer data on weather analysis and forecasting. *Bull. Amer. Meteorol. Soc.*, **82**, 1965–1990.

Atlas, R., S. Bloom, J. Ardizzone, E. Brin, J. Terry, and T.-W. Yu, 2001: The impact of Quikscat on weather analysis and forecasting, Preprint paper presented at 18th Conference on Weather Analysis and Forecasting and 14th Conference on Numerical Weather Prediction, 30 July–2 August 2001, Fort Lauderdale, Florida, USA, pp. 172–175.

Atlas, R., S. Bloom, R. Hoffman, E. Brin, J. Ardizzone, J. Terry, D. Bungato, and J. Jusem, 1999: Geophysical validation of NSCAT winds using atmospheric data and analysis. *J. Geophys. Res.*, **104** (C5), 11405–11424.

Andrew, P. L., and R. S. Bell, 1998: Optimizing the United Kingdom Meteorological Office data assimilation for ERS-1 scatterometer results. *Quart. J. Roy. Meteorol. Soc.*, **124**, 1831–1860.

Baker, W., R. Atlas, E. Kalnay, M. Halem, P. Woicesyhn, S. Peteherych, and D. Edelmann, 1984: Large-scale analysis and forecast experiments wind data from Seasat-A scatterometer. *J. Geophys. Res.*, **89** (D3), 4927–4936.

Courtier, P., E. Anerson, W. Heckley, J. Paileux, D. Vasiljevic, M. Hamrud, A. Holligsworth, F. Rabier, and M. Fisher, 1998: The ECMWF implementation of three-dimensional variational assimilation (3D-Var). I: Formulation. *Quart. J. Roy. Meteor. Soc.*, **124**, 1783–1807.

Cressman, G. P., 1959: An operational objective analysis scheme. *Mon. Wea. Rev.*, **87**, 367–374.

Derber, J., D. Parrish, and S. Lord, 1991: The new global operational analysis system at the National Meteorological Center. *Wea. Forecast.*, **6**, 538–547.

Duffy, D., and R. Atlas, 1986: The impact of Seasat-A scatterometer winds on the numerical prediction of the Queen Elizabeth II storm. *J. Geophys. Res.*, **91** (C2), 2241–2248.

Gandin, L. S., 1965: *Objective Analysis of Meteorological Fields*, Israel Program for Scientific Translation, Jerusalem, 242 pp.

Gemmill, W. H., T.-W. Yu, V. Krasnopolsky, C. Peters, and P. Woiceshyn, 1999: NCEP experience with "Real-time" ocean surface wind retrievals from satellites. NCEP Technical Note, OMB Contribution No. 164, 26 pp.

Goodberlet, M. A., C. T. Swift, and J. Wikerson, 1989: Remote sensing of ocean surface winds with the Special Sensor Microwave/Imager. *J. Geophys. Res.*, **94**, 547– 555.

Hoffman, R. N., 1993: A preliminary study of the impact of the ERS1 C-band scatterometer wind data on the ECMWF global data assimilation system. *J. Geophys. Res.*, **98** (C6), 10233–10244.

Hoffman, R. N., and C. Grassotti, 1996: A technique for assimilating SSM/I observations of marine atmospheric storms: tests with ECMWF analysis. *J. Appl. Meteorol.*, **35**, 1177–1187.

Hollingsworth, A. (2004): Performance of ECMWF global forecasts in terms of 500 mb geopotential height anomaly correlation statistics. Personal communication.

Ingleby, N. B., and R. A. Blomley, 1991: A diagnostic study of the impact of SEASAT scatterometer winds on numerical weather prediction. *Mon. Wea. Rev.*, **119**, 84–103.

Krasnopolsky, V., L. Breaker, and W. Gemmill, 1995a: A neural network as a nonlinear transfer function model for retrieving surface wind speeds from the Special Microwave Imager. *J. Geophys. Res.*, **100**, 11033–11045.

Le DiMet, F., and O. Talagrand, 1986: Variational algorithms for analysis and assimilation of meteorological observations: Theoretical aspects. *Tellus*, **38A**, 97–110.

Lord, S. (2004): Performance of NCEP global forecasts in terms of 500 mb geopotential height anomaly correlation statistics. Personal communication.

Parrish, D., and J. C. Derber, 1992: The National Meteorological Center's Spectral Statistical Interpolation analysis scheme. *Mon. Wea. Rev.*, **120**, 1747–1763.

Rabier, F., A. McNally, E. Anderson, P. Courtier, P. Unden, J. Eyer, A. Hollingsworth, and F. Bouttier, 1998: The ECMWF implementation of three-dimensional varaitional assimilation (3D-Var). II: Structure function. *Quart J. Roy. Meteorol. Soc.*, **124**, 1809–1829.

Rabier, F., J.-F. Mahfouf, M. Fisher, H. Haseler, A. Holligsworth, L. Isaksen, E. Klinker, S. Saarinen, C. Temperton, J.-N. Thepaut, P. Unden, and D. Vasiljevic, 1997: The ECMWF operational implementation of four-dimensional variational assimilation. ECMWF Research Development Tech. Memo. No. 240.

Sasaki, Y., 1970: Some basic formalisms in numerical variational analysis. *Mon. Wea. Rev.*, **98**, 875–883.

Sela, J. G., 1980: Spectral modeling at the National Meteorological Center. *Mon. Wea. Rev.*, **108**, 1279–1292.

Stoffelen, A., and G. J. Cats, 1991: The impact of Seasat-A scatterometer data on high-resoltuion analysis and forecasts· the development of the QE II storm. *Mon. Wea. Rev.*, **119**, 2794–2802.

Stoffelen, A., and D. Anderson, 1997: Ambiguity removal and assimilation of scatterometer data. *Quart. J. Roy. Meteor. Soc.*, **123**, 491–518.

Thepaut, J.-N., R. N. Hoffman, and P. Courtier, 1993: Interactions of dynamics and observations in a four-dimensional variational assimilation. *Mon. Wea. Rev.*, 3393–3414.

Yu, T.-W., and R. D. McPherson (1984): Global data assimilation experiments using SESAT-A scatterometer wind data. *Mon. Wea. Rev.*

Yu, T.-W., 1987: A technique for deducing wind direction from satellite microwave measurements of wind speed. *Mon. Wea. Rev.*, **115**, 1929–1939.

Yu, T.-W., and D. Deaven, 1991: Use of SSM/I wind speed data at NMC's GDAS. Preprints, 9th Conference on Numerical Weather Prediction, Denver, CO, USA, American Meteorological Society, pp. 416–417.

Yu, T.-W., W. Gemmill, and J. Wollen, 1992: Use of SSM/I wind speed in the operational numerical weather prediction system at NMC. Abstracts, NMC/NESDIS/DOD Conference on DMSP Retrieval Products, Washington DC, USA.

Yu, T.-W., P. Woiceshyn, W. Gemmill, and C. Peters, 1994: Analysis and forecast experiments using ERS-1 scatterometer wind measurements. Preprint paper presented at 7th Conference on Satellite Meteorology and Oceanography of the American Meteorological Society, June 6–10, Monterey, California, USA, pp. 600–601.

Yu, T.-W., and J. C. Derber, 1995: Assimilation experiments with ERS-1 winds: Part I — Use of backscatter measurements in the NCEP spectral statistical analysis system. NCEP/EMC/Ocean Modeling Branch Technical Note 116, 32 pp.

Yu, T.-W., 1995: Assimilation experiments with ERS-1 winds: Part II — Use of vector winds in NCEP spectral statistical analysis system. NCEP/EMC/Ocean Modeling Branch Technical Note 117, 29 pp.

Yu, T.-W., M. D. Iredell, and Y. Zhu, 1996: The impact of ERS-1 winds on NCEP numerical weather analysis and forecasts, Preprint paper presented at 11th Conference on Numerical Weather Prediction of the American Meteorological Society, August 19–23, 1996, Norfolk, Virginia, USA, pp. 276–277.

Yu, T.-W., M. Iredell, and D. Keyser, 1997: Global data assimilation and forecast experiments using SSM/I wind speed data derived from a neural network algorithm. *Wea. Forecast.*, **12**, 859–865.

Yu, T.-W., 2000: Assimilating satellite observations of ocean surface winds in the NCEP global data and regional atmospheric analysis systems. Proceedings of Third WMO Symposium on Assimilations of Observations in Meteorology and Oceanography, WMO/TD-No. 986, pp. 229–232.

Yu, T.-W., 2003: Operational use of Quikscat winds in NCEP GDAS. Preprint paper at 12th Conference on Satellite Meteorology of the American Meteorological Society, February 9–13, Long Beach, CA, USA.

Purdue Atmospheric Models and Applications

Wen-Yih Sun[*,**], Wu-Ron Hsu[†,**], Jiun-Dar Chern[‡], Shu-Hua Chen[$,**], Ching-Chi Wu[†],
Kate Jr-Shiuan Yang[*], Kaosan Yeh[‡], Michael G. Bosilovich[‡], Patrick A. Haines[¶], Ki-Hong Min[‖],
Tae-Jin Oh[*,**], B. T. MacCall[*,**], Ahmet Yildirim[*], Yi-Lin Chang[*], Chiao-Zen Chang[*]
and Yi-Chiang Yu[†]

Purdue University, West Lafayette, Indiana, USA
†National Taiwan University, Taipei, Taiwan
‡Goddard Space Fight Center/NASA, Greenbelt, Maryland, USA
$University of California, Davis, California, USA
¶Army Research Laboratory, White Sand, New Mexico, USA
‖Valparaiso University, Valparaiso, Indiana, USA
***Taiwan Typhoon and Flood Research Institute, Hsin-chu, Taiwan*
wysun@purdue.edu

This article summarizes our research related to geofluid dynamics and numerical modeling. In order
to have a better understanding of the motion in the atmosphere, we have been working on various
forms of the Navier–Stokes equations, including the linearized and nonlinear systems as well as
turbulence parametrization, cumulus parametrization, cloud physics, soil–snow parametrization,
atmospheric chemistry, etc. We have also been working on numerical methods in order to solve
the equations more accurately. The results show that many weather systems in the initial/growing
stage can be qualitatively described by the linearized equations; on the other hand, many developed
weather phenomena can be quantitatively reproduced by the nonlinear Purdue Regional Climate
Model, when the observational data or reanalysis is used as the initial and lateral boundary conditions. The model can also reveal the detailed structure and physics involved, which sometimes can
be misinterpreted by meteorologists according to the incomplete observations. However, it is also
noted that systematic biases/errors can exist in the simulations and become difficult to correct.
Those errors can be caused by the errors in the initial and boundary conditions, model physics and
parametrizations, or inadequate equations or poor numerical methods. When the regional model is
coupled with a GCM, it is required that both models should be accurate so as to produce meaningful
results. In addition to the Purdue Regional Climate Model, we have presented the results obtained
from the nonhydrostatic models, the one-dimensional cloud model, the turbulence-pollution model,
the characteristic system of the shallow water equations, etc. Although the numerical model is the
most important tool for studying weather and climate, more research should be done on data assimilation, the physics, the numerical method and the mathematic formulation in order to improve
the accuracy of the models and have a better understanding of the weather and climate.

1. Introduction

The motion of the atmosphere and ocean can be
represented by Navier–Stokes equations, which
should be solved numerically. However, the
evolution of a weather/climate system consists of motions on many different scales. At
the beginning, some of the small disturbances
may be described by the linearized equations.
When disturbances grow, nonlinear equations

become necessary. It is also noted that eigen values/vectors can be easily obtained and interpreted in a linearized system. Hence, for the past few decades, we have been working on linear instability and nonlinear numerical models to study meteorological phenomena, ranging from convection, turbulence, air pollution, cumulus clouds, cloud streets, symmetric instabilities, mountain waves, lee vortices, and land–sea breezes, to synoptic scale waves, barotropic instabilities, cyclones, fronts, and regional climate in East Asia and North America. In order to obtain accurate numerical results, we have also been developing aspects of the model such as the new diffusion equation; turbulence parametrization; snow–vegetation–soil and snow–sea ice packages; forward–backward, advection, and semi-Lagrangian schemes; the pressure gradient force; the multigrid method; the transport of dust and trace gases; and the interaction between aerosols and regional climate. The important components of our mesoscale model are shown in Fig. 1.

We developed the Purdue Regional Climate Model (PRCM). However, several mesoscale regional models exist, with varying approaches. They include the fifth-generation Penn State/ NCAR Mesoscale model (MM5; Grell *et al.*, 1995) and the Weather Research and Forecast (WRF) model (Skamarock *et al.*, 2005). Table 1 shows the important features that are used in the PRCM, MM5, and WRF.

A few topics will be briefly discussed here. Many of them require further study. The basic structure of the PRCM and applications will be discussed in Sec. 2, the nonhydrostatic models in Sec. 3, other topics in Sec. 4, and a summary is provided at the end.

2. Purdue Regional Climate Model

2.1. *Basic equations*

The PRCM is a hydrostatic primitive equation model that utilizes the terrain-following coordinate (σ_p) in the vertical direction. It has

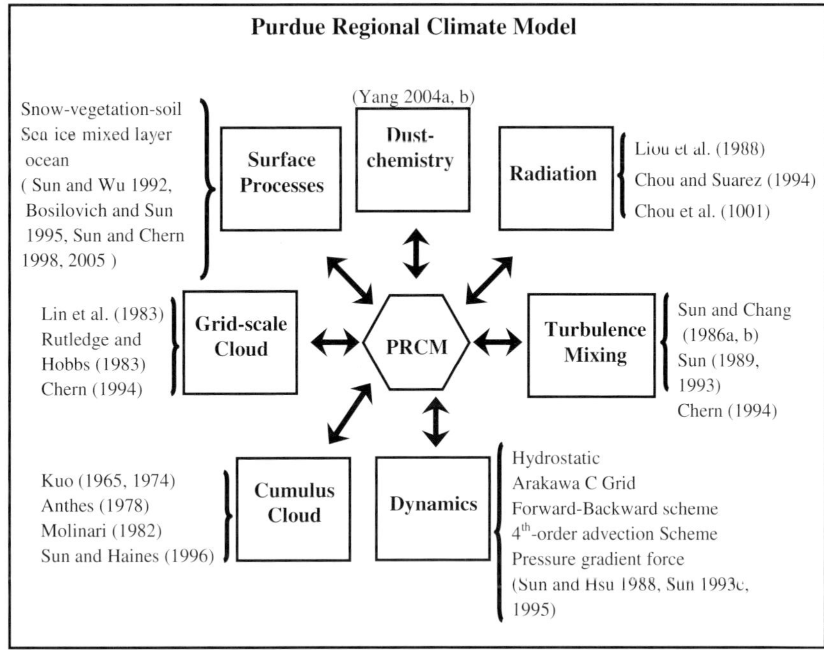

Figure 1. The schematic diagram of Purdue Regional Climate Model.

Table 1. The characteristics of PRCM, WRF, and MM5. WRF has three different dynamical cores. The one that mentioned here is Advanced Research WRF (ARW).

	PRCM	WRF (ARW)	MM5
Hydrostatic/Non-hydrostatic	Hydrostatic	Non-hydrostatic	Non-hydrostatic
Vertical Coordinate	σ_p	σ_p	σ_p
Equations	Advection form	Flux form	Advection form
Time differencing scheme	Forward-backward	Runge-Kutta 3rd order	Leap-frog
Space differencing scheme	4th order or Semi-Lagrangian advection	2nd to 6th order	2nd order
Time step	Timesplitting	Time splitting	Time splitting
Pressure gradient force	Local reference	Universal reference	Universal reference
Thermal variable	Ice equivalent temperature θ_{ei}	Potential temperature	Temperature
Turbulence parameterization	Based on θ_{ei} and total water substances	Based on temperature	Based on temperature

the prognostic equations for momentum; ice-equivalent potential temperature θ_{ei}; turbulent kinetic energy (TKE); surface pressure; all phases of water, including ice, liquid, snow, rain, graupel, and vapor (Lin *et al.*, 1983; Chern, 1994; Haines *et al.*, 1997; Chen and Sun, 2002); and multilayers of soil temperature and wetness, etc. (Sun and Wu, 1992; Bosilovich and Sun, 1995; Chern and Sun, 1998; Sun and Sun, 2004; Sun and Chern, 2005). Because θ_{ei} and total water substance, which are conserved without precipitation, are used as the prognostic variables, the model is able to include a comprehensive turbulence scheme, as discussed by Sun and Chang (1986a,b), Sun (1986, 1988, 1989), Sun (1993a,b), and Chern (1994). The PRCM also includes radiation parametrizations (Liou *et al.*, 1988, Chou and Suarez, 1994; Chou *et al.*, 2001) and cumulus parametrizations (Kuo, 1965, 1974; Anthes, 1977; Molinari 1982). The forward–backward scheme is applied in the Arakawa C grid to permit better computational accuracy and efficiency (Sun, 1980, 1984a). The fourth-order advection scheme (Sun, 1993c) is used to calculate the advection terms. A local reference is applied to calculate the pressure gradient force, which significantly reduces the error of the pressure gradient terms in the σ_p coordinate over

steep topography (Sun, 1995a). Recently, the transport of dusts and an atmospheric chemistry module have been added by Yang (2004a,b). We are incorporating the mass-conserved, positive-definite semi-Lagrangian scheme (Sun *et al.*, 1996; Sun and Yeh, 1997; Sun and Sun, 2004; Sun, 2007) and the sea-ice-mixed layer ocean model (Sun and Chern, 1998) into the PRCM, as shown in Fig. 1.

2.2. *Weather simulations*

The PRCM has been applied by Sun and Hsu (1988) to simulate cold air outbreaks over the East China Sea. They showed that the cloud, which has quite different properties than the cold air beneath the cloud base, was formed due to the warm and humid air being lifted by the cold air. This is in good agreement with observations. Hsu and Sun (1991) used the PRCM to simulate one of these cold air outbreak events, reproducing the three-dimensional mesoscale cellular convection, which has a horizontal wavelength of 20–30 km. The model has also been used to study air mass modification over Lake Michigan (Sun and Yildirim, 1989), baroclinic instability and frontogenesis (Sun, 1990a,b), as well as cyclogenesis and the life cycle of cyclones (Yildirim, 1994).

Sun and Wu (1992), applying the PRCM, were the first to successfully simulate the formation and diurnal oscillation of a dryline. Their results show that under a favorable combination of a strong soil moisture gradient, a terrain slope, and a vertical wind shear in Oklahoma and Texas in the late spring and early summer, the dryline can form within 12 hours in the absence of an initial atmospheric moisture gradient. Their results also show that the dryline moves eastward during the daytime, due to a strong mixing of the air near the dryline with the warm and dry westerly wind aloft (which descends from the Rocky Mountains) on the west side, quickly diluting the low level moist air coming from the southeast. At night, the cool, moist air on the east side continues moving into the deep, (still) hot, dry air on the west side, but is no longer vertically mixed and dissipated because the convection ceases to develop due to longwave radiative cooling at the surface. Hence, the dryline moves westward at night, as shown in Fig. 2. Their simulations (Fig. 3) also reproduce

a deeper intrusion of the moisture field far above the inversion of the potential temperature field, as observed [Fig. 9 of Schaefer (1974)]. A low-level jet, low-level convergence, and strong upward motion form along the dryline, which are consistent with the inland-sea breeze theory proposed by Sun and Ogura (1979; schematic diagram in Fig. 12 of their paper). These simulations provide an explanation for the frequent occurrence of a dryline in the Great Plains in the late spring and early summer, which becomes a favorable zone for storm development.

Chern (1994) has successfully simulated the two surface low pressures observed with severe winter storms (Fig. 4) in the US, which shut down the highways in the High Plains for a week but, at that time, were poorly predicted by the NWS model. With the PRCM, Haines *et al.* (1997) successfully simulated the ice and super-cooled liquid water observed by aircraft, the lee vortex observed by radar, and surface precipitation of the Denver basin's Valentine's Day storm (VDS) on 13 February 1990.

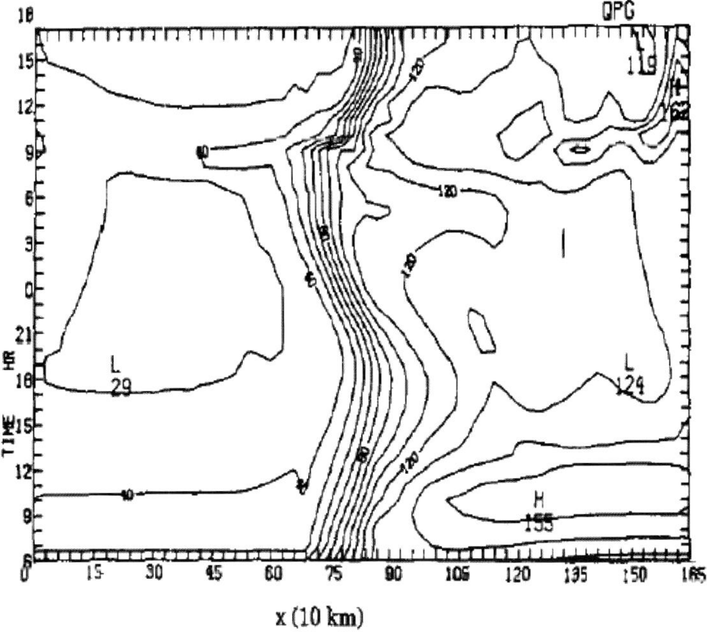

Figure 2.   Time-space variation of humidity q_v at $z \approx 10\,\mathrm{m}$, contour interval of $1\,\mathrm{g\,kg^{-1}}$ labels scaled by 10, dryline moves eastward during daytime but retreats at night (Sun and Wu, 1992).

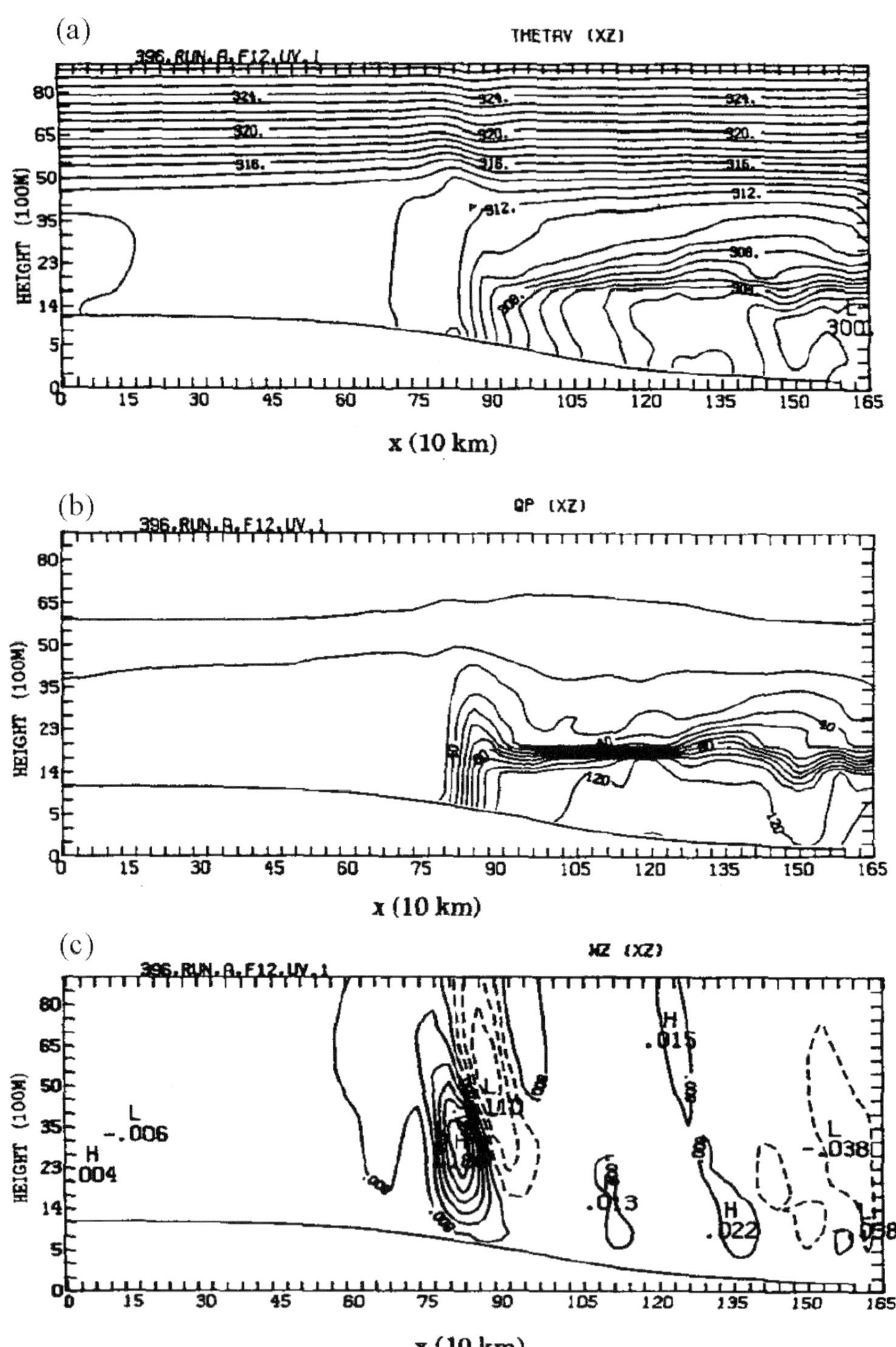

Figure 3. Cross section of the simulated (a) θ_v, (b) total water content q_w, (c) vertical velocity w. The intrusion of the moisture is much deeper than the potential temperature field along the dryline (Sun and Wu, 1992).

(a)

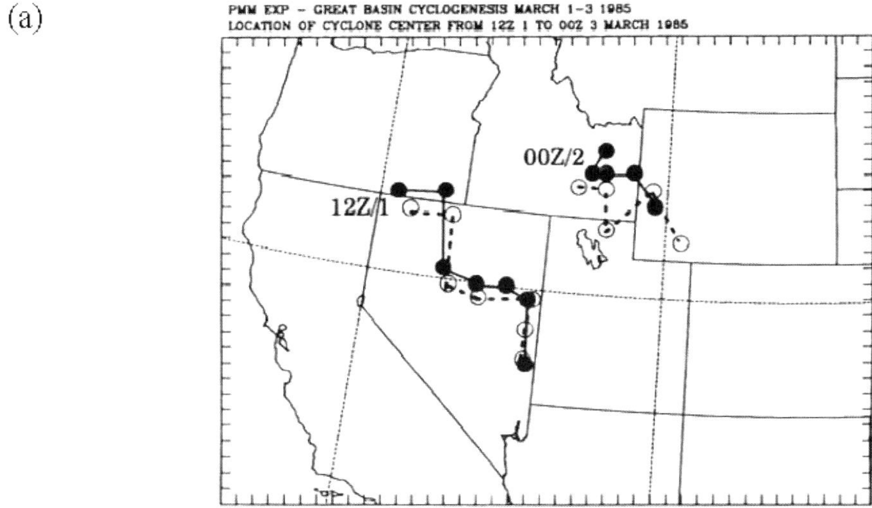

(b)

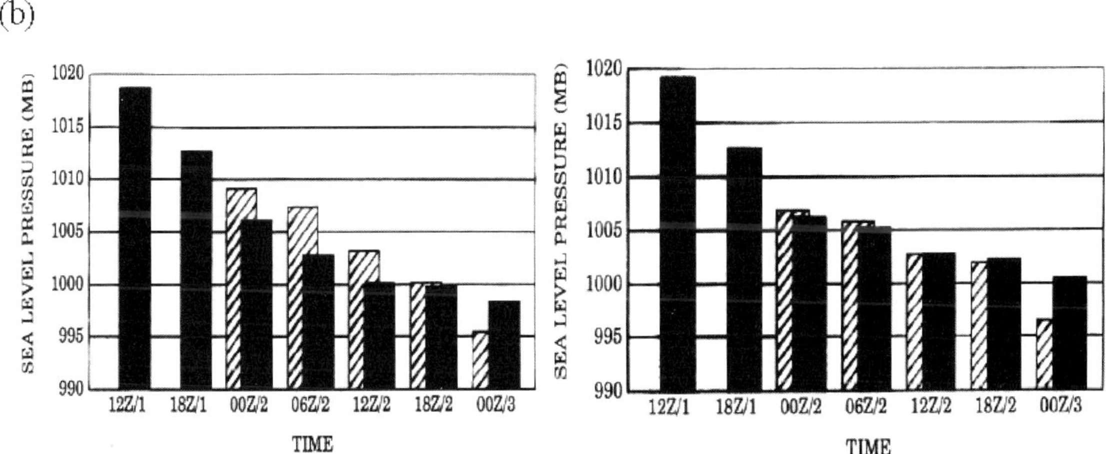

Figure 4. (a) Locations of the cyclones at 6-hr interval. Solid circles represent observation from ECMWF/TOGA analysis. Open from PRCM. The cyclones over Nevada (eastern Idaho) first appeared at 1200 UTC1 (0000 UTC 2) March 1985. (b) Observed (left) and simulated (right) surface pressure at low pressure centers in Nevada (filled bar) and Idaho (hatched bar) (Chern, 1994).

The results show that the PRCM is capable of reproducing the observed mesoscale/synoptic systems, clouds, and precipitation.

2.3. *Applications to Taiwan and East Asia*

Sun *et al.* (1991) applied the PRCM with a free-slip surface to generate a lee vortex and an area of strong wind in northwestern Taiwan under an easterly flow, and a lee vortex to the southeast of the island under a southwesterly flow. The budget of the vertical vorticity component showed that not only the tilting term, as proposed by Smolarkiewicz and Rotunno (1989), but also the stretching and friction terms are important to the formation of lee vortices. With detailed physics and surface parametrizations,

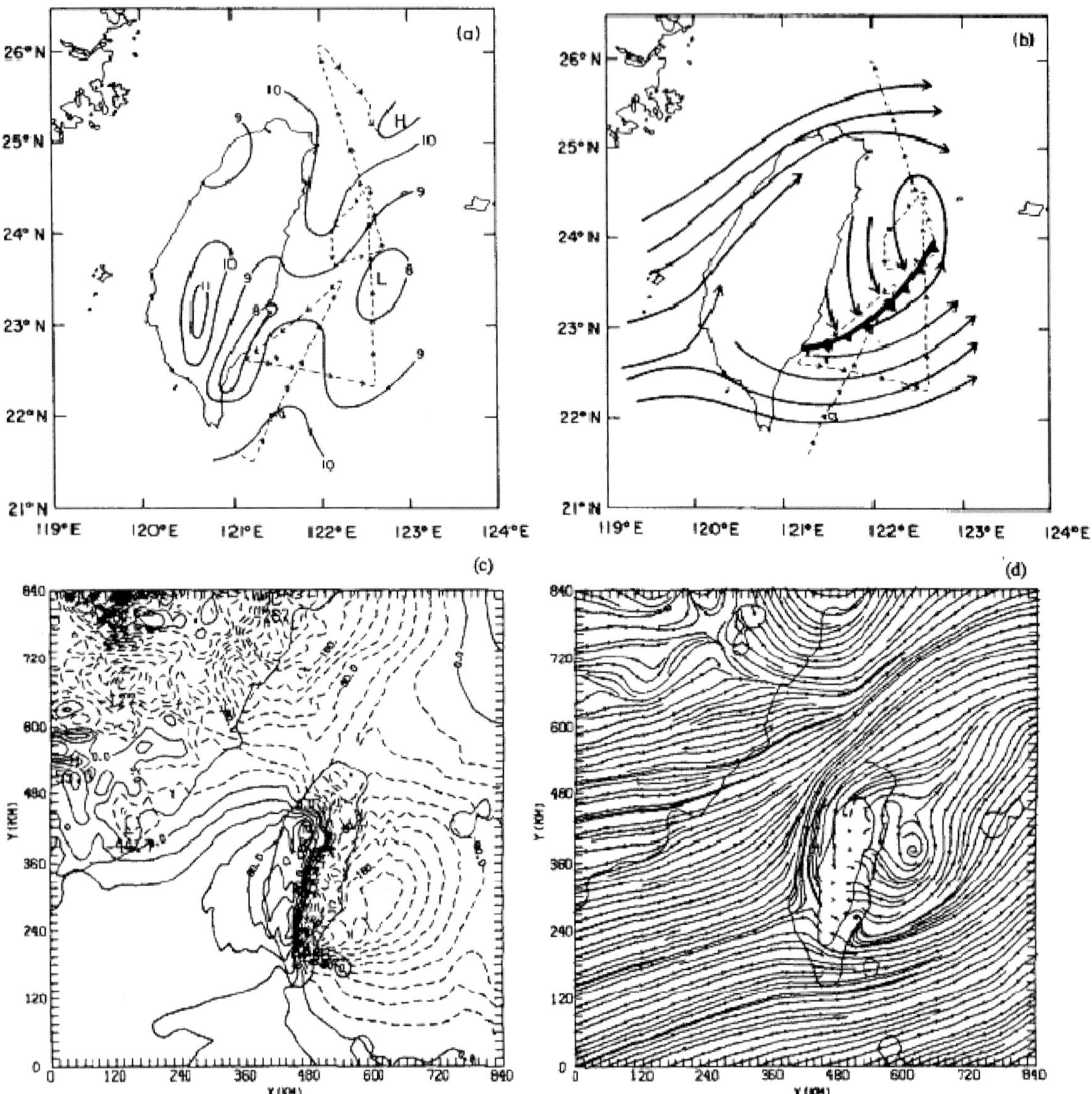

Figure 5. (a) Observed sea surface pressure, (b) observed streamline at 900 mb by airplane (from Kuo and Chen, 1990), (c) simulated sea surface pressure, and (d) simulated streamline at $z = 1$ km at 0200 LST 17 May 1987 (Sun and Chern, 1993).

Sun and Chern (1993) reproduced the observed mesolow, lee vortices, and downslope wind on the lee of the Central Mountain Range (CMR) during TAMEX (Taiwan Area Mesoscale Experiment) IOP(intensive observation period)-2. The lee vortex and the mesolow moved northeastward with time, and were located to the east of Taiwan at 0200LST, 17 May 1987, as shown in Fig. 5. It is also noted that the air moving over the CMR became drier and warmer than the surrounding environment and formed the mesolow to the south of the lee vortex. Hence, no meso-β scale front, which was suggested by Kuo and Chen (1990), developed around the vortex, as shown by smooth streamlines in Fig. 5(d). The model also captured the diurnal oscillation of the land–sea breeze and the formation of the clouds on both sides of the CMR, as observed during fair weather. Furthermore, the existence of mountain waves explains the observed light precipitation on the windward side with a clear sky near the peak of the CMR. Hence, the numerical model can provide plausible physical explanations of phenomena which have otherwise been misinterpreted according to incomplete observational data, such as the mesofront associated with a mesovortex and the distribution of rainfall in Taiwan in that particular situation.

Sun and Chern (1994) further investigated the formation of lee vortices and vortex shedding for low Froude number flow (the Froude number is defined as $\text{Fr} = U/NH$, where U is the mean flow in the upstream, N is the buoyancy frequency, and H is the height of the mountain in a rotational fluid. A pair of symmetric vortices forms on the lee side of the mountain in an irrotational, symmetric flow; but vortex shedding develops in a rotating system, because of the buildup of high pressure from the adiabatic cooling of the ascending motion on the windward side of the mountain, which enhances the anticyclonic circulation and produces a stronger wind on the left side (facing

downstream) of the mountain. The anticyclonic flow associated with the relative high on the windward side of the mountain was also observed by Sun and Chern (1993, 2006). Vortex shedding can also be developed by any asymmetry in the wind field, mountain shape, surface stress, etc.

The PRCM also faithfully reproduced the observed severe front during TAMEX IOP-2 (Sun *et al.*, 1995). The well-developed front extended from Japan and the Japan Sea through Taiwan into the South China Sea and Vietnam. The segment of the front between Taiwan and Japan is characteristic of midlatitude fronts, but it is a typical Mei-yu front in the lower latitudes, as reported by Hsu and Sun (1994) and many others: the southern segment has a large moisture gradient but a weak temperature gradient with an accompanying low-level jet on the moist side, which transports moisture from the warm ocean into the system.

In addition to the formation of a lee vortex, the front can be deformed by the CMR, as shown in many satellite images (Chen *et al.*, 2002) and surface maps (Chen and Hui, 1990). Figure 6(a) shows a simulated deformation of the cold front after a 30-hour integration (Sun and Chern, 2006), and Fig. 6(b) is the ECWMF

(a)

(b)

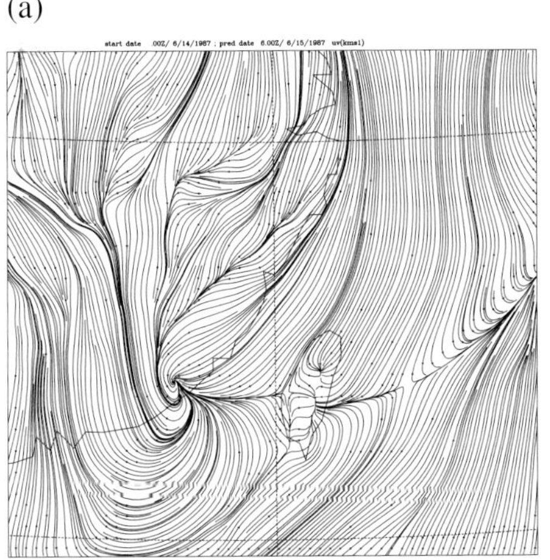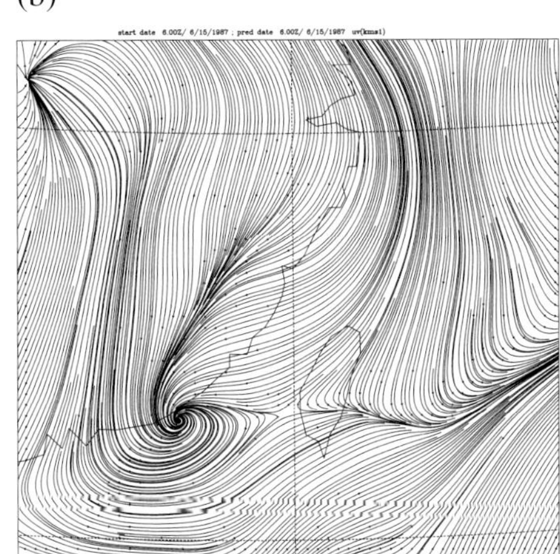

Figure 6. (a) Simulated streamline from the PRCM (after 30 h integration) at 0600 UTC15 June 1987 at $z \approx 25\text{m}$; (b) streamline from ECMWF reanalysis (from Sun and Chern, 2006).

reanalysis. The large-scale patterns are quite comparable, except that the deformation of the front by the CMR is missing in Fig. 6(b) due to a lack of observations in the Pacific and/or the coarse resolution used in the ECWMF. The momentum budget shows that the friction, ageostrophic forcing, and nonlinear advection terms are important for the propagation of the front, which has been frequently misinterpreted according to the theory of density current in the irrotational fluid (Chen and Hui, 1990) or trapped Kelvin waves. For more discussion, see Sun and Chern (2006).

2.4. *Regional climate studies*

The results discussed in the previous sections show that the PRCM is a useful tool for forecasting and studying short-term mesoscale and synoptic disturbances. The model has also been applied to study the monthly and seasonal variations of regional climate and the hydrological cycle. Bosilovich and Sun (1999a,b) applied the PRCM to study the 1993 summer flood in the Mississippi Valley. The model reproduced the observed precipitation associated with the transient synoptic waves in June and the Mesoscale Convective System (MCS) in July (Fig. 7) due to a change in atmospheric stability. Sensitivity tests showed that without the local surface source of water vapor, the flooded region's atmospheric hydrological

budget reduced to an approximate balance between precipitation and moisture flux convergence. Comparisons with the control simulation hydrology estimated that 12% and 20% of precipitation had a local source for June and July 1993, respectively (Bosilovich and Sun, 1999b).

Many scientists have applied numerical models to study the cause(s) of this drought. However, discrepancies exist among the different hypotheses. For example, Oglesby and Erikson (1989) and Oglesby (1991) demonstrated the persistence of an imposed soil moisture anomaly in the US using National Center for Atmospheric Research Community Climate Model 1 (NCAR-CCM1). However, also using NCAR-CCM1, Sun *et al.* (1997) carried out four experiments, including using the climatological SSTs (control case), 1988 SSTs, dry soil moisture anomaly (25% of soil moisture in May in the normal year between 35 and 50°N in the US), and 1988 SSTs and dry soil moisture anomaly, to study the effect of SST and soil moisture. For each experiment, three model simulations were performed and were initialized from arbitrary conditions. The results show that the 1988 SST did not cause the simulated weather pattern over the US to be drier than the control case with climatological SST. The ensemble mean with perturbed soil moisture experiment did show less precipitation than the control; however, due to the large variance in the data, the reduction in precipitation was not statistically significant.

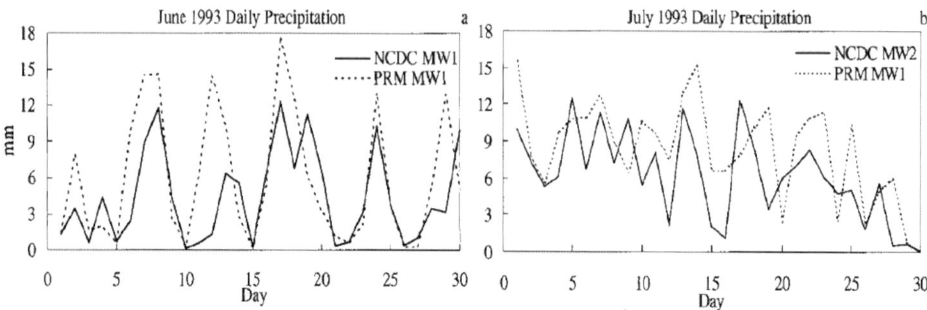

Figure 7. Simulated from PRCM and NCDC observation of daily integrated precipitation over flooding area in the midwestern US for: (a) June associated with synoptic waves and (b) July 1993 associated with mesoscale convective system (Bosilovich and Sun, 1999a).

The experiment with the soil moisture anomaly and 1988 SST obtained the most significant differences, specifically in the reduction of precipitation in the US. This may indicate that the 1988 severe drought may have been caused by a combination of dry soil and a special large scale weather pattern related to the SST, rather than by either the SST or dry soil alone.

Using the observed SST and the ECMWF analysis as initial and lateral boundary conditions, the PRCM (Sun *et al.*, 2004) reproduced a strong warm ridge at 500 hPa in North America over a hot, dry land. The monthly precipitation and soil moisture were far below the normal values in the Midwest and the Gulf states, but above normal in the Rocky Mountains, consistent with observations. It is noted that the PRCM reproduces not only the waves passing through the lateral boundaries, but also the development and decay of the disturbances inside the domain, which are not shown here. A sensitivity test also shows that monthly precipitation could significantly increase using the saturated soil moisture as the initial condition. However, the soil would dry up eventually, because the wet soil does not provide a positive feedback with the low-level jet. Hence, the large-scale weather pattern in the early summer of 1988 might have triggered the dry episode by forming a ridge in North America. This ridge gradually cut down precipitation in the Gulf states and Midwestern region. The soil became dry and hot, which further intensified the blocking of the warm ridge in the US. Hence, dry soil had a positive feedback on the severe drought in the summer of 1988. But the dry soil alone was not the cause of the drought, for in spite of even less soil moisture in early July, observed precipitation amounts returned to normal after mid-July 1988.

Min (2005) recently applied the PRCM with the new snow-land-surface module developed by Sun and Chern (2005) to study the 1997 spring flooding in Minnesota and the Dakotas due to spring snowmelt. With a wet soil, the north-central US experienced horrific conditions over the winter of 1996–97. An enormous snow pack was built up by blizzard after blizzard from the second half of November through January; many areas had more than 250 cm of snowfall. These amounts were as much as 2–3 times the normal annual amount. Although February and March were quite dry, frigid conditions throughout much of the winter ensured that as much as 25 cm of snow water equivalent remained on the ground when the spring melt period began in March. Early March brought temperatures below normal, delaying the onset of snowmelt. Significant melt of the deep snow started with particularly warm conditions at the end of March and in early April. At this time, many rivers in South Dakota, southern Minnesota, and southern North Dakota were rising, in some cases well above the flood stage. With a comprehensive snow–soil physics included in the PRCM, the simulated results of snow depth and precipitation rate can be dramatically improved (not shown in the figures). Our study revealed the importance of the detailed physics/parameters in a model, which can influence results in faraway regions. In addition, we also showed that the land properties can affect the flow pattern in the upper level as high as 200 hPa.

The PRCM has also been applied to study the East Asia Monsoon and short-term regional climate, by Sun and Chern (1999), Sun (2002), Hsu *et al.* (2003), Yu *et al.* (2004a,b), Hsu *et al.* (2004), etc. Figure 8 shows that the jet at 850 mb migrated northward over the summer of 1998 (Sun and Chern, 1999). Heavy precipitation occurred in Central China in midsummer, moving to Korea and northern and northeastern China in August. The front and the low-level jet dissipated after mid-August, as observed.

In simulations of the heavy flooding over Korea and China between 30 July and 18 August 1998, the model results show that the heavy rainfall along the Baiu/Mei-yu front was due to a combination of: (1) an anomalous

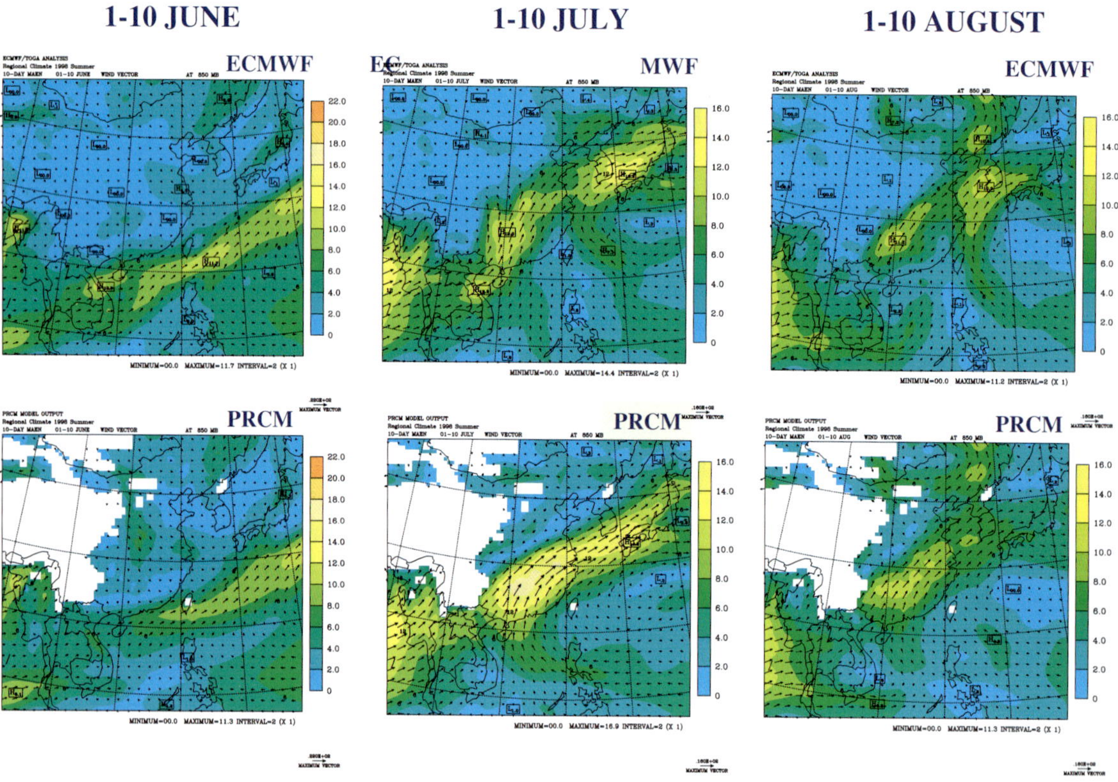

Figure 8. Ten day mean wind vector and isotachs (ms^{-1}) at 850 mb for the first 10 days of June, July, and August of 1998 (Sun and Chern, 1999).

850 hPa subtropical high, (2) a stronger baroclinicity around 40°N over eastern Asia and a low pressure located to the north of the front, and (3) excessive evaporation from the abnormally wet, warm land (Sun, 2002). The precipitation ended on 18 August, when the subtropical high retreated and the low pressure in Manchuria moved away from the Asian continent. The model reproduced well the observed baroclinic waves to the north, the subtropical high and low-level jet to the south, and the front with heavy precipitation extending from southern China and the Korean peninsula to Japan. High correlations (Table 2) were also found for most variables between the model simulation and the ECMWF reanalysis for the 20-day means. However, it is also noted that the simulated low pressure in Manchuria, which was to the north of the Mei-yu front, propagated more slowly than

observed, which may be responsible for the deepening of this low as well as for excessive heavy precipitation in the area (Fig. 9). It is also noted that the model generated a wet bias of precipitation in mountainous areas and a too strong subtropical high in the Pacific.

The PRCM also reproduced well the spatial distribution of mean surface temperature and mean sea level pressure during ten summers over East Asia as well as the temporal variation of mean vorticity over the South China Sea (Fig. 10; Yu *et al.*, 2004a,b; Hsu *et al.*, 2004).

2.5. *Simulations of dust and ozone*

Regional models coupling atmospheric chemistry have been applied with the PRCM to predict the transport and dispersion of trace gases and aerosols. Yang (2004a) has coupled the

Table 2.   20-day mean statistics of PRCM compared with ECMWF (Sun 2002).

FIELD	LEVEL	BIAS	RMS	COR
Mean Sea Level Pressure	SFC	$-4.80E+01$	$1.57E+02$	0.95
AirQv (kg/kg)		$2.45E-03$	$2.81E-03$	0.96
Temperature (K)		$7.33E-01$	$1.75E-00$	0.97
Wind vector		$7.81E-01$	$1.26E-00$	0.84
Height (m)	200 hPa	$3.70E+01$	$4.51E+01$	0.99
	500 hPa	$1.25E+01$	$1.77E+01$	0.98
	850 hPa	$2.54E-01$	$1.03E+01$	0.97
Temperature (K)	200 hPa	$7.44E-01$	$1.17E-00$	0.9
	500 hPa	$6.30E-01$	$8.55E-01$	0.98
	850 hPa	$1.32E-01$	$1.22E-00$	0.94
Qv (kg/kg)	500 hPa	$-1.34E-04$	$6.59E-04$	0.83
	700 hPa	$3.01E-04$	$1.22E-03$	0.82
	850 hPa	$1.13E-03$	$1.61E-03$	0.92
Wind vector	200 hPa	$-1.60E-00$	$3.98E-00$	0.91
	500 hPa	$-7.21E-01$	$2.14E-00$	0.8
	850 hPa	$1.28E-00$	$2.16E-00$	0.83

(a)   (b)

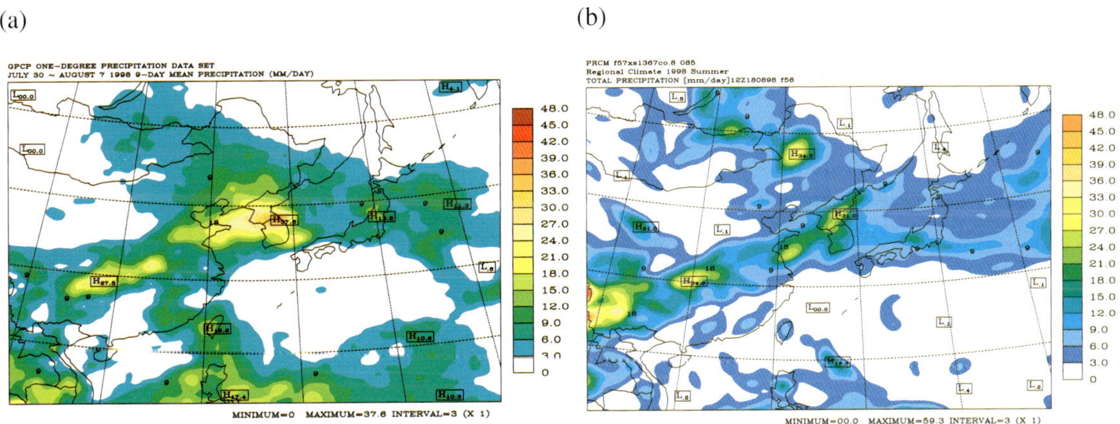

Figure 9.   (a) GPCP day mean precipitation; and (b) simulated day mean precipitation during July 30–August 18, flooding spread in Manchuria, Korea, and Yangtze River Valley. The model missed the tropical depression from southern boundary due to the coarse resolution (Sun, 2002).

PRCM and the dust/chemistry module as illustrated in Fig. 11 to simulate the distribution of aerosols and the interactions between aerosols and regional climate during April 1998.

The dust module consists of: (a) a dust source function derived by Ginoux et al. (2001) based on a 1°-by-1° terrain and vegetation data set derived from Advanced Very High Resolution Radiometer (AVHRR) data (DeFries and Townshend, 1994); (b) particle sizes, which are a function of the source region's soil properties (Tegen and Fung, 1994); seven-size bins (0.1–0.18 μm, 0.18–0.3 μm, 0.3–0.6 μm, 0.6–1 μm, 1–1.8 μm, 1.8–3 μm, and 3–6 μm, with corresponding effective radii of 0.15, 0.25, 0.4, 0.8, 1.5, 2.5, and 4 μm) are applied for dust sizes (Yang, 2004a); (c) the threshold friction velocity, defined as the horizontal wind velocity required to lift dust particles from the surface, is a function of particle diameter (Marticorena and

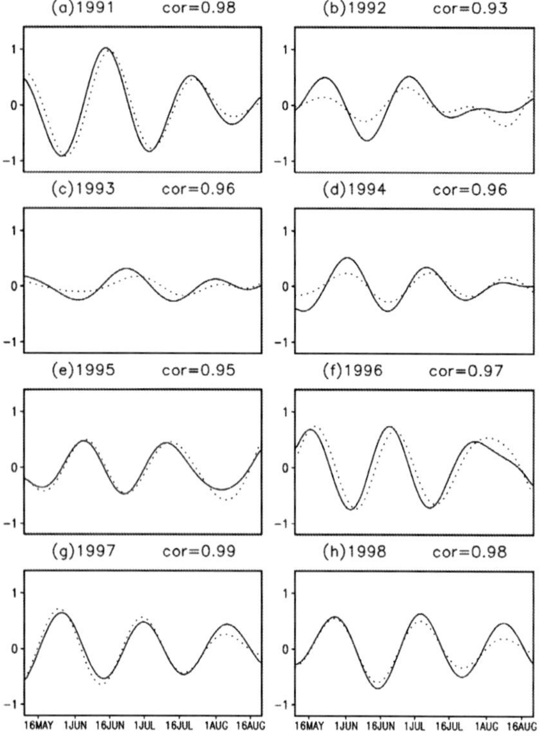

Figure 10. The mean vorticity over South China Sea (15–20°N; 110–120°E) during May–August from 1991–1998, solid lines are calculated from ECMWF analysis and dashed lines are PRCM results (Yu *et al.*, 2004a,b). Correlation coefficients between ECMWF analysis and PRCM results for individual years are also shown in each panel.

Bergametti, 1995) and soil wetness; (d) dust emission, which depends on source function, surface wind speed, threshold velocity, and the fraction of each size class; and (e) transport and removal processes — dust aerosols in the PRCM dust are transported by advection, dispersion, and subgrid cumulus convection, and are removed by wet and dry depositions. The dry deposition of dust aerosols is assessed through the gravitational settling for each model vertical layer and surface deposition velocity. The removal of dust aerosols by wet deposition is calculated using the model precipitation rate at all model levels, for both stratified and convective clouds. PRCM dust is capable of reproducing observed weather and satellite images of

dust during the 17-day continuous integration of April 1998 (not shown in the figures).

Yang (2004b) also included the chemistry mechanism of SAPRC97 (Statewide Air Pollution Research Center) (Carter *et al.*, 1995, 1997), which is a detailed mechanism for gas-phase atmospheric reactions of volatile organic compounds (VOCs) and oxides of nitrogen (NO_x) in urban and regional atmospheres. Coupling SAPRC97 and PRCM dust, Yang (2004b) showed that the model can reproduce the observed ozone concentration in the sky and at the surface in the US after a spin-up period of two days.

Wu *et al.* (2003) and Hsu (2001) applied the PRCM to study the transport of pollutants from a point source at the northern tip of Taiwan during four seasons in 1999. Figures 12(a) and 12(b) show the observed and simulated wind (after 48-hour integration) at 12Z on 31 October 1999; while the schematic diagram of pollutant transport and the vertically integrated concentration from the PRCM for the same time are shown in Figs. 12(c) and 12(d), respectively. They indicate that the plume moves around the western coast, the eastern coast, or even the entire island, depending on the detailed wind speed and direction under northerly wind.

3. Nonhydrostatic Models

3.1. *Basic equations*

In recent years, we have developed two non-hydrostatic models suited for studying a wide range of spatial scales of atmospheric systems. Both models are based on a fully compressible fluid in a terrain-following vertical coordinate. The National Taiwan University (NTU)/Purdue Nonhydrostatic Model (Hsu and Sun, 2001) is an explicit model in terms of its time integration scheme, while the other model (Chen and Sun, 2001), using a semi-implicit scheme, is an implicit model.

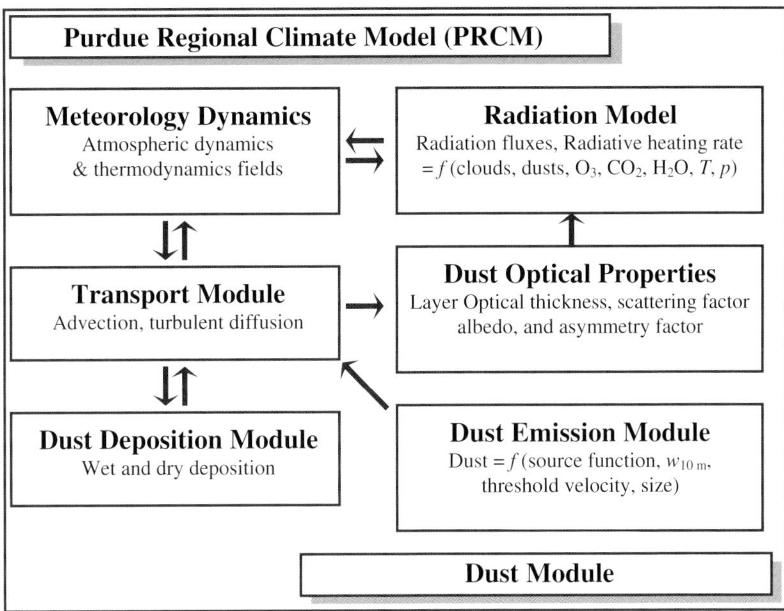

Figure 11. The schematic illustration of the components in the integrated PRCM-dust Model (Yang, 2004a).

The NTU/Purdue Nonhydrostatic Model uses a double forward–backward time integration procedure for treating both sound waves and internal gravity waves. The algorithm is stable, and does not generate computational modes. Although the integration time step is very small due to the CFL criterion imposed by treating high-frequency sound waves explicitly, this shortcoming is mitigated by using a time-splitting scheme and parallel computing. The explicit integration procedure is particularly well suited for parallel computing since a minimal amount of data transfer among CPUs is required with all partial differentiations calculated through local grid points. With the help of the National Center for High-performance Computing (NCHC) in Taiwan, the of has been written in Fortran 77 and Message Passing Interface (MPI) codes so that it can work on both supercomputers and PC clusters. The efficiency of the parallel processing depends on the number of grid points used in a simulation and the characteristics of the particular atmospheric circulation simulated. In one of the simulations

with no clouds, our model achieved over 95% efficiency for a 128-processor job (Hsu *et al.*, 2000). The high efficiency allows us to apply the model in studying resource-demanding problems, such as turbulence and local circulations.

In addition to its basic dynamic framework, the NTU/Purdue Nonhydrostatic Model takes into account many physical processes, such as land/ocean processes, cloud microphysics and atmospheric turbulence. The prognostic thermodynamic variables are the same as in the PRCM, with the addition of density in the continuity equation. The NTU/Purdue Nonhydrostatic Model is, thus, quite compatible with the PRCM. It will be possible in the future, with the improving computer technology, to nest the two models together for studying problems involving multiple-scale interactions.

Chen and Sun (2001) (CS) applied a multigrid solver in the Purdue Nonhydrostatic Model, which is a semi-implicit time integration scheme and can have a much larger time interval than that used in the NTU/Purdue

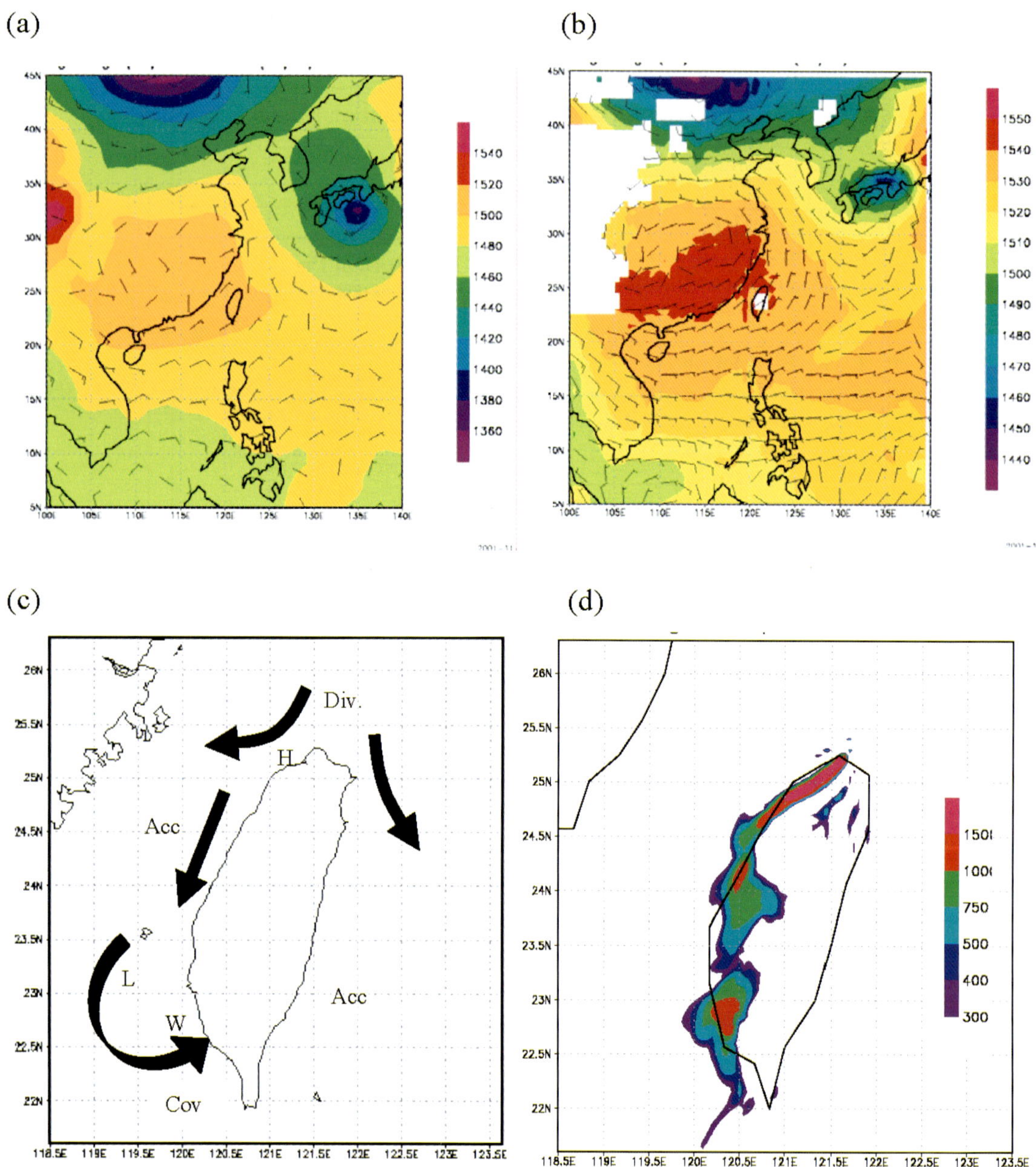

Figure 12. (a) ECWMF-reanalysis wind at 850 mb at 1200 UTC 31 Oct 1999, (b) same as (a) except for PRCM simulation (after 48 h integration), (c) schematic diagram of pollutant transport under northerly, and (d) vertical integrated concentration at 1200 UTC 31 Oct 1999 (after 48-h integration) (Wu *et al.*, 2003, Hsu, 2001) (Div: divergence, Acc: accelerateion, Con: convergence, W: westerly wind, H: high pressure, and L: low pressure).

Nonhydrostatic Model. In addition, a flexible hybrid coordinate in vertical was designed and used in that model. The two nonhydrostatic models produced consistent results for mountain waves, thermal convection, etc. The complicated multigrid approach has shown great potential for future development of nonhydrostatic models.

3.2. *Flow over mountains*

3.2.1. *Linear mountain waves*

Both (NTU/Purdue and CS) nonhydrostatic models have been validated with steady-state analytical solutions in several linear mountain wave situations, such as the two-dimensional, nonhydrostatic solution in Queney (1948); the two-layer, trapped-wave situation (Scorer, 1949); and the three-dimensional, nonhydrostatic situation (Smith, 1980). The model results are very close to the analytical solution in all cases. Figure 13 shows the horizontal distribution of the potential temperature anomaly, which is inversely proportional to the vertical displacement of airflow passing over a bell-shaped mountain at a height close the ground surface. The 2-hour simulation result is in good agreement with the analytical solution.

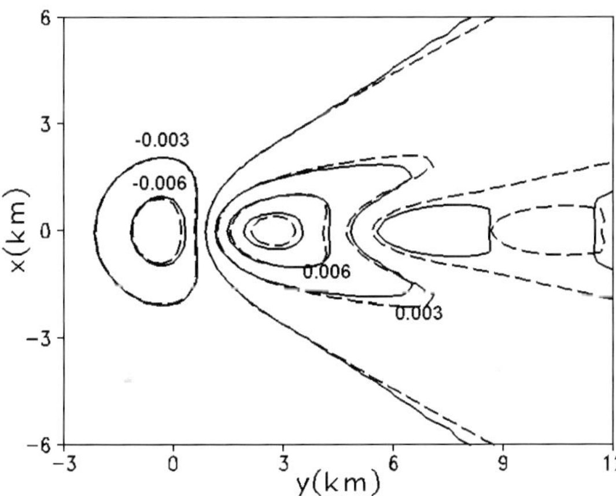

Figure 13. Horizontal distributions of the potential temperature anomaly for airflow passing over a bell-shaped mountain at the non-dimensional height of $Nz/U = \pi/4$. Solid contour lines correspond to the model result after 2 h. The grid interval is 300 m in all three directions. The dashed contour lines represent the analytical, steady-state solution in Smith (1980). The center of the bell-shaped mountain is located at the (0, 0) coordinate. The linear mountain waves is nonhydrostatic with $Na/U = 1$. The vertical static stability, mean wind speed and the half-width length of the mountain are $N = 0.01\,\mathrm{s}^{-1}$, $U = 10\,\mathrm{m\ s}^{-1}$ (from left to right in the figure), and $a = 1\,\mathrm{km}$, respectively. The contour interval is 0.003 K.

Hsu and Sun (2001) also found an error in Queney's classic 1948 paper. The surface pressure distribution in the two-dimensional, nonhydrostatic, linear mountain wave solution is off by about 100%, and the vertical displacement plotted in the original paper is not accurate either. The erroneous diagram has been cited repeatedly in many books (Gill, 1982; Smith, 1979) and papers over half a century. Our results, which have been confirmed by Chen and Sun (2001), can serve as a basis for future nonhydrostatic model development.

3.2.2. *11 January 1972 Boulder windstorm*

Both nonhydrostatic models have been used to simulate the famous 11 January 1972 Boulder windstorm (Lilly, 1978). The models reproduced the strong downslope wind and hydraulic jump for both free-slip surface and viscous surface simulations. With a free-slip boundary, the simulated hydraulic jump propagates downstream; however, the jump becomes stationary with a more realistic no-slip boundary condition (Fig. 14). The maximum wind also differs, depending on the lower boundary. For a viscous surface, it is around $45\text{--}50\,\mathrm{m\ s}^{-1}$, compared with $70\,\mathrm{m\ s}^{-1}$ or more for a free-slip surface (Sun and Hsu, 2005).

3.2.3. *White Sands simulation*

The NTU/Purdue model was applied to several well-observed real terrain cases for local circulations over the Organ Mountains in White Sands, New Mexico, USA, using 1 km grid resolution. The model successfully simulated strong winds over the valley, lee vortices, and downdrafts on the lee side for a weak prevailing wind situation (not shown in the figure). On the other hand, waves of strong–weak surface winds were simulated on the lee under a strong prevailing wind. The simulated wind patterns have been confirmed by upper air soundings and surface observations (Haines *et al.*, 2003). Oh (2003) also

(a)   (b)

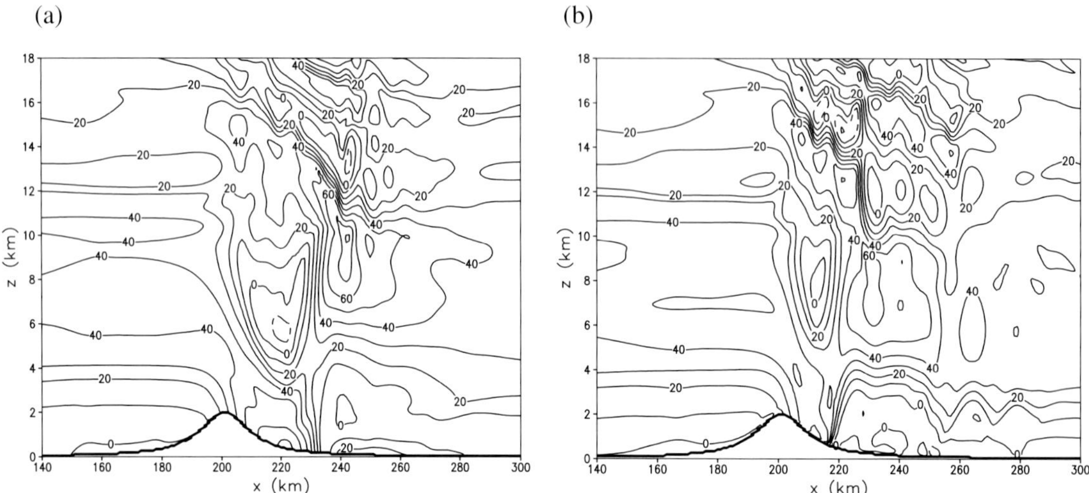

Figure 14. Simulated u-component wind for a 2-km mountain after 2.5 h of integration: (a) free-slip surface and (b) rigid surface (Hsu and Sun, 2001, and Sun and Hsu, 2005).

applied this model to study the flow over an idealized, bell-shaped mountain under different environments.

3.3. *Large eddy simulation (LES)*

With the use of an open lateral boundary and a very large number of grid points in the NTU/Purdue Nonhydrostatic Model, Hsu *et al.* (2004) were able to simulate the development of stratocumulus in a heterogeneous environment during a cold air outbreak event. A convective boundary layer (CBL) develops as very cold air originating from Siberia and China flows over the Japan Sea, the Yellow Sea, and the East China Sea during winter seasons. The CBL quickly deepens away from the coastline, with increasing fetch length and sea surface temperature. As the depth of the CBL changes, the embedded roll vortices (cloud streets) grow in size. The convection eventually becomes three-dimensional in the downstream region. The simulated convection also shows both 2D rolls and 3D cells in our simulation (Fig. 15). The increase of the CBL's depth and strength may result in changing cloud shapes.

4. Other Topics

4.1. *Turbulence, pollution and PBL*

In a higher-order turbulence parametrization, Sun and Ogura (1980) introduced a turbulence length scale in the stable layer that is related to the atmospheric stratification. They also included the equation for the potential temperature–humidity covariance $(\overline{\theta'q'})$ and were able to simulate the observed $(\overline{\theta'q'})$ and some third-order terms. Sun and Chang (1986a,b) and Sun (1986, 1988, 1989) extended Sun and Ogura's work by including the temperature-concentration covariance and using the observed turbulence length scale in the CBL (Caughey and Palmer, 1979) as a length scale in the model, successfully simulating the transport and dispersion of plumes in a convective boundary layer. Simulated results (Sun, 1988, 1989) were compared with the laboratory study by Deardorff and Willis (1975) for pollution released from point sources at different heights. Both laboratory and model simulations show that the plumes move upward from the point source near the surface, but the plumes from the elevated point sources descend quickly and form a high concentration zone near the

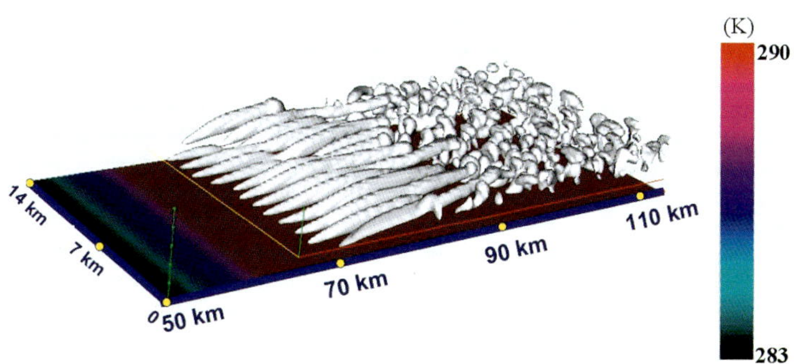

Figure 15. The simulated stratocumulus after 2-h integration during a cold air outbreak over a warm ocean. The background color shows the sea surface temperature distribution (scale indicated to the right of the figure; units in K). The thin red and orange lines identify the lowest cloud level at $z = 300\,\mathrm{m}$. Cold air with temperature of $280\,\mathrm{K}$ near surface comes from the left with wind speed of $10\,\mathrm{m\,s^{-1}}$. Cloud streets broke into three-dimensional cells in the downstream region. The grid intervals in x, y and z directions are 200, 100, and $50\,\mathrm{m}$, respectively (Hsu *et al.*, 2004).

surface before they move upward. The modeled mean plume height and the surface concentration were also comparable with observations (Willis and Deardorff, 1978, 1981). Sensitivity tests show that the temperature-concentration covariance is crucial to turbulence-pollution modeling, which had been ignored before. The simplified version of this turbulence parametrization is also used in the PRCM (Chern, 1994). Currently, MacCall (2006) is working on the third-order turbulence closure to study the turbulence in a stable boundary layer.

4.2. *Snow–vegetation–soil*

A comprehensive snow–vegetation–soil module has been developed for the PRCM (Wu and Sun, 1990a,b; Sun and Wu, 1992; Sun, 1993a,b; Bosilovich and Sun, 1995; Sun and Bosilovich, 1996; Bosilovich and Sun, 1998; Chern and Sun, 1998; Sun and Chern, 2005). It handles diffusion and transport of heat and water substance inside snow, vegetation, and soil as well as the processes of melting and freezing and the fluxes at the interfaces, etc. Figure 16 shows a comparison between the observations and model simulations of the soil temperature during summer. Sun and Chern (2005) simulated the changes

of the snow depth and soil temperature in the Sleepers Watershed Experiment during 1969–1974. The simulations are in good agreement with observations.

4.3. *Cloud streets and symmetric instability*

Kuo (1963) applied the linear theory to explain the cloud streets observed in a trough in the easterly wave (Malkus and Riehl, 1964). He successfully simulated the cloud streets forming along the wind shear in a dry, unstable atmosphere with a constant lapse rate, but failed to produce the larger cloud streets forming perpendicular to the wind shear. Using the real atmospheric stratification and including the effect of latent heat release, Sun (1978) demonstrated that observed clouds may be explained by the coexistence of two different types of clouds: the shallow convective-type cloud streets form along the wind shear, and the deep wave-type cloud streets develop perpendicular to the wind shear.

Sun and Orlanski (1981a,b) solved both linearized and nonlinear equations as initial value problems and confirmed that the two-day waves can be easily excited by the diurnal

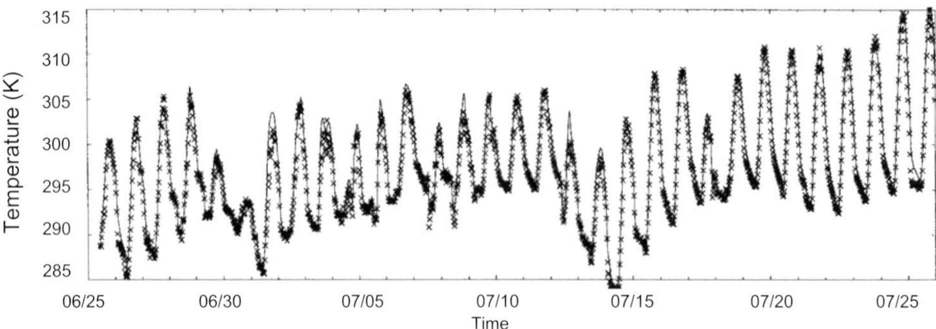

Figure 16. Comparison of model-simulated surface temperature (solid line) and observations (symbol x) during FIFE campaign, 25 June–25 July 1987 (Bosilovich and Sun, 1998).

oscillation of the land–sea contrast at lower latitudes ($<15°$). On the other hand, one-day and two-day waves may coexist at latitudes of up to 30°. These waves may correspond to the mesoscale cloud bands observed along coastlines with a space interval of a few ten to a few hundred kilometers (Fig. 1 in Sun and Orlanski, 1981a).

Integrating the linearized equations as a wet-symmetric instability problem, Sun (1984b, 1987) showed that the small-sized rainbands can organize into larger ones with strong narrow upward motions interspersed with weak and widespread downward motion, because the ascending air should be warmer than the descending air parcel to maintain the positive circulation. Those rain bands also propagated toward the warm side, which is the source of moisture in a rotating fluid. Sun (1995b) further proved that the y component of the earth rotation ($2\Omega\cos\phi$; where ϕ is the latitude) can significantly impact the growth rate of symmetric instability in the lower latitudes, because the x component wind u can either enhance or decrease the z component acceleration through the $u\,2\Omega\cos\phi$ term in the momentum equation.

4.4. *Cumulus parametrization scheme (CPS) and cloud models*

Haines and Sun (1994) developed a simple steady 1D cloud model to be used in the cumulus

parametrization scheme (Sun and Haines, 1996). This CPS also includes the Weisman and Klemp (1984) storm intensity according to the wind shear and buoyancy environment and allows three sizes, of clouds: small, medium, and large sizes, with different radius and cloud depth. The population of each cloud type is determined by the conservation of mass and moisture flux, and the ability to generate the fastest heating rate from the combined clouds. The CPS has been tested against the SESAME V storm-scale analysis from 2000 UTC to 2300 UTC on 20 May 1979, when the precipitation was almost exclusively convective in nature. They are in good agreement in the apparent heat source (Q1) and the apparent moisture sink (Q2), as shown in Fig. 17 for both strong and weak convections. The CPS was also successful in diagnosing the apparent heat source (Q1) and the apparent moisture sink (Q2) from purely convective sources even later in the SESAME V case, when a significant amount of stable type precipitation developed. The CPS has been incorporated in the PRCM to simulate the squall line observed during SESAME V. The simulated squall line formed about when and where the observed squall line did. Additionally, later deep convection to the south along the dry line was successfully simulated. However, it is very important and challenging to develop a better CPS with more rigorous assumptions and formation.

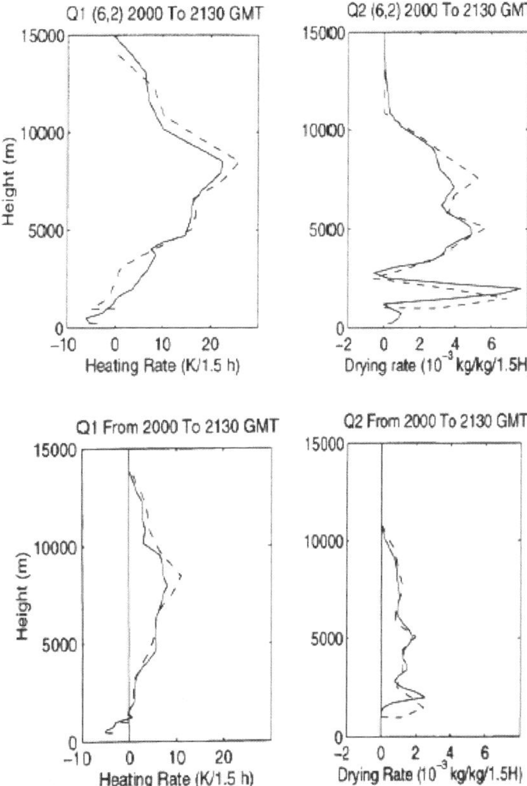

Figure 17. Comparison of observed (solid) and CPS diagnosed (dashed) from 2000–2130 UTC, 20 May 1979 for Q1 and Q2 at two locations during SESAME V experiment (Sun and Haines, 1996).

In order to avoid using a steady state cloud model as in Haines and Sun (1994), Chen and Sun (2002) have developed a time-dependent one-dimensional cloud model, which represents well the average properties of the clouds generated by the WRF model (Fig. 18). Furthermore, Chen and Sun (2004) developed a one-dimensional time-dependent tilting cloud model to represent the effect of wind shear. Hopefully, those cloud models can be incorporated into the CPS in the near future.

4.5. *Barotropic model and shear instability*

Sun and Chang (1992) applied a nonlinear barotropic model to study the barotropic instability (Kuo, 1949) of the modified hyper-tangent shear flows, which may represent the flow along a Mei-yu front or a cyclone family in the middle-to-high latitudes. In a β plane, the waves initially propagate as Rossby waves through the various basic states following linearized equations. Those waves move into the most unstable zone and, thus, grow more rapidly than in the weak shear zone. In the later stage, the nonlinear wave interaction becomes dominant and the transition among the modes becomes very difficult to predict. The simulations confirm that the results are very sensitive to external forcing, the basic wind field, the effect of β, as well as the initial perturbation field. This may suggest that predictability is very limited even in this simple 2D barotropic flow. The evolution of 2D shear flow has also been studied by Oh (2007) using the semi-Lagrangian scheme on the characteristic shallow water equations.

4.6. *Numerical schemes*

The results of a numerical prediction model depend upon the model equations, the physics, the initial and boundary conditions, and the numerical methods. Therefore, we have also been working on numerical schemes in order to provide more accurate results. In addition to diffusion equations (Sun, 1982) and a forward–backward scheme for inertial–internal gravity waves (Sun, 1980, 1984a), Sun developed an advection scheme which is more accurate than the popular Crowley fourth-order advection scheme (Crowley, 1969). Sun *et al.* (1996), Sun and Yeh (1997), and Sun and Sun (2004) also developed a mass-conserved, positive-definite semi-Lagrangian scheme. A forward trajectory and simple mass correction are applied in this scheme. The procedure includes: (a) constructing the Lagrangian network induced by the motion of the fluid from the Eulerian network and finding the intersections of the networks by a general interpolation from the irregularly distributed Lagrangian grid to the regularly distributed Eulerian grid,

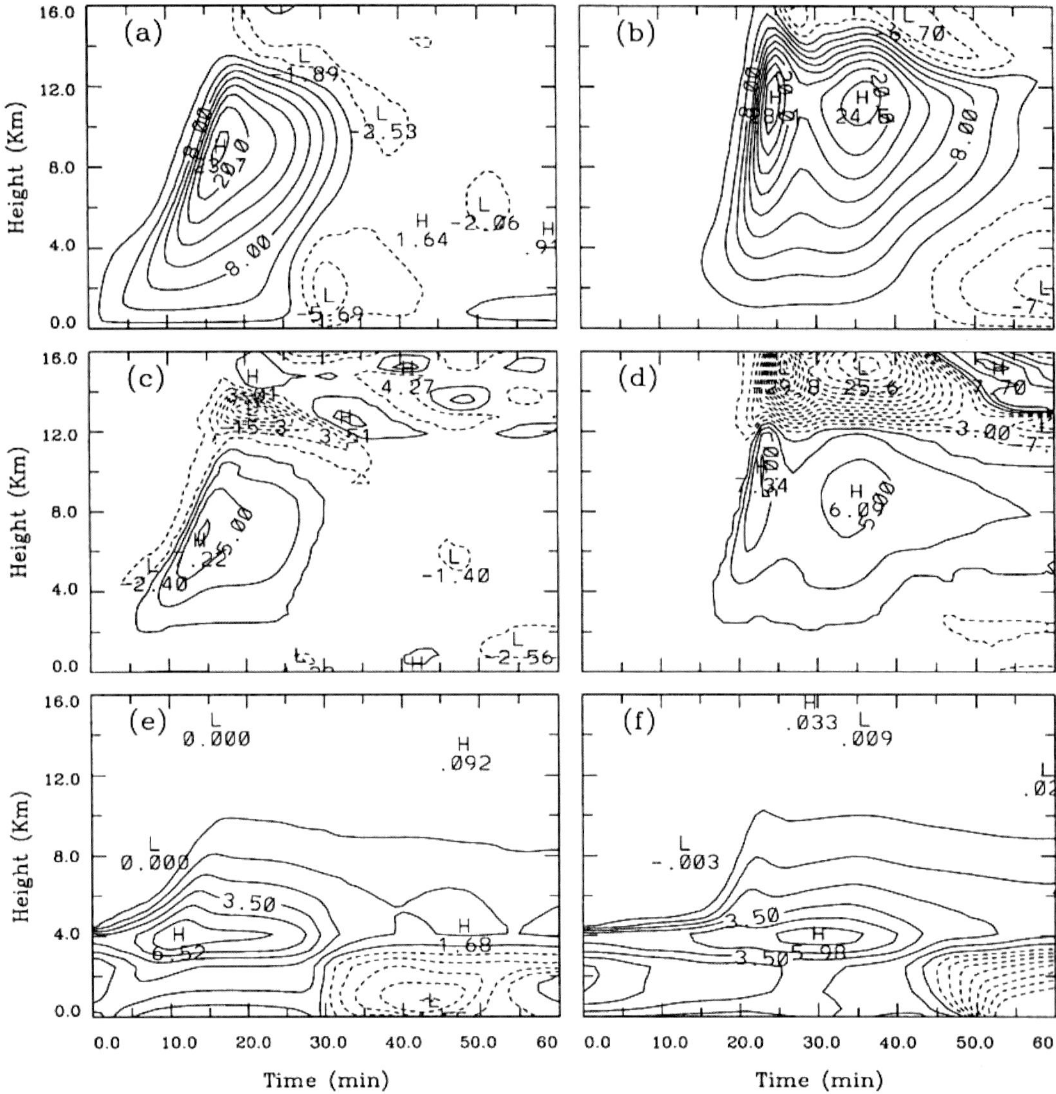

Figure 18. Time evolutions of the (a) vertical velocity (ms^{-1}), (c) potential temperature anomaly (K), and (e) moisture anomaly (kg kg^{-1}) from the one-dimensional cloud model, and the averaged (b) vertical velocity (m s^{-1}), (d) potential temperature anomaly (K), and (f) moisture anomaly (kg kg^{-1}) within the radius of 5000 m cloud from the WRF model. The values in (e) and (f) are multiplied by 1×103 (Chen and Sun, 2002).

(b) applying a spatial filter to remove the unwanted shortwaves and the values beyond the constraints, and finally (c) introducing a polynomial or sine function as the correction function that conserves mass, while inducing the least modification to the results obtained from (a) and (b). Numerical simulations of pure advection, rotation, and idealized cyclogenesis show that the scheme is very accurate compared with analytic solutions, as shown in Fig. 19 (Sun and Sun, 2004). With the variational techniques, this scheme is capable of reproducing the positive-definite results and conservation of total mass and total energy in the shallow water equations in both rotational and irrotational frames (Sun, 2007).

(a) (b)

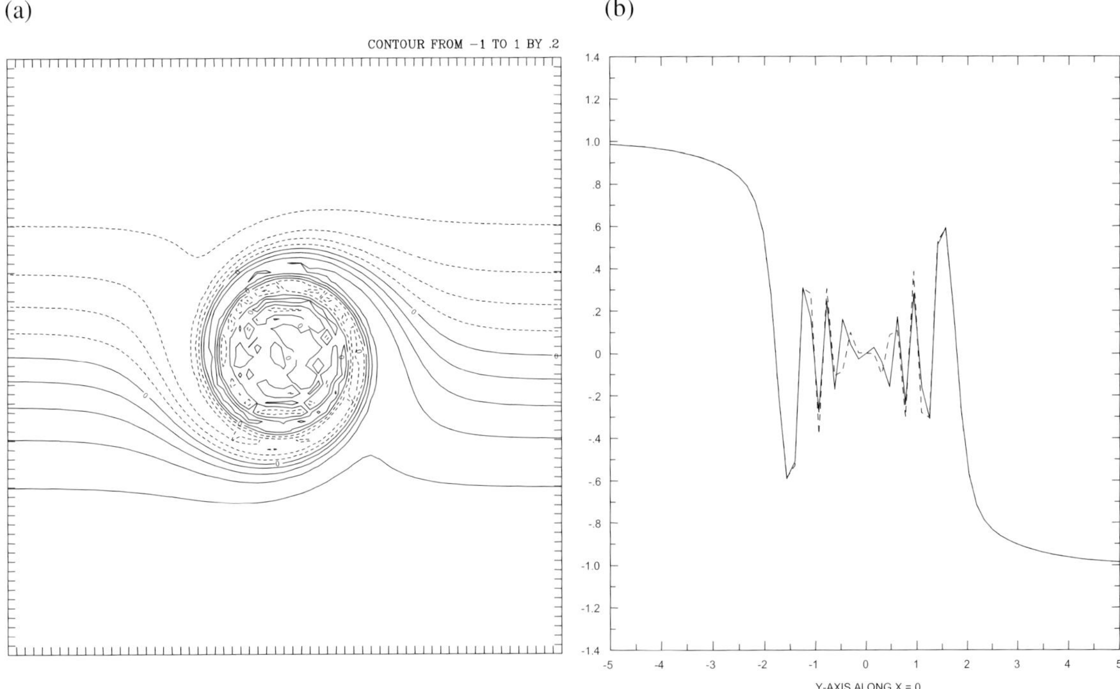

Figure 19. (a) Horizontal plane of numerical simulation and (b) vertical profiles of analytic solution (dashed line) and numerical result (solid line) of Doswell's idealized cyclogenesis after 16 time steps with courant number = 4.243, revolution = 4.386 and δ (transition width) = 2.0 with mass correction but without internet filter (Sun and Sun, 2004).

4.7. *2D shallow water equations*

Oh (2007) applied the semi-Lagrangian integration method (Sun *et al.*, 1996; Sun and Yeh, 1997) combined with the method of characteristics to the 2D shallow water equations to study the geostrophic adjustment problem, the shear instability, the collapse of a circular dam, the interaction of a vortex with the terrain, and the merging of vortices. His results show that the scheme is highly accurate in simulating flows involving a sharp gradient, such as the collapse of the cylindrical dam and the merging of two vortices. The characteristic approach is easier to interpret. Figure 20(a) shows the height at different times, and Fig. 20(b) the PV (potential vorticity), revealing that the height (or kinetic energy) cascades to the larger wavelength while the vorticity field cascades to the smaller wavelength in a 2D flow.

5. Summary

For the past several decades, numerical models have substantially improved in simulating short-term weather and long-term climate change. However, tremendous work is still required to improve the basic equations, numerical methods, resolvable and subgrid scale physical parametrizations, and initial and boundary conditions (Sun, 2002). More observational data is also required to provide better initial and boundary conditions. The reanalysis data, which has been used for initial and boundary conditions in weather/climate simulations, is a combination of model output and observations. More observations are also needed to validate the model results. Finally, we need better computational resources (both hardware and software) in order to develop new models with finer resolution and comprehensive

(a)

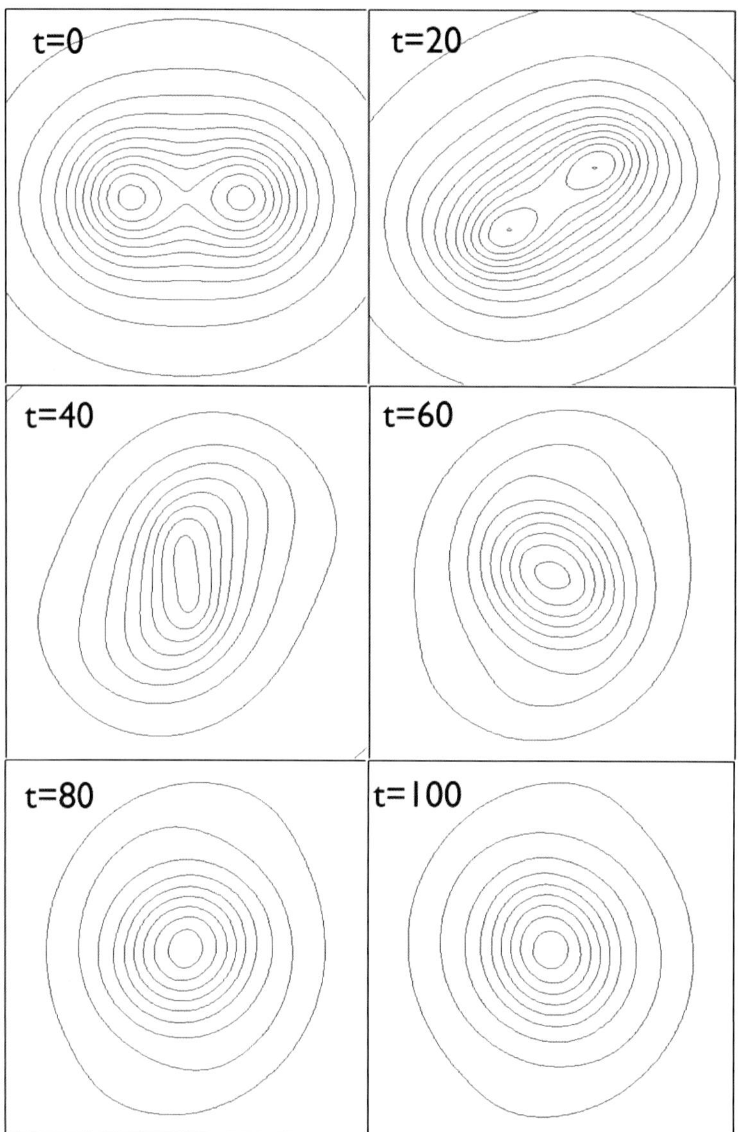

Figure 20.   Simulated (a) height field and (b) potential vorticity of a merger of two vortices. Results are shown in 20 s time intervals.

physics/parametrizations, as well as to display model results.

Numerical modeling is an exciting and challenging field. With the appropriate equations, numerical methods, and initial and boundary conditions, the models can reveal the spatial and temporal evolution of the processes involved. Models can also be used to simulate climate/environment in the past or in the future, which cannot be carried out in the laboratory or in field experiments. Hence, numerical models are the most important tool in weather forecasting and climate study. However, we should also be aware that uncertainty

(b)

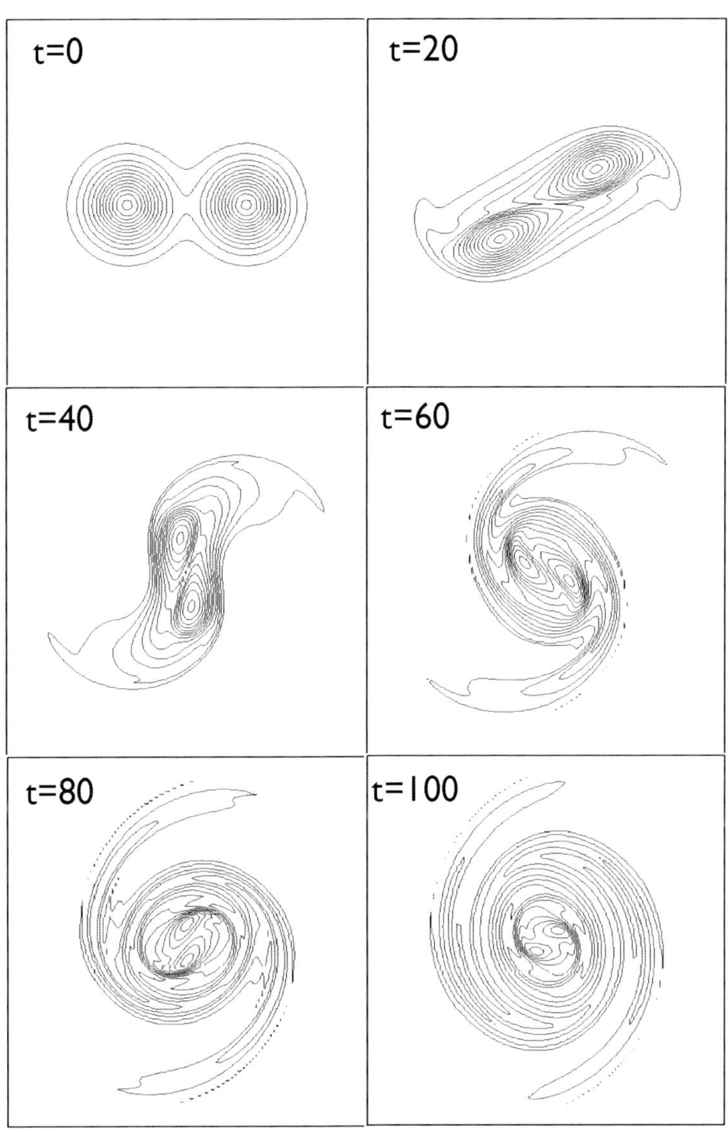

Figure 20. *(Continued)*

always exists in weather/climate models, which can be caused by errors in equations, physical parametrizations, resolution, initial and boundary conditions, numerical methods, etc. (Sun, 2006a,b).

[Received 19 January 2007; Revised 19 October 2007; Accepted 23 October 2007.]

References

Anthes, R. A., 1977: A cumulus parameterization scheme utilizing a one-dimensional cloud model. *Mon. Wea. Rev.*, **105**, 270–286.

Bosilovich, M. G., and W. Y. Sun, 1995: Formation and verification of a land surface parameterization for atmospheric models. *Boundary-Layer Meteor.*, **73**, 321–341.

Bosilovich, M. B., and W. Y. Sun, 1998: Monthly simulation of surface layer fluxes and soil properties during FIFE. *J. Atmos. Sci.*, **55**, 1170–1183.

Bosilovich, M. G., and W. Y. Sun, 1999a: Numerical simulation of the 1993 Midwestern flood: local and remote sources of water. *J. Geophys. Res.*, **104**, D1, 19415–19423.

Bosilovich, M. G., and W. Y. Sun, 1999b: Numerical simulation of the 1993 Midwestern flood: land–atmosphere interactions. *J. Climate*, **12**, 1490–1505.

Carter, W. P. L., J. A. Pierce, D. Luo, and I. L. Malkina, 1995: Environmental chamber studies of maximum incremental reactivities of volatile organic compounds. *Atmos. Environ.*, **29**, 2499–2511.

Carter, W. P. L., D. Lu, and I. L. Malkina, 1997: Environmental chamber studies for development of an updated photochemical mechanism for VOC reactivity assessment. Draft final report to California Air Resources Board Contract 92–345, Coordinating Research Council Project M-9, and National Renewable Energy Laboratory Contract ZF-2-12252-07.

Caughey, S. J., and S. G. Palmer, 1979: Some aspects of turbulence structure through the depth of convective layer. *Quart. J. Roy. Meteor. Soc.*, **105**, 811–827.

Chen, S. H., and W. Y. Sun, 2001: The applications of the multigrid method and a flexible hybrid coordinate in a nonhydrostatic model. *Mon. Wea. Rev.*, **129**, 2660–2676.

Chen, S. H., and W. Y. Sun, 2002: A one-dimensional time dependent cloud model. *J. Meteor. Soc. Japan*, **80**, 99–118.

Chen, S. H., and W. Y. Sun, 2004: An explicit one-dimensional time-dependent tilting cloud model. *J. Atmos. Sci.*, **61**, 2797–2816.

Chen, Y. L., and N. B. F. Hui, 1990: Analysis of a shallow front during the Taiwan Area Mesoscale Experiment. *Mon. Wea. Rev.*, **118**, 2649–2667.

Chen, T.-C., W.-R Huang, and W. A. Gallus Jr., 2002: An East Asian cold surge: case study. *Mon. Wea. Rev.*, **130**, 2271–2290.

Chern, J. D., 1994: Numerical simulation of cyclogenesis over the western United States. PhD thesis (Department of Earth and Atmospheric Sciences, Purdue University, W. Lafayette, IN, USA), pp. 178.

Chern, J. D., and W. Y. Sun, 1998: Formulation and validation of a snow model. A11A-06, 1998 Fall Meeting, AGU, supplement to EOS, AGU, 79, 45. F96. 10 Nov. 1998.

Chou, M. D., and M. J. Suarez, 1994: An efficient thermal infrared radiation parameterization for use in general circulation models. NASA Tech. Memo. 104606, Vol. 3, pp. 85.

Chou, M. D., M. J. Suarez, X.-Z. Liang, and M. M.-H. Yan, 2001: A thermal infrared radiation parameterization for atmospheric studies. NASA/ TM-2001-104606, NASA GSFC, Greenbelt USA, MD, USA.

Crowley, W. P., 1969: Numerical advection experiment. *Mon. Wea. Rev.*, **96**, 1–11.

Deardorff, J. W., and G. E. Willis, 1975: A parameterization of diffusion into mixed layer. *J. Appl. Meteor.*, **14**, 1451–1458.

DeFries, R. S., and J. R. G. Townshend, 1994: NDVI-derived land cover classification a global scale. *Int. J. Remote Sens.*, **15**, 3567–3586.

Doswell, C. A., III, 1984: A kinematic analysis associated with a nondivergent vortex. *J. Atmos. Sci.*, **41**, 1242–1248.

Gill, A. E., 1982: *Atmosphere–Ocean Dynamics.* (Academic, New York), pp. 662.

Ginoux, P., M. Chin, I. Tegen, J. M. Prospero, B. Holben, O. Dubovik, and S.-J. Lin, 2001: Sources and distributions of dust aerosols simulated with the GOCART model. *J. Geophys. Res.*, **106**, D17, 20255–20273.

Grell, G. A., J. Dudhia, and D. R. Stauffer, 1995: A description of the fifth-generation Penn State/ NCAR mesoscale model (MM5). NCAR/TN-398 + STR, (National Center for Atmospheric Research, Boulder, CO, USA), pp. 122.

Haines, A. P., and W. Y. Sun, 1994: A convective cloud model for use in a cumulus parameterization scheme. *Mon. Wea. Rev.*, **122**, 165–182.

Haines, P. A., J. D. Chern, and W. Y. Sun, 1997: Numerical simulation of the Valentine's Day storm during WISP 1990. *Tellus*, **49A**, 595–612.

Haines, P. A., D. J. Grove, W. Y. Sun, W. R. Hsu, and S. C. Tcheng, 2003: High resolution results and scalability of numerical modeling of wind flow at White Sand missile range. BACIMO (DOD atmospheric science and effects) conference (Sept. 2003, Monterey, California).

Hsu, K. J., 2001: Numerical simulations of local atmospheric circulation and pollution tracks in northern Taiwan. Taipei County Environment Protection Bureau Report, 235 pp.

Hsu, H. H., Y. C. Yu, W. S. Kau, W. R. Hsu, and W. Y. Sun, 2004: Simulation of the 1998 East Asian summer monsoon using Purdue regional model. *J. Meteor. Soc. Japan*, **82**, 1715–1733.

Hsu, H. H., W. R. Hsu, W. S. Kau, Y. C. Yu, and W. Y. Sun, 2003: Simulation of the 1998

East Asian Summer Monsoon using Purdue Regional Climate Model. The Second Workshop on Regional Climate Modeling for Monsoon System. (Mar. 4–6, 2003; Yokohama, Japan).

Hsu, W. R., and W. Y. Sun, 1991: Numerical study of mesoscale cellular convection. *Boundary-Layer Meteor.*, **57**, 167–186.

Hsu, W. R., and W. Y. Sun, 1994: A numerical study of a low-level jet and its accompanying secondary circulation in a Mei-yu system. *Mon. Wea. Rev.*, **122**, 324–340.

Hsu, W. R., and W. Y. Sun, 2001: A time-split, forward–backward numerical model for solving a nonhydrostatic and compressible system of equations. *Tellus*, **53A**, 279–299.

Hsu, W. R., W. Y. Sun, S. C. Tcheng, and H. Y. Chang, 2000: Parallel of the NTU–Purdue non-hydrostatic numerical model: simulation of the 11 January 1972 Boulder windstorm. The Seventh National Computational Fluid Dynamics Conference. (Aug. 2000, Kenting, Taiwan).

Hsu, W. R., J. P. Hou, C. C. Wu, and W. Y. Sun, 2004: Large-eddy simulation of cloud streets over the East China Sea during cold-air outbreak events. 16th Symposium on Boundary Layer and Turbulence, (Aug. 9–13, 2004; Portland, Maine, USA).

Kuo, H. L., 1949: Dynamically instability of two-dimensional nondivergent flow in a barotropic atmosphere. *J. Meteor.*, **6**, 105–122.

Kuo, H. L., 1963: Perturbations of plane Couetter flow. I: A stratified fluid and origin of cloud streets. *Phys. Fluid*, **6**, 195–211.

Kuo, H. L., 1965: On formation and intensification of tropical cyclones through latent heat release by cumulus convection. *J. Atmos. Sci.*, **22**, 40–63.

Kuo, H. L., 1974: Further studies of the parameterization of the influence of cumulus convection on large-scale flow. *J. Atmos. Sci.*, **31**, 1232–1240.

Kuo, Y. H., and G. T. J. Chen, 1990: Taiwan Area Mesoscale Experiment: an overview. *Bull. Amer. Meteor. Soc.*, **71**, 488–503.

Lin, Y.-L., R. D. Farley, and H. D. Orville, 1983: Bulk parameterization of the snow field in a cloud model. *J. Climate Appl. Meteor.*, **22**, 1065–1092.

Lilly, D. K., 1978: A severe downslope windstorm and aircraft turbulence event induced by a mountain wave. *J. Atmos. Sci.*, **35**, 59–77.

Liou, K. N, Q. Fu, and T. P. Ackerman, 1988: A simple formulation of the delta-four-stream approximation for radiative transfer parameterizations. *J. Atmos. Sci.*, **45**, 1940–1947.

MacCall, B. T., 2006: Application of Reynolds' stress closure to the stable boundary layer. PhD thesis, Department Earth and Atmospheric Sciences, Purdue University, W. Lafayette, IN USA.

Malkus, J. S., and H. Riehl, 1964: *Cloud Structure and Distribution over the Tropical Pacific Ocean* (University of California Press, Berkeley, Los Angeles), pp. 229.

Marticorena, B., and G. Bergametti, 1995: Modeling the atmospheric dust cycle. I: Design of a soil-derived dust emission scheme. *J. Geophys. Res.*, **100**(D8), 16415–16430.

Min, K.-H., 2005: Coupled cryosphere model development for regional climate study and the initialization of Purdue Regional Model with the land data assimilation system. PhD thesis (Department Earth and Atmospheric Sciences, Purdue University, W. Lafayette, IN, USA) pp. 149.

Molinari, J., 1982: A method for calculating the effects of deep cumulus convection in numerical models. *Mon. Wea. Rev.*, **110**, 1527–1534.

Oglesby, R. J., and D. Erikson, 1989: Soil moisture and persistence of North American drought. *J. Climate*, **2**, 1362–1380.

Oglesby, R. J., 1991: Springtime soil moisture, natural climatic variability and North American drought as simulated by the NCAR Community Climate Model 1. *J. Climate*, **4**, 890–897.

Oh, T.-J., 2003: Numerical study of flow past an idealized bell shape mountain with varying Froude number. MS thesis (Department of Earth and Atmospheric Sciences, Purdue University, W. Lafayette, IN, USA) pp. 106.

Oh, T.-J., 2007: Development and testing of characteristic-based semi-Lagrangian two-dimensional shallow water equations development and testing of characteristic-based semi-Lagrangian two-dimensional shallow water equations model. PhD thesis, department of Earth and Atmospheric Sciences, Purdue University, W. Lafayette, IN, USA.

Queney, P., 1948: The problem of airflow over mountains: a summary of theoretical studies. *Bull. Amer. Meteor. Soc.*, **29**, 16–26.

Schaefer, J. T., 1974: The life cycle of the dryline. *J. Appl. Meteor.*, **13**, 444–449.

Scorer, R. S., 1949: Theory of waves in the lee of mountains. *Quart. J. Roy. Meteor. Soc.*, **75**, 41–56.

Skamarock, W., J. Klemp, J. Dudhia, D. Gill, D. Barker, W. Wang, and J. Powers, 2005:

A description of the Advanced Research WRF version 2. NCAR technical note, 100 pp.

Smith, R. B., 1979: The influence of mountains on the atmosphere. In *Advances in Geophysics*, Vol. 21 ed. B. Saltzman, (Academic), pp. 87–230.

Smith, R. B., 1980: Linear theory of stratified hydrostatic flow past an isolated mountain. *Tellus*, **32**, 348–364.

Smolarkiewicz P. K., and R. Rotunno., 1989: Low Froude number flow past three-dimensional obstacles. Part I: Baroclinically generated Lee vortices. *J. Atmos. Sci.*, **46**, 1154–1164.

Sun, W. Y., 1978: Stability of deep cloud streets. *J. Atmos. Sci.*, **35**, 466–483.

Sun, W. Y., 1980: A forward–backward time integration scheme to treat internal gravity waves. *Mon. Wea. Rev.*, **108**, 402–407.

Sun, W. Y., 1982: A comparison of two explicit time integration schemes applied to the transient heat equation. *Mon. Wea. Rev.*, **110**, 1645–1652.

Sun, W. Y., 1984a: Numerical analysis for hydrostatic and nonhydrostatic equations of inertial–internal gravity waves. *Mon. Wea. Rev.*, **112**, 259–268.

Sun, W. Y., 1984b: Rainbands and symmetric instability. *J. Atmos. Sci.*, **41**, 3412–3426.

Sun, W. Y., 1984c: Pollution in the atmosphere. *NATPA Bull.*, **4**, 30–32.

Sun, W. Y., 1986: Air pollution in a convective boundary. *Atmos. Environ.*, **20**, 1877–1886.

Sun, W. Y., 1987: Mesoscale convection along the dryline. *J. Atmos. Sci.*, **44**, 1394–1403.

Sun, W. Y., 1988: Air pollution in a convective atmosphere. In *Library of Environmental Control Technology* ed. Cheremisinoff (Gulf), pp. 515–546.

Sun, W. Y., 1989: Diffusion modeling in a convective boundary layer. *Atmos. Environ.*, **23**, 1205–1217.

Sun, W. Y., 1990a: Modified Eady waves and frontogenesis. Part I: Linear stability analysis. *Terrestrial, Atmospheric and Oceanic Sciences*, **1**, 363–378.

Sun, W. Y., 1990b: Modified Eady waves and frontogenesis. Part II: Nonlinear Integration. *Terrestrial, Atmospheric and Oceanic Sciences*, **1**, 379–392.

Sun, W. Y., 1993a: Numerical simulation of a planetary boundary layer. Part I. Cloud-free case. *Beitr. Phys. Atmos.*, **66**, 3–16.

Sun, W. Y., 1993b: Numerical simulation of a planetary boundary layer. Part II. Cloudy case. *Beitr. Phys. Atmos.*, **66**, 17–30.

Sun, W. Y., 1993c: Numerical experiments for advection equation. *J. Comput. Phys.*, **108**, 264–271.

Sun, W. Y., 1995a: Pressure gradient in a sigma coordinate. *Terrestrial, Atmospheric and Oceanic Sciences*, **4**, 579–590.

Sun, W. Y., 1995b: Unsymmetrical symmetric instability. *Quart. J. Roy. Meteor. Soc.*, **121**, 419–431.

Sun, W. Y., 2002: Numerical modeling in the atmosphere. *J. Korean Meteorol. Soc.*, **38**, 3, 237–251.

Sun, W. Y., 2006a: Uncertainties in climate models (To appear in 10th edition of *McGraw-Hill Encyclopedia of Sciences & Technology*).

Sun, W. Y., 2006b: Regional climate modeling. In *McGraw-Hill 2006 Yearbook of Science & Technology*.

Sun, W. Y., 2007: Conserved semi-Lagrangian scheme applied to one-dimensional shallow water equations. *Terrestrial, Atmospheric and Oceanic Science*, **18**, 777–803.

Sun, W. Y., and M. G. Bosilovich, 1996: Planetary boundary layer and surface layer sensitivity to land surface parameters. *Boundary-Layer Meteor.*, **77**, 353–378.

Sun, W. Y., and C. Z. Chang, 1986a: Diffusion model for a convective layer. Part 1: Numerical simulation of a convective boundary layer. *J. Climate Appl. Meteor.*, **25**, 1445–1453.

Sun, W. Y., and C. Z. Chang, 1986b: Diffusion model for a convective layer. Part II: Plume released from a continuous point source. *J. Climate Appl. Meteor.*, **25**, 1454–1463.

Sun, W. Y., and Y. L. Chang, 1992: Barotropic instability of the modified hyper-tangent shear flows. *Beitr. Phys. Atmos.* **65**, 193–210.

Sun, W. Y., and J. D. Chern, 1993: Diurnal variation of lee-vortexes in Taiwan and surrounding area. *J. Atmos. Sci.*, **50**, 3404–3430.

Sun, W. Y., and J. D. Chern, 1994: Numerical experiments of vortices in the wake of idealized large mountains. *J. Atmos. Sci.*, **51**, 191–209.

Sun, W. Y., and J. D. Chern, 1998: Air-sea-ice model applied in the Arctic region. A72D-06, 1998 Fall Meeting, AGU, supplement to EOS, AGU, 79, 45. F94. 10 Nov. 1998.

Sun, W. Y., and J. D. Chern, 1999: Numerical simulation of southeast Asia weather in May–June of 1998. Preprint, Workshop on Regional Climate (11 Mar 1999; The Center of Ocean Research, National Taiwan University, Taipei, Taiwan) p. 12.

Sun, W. Y., and J. D. Chern, 2005: One-dimensional snow-land surface model applied to

Sleepers experiment. *Boundary-Layer Meteor.*, **116**, 95–115.

Sun, W. Y., and J. D. Chern, 2006: A numerical study of influence of central mountain ranges on a cold front. *J. Meteorol. Soc. Japan*, **84**, 27–46.

Sun, W. Y., and P. A. Haines, 1996: Semi-prognostic tests of a new mesoscale cumulus parameterization scheme. *Tellus*, **48A**, 272–289.

Sun, W. Y., and W. R. Hsu, 1988: Numerical study of cold air outbreak over the warm ocean. *J. Atmos. Sci.*, **45**, 1205–1227.

Sun, W. Y., and W. R. Hsu, 2005: Effect of surface friction on downslope wind and mountain waves. *Terrestrial, Atmospheric and Oceanic Science*, **16**, 393–418.

Sun, W. Y., and Y. Ogura, 1980: Modeling the evolution of the convective planetary boundary layer. *J. Atmos. Sci.*, **37**, 1558–1572.

Sun, W. Y., and Y. Ogura, 1979: Boundary-layer forcing as a possible trigger to a squall-line formation. *J. Atmos. Sci.*, **36**, 235–254.

Sun, W. Y., and I. Orlanski, 1981a: Large mesoscale convection and sea breeze circulation. Part I: linear stability analysis. *J. Atmos. Sci.*, **38**, 1675–1693.

Sun, W. Y., and I. Orlanski, 1981b: Large mesoscale convection and sea breeze circulation. Part 2: Non-linear numerical model. *J. Atmos. Sci.*, **38**, 1694–1706.

Sun, W. Y., and M. T. Sun, 2004: Mass correction applied to semi-Lagrangian advection scheme. *Mon. Wea. Rev.*, **132**, 975–984.

Sun, W. Y., and R. Y. Sun, 2004: Relationship between canopy and climate chap. 2 of *Protected Horticulture Science*, ed. C. Tsay (Research Institute of Agriculture, Taiwan), pp. 11–31.

Sun, W. Y., and C. C. Wu, 1992: Formation and diurnal oscillation of dryline. *J. Atmos. Sci.*, **49**, 1606–1619.

Sun, W. Y., and K. S. Yeh., 1997: A general semi-Lagrangian advection scheme employing forward trajectories. *Quart. J. Roy. Meteor. Soc.*, **123**, 2463–2476.

Sun, W. Y., and A. Yildirim, 1989: Air mass modification over Lake Michigan. *Boundary-Layer Meteor.*, **48**, 345–360.

Sun, W. Y., J. D. Chern, C. C. Wu, and W. R. Hsu, 1991: Numerical simulation of mesoscale circulation in Taiwan and surrounding area. *Mon. Wea. Rev.*, **119**, 2558–2573.

Sun, W. Y., W. R. Hsu, and J. D. Chern, 1995: Numerical simulation of an active Mei-yu front.

International Workshop on Heavy Rainfall in East Asia, (27–29 Seoul, South Korea, Mar. 1995), pp. 108–113.

Sun, W. Y., K. S. Yeh, and R. Y. Sun, 1996: A simple semi-Lagrangian scheme for advection equation. *Quart. J. Roy. Meteor. Soc.*, **122**, 1211–1226.

Sun, W. Y., M. G. Bosilovich, and J. D. Chern, 1997: CCM1 regional response to anomalous surface properties. *Terrestrial, Atmospheric and Oceanic Sciences*, **8**, 271–288.

Sun, W. Y., J. D. Chern, and M. Bosilovich, 2004: Numerical study of the 1988 drought in the United States. *J. Meteorol. Soc. Japan*, **82**, 1667–1678.

Tegen, I., and I. Fung, 1994: Modeling of mineral dust in the atmosphere: sources, transport, and optical thickness. *J. Geophys. Res.*, **99**(D11), 22897–22914.

Weisman, M. L, and J. B. Klemp, 1984: The structure and classification of numerically simulated convective storms in directionally varying wind shears. *Mon. Wea. Rev.*, **112**, 2479–2498.

Willis, G. E., and J. W. Deardorff, 1978: A laboratory study of dispersion from an elevated source within a modeled convective planetary boundary layer. *Atmos. Environ.*, **12**, 1305–1311.

Willis, G. E., and J. W. Deardorff, 1981: A laboratory study of dispersion from a source in the middle of the convective planetary boundary layer. *Atmos. Environ.*, **15**, 109–117.

Wu, C.-C., and W. Y. Sun, 1990a: Diurnal oscillation of convective boundary layer. Part I: Cloud-free atmosphere. *Terrestrial, Atmospheric and Oceanic Sciences*, **1**, 23–43.

Wu, C.-C., and W. Y. Sun, 1990b: Diurnal oscillation of convective boundary layer. Part II: Cloudy atmosphere. *Terrestrial, Atmospheric and Oceanic Sciences*, **2**, 157–174.

Wu, C. C., Y. C. Yu, W. R. Hsu, K. J. Hsu, and W. Y. Sun, 2003: Numerical study on the wind fields and atmospheric transports of a typical winter case in Taiwan and surrounding area. *Atmos. Sci.*, **31**, 29–54 (in Chinese).

Yang, K. J.-S., 2004a: Numerical simulations of Asian aerosols and their regional climatic impacts: dust storms and bio-mass burning. MS thesis, (Department of Earth and Atmospheric Sciences, Purdue University, W. Lafayette, IN, USA), pp. 141.

Yang, K. J.-S., 2004b: Regional climate-chemistry model simulations of ozone in the lower troposphere and its climatic impacts. PhD thesis,

(School of Civil Engineering, Purdue University, W. Lafayette, IN, USA), pp. 157.

Yildirim, A., 1994: Numerical study of an idealized cyclone evolution and its sub-synoptic features. PhD thesis (Department of Earth and Atmospheric Sciences, Purdue University, W. Lafayette, IN USA), pp. 145.

Yu, Y. C., H. H. Hsu, W. S. Kau, T. H. Tsou, W. R. Hsu, and W. Y. Sun, 2004a: An evaluation of the East Asian summer monsoon simulation by the Purdue Regional Model . *J. Environ. Protect Soc. Taiwan*, **27**, 41–56.

Yu, Y. C., H.-H. Hsu, W. S. Kau, T. H. Tsou, W. R. Hsu, W. Y. Sun, and W. N. Yu, 2004b: An evaluation of the east Asian summer monsoon simulation using the Purdue Regional Model. Third Workshop on Regional Climate Modeling for Monsoon System (Hawaii, 17 Feb. 2004).

PART 2

Atmospheric Physics and Chemistry

Interaction between Aerosols and Clouds: Current Understanding

Jen-Ping Chen, Anupam Hazra, Chein-Jung Shiu, I-Chun Tsai and Hsiang-He Lee

Department of Atmospheric Sciences, National Taiwan University, Taipei, Taiwan

jpchen@as.ntu.edu.tw

This article reviews our current understanding of the interactive processes between aerosols and clouds pertaining to the issues of climate change and the hydrological cycle. We first introduce the aerosol effects on clouds by the classification of hygroscopic aerosols, carbonaceous aerosols and mineral dust according to the aerosol chemical contents. Following that are discussions on how clouds influence aerosols via microphysical and chemical mechanisms, scavenging by precipitation, and vertical transport of cloud venting, as well as some indirect effects. The main topics covered here include cloud microphysics, cloud chemistry, photochemistry and gas-to-particle conversion, radiation and climate impact, and interactions with the biosphere. The examples given focus more on the work conducted by the Cloud and Aerosol Research Group in the Department of Atmospheric Sciences, National Taiwan University.

1. Introduction

Aerosols and clouds are both colloidal entities that contain suspended particles in the atmosphere. The two are mainly distinguished by whether water vapor may continuously condense/deposit on the particles, which also means that they may interchange by condensation and evaporation of water. Such transformations are frequent and have important consequences for their internal metamorphosis, as well as impacts on the environment.

Aerosol and cloud processes have strong influences on the chemical, hydrological, and radiative budgets of the atmosphere, and through these processes human activities may significantly alter the balances of our climate system. In particular, aerosols may produce climate forcing through their interaction with clouds, which are called the indirect effect of aerosols (Twomey, 1974; Albrecht, 1989; Haywood and Boucher, 2000). This forcing may proceed by enhanced cloud albedo due to more numerous cloud drops (commonly termed the *first indirect effect* or *cloud albedo effect*) or by a prolonged cloud lifetime due to reduced cloud drop size and precipitation efficiency (commonly termed the *second indirect effect* or *cloud lifetime effect*) if the aerosol number concentrations are increased (IPCC, 2001).

Many observational and modeling results support the first indirect effect (Charlson *et al.*, 1992, Han *et al.*, 1994; Ackerman *et al.*, 2000; Rosenfeld, 2000), but rather limited evidence has been shown to verify the second indirect effect. For both effects a quantitative assessment is still difficult to achieve. The major issue is that there are still no convincing understandings of the aerosol–cloud interactions, including the interaction between aerosols and cloud condensation nuclei (CCNs), CCNs and cloud droplets, and cloud droplets and cloud albedo (IPCC, 2001). In fact, aerosols' effects, particularly the indirect effects, represent the largest source of uncertainty in the current estimates of global climate forcing.

One of the major complications stems from the diversified chemical compositions

"

of aerosols. Different types (composition) of aerosols act differently in climate forcing. For instance, acidic (sulfate, nitrate) aerosols tend to cool the surface, while soot has the additional effect of warming the atmosphere, which may lead to changes in large-scale atmospheric circulation (Hansen *et al.*, 1997; Hansen and Sato, 2001) and even "burning out" the clouds (Ackerman *et al.*, 2000). Acidic particles and sea salt particles are both good condensation nuclei, but the increase of the former tends to reduce precipitation (Twomey, 1974; Rosenfeld, 2000), while the latter tend to enhance precipitation by acting as a rain embryo (Woodcock, 1971; Rosenfeld, 2002). Soot particles may also influence cloud formation, either acting as condensation nuclei or as ice nuclei (Motoi, 1951; Bigg, 1990). Mineral dust, on the other hand, may be the main source of ice nuclei (Motoi, 1951; Georgii and Kleinjung, 1967), which are of great importance in precipitation development, particularly in the mid-to-high latitudes. Moreover, different particulate chemicals may coexist in a specific air parcel by external or internal mixing, the numeral combination of which further muddles up their role in the direct and indirect forcing (Jacobson, 2001; Nenes *et al.*, 2002). Therefore, the studies of aerosols, by either measurement or modeling, necessarily become utterly sophisticated.

The problem becomes even more challenging when aerosols interact with clouds. Aerosols may affect clouds directly through their activation into cloud drops or heterogeneous nucleation into ice crystals, and indirectly through the cloud burning effect by soot or dust particles. Clouds may remove aerosols by wet scavenging (collection and washout), or augment aerosols by aqueous chemical reactions and then deactivation of cloud drops. Clouds may also interact indirectly with aerosols through the alteration of actinic flux and photochemistry, which in turn produce condensable vapor of trace chemicals that may form new

aerosols by nucleation or grow on existing ones. The interplay between aerosols and clouds also strongly influences the precipitation processes; therefore, much attention has also been paid to the impact of air pollution on the hydrological cycle, as well as the removal of pollutants by the cloud-cleansing mechanism.

Because of the crucial roles that aerosols and clouds play in the complex atmospheric system, coupled aerosol–cloud models with high accuracy and efficiency have become increasingly important, particularly in the absence of sufficient observational data. Here, we identify some of the most important mechanisms involved in the interaction of aerosols and clouds, as well as review current progress of the theoretical and technical development of the modeling aspect. The examples given are biased toward our research efforts.

2. Aerosol Effect on Clouds

Most atmospheric aerosols tend to have a mixed chemical composition, including a variety of inorganic and organic species. In the lower atmosphere, particulate matter is composed of highly water-soluble inorganic salts, insoluble mineral dust and carbonaceous material. This section discusses the effects of aerosols on clouds according to their main chemical compositions. Caution needs to be taken — natural aerosols are often not pure and may coexist by internal or external mixing.

2.1. *Hygroscopic aerosols*

Aerosol particles that are capable of initiating the formation of cloud drops and raindrops are called cloud condensation nuclei (CCNs). The ability of CCNs to activate into cloud drops is strongly related to the mass and composition of their water-soluble component. The most common of them include the gaseous and aqueous chemical conversions of their precursors (such as the oxidation of SO_2 and NO_x into sulfate and nitrate), as well as a direct emission from the Earth's surface (such as ammonia or sea salt). Whether an aerosol particle may be activated into cloud drops can be decided from its *Köhler curve*, which describes the equilibrium saturation ratio S_{eq} of a solution drop containing a fixed mass of dissolved salt as a function of the drop radius r_{eq}. Figure 1 gives an example of the Köhler curve family for ammonium sulfate aerosols. For each droplet the solution (Raoult) effect reduces the surface vapor pressure while the curvature (Kelvin) effect does the opposite, and the combination gives a Köhler curve that has a peak in S_{eq} (commonly called the *critical* saturation ratio). If the environmental saturation ratio never exceeds the critical S_{eq}, this aerosol will not be activated[a] into a cloud drop to grow further. From Fig. 1, it can be seen that larger aerosols

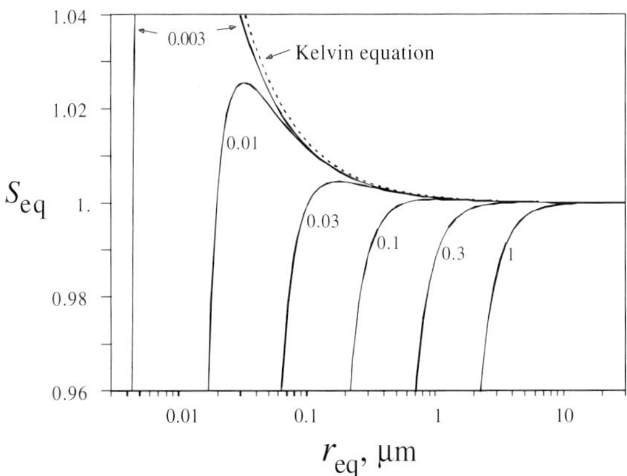

Figure 1. Köhler curves: variation of the equilibrium drop size with the saturation ratio. All curves are for droplets formed from ammonium sulfate particles, with the dry radius labeled in units of micrometers. The dashed curve represents the curvature effect for pure water droplets. (From Chen, 1994.)

have a lower critical S_{eq}, and thus the advantage of easier activation.

It is important to recognize that hygroscopic aerosols have a broad size range, and very large CCNs may play very different roles in cloud and precipitation formation than the smaller ones. In addition, the sizes of aerosols might be much more important than their chemical composition, which normally has smaller variability, in determining their ability to activate into cloud drops (Rosenfeld, 2006; Dusek *et al.*, 2006), unless the aerosols have large insoluble fractions.

2.1.1. *Twomey's indirect effects*

Twomey in 1974 first pointed out that increasing anthropogenic pollution would result in higher CCN and cloud drop number concentrations. Note that traditionally we discuss Twomey's effects with a focus on hygroscopic CCNs, but other types of aerosols may also contribute to such effects. The relationship between the number concentrations of CCNs and cloud drops during the activation process (at the cloud base) can be seen through the calculations from a detailed parcel model, as shown in Fig. 2, in which the CCN size distribution assumes the empirical expression $N_{CCN} - C\Delta S^k$, as derived by Twomey (1959), where N_{CCN} is the number concentration of CCNs (outside the cloud) or cloud drops (inside the cloud), C is a coefficient that roughly represents the total number of condensation nuclei, ΔS is the supersaturation, and k represents the gradient by which the number density (in log scale) varies with the CCN size (note: CCN size is connected with the supersaturation ΔS, following the Köhler theory). Observation evidences also concur with

---

[a]The process of aerosols turning into cloud drops was often termed "cloud nucleation." However, hygroscopic aerosols usually become wet (a process called "deliquescence"; see Subsec. 2.1.3 for details) long before reaching the cloud state. So, the formation of cloud drops from hygroscopic aerosols usually does not involve new-phase formation. Therefore, the proper term is "activation," instead of "nucleation."

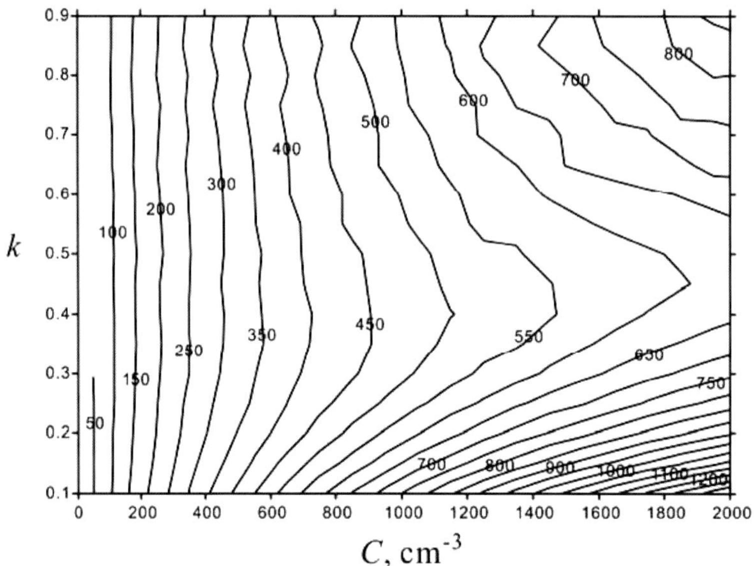

Figure 2. Model-calculated number concentration of cloud drops activated from a CCN distribution of $N_{\rm CCN} = C\Delta S^k$, expressed as a function of the coefficients C and k. The updraft speed used for this calculation is $1\,{\rm m\ s^{-1}}$. (From Chen and Liu, 2004.)

such dependence (e.g. Hegg *et al.*, 1991; Leaitch *et al.*, 1992). In the previous section we quoted from Rosenfeld (2006) that sizes may be a more important factor than chemical composition for the activation capability of aerosols, but in terms of Twomey's indirect effects, the number concentration is certainly an even more critical attribute of aerosols.

More aerosols means more cloud drops, and more cloud drops tends to enhance the reflectance of clouds and therefore cool the atmosphere by reflecting more sunlight. A vivid illustration of such an effect is the co-called ship tracks, where the marine boundary layer clouds that are contaminated with particles of ship exhaust have brighter reflectivity (Coakley *et al.*, 1987). Other examples are the reduced cloud particle size and even suppressed precipitation originating from major urban areas or industrial facilities (Rosenfeld, 2000; Jirak and Cotton, 2006) and from downwind of biomass burning (Warner and Twomey, 1967; Rosenfeld, 1999). Twomey and others demonstrated theoretically that the cooling effect has

a magnitude comparable, but opposite in sign, to that of greenhouse warming due to anthropogenic trace gases (Twomey *et al.*, 1984; Charlson *et al.*, 1992; Boucher and Lohmann, 1995). Using results from a mesoscale meteorological model and a regional air pollution model, Tsai (2001) diagnosed the indirect effect of aerosols at the Pacific Rim and found that anthropogenic aerosols significantly affect near-continent oceans that are located downwind of the major emission source under northeasterly monsoon winds. The cloud albedo averaged over the whole domain change (including areas without clouds) from 0.08 for natural conditions to 0.13 for polluted conditions. Apparently, anthropogenic aerosols of East Asia may significantly increase the CDNC, cloud optical thickness, and cloud albedo over the western Pacific Ocean and the South China Sea during the winter monsoon.

However, the relationship between the numbers of aerosols and cloud drops might not be linear or monotonic, but rather "jumpy," if a more complete microphysical cycle was considered. Baker and Charlson (1990) and

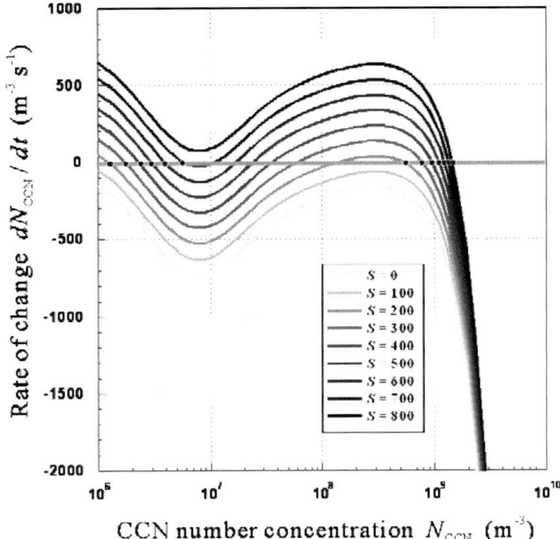

Figure 3. Variations of the rate change of N_{CCN} as a function of N_{CCN} for different production strengths. The zero intercept represents an equilibrium condition, but only those marked by black dots are stable modes where any deviation from that N_{CCN} value will eventually return to the stable conditions. (From Fu, 2002.)

Baker (1993) proposed an interesting bistability hypothesis to describe the possible relationship between CCNs and cloud drops. With a mixed-layer approach, she analyzed theoretically the variation of CCN and cloud drop number concentrations in marine stratiform clouds. Figure 3 is a reproduction of her analysis, but using a newer microphysical parameterization (Fu, 2002). Each curve represents the overall rate change of N_{CCN} as a function of N_{CCN} itself under a particular strength of CCN production, S. The microphysical processes included are Brownian coagulation of CCNs, self-collection of cloud drops, conversion of cloud drop to raindrop (autoconversion), precipitation fallout, and scavenging of CCNs by rain. On each curve one can see 1–3 intercepts at zero which represent the equilibrium (steady) states. However, only the intercepts at the right or left (marked by the red dots) are stable modes where any perturbation of N_{CCN} from that value will cause a restoring force to bring it back to equilibrium. This means that the atmosphere

has the capability of cleansing itself, but only when it is not heavily polluted (to the left of the central intercept). If all of a sudden too much CCN is put into the air (the value of N_{CCN} greater than that at the central intercept), then N_{CCN} will shift toward the other stable mode such that the polluted conditions will persist, and does not return to the clean mode unless the production strength S is reduced.

Furthermore, S must be reduced drastically to revert to the clean mode due to the hysteresis effect shown in Fig. 4(a). Because the two stable modes do not coexist all the time, the condition of jumping from clean to dirty modes occurs at a different production strength than that of jumping from dirty to clean modes. For example, the clean conditions may be sustained until S reaches about $700 \, \mathrm{m}^{-3} \, \mathrm{s}^{-1}$; but once it jumps to a dirty condition, aerosol production must be reduced to about $200 \, \mathrm{m}^{-3} \, \mathrm{s}^{-1}$ in order to return to the clean mode. Cloud albedo also exhibits such a hysteresis cycle, jumping between the clean value of about 0.2 and the dirty value of about 0.7 [Fig. 4(b)]. However, the simulation of Ackerman *et al.* (1994) did not show such a hysteresis effect. Fu (2002) pointed out that the discrepancy could be due to the transient effect as the time to reach equilibrium often takes several days, which might be much longer than the time of simulation spent by Ackerman *et al.* (1994). In addition, during the long time required to approach equilibrium, the properties of the cloud, such as the liquid water content and cloud depth, might have changed due to precipitation or dynamic adjustment, and yet Baker (1993) held these parameters as constant. Although the bistability hypothesis needs further verification, one may nevertheless think of the two stable modes as attractors that the numbers of cloud drops and CCNs tend to approach. As the stratocumulus deck in the marine boundary layer (MBL) has a major impact upon the Earth's radiation budget by reflecting solar radiation, Baker's bistability hypothesis has important implications for the

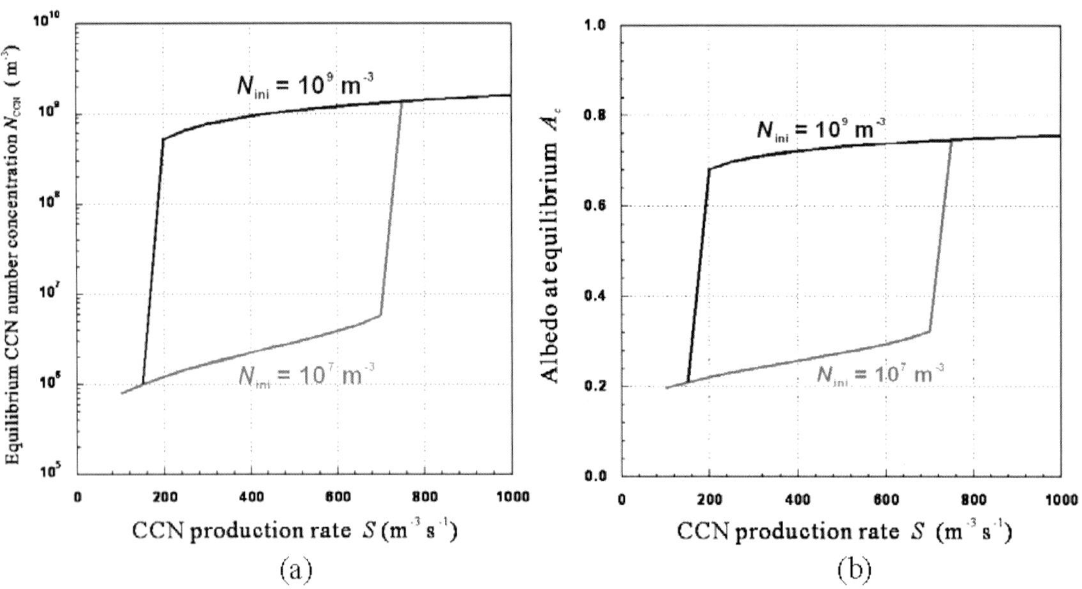

Figure 4. (a) Equilibrium N_{CCN} as a function of the production strength S. (b) Cloud albedo that corresponds to the equilibrium N_{CCN}. The lower curves represent the scenario of shifting from clean to dirty conditions by sweeping S from low to high values for an initially low N_{CCN}, whereas the upper curve is just the opposite. (From Fu, 2002.)

limitation of atmospheric self-cleansing capability pertinent to the injection of anthropogenic pollutants.

The effect of aerosols on cloud drop concentration may occur due to not only anthropogenic pollution but also natural biogenic production. One of the main sources of natural condensation nuclei (CNs) over the ocean is believed to be dimethylsulfide (DMS) excreted by phytoplankton and liberated into the atmosphere, where it is photochemically oxidized to form sulfate particles and become CCNs. By providing CCNs, DMS may determine the properties of the cloud. Thus, the production of DMS in oceans and its subsequent transformation into CCNs act as a negative feedback mechanism to counteract global warming (Shaw, 1983; Charlson *et al.*, 1987; Easter and Peters, 1994), which is called the DMS–cloud–climate hypothesis. It is speculated that the productivity of DMS is temperature-dependent, such that any increase in the ocean surface temperature (expected outcome of greenhouse warming) would enhance CCN production,

which, in turn, would increase cloud reflectance (albedo) and cause a cooling effect similar to the Twomey effect. Thus DMS (or anthropogenic SO_2, for that matter) is sometimes referred to as the "anti greenhouse gas." The cooling of the sea surface or the blocking of sunlight by denser clouds could reduce the productivity of phytoplankton, thus forming a negative feedback loop. Recently, Meskhidze and Nenes (2006) analyzed satellite data and showed that cloud drop number concentration over the regions of the phytoplankton bloom was twice as high as that away from the bloom. The resulting change of short-wave radiative flux at the top of the atmosphere may be comparable to the aerosol indirect effect over highly polluted regions. However, the authors proposed that the changes were caused by secondary organic aerosols (also hygroscopic) formed from isoprene that was released by the phytoplankton. Yet, the DMS–cloud–climate hypothesis still cannot be discarded.

CCNs not only affect cloud-physical properties and albedo but may also influence precipitation formation. As cloud drops get smaller due to there being more CCNs, they collide less efficiently, such that the warm rain formation mechanism may be retarded. The sensitivity test of Teller and Levin (2006) nicely shows the decrease of total precipitation with increasing CCN concentration. Cheng et al. (2007) also showed that in more polluted conditions (i.e. higher CCN concentration) less rain is formed, whereas more cloud water is retained in the cloud (Fig. 5). Their results exemplify both the first and the second indirect effect of Twomey. Note that the influence of CCNs on precipitation might deviate from those discussed above when ice-phase processes are involved, because there could be stronger growth by riming and latent heat release in the cloud when more liquid water is retained. Giant CCNs discussed in the next section are an even clearer example of aerosols enhancing precipitation.

Quantitative estimates of indirect aerosol effects were provided in numerous articles (e.g. Lohmann and Feichter, 2005; IPCC, 2001, 2007). However, care must be taken that there exist large uncertainties in the estimation or verification of aerosol indirect effects. Besides our limited knowledge about the processes involved, a few other factors may contribute to the uncertainties. For example, for those models that are able to diagnose the mean cloud drop radius (usually volume-weighted), they also use it to represent the effective radius in determining the cloud radiative effects. Yet, these two radii may differ significantly due to the dispersive nature of the drop size distribution (Martin et al., 1994; Liu and Duam, 2002). Liu and Duam (2002) suggested that Twomey's first indirect effect of aerosols may be partially offset due to such dispersive effect, whereas Chen and Liu (2004) further suggested that the fraction of the offset should be about one-ninth, which is supported by the calculations of Cheng et al. (2007).

Twomey's second indirect effect seems to be a sound hypothesis. Yet, there are numerous examples where increases in CCNs do not lead to higher liquid water content clouds and longer lifetimes via the second indirect effect. For instance, Ackerman et al. (2004) demonstrated that the entrainment of overlying dry air above boundary layer stratocumulus may mitigate the second indirect effect and sometimes even results in cloud water decreases when the humidity of the overlying air is very low. In fact, more and smaller cloud drops may enhance evaporation at the downdraft region of cloud edges, which leads to a lower cloud fraction, cloud size, and depth (Teller and Levin, 2006; Xue and Feingold, 2006). Guo et al. (2007) also showed such a positive indirect effect due to stronger entrainment when large-scale subsidence is weak. Tao et al. (2007) pointed out another effect of enhanced evaporation. From simulations using a two-dimensional cloud resolving model, they found that in a tropical cloud more aerosols caused more but smaller raindrops, which evaporate faster and cause a stronger downdraft, in the stratiform region of the clouds. The stronger cold pool associated with the enhanced downdraft induced stronger low-level convergence and thus stronger convection, and this resulted in greater precipitation. But, for a continental convective cloud, increasing CCNs does result in less precipitation. More and smaller cloud drops may also suppress low-level precipitation, elevating the release of latent heat at the upper levels and the onset of precipitation, which result in more intense convection (Rosenfeld and Woodley, 2000). These examples caution us that the effects of aerosols on clouds and precipitation are not so straightforward. Whether the response follows Twomey's indirect effects depends on many intricate microphysical factors as well as dynamic feedbacks.

2.1.2. Giant CCNs

Some of the hygroscopic aerosols are inherently large in size, such as sea salt generated

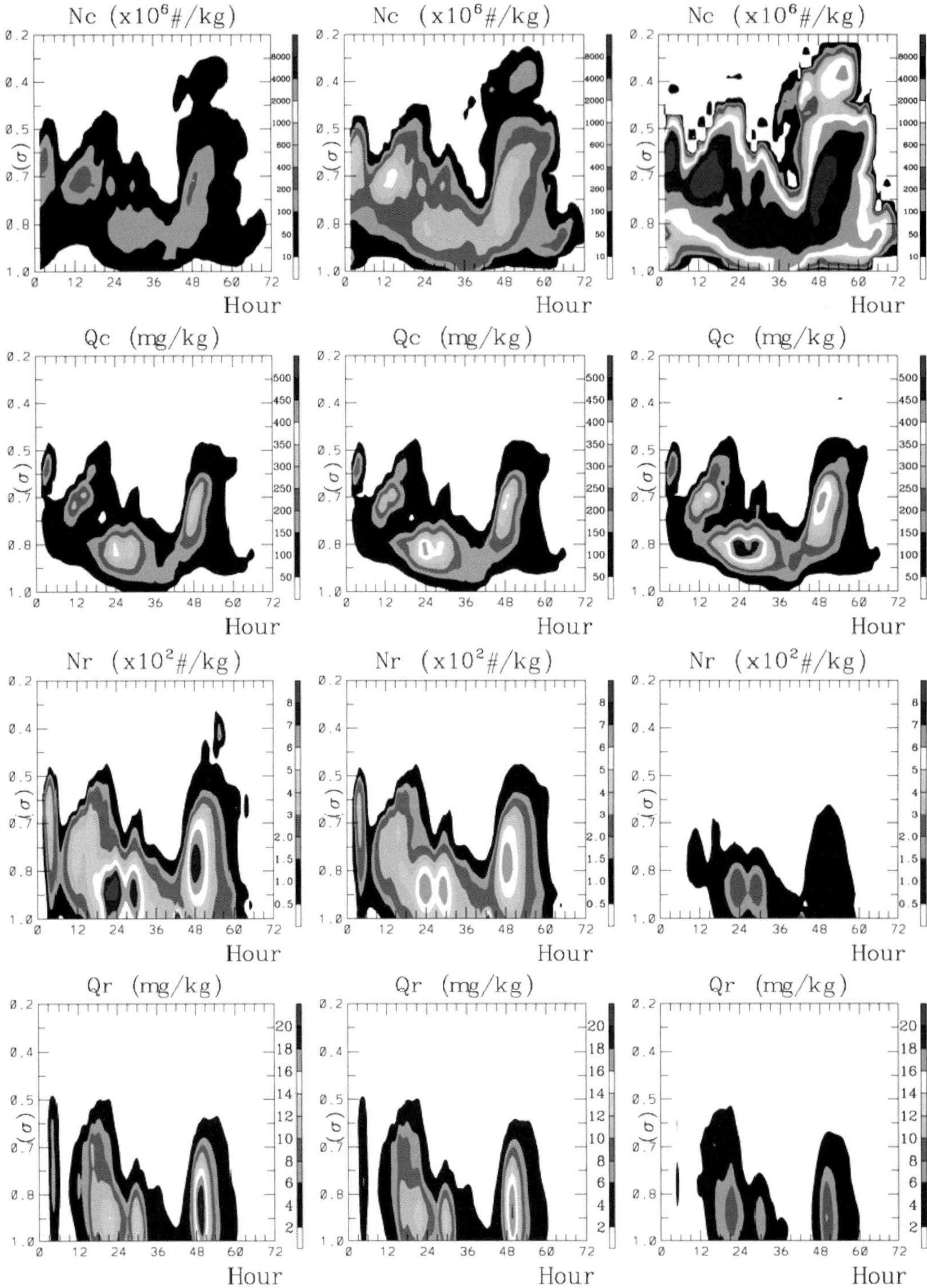

Figure 5.	Time series of the simulated vertical profiles (in σ coordinates) of cloud drop fields averaged over the model domain. The rows from the top down are the number concentration N_c and the mixing ratio Q_c of cloud drops, and the number concentration N_r and the mixing ratio Q_r of raindrops. The three columns, from left to right, are for initial aerosol types — clean continental, averaged continental, and urban. (From Cheng *et al.*, 2007.)

by the breakup of air bubbles in the ocean or by wave tearing. These giant CCNs, with radii of a few microns or more, play very different roles than common CCNs in either the hydrological cycle or Twomey's indirect effects, mentioned in the previous subsection. They not only activate more readily into cloud drops but may also act as rain embryos to initiate the warm rain process because they grow, by condensation, rather quickly to exceed the so-called "Hocking limit" for collision–coalescence. Large, insoluble particles such as mineral dust may exhibit similar properties when coated with hygroscopic materials.

The role of rain embryo that giant CCNs play in cloud processes has been demonstrated to be important for warm rain initiation (Johnson, 1982; Feingold *et al.*, 1999; Lasher-Trapp *et al.*, 2001). This is particularly true for the statocumulus clouds in the marine boundary layer (MBL), where drizzle is the main form of precipitation. Studies have shown that sea salt particles are often present in concentrations similar to those of drizzle drops, implying that these giant CCNs may be the main source of drizzle production (see O'Dowd *et al.*, 1997). The stratocumulus clouds in the MBL are particularly worthy of understanding, because they play an important role in the radiative balance of the Earth (Albrecht, 1989). It has been suggested that drizzle can significantly affect the thermodynamics and energetics of the MBL, leading to profound changes in the cloud amount and cloud structure (Paluch and Lenschow, 1991; Stevens *et al.*, 1998). Strong drizzle may even lead to the collapse of the marine stratocumulus, as suggested by the observational evidence shown by Stevens *et al.* (2004).

Drizzle may be produced either via the so-called "autoconversion" process or from rain embryos. The former produces rain-sized droplets by a gradual collision–coalescence among cloud drops, whereas the latter immediately introduce droplets with sizes larger than

the Hocking limit to initiate fast collision. Traditional bulkwater (also known as "Kessler type") microphysical schemes do not consider the effect of giant CCNs. The only way for them to produce rain in warm clouds is through the autoconversion process. Note that without considering aerosols, the traditional (one-moment) bulkwater schemes cannot simulate Twomey's indirect effects, not to mention the giant CCN effect.

However, whether giant CCNs are crucial to the initiation of drizzle or not depends on how high the cloud drop number concentration (CDNC) is, which in turn is driven largely by variability in the concentration of the smaller (e.g. accumulation mode) aerosol particles, as discussed previously. Under low CDNC, cloud drops are able to grow large enough by condensation to enable efficient autoconversion. Note that when the CDNC is too low, clouds could become optically thin, so that cloud-top radiative cooling could no longer drive vertical mixing, which is another possible cause of the collapse of the stratiform clouds in the MBL (Ackerman *et al.*, 1993). Also note that low CCN and thus low CDNC may cause a stronger autoconversion process that may possibly contribute to the collapse of the marine stratocumulus deck (Khairoutdinov and Kogen, 2000). Under high CDNC (thus small droplet sizes), the autoconversion process tends to shut down due to the Hocking limit restriction, and so the formation of drizzle depends on the amount of giant CCNs present in the air. Feingold *et al.* (1999) found that giant CCNs are effective in enhancing precipitation in clouds with a CDNC of above $50\,\mathrm{cm}^{-3}$ but are ineffective at a lower CDNC. Of course, there must be a sufficient amount of giant CCNs to start with. Figure 6 shows a similar effect of giant CCNs from simulations using the Mesoscale Meteorological Model, version 5 (MM5), with an improved bulkwater scheme that takes aerosols into account (Cheng *et al.*, 2007). The setup of the case presented is the same as that for Fig. 6. One can see that with

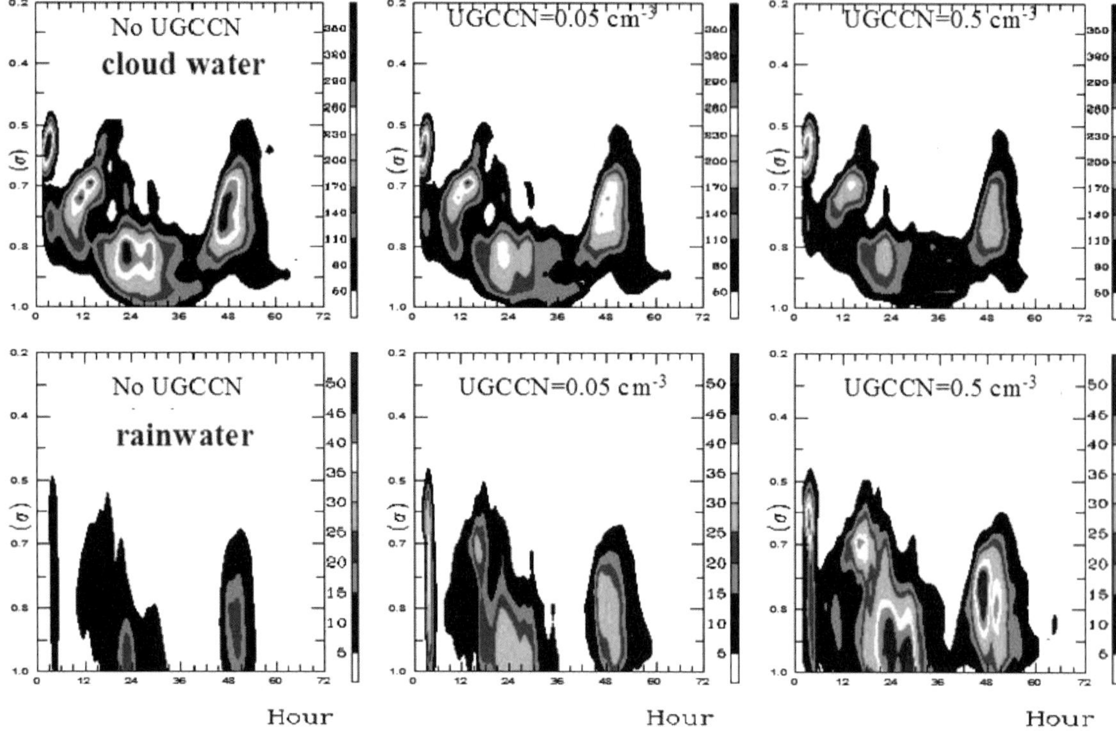

Figure 6. Distribution of cloud water (*top*) and rainwater (*bottom*) under various amounts of ultragiant CCNs, (radius larger than $10\,\mu$m). The amount of giant CCNs, from left to right is 0, 0.05 and 0.5 cm^{-3}, respectively. A greater amount of giant CCNs, causes more rains and retains less water in the cloud.

the addition of giant CCNs, progressively more rain is produced while less water is retained in the cloud. Such an effect is particularly strong for polluted conditions where autoconversion is largely inhibited.

However, at least for not-very-polluted situations, the actual cause of such rain enhancement is not just for having more rain embryos. In fact, the introduction of giant CCNs may also enhance the autoconversion process. Because of their strong water-absorbing capabilities, giant CCNs consume water vapor during the activation stage, thus decreasing the maximum supersaturation and reducing the chance for smaller aerosols to be activated into cloud drops (Ghan *et al.*, 1998; Cheng *et al.*, 2007). As less cloud drops are activated, they can grow larger and have a higher efficiency of collision, thus enhancing autoconversion. Furthermore, the rain embryos may reduce the CDNC by accretion, as shown by Cheng *et al.*, (2007). This has the long-term effect of reducing aerosols that can be recycled during cloud evaporation, and is particularly important for marine stratocumuli that normally go through many cloud cycles in a relatively short time. These concepts have actually been adopted for warm cloud seeding, where giant nuclei (typically composed of $CaCl_2$) are artificially introduced into the updraft region below the cloud base to not only promote coalescence but also reduce cloud drops and cause those activated to grow bigger (Cooper *et al.*, 1997; Mather *et al.*, 1997). It is also worthwhile to note that giant CCNs cause an anti-Twomey's first indirect-effect by reducing CDNC, and an anti-Twomey's-second-indirect-effect by enhancing precipitation. So, giant CCNs play an exactly opposite role to that of typical CCNs in the context of aerosol–cloud–climate interactions.

2.1.3. *Effects on cloud ice*

Precipitation formation through ice-phase mechanisms is much more efficient than the warm rain process, provided that a sufficient amount of ice crystals can be formed in the cloud. Ice crystals can be formed most effectively by heterogeneous nucleation with the help of ice nuclei (INs). When there is a lack of INs, ice crystals can also be formed through homogeneous freezing of liquid droplets, but only at low temperatures (typically $\leq -35°C$). It is also theoretically possible for water vapor to nucleate directly into ice crystals, but this would require extremely high supersaturations that are rarely found in the atmosphere. Due to the Raoult effect of freezing point depression, soluble compounds tend to inhibit the freezing of liquid aerosols and cloud droplets, and thus the formation of ice crystals. Yet, liquid droplets

normally cannot form without these soluble chemicals. So, when the atmosphere is short of INs, the only viable way to form ice is to freeze the water in haze or cloud drops. In fact, ice clouds formed from homogeneous nucleation are very common, such as the cirrus formed by slow lifting of air or in orographic wave flow at high altitudes (Heymsfield and Sabbin, 1989; Heymsfield and Miloshevich, 1993), as well as the cirrus anvil generated in convective storms (Knollenberg *et al.*, 1993). These ice clouds cover a large portion of the Earth's surface and have a major influence on the global energy budget (Liou, 1986). Simulations of cirrus anvils from convective outflow performed by Chen *et al.* (1997) using a column model showed that aerosols have strong influences on not only the microphysical structure but also the extent and lifetime of cirrus. As shown in Fig. 7, with five times the aerosol

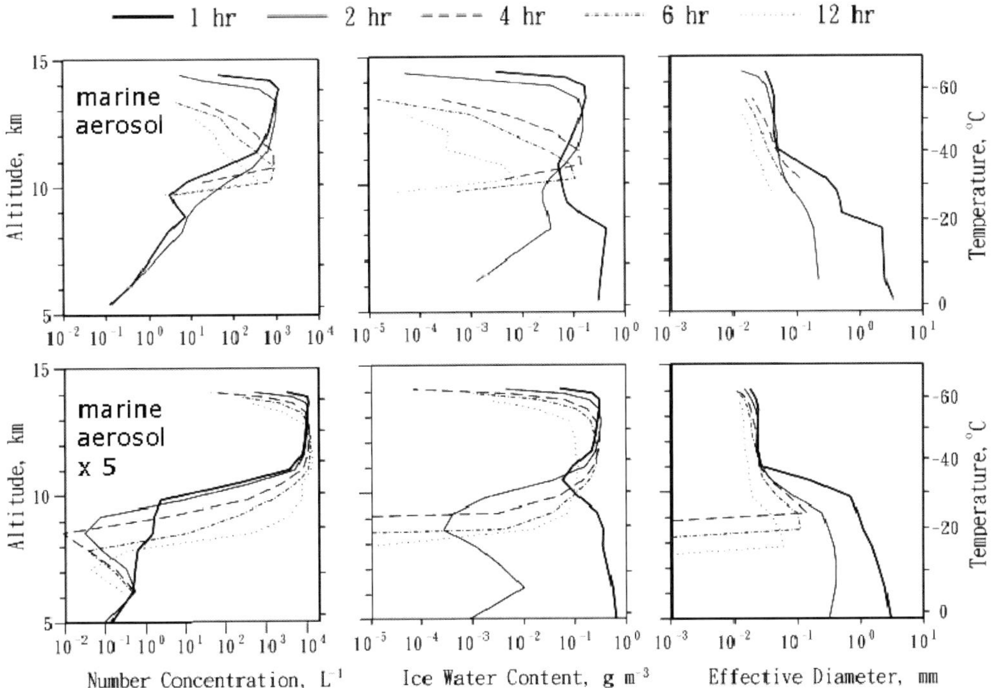

Figure 7. Simulation of ice microphysics in a tropical convective outflow anvil. The top panels show results of applying typical maritime aerosol size distribution, and the bottom panels show results with five times the aerosol concentration. The left, center and right panels display the number concentration, mass concentration and effective radius of ice particles. Different curves represent the vertical profiles at different times of evolution. (From Chen *et al.*, 1997.)

concentration, the number concentration of ice particles may increase by more than one order of magnitude. With more aerosols, there will be more activation to form more and smaller cloud drops; smaller cloud drops also cause a reduction of droplet coalescence or accretion by ice, and thus more droplets will remain for eventual freezing into ice particles. As a consequence, more ice water but smaller effective radii are formed in the anvil, and the anvil becomes thicker and lasts longer. At 12 hours of simulation time, the vertical extent (as defined by the ice water content) increases from 3 km to 5.5 km as a result of the increase of aerosols.

The evolution from an aerosol particle to an ice crystal is illustrated in Fig. 8. Dry hygroscopic aerosols first obtain water when the humidity reaches a particular value called the deliquescence point, which is actually a discontinuity of the Köhler curve caused by the solubility limitation, as shown in Fig. 9 (Chen, 1994). Once deliquesced, the aerosol reaches an equilibrium size determined by the traditional Köhler curve and stays in a haze state. Further lifting and cooling of the air causes the humidity to rise and the aerosol droplet to swell. If the air is not cold enough (but still colder than $-35°C$), it needs to reach over

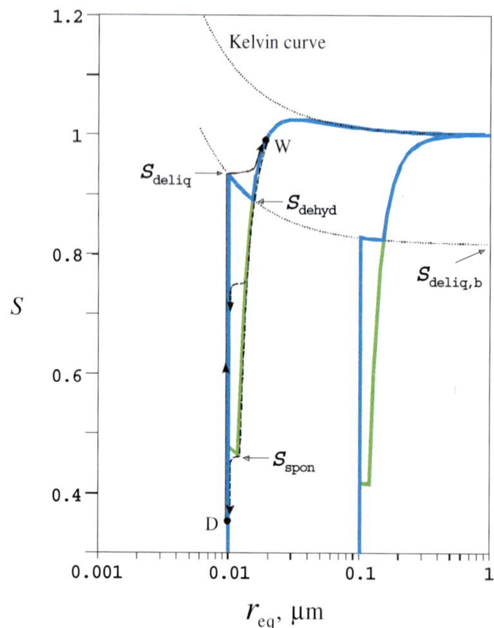

Figure 9. Modified Köhler curves and the hysteresis cycle. The blue curves are the modified Köhler curves; the green curves are the traditional Köhler curves, except below the spontaneous dehydration point, S_{spon}; the upper dotted line indicates the Kelvin effect, and the lower dotted line connects the deliquescence part of each curve, the end of which is the common or "bulk" deliquescence point, $S_{\text{deliq,b}}$. The thin solid line shows the growth path of dry aerosols (from point D) in moisturizing air; it acquires water only when the humidity reaches the deliquescence point, S_{deliq}. The thin dashed lines show the growth path of a wet aerosol (from point W) in drying air; it will suddenly lose all the water at the dehydration point, S_{dehyd}, if under thermodynamic equilibrium, at S_{spon} if the salt core appears by homogeneous nucleation, or somewhere in between if the salt core appears by heterogeneous nucleation. (See Chen, 1994, for details.)

100% relative humidity with respect to water such that the aerosols may activate into cloud drops. Only then can the strong uptake of water by the cloud drop dilute the solute effect to allow droplets to freeze. However, if the air is cold enough (significantly lower than $-40°C$) to overcome the suppression of the freezing point, the haze particles themselves can directly freeze to form ice. Note that these processes are time-dependent and self-limiting, as the ice crystals already formed will grow and deplete

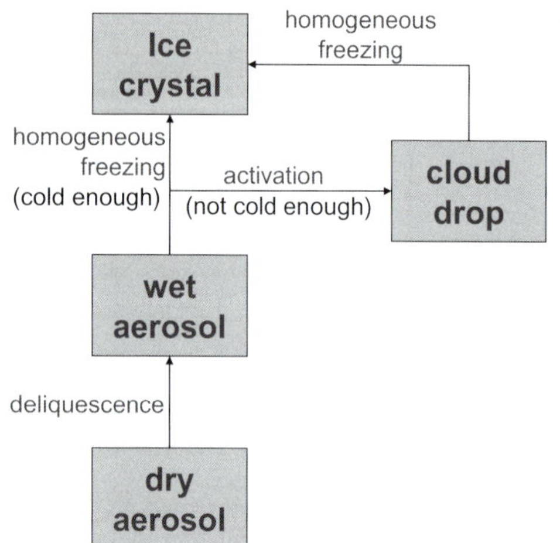

Figure 8. Schematics of the processes where aerosols are converted into ice crystals in ascending air.

water vapor rather quickly. Then the lowering of relative humidity forces the unnucleated cloud drops and aerosols to evaporate and become more concentrated, such that they are more difficult to freeze. So, the microphysical properties of high-altitude ice clouds depend very much on the chemical and physical properties of the aerosols.

Besides the freezing mechanism, hygroscopic aerosols may aid the formation of ice by taking on the role of INs. Abbatt *et al.* (2006) pointed out that dry ammonium sulfate particles may serve as INs to provide another pathway for cirrus formation. As indicated by Chen (1994), the deliquescence point becomes higher at lower temperatures, so at high altitudes aerosols more easily stay in a dry state and serve as INs. Note that the role of ammonium becomes important because if the aerosol loses its ammonium, the deliquescence point becomes lower and the aerosol will "get wet" more easily. Another possible mechanism for ice nucleation from hygroscopic aerosols involves the physical chemistry of solutions. Hazra *et al.* (2003) showed that hydrate crystals may form in concentrated ammonium sulfate solution, then trigger ice nucleation. This process could occur at rather high temperatures. They also pointed out that a solid mixture of ammonium sulfate and water may form below the eutectic temperature, which is around $-18°C$ (see Fig. 5 and Chen, 1994). Such solids, just like the dry aerosols or the hydrate crystals, may initiate heterogeneous ice nucleation and formation of cirrus clouds.

Another effect that aerosols may have on ice clouds is the influence of solute coating on the evaporation of ice crystals. Chen and Crutzen (1994) hypothesized that nonvolatile solutes, which enter the ice through surface chemistry or the processes shown in Fig. 8, will remain on the surface of the ice when the water containing them evaporates. The solute interacts with the quasi-liquid layer on the surface of the ice and forms a solute coating that may

reduce the surface vapor pressure and retard the evaporation of ice crystals. Such a mechanism has important implications for prolonging the lifetime of cirrus clouds. In addition, it may explain the unusually long survival time of cirrus crystal trails that fall a long distance to seed and glaciate low-level clouds, as observed by Braham (1967) and Braham and Spyers-Duran (1967).

The physical chemistry of aerosols is also important to the formation of polar stratospheric clouds (PSCs), which are responsible for the formation of the ozone hole (Turco *et al.*, 1989). As shown in Fig. 10, it starts with the formation of stratospheric aerosols composed of sulfuric acid and water. Then, during the polar night, as the temperature in the stratosphere decreases to below $210\,K$, the solution droplet has the chance to freeze into a sulfuric acid tetrahydrate (SAT) crystal. Further cooling to temperatures less than about $200\,K$ allows the simultaneous deposition of nitric acid vapor and water vapor to form nitric acid trihydrate (NAT). Continuous growth with NAT will result in Type I PSCs. If the cooling continues to reach the frost point (generally around $-185\,K$ in stratospheric conditions), then the crystals may grow by deposition of pure water vapor and form Type II PSCs.

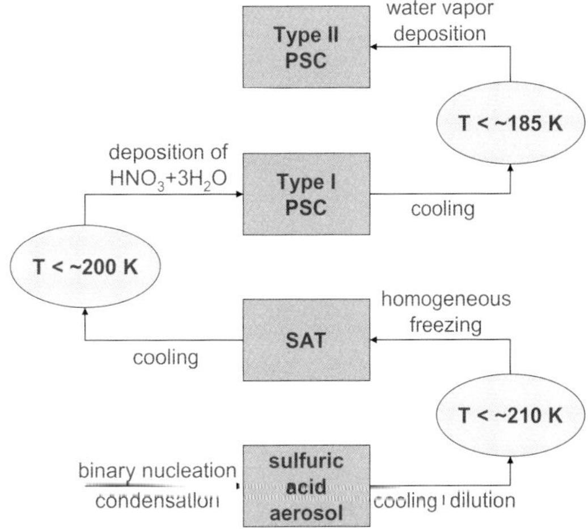

Figure 10. Pathway for the formation of Type I and Type II polar stratospheric clouds.

2.1.4. *Effects on cloud chemistry*

Soluble ingredients contained in aerosols are brought along into cloud water when they are activated into cloud drops. Some of them have already been dissociated into ions in the haze state, but some dissociate only when the drops are diluted under fast condensation growth after activation. Either way, they are involved in aqueous chemical reactions that may be important to many atmospheric phenomena. The most common soluble ingredients in aerosols are sulfate, ammonium, nitrate, sea salt (Na^+ and Cl^-), calcium, potassium, magnesium, organic acids, and transition metals. Because of the dominant role of sulfate and sometimes nitrate in controlling the ionic balance, most of the aerosols, as well as the cloud drops forming on them, are acidic, except over areas where the atmosphere has high loading of alkaline material such as mineral dust or sea salt.

Aerosol compositions may first affect cloud chemistry by controlling the acidity of the cloud water, because many aqueous phase reactions are strongly dependent on the pH value. For instance, the solubility of SO_2 and many other species that dissociate upon dissolution in water

is stronger at higher pH. Furthermore, the oxidation of dissolved SO_2 by O_3 or by O_2 (with Fe^{3+} or Mn^{2+} as catalyst) also increases with pH under typical conditions (see Seinfeld and Pandis, 2006; p. 317). So the presence of alkaline material tends to enhance the conversion of atmospheric SO_2 into sulfate, whereas the acidic materials do just the opposite.

Less noticed is that the size distribution of aerosols could also affect cloud chemistry. Cloud drops are formed from CCNs with different sizes, which means that they inherently contain different amounts of solutes after activation. With the same chemical compositions, the concentration of solutes in haze drops varies only slightly with the particle size, being higher in smaller drops due to the curvature effect (curves at time 0 in Fig. 11). But when the particles are activated into cloud drops, the solutes in them are quickly diluted because of the water uptake by condensation. Yet, the degree of dilution varies drastically with particle size, as shown in Fig. 11. Large drops dilute much slower than smaller drops, because the rate increase of water mass (i.e. $d\ln m/dt$) is roughly proportional to $1/r^2$, where m is the water mass and r is the

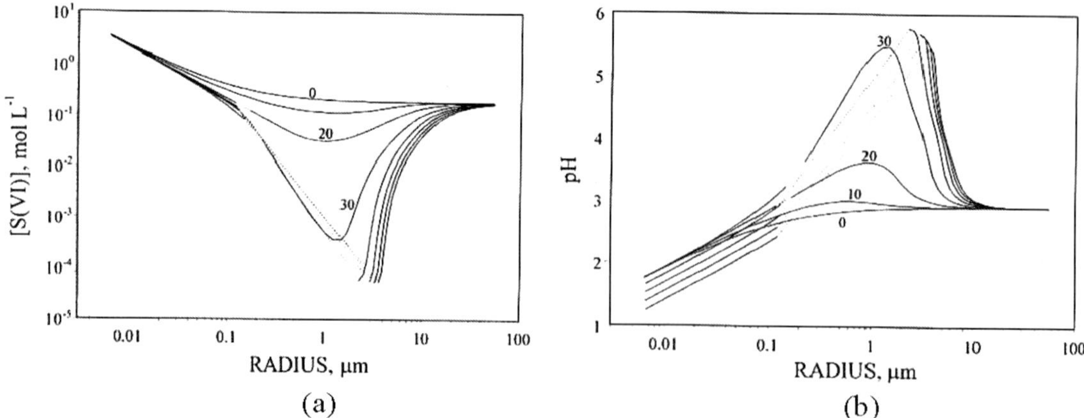

(a)   (b)

Figure 11. Time evolution of the concentration spectrum of aerosol and cloud droplets in an ascending air parcel considering only the condensation process. (a) Sulfate concentration; (b) pH value. Each curve represents the spectrum at a different time of cloud formation, and the labels indicate simulation time in seconds. At time zero, the air parcel is not saturated, so all particles are in their haze state. The air parcel becomes saturated after about 20 seconds, and a gap appears in the spectrum, which separates those activated into cloud drops and those remaining as interstitial haze particles.

drop radius. From Fig. 11(a) one can see that the solute concentrations in the smallest cloud drops decrease by more than 4 orders of magnitude in less than 1 minute of cloud formation, whereas those in the largest drops (giant CCNs) hardly change at all. Similar changes are found for pH values, with those in the smallest cloud drops rising from less than 2 to near 6 in just 1 minute [Fig. 11(b)]. In a relatively narrow size range, the pH values may differ by 3 (3 orders of magnitude in hydronium concentrations) between large and small cloud drops. Such a drastic variation in solute concentration and pH values has profound influence on the cloud chemistry. Again, take SO_2 chemistry as an example; most of the conversion into sulfate would occur in the smaller and diluted drops, which were formed from the smaller CCNs, unless the larger cloud drops were formed on alkaline aerosols such as sea salt or mineral dust.

The calculation shown in Fig. 11 considers only the condensation process. More complicated spectral distribution of solute concentration will result from the collision between cloud drops. In Fig. 12 is a two-dimensional particle framework for aerosols and cloud drops,

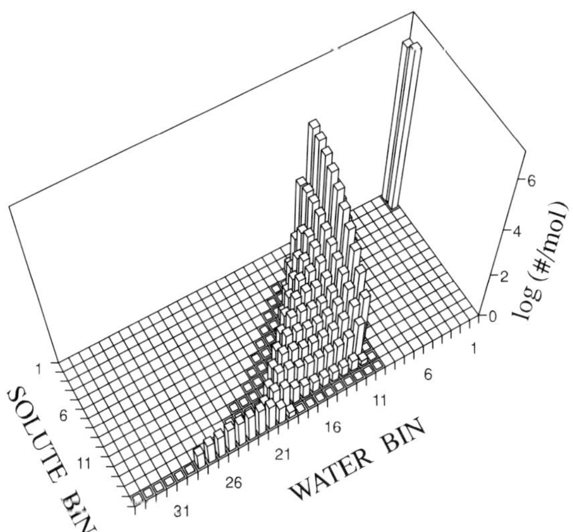

Figure 12. Evolution of the drop spectrum in the two-dimensional particle framework due to the condensation and coalescence processes. (From Chen and Lamb, 1992.)

which was developed by Chen and Lamb (1992) to describe the variation of drop composition in terms of both the water and solute contents. These aerosols are activated into cloud drops, then grow by vapor condensation and collision–coalescence, as simulated in a parcel model. Note that the first two solute bins are not activated due to a limited supersaturation acquired in the ascending air. One can see that the initially narrow spectrum evolved into a broad two-dimensional spectrum, in which droplets of the same water content may have very different solute concentrations and vice versa. Purdue and Beck (1988), Ogren and Charlson (1992), and as Pandis and Seinfeld (1991) all pointed out that the chemistry in mixed drops behaves very differently than while they are separated. Mixing of droplets with different solute concentrations will normally result in significant outgassing of dissolved volatile chemicals. This also means that droplets with solutes distributed nonuniformly among them may dissolve more trace gases than droplets of homogeneous concentrations, and thus relevant chemical reactions in the former proceed faster than in the latter. Traditional detailed (binned) cloud-microphysical and chemical models that do not consider the extra solute component must assume that droplets of the same size contain the same amount of solutes, and this leads to large errors in cloud chemistry calculation. How the final cloud drop spectrum is distributed in the two-dimensional framework as shown in Fig. 12 depends largely on the aerosol size distribution and composition. Thus one may conclude that aerosols have a strong influence on cloud chemistry through not only their chemical contents but also their effect on cloud microphysics.

2.2. Carbonaceous aerosols

Carbonaceous compounds constitute a significant fraction of aerosol mass, particularly in the area with heavy fossil fuel and biofuel burning (Nunes and Pio, 1993; Novakov *et al.*, 2000; Andreae and Merlet, 2001). The two main classes of anthropogenic carbonaceous compounds are organic carbon (OC) and black carbon (BC). These carbonaceous ingredients play crucial roles in cloud-microphysical and radiative properties. BC has been know to influence atmospheric radiation either directly by absorbing shortwave radiation or indirectly by retarding the growth of droplets by the mechanism of radiative heating (Conant *et al.*, 2002; Nenes *et al.*, 2002), whereas organic chemicals may affect the capability of CCNs to activate into cloud drops in many ways. Soluble organic compounds decrease the water activity as well as the surface tension of droplets, both resulting in an increase of drop growth. (Facchini *et al.*, 1999a; Abdul-Razzak and Ghan, 2004). Natural vegetation also emits an ample amount of organic gases into the atmosphere, some of which may turn into an aerosol or part of an aerosol. These organics generally behave similarly to the anthropogenic OC. Another set of carbonaceous aerosols are biogenic — those directly emitted by organisms such as pollen, fungi, or even the organisms themselves, as in bacteria and viruses. These bioaerosols may play a particular role in the formation of ice in clouds.

2.2.1. *Soluble organic aerosols*

Organic particulate matter can be represented as a complex mixture of OC of biogenic and/or nonbiogenic origin (Andrews *et al.*, 1997). In the early 1980s, Likens *et al.* (1983) reported the existence of organic compounds in precipitation. Recently, a new class of macromolecular polycarboxylic acids has been detected in aerosol samples (Mukai and Ambe, 1986), accounting for a significant fraction of the water-soluble organic carbon (WSOC) aerosols (Facchini *et al.*, 1999b). This class of macromolecular compounds has physical and chemical properties similar to those of humic acids, the main constituent of dissolved OC in natural water (Stumm

and Morgan, 1981), and for this reason they are sometimes referred to as HULIS (humic-like substances) in the literature. The possible sources of HULIS (Mukai and Ambe, 1986) are particularly the process of biomass combustion. It is noted that the molecular forms of the oxygenated, water-soluble organics (WSOC) and volatile organic carbon (VOC) are also produced in biomass combustion processes (Falkovich *et al.*, 2004). Chemical characterization of atmospheric aerosols has revealed that OC is usually the second-most-abundant component of fine aerosols (0.01–1μm), after sulfates, around the globe (Novakov and Penner, 1993). Hence, organic aerosols are an important part of the global CCN budget.

Organic aerosols can be classified into directly emitted species (primary) and those formed by chemical conversion in the atmosphere (secondary). Both contribute to the atmospheric population of CCNs. The natural primary organic aerosol originates from a wide range of primary emissions of anthropogenic and biogenic activities, as well as the burning of open biomass due to natural fires and land-use practices. Based on the energy statistics for the year 1996, Bond *et al.* (2004) estimated that emissions of primary OC are 2.4, 5.8, 25 Tg yr^{-1} from combustion of fossil fuels, combustion of biofuels and open biomass burning, respectively. The main pathway of the anthropogenic emissions is vehicular exhaust, which is the primary source of dicarboxylic acid and monocarboxylic acid (Yao *et al.*, 2003). The biosphere is another major source of primary organic aerosols. Primary bioaerosols play a very special role in cloud processes, so a separate subsection (2.2.4) is devoted to giving a more detailed introduction. Secondary organic aerosols are formed due to the oxidation of VOCs emitted from biological organisms. Biogenic emissions are driven by temperature, light, and vegetation. Photo-oxidation products of monoterpenes (e.g. α- and β-pinene) (Hoffman *et al.*, 1997), which are biogenic VOCs and are emitted mainly by terrestrial vegetation, also contribute to the aerosol budget (Kavouras *et al.*, 1998).

Traditional cloud activation theory is commonly applied to CCNs that are composed of highly soluble inorganic salts. Yet, there are many highly or slightly soluble organic compounds that also can be considered as cloud-active nuclei (Kulmala *et al.*, 1996). The importance of organics as CCNs was first noted during the field studies made by Desalmand *et al.* (1985), who found a positive correlation between the concentration of water-soluble organics and the number of CCNs. They suggested that vegetation can produce CCNs. Novakov and Penner (1993) suggested that organic aerosol particles may make up a significant portion of CCNs, comparable perhaps with the sulfate aerosol contribution to CCNs. They also indicated that 37% of the CCN number concentration measured at a marine site were sulfate particles, while the remaining 63% were attributed to organic aerosols. The CCN properties of highly soluble organic acids like oxalic, malonic and glutaric acids have also been identified (Kumar *et al.*, 2003). Novakov and Penner (1993) found that organic aerosol mass dominated the total mass in the nucleation mode at a marine site in Puerto Rico. The presence of organics in small aerosols may be the result of nucleation from organic vapors. Therefore, identifying sources of the gas phase precursors is important to determining the origins of the organic CCNs.

Just as with the inorganic salts, there are two effects involved for organic aerosols to serve as CCNs: the solute effect, which lowers the activity of water, and the effect on surface tension. The modeling results of Anttila and Kerminen (2002) suggested that soluble organics indeed influence the activation process with both effects, but the slightly soluble compounds play only a minor role. Raymond and Pandis (2002) performed a laboratory study to show that organic species with solubility less than

$0.01\,\mathrm{g\,cm^{-3}}$ can still be a good source of CCNs. They also found that the traditional Köhler theory (see Fig. 1) works well in predicting the activation of soluble organic aerosols, but needs modification when dealing with slightly soluble organic species. Note that the great majority of organic compounds identified in aerosols are semivolatile (Rogge *et al.*, 1993), so when treating the activation and condensation growth of these particles one also needs to consider the mass transfer of these organics from the gas phase.

2.2.2. *Soot*

While aerosols of any composition reflect sunlight, only a few can also have absorption. These absorbing aerosols include BC or soot, desert dust (Sokolik and Toon, 1996) and some organic carbon species (Bond, 2001). They may have a warming effect, opposed to the cooling by scattering aerosols (Charlson and Pilat, 1969; Schneider, 1971). Soot, also called carbon black or black carbon, is residue from the combustion of carbon-rich organic fuels in the lack of sufficient oxygen. Fossil fuel, biofuel or biomass combustions often release soot particles in large quantities into the atmosphere; therefore, soot has become a major component of aerosols in polluted regions. Past studies of soot focused mostly on their direct effect on climate, such as the influence on large-scale circulation (Hansen *et al.*, 1997), as well as atmospheric stability and convection activities (Ackerman *et al.*, 2000). Much less attention has been paid to the effect of soot, also due to solar heating, on the growth of cloud drops. As demonstrated by Conant *et al.* (2002), when BC-containing CCNs are activated into cloud drops, the absorption of sunlight will raise droplet temperature and thus surface vapor pressure. Such heating is in effect increasing the Köhler curve saturation ratio S_d shown in Fig. 1. As derived by Chen and Hsieh (2004), S_d is modified by an exponential term under BC heating:

$$S'_d = S_d \cdot \exp\left(\frac{L_v}{R_v}\frac{\Delta T}{T_a(T_a + \Delta T)}\right)$$

$$\approx S_d \cdot \left(1 + \frac{L_v}{R_v}\frac{\Delta T}{T_a^2}\right), \qquad (1)$$

where L_v is the latent heat of condensation, R_v is the gas constant of water vapor, T_a is air temperature, $\Delta T = Q/4\pi r k_a$ is the temperature rise under a BC heating rate of Q, r is the drop radius and k_a is the thermal conductivity of air ($= 0.0238\,\mathrm{W\,m^{-1}\,K^{-1}}$ at $0°\mathrm{C}$) corrected with the gas kinetic theory. With a 50% BC content, ΔT is about $0.02\,\mathrm{K}$ for $1\,\mu\mathrm{m}$ aerosols and $0.1\,\mathrm{K}$ for $10\,\mu\mathrm{m}$ aerosols. Note that, as will be mentioned later, the heating rate is significantly stronger if the aerosols are wet. The heating effect may cause a change in Köhler curves because of the rise in drop surface vapor pressure, and this not only retards the condensation growth but also hinders the activation process because the critical saturation is increased. Chen and Hsieh (2004) showed that condensation growth (see Pruppacher and Klett, 1997) may be retarded by a factor $(1 - B)$:

$$\frac{dm'}{dt} \approx \frac{dm}{dt} \cdot (1 - B), \qquad (2)$$

where m is droplet mass. dm' is mass change under BC heating,

$$B = Q \cdot \frac{L/T_v T_a^{-1}}{(S_a - S_d)4\pi r k_a f_h T_a}, \qquad (3)$$

S_a is the saturation ratio of ambient air and f_h is the heat ventilation coefficient. Note that this equation can be rearranged into a form similar to that given by Conant *et al.* (2002):

$$\frac{dm'}{dt} \approx \frac{dm}{dt}\frac{S_a - S'_d}{S_a - S_d}. \qquad (4)$$

The overall heating from individual droplets causes two types of group effects. The first is that the reduced overall condensation [see Eq. (2)] will raise the ambient saturation S_a. So even though the critical S_d of each aerosol has been elevated, the enhanced ambient saturation S_a may increase the chance of activation,

particularly for the smaller aerosols whose critical S_d is less affected. The second effect is the heating of air by energy transferred from the heated droplets. An increase in air temperature will depress S_a, thus reducing the amount of water that can be condensed. Such a "cloud-burning effect" (Ackerman *et al.*, 2000) may have a significant influence on cloud formation and climate in polluted regions. In addition, a depressed S_a reduces the chance of aerosol activation into cloud drops, particularly for smaller aerosols which require a higher S_a to be activated. In summary, BC heating has three effects: (1) It increases the critical S_a of individual droplets, (2) it increases the ambient S_a due to retarded condensation growth on heated droplets, and (3) it decreases the ambient S_a due to heating on ambient air. How these effects combine to affect the CDNC requires a detailed cloud-microphysical model to explain.

Following the approach of Nenes *et al.* (2002), Chen and Hsieh (2004) applied the microphysical model of Chen and Lamb (1994) to evaluate the overall BC heating effect. The radiative heating algorithm of Toon and Ackerman (1981) was adopted to calculate BC heating. Figure 13 shows their results for three different heating scenarios, compared with the situation without BC heating (the curve NoHeat). If the heating occurs only on the droplet and the heat does not conduct to the ambient air (the curve Drop), then the peak S_a is raised due to retarded condensation growth. This led to more cloud drop activation (higher CDNC) even though the critical S_d of individual aerosols was also raised. For the externally mixed situation (the curve XM), where BC particles do not reside within the hygroscopic material, the heating is only on the air, but not the cloud drops, such that the peak S_a and CDNC are significantly reduced. If the BC and hygroscopic materials are internally mixed (the curve IM), the heating will be on both the cloud drops and the ambient air, and then the air-heating effect is mostly offset by

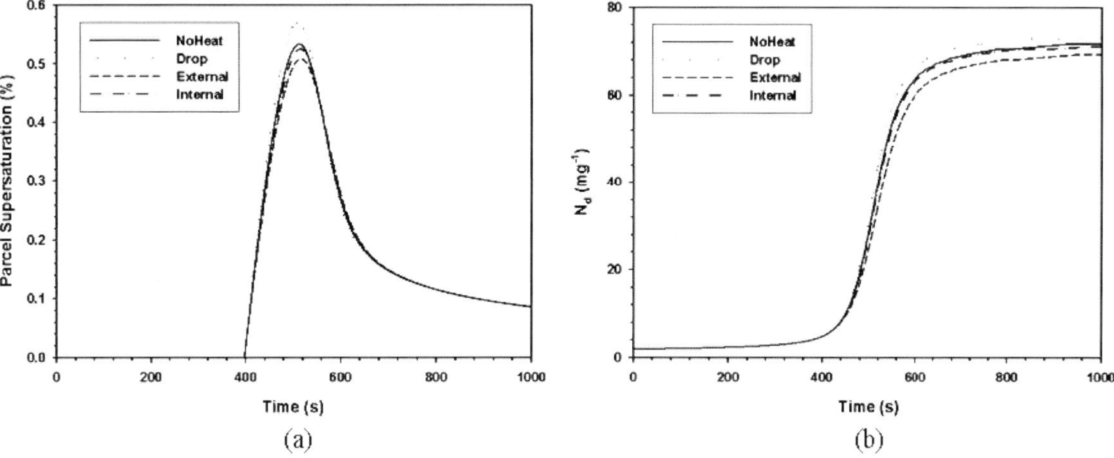

Figure 13. Time evolution of (a) supersaturation and (b) cloud drop number concentration in different heating scenarios. Marine aerosol distribution is applied with a BC mass content of 50%. The updraft velocity is set as 0.25 m/s. (From Hsieh, 2004.)

the drop-heating effect such that there is only a slight decrease in the peak S_a and CDNC. Note that heating of BC with a thick liquid shell is stronger than that without, a mechanism included in the above calculation.

The relative strength of saturation enhancement or reduction by the drop-heating effect and that by the air-heating effect depends on the total mass fraction of the BC and its distribution among aerosols. The changes in the peak S_a become less obvious when the aerosol distribution is of the continental type, because many more droplets are activated to quickly consume the excess water vapor. Chen and Hsieh (2004) found that the maximum S_a and CDNC can be either enhanced or depressed, depending on the types of aerosol distribution and BC distribution, a conclusion different from that of Nenes *et al.* (2002). Nenes *et al.* (2002) also indicated that soot heating may prohibit giant CCNs from reaching threshold sizes to initiate drizzle. This mechanism, most efficient at strong updrafts, tends to increase the CDNC, because there will be less accretion of cloud drops by drizzles.

Being insoluble, soot particles may also act as nuclei for heterogeneous ice nucleation. This subject is of particular interest with

regard to the contrail formation from aircraft exhausts. Demott (1990) found that soot can enhance ice nucleation by the immersion-freezing mechanism, and the activity increases with particle size. For the temperature range of -5 to $-20°$C, Gorbunov *et al.* (2001) found that the ice-forming activity of soot particles increases with decreasing temperature and increasing size, as well as the degree of oxidization of the particle surface (which helps in forming hydrogen bonds with water molecules). Möhler *et al.* (2005) found that at temperatures below $-38°$C, soot particles acted as deposition nuclei at very low ice saturation ratios, between 1.1 and 1.3. At higher temperatures, ice nucleation occurred only after approaching liquid saturation. They further found that a coating of sulfuric acid elevates the ice nucleation thresholds. This result shows the importance of knowing the mixing state of soot and sulfuric acid aerosol particles. Note that freshly emitted soot is extremely hydrophobic, but during aging soot becomes less so. Decesari *et al.* (2002) found that the soot oxidation process causes the formation of water-soluble polycarboxylic compounds, which might cause soot to become effective CCNs.

2.2.3. *Surfactants*

One particular type of OC — called "surfactants" — consists of polar (hydrophilic) and nonpolar (hydrophobic) segments. Surfactants can form a monolayer surface film when exposed to water or droplets, and therefore are also known as "film-forming compounds" (FFCs). The surface film may cause a significant decrease in the water accommodation coefficient, and thus it retards the condensation or evaporation over a water surface. This phenomenon was noticed as early as in the 18th century by Benjamin Franklin, and relevant experimental work extended through the late 19th and the early 20th century until Irvin Langmuir in 1917 decisively determined many basic properties of the surface film (see La Mer, 1962).

Natural FFCs have been observed widely in the marine atmosphere (Blanchard, 1964; Goetz, 1965; Barger and Garrett, 1970). Blanchard (1963) and Garret (1967) suggested that FFCs may be injected into the atmosphere by bubble bursting over the oceans. Over the land, hydrocarbons emitted by plants may be transformed into polar species due to oxidation or polymerization by photocatalytic processes (Garrett, 1978). Seidl (2000) showed that leaf abrasion or biomass burning may produce significant fatty acids that form a dense surface film on aerosol particles. Husar and Shu (1975) found direct evidence from electron micrographs that organic coating does exist on urban haze particles, which might be related to the persistence of smog in the Los Angeles basin (Husar *et al.*, 1976). FFCs are also common ingredients in fog, rain and snow, particularly near polluted or forested regions (Lunde *et al.*, 1977; Meyers and Hites, 1981; Capel *et al.*, 1990; Facchini *et al.*, 1999b). Graedel *et al.* (1983) suggested that typical mass fractions of surface-active compounds in aerosol particles are on the order of 10%. This might not seem much, but if this amount is present entirely on the surface, the surface film formed may cover a large fraction of the droplet surface because

it is only one molecule thick (Langmuir and Langmuir, 1927).

Derjaguin *et al.* (1985) considered the FFC effect using a one-dimensional model to evaluate surfactant vapor on the spectrum of cloud drops forming in the process of condensation growth. Their results indicated that as the small droplets are passivated, the number of growing droplets increases, and the growth of large droplets is accelerated. Feingold and Chuang (2002) also evaluated the potential influence of FFCs on droplet growth by using a parcel model, and provided an alternative explanation for droplet spectral broadening resulting from the presence of FFCs in CCNs. They summarized that the ability of FFCs on droplet spectral broadening is a function of both the total mass of FFCs and how it is distributed among the particles. In the following, a more detailed analysis is given to elucidate the mechanisms involved and the impacts of FFCs on cloud properties.

The basic condensation growth equation [i.e. dm/dt in Eq. (2)] can be transformed into a form commonly seen in textbooks (see Pruppacher and Klett, 1997; p. 511):

$$\frac{dm}{dt} \approx \frac{4\pi r}{A_D + A_K} \cdot (S_a - S_d), \qquad (5)$$

with

$$A_D = \frac{R_v T_a}{D'_v f_v e_{sw}(T_a)}, \qquad (6)$$

$$A_K = \left(\frac{L}{R_v T_a} - 1\right) \frac{L}{k'_a f_h T_a}, \qquad (7)$$

where e_{sw} is the saturation vapor pressure and D'_v is the diffusion coefficient D_v corrected with the gas kinetic effect:

$$D'_v = D_v \left/ \left[\frac{D_v}{\alpha \cdot r \cdot (\nu/4)} + \frac{r}{r + \lambda} \right] \right., \qquad (8)$$

in which α is the mass accommodation coefficient, and ν and λ are the mean thermal velocity and the mean free path of water vapor, respectively. The gas kinetic effect (also called the "vapor jump effect") arises because within one mean free path close to the droplet surface, the transport of water vapor (as well as heat conduction) is considered to proceed not by diffusion but by the gas kinetic process. FFCs retard droplet growth through the influence on α, which represents the probability of a water vapor molecule staying when it hits the droplet surface. From Eq. (8) one can see that the effect of α on diffusion is most significant when either the droplet size or α is small, and thus the FFC is able to retard condensation/evaporation by drastically reducing the value of α.

Past studies (e.g. Derjaguin *et al.*, 1985; Feingold and Chuang, 2002) often assumed that the effect of FFCs on α occurs only when the surface coverage by the FFCs exceeds a threshold value. So the value of α is either 0.035 for pure water (α_w) or $\sim 10^{-5}$ for full FFC coverage (α_H). However, Chen and Hsieh (2004) re-examined the experimental data in relevant articles (e.g., Archer and La Mer, 1955; Rubel and Gentry, 1984; Seaver *et al.*, 1992) and suggested that the effect should still exist under partial coverage, as shown in Fig. 14. Based on molecule flux theory, they developed a new method to describe the relationship between α and FFC coverage, then fitted the experimental data of α with extrapolation to that of pure water at zero coverage.

The above parametrization was incorporated into a detailed microphysical parcel model of Chen and Lamb (1994) to simulate the FFC effects. For a mass fraction more than just 1%, FFCs may drastically retard the condensation growth and elevate S_a. The more FFCs there are, the higher S_a may reach. However, CDNC does not respond so linearly. Intuitively, a higher value of the peak S_a should allow more smaller aerosols to be activated. Yet, the retardation of condensation alters (lengthens) the characteristic time of growth such that some of the aerosols whose critical S_d is below the peak S_a (and may be activated) do not have time to grow to their critical size before S_a starts to taper off. An analogy is that someone buys an airplane ticket but does not arrive before the gate closes.

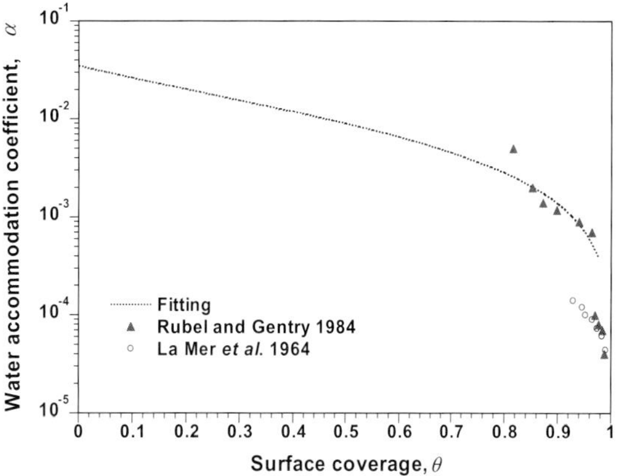

Figure 14. Parametrization of the overall mass accommodation coefficient as a function of FFC coverage. Filled triangles are the experimental data from Rubel and Gentry (1984), which show a sharp transition of FFCs from a liquid film to a solid film under surface coverage exceeding about 97.5%; whereas circles are data from La Mer *et al.* (1964), which show the state of the solid film extending to lower coverage. The dotted curve is a fitting of the overall α for FFC coverage up to 97.5%, above which we assume that α is at a solid state and $\alpha_H = 2 \times 10^{-5}$.

The above effect can be seen from the evolution of the droplet size spectrum shown in Fig. 15. For normal conditions (i.e. without the FFC effect), the initially continuous size spectrum breaks up into two segments with distinct gaps during and after the activation stage [Fig. 15(a)]. On the left are the unactivated haze particles, which swell only a little bit initially; on the right are those activated into cloud drops, which continue to grow with time. The cutoff size of activation here is around 0.1 μm. For the case with 1% FFCs [Fig. 15(b)], the cutoff size is reduced to 0.07 μm because of the elevated S_a, and so more smaller droplets are activated. However, taking the spectrum at 600 s for example, the largest droplets do not seem to grow at all, let alone become activated. But why can the smaller droplets grow under the influence of FFCs? The reason is that the dilution rate (mass acquired compared to existing mass) is faster for smaller droplets. Also, the amount of the FFC in each droplet is

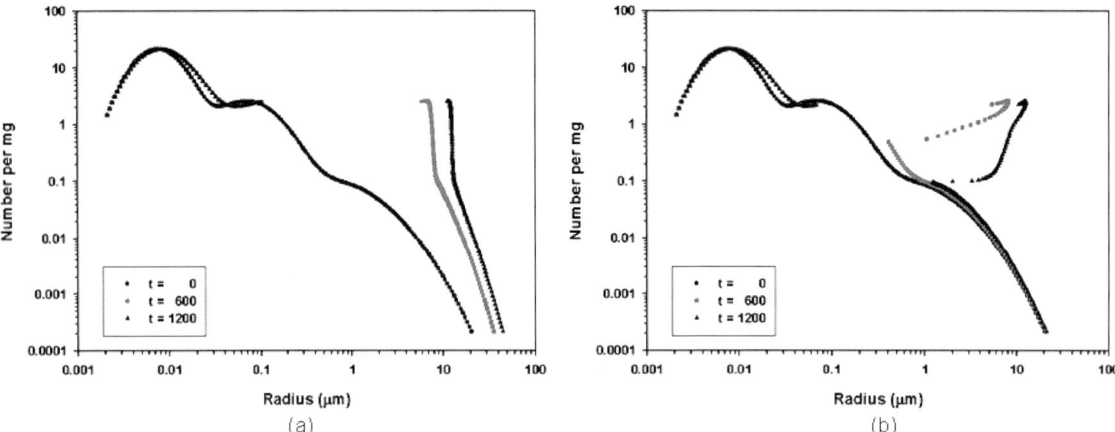

Figure 15. The evolution of droplet size distribution (DSD) as a function of the corresponding droplet radius for (a) no FFC effect and (b) an FFC mass fraction of 1%. Each dot represents the number concentration of droplets in each size bin. The black circle, light-gray square and dark triangle represent the spectra at the initial time, 600 s and 1200 s of simulation.

fixed, and initially is enough to form more than one layer of surface film for this case. At some point the slow swelling will eventually break off the FFC coverage and induce stronger condensation and thus accelerated dilution. As the dilution process is faster for smaller droplets, only the smaller aerosols are able to grow. Given a longer time, as shown in the spectrum at 1200s simulation time, the smaller of the previously un-activated larger droplets were able to break through the FFC shielding and try to catch up in growth. Some of the remaining ones will never have the chance of activation, because the S_a falls off rather fast to a value that is below their critical S_d. Note that FFCs may enter hygroscopic aerosols by deposition from the gas phase or by coagulation. So, theoretically, their mass fraction may be proportional to the surface area or mass of the aerosol, respectively, and the calculations for Fig. 15 assume the former.

The FFC effect on the CDNC can be summarized as: (1) the retarded condensation growth elevates the ambient supersaturation, thus reducing the activation cutoff size and increasing the CDNC; (2) large aerosols cannot break off the FFC shielding to grow into cloud drops, causing a decrease in the CDNC. Which effect is stronger depends on the FFC content

and aerosol size distribution, as well as various environment parameters. Chen and Hsieh (2004) performed a comprehensive calculation of the change of CDNC as a function of FFC content and updraft speed for different aerosol size distributions. In general, for maritime aerosols containing FFCs, the influence on the CDNC is negative for high FFC contents, and positive for low FFC contents. The maximum change is a-few-ten percent. For urban aerosol distributions, the changes in the CDNC are all positive, reaching over 200% in some situations.

Besides modifying the CDNC by directly influencing the activation process, there are other indirect effects that FFCs may have on cloud microphysics and chemistry. First of all, they change the size distribution of cloud drops. From Fig. 15 one may notice the significant reduction of large cloud drops when the FFC is in operation. Those large cloud drops are important to the initiation of rain. They also have the function of decreasing the CDNC by accretion. In this regard, the FFC yields consequences equivalent to Twomey's first and second indirect effects. From the figure one may also realize that, with the FFC effect, smaller CCNs will replace large CCNs to activate into cloud drops. This means that there are less hygroscopic materials in cloud water, which may have a substantial effect on the cloud chemistry. Furthermore, the FFC retards not only the condensation and evaporation of cloud drops but also the mass exchange (absorption/desorption) of other gases. This again has potential influences on cloud chemistry and acid rain formation.

Surfactants can also affect cloud drop activation through the influence on water activity (Raymond and Pandis, 2002) or surface tension (Facchini *et al.*, 1999a; Anttila and Kerminen, 2002). Charlson *et al.* (2001) reported measurements on samples of cloud water, which showed a large decrease in the surface tension due to surface-active organics. They suggested that if a large surface tension depression occurs in cloud droplets near the critical size for activation, it will lead to an increase in droplet concentration and hence in cloud albedo. They estimated that if there is such an effect on all stratus clouds, a global mean forcing of almost $-1\,\mathrm{W\,m^{-2}}$ will arise. The change of surface tension may also affect other microphysical processes. Garrett (1978) suggested that partially covered surface film tends to concentrate at the downwind portion of the free-falling droplet surface and thus gives rise to a surface tension gradient, which then influences internal circulation. This is another way to affect mass transfer across the phase boundary and the chemical reactions inside the droplets. The influence of surfactants on surface tension may also affect the dynamics of droplet coalescence and breakup during collision (Ryan, 1976), processes that are very important to rain formation.

2.2.4. *Bioaerosols*

Primary biogenic organic aerosols (hereafter called "bioaerosols") include whole organisms (e.g. bacteria, fungi and phytoplankton), reproductive materials (e.g. pollen) and fragments (e.g. plant waxes). Air masses that are influenced in rural areas generally have large

amounts of giant biological particles such as pollen spores, whereas in urban and industrial air there tend to be higher concentrations of smaller biological particles such as bacteria (Matthias-Maser and Jaenicke, 1995). Over the ocean, primary production results from the ejection and dispersion of saltwater droplets from bursting bubbles at the sea surface (Woodcock, 1953). Matthias-Maser *et al.* (2000) suggested that the proportion by volume of atmospheric particles made up by biological material in remote continental, populated continental and remote maritime environments is respectively 28%, 22% and 10%.

Bioaerosols may lead to the formation of ice crystals by serving as INs (Schnell and Vali, 1973; Hazra *et al.*, 2006), among which bacteria are possibly the most abundant and effective in ice nucleation. Decaying plant leaves can also act as INs, but the ice nucleation has been identified to be of bacterial origin. In fact, some plant frost injury has been shown to involve an interaction of certain leaf-surface bacteria that cause the frost-sensitive plants on which they reside to become more susceptible to freezing damage (Lindow, 1983). The worldwide availability of such nuclei was established by finding ice-forming nuclei in plant litters collected in different climatic zones. The studies by Vali *et al.* (1976), Levin *et al.* (1987) and Hazra *et al.* (2004) showed that biogenic INs might be released from the Earth's surface to the atmosphere and be active in initiating ice formation at temperatures as high as $-2°\mathrm{C}$. Interestingly, some bacteria also serve as CCNs (Bauer *et al.*, 2003), and have indeed been observed in the active form in clouds (Sattler *et al.*, 2001).

Here, we demonstrate the effect of bacteria on precipitation formation using the model of Cheng *et al.* (2007), but add in the mechanism of ice nucleation from bacteria. The rate equation of bacteria ice nucleation is derived from the experimental data of Hazra *et al.* (2004) for a common bacterium — *Pseudomonas aeruginosa*. A typical bacteria concentration of

254   Jen-Ping Chen *et al.*

20 per liter is applied to the whole model domain. For comparison, we also apply the heterogeneous (deposition–condensation) nucleation rate from Hoffman (1973) to represent natural ice nucleation. A springtime cold-front system passing through northern Taiwan is selected for simulation. Since bacteria INs are effective at high subzero temperatures, one would expect more ice particles to be formed in the cloud. But there seems to be less snow and graupel in the case of bacteria INs (not shown). In fact, the formation of ice from bacteria is so effective that ice particles formed from them largely fall and melt to form raindrops. Interestingly, although bacteria cause more rainwater in the air, the precipitation intensity on the ground is actually weaker before the middle of the second day, as shown in Fig. 16. This is because there are so many raindrops formed in the bacteria IN case that each raindrop becomes much smaller. Smaller raindrops not only fall slower (hence lower precipitation intensity) but also evaporate faster. Nevertheless, bacteria INs seem to be able to produce precipitation with a similar order of magnitude to that by natural INs. Furthermore, toward the end of the second day, rain formation in the bacteria INs case seems to pick up, and produces more rain on the ground. The 48-hour cumulative rainfall of 67.0 mm is, in fact, slightly

higher than the 65.5 mm in the natural INs case. Note that we also run the simulation with all ice processes turned off, and find that warm rain processes are also effective but less prominent (49.2 mm rainfall in 48 hours) than ice processes.

Because of their strong ice-nucleating capability and large abundance, bioaerosols may play an important role in exerting control over cloud development and precipitation. In turn, the quantity and the type of bioaerosols are strongly influenced by clouds, which alter the availability of solar radiation for photosynthesis and surface temperature, as well as precipitation that provides the main water supply. Such a feedback process is very complicated and highly uncertain, but the potential importance cannot be ignored. As summarized by Barth *et al.* (2005), an investigation into this coupled cycle is necessary for a better understanding of the Earth system, including weather and climate change, regional and global atmospheric chemistry, as well as changes in land cover ecology.

The concentration of bacteria in the atmosphere depends on transportation from the surface boundary layer. The bacteria flux is closely connected with the reproduction rate of bacteria. Living and dead bacteria, including ice-nucleating species, have been found in clouds and fogs (Fuzzi *et al.*, 1997), raindrops (Maki and Willoughby, 1978) and hailstones (Mandrioli *et al.*, 1973). Bacteria have been observed in the boundary layer, in the upper troposphere (Lindemann *et al.*, 1982), and even in the stratosphere at altitudes of up to 41 km above the sea surface (Wainwright *et al.*, 2003). The flux of bacteria entering the atmosphere is recognized as originating from two types of temperate vegetation zones: (a) high-primary-production row crop areas and (b) relatively-low-production desert areas. The highest flux was 1.95×10^6 colony-forming units (CFU) $m^{-2} h^{-1}$, measured in the crop area (Lindemann *et al.*, 1982) and 1.7×10^3 CFU $m^{-2} h^{-1}$ over

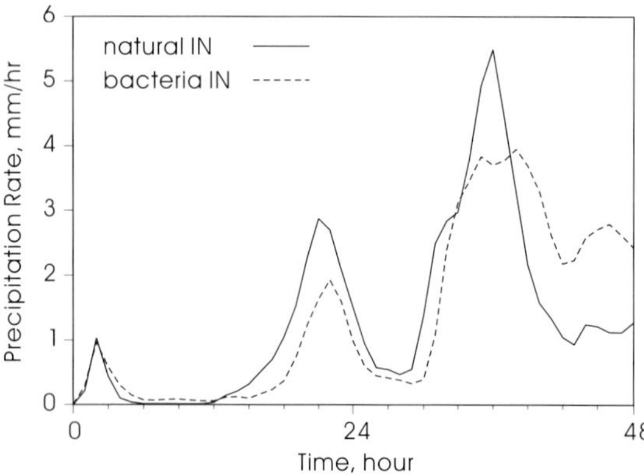

Figure 16. Time series of surface precipitation intensity simulated with natural INs (solid) and bacteria INs (dashed).

the desert area (Lighthart and Sharffer, 1994). Interestingly, bacteria not only have capabilities as CCNs and INs, but may also play a role in the modification of other atmospheric OCs that act as CCNs. Herlihy *et al.* (1987) showed that the bacterial utilization of formic and acetic acid is in rainwater. Bacteria and fungi transformed dicarboxilic acid (DCA) efficiently in the boundary layer, with estimated degradation lifetimes comparable to those of major atmospheric oxidants (Ariya and Amyot, 2004). The authors also showed that different fungus species drive microbiological degradation of several atmospherically active OCs at different rates. This degradation is also a function of several environmental factors, namely pH, temperature and nutrient levels.

2.3. *Mineral dust*

Atmospheric mineral dusts originate mainly from deserts, semiarid areas and, to a lesser extent, cultivated lands, sandy seashores, riverbanks and grasslands. These airborne dust particles (which are often termed "mineral aerosols," as well as "yellow dust" or "kosa" in some regions) have potential effects on cloud and precipitation formation, atmospheric radiative transfer, air quality and atmospheric chemistry, as well as ecology over land and in the ocean (Eppley, 1980; Duce, 1986; Guieu *et al.*, 2002). Besides the air quality and ecological issues, a lot of dust studies emphasize the climate forcing aspect of mineral aerosols due to their ability to scatter and absorb solar radiation. The report of the IPCC (2001) specifically points out that the influence of mineral dust on radiative forcing can be both positive and negative, and ranges from -0.6 to $+0.4\,\mathrm{W\,m^{-2}}$. However, these are just the direct effects. Evidence is mounting that mineral dust may influence the climate indirectly by influencing cloud processes. There are also influences on the hydrological cycle through similar processes mentioned in previous subsections.

2.3.1. *Roles as CCNs and INs*

Mineral dust particles from the arid regions of the Asiatic continent have been implicated as possible heterogeneous INs, more than 40 years ago (Isono *et al.*, 1959). More recently, emerging evidence has indicated that mineral particles may reach the upper troposphere, where they may serve as INs for cirrus cloud formation. In an aircraft campaign conducted over the Alps, Heintzenberg *et al.* (1996) found that minerals were common constituents in cirrus crystal residues. Upper-tropospheric INs activated in a diffusion chamber and subsequently collected by impaction (Chen *et al.*, 1998) also had increased number fractions of crustal particles when compared with the ambient aerosol population. A lidar study by Sassen (2002) suggested that cirrus-like ice clouds form at the top of layers of transported Asian aerosols, at temperatures considerably warmer than the climatological mean temperature for midlatitude cirrus formation. Sassen *et al.* (2003) documented glaciations of altocumuli forming at the top of the dust layer at $-8°C$ during one of the same episodes. DeMott *et al.* (2003) detected concentrations of heterogeneous INs exceeding $1-3\,\mathrm{cm^{-3}}$, up to 100 times higher than typical background values in and above the marine boundary layer in Florida during Saharan dust episodes. These INs were within the air layer that feeds thunderstorm development and subsequent cirrus anvil formation. Measurements of high mineral dust fractions in residual particles from anvil cirrus during that study (Cziczo *et al.*, 2004) and cloud model simulations of these cases (Van den Heever *et al.*, 2006) all supported the hypothesis that the dust modified the cloud microphysics and dynamics.

Mineral dust particles may trigger ice nucleation via either the deposition mode or the freezing mode. The former mode forms ice by direct deposition of water vapor, while the latter mode usually occurs on dust particles coated with solutes, which absorb water and form a solution coating. Quantitative evidence of such ice nucleation comes best from laboratory efforts. Hung *et al.* (2003) studied ice freezing of aqueous ammonium sulfate particles containing mineral dust cores, and found that the heterogeneous nucleation rates vary from 10^2 to $10^6\,\mathrm{cm}^{-2}\,\mathrm{s}^{-1}$, depending on the diameter of the dust core, the temperature and the solute concentration. They also rationalized the nucleation rates by the equations of classical heterogeneous nucleation theory. Archuleta *et al.* (2005) measured the ice nucleation behavior of metal oxides coated with sulfuric acid and found that heterogeneous freezing occurs at lower relative humidities than those calculated for homogeneous freezing of the diluted particle coatings. In addition, for all dust types the heterogeneous freezing rates increased with particle size for the same thermodynamic conditions, whereas for the same particle size, natural mineral dust particles were the most effective INs. Salam *et al.* (2006) reported the ice nucleation results of two dust types that are of potential atmospheric relevance. They showed that both kaolinite and montmorillonite act as efficient INs in the deposition/condensation nucleation mode, but the latter is somewhat more efficient. Also, the fraction of active ice nuclei increases with decreasing temperature and with increasing relative humidity. Mohler *et al.* (2006) analyzed mineral dust particles from three deserts — Arizona, Taklamakan and Sahara — and found that deposition ice nucleation was most efficient on Arizona dust particles. For all samples the ice-activated aerosol fraction could be approximated by an exponential equation as a function of S_i, but more experimental work is needed to quantify the variability of the ice activation spectra as functions of the temperature and dust particle properties. Field *et al.* (2006) collected and analyzed desert dust samples from the Sahara and Asia, and found two nucleation events. The primary nucleation event occurred at ice saturation ratios of 1.1–1.3, and is likely to have been in the deposition nucleation mode. The secondary nucleation event occurred at ice saturation ratios between 1.35 and 1.5, and is likely to have occurred via condensation mode nucleation. The activated fractions of desert dust ranged from 5–10% at $-20°\mathrm{C}$ to 20–40% at temperatures colder than $-40°\mathrm{C}$. But the authors did not find an obvious difference between the nucleation behaviors of the two dust samples.

The knowledge of heterogeneous ice nucleation in precipitation formation has been utilized to artificially enhance precipitation. As early as World War II, Findeisen attempted a cloud-seeding flight using sand as an ice-nucleating agent, but apparently without success (Schaefer, 1951), which indicates the limited ice-nucleating ability of natural mineral dust. More effective agents were later developed and applied in weather modification; they include the commonly used silver iodine (Schaefer, 1946) and rarely applied but perhaps more effective organic agents such as metaldehyde (Fukuta, 1963) and phloroglucinol (benzene-1,3,5-triol) (Langer *et al.*, 1963). Even bacteria have been suggested as artificial seeding agents because of their superb nucleation efficiency (see Subsec. 2.2.4).

Besides acting as INs, mineral dust may take on the role of CCNs. It is now more clearly known that mineral dust reacts with various trace gases, and through these reactions the cycles of various chemical constituents and mineral dust become linked (Dentener *et al.*, 1996). Such processes may also affect cloud formation. For example, dust particles provide reactive sites for heterogeneous chemistry, which enhance the uptake of nitrate onto the dust particles (Phadnis and Carmichael, 2000; Grassian, 2002; Bauer *et al.*, 2004; Krueger *et al.*, 2004). The addition of nitrate gives dust particles a hygroscopic coating, which turns them into effective CCNs and thus impacts cloud formation. Hygroscopic coating on dust may also be attained through cloud-chemical and -microphysical processes, details of which are given in Subsec. 3.1. Rosenfeld *et al.* (2001)

noted the effect of Saharan dust in reducing precipitation in shallow convective clouds near the source. This was hypothesized to be due to lowering of the coalescence efficiency of clouds resulting from increases in dust particles that act as cloud condensation nuclei.

2.3.2. *Effects on cloud chemistry*

Although mineral dust aerosols are considered insoluble, they may still dissolve to a certain degree and release trace ingredients that affect chemical processes in cloud water and aerosol water. The most interesting ingredients include calcium and transition metals such as iron and magnesium. Calcium and minerals alike are base substances that can neutralize acid rain in a manner similar to the way antacids counteract excess acid in an upset stomach. Barnard *et al.* (1986) proposed that dust from unpaved roads, which contribute more than 90% of all open source calcium and magnesium in most of the US states, may be of potential importance to acid rain neutralization. Inoue *et al.* (1991) also suspected that calcite in the eolian dust may have been neutralizing acids in rainwater and snow. More quantitative evaluations were provided by Wang *et al.* (2002), who showed that the yellow sand in East Asia can strongly neutralize the acidic precipitation, with the highest seasonal increase of pH values by 0.1–0.4 in Japan, 0.5–1.5 in Korea, and more than 2 in northern China. Note that the elevation of pH values in turn affects many aqueous phase chemical reactions, as mentioned in Subsec. 3.1 and discussed below.

Mineral dust particles also impact the ecology of land and sea via the release of their trace elements. They may provide iron, which is known to exert a controlling influence on biological productivity in surface waters over large areas of the ocean, and may have been an important factor in the variation of the concentration of atmospheric carbon dioxide over the glacial cycle (Martin, 1990; Boyd *et al.*, 2000).

The enhancement of phytoplankton activity through iron fertilization may also exert an influence on the DMS–cloud–climate interactions mentioned in Subsec. 2.1.1.

Dissolved iron (Fe^{3+}) and magnesium (Mn^{2+}) may also affect aqueous chemistry in cloud water by acting as a catalyst in the oxidation of SO_2 by O_2 to produce sulfate (see Seinfeld and Pandis, 2006; p. 314). There are further interesting links between iron and sulfate. Iron in mineral dust mostly exists in the form of Fe(III), whose low solubility limited its involvement in either cloud chemistry or marine biological fertilization. There are a series of reactions cycling iron between the Fe(III) and Fe(II) states that are pH-dependent (Zhuang *et al.*, 1992; Deutsch *et al.*, 2001). In fact, iron can be significantly mobilized below a threshold pH of $\sim$3.6 or above a pH value of 7.1 (Mackie *et al.*, 2005). In summary, Fe(III) affects the oxidation of SO_2 into sulfate, the production of sulfate decreases pH, and pH affects the cycling of iron as well as the oxidation of SO_2. The main source of SO_2 in the marine area is the production of DMS by phytoplankton, whose growth may be accelerated by the input of Fe(II) which is mobilized from Fe(III) that is contained in mineral dust. If dust is the one that provides iron, its alkaline contents (e.g. calcium and magnesium) also involve the Fe–sulfur coupling by affecting pH. Such iron–sulfur cycles are illustrated in Fig. 17.

3. Cloud Effect on Aerosols

Clouds affect aerosols in many intricate ways. They can redistribute aerosols by processes such as cloud convection and scavenging. They also provide sites for chemical reactions that would not otherwise occur in a clear sky, thus influencing the production and destruction of many trace gases and aerosol particles. Clouds can also reflect or diffuse the solar actinic flux that drives many photochemical reactions that produce precursors of aerosols. Lightning generated from

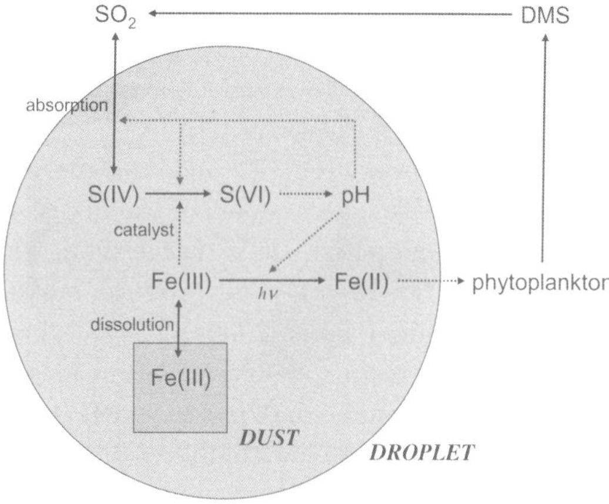

Figure 17. Schematics of the interactions between iron and sulfur cycles. Solid arrows indicate conversions, while dotted arrows indicate influences.

clouds may induce similar effects by influencing NO_x-related chemistry.

3.1. *Microphysical and chemical modifications*

Cloud water is an effective medium for chemical reactions. It can absorb trace gases, and dissociate them into ionic species, which may then react to form new species. Some of the aqueous phase reactions dominate the production or destruction of certain important chemical species in the atmosphere. Aqueous phase oxidation of sulfur and nitrogen species adds to the nonvolatile solute mass in cloud drops, such that larger and more efficient CCNs can be produced if they are not removed through rainout. Lelieveld (1993, 1994) estimated that up to 80–90% of the atmospheric SO_2 is converted into sulfate within clouds, while the rest of the conversion proceeds mainly via reaction with OH radicals in the gas phase. Most of the condensed water in clouds does not precipitate out, and the remaining droplets unavoidably evaporate and return the remnants to the aerosol population. So, aqueous phase chemical production may increase the chemical mass load of aerosols. However, the increase in particle size also leads to a greater deposition velocity. So clouds are

effective in shortening the lifetimes of atmospheric chemicals not only directly by in-cloud chemical conversion and wet deposition but also indirectly by enhancing dry deposition.

Cloud processes may modify the properties of mineral dust. Wurzler *et al.* (1997) measured aerosol composition in the Mediterranean and found that mineral dust particles often get coated with sulfate and other soluble substances. It has been suggested that the formation of the sulfate coating is related to the scavenging of aerosol particles by cloud droplets and the subsequent impaction scavenging of mineral dust particles. Certainly, some of the sulfate may also be produced within the cloud drops via the aqueous phase chemistry discussed above. Wurzler *et al.* (1997) also found that the processed dust particles may become effective giant CCNs which, upon entering subsequent clouds, enhance the development of precipitation through the collision–coalescence process. Yan *et al.* (2002), on the other hand, suggested that the sulfate-coated dust particles become effective CCNs after passing through a convective cloud, and these particles increase the concentration of the activated drops and widen the drop size distribution while entering the next cloud. The widening of drop size distribution also accelerates precipitation formation.

The dissolution of gaseous SO_2 into cloud water and its oxidation into sulfate are positively dependent on the pH, which means that the conversion of SO_2 into sulfate (a process that decreases the pH) is generally self-limiting. However, the ammonia or even the ammonium in aerosols may elevate the pH and boost the sulfur chemistry. During the condensation growth, the dilution of cloud drops causes a departure from Henry's equilibrium between the air and the droplet. So, ammonia in the air is immediately drawn into the cloud drops. As soon as the ammonia concentration in the air is reduced, the disequilibrium is relayed to the interstitial aerosols, which do not swell as much as cloud drops do. As a result, ammonia will be drawn from the interstitial aerosols into the air and, eventually, into the cloud drops. Such a redistribution of ammonia is in fact the main reason why in Fig. 13 the pH values of the interstitial aerosols decreased during the cloud formation shown. This ammonia input helps to fix more sulfate and other acidic materials in the cloud drops; if the acidic materials (such as sulfate) are nonvolatile, they also help to fix ammonia in the particle when cloud drops evaporate. In this way, cloud processes affect the chemical distribution among aerosols.

Aqueous phase oxidation in cloud water is not the only mechanism that increases the mass of nonvolatile chemicals in aerosols (when the cloud evaporates). The absorption of ammonia, an alkaline chemical as discussed above, also helps to absorb nitric acid in the air and fix it in the droplet. Figure 18 shows the partition of two acids in the gas and liquid phases. Sulfuric acid has very high solubility; thus, it mostly stays in haze particles under typical atmospheric liquid water contents (LWCs) and pH values [see Fig. 18(a)]. But nitric acid has relatively weak solubility in aerosols, particularly when the aerosol particle contains sulfuric acid, which makes the aerosol very acidic. From Fig. 18(b)

one can see that under typical clear air LWCs of below $10^{-5}\,\mathrm{g\,kg^{-1}}$ and aerosol pH of around 2 (see Fig. 11), about 90% of the nitric acid exists in the gas phase (point A). When clouds are formed, both the LWCs and the droplet pH increase significantly, and essentially all gas phase nitric acid will enter cloud drops (point B). As pointed out by Kulmala *et al.* (1993), Xue and Feingold (2004) and Romakkaniemi *et al.* (2004), the intake of nitric acid that reserved in the gas phase may enhance the activation of aerosols and the growth of cloud drops. Furthermore, the ammonia intake mechanism mentioned above can fix such nitric acid in the particle phase and alter the composition of aerosols.

Other alkaline substances that may enhance the mass production of aerosols through cloud processes include sea salt and mineral dust, as mentioned in Subsec. 2.3.2. These aerosols tend to induce stronger modification by cloud chemistry. Hegg *et al.* (1992) pointed out that alkaline sea salt particles significantly impact the sulfate production and modify the size distribution, thus enhancing the light-scattering efficiency of aerosols. Yuen *et al.* (1994) then suggested a different behavior from continental

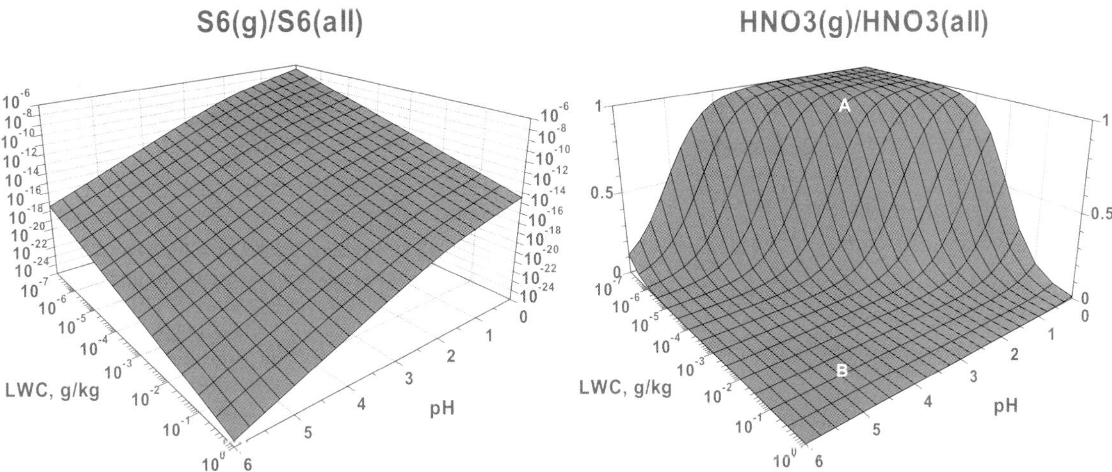

Figure 18. Partition of sulfuric acid and nitric acid between the gas phase and the liquid phase as a function of atmospheric liquid water content and droplet pH. The vertical axis is the fraction of these acids that exists in the gas phase.

type aerosols, from which sulfate was produced in particles but outside of the light-scattering size range.

In the above we discussed the modification of aerosol composition and size distribution through the combination of cloud chemistry and two microphysical processes — activation and condensation. Other cloud-microphysical processes may modify aerosols without the involvement of chemistry, such as droplet collision, riming or sedimentation. When droplets collide and coalesce, they not only combine water mass but also solute mass, as shown previously in Fig. 12. When these coalesced droplets evaporate, they leave behind much larger aerosols, but with lower number concentration. This mechanism is a possible reason why some marine stratiform clouds suddenly develop drizzle. It is thought that the slow collision–coalescence occurring in these clouds gradually modifies the CCN size distribution and produces more and more giant CCNs. After a few cloud cycles, there might be enough giant CCNs to initiate strong drizzle (Feingold, 1999). To the extreme, this may lead to a sudden breakdown of marine stratiform clouds (see Subsecs. 2.1.1 and 2.1.2). Mixed-phase collision (e.g. riming) may have similar effects, and this often leads to precipitation — hence the removal of the potential aerosols. Note that traditional spectral (bin) microphysical models such as those of Feingold *et al.* (1999) or Lynn *et al.* (2005), although very detailed in resolving the size spectrum of cloud particles, cannot accurately simulate aerosol recycling through clouds as discussed above, because they do not resolve the additional solute component shown in Fig. 12.

3.2. *Precipitation scavenging*

Cloud and precipitation scavenging is an important factor in determining the lifetimes of many atmospheric trace chemicals, including aerosols. Aerosols may be incorporated into cloud drops and ice crystals either by activation/nucleation or by collections (gravitational collection, Brownian collection and phoretic collection), then removed along with the precipitation — a process often referred to as in-cloud scavenging or rainout. The collection process may also occur below the cloud base, which is referred to as below-cloud scavenging or washout. The fraction of aerosols (as well as trace gases) being scavenged depends on the microphysical processes through which they are incorporated into cloud particles. In other words, the scavenging of aerosols by clouds is essentially determined by how the precipitation is formed and cannot be easily estimated by the amount of the precipitation fall on the surface. This is in fact one of the most difficult issues in the modeling of aerosols in regional and global models, which often lacks detailed description of cloud microphysics. The interaction between microphysics and chemistry which was discussed earlier further complicates the problem.

Recognizing the importance of cloud microphysics for the removal of atmospheric trace chemicals, Chen and Lamb (1994, 1999) developed a mixed phase cloud-microphysical model coupled with aqueous phase and ice phase chemistry to investigate the complex scavenging mechanisms. This multicomponent model allows simultaneous and independent changes of various physical and chemical properties of the cloud particles. For the results given below, they applied two bin components (water mass and major solute mass) for the liquid phase framework (see Subsec. 2.1.4) and three bin components (water mass, major solute mass, and aspect ratio — defined as the ratio of the c-axis length to the a-axis length of crystals) for the ice-phase framework. The simultaneous consideration of water and solute mass contents allows this model to resolve truthfully the aerosol–cloud interactive processes, such as the size-dependent aqueous chemistry and the recycling of aerosols after cloud dissipation. The liquid phase microphysical processes considered in the model are the activation of condensation nuclei into cloud drops, the subsequent condensational growth, and collision–coalescence and breakup. The ice phase processes included are heterogeneous deposition/condensation nucleation, heterogeneous freezing, contact freezing, homogeneous freezing, diffusional growth, accretional growth, rime-splintering, melting, shedding due to melting and wet riming, and the aggregation of snow crystals. Besides the acquisition of water and solute, the change of shape (aspect ratio) due to all these processes is calculated explicitly following the adaptive crystal shape scheme of Chen and Lamb (1994a). To facilitate discussions, ice particles in the three-dimensional (water mass, solute mass, shape) ice framework are classified, following conventional terms, into cloud ice, planner ice, columnar ice, rimed crystals (i.e. graupel) and crystal aggregates according to their shape, chemical content and density. The aqueous phase chemistry considered includes the absorption/desorption of NH_3, H_2SO_4, HNO_3, SO_2, O_3, H_2O_2, CO_2, their ionic dissociations, and the oxidation of sulfite by O_3 and H_2O_2. The ice phase chemical processes included are the sorption/desorption of SO_2 and H_2O_2 onto the ice surface, as well as the entrapment of SO_2 inside the ice during riming.

The example given here is a case of wintertime orographic cloud formation over the Sierra Nevada Crest and Carson Range in the western US on 18 December 1986. Figure 19 shows the simulated liquid phase and ice-phase cloud fields over the mountains. The LWCs in both clouds are mostly below $0.2\,g\,m^{-3}$, which is typical of such orographic clouds. The aerosol type applied in this simulation is of remote continental conditions with a total aerosol concentration of about $6{,}000\,cm^{-3}$. This relatively clean condition plus a low updraft speed of the upslope wind result in a low CDNC, which has the highest value near the cloud base, ranging from 106 to $274\,cm^{-3}$, then gradually decreases toward the upslope because of collision-coalescence and collection by ice crystals. Light rains develop near and below the base of the first

cloud, and these raindrops are not formed by warm cloud processes, but by melted ice falling from aloft. In Figs. 19(d) and 19(f) one may observe two zones of planner ice and two zones of columnar ice that are in accordance with the growth habit regimes reported by Nakaya (1954) and Hallet and Mason (1958). Some of the earliest-formed plates and columns grow big enough to commence accretion of cloud drops and turn into rimed crystals [Figs. 19(g) and 19(h)], which are the main source of raindrops at the lower levels. After the depletion of larger cloud drops by the riming process, ice crystals grew mainly by vapor deposition. Some of these pristine crystals collide and coagulate into snow aggregates [Figs. 19(i) and 19(j)], becoming the main form of precipitation at the latter stage of cloud development. Note that all liquid droplets evaporate over the downslope region between the two mountain peaks, but ample ice crystals survive and re-enter the second cloud.

The types of precipitation falling on the ground are shown in Fig. 20(a). One can see that rainfall first develops between about 30 and 70 km. Beyond about 70 km, there is still a minor amount of "rainfall", but it is actually a deposition of cloud droplets because by then the cloud is in contact with the ground. As mentioned earlier, the rain forms mainly from melted graupel. Since the temperature becomes lower while going uphill, the portion of graupel that remains unmelted becomes greater, and reaches a peak at about the 70 km distance. After that, snow aggregates started to appear and become the main precipitation type for a short while before columnar ice emerges. Such a sequential change of precipitation types is commonly observed in cold orographic or stratiform clouds. In the second cloud (over the second mountain peak), because there are remnant ice crystals from the first cloud to feed in, aggregates form rather early as compared to the first cloud. In addition, the early presence of ice crystals prohibits (due to the Bergeron–Findeison process) cloud drops to grow large enough

262   Jen-Ping Chen *et al.*

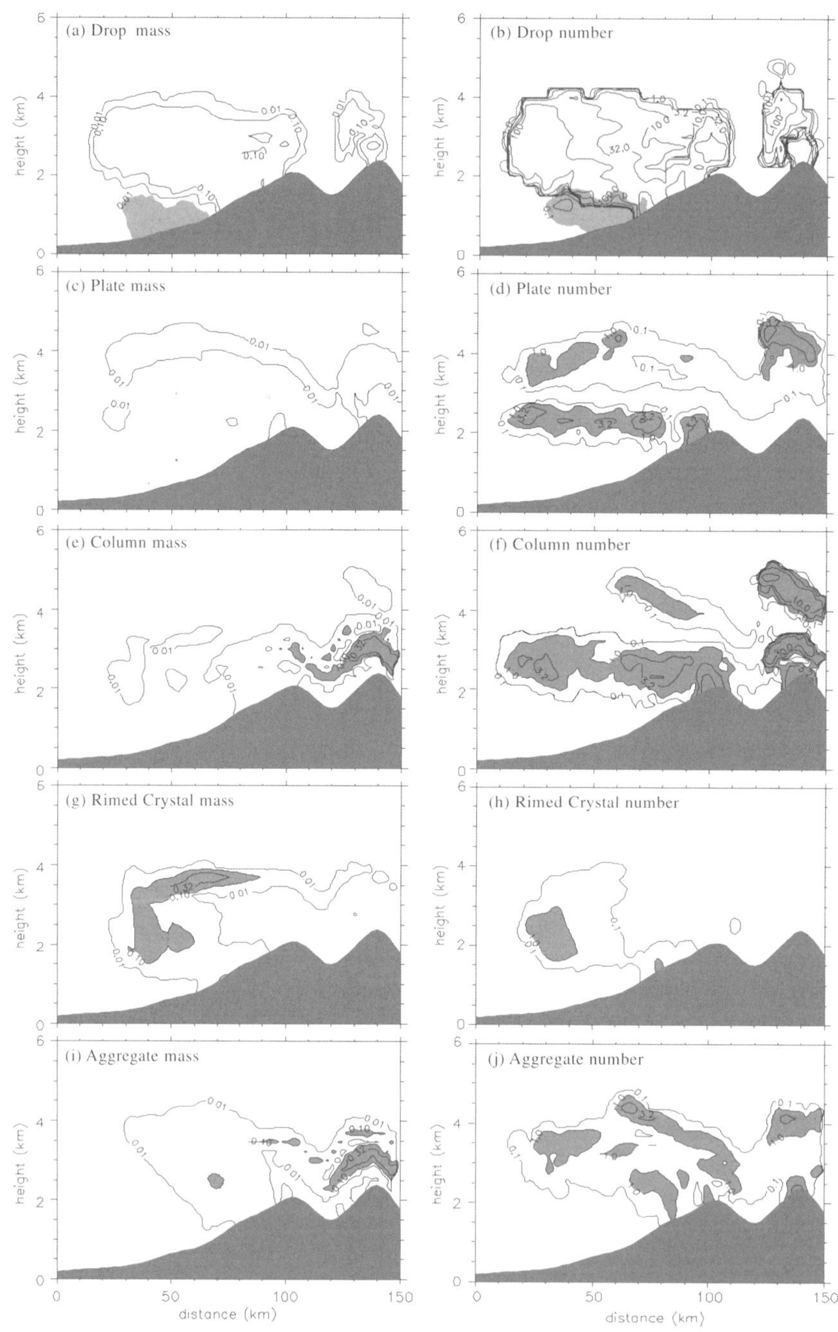

Figure 19. Simulated distribution of various cloud condensates over the Sierra Nevada Crest and Carson Range. The left and the right panel show are mixing ratios of condensed water and their number concentrations, respectively. From the top down are different condensate types of droplets including cloud drops (light shading) and raindrops (dark shading), planner ice (plates), columnar ice (columns), rimed ice (graupel) and snow aggregates. The mixing ratios are in g m^{-3}, while the number concentrations are in cm^{-3} for cloud drops and in L^{-1} for raindrops and ice crystals. The minimum contour of the mixing ratio $\langle$number concentration$\rangle$ is $0.01\langle 0.1\rangle$ and increases by a half-decade for each additional interval except for the $0.032\ \langle 0.32\rangle$ contours that are omitted; shadings indicate values $> 0.1\langle > 1\rangle$ for cloud drops and ice crystals, and $> 0.01\langle > 1\rangle$ for raindrops. (From Chen and Lamb, 1999.)

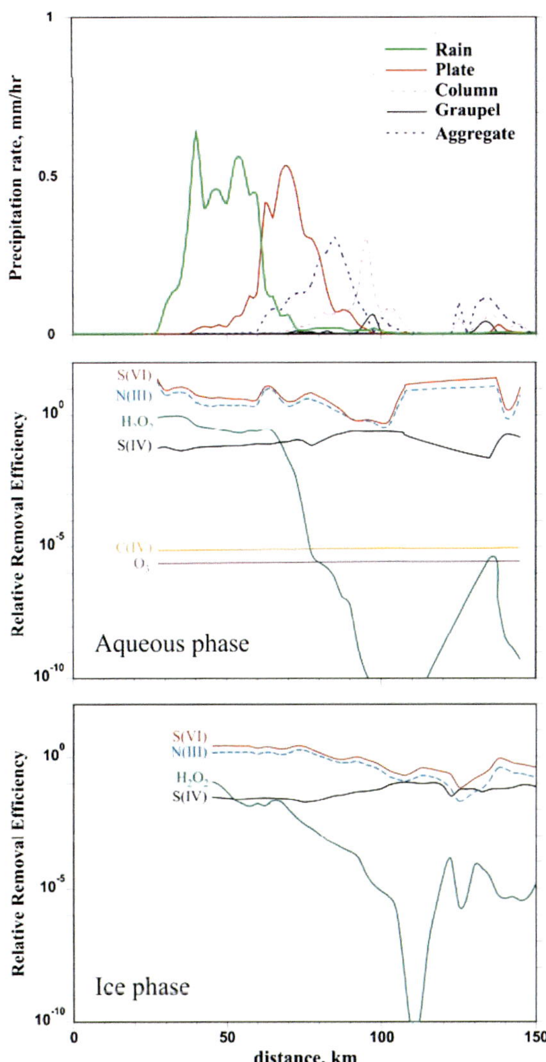

Figure 20. Simulated surface precipitation (*top*) and column-integrated relative removal efficiencies of various trace chemicals by liquid precipitation (*middle*) and ice precipitation (*bottom*) from the clouds shown in Fig. 19.

to be accreted by rimed ice. That is why not much graupel-type precipitation develops from the second cloud.

The types of precipitation (and the formation mechanisms) strongly influence the removal of trace chemicals. As the "precipitation rate of water" is quite a common scientific notion, we shall express the strength of chemical scavenging (i.e. precipitation rate of chemicals)

using a relative concept. We define the *relative removal efficiency, E_R*, which represents the proportion of chemicals in the air being removed by precipitation compared to that of water in the air, with the following formulation (Lamb and Chen, 1990):

$$E_R = \frac{k_i}{k_{H_2O}} = \frac{[i]/C_i}{[H_2O]/C_{H_2O}}, \qquad (9)$$

where k_i and k_{H_2O} are the precipitation frequency (or rate) of a chemical species i and water, respectively; variables in square brackets denote the species concentration in the precipitation; and C represents the concentrations in the air. Figures 20(b)–20(c) show the E_R of various trace chemicals due to scavenging by liquid phase and ice phase precipitation, respectively, from the orographic clouds shown in Fig. 19. Note that since essentially all sulfate exists in aerosol or cloud droplets (see Fig. 18), and sulfate in cloud drops will return to aerosols when clouds dissipate, the curves labeled S(VI) may represent the E_R of aerosols. One can see that the aerosol E_R is the highest value among all chemicals. Furthermore, the aerosol E_R in rainfall [Fig. 20(b)] is significantly higher than unity, which means that they can be removed more efficiently than water from the air. Between 25 and 60 km, the aerosol E_R of the aqueous phase decreases gradually, due to two factors: (1) the rimed crystals accrete smaller and smaller cloud drops containing progressively less solute, and (2) the less-rimed ice actually relies more on depositional growth and obtains more clear water. At 60–65 km, the main form of aqueous phase precipitation switches from melted ice to intercepted cloud water, so the aerosol E_R is elevated somewhat. The only place in Fig. 20(b) where the aerosol E_R is lower than unity is at 90–100 km; at these locations the "rain" is actually ground interception of cloud drops, and these cloud drops were activated from smaller CCNs which contain less sulfate. Cloud drops activated from larger CCNs tend to grow larger and get accreted more easily by ice crystals. Ammonia is highly, but not completely, condensable in aerosol and cloud drops, so it also has a rather high E_R. S(IV) has a moderate E_R, and the value is higher further up the first hill because the intercepted cloud drops are more diluted and have higher pH values to allow high solubility of SO_2.

The atmospheric chemistry model and air pollution models often neglect the scavenging of chemicals by ice phase precipitation, which is commonly regarded as a "clean" form of falling water. Yet, from Fig. 20(c) one can see that ice phase precipitation does contain chemicals. In fact, the removal efficiency of aerosols by ice phase precipitation can be several times higher than that of water, shown for example by the high E_R at distances before 80 km. This, of course, is due to the riming process through which ice crystals obtain solutes in cloud drops. Note that the E_R of S(IV) may approach that of aerosols (between 105 and 135 km), as S(IV) can be incorporated into ice by entrapment during riming and by sorption during vapor deposition (see Chen and Lamb, 1990).

From the above discussions we may summarize that E_R is determined by solubility and microphysical mechanisms by which the chemicals are incorporated into the precipitation. Furthermore, precipitation is very effective in removing aerosols from the atmosphere even in a weak-lifting orographic cloud. In the previous section we saw that clouds can also be strong sources of aerosols. The tug of war between in-cloud production and removal by precipitation causes strong variability of atmospheric CCNs.

3.3. *Cloud venting*

Cloud venting is the process of exchanging mixed-layer air masses with the free atmosphere through updrafts and downdrafts associated with cloud activities. It plays a major role in the vertical redistribution of atmospheric trace chemicals, including water vapor and aerosols. Trace chemicals produced near

the Earth's surface, if not already scavenged, can reach the upper troposphere rather quickly with the assistance of strong updrafts. Strong convection may even significantly contribute to cross-tropopause transport into the stratosphere (e.g. Wang, 2003; Sherwood and Dessler, 2003). Stratospheric chemicals can also be transported downward by cloud turbulence and precipitation-induced downdrafts.

Cloud venting may proceed not only through cloud scale convections but also through synoptic cloud systems. For instance, the warm conveyor belts (WCBs), which are ascending airstreams ahead of cold fronts, are the primary mechanism for rapidly transporting air pollution from the continental planetary boundary layer to the upper troposphere and from one continent to another. Cooper *et al.* (2004) estimated in a case study that 8% of the WCB mass originated in the stratosphere and 44% passed through the lower troposphere. Cotton *et al.* (1995) performed a global analysis and found that the extratropical cyclone (e.g. WCBs) has the highest boundary layer mass flux of all cloud venting systems, followed by the general class of mesoscale convective systems, ordinary thunderstorms, tropical cyclones and mesoscale convective complexes. According to their estimation, these cloud ventings may pump the entire atmospheric boundary layer into the free troposphere 90 times a year. Note that the proportion of chemicals being transported by clouds varies from species to species because of chemical fractionation in scavenging processes. So, cloud venting and precipitation scavenging are actually two sides of the same coin.

Some of the cloud systems are too small to be represented in global scale models. So, cloud venting of trace chemicals is parametrized in essentially all current global chemical transport models (e.g. Jacob *et al.*, 1997), in which the vertical transport of soluble gases is parametrized by normalizing against the transport of tracers such as water. However, as pointed out by Yin *et al.* (2001), there is no straightforward scaling factor, particularly if small concentrations of highly soluble gases in the upper troposphere need to be defined. The transport of aerosols is no exception. In fact, the amount of chemicals being transported vertically depends not only on the strength of convection but also on the scavenging processes involving many cloud-microphysical mechanisms, as mentioned in the previous section. So the parametrization of aerosol scavenging and venting in large-scale models are particularly difficult, because these models do not truly resolve the cloud microphysical mechanisms.

3.4. *Influences on aerosol production*

In Subsec. 2.1.1 we mentioned the DMS–cloud–climate cycle, in which clouds may indirectly influence the formation of aerosols by impacting marine biological activities. Similarly, clouds may influence the productivity of land plants and the generation of ice nucleation bioaerosols (see Subsec. 2.2.4). The generation of atmospheric sea salt particles and of mineral dust are also related to clouds, as they favor the conditions of strong gust winds associated with cloud activities. All these indirect influences have strong feedback cycles, and the cycles may link to each other, forming a convoluted process network. In this Subsection we will introduce only a few additional mechanisms, with focus on aerosol nucleation, to exemplify the complexity of aerosol–cloud interactions.

3.4.1. *Actinic flux effect*

Although clouds may increase the total aerosol mass by aqueous phase chemical production, they normally reduce the number of aerosols either by the coalescence and accretion processes or by precipitation scavenging. Yet, some observational studies have found evidence of strong new aerosol production near clouds (e.g. Dinger *et al.*, 1970; Hegg *et al.*, 1990; Frick and Hoppel,

1993; Saxena and Gravenstein, 1994; Wiedensohler *et al.*, 1997). Earlier postulations of the causes of such particle production included the shattering of salt crystals formed by rapid evaporation of cloud drops (e.g. Dessens, 1949; Twomey and McMaster, 1955; Radke and Hegg, 1972), but Mitra *et al.* (1992) disproved this mechanism with wind tunnel experiments. More advanced aerosol measurements later revealed that the new particles are actually quite small in the nuclei mode size range. A more plausible cause of the generation of such nuclei mode aerosols is the homogeneous nucleation from the gas phase, which involves photochemical reactions. The remaining question is: How do clouds get involved in the process?

Homogeneous nucleation of atmospheric particles usually proceeds via the H_2O–H_2SO_4 binary interaction. Ternary nucleation with the participation of an additional gas such as ammonia or nitric acid may also enhance the process. The main conditions favorable for binary or ternary homogeneous nucleation are a high saturation ratio of water vapor S_w (relative humidity) and a high saturation ratio of sulfuric acid vapor S_a (relative acidity). High S_w and S_a may be caused by either increasing the vapor concentration or lowering the temperature. The increase of sulfuric acid concentration is particularly complex, because the amount of sulfuric acid in the gas phase is normally very low (sec Fig. 18). It requires either a strong production (such as via photochemical reactions) or a low concentration of existing aerosols which may otherwise consume the vapor quickly. Shaw (1989) postulated that only in very clean air could binary nucleation of H_2O–H_2SO_4 particles occur, because the existing aerosols may consume sulfuric acid vapor and inhibit high S_a. Hegg *et al.* (1990) then pointed out that high water vapor concentrations detrained from the cloud coupled with high actinic flux due to backscatter from the cloud droplets are additional controlling factors. A high actinic flux may boost photochemical reactions that lead to sulfuric acid production and high relative acidity. Crawford *et al.* (2003) measured enhancements in UV actinic flux of up to 40% over clear-sky values, whereas Kylling *et al.* (2005) found a 60–100% increase. Los *et al.* (1997) even estimated that photolysis rate coefficients can be 300% higher at cloud tops than in clear-sky conditions. Perry and Hobbs (1994) further drew the picture that convective clouds bring up the precursor of sulfuric acid, SO_2, by cloud venting and remove a large portion of the existing aerosols by cloud and precipitation scavenging. Then, high relative humidity and reflected sunlight near the clouds together form a favorable nurturing environment for new particles.

Reflectance of actinic flux by clouds may enhance particle nucleation, but only on the sunward side of the cloud. Behind the cloud, an opposite effect would occur. Figure 21 demonstrates such influences by cloud shadows. The results are from a parcel simulation using a detailed (binned) aerosol microphysical model with a multicomponent particle framework similar to that of Chen and Lamb (1994, 1999) for cloud microphysics (see Subsec. 3.2). It includes detailed treatments on binary nucleation and particle aggregation. For the production of sulfuric acid through photochemical reactions, a simplified production rate is used by applying the typical SO_2 concentration and

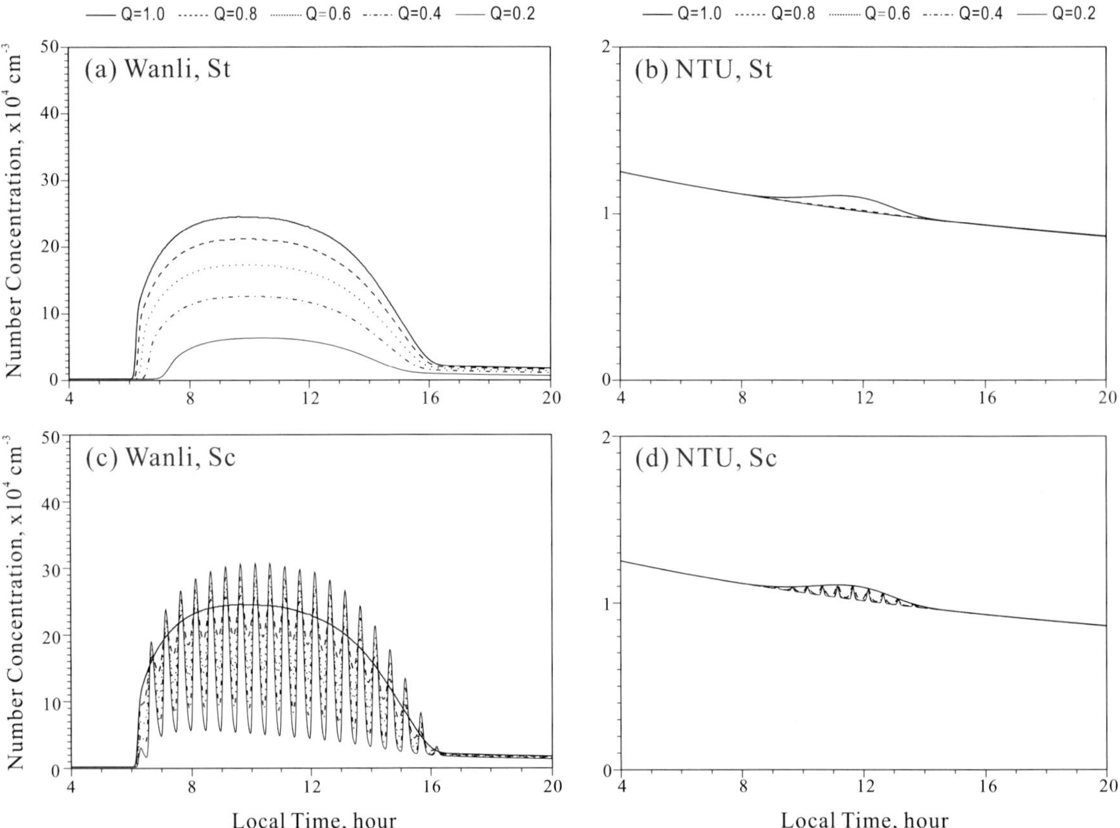

Figure 21. Simulated evolutions of aerosol number concentration under the influence of cloud cover. The top and the bottom panels are with stratus (St) and stratocumulus (Sc) cloud cover, respectively. The left panels are initialized with the Wanli aerosol type, and the right panels apply the NTU aerosols. Five values of fixed transmissivity are applied for the St clouds, and five values of minimum transmissivity for the Sc clouds. Note that the smooth curve in (c) is the same as the curve with $Q = 1$ in (a), and likewise for (b) and (d).

diurnally varying reaction rate. Two types of initial aerosols are considered to demonstrate the effect of existing aerosols: (1) the aerosol size distribution measured at the Wanli station located at the northern tip of Taiwan, and its total number concentration of $2200\,\mathrm{cm}^{-3}$ representing a relatively clean condition; (2) the aerosol size distribution measured at the National Taiwan University (NTU), and the $12500\,\mathrm{cm}^{-3}$ total number concentration representing an average urban condition in Taipei.

We consider two types of cloud influences: stratus clouds (St), which have uniform influence on the actinic flux; and stratocumulus clouds (Sc), which appear and attenuate sunlight periodically. Their influence on the actinic flux is represented by a factor of atmospheric transmissivity, Q. For St clouds, fixed values of Q are imposed to modify the diurnally varying clear sky actinic flux. Similar treatment is applied to the Sc clouds, except that the transmissivity varies periodically between unity and a minimum value. Figure 21(a) shows a drastic increase in aerosol concentration after sunrise due to photochemical production of sulfate for the rural (Wanli) condition. The number of aerosols may increase by two orders of magnitude under a clear sky condition. Lower Q values result in an obvious reduction of aerosol production, and the time of the particle burst is somewhat delayed. For the urban (NTU) aerosol condition, however, particle production is much weaker. In fact, the production is discernible only for the clear-sky ($Q = 1$) condition. This result further confirms that particle nucleation is stronger in cleaner environments. Note that the general decreasing trend in Figs. 21(b) and (d) is due to particle aggregation.

Under the stratocumulus condition, as shown in Fig. 21(c), particles are generated during cloud breaks but quickly diminish due to Brownian coagulation under cloud cover (which indicates that most of the increase is in ultrafine particles). One may notice that the maximum particle concentration can be higher

in the cloudy conditions than in the clear-sky situation. This again shows the effect of a lower existing aerosol condensation sink at the beginning of each burst. In the urban case [Fig. 21(d)], particle production is still much weaker than in the rural case. But unlike the St case [Fig. 21(b)], the fluctuation is discernible even for the lowest Q value of 0.2.

The above examples have demonstrated that aerosol nucleation is very sensitive to the changes of actinic flux and the amount of existing aerosols. As both factors are often highly variable, no wonder that aerosol particles are found to distribute rather unevenly in both space and time. But to what extent clouds contribute to such variations remains an unresolved issue.

3.4.2. *Turbulence effect*

A factor overlooked by earlier studies is the turbulent condition in the atmospheric boundary layer (ABL), in which the air parcels rise and sink repeatedly with the so-called large eddy circulation. Clouds often form at the top of the large eddies, as in the circumstances of stratiform clouds in the marine boundary layer. In the rising part of the large eddy, the expanding air cools down adiabatically, and the lowering temperature necessarily causes an increase in both S_w and S_a. This is another way of increasing the nucleation rate, besides photochemical production of sulfuric acid as just mentioned above. Figure 22 shows the simulation of induced particle nucleation in an air parcel that goes up and down with the large eddies. One can see a pulsating production of aerosols similar to that shown in Fig. 21(c) under Sc cloud covers, but with significantly stronger amplitudes. The particle burst is strong even for the urban aerosol scenario.

However, the cause of the pulsating particle burst is not as simple as the cooling effect. Looking closely at each pulse, one may find that the variations are not as smooth as that shown

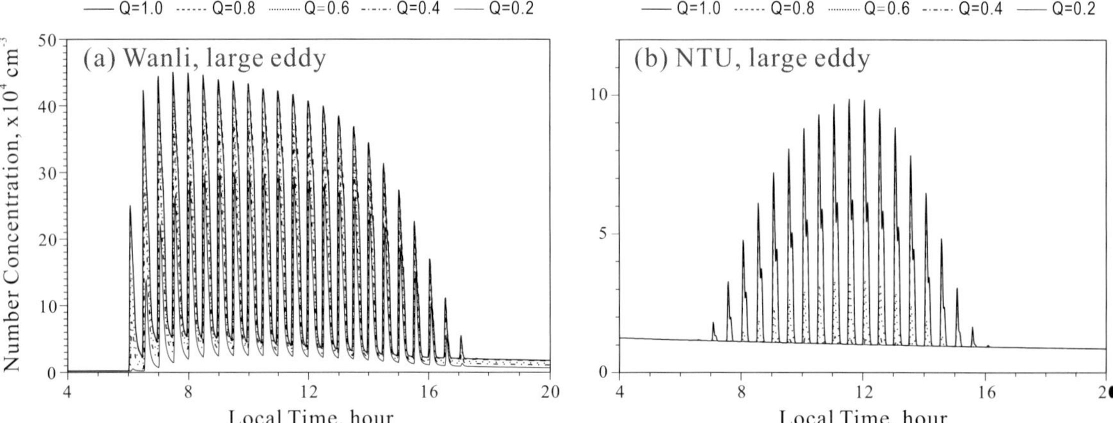

Figure 22. Simulation of new particle production in an air parcel moving in the form of a large eddy circulation within the atmospheric boundary layer. The amplitude and period of the large eddy are set as 250 m and 30 minutes, respectively. The setup of the initial aerosol type is the same as those in Fig. 21.

in Fig. 21(c), particularly for the urban situation where two peaks are visible. So, in Fig. 23, we zoom in on a couple of the cycles to show the details. Note that the left panels show the same results as for Fig. 21, and we put an additional simulation in the right panels that represents a moister condition (5% higher in S_w) in order to inspect also the effect of S_w.

The ups and downs of S_w in the top two panels of Fig. 23 reflect the imposed periodical vertical motion (with a period of 30 minutes and an amplitude of 250 m). In Fig. 23(b) one may also notice the formation of clouds as indicated by the development of supersaturation at the tops of the circulation. The expansion cooling should in principle cause an increase in S_a if the mixing ratio of sulfuric acid vapor is kept constant, but in reality the aerosol (haze) drops swell upon increasing S_w (cf. Köhler curves in Subsec. 2.1.1), and thus quickly suck up sulfuric acid vapor and reduce S_a. An even more interesting phenomenon occurs during the downward motion period for the NTU aerosol scenario, where the compression warming causes haze drops to evaporate and become more and more concentrated, such that the sulfuric acid in them is forced out into the gas phase to reach a new Henry equilibrium. The outgassing from the evaporating haze is more than enough to

offset the warming effect on S_a during the fastest descending period. Later, the air gets too warm such that S_a eventually falls off. As a consequence, every cycle of circular motion (and S_w) is accompanied by two peaks of S_a evolution. Such a phenomenon is less obvious in the Wanli aerosol scenario, because less aerosol sulfate is present.

The out-of-phase variations of S_w and S_a result in two peaks in the binary nucleation rate for every large-eddy cycle. So, as shown in Figs. 23(c)–23(f), the evolution of aerosol concentrations tends to exhibit a double-peak feature. Note that the two peaks are much more pronounced in the NTU aerosol scenario. The peak of the descending (evaporating) period is less prominent than that of the ascending period, particularly for the case of Wanli aerosols with lower relative humidity, where the second peak is almost indiscernible. Also worth noting is the stronger aerosol production with the Wanli aerosol scenario, which indicates that in a cleaner (less existing aerosols) environment more of the photochemically produced sulfuric acid can be retained in the gas phase, thus resulting in higher S_a, as shown in Figs. 23(a) and 23(b). As for the humidity effect, one can find higher aerosol production in higher relative humidity for the Wanli aerosol

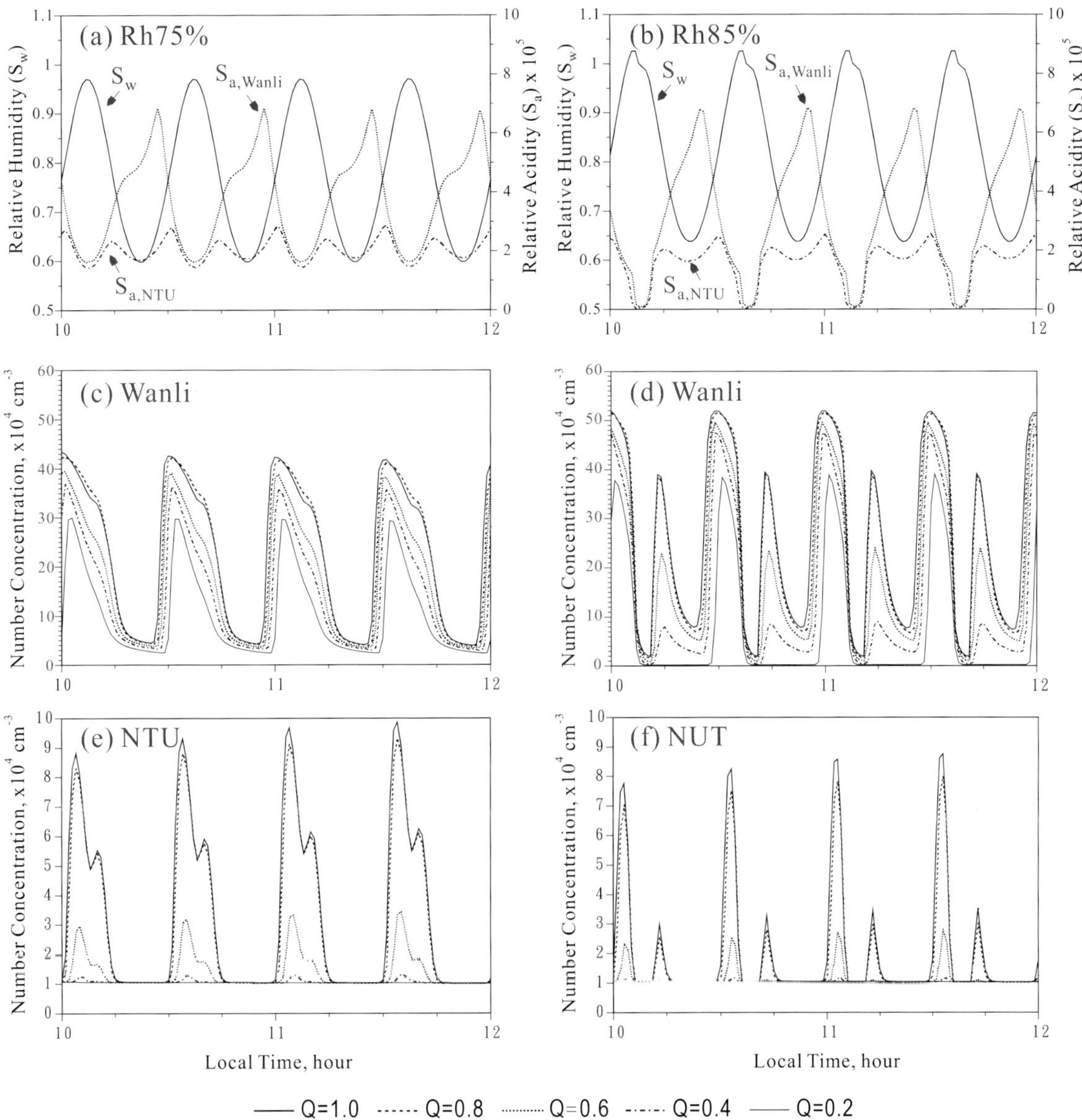

Figure 23. Enlargements of the results shown in Fig. 22. The left panels are from the same simulation in Fig. 22, which takes a midlevel relative humidity of 75%, whereas for the right panels the relative humidity is increased by 5%. Top two panels show the evolution of relative humidity (solid lines) and relative acidity in the Wanli aerosol (dashed lines) and NTU aerosol (dotted–dashed lines) scenarios. The middle and bottom panels show the evolution of aerosol number concentration in the Wanli and NTU aerosol scenarios, respectively.

scenario, but just the opposite in the NTU aerosol scenario. This result indicates that the variation of S_a is a stronger controlling factor than that of S_w.

The above example shows that the source of sulfuric acid for binary nucleation may come from not only the gas phase photochemical production but also the outgassing from existing aerosols. This leads to an interesting question: whether in-cloud aqueous phase production of sulfate (see Subsec. 3.1) may enhance the above mechanism. On the other hand, there may be other cloud processes that may enhance the gas phase photochemical production of sulfuric acid,

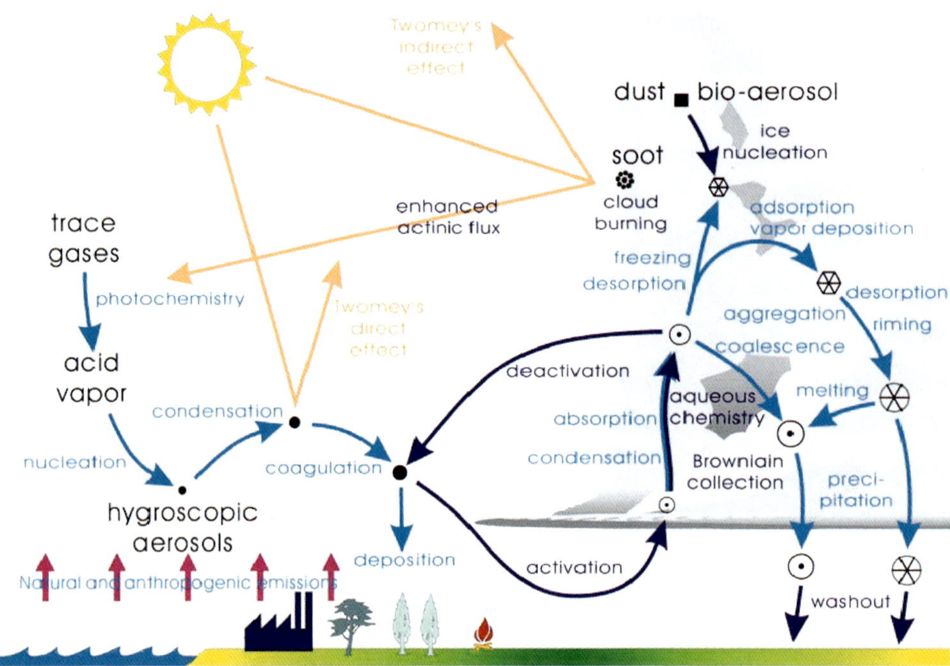

Figure 24. Schematics of the complicated aerosol and cloud processes (light blue), their interactions (dark blue), and their impacts on atmospheric radiation (brown).

such as the lightning activity which may produce NO_x and involve in the photochemical reactions.

4. Summary

In the final diagram (Fig. 24), we summarize the main processes regarding the interactions between aerosols and clouds. It reveals the convoluted nature of this complex aerosol–cloud coupling system. This system not only plays very important roles in the Earth's climate change and hydrological cycle but also interacts strongly with the biosphere and influences the biogeochemical cycles.

The discussions given in the previous sections may lead us to believe that we already know plenty about the important mechanisms involved. But, in fact, a lot of details are absent, such as how the internal mixing of different chemical compositions alters the properties of aerosols, what is the precise rate of each microphysical and chemical process, and what are the spatial and temporal variations of different kinds of CCNs and natural ice nuclei. Worse yet, there is no comprehensive understanding of the combined effects on either the regional or the global scale. The most versatile and effective tools for studying the aerosol and cloud effects on such scales are probably numerical models. But from the standpoint of our current knowledge (which is, by the way, far from complete), their treatments of aerosol–cloud interactions are very primitive. One of the main difficulties lies in the inability to link microscale processes of aerosols and clouds with the large-to-global-scale dynamic and radiative processes.

The third and fourth IPCC assessment reports (2001, 2007) listed the effect of aerosols and clouds on global climate as potentially important, with their radiative forcing nearly canceling out the CO_2 greenhouse forcing (which accounts for about two-thirds of the total greenhouse forcing). However, the levels of scientific understanding of these aerosol and cloud effects are quite low. Although the level of scientific understanding of the aerosol direct effect has improved from very low (IPCC, 2001) to median low (IPCC, 2007), and that of both the first and the second indirect aerosol effect (through interaction with clouds) has improved from very low to low, they still give the largest uncertainties in our estimation and projection of climate change. As summarized by Lohmann *et al.* (2007), future studies should emphasize "synergetic approaches involving modeling and observational evidence at different spatial and temporal scales."

Besides the climate change issue, the influence of aerosol–cloud interactions on the hydrological cycle is also poorly understood. Among the various meteorological parameters, precipitation is perhaps the least accurate prognostic variable in global climate models. Even with mesoscale meteorological models using the explicit (but still highly parametrized) cloud-microphysical scheme, quantitative precipitation forecasts are very often far from satisfactory compared to the prediction of other parameters. One of the common causes of these models' deficiency in precipitation calculation is the inability to include cloud–aerosol interactions properly. Furthermore, it is hard to believe that the microphysical processes can be simulated correctly when very little CCN observation and essentially no IN observation is available for model initialization. Thus, many tasks remain to be accomplished in understanding the detailed mechanisms of microscale physics and chemistry via diligent observational, laboratory and theoretical research. It is equally important to convert obtained knowledge into mathematical formulas and aggregate these into a robust, physically based model. With unrelenting efforts, the uncertainties in our estimation of climate change or precipitation forecasts may be reduced to an acceptable level.

[Received 17 April 2007; Revised 19 December 2007; Accepted 23 December 2007.]

References

Abbatt, J. P. D., S. Benz, D. J. Cziczo, Z. Kanji, U. Lohman, and O. Mohler, 2006: Solid ammonium sulfate aerosols as ice nuclei: a pathway for cirrus cloud formation. *Science,* **313**, 5794, 1770–1773.

Abdul-Razzak, H., and S. J. Ghan, 2004: Parameterization of the influence of organic surfactants on aerosol activation. *J. Geophys. Res.,* **109**, D03205, doi: 10.1029/2003JD004043.

Ackerman, A. S., O. B. Toon, and P. V. Hobbs, 1993: Dissipation of marine stratiform clouds and collapse of the marine boundary layer due to the depletion of cloud condensation nuclei by clouds. *Science,* **262**, 226–229.

Ackerman, A. S., O. B. Toon, and P. V. Hobbs, 1994: Reassessing the dependence of cloud condensation nucleus concentration on formation rate. *Nature,* **367**, 445–447.

Ackerman, A. S., O. B. Toon, D. E. Stevens, A. J. Heymsfield, V. Ramanathan, and E. J. Welton, 2000: Reduction of tropical cloudiness by soot. *Science,* **288**, 1042–1047.

Ackerman, A. S., M. P. Kirkpatrick, D. E. Stevens, O. B. Toon, 2004: The impact of humidity above stratiform clouds on indirect aerosol climate forcing. *Nature,* **432**, 1014–1017.

Albrecht, B. A., 1989: Aerosols, cloud microphysics, and fractional cloudiness. *Science,* **245**, 1227–1230.

Andreae, M. O., and P. Merlet, 2001: Emission of trace gases and aerosols from biomass burning. *Global Biogeochem. Cycles,* **15**, 955–966.

Andrews, E., S. M. Kreidenweis, J. E. Penner and S. M. Larson, 1997: Potential origin of organic cloud condensation nuclei observed at marine site. *J. Geophys. Res.,* **102**, 21997–22012.

Anttila, T., and V.-M. Kerminen, 2002: Influence of organic compounds on the cloud droplet activation: a model investigation considering the volatility, water solubility, and surface activity of organic matter. *J. Geophys. Res.,* **107**, AAC 12–1, doi 10.1029/2001JD001482.

Archer, R. J., and V. K. La Mer, 1955: The rate of evaporation of water through fatty acid. *J. Phys. Chem.,* **59**, 200–208.

Archuleta, C. M., P. J. DeMott, and S. M. Kreidenweis, 2005: Ice nucleation by surrogates for atmospheric mineral dust and mineral dust/sulfate particles at cirrus temperatures. *Atmos. Chem. Phys.,* **5**, 2617–2634.

Ariya, P. A., and M. Amyot, 2004: Bioaerosols: impact on physics and chemistry of the atmosphere. *Atmos. Environ.*, **38**, 1231–1233.

Baker, M. B., and R. J. Charlson, 1990: Bistability of CCN concentrations and thermodynamics in the cloud-topped boundary layer. *Nature*, **345**, 142–145.

Baker, M. B., 1993: Variability in concentrations of cloud condensation nuclei in the marine cloud-topped boundary layer. *Tellus*, **45**, 458–472.

Barger, W. R., and W. D. Garrett, 1970: Surface active organic material in the marine atmosphere. *J. Geophys. Res.*, **75**, 4561–4566.

Barnard, W. R., G. J. Stensland, and D. F. Gatz, 1986: Alkaline materials flux from unpaved roads: source strength, chemistry and potential for acid rain neutralization. *Water, Air, & Soil Pollution*, **30**, 285–293.

Barth, M. C., P. J. Rasch, J. T. Kiehl, C. M. Benkovitz, and S. E. Schwartz, 2000: Sulfur chemistry in the National Center for Atmospheric Research Community Climate Model: description, evaluation, features, and sensitivity to aqueous chemistry. *J. Geophys. Res.*, **105**, 1387–1415.

Barth, M. C., *et al.*, 2005: Coupling between land ecosystems and the atmospheric hydrologic cycle through biogenic aerosol pathways. *Bull. Amer. Meteorol. Soc.*, **86**, 1738–1742.

Bauer, H., H. Giebl, R. Hitzenberger, A. Kasper-Giebl, G. Reischl, F. Zibuschka, and H. Puxbaum, 2003: Airborne bacteria as cloud condensation nuclei. *J. Geophys. Res.*, **108**, AAC2/1–AAC2/5.

Bauer, S. E., Y. Balkanski, M. Schulz, D. A. Hauglustaine, and F. Dentener, 2004: Global modeling of heterogeneous chemistry on mineral aerosol surfaces: influence on tropospheric ozone chemistry and comparison to observations. *J. Geophys. Res.*, **109**, D02304.1–D02304.17.

Bigg, E. K., 1990: Measurements of concentrations of natural ice nuclei. *Atmos. Res.*, **25**, 397–408.

Binkowski, F. S., and U. Shankar, 1995: The regional particulate matter model. Part. I: Model description and preliminary results. *J. Geos. Res.*, **100**, 191–209.

Blanchard, D. C., 1963: The electrification of the atmosphere by particles from bubbles in the sea. *Prog. Oceanogr.*, **1**, 71–202.

Blanchard, D. C., 1964: Sea to air transport of surface active material. *Science*, **146**, 396–397.

Bond, T. C., 2001: Spectral dependence of visible light absorption by carbonaceous particles emitted from coal combustion. *Geophys. Res. Lett.*, **28**, 4075–4078.

Bond, T. C., D. G. Streets, K. F. Yarber, S. M. Nelson, J. H. Woo, and Z. Klimont, 2004: a technology-based global inventory of black and organic carbon emissions from combustion. *J. Geophys. Res.*, **109**, D14203.

Boucher, O., and U. Lohmann, 1995: The sulfate–CCN–cloud albedo effect: a sensitivity study with two general circulation models. *Tellus*, **47B**, 281–300.

Boyd, P. W., *et al.*, 2000: A mesoscale phytoplankton bloom in the polar Southern Ocean stimulated by iron fertilization. *Nature*, **407**, 695–702.

Braham, R. R., 1967: Cirrus cloud seeding as a trigger for storm development. *J. Atmos. Sci.*, **24**, 311–312.

Braham, R. R., and P. Spyers-Duran, 1967: Survival of cirrus crystals in clear air. *J. Appl. Meteor.*, **6**, 1053–1061.

Capel, P. D., R. Gunde, F. Zürcher, and W. Giger, 1990: Carbon speciation and surface tension of fog. *Environ. Sci. Tech.*, **24**, 722–727.

Chang, K.H., *et al.*, 2000: Modeling of long-range transport on Taiwan's acid deposition under different weather conditions. *Atmos. Environ.*, **34**, 3281–3295.

Charlson, R. J., and M. J. Pilat, 1969: Climate: the influence of aerosols. *J. Appl. Meteorol.*, **8**, 1001–1002.

Charlson, R. J., J. E. Lovelock, M. O. Andreae, and S. G. Warren, 1987: Oceanic phytoplankton, atmospheric sulphur, cloud albedo and climate. *Nature*, **326**, 655–661; doi: 10.1038/326655a0.

Charlson, R. J., S. E. Schwartz, J. M. Hales, R. D. Cess, J. A. Coakley, Jr., J. E. Hansen, and D. J. Hofmann, 1992: Climate forcing by anthropogenic aerosols. *Science*, **255**, 423–430.

Charlson, R. J., J. H. Seinfeld, A. Nenes, M. Kulmala, A. Laaksonen, and M. C. Facchini, 2001: Reshaping the theory of cloud formation. *Science*, **292**, 2025–2026.

Chen, J.-P., 1994: Theory of deliquescence and modified Köhler curves. *J. Atmos. Sci.*, **51**, 3505–3516.

Chen, J.-P., and D. Lamb, 1990: The role of precipitation microphysics in the selective filtration of air entering the upper troposphere. Preprints, *24th Conf. on Cloud Physics* (July 23–27, San Francisco, CA, USA), pp. 479–484.

Chen, J-.P., and D. Lamb, 1992: The effect of cloud microphysics on the composition of rain.

In S. E. Schwartz and W. G. N. Slinn (eds.), *Precipitation Scavenging and Atmosphere–Surface Exchange* (Hemisphere, Washington), pp. 51–62.

Chen, J-.P., and D. Lamb, 1994a: The theoretical basis for the parameterization of ice crystal habits: growth by vapor deposition. *J. Atmos. Sci.*, **51**, 1206–1221.

Chen, J-.P., and D. Lamb, 1994b: Simulation of cloud microphysical and chemical processes using a multicomponent framework. Part I: Description of the microphysical model. *J. Atmos. Sci.*, **51**, 2613–2630.

Chen, J-.P., and P. J. Crutzen, 1994: Solute effects on the evaporation of ice particles. *J. Geophys. Res.*, **99**, 18847–18859.

Chen, J.-P., G. M. McFarquhar, A. J. Heymsfield, and V. Ramanathan, 1997: A modeling and observational study of the detailed microphysical structure of tropical cirrus anvils. *J. Geophys. Res.*, **102**, 6637–6653.

Chen, J-.P., and D. Lamb, 1999: Simulation of cloud microphysical and chemical processes using a multicomponent framework. Part II: Microphysical evolution of a wintertime orographic cloud. *J. Atmos. Sci.*, **56**, 2293–2312.

Chen, J.-P., and S.-T. Liu, 2004: Physically-based two-moment bulk-water parameterization for warm cloud microphysics. *Q. J. Royal Meteor. Soc.*, **130**, Part A, 51–78.

Chen, J.-P., and W.-C. Hsieh, 2004: Reexamination of the influences of carbonaceous chemicals on cloud drop activation. In *Proc. 14th Int. Conf. Cloud and Precipitation* (July 18–23, Bologna, Italy).

Chen, Y., S. M. Kreidenweis, L. M. McInnes, D. C. Rogers, and P. J. DeMott, 1998: Single particle analysis of ice nucleation aerosols in the upper troposphere and lower stratosphere. *Geophys. Res. Lett.*, **25**, 1391–1394.

Cheng, C.-T., W.-C. Wang, and J.-P. Chen, 2007: A modeling study of aerosol impacts on cloud radiative properties and precipitation. *Q. J. Royal Meteor. Soc.*, **133**, Part B, 283–297.

Chin, M., D. J. Jacob, G. M. Gardner, M. S. Foreman-Fowler, P. A. Spiro, and D. L. Savoie, 1996: A global three-dimensional model of tropospheric sulfate. *J. Geophys. Res.*, **101**, 18667–18690.

Chin, M., D. L. Savoie, B. J. Huebert, A. R. Bandy, D. C. Thornton, T. S. Bates, P. K. Quinn, E. S. Saltzman, and W. J. De Bruyn, 2000: Atmospheric sulfur cycle simulated in the global model GOCART: comparison with field observations and regional budgets. *J. Geophys. Res.* **105**, 24689–24712.

Chou, C., 2003: Land–sea heating contrast in an idealized Asian summer monsoon. *Clim. Dyn.*, **21**, 11–25.

Chou, Chia, J. D. Neelin, U. Lohmann, and J. Feichter, 2005: Local and remote impacts of aerosol climate forcing on tropical precipitation. *J. Climate*, **18**, 4621–4635.

Coakley, J. A., Jr., R. L. Bernstein, and P. A. Durkee, 1987: Effect of ship-stack effluents on cloud reflectivity. *Science*, **237**, 4818, 1020–1022.

Conant, W. C., A. Nenes, and J. H. Seinfeld, 2002: Black carbon radiative heating effects on cloud microphysics and implications for the aerosol indirect effect. 1: Extended Köhler theory. *J. Geophys. Res.*, **107**, doi:10.1029/2002JD002094.

Cooper, W. A., R. T. Bruintjes, and G. K. Mather, 1997 Some calculations pertaining to hygroscopic seeding with flares. *J. Appl. Meteorol.*, **36**, 1449–1469.

Cooper, O. R., C. Forster, D. Parrish, M. Trainer, E. Dunlea, T. Ryerson, G. Hübler, F. Fehsenfeld, D. Nicks, J. Holloway, J. de Gouw, C. Warneke, J. M. Roberts, F. Flocke, and J. Moody, 2004: a case study of transpacific warm conveyor belt transport: influence of merging airstreams on trace gas import to North America. *J. Geophys. Res.*, **109**, D23S08, doi: 10.1029/2003JD003624.

Cotton, W. R., G. D. Alexander, R. Hertenstein, R. L. Walko, R. L. McAnelly, and M. Nicholls, 1995: Cloud venting: a review and some new global annual estimates. *Earth-Sci. Rev.*, **39**, 169–206.

Crawford, J., R. E. Shetter, B. Lefer, C. Cantrell, W. Junkermann, S. Madronich, and J. Calvert, 2003: Cloud impacts on UV spectral actinic flux observed during the International Photolysis Frequency Measurement and Model Intercomparison (IPMMI). *J. Geophys. Res.*, **108**(D16), 8545, doi: 10.1029/2002JD002731.

Cziczo, D. J., D. M. Murphy, P. K. Hudson, and D. S. Thomson, 2004: Single particle measurements of the composition of cirrus ice residue during CRYSTAL-FACE. *J. Geophys. Res.*, **109**, D04201, doi: 10.1029/2004JD004032.

Decesari, S., M. C. Facchini, E. Matta, M. Mircea, S. Fuzzi, A. R. Chughtai, and D. M. Smith, 2002: Water-soluble organic compounds formed by oxidation of soot. *Atmos. Environ.*, **36**, 1827–1832.

DeMott, P. J., 1990: An exploratory study of ice nucleation by soot aerosols. *J. Appl. Meteorol.*, **29**, 1072–1079.

DeMott, P. J., K. Sassen, M. Poellot, D. Baumgardner, D. C. Rogers, S. D. Brooks, A. J. Prenni, and S. M. Kreidenweis, 2003a: African dust aerosols as atmospheric ice nuclei. *Geophys. Res. Lett.*, **30**, No. 14, 1732, doi: 10.1029/2003GL 017410.

DeMott, P. J., D. J. Cziczo, A. J. Prenni, D. M. Murphy, S. M. Kreidenweis, D. S. Thomson, R. Borys, and D. C. Rogers, 2003b: Measurements of the concentration and composition of nuclei for cirrus formation. *Proc. Natl. Acad. Sci.*, **100**, No. 25, 14655–14660.

Dentener, F. J., G. R. Carmichael, Y. Zhang, J. Lelieveld, and P. J Crutzen, 1996: Role of mineral aerosol as a reactive surface in the global troposphere. *J. Geophys. Res.*, **101**, 22869–22890, doi: 10.1029/96JD01818.

Desalmand, F., R. Serpolay, and J. Podzimek, 1985: Some specific features of the aerosol particle concentrations during the dry season and during a bushfire event in West Africa. *Atmos. Environ.*, **19**, 1535–1543.

Dessens, H., 1949: The use of spider's threads in the study of condensation nuclei. *Quart. J. Roy. Meteorol. Soc.*, **75**, 23–27.

Derjaguin, B. V., Yu. S. Kurghin, S. P. Bakanov, and K. M. Merzhanov, 1985: Influence of surfactant vapor on the spectrum of cloud drops forming in the process of condensation growth. *Langmuir*, **1**, 278–281.

Deutsch, F., P. Hoffmann, and H. M. Ortner, 2001: Field experimental investigations on the Fe(II)- and Fe(III)-content in cloudwater samples. *J. Atmos. Chem.*, **40**, 87–105, doi: 10.1023/ A:1010684628804.

Dinger, J. E., H. B. Howell, and T. A. Wojciechowski, 1970: On the sources and composition of cloud condensation nuclei in the subsidence air mass over the North Atlantic. *J. Atmos. Sci.*, **27**, 791–797.

Duce, R. A., C. K. Unni, and B. J. Bay, 1980: Long-range atmospheric transport of soil dust from Asia to the Tropical North Pacific: temporal variability. *Science*, **209**, 1522–1423.

Duce, R. A. 1986: The impact of atmospheric nitrogen, phosphorus, and iron species on marine biological productivity. In: P. Buat-Menard (ed.), *The Role of Air–Sea Exchange in Geochemical Cycling* (Redel), pp. 495–529.

Dusek, U., *et al.*, 2006: Size matters more than chemistry for cloud-nucleating ability of aerosol particles. *Science*, **312**, 1375–1378.

Easter, R. C., and L. K. Peters, 1994: Binary homogeneous nucleation: temperature and relative humidity fluctuations, nonlinearity, and aspects of new particle production in the atmosphere. *J. Appl. Meteorol.*, **33**, 775–784.

Endresen, Ø., E. Sørgård, J. K. Sundet, S. B. Dalsøren, I. S. A. Isaksen, T. F. Berglen, and G. Gravir, 2003: Emission from international sea transportation and environmental impact. *J. Geophys. Res.*, **108**, 4560, doi: 10.1029/ 2002JD002898.

Eppley, R. W., 1980: Estimating phytoplankton growth rates in the central oligotrophic oceans. In *Primary Productivity in the Sea*, ed. P. G. Falkowski (Plenum, New York), pp. 231–242.

Eyring, V., N. R. P. Harris, M. Rex, T. G. Shepherd, D. W. Fahey, G. T. Amanatidis, J. Austin, M. P. Chipperfield, M. Dameris, P. M. De F. Forster, A. Gettelman, H. F. Graft, T. Nagashima, P. A. Newman, S. Pawson, M. J. Prather, J. A. Pyle, R. J. Salawitch, B. D. Santer, and D. W. Waugh, 2005: A strategy for process-oriented validation of coupled chemistry–climate models. *Bull. Ameri Meteor Soc.*, 1117–1133.

Facchini, M. C., M. Mircea, S. Fuzzi, and R. J. Charlson, 1999a: Cloud albedo enhancement by surface-active organic solutes in growing droplets. *Nature*, **401**, 257–259.

Facchini, M. C., S. Fuzzi, S. Zappoli, A. Andracchio, A. Gelencsér, G. Kiss, Z. Krivácsy, E. Mészáros, H.-C. Hansson, T. Alsberg, and Y. Zebühr, 1999b: Partitioning of the organic aerosol component between fog droplets and interstitial air. *J. Geophys. Res.*, **104**, 26821–26832.

Falkovich, A. H., E. R. Graber, G. Schkolnik, Y. Rudich, W. Maenhaut, and P. Artaxo, 2004: Low molecular weight organic acids in aerosol particles from Rondônia, Brazil, during the biomass-burning, transition and wet periods. *Atmos. Chem. Phys. Discuss.*, **4**, 6867–6907.

Feichter, J., E. Roeckner, U. Lohmann, and B. Liepert, 2004: Nonlinear aspects of the climate response to greenhouse gas and aerosol forcing. *J. Climate*, **17**, 2384–2398.

Feingold, G., W. R. Cotton, S. M. Kreidenweis, and J. T. Davis, 1999: The impact of giant cloud condensation nuclei on drizzle formation in stratocumulus: implication for cloud radiative properties. *J. Atmos. Sci.*, **56**, 4100–4117.

Feingold, G., and P. Y. Chuang, 2002: Analysis of the influence of film-forming compounds on droplet growth: implications for cloud microphysical processes and climate. *J. Atmos. Sci.*, **59**, 2006–2018.

Fiedler, F., 1982: Atmospheric circulation. In *Chemistry of the Unpolluted and Polluted Troposphere*, eds. H. W. Georgii and W. Jaeschke (Reidel Dordrecht), p. 509.

Field, P. R., O. Möhler, P. Connolly, M. Krämer, R. Cotton, A. J. Heymsfield, H. Saathoff, and M. Schnaiter, 2006: Some ice nucleation characteristics of Asian and Saharan desert dust. *Atmos. Chem. Phys.*, **6**, 2991–3006.

Frick, G. M., and W. A. Hoppel, 1993: Airship measurements of aerosol size distribution, cloud droplet spectra, and trace gas concentrations in the marine boundary layer. *Bull. Amer. Meteor. Soc.*, **74**, 2195–2202.

Fu, T. M., 2002: Microphysical stability analyses in the stratocumulus-topped marine boundary layer. (master's thesis, National Taiwan University), 123 pp.

Fukuta, N., 1963: Ice nucleation by metaldehyde. *Nature*, **199**, 475–476.

Fuzzi, S., P. Mandrioli, and A. Perfetto, 1997: Short communication: fog droplets — an atmospheric source of secondary biological aerosol particles. *Atmos. Environ.*, **31**, 287–290.

Ghan, S. J., G. Guzman, and H. Abdul-Razzak, 1998: Competition between sea salt and sulfate particles as cloud condensation nuclei. *J. Atmos. Sci.*, **55**, 3340–3347.

Garrett, W. D., 1967: The organic chemical composition of the ocean surface. *Deep-Sea Res.*, **14**, 221–227.

Garrett, W. D., 1978: The impact of organic material on cloud and fog processes. *Pageoph.*, **116**, 316–334.

Gelbard, F., Y. Tambour, and J. H. Seinfeld, 1980: Sectional representations for simulating aerosol dynamics. *J. Colloid Interface. Sci.*, **76**, 541–556.

Georgii, H. W., and E. Kleinjung, 1967: Relations between the chemical composition of atmospheric aerosol particles and the concentration of natural ice nuclei. *J. Rech. Atmos.*, **3**, 145–156.

Ghan, S. J., R. C., Easter, J., Hudson, and F.-M. Breon, 2001: Evaluation of aerosol indirect radiative forcing in MIRAGE. *J. Geophys. Res.*, **106**, 5317–5334.

Goetz, A., 1965: The constitution of aerocolloidal particulates above the ocean surface. In *Proc. Int. Conf. Cloud Physics* (Jap. Geophys. Union, Tokyo–Sapporo), p. 42.

Gorbunov, B., A. Baklanov, N. Kakutkina, H. L. Windsor, and R. Toumi, 2001: Ice nucleation on soot particles. *J. Aerosol. Sci.*, **32**, 199–215.

Graedel, T. E., P. S. Gill, and C. J. Weschler, 1983: Effects of organic surface films on the scavenging of atmospheric gases by raindrops and aerosol particles. In *Precipitation Scavenging, Dry Deposition, and Resuspension*, eds. Pruppacher *et al.*, pp. 417–430.

Grassian, V. H., 2002: Chemical reactions of nitrogen oxides on the surface of oxide, carbonate, soot, and mineral dust particles: implications for the chemical balance of the troposphere. *J. Phys. Chem. A*, **106**, 860–877.

Gu, Y., K. N. Liou, Y. Xue, C. R. Mechoso, W. Li, and Y. Luo, 2006: Climatic effects of different aerosol types in China simulated by the UCLA GCM. *J. Geophy. Res.*, **111**, D15201, doi: 10.1029/2005JD006312.

Guieu, C., Y. Bozec, S. Blain, C. Ridame, G. Sarthou, and N. Leblond, 2002: Impact of high Saharan dust inputs on dissolved iron concentrations in the Mediterranean Sea. *Geophys. Res. Lett.*, **29**, 1911, doi: 10.1029/2001GL014454.

Guo, H., J. E. Penner, M. Herzog, and H. Pawlowska, 2007: Examination of the aerosol indirect effect under contrasting environments during the ACE-2 experiment. *Atmos. Chem. Phys.*, **7**, 535–548.

Hallett, J., and B. J. Mason, 1958: The influence of temperature and supersaturation on the habit of ice crystals grown from the vapor. *Proc. R Soc. London*, **A247**, 440–453.

Han, Q., W. B. Rossow, and A. A. Lacis, 1994: Near-global survey of effective droplet radii in liquid water clouds using ISCCP data. *J. Climate*, **7**, 465–497.

Hansen, J. E., and M. Sato, 2001: Trends of measured climate forcing agents. *Proc. Natl. Acad. Sci.*, **98**, 14778–14783, doi: 10.1073/pnas.261553698.

Hansen, J., M. Sato, and R. Ruedy, 1997: Radiative forcing and climate response. *J. Geophys. Res.*, **102**, 6831–6864.

Haywood, J., and O. Boucher, 2000: Estimates of the direct and indirect radiative forcing due to tropospheric aerosol: a review. *Rev. Grophys.*, **38**, 4, 513–543.

Hazra, A., S. Paul, U. K. De, S. Bhar, and K. Goswami, 2003: Investigation on ice nucleation/hydrate crystallization by aqueous solution of ammonium sulfate. *Progress in*

Crystal Growth and Characterization of Materials, **47**, 1, 45–61.

Hazra, A., M. Saha, U. K. De, J. Mukherjee and K. Goswami, 2004: Study of ice nucleating characteristics of *Pseudomonas aeruginosa. J. Aeros. Sci.*, **35**, 1405–1414.

Hazra, A., U. K. De and K. Goswami, 2006: Nucleating characteristics of *Pseudomonas aeruginosa* above 0°C and the role of L-leucine. *J. Cryst. Growth*, **286**, 114–120.

Heintzenberg, J., K. Okada, and J. Ström, 1996: On the composition of non-volatile material in upper tropospheric aerosols and cirrus crystals. *Atmos. Res.*, **41**, 81–88.

Hegg, D. A., L. F. Radke, and P. V. Hobbs, 1990: Particle production associated with marine clouds. *J. Geophys. Res.*, **95**, 917–926.

Hegg, D. A., L. F. Radke, and P. V. Hobbs, 1991: Measurements of Aitken nuclei and cloud condensation nuclei in the marine atmosphere and their relation to the DMS–cloud–climate hypothesis. *J. Geophys. Res.*, **96**, 18727–18733.

Hegg, D. A., P.-F. Yuen, and T. V. Larson, 1992: Modeling the effects of heterogeneous cloud chemistry on the marine particle size distribution. *J. Geophys. Res.*, **97**, 12927–12933.

Herlihy, L. J., J. N. Galloway, and A. L. Mills, 1987: Bacterial utilization of formic and acetic acid in the rainwater. *Atmos. Environ.*, **21**, 2397–2402.

Heymsfield, A. J., and L. M. Miloshevich, 1993: Homogeneous ice nucleation and supercooled liquid water in orographic wave clouds. *J. Atmos. Sci.*, **50**, 2335–2353.

Heymsfield, A. J., and R. M. Sabbin, 1989: Cirrus crystal nucleation by homogeneous freezing of solution droplets. *J. Atmos. Sci.*, **46**, 2252–2264.

Hoffman, P. J., 1973: Supersaturation spectra of AgI and natural ice nuclei. *J. Appl. Meteor.*, **12**, 1080–1082.

Hoffman, T., J. R. Odum, F. Bowman, D. Collins, D. Klockow, R. C. Flagan, and J. H. Seinfeld, 1997: Formation of organic aerosols from the oxidation of biogenic hydrocarbons. *J. Atmos. Chem.*, **26**, 189–222.

Hsieh, W.-C., 2004: The impacts of carbonaceous chemicals on cloud formation: cloud burning effect and surface film effect. (master's thesis, National Taiwan University), 105 pp.

Hung, H.-M., A. Malinowski, and S. T. Martin, 2003: Kinetics of heterogeneous ice nucleation on the surfaces of mineral dust cores inserted into aqueous ammonium sulfate particles. *J. Phys. Chem. A*, **107**, 1296–1306.

Husar, R. B., and W. R. Shu, 1975: Thermal analysis of the Los Angeles smog aerosol. *J. Appl. Meteorol.*, **14**, 1558–1565.

Husar, R. B., W. H. White, and D. L. Bluementhal, 1976: Direct evidence of heterogeneous aerosol formation in Los Angeles smog. *Envrion. Sci. Tech.*, **10**, 490–191.

Inoue, K., H. Satake, T. Shima, N. Yokota, 1991: Partial neutralization of acid rain by Asian eolian dust transported over a long distance. *Soil Science and Plant Nutrition*, **37**, 83–91.

IPCC, 2001: Climate change 2001: the scientific basis. Contribution of Working Group I to the Third Assessment Report of the Intergovernmental Panel on Climate Change. eds. J. T. Houghton, Y. Ding, D. J. Griggs, M. Noguer, P. J. van der Linden, X. Dai, K. Maskell, and C. A. Johnson (Cambridge University Press, Cambridge, New York), 881 pp.

IPCC, 2007: Climate Change 2007: The physical science basis, summary for policymakers. Contribution of Working Group I to the Fourth Assessment Report for the Intergovernmental Panel on Climate Change. eds. J. Arblaster *et al.*

Isaksen, I. S. A., and H. Rodhe, 1978: A two-dimensional model for the global distribution of gases and aerosol particles in the troposphere. Report AC-47, Department of Meteorology, University of Stockholm.

Isono, K., M. Kombayasi, and A. Ono, 1959: The nature and the origin of ice nuclei in the atmosphere. *J. Met. Soc. Japan*, **37**, 211.

Jacob, D. J., *et al.*, 1997: Evaluation and intercomparison of global atmospheric transport models using Rn-222 and other short-lived tracers. *J. Geophys. Res.*, **102**, 5953–5970.

Jacobson, M. Z., 2001: Global direct radiative forcing due to multicomponent anthropogenic and natural aerosols. *J. Geophys. Res.*, **106**, 1551–1568, doi: 10.1029/2000JD900514.

Jirak, I. L., and W. R. Cotton, 2006: Effect of air pollution on precipitation along the front range of the Rocky Mountains. *J. Appl. Meteor.*, **45**, 236–245.

Johnson, D. B., 1982: The role of giant and ultra-giant aerosol particles in warm rain initiation. *J. Atmos. Sci.*, **39**, 448–460.

Jones, A. and Slingo, A., 1996: Predicting cloud-droplet effective radius and indirect sulphate aerosol forcing using a general circulation model. *Quart. J. Roy. Meteorol. Soc.*, **122**, 1573–1595.

Kavouras, I. G., N. Mihalopoulos, and E. G. Stephanou, 1998: Formation of atmospheric particles from organic acids produced by forests. *Nature*, **395**, 683–686.

Khairoutdinov, M., and Y. Kogen, 2000: A new cloud physics parameterization in a large-eddy simulation model of marine stratocumulus. *Mon. Wea. Rev.*, **128**, 229–243.

Knollengerg, R. G., K. Kelly, and J. C. Wilson, 1993: Measurements of high number densities of ice crystal populations in the tops of tropical cumulonumbus. *J. Geophys. Res.*, **98**, 8639–8664.

Koch, D., D. J. Jacob, I. Tegen, D. Rind, and M. Chin, 1999: Tropospheric sulfur simulation and sulfate direct radiative forcing in the Goddard Institute for Space Studies general circulation model. *J. Geophys. Res.*, **104**, 23799–23822.

Krueger, B. J., V. H. Grassian, J. P. Cowin, and A. Laskin, 2004: Heterogeneous chemistry of individual mineral dust particles from different dust source regions: the importance of particle mineralogy. *Atmos. Environ.*, **38**, 6253–6261.

Kulmala, M., A. Laaksonen, P. Korhonen, T. Vesala, T. Ahonen, and J. C. Barrett, 1993: The effect of atmospheric nitric acid vapor on cloud condensation nucleus activation. *J. Geophys. Res.*, **98**, 22949–22958.

Kulmala, M., P. Korhonen, T. Vesala, H.-C. Hansson, K. Noone, and B. Svenningsson, 1996: The effect of hygroscopicity on cloud droplet formation. *Tellus. Ser. B*, **48**, 347–360.

Kumar, P. P., K. Broekhuizen, and J. P. D. Abbatt, 2003: Organic acids as cloud condensation nuclei: laboratory studies of highly soluble and insoluble species. *Atmos. Chem. Phys.*, **3**, 509–520.

Kylling, A., *et al.*, 2005: Spectral actinic flux in the lower troposphere: measurement and 1-D simulations for cloudless, broken cloud and overcast situations. *Atmos. Chem. Phys. Discuss.*, **5**, 1421–1467.

Lamb, D., and J.-P. Chen, 1990: A modeling study of the effects of ice-phase microphysical processes on trace chemical removal efficiencies. *Atmos. Res.*, **25**, 31–51.

La Mer, V. K. (ed.), 1962: *Retardation of Evaporation by Monolayers: Transport Processes* (Academic, New York, London), 277 pp.

La Mer, V. K., T. W. Healy, and L. A. G. Aylmore, 1964: The transport of water through monolayers of long-chain n-paraffinic alcohols. *J. Colloid Sci.*, **19**, 673–684.

Langer, G., J. Rosinski, and S. Bernsen, 1963: Organic crystals as icing nuclei. *J. Atmos. Sci.*, **20**, 557–562.

Langmuir, I., and D. B. Langmuir, 1927: The effect of monomolecular films on the evaporation of ether solutions. In *The Collected Works of Irvin Langmuir*, Vol. 9, pp. 472–485.

Lasher-Strapp, S. G., C. A. Knight, and J. M. Straka, 2001 Early radar echoes from ultragiant aerosol in a cumulus congestus: modeling and observations. *J. Atmos. Sci.*, **58**, 3545–3562.

Leaitch, W. R., G. A. Isaac, J. W. Strapp, and C. M. Banic, and H. A., Wiebe, 1992: The relationship between cloud droplet number concentrations and anthropogenic pollution: observations and climatic implications. *J. Geophy. Res.*, **97**, D2, 2463–2474.

Leleiveld, J., 1993: Multi-phase processes in the atmospheric sulfur cycle. In *Interactions of C, N, P and S Biogeochemical Cycles and Global Change*, eds. Wollast *et al.*, NATO ASI Series, Vol. 14 (Springer-Verlag, Berlin, Heidelberg).

Leleiveld, J., 1994: Modeling of heterogeneous chemistry in the global troposphere. In eds. G. Bidoglio, and W. Stumm *Chemistry of Aquatic Systems: Local and Global Perspectives*, pp. 73–95.

Levin, Z., S. A. Yankofsky, D. Pardes, and N. Magal, 1987: Possible application of bacterial condensation freezing to artificial rainfall enhancement. *J. Clim. Appl. Meteorl.*, **26**, 1188–1197.

Lighthart, B., and B. T. Sharffer, 1994: Bacterial flux from chaparral into the atmosphere in the mid-summer at a high desert location. *Atmos. Environ.*, **28**, 1267–1274.

Likens, G. E., E. S. Edgerton, and J. N. Galloway, 1983: The composition and deposition of organic carbon in precipitation. *Tellus. Ser. B*, **35**, 16–24.

Lindemann, J., H. A. Constantinidou, W. R. Barchet, and C. D. Upper, 1982: Plants as sources of airborne bacteria including ice nucleation active bacteria. *Appl. Environ. Microbiol.*, **44**, 1059–1063.

Lindow, S. E., 1983: The role of bacterial ice nucleation in frost injury to plants. *Ann. Rev. Phytopathol.*, **21**, 363–384.

Liou, K.-N., 1986: Influence of cirrus clouds on weather and climate processes: a global perspective. *Mon. Wea. Rev.*, **114**, 1167–1199.

Liu, Y., and P. H. Daum, 2002: Indirect warming effect from dispersion forcing. *Nature*, **419**, 580–581.

Lohmann, U., and J. Feichter, 2005: Global indirect aerosol effects: A review. *Atmos. Chem. Phys.*, **5**, 715–737.

Lohmann, U., J. Quaas, S. Kinne, and J. Feichter, 2007: Different approaches for constraining global climate models of the anthropogenic indirect aerosol effect. *Bull. Amer. Meteorol. Soc.*, **88**, 243–249.

Los, A., M. van Weele, and P. G. Duynkerke, 1997: Actinic fluxes in broken cloud fields. *J. Geophys. Res.*, **102**, 4257–4266.

Lunde, G., J. Gether, N. Gjos, and M. B. S. Lande, 1977: Organic micropollutants in precipitation in Norway. *Atmos. Environ.*, **11**, 1007–1014.

Lynn, B., A. Khain, J. Dudhia, D. Rosenfeld, A. Pokrovsky, and A. Seifert 2005: Spectral (bin) microphysics coupled with a mesoscale model (MM5). Part 1: Model description and first results. *Mon. Wea. Rev.*, **133**, 44–58.

Mackie, P. W., K. A. Hunter, and G. H. McTainsh, 2005: Simulating the cloud processing of iron in Australian dust: pH and dust concentration. *Geophys. Res. Lett.*, **32**, L06809, doi: 10.1029/2004GL022122.

Maki, L. R., and K. J. Willoughby, 1978: Bacteria as biogenic sources of freezing nuclei. *J. Appl. Meteorol.*, **17**, 1049–1053.

Mandrioli, P., G. K. Puppi, N. Bagni, and F. Prodi, 1973: Distribution of microorganisms in hailstones. *Nature*, **246**, 416–417.

Martin, J., 1990: Glacial–interglacial CO_2 change: the iron hypothesis. *Paleoceanography*, **5**, 1–13.

Martin, G. M., D. W. Johnson, and A. Spice, 1994: The measurement and parameterization of effective radius of droplets in warm stratocumulus clouds. *J. Atmos. Sci.*, **51**, 1823–1842.

Mather, G. K., D. E. Terblanche, F. E. Steffens, and L. Fletcher, 1997: Results of the South African cloud seeding experiments using hygroscopic flares. *J. Appl. Meteorol.*, **36**, 1433–1447.

Matthias-Maser, S., and R. Jaenicke, 1994: Examination of atmospheric bioaerosol particles with radii 0.2 μm. *J. Aero. Sci.*, **25**, 1605–1613.

Matthias-Maser, S., and R. Jaenicke, 1995: The size distribution of primary biological aerosol particles with radii 0.2 μm in an urban/rural influenced region. *Atmos. Res.*, **39**, 279–286.

Matthais-Maser, S., V. Obolkin, T. Khodzer, and R. Jaenicke, 2000: Seasonal variation of primary biological aerosol particles in the remote continental region of Lake Baikal/Siberia. *Atmos. Environ.*, **34**, 3805–3811.

Meskhidze, N., and A. Nenes, 2006: Phytoplankton and cloudiness in the Southern Ocean. *Science*, **314**, 1419–1423, doi: 10.1126/science.1131779.

Mitra, S. K., J. Brinkmann, and H. R. Pruppacher, 1992: A wind tunnel study on the drop-to-particle conversion. *J. Aerosol. Sci.*, **23**, 245–256.

Möhler, O., *et al.*, 2005: Effect of sulfuric acid coating on heterogeneous ice nucleation by soot aerosol particles. *J. Geophys. Res.*, **110**, D11210, doi: 10.1029/2004JD005169.

Möhler, O., P. R. Field, P. Connolly, S. Benz, H. Saathoff, M. Schnaiter, R. Wagner, R. Cotton, M. Krämer, A. Mangold, and A. J. Heymsfield, 2006: Efficiency of the deposition mode ice nucleation on mineral dust particles. *Atmos. Chem. Phys. Discuss.*, **6**, 1539–1577.

Motoi, K., 1951: Electron-microscope study of snow crystal nuclei. *J. Meorol.*, **8**, 151–156.

Mukai, H., and Y. Ambe, 1986: Characterization of a humic acidlike brown substance in airborne particulate matter and tentative identification of its origin. *Atmos. Environ.*, **20**, 813–819.

Meyers, P. A., and R. A. Hites, 1982: Extractable organic compounds in midwest rain and snow. *Atmos. Environ.*, **16**, 2169–2175.

Nakaya, U., 1954: *Snow Crystals: Natural and Artificial.* (Harvard University Press), 510 pp.

Nenes, A., W. C. Conant, and J. H. Seinfeld, 2002: Black carbon radiative heating effects on cloud microphysics and implications for the aerosol indirect effect. 2: Cloud microphysics. *J. Geophys. Res.*, 107, doi: 10.1029/2002JD 002101.

Novakov, T., and J. E. Penner, 1993: Large contributions of organic aerosols to cloud-condensation nuclei concentrations. *Nature*, **365**, 823–825.

Novakov, T., M. O. Anfreae, R. Gabriel, T. W. Kirchstetter, O. L. Mayol-Bracero, and V. Ramanathan, 2000: Origion of carbonaceous aerosols over the tropical Indian Ocean: biomass burning or fossil fuels? *Geophys. Res. Lett.*, **27**(24), 4061–4064.

Nunes, T. V., and Pio, C. A., 1993: Carbonaceous aerosols in industrial and costal atmospheres. *Atmos. Environ.*, **27**, 1339–1346.

O'Dowd, C. D., M. H. Smith, I. E. Consterdine, and J. A. Lowe, 1997: Marine aerosol, sea-salt, and the marine sulphur cycle: a short review. *Atmos. Environ.*, **31**, 73–80.

Ogren, J. A., and R. J. Charlson, 1992: Implications for models and measurements of chemical inhomogeneities among cloud drops. *Tellus*, **44B**, 208–225.

Paluch, I. R., and D. H. Lenschow, 1991: Stratiform cloud formation in the marine boundary layer. *J. Atmos. Sci.*, **48**, 2141–2158.

Pandis, S. N., and J. H. Seinfeld, 1991: Should bulk cloudwater or fogwater samples obey Henry's law? *J. Geophys. Res.*, **96**, 10791–10798.

Perry, K. D., and P. V. Hobbs, 1994: Further evidence for particle nucleation in clear air adjacent to marine cumulus clouds. *J. Geophys. Res.*, **99**, 22803–22818.

Phadnis, M. J., and G. R. Carmichael, 2000: Numerical investigation of the influence of mineral dust on the tropospheric chemistry of East Asia. *Earth and Environmental Science*, **36**, doi: 10.1023/A:1006391626069.

Pruppacher, H. R., and J. D. Klett, 1997: *Microphysics of Clouds and Precipitation* (Kluwer).

Purdue, E. M., and K. C. Beck, 1988: Chemical consequences of mixing atmospheric droplets of varied pH. *J. Geophys. Res.*, **93**, 691–698.

Raymond, T. M., and S. N. Pandis, 2002: Cloud activation of single-component organic aerosol particles. *J. Geophys. Res.*, **107**, 4787, doi: 10.1029/2002JD002159.

Radke, L. F., and D. A. Hegg, 1972: The shattering of saline droplets upon crystallization. *J. Rech. Atmos.*, **6**, 447–455.

Rogge, W. F., M. A. Mazurch, L. M. Hildmann, G. R. Cass, and B. R. T. Simoneit, 1993: Quantification of urban organic aerosols at a molecular level: identification, abundance and seasonal variation. *Atmos. Environ.*, Part A, **27**, 1309–1330.

Romakkaniemi, S., H. Kokkola, K. E.J. Lehtinen, and A. Laaksonen, 2004: The influence of nitric acid on the cloud processing of aerosol particles. *Atmos. Chem. Phys.*, **6**, 1627–1634.

Rosenfeld, D., 1999. TRMM observed first direct evidence of smoke from forest fires inhibiting rainfall. *Geophys. Res. Lett.*, **26**, 3105–3108.

Rosenfeld, D., 2000: Suppression of rain and snow by urban and industrial air pollution. *Science*, **287**, 5459, 1793–1796.

Rosenfeld, D. Y. Rudich, and R. Lahav, 2001: Desert dust suppressing precipitation: a possible desertification feedback loop. *PNAS*, **98**, 5975–5980.

Rosenfeld, D., and W. L. Woodley, 2000: Deep convective clouds with substained supercooled liquid water down to −37.5°C. *Nature*, **405**, 440–442.

Rosenfeld, D, 2006: Aerosols, clouds and climate. *Science*, **312**, 1323–1325.

Rubel, G. O. and J. W. Gentry, 1984: Measurement of the kinetics of solution droplets in the presence of absorbed monolayers: determination of water accommodation coefficients. *J. Phys. Chem.*, **88**, 3142–3148.

Ryan, R. T., 1976: The behavior of large, low-surface-tension water drops falling at terminal velocity in air. *J. Appl. Meteorol.*, **15**, 157–165.

Salam, A., U. Lohmann, B. Crenna, G. Lesins, P. Klages, D. Rogers, R. Irani, A. MacGillivray, and M. Coffin, 2006: Ice nucleation studies of mineral dust particles with a new continuous flow diffusion chamber. *Aerosol Sci. Tech.*, **40**, 134–143.

Sassen, K., 2002: Indirect climate forcing over the western US from Asian dust storms. *Geophys. Res. Lett.*, **29**(10), doi: 10.1029/2001GL014051.

Sassen, K., P. J. DeMott, J. Prospero, and M. R. Poellot, 2003: Saharan dust storms and indirect aerosol effects on clouds: CRYSTAL-FACE results. *Geophys. Res. Lett.*, **30**, No. 12, 1633, doi: 10.1029/2003GL017371.

Sattler, B., H. Puxbaum, and R. Psenner, 2001: Bacterial growth in supercooled cloud droplets. *Geophys. Res. Lett.*, **28**, 239–242.

Saxena, V. K., and J. D. Grovenstein, 1994: The role of clouds in the enhancement of cloud condensation nuclei concentrations. *Atmos. Res.*, **31**, 71–89.

Saxena, V. K., 1996: Bursts of cloud condensation nuclei (CCN) by dissipating clouds at Palmer Station, Antarctica. *Geophys. Res. Lett.*, **23**, 69–72.

Schaefer, V. J., 1946: The production of ice crystals in a cloud of supercooled water droplets. *Science*, **104**, 457–459.

Schaefer, V. J., 1951: Snow and its relationship to experimental meteorology. In *Compendium of Meteorology*, ed. T. F. Malone (Am. Meteorol. Soc., Boston), pp. 221–234.

Schneider, S. H., 1971: A comment on "Climate: the influence of aerosols." *J. Appl. Meteorol.*, **10**, 840–841.

Schnell, R. C., and G. Vali, 1973: World-wide source of leaf derived freezing nuclei. *Nature*, **246**, 212–213.

Seaver, M., J. R. Peele, T. J. Manuccia, G. O. Rubel, and G. Ritchie, 1992: Evaporation kinetics of ventilated waterdrops coated with octadecanol monolayers. *J. Phys. Chem.*, **96**, 6389–6394.

Seidl, W., 2000: Model for a surface film of fatty acids on rain water and aerosol particles. *Atmos. Environ.*, **34**, 4917–4932.

Seinfeld, J. H., and S. N. Pandis, 2006: *Atmospheric Chemistry and Physics — From Air Pollution to Climate Change*, 2nd edn. (Wiley-Interscience), 1203 pp.

Shaw, G. E., 1983: Bio-controlled thermostasis involving the sulfur cycle. *Climate Change*, **5**, 297–303.

Shaw, G. E., 1989: Production of condensation nuclei in clean air by nucleation of H_2SO_4. *Atmos. Environ.*, **23**, 2841–2846.

Sherwood, S. C., and A. E. Dessler, 2003: Convective mixing near the tropical tropopause: insights from seasonal variations. *J. Atmos. Sci.*, **60**, 1847–1856.

Sokolik, I. N., and O. B. Toon, 1996: Direct radiative forcing by anthropogenic airborne mineral aerosols. *Nature*, **381**, 681–683.

Stevens, B., W. R. Cotton, G. Feingold, and C.-H. Moeng, 1998: Large-eddy simulations of strongly precipitating, shallow, stratocumulus-topped boundary layers. *J. Atmos. Sci.*, **55**, 3616–3638.

Stevens, B., G. Vali, K. Comstock, R. Wood, M. C. van Zanten, P. H. Austin, C. S. Bretherton, and D. H. Lenschow, 2004: Pockets of open cells (POCs) and drizzle in marine stratocumulus. *Bull. Amer. Meteor. Soc.*, **86**, 51–57.

Stumm, W., and J. J. Morgan, 1981: *Aquatic Chemistry*. (John Wiley & Sons, New York).

Tao, W.-K., X. Li, A. Khain, T. Matsui, S. Lang, and J. Simpson, 2007: The role of atmospheric aerosol concentration on deep convective precipitation: cloud-resolving model simulations (Submitted to *J. Geophys. Res.*)

Teller, A., and Z. Levin, 2006: The effects of aerosols on precipitation and dimensions of subtropical clouds: a sensitivity study using a numerical cloud model. *Atmos. Chem. Phys.* **6**, 67–80.

Toon, O. B., and T. P. Ackerman, 1981: Algorithms for the calculation of scattering by stratified spheres.*Appl. Opt.*, **20**, 3657–3660.

Tsai, I.-C., 2001: The transport of anthropogenic aerosols and their impact on cloud microphysics over East Asia (master's thesis, National Taiwan University), 97 pp. (in Chinese).

Turco, R. P., O. B. Toon, and P. Hamill, 1989: Heterogeneous physicochemistry of the polar ozone hole. *J. Geophys. Res.*, **94**, 16493–16510.

Twomey, S., 1959: The nuclei of natural cloud formation. *Geofis. Pura Appl.*, **43**, 227–249.

Twomey, S., 1974: Pollution and the planetary albedo. *Atmos. Environ.*, **8**, 1251–1256.

Twomey, S., and K. McMaster, 1955: The production of condensation nuclei by crystallizing salt particles. *Tellus*, **7**, 458–461.

Twomey, S. A., M. Piepgrass, and T. L. Wolfe, 1984: An assessment of the impact of pollution on global cloud albedo. *Tellus*, **36B**, 356–366.

Vali, G., M. Christensen, R. W. Fresh, E. L. Galyan, L. R. Maki, and R. C. Schnell, 1976: Biogenic ice nuclei. Part II: Bacterial sources. *J. Atmos. Sci.*, **33**, 1565–1570.

Van den Heever, S. C., G. Carrio, W. R. Cotton, P. J. DeMott, and A. J. Prenni, 2006: Impacts of nucleating aerosol on Florida convection. Part I: Mesoscale simulations. *J. Atmos. Sci.*, **63**, 1752–1775.

Wainwright, M., N. C. Wickramasinghe, J. V. Narlikar, and P. Rajaratnam, 2003: Microorganisms cultured from stratospheric air samples obtained at 41 km. *FEMS Microbiol. Lett.*, **218**, 161–165.

Wang, P. K., 2003: Moisture plumes above thunderstorm anvils and their contributions to cross-tropopasue transport of water vapor in midlatitudes. *J. Geophys. Res.*, **108**, doi: 10.1029/2002JD002581.

Warner, J., and S. Twomey, 1967: The production of cloud nuclei by cane fires and the effect on cloud droplet concentration. *J. Atmos. Sci.*, **24**, 704–706.

Wang, Z., H. Akimoto, and I. Uno, 2002: Neutralization of soil aerosol and its impact on the distribution of acid rain over east Asia: observations and model results. *J. Geophys. Res.*, **107**, 4389, doi: 10.1029/2001JD001040.

Wexler, A. S., F. W. Lurmann, and J. H. Seinfeld, 1994: Modeling urban and regional aerosols. I: Model development. *Atmos. Environ.*, **28**, 531–546.

Wiedensohler, A., *et al.*, 1997: Night-time formation and occurrence of new particles associated with orographic clouds. *Atmos. Environ.*, **31**, 2545–2560.

Whitby, K. T., 1978: The physical characteristics of sulfur aerosols. *Atmos. Environ.*, **12**, 135–159.

Whitby, E. R., P. H. McMurry, U. Shankan, and F. S. Binkowski, 1991: Modal aerosol dynamics modeling. Rep. 600/3–91/020, Atmospheric Research and Exposure Assessment Laboratory, US Environmental Protection Agency, Research Triangle Park, NC (NTIS PB91-161729/AS).

Woodcock, A. H., 1953: Salt nuclei in marine air as a function of altitude and wind force. *J. Meteorol.*, **10**, 362–371.

Wurzler, S. C., Z. Levin, and T. G. Reisin, 1997: Cloud processing of dust particles and subsequent effects on drop size distributions. *J. Aerosol Sci.*, **28**, 427–428.

Xue, H., and G. Feingold, 2004: A modeling study of the effect of nitric acid on cloud properties. *J. Geophys. Res.*, **109**, D18204, doi:10.1029/2004JD004750.

Xue, H., and G. Feingold, 2006: Large eddy simulations of tradewind cumuli: investigation of aerosol indirect effects. *J. Atmos. Sci.*, **63**, 1605–1622.

Yan, Y., Wurzler, S., G. T. Reisin, and Z. Levin, 2002: Interactions of mineral dust particles and clouds: effects on precipitation and cloud optical properties. *J. Geophys. Res.*, **107**, doi:10.1029/2001JD001544.

Yao, X., M. Fang, C. K. Chan, and M. Hu, 2003: Formation and size distribution characteristics of ionic species in atmospheric particulate matter in Beijing, China: (2) dicarboxylic acids. *Atmos. Environ.* **37**, 3001–3007.

Yin, Y., D. J. Parker, and K. S. Carslaw, 2001: Simulation of trace gas redistribution by convective clouds — liquid phase processes. *Atmos. Chem. Phys. Discuss.*, **1**, 125–166.

Yuen, P.-F., D. A. Hegg, and T. V. Larson: The effects of in-cloud sulfate production on light-scattering properties of continental aerosol. *J. Appl. Meteorol.*, **33**, 848–854.

Zhuang, G., Z. Yi, R. A. Duce, and P. R. Brown, 1992: Link between iron and sulphur cycles suggested by detection of Fe(II) in remote marine aerosols. *Nature*, **355**, 537–539.

Application of Satellite-Based Surface Heat Budgets to Climate Studies of the Tropical Pacific Ocean and Eastern Indian Ocean

Ming-Dah Chou, Shu-Hsien Chou* and Shuk-Mei Tse

Department of Atmospheric Sciences, National Taiwan University, Taipei, Taiwan
mdchou@as.ntu.edu.tw

In tropical oceanic regions, surface heat fluxes influence the climate at all spatial and temporal scales, ranging from diurnal variations of the sea surface temperature (SST) and clouds to intraseasonal variations of atmospheric disturbances, as well as interannual variations of the El Niño/Southern Oscillation. In the past, research on the impact of sea surface heat fluxes on climate was limited by the lack of quality global data sets with high temporal and spatial resolutions. More recently, good quality sea surface heat flux data sets with a temporal resolution of 1 day and a spatial resolution of $1° \times 1°$ latitude–longitude have been produced, based on satellite radiation measurements. We have applied these data sets to study climate in the tropical Pacific Ocean and eastern Indian Ocean. These studies include the climatology of surface heat budgets (SHB's), the role of SHB's in the regulation of Pacific warm pool temperature, the correlation between local SHB's and the rate of change of the SST, and the correlation between the interannual variations of the tropical Pacific basin-wide mean SHB and SST. We review the results of those studies in this article.

1. Introduction

The Earth–atmosphere system reflects $\sim$30% of the incoming solar (or shortwave, SW) radiation back to space, and absorbs $\sim$20% in the atmosphere. The rest, $\sim$50%, is absorbed at the Earth's surface (Kiehl and Trenberth, 1997). Most of the SW radiation absorbed at the surface occurs over the tropical ocean, and is then transferred to the atmosphere primarily in the form of latent heat. The latent heat released in the atmosphere is a major force driving atmospheric circulation. Atmospheric circulation, in turn, provides wind stress for driving the ocean circulation. Thus, the heat exchanges at the tropical sea–air interface are among the most important mechanisms affecting Earth's climate.

The warm pool of the tropical western Pacific Ocean and Indian Ocean, with a monthly mean temperature $>28°C$, is a climatically important region, characterized by the warmest sea surface temperature (SST), frequent heavy rainfall, strong atmospheric heating, and weak mean winds with highly intermittent westerly wind bursts (Webster and Lukas, 1992). The large amount of heat contained in the warm pool drives the global climate, and plays a key role in the El Niño–Southern Oscillation (ENSO) and the Asian–Australian monsoon (Webster *et al.*, 1998). Small changes in the SST of the Pacific warm pool by $\sim$1°C, associated with the eastward shift of the warm pool during the ENSO events, have been shown to affect the global climate (Palmer and Mansfield, 1984).

---

*Deceased.

In order to enhance our understanding of the air–sea interaction and other tropical dynamical and physical processes in the warm pool, the Tropical Ocean–Global Atmosphere (TOGA) Coupled Ocean–Atmosphere Response Experiment (COARE) Intensive Observing Period (IOP) was conducted from November 1992 to February 1993 (Webster and Lukas, 1992).

The SST undergoes diurnal variation especially on clear days with weak surface winds due to strong diurnal SW radiation, weak evaporative cooling, and weak mixing of water in the surface layer. Webster *et al.* (1996) investigated the diurnal SST variation in the tropical western Pacific using a one-dimensional ocean mixed layer model. They suggested that because of the nonlinear relationships between clouds, winds, surface fluxes, and the SST, information on the SST diurnal variation may be important for simulation of large scale cloud systems and feedback to the large scale atmospheric dynamics. Using data collected from different observation platforms during the COARE IOP, Sui *et al.* (1997) investigated the diurnal variations of cumulus convection. They found that the tropical oceanic rainfall can be classified into three stages: morning cumulus rainfall, afternoon convective showers, and nocturnal rainfall. These stages of rainfall are related to the diurnal SW heating of the ocean and the infrared (or longwave, LW) cooling at the cloud top.

During El Niño events, the central equatorial Pacific and the eastern Pacific are anomalously warm, and the convection center in the western Pacific warm pool shifts eastward by $\sim 40°$ latitude to the central equatorial Pacific, which can be seen from the satellite measurements in the $11\,\mu$m IR window channel (Chou *et al.*, 2001). Also associated with the El Niño events are more frequent and intensive eastward propagating intraseasonal disturbances, known as the Madden–Julian Oscillation (MJO), with a period of 30–60 days. These disturbances are often followed by strong westerly wind bursts (Sui and Lau, 1992). Lau and Sui (1997)

investigated the mechanisms of short term SST variation associated with the MJO using data measured during the TOGA COARE IOP. Their study demonstrated the importance of the MJO in regulating the short term SST through the induced changes in surface heat fluxes. They suggested that the MJO-induced SST variation might feed back to influencing the MJO. In a study of the mechanisms of the 1997–98 El–La Niña, Picaut *et al.* (2002) found that successive westerly wind bursts excited equatorial Kelvin waves and advected warm water eastward, which triggered the strong 1997 El Niño. They also found that the existing theories of the ENSO were all legitimized at various stages during the development of the 1997 El Niño. In an analysis of the heat sources and sinks of the 1986–87 El Niño, Sun (2000) suggested that the ENSO system behaves like a heat pump: the equatorial ocean absorbs heat during the cold phase, and transports that heat out of the tropics. It implies that the heat fluxes at the ocean surface may be a driving force for El Niño.

Surface heat fluxes over global oceans can best be derived from satellite-inferred atmospheric and surface parameters. Currently, there are several satellite-based sea surface heat fluxes available for studying the relationships between SST, convection, cloud, and atmospheric and oceanic circulation. These data sets include the GSSTF2 (Chou *et al.*, 2003), HOAPS (Grassl *et al.*, 2000), and J-OFURO (Kubota *et al.*, 2002) for turbulent heat fluxes and GSSRB (Chou *et al.*, 2001) and NASA SRB Rel2 (available from http://daac.gsfc.nasa.gov/ precipitation/gssrb.shtml) for radiative fluxes. We have applied these satellite-based data sets to study the interactions between the surface heat fluxes, SST, and climate in the tropical Pacific and eastern Indian Ocean. The results were published in a series of papers, which included the study of the region in the TOGA COARE IOP (Chou *et al.*, 1998; Chou *et al.*, 2000), the tropical western Pacific Ocean and Indian Ocean (Chou *et al.*, 2001; Chou *et al.*, 2004;

Chou *et al.*, 2005a), and the entire tropical Pacific (Chou *et al.*, 2005b). Those studies are reviewed in this article.

2. Data Sources and Satellite Retrievals of Surface Heat Fluxes

Data used in this study include SST, surface wind, surface humidity, high level clouds, outgoing longwave radiation (OLR), and the surface radiative and turbulent heat fluxes. All the data have a spatial resolution of $1° \times 1°$ latitude–longitude and a temporal resolution of 1 day except for the SST, which has a temporal resolution of 1 week. The SST was taken from the National Centers for Environmental Prediction (NCEP) data archive (Reynolds and Smith, 1994), and the surface wind was retrieved from radiance measurements (Wentz, 1997) of the Special Sensor Microwave Imager (SSM/I). The high level cloud cover was inferred from the brightness temperature measured in the 11μm channel of Japan's Geostationary Meteorological Satellite-5 (GMS-5) using a threshold brightness temperature of $260 \, \mathrm{K}$. The GMS pixels have a spatial resolution of $5 \, \mathrm{km}$ and a temporal resolution of $1 \, \mathrm{hr}$. The high level cloud cover was inferred at these high spatial and temporal resolutions, and then averaged to a spatial resolution of $1° \times 1°$ latitude–longitude and a temporal resolution of 1 day (Chou *et al.*, 2001). The OLR inferred from the NOAA Advanced Very High Resolution Radiometer (Gruber and Winston, 1978) was also used as a proxy for cloudiness in the tropical Pacific. The surface latent and sensible heat fluxes were taken from the GSSTF2 data archive of Chou *et al.* (2003) and the HOAPS-II data archive of Grassl *et al.* (2000). Two surface radiation data sets were used. One was the GSSRB of Chou *et al.* (2001), which covers the tropical western Pacific Ocean and eastern Indian Ocean, and the other was the NASA SRB Rel2 global data archive (available from http://eosweb.larc.nasa.gov/PRODOCS/srb/table_srb.html).

2.1. *The GSSRB surface radiation*

A common method for deriving surface radiation budgets is to couple satellite retrievals of cloud, surface, and atmospheric parameters to radiation model calculations. This method can be grouped into two types. One is to derive cloud parameters from satellite radiation measurements, and to compute fluxes using a radiation model (e.g. Chou, 1994; Zhang *et al.*, 1995). The other is to derive a predetermined relationship between the radiation at the top of the atmosphere and the surface (e.g. Pinker and Laszlo, 1992; Li and Leighton, 1993). The predetermined relationship is derived from radiation model calculations with climatological temperature, humidity, clouds, and aerosols as input to the radiation model. Both approaches are attractive because the physical processes that affect the radiative transfer are explicitly included. Nevertheless, the accuracy of the calculated fluxes is inherently limited by inadequate information on the absorption and scattering properties of atmospheric gases and particulates (clouds and aerosols), as well as on the temporal and spatial distributions of clouds and aerosols. One cannot with absolute certainty define either the accuracy of cloud and aerosol parameters retrieved from satellite observations, or the accuracy of radiative transfer models in computing fluxes in both clear and cloudy atmospheres. Another method for deriving the downward surface SW and LW fluxes is to empirically relate the surface flux measurements to the satellite radiance measurements. This method reduces errors introduced by the uncertainties in cloud and aerosol retrievals, and in radiative transfer models.

Based on GMS radiance measurements and surface SW and LW radiative flux measurements at six sites on islands, research ships, and a buoy during the TOGA COARE IOP from November 1992 to February 1993, Chou *et al.* (1998) derived the relationship between the GMS radiance measurements and the surface SW and LW radiative fluxes. The surface

SW radiative flux was empirically related to the satellite-measured radiance in the visible channel and the solar zenith angle, whereas the surface LW radiative flux was related to the temperature and humidity near the surface and the satellite-measured radiance in the 11 μm window channel.

Chou *et al.* (2001) applied these relationships to empirically derive surface SW and LW radiative fluxes from the GMS-5 radiance measurements in the tropical western Pacific and eastern Indian Ocean for the period of October 1997–December 2000. The resultant surface radiation was validated against the observations at the US Department of Energy's Atmospheric Radiation Measurement (ARM) site on Manus Island (2.06°S, 147.43°E) (Mather *et al.*, 1998). Figure 1 shows daily variations of the observed (solid curve) and retrieved (dashed curve) downward SW radiation at the Manus site during the period of December 1999–November 2000. The retrieved SW radiation is in agreement with the measured SW radiation. Averaged over the entire GSSRB period (October 1997–December 2000), the retrieved daily-mean downward surface SW flux has a positive bias of $6.7\,\mathrm{W\,m^{-2}}$ and a standard deviation (SD) error of $28.4\,\mathrm{W\,m^{-2}}$. In the GSSRB data archive, this small bias at the Manus site was assumed to be universal and subtracted from the retrievals over the entire domain. The error of the retrieved LW radiation is small.

The surface radiation was originally retrieved hourly with a spatial resolution of 25 km and subsequently averaged to daily values with a spatial resolution of $0.5° \times 0.5°$ latitude–longitude. The GSSRB covers the domain 40°S–40°N, 90°E–170°W and the period of October 1997–December 2000. There is also a subset of GSSRB that covers the TOGA COARE IOP from November 1992 to February 1993. We used the GSSRB surface radiation to study the relationship between SST and surface heat budget (SHB) in the tropical western Pacific and

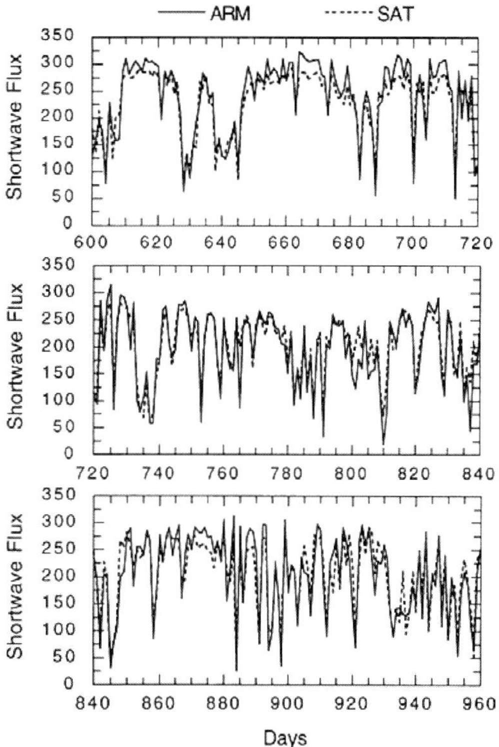

Figure 1. Daily variations of the downward surface SW flux measured at the ARM Manus site (2.06°S, 147.43°E) and retrieved from GMS-5 albedo measurements. Comparison is shown only for the period of December 1999–November 2000. Day 600 corresponds to December 1, 1999. Units of flux are $\mathrm{W\,m^{-2}}$. (After Chou *et al.*, 2004.)

eastern Indian Ocean. The GSSRB data archive is available at http://daac.gsfc.nasa.gov/precipitation/gssrb.shtml.

2.2. *The NASA SRB_Rel2 surface radiation*

The global surface radiation of the NASA SRB_Rel2 was derived using the Pinker and Laszlo algorithm (Pinker and Laszlo, 1992) for SW radiation, and a modified version of the Fu *et al.* (1992) algorithm for LW radiation. The cloud data input to these algorithms is taken from International Satellite Cloud Climate Project (ISCCP; Rossow and Schiffer, 1999) D series, and the temperature and humidity profiles are taken from the version 1 of the NASA Goddard Earth Observing System model (GEOS-1) 4D data assimilation. The radiative fluxes are derived at a three-hourly resolution and subsequently averaged to daily and monthly resolutions. The latitudinal spatial resolution is $1°$ globally, whereas the longitudinal spatial resolution decreases from $1°$ in the tropics to $120°$ at the poles. We interpolated the data into $1° \times 1°$ latitude–longitude grids. The data we used cover the period from July 1987 to December 2000, which is the same period as the GSSTF2 turbulent flux data set. We use the NASA SRB_Rel2 surface radiation in the study of sea surface heating and interannual SST variation in the entire tropical Pacific.

2.3. The GSSTF2 surface turbulent fluxes, temperature, and humidity

A few data sets of sea surface turbulent fluxes derived from satellite observations using bulk flux models are available (Curry *et al.*, 2004, and references therein). The latent heat (LH) and sensible heat (SH) fluxes using bulk flux models are computed from

$$F_{\mathrm{LH}} = \rho_a L_v C_E U (Q_s - Q_a), \qquad (1)$$

$$F_{\mathrm{SH}} = \rho_a C_p C_H U (T_s - T_a), \qquad (2)$$

where ρ_a is the surface air density, L_v the latent heat of vaporization, C_p the isobaric specific heat, U the surface wind speed, $(Q_s - Q_a)$ the sea–air humidity difference and $(T_s - T_a)$ the sea–air temperature difference, and C_E and C_H are transfer coefficients which are dependent on the reference height, stability of the air, and surface roughness.

The SSM/I on board a series of the Defense Meteorological Satellite Program (DMSP) spacecraft have provided global radiance measurements for both the atmosphere and the surface. Chou *et al.* (2003) derived the GSSTF2 surface turbulent fluxes over global oceans based on the surface wind and humidity inferred from SSM/I radiance measurements. The GSSTF2 covers the 13–year period from July 1987 to December 2000 with a spatial resolution of $1° \times 1°$ latitude–longitude and a temporal resolution of 1 day. The GSSTF2 flux model is a bulk aerodynamics algorithm based on the Monin–Obukhov similarity theory including the salinity and cool-skin effects. The daily mean input parameters for turbulent flux calculations include the wind speed at $10\,\mathrm{m}$ height ($U_{10\mathrm{m}}$) derived from SSM/I radiance measurements (Wentz, 1997), the $10\,\mathrm{m}$ specific humidity ($Q_{10\mathrm{m}}$) derived from the water amount of the entire atmospheric column, and the $500\,\mathrm{m}$ layer near the surface (Chou *et al.*, 1997). The input parameters also include the $2\,\mathrm{m}$ air temperature ($T_{2\mathrm{m}}$) and the SST of the NCEP–NCAR reanalysis (Kalnay *et al.*, 1996).

Compared to 134 daily turbulent fluxes of five tropical field experiments conducted by the NOAA/Environmental Technology Laboratory (ETL), Chou *et al.* (2003) found that F_{LH} of GSSTF2 had a negative bias of $-2.6\,\mathrm{W\,m^{-2}}$, an SD error of $29.7\,\mathrm{W\,m^{-2}}$, and a correlation coefficient of 0.80. F_{SH} had a positive bias of $7.0\,\mathrm{W\,m^{-2}}$, an SD error of $6.2\,\mathrm{W\,m^{-2}}$, and a correlation coefficient of 0.45. The GSSTF2 data archive can be found at http://daac.gsfc.nasa.gov/precipitation/gsstf2.0.shtml.

2.4. The HOAPS-II surface turbulent fluxes, temperature, and humidity

The HOAPS-II data archive (Grassl *et al.*, 2000) covers the global ice-free oceans from July 1987 to December 2002 with a spatial resolution of $0.5° \times 0.5°$ latitude–longitude and a temporal resolution of 5 days. The data set includes the SST, wind speed, air–sea humidity difference, and turbulent heat fluxes. The SST is retrieved from the Advanced Very High Resolution Radiometer (AVHRR) radiance measurements (Kilpatrick *et al.*, 2001), wind speed and humidity are retrieved from the SSM/I radiance measurements, the turbulent heat fluxes are derived by a bulk formula and parametrization scheme of Fairall *et al.* (1996). The air–sea humidity difference is calculated using the approach of Bentamy *et al.* (2003) without a fixed height. Pentad data are degraded into monthly averages with a spatial resolution of $1° \times 1°$ latitude–longitude.

3. Tropical Warm Pool Surface Heat Budget (SHB)

The downward net heat flux, F_{net}, at the surface is given by

$$F_{\mathrm{net}} = F_{\mathrm{SW}} - (F_{\mathrm{LW}} + F_{\mathrm{LH}} + F_{\mathrm{SH}}), \quad (3)$$

where F_{SW} is the downward net short wave flux (solar heating), F_{LW} the upward net long wave flux (IR cooling), F_{LH} the upward latent heat flux (evaporative cooling), and F_{SH} the upward sensible heat flux (sensible cooling).

Chou *et al.* (2004) investigated the annual-mean, seasonal, and interannual variations of the SHB over the tropical eastern Indian Ocean and western Pacific Ocean (30°S–30°N, 90°E–170°W) using the surface SW and LW radiative fluxes taken from GSSRB and the surface latent and sensible fluxes taken from GSSTF2. Limited by the length of the GSSRB surface radiation data, it covers only the three-year period from October 1997 to September 2000. During this period, there was a strong El Niño in the winter of 1997–98 with rapid onset and decay (Bell *et al.*, 1999). The El Niño was followed by a moderate and long-lasting La Niña from 1998 to 2000 (Bell *et al.*, 2000).

3.1. *Annual mean*

The spatial distributions of the three-year mean F_{SW}, F_{LW}, F_{LH}, and F_{net}, from October 1997 to September 2000, are shown in Fig. 2. F_{SH} is small ($\sim$5–15 W m^{-2}) and is not shown in the figure. The solar heating F_{SW} varies from 180 to 240 W m^{-2} with a range of 60 W m^{-2}, and the evaporative cooling F_{LW} varies from 80 to 190 W m^{-2} with a range of 110 W m^{-2}.

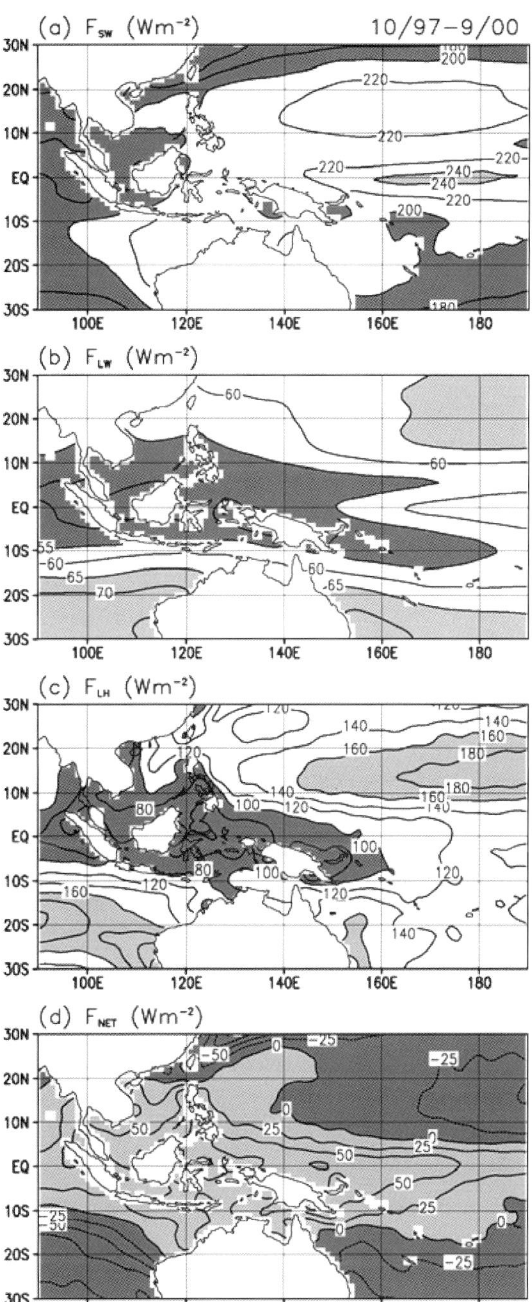

Figure 2. Surface (a) SW heating, (b) LW cooling, (c) evaporative cooling, and (d) net heating, averaged over the three-year period of October 1997–September 2000. Regions of large values are marked with light shading, and regions of small values are marked with dark shading. (After Chou *et al.*, 2004.)

The magnitude and spatial variation of F_{LW} are both small. Among the heat flux components, F_{SW} has the largest magnitude, and F_{LH} has the largest spatial variation. As a result, the spatial variation of F_{net} is primarily determined by F_{LH}. Because the spatial variation of evaporative cooling is larger than that of solar heating, regions of the maximum net surface heating are regions of minimum evaporative cooling, which correspond to regions of minimum air–sea humidity difference and minimum wind speed. These regions also have maximum cloud cover. Data from the NCEP/NCAR (National Center for Atmospheric Research) reanalysis (Kalnay et al., 1996) for the same three-year period as show the same result, that the maximum surface heating coincides with the maximum cloud cover, owing to the reduced evaporative cooling. The reduced evaporative cooling in the convective regions was also shown by Ramanathan and Collins (1991). For the three-year mean condition within $\sim 10°$ of the equator, the oceans gain heat of up to ~ 50–$75\,\mathrm{W\,m^{-2}}$, owing to the large solar heating and small evaporative cooling. Poleward of this region the heat loss by the ocean increases with latitudes, primarily owing to an increase of evaporative cooling.

In the tropical western Pacific, the SST is high. It is in the ascending branch of both the Hadley and Walker circulations. Trade winds that converge in this region induce strong, deep convection and produce large amounts of clouds. Figure 3 shows the OLR, SST, air–sea humidity difference (Q_S–Q_{10m}), and surface wind speed (U_{10m}) averaged over the three-year period. The deep convective cloudiness, as indicated by OLR $<240\,\mathrm{W\,m^{-2}}$, occurs in the maritime continent, the intertropical convergence zone (ITCZ; centered at $\sim 7°$N), and the southern Pacific convergence zone (SPCZ; centered at $\sim 10°$S). Regions of high SST cover most of the western Pacific and eastern Indian Ocean in the tropics. They have a large cloud cover (or small OLR), a small air–sea humidity difference, and weak wind.

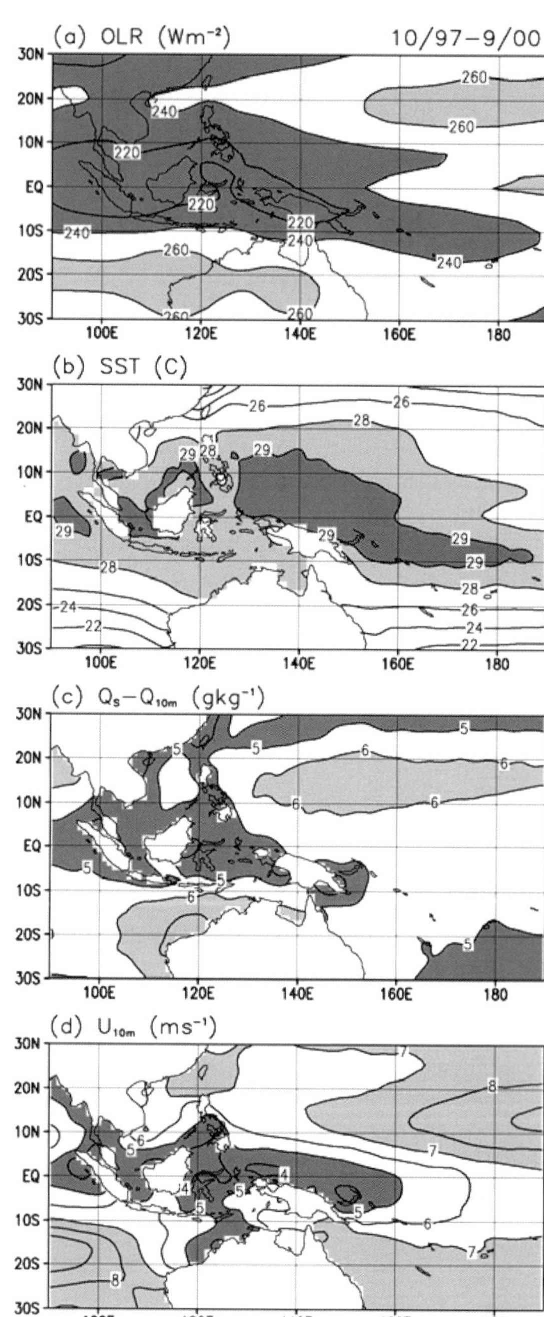

Figure 3. (a) OLR, (b) SST, (c) sea–air humidity difference, and (d) 10 m wind speed, averaged over the three-year period of October 1997–September 2000. Regions of large values are marked with light shading, and regions of small values are marked with dark shading, except for the SST shown in (b). (After Chou et al., 2004.)

The solar heating depends not only on clouds but also on the seasonal and latitudinal variations of insolation at the top of the atmosphere. Averaged over three years, the spatial distribution of the solar heating is rather homogeneous, with a range of 200–240 W m^{-2} within 20° latitude of the equator. As a result, the spatial distribution of annual-mean solar heating [Fig. 2(a)] only slightly resembles that of clouds [or OLR, Fig. 3(a)]. The longwave (LW) cooling depends on clouds, the SST, and the temperature and humidity of the atmospheric boundary layer. Although the range of LW cooling is small [Fig. 2(b)], its spatial distribution closely resembles that of clouds (or OLR). Figure 3(c) shows that the annual-mean air–sea humidity difference (Q_s-Q_{10m}) is large, ~5–6 g kg^{-1}, but the spatial variation is small. On the other hand, the spatial variation of the annual-mean surface wind speed (U_{10m}) is large, varying by a factor >2 from 4 m s^{-1} to 9 m s^{-1} [Fig. 3(d)]. The surface wind speed has a minimum in the equatorial convergence region and a maximum in the trade wind regions. The spatial distribution of the evaporative cooling [Fig. 2(c)] is determined primarily by wind speed and secondarily by Q_s-Q_{10m}. The minimum evaporative cooling of 80–100 W m^{-2} in the maritime continent coincides with the minimum U_{10m} (4–5 m s^{-1}) and Q_s-Q_{10m}, (~5 g kg^{-1}). The maximum evaporative cooling of 180 W m^{-2} occurs in the trade wind regions, where U_{10m} (8–9 m s^{-1}) and Q_s-Q_{10m} (~6 g kg^{-1}) are both large.

3.2. *Seasonal variation*

Seasonal variations of F_{net} over the tropical eastern Indian Ocean and western Pacific Ocean were investigated for the boreal winter (December, January, February; DJF), spring (March, April, May; MAM), summer (June, July, August; JJA) and fall (September, October, November; SON) of the three-year

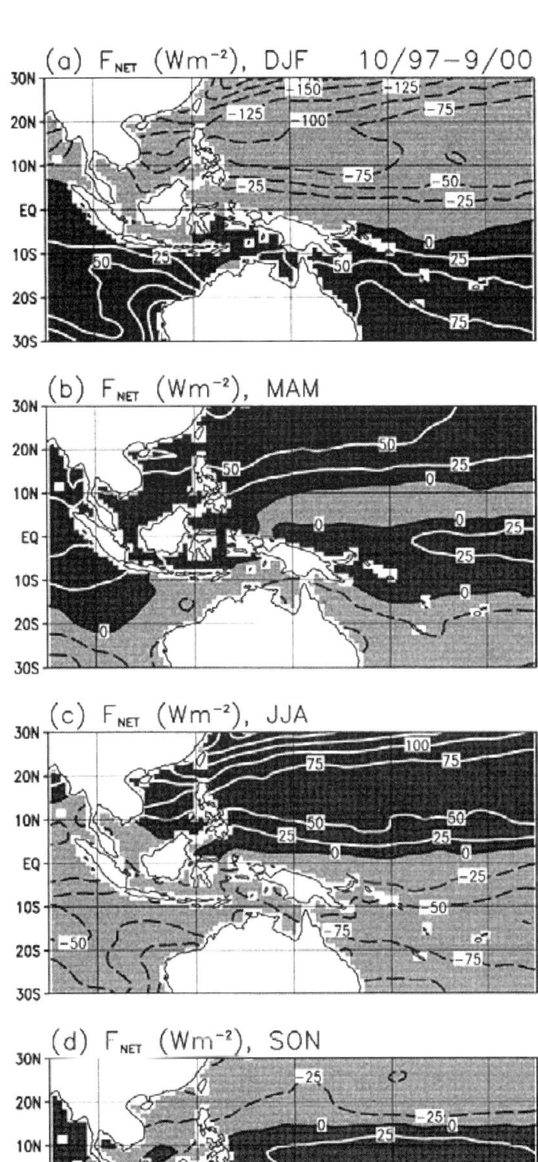

Figure 4. Deviations of seasonal-mean net surface heating from the three-year mean for the boreal (a) winter, DJF, (b) spring, MAM, (c) summer, JJA, and (d) fall, SON, of October 1997–September 2000. Positive values are marked with dark shading, and negative values are marked with light shading. (After Chou *et al.*, 2004.)

period from October 1997 to September 2000. Figure 4 shows the spatial distributions of the deviation of seasonal mean F_{net} from the three-year mean. The ITCZ follows the Sun to the summer hemisphere, and the trade winds are stronger in the winter hemisphere than in the summer hemisphere. The weak solar heating and the strong evaporative cooling cause a strong wintertime cooling of the ocean, whereas the weak evaporative cooling and the strong solar heating cause a strong summertime heating [Figs. 4(a) and 4(c)]. The contrast between summertime heating and wintertime cooling is very large, $\sim 150\,\mathrm{W\,m^{-2}}$ at $20°$ latitude in both hemispheres. F_{net} increases during the spring and summer of both hemispheres but decreases during the fall and winter, except for the equatorial region. The magnitude of the seasonal deviation of F_{net} from the annual mean increases poleward as the seasonal variations of both F_{SW} and F_{LH} increase poleward. The seasonal variation of F_{net} is larger in the northern hemisphere than in the southern hemisphere, mainly owing to a larger seasonal variation of F_{LH} in the northern hemisphere (not shown in the figures).

3.3. *Interannual variation*

The central and eastern equatorial Pacific warmed significantly during May 1997–May 1998, but cooled slightly from July 1998 through 2000 (Bell *et al.*, 1999, 2000). The former is a strong El Niño, and the latter is a moderate La Niño. To consider the bigger picture of the changes in the SST and atmospheric circulation over the tropical Pacific and Indian Oceans during El Niño and La Niña episodes, the differences in SST, OLR, zonal wind stress and surface wind speed between the boreal winter (October–March) of 1997–98 and 1998–99 are shown in Fig. 5 for a large domain of 30°S–30°N, 40°E–120°W. Compared to the La Niña (1998–99), the SST during the El Niño (1997–98) increases by a maximum of 5°C in the central and eastern equatorial Pacific and by $\sim$1–2°C in the tropical western Indian

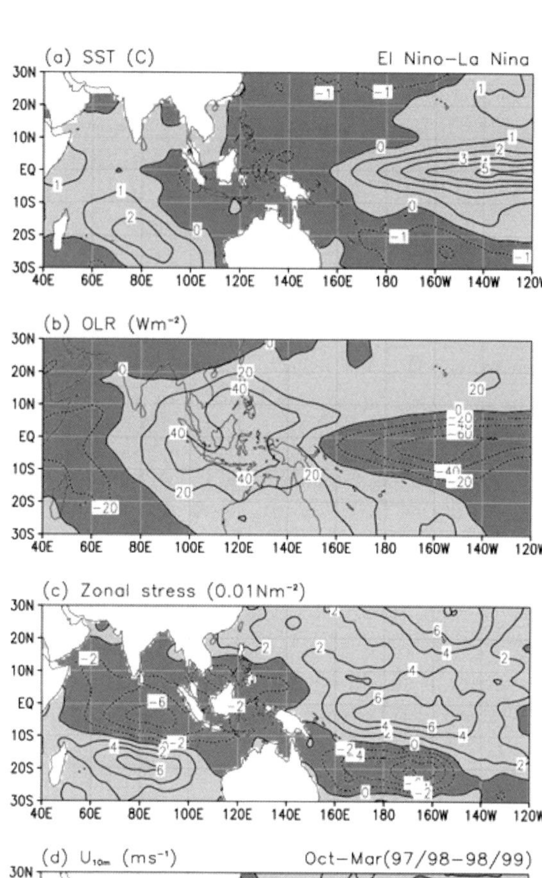

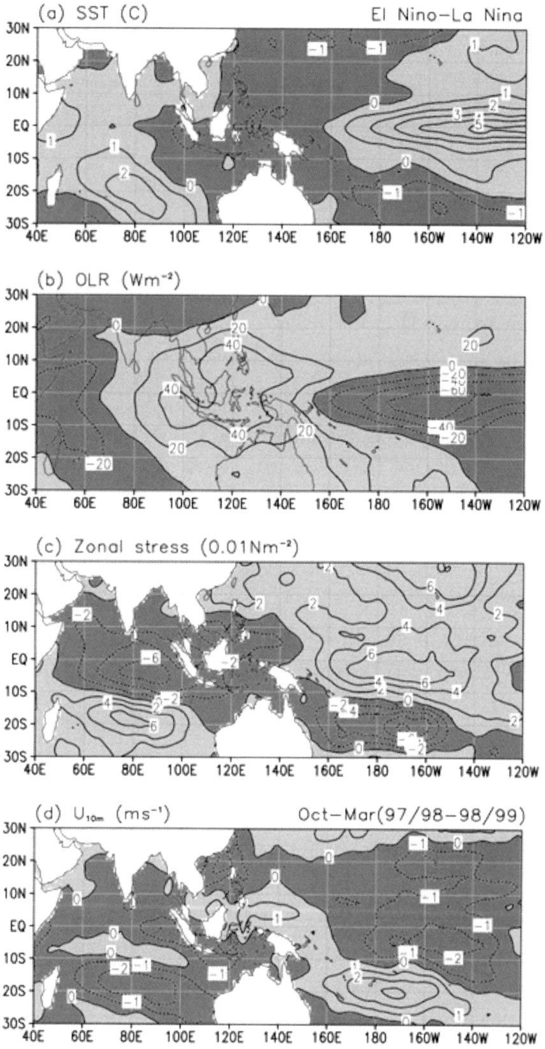

Figure 5. 1997/98 El Niño–1998/99 La Niña differences of (a) SST, (b) OLR, (c) zonal wind stress, and (d) 10 m wind speed for the boreal winter of October–March. Positive values are marked with light shading, and negative values are marked with dark shading. (After Chou *et al.*, 2004.)

Ocean, but decreases slightly by $<$1°C in the maritime continent [Fig. 5(a)]. Corresponding to the change in the SST, the OLR decreases by 40–60 $\mathrm{W\,m^{-2}}$ and 20 $\mathrm{W\,m^{-2}}$, respectively, in the central equatorial Pacific and the tropical western Indian Ocean, but increases by 20–40 $\mathrm{W\,m^{-2}}$ in the maritime continent and the tropical eastern Indian Ocean [Fig. 5(b)]. This is an indication of shifting the convection center from the maritime continent eastward to the central equatorial Pacific and westward to the western equatorial Indian Ocean.

Concurrent with the shift of convection centers is the shift of zonal winds. In the equatorial region during the El Niño, the zonal wind stress increases (more westerly as trade winds weaken) east of $\sim$150°E but decreases (more easterly as trade winds strengthen) west of $\sim$150°E [Fig. 5(c)]. In the subtropical regions, the zonal wind stress increases in the South Indian Ocean and the North Pacific during the El Niño, but decreases in the southwestern Pacific near the SPCZ. The latter is due to the northward shift of the SPCZ, as indicated by the OLR change. The wind stress change in the tropical southern Indian Ocean is a result of the equatorward shift of trade wind belts during the El Niño Niño (Yu and Rienecker, 2000). The wind stress change in the South China Sea is an indication of increased subtropical high during the El Niño, which is consistent with the OLR change there. Compared to the La Niña, the surface wind speed generally decreases by $\sim$1–2 m s^{-1} during the El Niño, except for the regions extending from the southern section of the South China Sea to the maritime continent and further to the SPCZ [Fig. 5(d)].

Figure 6 shows the El Niño–La Niña differences in F_{SW}, F_{LH}, F_{net}, and the SST tendency for the region 30°S–30°N, 90°E–170°W. In accordance with the change in cloudiness [or OLR, Fig. 5(b)], F_{SW} decreases in the equatorial central Pacific but increases in the rest of the domain during the El Niño. The pattern of the change of F_{LH} is similar to that of the surface wind. The spatial variation of the F_{LH} change is much greater than that of the F_{SW} change, and hence it dominates the variation of the F_{net} change. In the central equatorial Pacific, the cloudiness increases but wind decreases during the El Niño. Correspondingly, both F_{SW} and F_{LH} decrease with an increasing SST. The

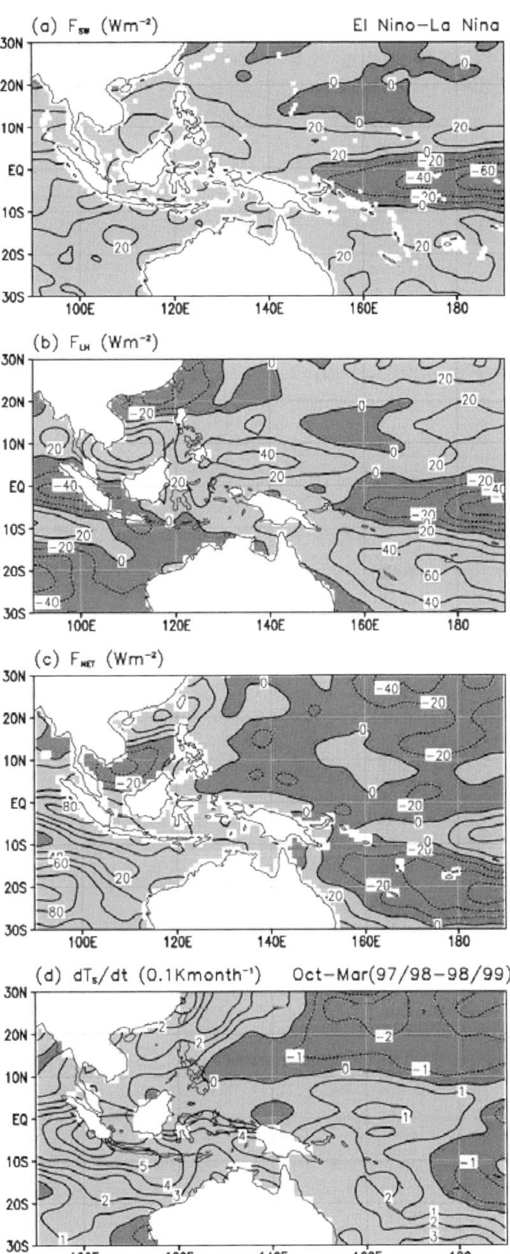

Figure 6. Same as Fig. 5, except for surface (a) solar heating, (b) evaporative cooling, (c) net heating, and (d) SST tendency. (After Chou *et al.*, 2004.)

decrease of F_{LH} surpasses the decrease of F_{SW}, leading to an increase of F_{net}. Thus, the enhancement of convection and cloudiness in the central equatorial Pacific is associated with a higher SST that leads to an enhancement of surface heating, with a maximum of $50\,\mathrm{W\,m^{-2}}$ at 8°S, 175°W.

Compared to the La Niña, the southern section of the SPCZ during the El Niño is less cloudy, more windy, and dryer (greater $Q_s\text{-}Q_{10\mathrm{m}}$) as the SPCZ shifts northward. The first factor enhances the solar heating, but the last two enhance the evaporative cooling. The change of evaporative cooling exceeds the change of solar heating, resulting in a larger net surface cooling during the El Niño. The region of reduced F_{net} covers a large area, with a maximum of $56\,\mathrm{W\,m^{-2}}$ at 22°S, 175°W.

In the South China Sea (SCS), the change of F_{SW} is positive. The change of F_{net} is large in both the northern and southern sections but with opposite signs, dominated by F_{LH}. The maximum change of heating in the northern section of the SCS and the maximum change of cooling in the southern section of the SCS are $\sim$50 and $40\,\mathrm{W\,m^{-2}}$, respectively.

In the eastern Indian Ocean, the F_{net} change is also dominated by the F_{LH} change. The large increases of F_{net} in the tropical southeastern Indian Ocean, $\sim$110 $\mathrm{W\,m^{-2}}$, and in the equatorial eastern Indian Ocean, $\sim$95 $\mathrm{W\,m^{-2}}$, are due to a large reduction of F_{LH} associated with weakened wind and a reduced air–sea humidity difference. The enhanced solar heating due to reduced cloudiness during the El Niño [Fig. 5(b)] also contributes to the large increase of the net surface heating.

The change of F_{net} between the 1997–98 El Niño and the 1998–99 La Niña is significantly larger in the tropical eastern Indian Ocean than in the tropical western Pacific [Fig. 6(c)]. For the eastern Indian Ocean, the reduced evaporative cooling arising from weakened winds during the El Niño is generally associated with enhanced solar heating due to reduced cloudiness, leading to large interannual variability of F_{net}. For the western Pacific, the reduced evaporative cooling due to weakened winds is generally associated with the reduced solar heating

arising from increased cloudiness, and vice versa. Consequently, the interannual variability of F_{net} is reduced.

4. Tropical Warm Pool SHB and SST Tendency

The rate of change of the ocean mixed layer temperature is determined by the heat convergence and the depth of the mixed layer. The heat convergence involves the net heating at the surface and advection of heat by ocean currents, as well as the transmission of solar radiation and the entrainment of cold water through the bottom of the mixed layer. In a study of the depth of the ocean mixed layer with a vertically uniform density in the Pacific warm pool, Lukas and Lindstrom (1991) found that the mixed layer in this region was very shallow, with a mean of $\sim$29 m. The shallow mixed layer enhances the solar radiation penetrating through the bottom of the layer (e.g. Paulson and Simpson, 1977; Lewis *et al.*, 1990; Siegel *et al.*, 1995). Between the bottom of the mixed layer and the top of the thermocline, there is a barrier layer characterized by nearly constant temperature with stable stratification due to downward increase of salinity (Lukas and Lindstrom, 1991). This stable barrier layer inhibits the entrainment of cold water from the deeper thermocline except during very strong westerly wind bursts and is very important to the ocean mixed layer heat budget (Sprintall and Tomczak, 1992; Ando and McPhaden, 1997). The horizontal advection of heat in the warm pool has generally been estimated to be very small, due to a small SST gradient and weak currents (Niiler and Stevenson, 1982; Enfield, 1986). However, some studies have suggested that the advection of heat may be important to the warm pool heat budget during strong wind events (Feng *et al.*, 1998; Godfrey *et al.*, 1998).

Assuming that the horizontal advection of heat and entrainment of cold water from below are negligible, the heat budget of the mixed layer is given by

$$h\rho C \left(\frac{dT_s}{dt}\right) = F_{\text{net}} - f(h)F_{\text{SW}}, \qquad (4)$$

where h is the mixed layer depth, ρ the density of sea water, C the heat capacity of sea water, dT_s/dt the rate of SST change, and $f(h)$ the fraction of F_{SW} penetrating through the depth h. Using the water "type 1A" data of Jerlov (1968), Paulson and Simpson (1977) derived $f(h)$ from

$$f(h) = \gamma e^{-\alpha h} + (1-\gamma)e^{-\beta h}, \qquad (5)$$

where $\gamma = 0.38$, $\alpha = 0.05\,\text{m}^{-1}$, and $\beta = 1.67\,\text{m}^{-1}$. In the above equation, the solar spectrum is divided into two regions, one corresponding to the visible region and the other corresponding to the near-infrared region. The fractions of the solar radiation contained in these two spectral regions are γ and $1-\gamma$, and the absorption coefficients of sea water are α and β, respectively. Siegel et $al.$ (1995) studied the solar radiation penetrating at various depths in the upper ocean using the measurements made from the research vessel $Vickers$ during the COARE IOP. We have found that the data given by Siegel et $al.$ closely correspond to Eq. (5). Using daily values of dT_s/dt, F_{net}, and F_{SW}, we solve Eqs. (4) and (5) for h and $f(h)F_{\text{SW}}$.

4.1. *SHB and SST tendency during the TOGA COARE IOP*

Extensive measurements were carried out during the TOGA COARE IOP, from November 1992 to February 1993, with the primary goal of better understanding various physical processes responsible for the SST variation (Webster and Lukas, 1992). Chou et $al.$ (2000) studied the SHB and SST in the equatorial Pacific warm pool region (10°S–10°N; 135°E–175°E) during the COARE IOP. The COARE IOP is the boreal winter. Trade winds converge to the south of the equator. The northern warm pool has strong northeasterly trade wind, and the southern warm pool has weak converging

wind. During the COARE IOP, there were two Madden–Julian oscillations (MJO's) propagating eastward from the Indian Ocean to the central Pacific (Gutzler et $al.$, 1994; Sui et $al.$, 1997). These two MJO's passed through the warm pool just south of the equator. A super cloud cluster followed by low level westerly wind bursts lasting for about 2–3 weeks was associated with each of the two MJO's. These super cloud clusters cause significant variability of F_{SW} in the southern warm pool, reaching $50\,\text{W}\,\text{m}^{-2}$ on the monthly scale. The westerly wind bursts following the super cloud clusters cause a large increase of F_{LH}, but the weak winds ahead of the cloud clusters cause a large decrease of F_{LH} (Lau and Sui, 1997). When averaged over the southern warm pool, the impact of MJO's on F_{LH} is small, $<10\,\text{W}\,\text{m}^{-2}$ on the monthly scale. Thus, the impact of the two MJO's on F_{SW} is significantly larger than the impact on F_{LH}. In the northern warm pool, on the other hand, the northeasterly trade wind averaged over the IOP is very strong, with a maximum $>8\,\text{m}\,\text{s}^{-1}$. The net surface heating F_{net} is dominated by the large F_{LH} owing to a strong northeasterly trade wind.

The spatial distributions of F_{net} and dT_s/dt in the equatorial Pacific warm pool region averaged over the COARE IOP (November 1992–February 1993) are shown in Figs. 7(a) and 7(b), respectively. The ocean gains heat in the summer (southern) hemisphere while losing heat in the winter (northern) hemisphere, with the net heat flux ranging from $-80\,\text{W}\,\text{m}^{-2}$ in the northeastern section of the equatorial warm pool to $40\,\text{W}\,\text{m}^{-2}$ north of New Guinea. Averaged over the warm pool, the net heating is only $0.7\,\text{W}\,\text{m}^{-2}$. This result is consistent with the study of Godfrey et $al.$ (1991), which demonstrated that the mean surface net heat flux into the warm pool is small ($<10\,\text{W}\,\text{m}^{-2}$). On the other hand, Fig. 6(b) shows that, except near the southern edge, the warm pool cools during the IOP. The cooling increases northward and reaches a maximum $>0.5°\text{C}\,\text{month}^{-1}$. Averaged

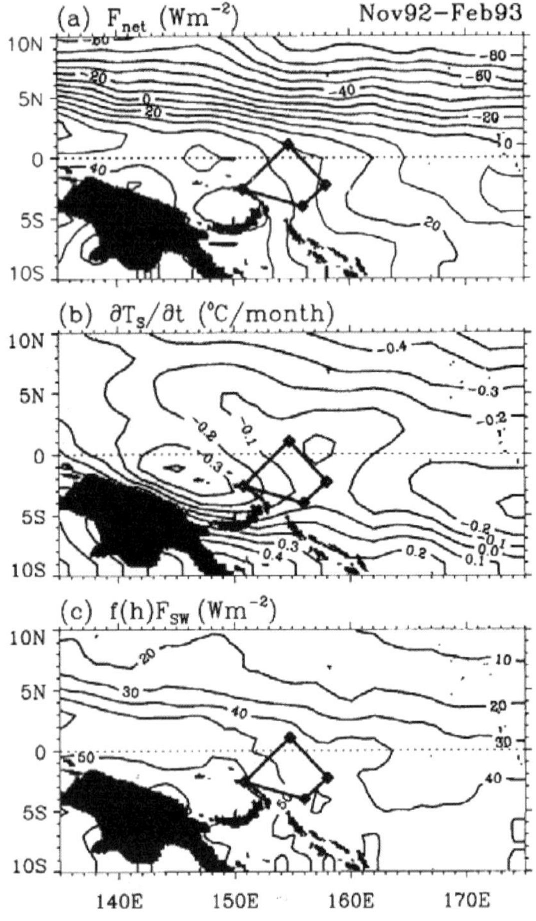

Figure 7. Spatial distributions of (a) surface net heat flux, (b) dT_s/dt, and (c) penetration of solar radiation through the bottom of the mixed layer, $f(h)F_{\rm SW}$, averaged over the COARE IOP. (After Chou *et al.*, 2000.)

over the warm pool the SST cools at a rate of $0.14°$C month^{-1}, which does not agree with the surface heating of $0.7\,$W m^{-2}. The discrepancy is particularly large in the southern warm pool. For example, the net surface heat flux north of New Guinea has a maximum $>40\,$W m^{-2}, but the SST cools at a rate of $0.3°$C month^{-1}. Therefore, much of the solar flux incident at the surface penetrates through the bottom of the mixed layer, and is not available for heating the layer.

Figure 7(c) shows the estimated solar flux penetrating through the bottom of the mixed layer, $f(h)F_{\rm SW}$, averaged over the IOP. The

solar radiation lost through the bottom of the mixed layer is $\sim 26\,$W m^{-2} for the northern warm pool and $\sim 45\,$W m^{-2} for the southern warm pool. The domain-averaged value of $35\,$W m^{-2} is significantly larger than the net surface heat flux of $0.7\,$W m^{-2}. The cooling of the southern warm pool during the IOP is primarily due to this loss of solar radiation, even though the westerly wind bursts may deepen the mixed layer to the top of the thermocline, and entrain cold water from below.

4.2. *Seasonal variation of SHB and SST tendency*

The seasonal-mean SST tendency, dT_s/dt, averaged over the three-year period of October 1997–September 2000, is shown in Fig. 8. Consistent with the seasonal variation of $F_{\rm net}$ (Fig. 4), the SST increases during the spring and summer of both hemispheres, but decreases during the fall and winter — except for the equatorial region. The seasonal variation of dT_s/dt is larger in the northern hemisphere than in the southern hemisphere, which is also consistent with the seasonal variation of $F_{\rm net}$. The magnitude of dT_s/dt increases poleward as the seasonal variation of $F_{\rm net}$ increases poleward.

Figure 9 shows the correlation coefficient, r, of monthly dT_s/dt, $F_{\rm net}$, $F_{\rm SW}$, and $-F_{\rm LH}$ for the three-year period. The regions where the correlation is significant at the 95% level are light-shaded. To determine the significance of the correlation, the standard deviation of the correlation coefficient, SD, is derived following the approach of Lau and Chan (1983). The correlation is considered significant if $r > 2$ SD, which has a 95% confidence level. Regions of $r > 2$ SD generally coincide with regions of $r > 0.5$–0.6 for all three cases of correlation shown in the figure. Figure 9(a) shows that the correlation between monthly dT_s/dt and $F_{\rm net}$ increases poleward and is generally very high, $r \sim 0.8$–0.9 poleward of $15°$–$20°$ latitude. The correlation is generally lower ($r \sim 0.4$–0.6) in

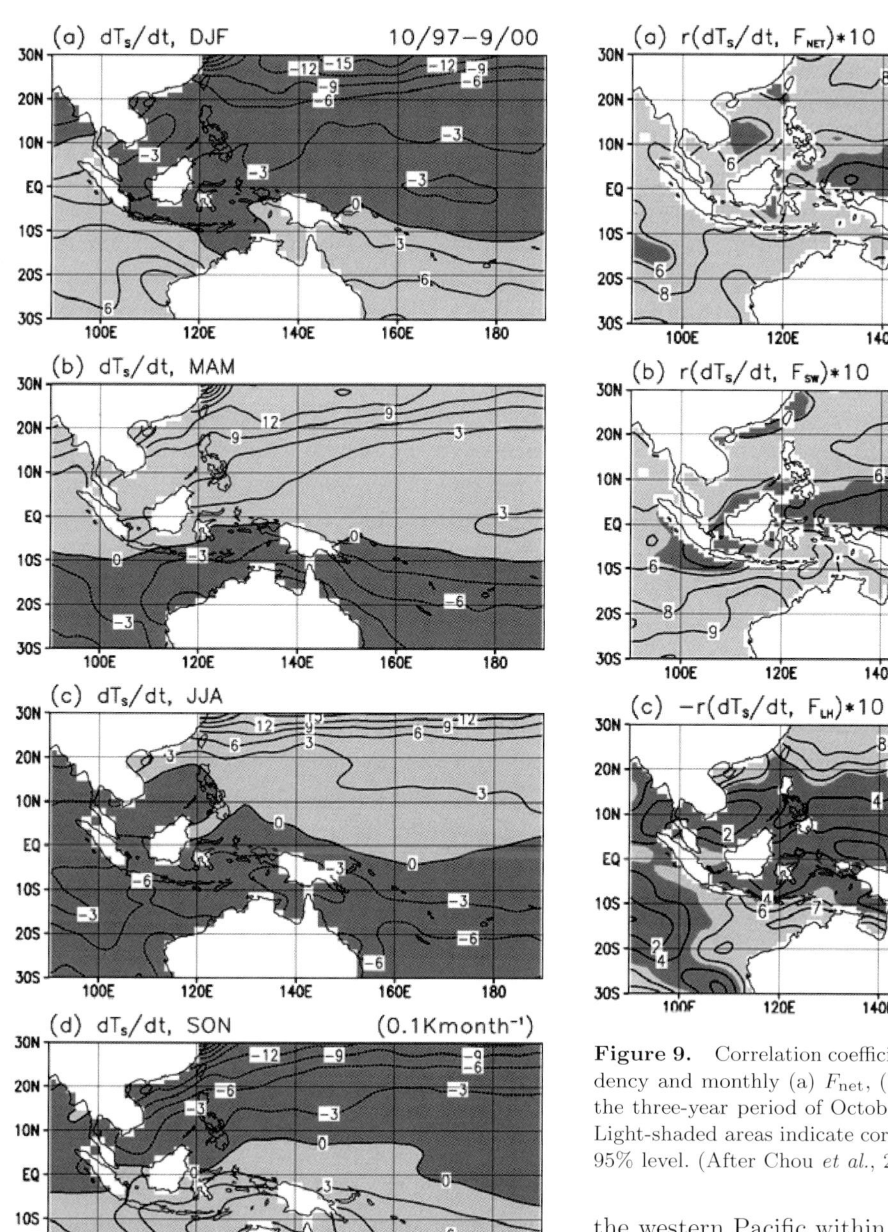

Figure 8. Deviations of seasonal-mean SST tendency from the three-year mean (October 1997–September 2000) for the boreal (a) winter, DJF, (b) spring, MAM, (c) summer, JJA, and (d) fall, SON, of October 1997–September 2000. Positive values are marked with light shading, and negative values are marked with dark shading. Units are 0.1 K month^{-1}. (After Chou et al., 2004.)

Figure 9. Correlation coefficients of monthly SST tendency and monthly (a) F_{net}, (b) F_{SW}, and (c) F_{LH} for the three-year period of October 1997–September 2000. Light-shaded areas indicate correlation significant at the 95% level. (After Chou et al., 2004.)

the western Pacific within $\sim 10°$ of the equator. Thus, the seasonal variations of dT_s/dt and F_{net} are significantly correlated at a confidence level of $\sim 95\%$, except for some small areas in the equatorial western Pacific. For the North Indian Ocean, Loschnigg and Webster (2000) found that the seasonal variation of ocean heat storage was nearly balanced by that of the oceanic heat transport through the equator, and that the correlation between seasonal variations of net surface heating and ocean heat storage was poor.

Our results do not support their conclusion about the northeastern Indian Ocean, as the correlation (>0.6) is significant there.

Figure 9(b) shows that the correlation between monthly dT_s/dt and F_{SW} increases poleward and is generally very high, ~ 0.8–0.9 poleward of $10°$ latitude, but the correlation within $\sim 10°$ of the equator is weak, ranging from -0.2 to 0.6. Thus, the seasonal variations of dT_s/dt and F_{SW} are significantly correlated at a confidence level of $\sim 95\%$, except for the equatorial region. The correlation between monthly dT_s/dt and $-F_{LH}$ is shown in Fig. 9(c). Comparing Figs. 9(b) and 9(c), it can be seen that dT_s/dt has a higher correlation with F_{SW} than $-F_{LH}$, except for the western equatorial Pacific, $\sim 5°S$–$5°N$, $150°E$–$170°W$. In the equatorial regions, the weak correlation of dT_s/dt with F_{SW} and F_{net} is likely due to a large SW radiation penetration through the shallow ocean mixed layer associated with weak winds, together with the small seasonal variations of F_{SW} and F_{net}. When the mixed layer reaches the thermocline during intermittent westerly wind bursts, the entrainment of cold water from the thermocline can also weaken the correlation of dT_s/dt with F_{SW} and F_{net}.

4.3. *Interannual variations of SHB and SST*

The El Niño–La Niña differences in dT_s/dt, as well as the differences in F_{SW}, F_{LH}, and F_{net}, for the region $30°S$–$30°N$, are shown in Fig. 6. The correlation between interannual variations of F_{net} and dT_s/dt is very weak [Figs. 6(c) and 6(d)]. In particular, the strong meridional dipole-like change of F_{net} in the South China Sea and the tropical western Pacific south of the equator is not consistent with the much smoother change of dT_s/dt. In the eastern Indian Ocean, the spatial pattern of the F_{net} change is not consistent with the dT_s/dt change. Furthermore, the anomalous correlation between the monthly SST tendency

and monthly net surface heating computed for the three-year period, although too short for studying interannual variability, is generally ~ 0.2–0.5 and is not significant at the 95% level (not shown). These results indicate that the change in ocean dynamics, other than the change in surface heating, plays an important role in the interannual variation of dT_s/dt. The change in ocean dynamics may include the variations of SW radiation penetrating through the ocean mixed layer, upwelling of cold thermocline water, Indonesian throughflow for transporting heat from the Pacific to the Indian Ocean, and interhemispheric transport in the Indian Ocean as inferred from the changes in zonal wind stress and surface wind speed [Figs. 5(c) and 5(d)]. Wang and McPhaden (2001) found that all terms in the heat balance of the oceanic mixed layer contributed to the SST variation in the equatorial Pacific during the 1997–98 El Niño and 1998–99 La Niño. The low correlation between the interannual variations of F_{net} and dT_s/dt shown in Figs. 6(c) and 6(d) is consistent with their finding.

5. Regulation of the Pacific Warm Pool SST

The tropical western Pacific is warm and cloudy. The SST between $20°S$ and $20°N$ is nearly homogeneous, with a magnitude of $28°C$–$30°C$. Regions of SST $>30°C$ are rare and are associated with a stable and clear atmosphere with descending air (Waliser and Graham, 1993). Ramanathan and Collins (1991) postulated that the increase in thick cirrus clouds in response to an increase in the SST had an effect of reducing the solar heating of the ocean and limiting the SST to the observed high values. Subsequently, various mechanisms were proposed by several investigators that regulate the SST in the Pacific warm pool (e.g. Hartmann and Michelsen, 1993; Lau *et al.*, 1994; Pierrehumbert, 1995; Sud *et al.*, 1999; Sun and Liu, 1996; Wallace, 1992). Most of these studies stressed the importance of large scale circulation in distributing heat. Some also stressed the importance of the high sensitivity of evaporation to high temperature and the ocean dynamic in regulating the SST. In addition to the magnitude of the SST, a relevant question can also be raised as to the mechanisms that cause the SST in the warm pool to be homogeneous. Wallace (1992) suggested that the homogeneous tropospheric temperature in the tropics is a result of the efficient heat transport by large scale circulation. In addition to viewing the issue from a large scale perspective, analysis of the SHB can provide information on the causes that lead to a homogeneous SST distribution in the Pacific warm pool. Chou *et al.* (2005a) used the satellite-based SHB to study the influence of transient atmospheric circulation on the SST of the western Pacific warm pool.

The high SST regions in the western Pacific have large amounts of clouds and weak converging wind, corresponding to weak surface solar heating and evaporative cooling. Because the spatial variation of solar heating is considerably smaller than that of evaporative cooling (see Sec. 3), the high SST regions have a maximum surface heating. With the maximum heating occurring at high SST regions, one might wonder what the physical mechanisms are that would prevent the SST from increasing further. In addition to the SW radiation penetrating through the bottom of the ocean mixed layer, the transient nature of the solar radiation and the atmospheric circulation are among the major mechanisms that limit the SST to below $\sim 30^\circ$C and cause the warm pool SST to be homogeneous. The surface heating and cooling shown in Fig. 2 are for an annual mean over three years (1998–2000). Owing to the seasonal variations of the insolation at the top of the atmosphere (TOA) and the corresponding shifts in the ITCZ and trade winds, the maximum heating as shown in the lower panel of Fig. 2 does not occur in all seasons.

For a given season, regions of maximum cloudiness (i.e. ITCZ) follow the maximum SST,

minimum winds, and maximum insolation at the TOA. As the season advances, the Sun moves away from this maximum SST region to the neighboring regions, which are less cloudy. The neighboring regions then receive a great deal of solar heating. The SST increases, the convergence zone follows, and the wind speed decreases. The decrease in wind speed causes a reduction in evaporative cooling, and further enhances the SST and convection in the neighboring regions, while the region of maximum surface heating in the previous season experiences a decrease clouds and an increase of winds, which enhances the evaporative cooling. The high SST region in the previous season is then cooled down as the Sun moves away and the strong trade winds set in.

The processes outlined above are demonstrated in Fig. 10, which shows time series of the SST, the TOA insolation (S_o), the surface solar heating (SW), and the evaporative cooling (LH) averaged over the warmest region of the Pacific (5°N–15°N, 130°E–160°E). The insolation reaches a minimum in December and

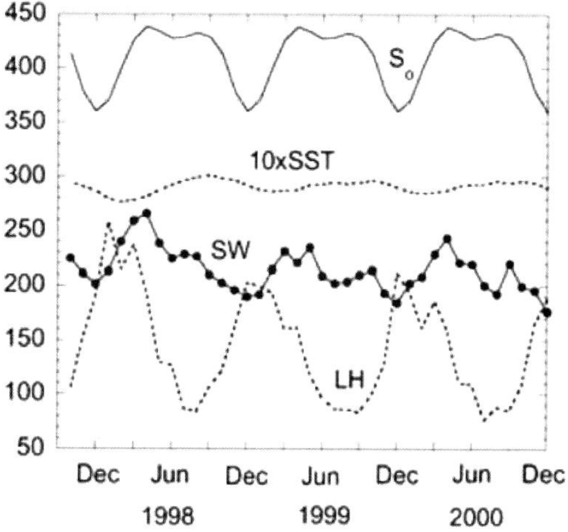

Figure 10. The monthly-mean SST, the insolation at the top of atmosphere, S_o, the surface solar heating, SW, and the evaporative cooling, LH, averaged over the domain 5°N–15°N, 130°E–160°E. Units are $^\circ$C for the SST and W m^{-2} for heat fluxes. (After Chou *et al.*, 2005a.)

increases until April. It remains nearly constant from April through September and decreases afterward as the Sun moves southward. From April to September, the S_o remains nearly constant but the SW decreases, indicating an increase in cloudiness. Simultaneously, the LH also decreases but at a much larger rate than the SW, indicating a greatly reduced wind speed in the convective and high cloudiness region. After August–September, the Sun moves southward and the SW continues to decrease, but the evaporative cooling increases when the strong northeast trade wind sets into this region. The SST follows S_o with a lag of ~2 months.

To demonstrate the seasonal changes over the tropical western Pacific and the tropical eastern Indian Ocean, differences in the SST, the cloud cover, and the surface wind speed between the months of December, January, and February (DJF) and the months of June, July, and August (JJA) for the three years (1987–2000) are shown in Fig. 11. Following the position of the Sun, the SST increases in the summer and decreases in the winter. The ITCZ moves from the northern hemisphere in JJA to the southern hemisphere in DJF. Consistent with the position of the ITCZ, the strong southeast trade wind south of the equator during JJA is replaced by converging weak winds during DJF. Similarly, the weak wind north of the equator during JJA is replaced by the strong northeast trade wind during DJF. Figure 12 shows the responses of the surface heating to the seasonal changes in the SST, clouds, and winds. In spite of a larger cloudiness, the surface solar heating increases from to JJA to DJF in the southern hemisphere owing to an increase in the TOA insolation, whereas the evaporative cooling decreases due to a reduced wind speed. The result is a large increase in the total surface heating by ~100 W m^{-2} at 10°S latitude. This large surface heating and the associated increase in the SST is not sustainable over a season, because the Sun and the ITCZ will soon move away from this maximum heating region.

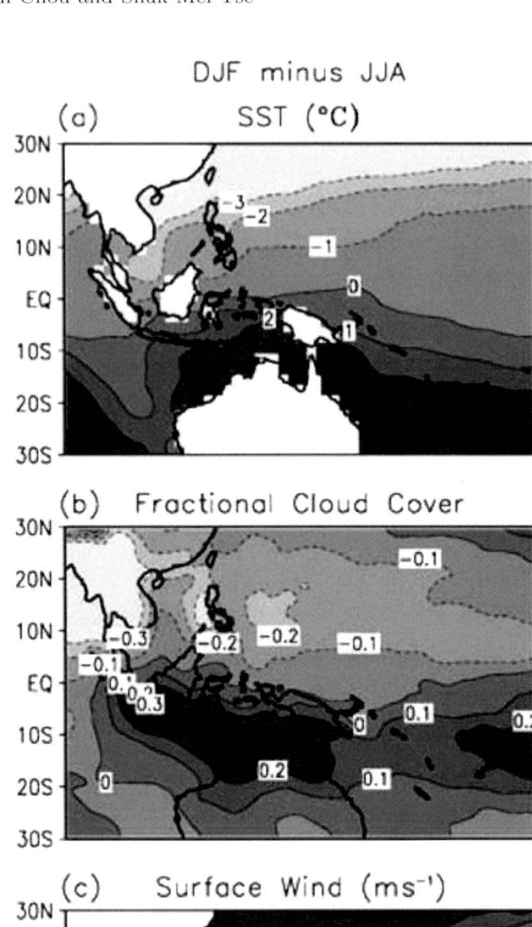

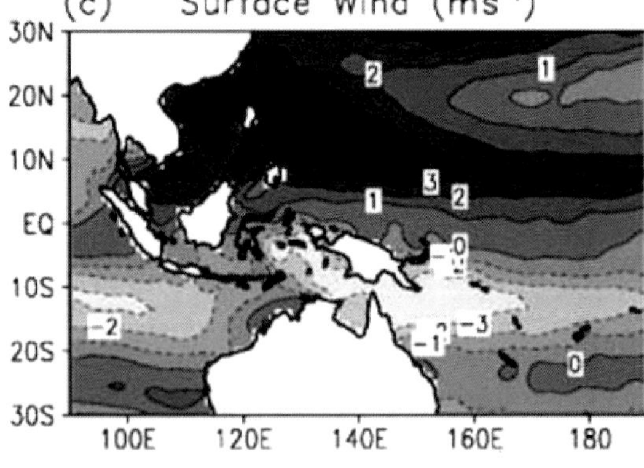

Figure 11. The difference of (a) SST, (b) high level cloud cover, and (c) wind speed between the months of December, January, February (DJF) and the months of June, July, August (JJA). (After Chou *et al.*, 2005a.)

Our study covered the entire tropical western Pacific in the period of 1998–2000. The SST, cloud, and ITCZ follow the seasonal march of the Sun. Within the ITCZ, enhanced surface evaporation might accompany the passage of

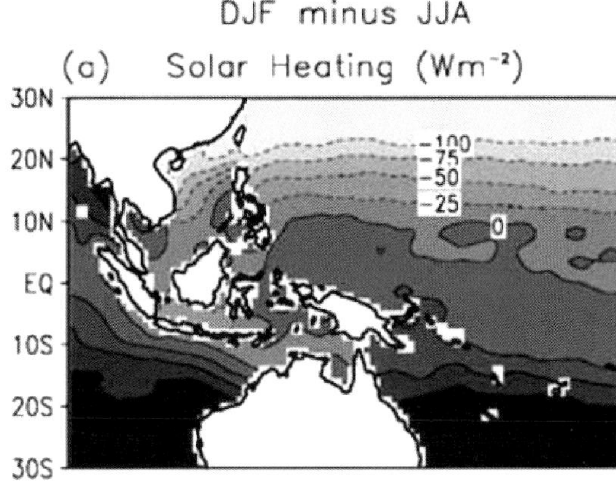

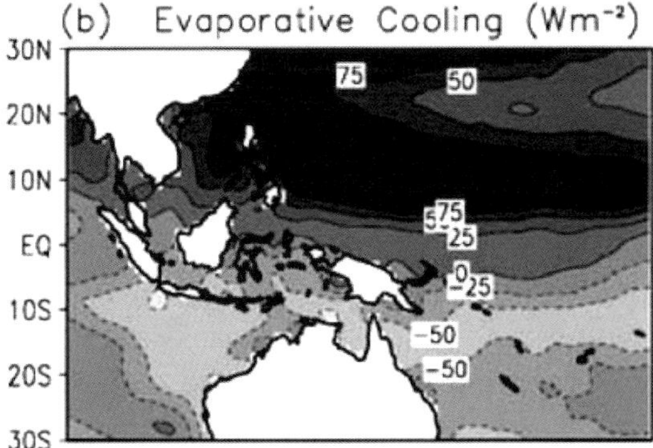

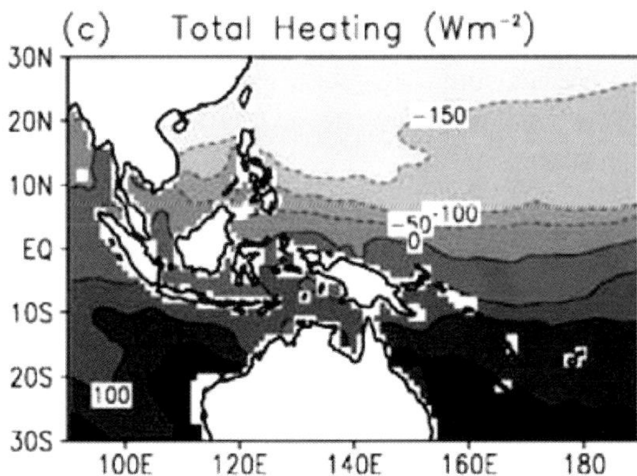

Figure 12. The difference of (a) surface solar heating, (b) evaporative cooling, and (c) total heating between the months of December, January, February (DJF) and the months of June, July, August (JJA). (After Chou *et al.*, 2005a.)

disturbances associated either with Madden–Julian oscillations (MJO's) or with easterly waves. For the ITCZ as a whole, however, both the reduced speed of converging winds and the reduced sea–air humidity difference contribute

to the large reduction of surface evaporative cooling. We can conclude that the seasonal variation of evaporative cooling associated with the seasonal variation of trade and monsoonal winds is one of the major mechanisms that cause the SST of the tropical western Pacific to be homogeneous and limited to the observed high values.

6. Correlation between the SST and the SHB in the Tropical Pacific

Limited by the domain and period of the GMS-retrieved surface radiation data, the work of Chou *et al.* (2004) only covered the tropical eastern Indian Ocean and western Pacific for three years (October 1998–December 2000). The NASA/GEWEX SRB_Rel2 surface radiation data sets cover the entire globe and a much longer period. Using the SRB Rel2 surface radiation, Chou *et al.* (2005b) extended the work of Chou *et al.* (2004) to investigate the correlation between the surface heating and the SST of the entire tropical Pacific for a period of seven years.

The surface solar, latent heat, and net heat fluxes averaged over the tropical Pacific in the seven-year period from January 1988 to December 1994 are shown in Fig. 13. The net surface flux is computed from Eq. (3), with radiative fluxes taken from the SRB_Rel2 and turbulent fluxes taken from the GSSTF2. It can be seen that the maximum solar heating and evaporative cooling is in the subtropical high pressure regions, where the atmosphere is dry and clear, the trade winds are strong, and the air–sea humidity difference is large. The region of minimum evaporative cooling is near the equator, where the winds are converging and weak. On the other hand, the region of the minimum solar heating is just north of the equator, where the mean ITCZ is located. The bottom panel of the figure shows that the maximum net surface heating is located at the equator, where solar heating is strong and

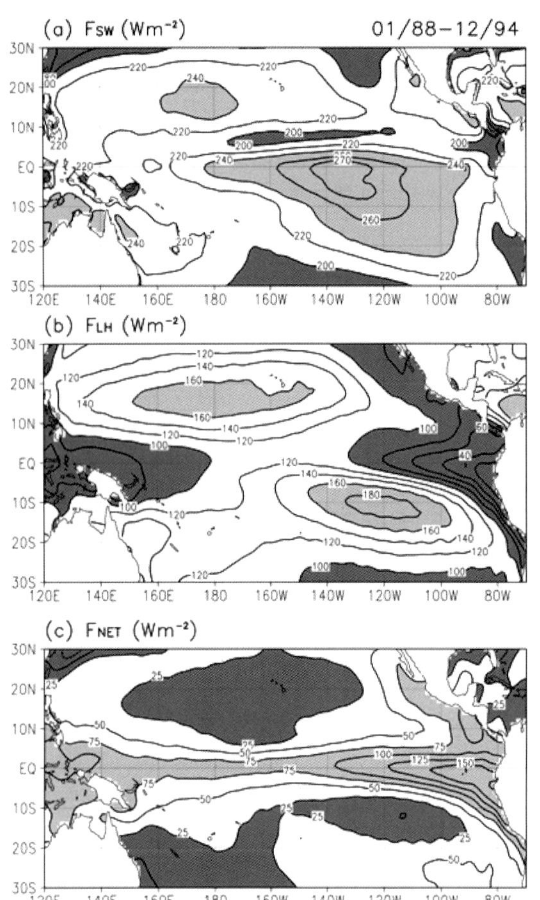

Figure 13. Surface solar heating, evaporative cooling, and net heating, averaged over the seven-year period of January 1988–December 1994. In computing the net heating, turbulent fluxes are taken from GSSTF2 and radiative fluxes are taken from NASA SRB_Rel2.

evaporative cooling is weak. The minimum net surface heating is located at the subtropical high pressure regions, where the strong solar heating and evaporative cooling offset one or the other to a large degree. Over the entire tropical Pacific, the magnitude of the solar heating $(\sim 270\,\mathrm{W\,m^{-2}})$ is significantly larger than that of the evaporative cooling $(\sim 180\,\mathrm{W\,m^{-2}})$, but the spatial variation of the former $(\sim 70\,\mathrm{W\,m^{-2}})$ is smaller than that of the latter $(\sim 130\,\mathrm{W\,m^{-2}})$. As a result, the spatial variation of the net surface heating is predominantly determined by the evaporative cooling.

In the search for the origin of the tropical intraseasonal and interannual oscillations, it is important to understand the correlation between the surface heating and the SST in determining the degree to which the surface heating affects the SST and, conversely, the surface heating responses to the SST. In Fig. 14 we show the time series of monthly F_{net} and SST anomalies (deviation from the eight-year mean annual cycle) over the Niño 3.4 region (5°S–5°N, 120°W–170°W) for the period of July 1987–August 1995. Four groups of data are shown in the figure; two are based on satellite observation and the other two are reanalysis. For the satellite-based data group (GSSTF2/SRB_Rel2, HOAPS-II/SRB_Rel2), the SRB+Rel2 radiative fluxes are used in computing the net surface heating. It appears that the F_{net} anomaly and the negative SST anomaly follow each other rather well for all the four data sets. At the extreme sea surface temperature in January 1989, January 1992, and December 1994, the F_{net} anomalies are ~ 30, -40, and $-25\,\mathrm{W\,m^{-2}}$ for all the four data sets, except for the ERA-40 in January 1992, when the F_{net} anomaly is $-80\,\mathrm{W\,m^{-2}}$.

The lag correlation between the SST and the surface heat fluxes is shown in Fig. 15 for the four data groups. For all the data sets, there do not seem to be any significant lags. The correlation coefficient is a maximum at zero lag and nearly symmetric relative to the zero lag. The correlation is significantly higher for the reanalyses than the correlation for the observation-based data sets. In the reanalyses, the SST is given as an input data for model simulation, and the surface heat fluxes are a response to the specified SST field. It is reasonable to expect a high correlation between the SST and the surface heating for the reanalyses. On the other hand, the observation-based surface heat fluxes are derived from many parameters independently retrieved, and the SST

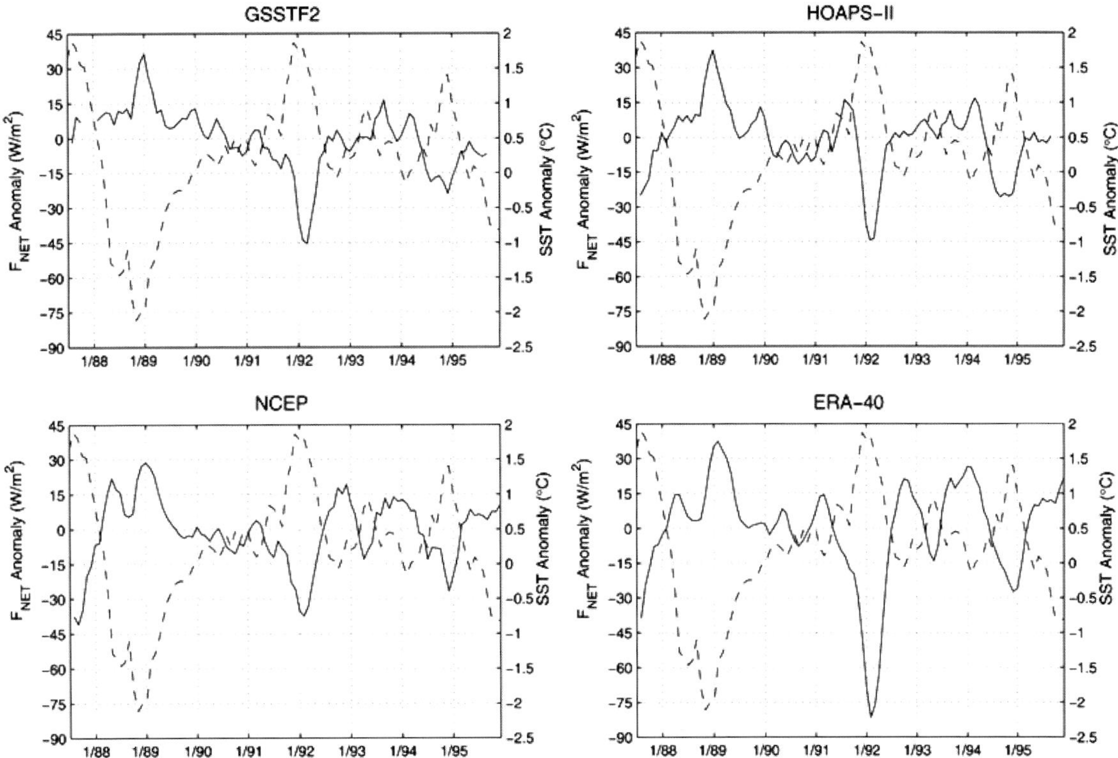

Figure 14. Monthly anomaly of the SST of the Niño-3.4 and net surface heating of the region 5°S–5°N; 120°W–170°W. For the GSSTF2 and HOAPS-II, the SRB_Rel2 radiative fluxes are used in computing the surface net heating.

is only one of those parameters. The correlation between the SST and the surface heating is not necessarily high. This is especially true for the GSSTF2/SRB_Rel2, where the correlation between the SST and F_{LH} is weak. The maximum correlation between the SST and F_{net} at zero lag is only 0.38. If the change in the SST is forced by F_{net}, the correlation should be positive. The negative correlation between the SST and F_{net} implies that a decrease of F_{net} is caused by an increase of the SST. Because of the large heat capacity of the ocean, the SST lags F_{net} by ~2 months if the change of the SST is due to surface heating. The zero lag correlation further implies the fast response of the atmosphere to the change of the SST. As can be seen from Fig. 14, this negative, zero lag correlation is especially clear during El Niño and La Niña events. Thus, the interannual variation of the SST is not caused directly by

surface heating or cooling. It must be caused by the ocean dynamics, either directly through the delayed action oscillator hypothesis of Suarez and Schopf (1988) or through the recharge–discharge mechanism proposed by Jin (1997), or indirectly through the heat pump mechanism suggested by Sun (2000).

7. Summary

Heat fluxes at the sea surface are among the major forces driving the atmospheric and oceanic circulations. During the past 10 years, we have published a series of papers on the derivations of the sea surface radiative and turbulent heat fluxes based on satellite radiance measurements, the intraseasonal and interannual variations of the sea surface heat budget (SHB), the influence of transient atmospheric circulation on the surface heating of the western Pacific

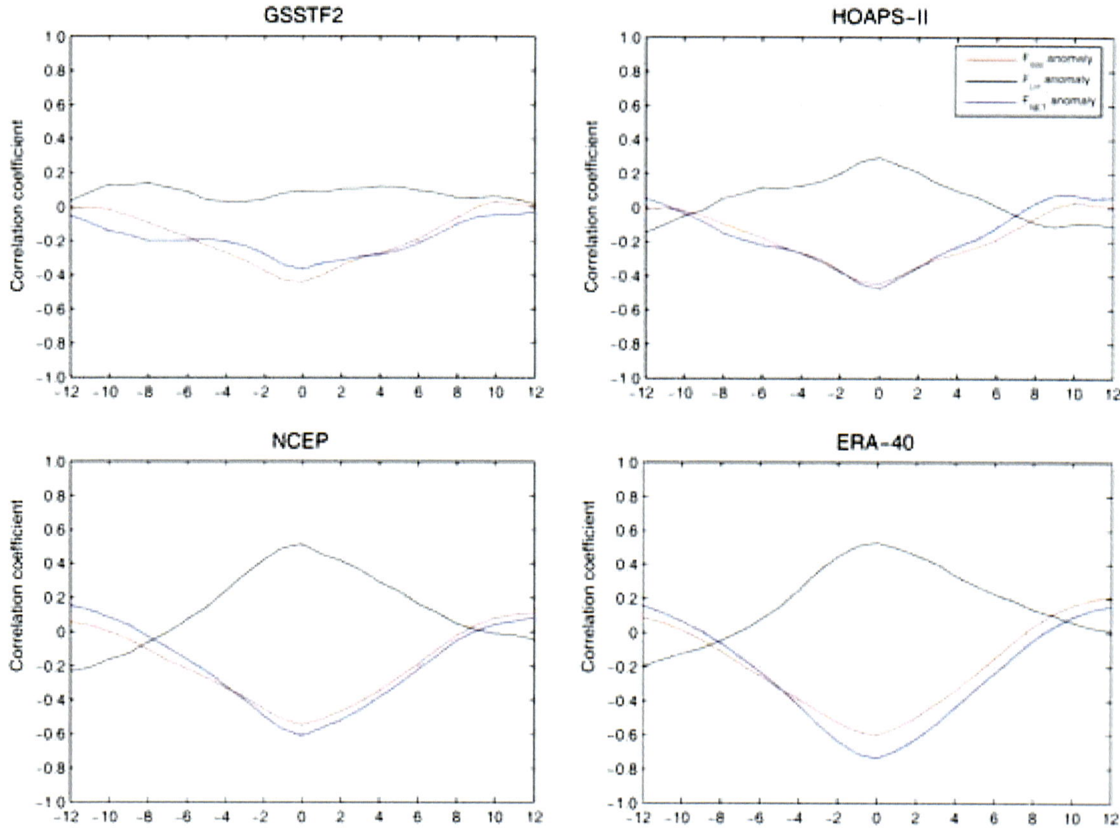

Figure 15. Coefficient of lag correlation between the SST and the surface solar, latent heat, and net heat fluxes. A positive number of months indicates SST leading heat fluxes.

warm pool, and the correlation between the SST and the SHB in the tropical Pacific. We have reviewed those studies in this article.

Monthly variations of the net surface heating (F_{net}) and its relationship to the SST tendency (dT_s/dt) in the equatorial western Pacific warm pool region (10°S–10°N; 135°E–175°E) are studied for the TOGA COARE IOP, November 1992–February 1993, using the satellite-based surface radiative and turbulent heat fluxes. The characteristics of surface heat fluxes are very different between the southern and the northern warm pool. In the southern warm pool during the northern hemispheric winter, F_{net} is dominated by solar radiation, which is, in turn, modulated by the two Madden–Julian oscillations. Winds are generally weak, except during short episodes of westerly wind bursts.

The weak surface winds leads to a shallow ocean mixed layer, and the solar radiation penetrating through the bottom of the mixed layer is significant. Therefore, the temperature tendency dT_s/dt during the four IOP months does not follow F_{net}. In the northern warm pool, the northeasterly trade wind is strong, and undergoes significant monthly (seasonal) variation, and the variation of F_{net} is dominated by evaporation. The two westerly wind bursts associated with the Madden–Julian oscillations have little effect on F_{net} of the northern warm pool. The ocean mixed layer is deep, and the solar radiation penetrating through the bottom of the mixed layer is small. As opposed to the southern warm pool, the dT_s/dt in the northern warm pool during the IOP is consistent with F_{net}.

Seasonal to interannual variations of F_{net} and its relationship to dT_s/dt in the tropical eastern Indian and western Pacific Oceans are studied for the three-year period October 1997–September 2000. The magnitude of solar heating is larger than that of evaporative cooling, but the spatial variation of the latter is significantly larger than that of the former. As a result, the spatial patterns of the seasonal and interannual variability of F_{net} are dominated by the variability of evaporative cooling. Seasonal variations of F_{net} and dT_s/dt are significantly correlated, except for the equatorial western Pacific. This high correlation is augmented by the high negative correlation between solar heating and evaporative cooling.

The change of F_{net} between the 1997/98 El Niño and the 1998/99 La Niña is significantly larger in the tropical eastern Indian Ocean than in the tropical western Pacific. For the former region, convection is reduced during enhanced subsidence of air during the El Niño. The reduced evaporative cooling arising from weakened winds during the El Niño is generally associated with enhanced solar heating due to reduced cloudiness, leading to enhanced interannual variability of F_{net}. For the latter region, reduced evaporative cooling due to weakened winds is generally associated with reduced solar heating arising from increased cloudiness during the El Niño. Consequently, the interannual variability of F_{net} is reduced. The correlation between interannual variations of F_{net} and dT_s/dt is weak in the tropical western Pacific and eastern Indian Oceans, indicating the importance of ocean dynamics in affecting the interannual SST variation.

Various mechanisms that regulate the western Pacific SST have been extensively investigated in many studies. They include the surface SW heating, evaporative cooling, and large scale atmospheric ventilation. In view of the large heat capacity of the ocean mixed layer and the seasonal variation SW radiation, the transient nature of SW radiation and the associated seasonal variation of winds and clouds can be expected to have a significant contribution to the homogeneity of the SST in the Pacific warm pool. Trade and monsoonal winds converge in regions of the highest SST in the equatorial western Pacific. These regions have the largest cloud cover and smallest wind speed. Both surface solar heating and evaporative cooling are weak. Data on surface heat fluxes show that the reduced evaporative cooling due to weakened winds exceeds the reduced solar heating due to enhanced cloudiness. The result is a maximum surface heating in the strong convective and high SST regions, which follow the seasonal march of the Sun. As the Sun moves away from a convective region, the strong trade and monsoonal winds set in, and the evaporative cooling is enhanced, resulting in a net cooling of the surface. The seasonal variation of evaporation associated with the seasonal variation of trade and monsoonal winds is one of the major factors that modulate the SST distribution of the Pacific warm pool.

The heating of the entire tropical Pacific and its correlation with the SST is investigated for the period of 1987–1995 using the radiative and turbulent fluxes derived from satellite observations and reanalyses. The data show that during an El Niño year, when the central and eastern equatorial Pacific is abnormally warm, the surface solar heating decreases owing to enhanced cloudiness, but the surface evaporative cooling increases are due primarily to increases in the air–sea humidity difference. The combined effect is a reduction of surface heating during a warm El Niño episode. On the other hand, during a La Niña year, when the central and eastern equatorial Pacific regions are anomalously cool, the surface solar heating increases and the evaporative cooling decreases. The result is an enhanced net heating of the equatorial Pacific. Thus, the surface heating has a stabilizing effect on the SST; the surface heating is anomalously low during a warm episode and anomalously high during a cool episode.

Taiwan is surrounded by oceans, and the weather and climate are highly influenced by the heat and water exchanges at the air–sea interface. During the winter monsoon seasons, the dry and cold continental air mass emanating from the Asian continent undergoes enormous transformation by receiving a large amount of heat and water vapor from the ocean when passing over the East China Sea. The wintertime weather of northern Taiwan is dominated by this ocean-transformed continental air mass with prevailing northeasterly winds. During the summer monsoon seasons, the southwesterly wind carries humidity and heat from the South China Sea, and largely determines the summertime weather of Taiwan. Although it is commonly acknowledged that the heat and water vapor (evaporation) fluxes at the air–sea interface have a large impact on the weather and climate of Taiwan, detailed studies of these topics are sparse. Finally, Taiwan suffers from severe floods caused by typhoons in summer and autumn, and typhoon forecasting is very important in mitigating the damage by flood. It has been shown by Lin *et al.* (2005) and Pun *et al.* (2007) that the temperature of the upper ocean layer has a large impact on the strength of typhoons, indicating the importance of heat and water vapor fluxes at the sea surface in influencing the strength of typhoons. We can conclude that future research on the air–sea interaction is essential to improving the accuracy of weather forecasts in Taiwan.

Acknowledgements

This work was supported by the National Science Council, Taiwan.

[Received 31 August 2006; Revised 31 July 2007; Accepted 31 July 2007.]

References

Ando, K., and M. J. McPhaden, 1997: Variability of surface layer hydrography in the tropical Pacific Ocean. *J. Geophys. Res.*, **102**, 23063–23078.

Bell, G. D., and coauthors, 1999: Climate assessment for 1998. *Bull. Amer. Meteor. Soc.*, **80**, S1–S48.

Bell, G. D., and coauthors, 2000: Climate assessment for 1999. *Bull. Amer. Meteor. Soc.*, **81**, S1–S50.

Bentamy A., K. B. Katsaros, A. M. Mestas-Nuñez, W. M. Drennan, E. B. Forde, and H. Roquet, 2003: Satellite estimates of wind speed, and latent heat flux over the global oceans. *J. Climate*, **16**, 637–656.

Chou, M.-D., 1994: Radiation budgets in the western Pacific. *J. Climate*, **7**, 1958–1971.

Chou, M.-D., W. Zhao, and S.-H. Chou, 1998: Radiation budgets and cloud radiative forcing in the Pacific warm pool during TOGA COARE. *J. Geophys. Res.*, **103**, 16967–16977.

Chou, M.-D., P.-K. Chan, and M. M.-H. Yan, 2001: A sea surface radiation data set for climate applications in the tropical western Pacific and South China Sea, *J. Geophys. Res.*, **106**, 7219–7228.

Chou, M.-D., S.-H. Chou, and P.-K. Chan, 2005a: Influence of transient atmospheric circulation on the surface heating of the western Pacific warm pool. *Geophys. Res. Lett.*, **32**, L08705, doi: 10.1029/2005GL022359.

Chou, M.-D., S.-H. Chou, and S.-M. Tse, 2005b: Contrast of surface heating between El Niño and La Niña. *Proc., Conference on Weather Analysis and Forecasting*, October 18–20, 2005, Central Weather Bureau, Taipei, Taiwan.

Chou, S.-H., C.-L. Shie, R. M. Atlas, and J. Ardizzone, 1997: Air–sea fluxes retrieved from Special Sensor Microwave Imager data. *J. Geophys. Res.* **102**, 12705–12726.

Chou, S.-H., W. Zhao, and M.-D. Chou, 2000: Surface heat budgets and sea surface temperature in the Pacific warm pool during TOGA COARE, *J. Climate*, **13**, 634–649.

Chou, S.-H., E. Nelkin, J. Ardizzone, R. M. Atlas, and C.-L. Shie, 2003: Surface turbulent heat and momentum fluxes over global oceans based on the Goddard satellite retrievals, version 2 (GSSTF2). *J. Climate*, **16**, 3256–3273.

Chou, S.-H., M.-D. Chou, P.-K. Chan, P.-H. Lin, and K.-H. Wang, 2004: Tropical warm pool surface heat budgets and temperature: contrasts between 1997–98 El Niño and 1998–99 La Niña. *J. Climate*, **17**, 1845–1858.

Curry, J. A., A. Bentamy, M. A. Bourassa, D. Bourras, E. F. Bradley, M. Brunke, S. Castro, S.-H. Chou, C. A. Clayson, W. J. Emery, L. Eymard, C. W. Fairall, M. Kubota, B. Lin, W. Perrie, R. A. Reeder, I. A. Renfrew, W. B. Rossow, J. Schulz, S. R. Smith, P. J. Webster, G. A. Wick, and X. Zeng, 2004: SEAFLUX. *Bull. Amer. Meteor. Soc.*, **85**, 409–424.

Enfield, D. B., 1986: Zonal and seasonal variations of the near-surface heat balance of the equatorial Pacific Ocean. *J. Phys. Oceanogr.*, **16**, 1038–1054.

Fairall, C. W., E. F. Bradley, D. P. Rogers, J. B. Edson, G. S. Young, 1996: Bulk parameterization of air–sea fluxes for tropical ocean–global atmosphere coupled-ocean atmosphere response experiment. *J. Geophys. Res.*, **101**, 3747–3764.

Feng, M., P. Hacker, and R. Lukas, 1998: Upper ocean heat and salt balances in response to a westerly wind burst in the western equatorial Pacific during TOGA COARE. *J. Geophys. Res.*, **103**, 10289–10311.

Fu, Q., K. N. Liou, M. C. Cribb, T. P. Charlock, and A. Grossman, 1992: Multiple scattering parameterization in thermal infrared radiative transfer. *J. Atmos. Sci.*, **54**, 2799–2812.

Godfrey, J. S., M. Nunez, E. F. Bradley, P. A. Coppin, and E. J. Lindstrom, 1991: On the net surface heat flux into the western equatorial Pacific. *J. Geophys. Res.*, **96**, 3343–3357.

Godfrey, J. S., R. A. House Jr., R. H. Johnson, R. Lukas, J.-L. Redelsperger, A. Sumi, and R. Weller, 1998: Coupled Ocean–Atmosphere Response Experiment (COARE): An interim report. *J. Geophys. Res.*, **103**, 14395–14450.

Grassl H., V. Jost, R. Kumar, J. Schulz, P. Bauer and P. Schluessel, 2000: The Hamburg Ocean–Atmosphere Parameters and Fluxes from Satellite Data (HOAPS): a climatological atlas of satellite-derived air–sea-interaction parameters over the oceans. *Max Planck Institute for Meteorology*, Rep. **312**, Hamburg, 130 pp.

Gruber, A., and J. S. Winston, 1978: Earth–atmosphere radiative heating based on NOAA scanning radiometer measurements. *Bull. Amer. Meteorol. Soc.*, **59**, 1570–1573.

Gutzler, D., G. N. Kiladis, G. A. Meehl, K. M. Weickman, and M. Wheeler, 1994: The global climate of December 1992–February 1993. Part II: Large-scale variability across the tropical western Pacific during TOGA COARE. *J. Climate*, **7**, 1606–1622.

Hartmann, D. L., and M. L. Michelsen, 1993: Large-scale effects on the regulation of tropical sea surface temperature. *J. Climate*, **6**, 2049–2062.

Jerlov, N. G., 1968: *Optical Oceanography*. Elsevier, 194 pp.

Jin, F.-F., 1997: An equatorial ocean recharge paradigm for ENSO. Part I: Conceptual model. *J. Atmos. Sci.*, **54**, 811–829.

Kalnay, E., and coauthors, 1996: The NCEP/NCAR 40-year reanalysis project. *Bull. Amer. Meteorol. Soc.*, **77**, 437–471.

Kiehl, J. T., and K. E. Trenberth, 1997: Earth's annual global mean energy budget. *Bull. Amer. Met. Soc.*, **78**, 197–208.

Kilpatrick, K. A., G. P. Podestá, and R. Evans, 2001: Overview of the NOAA/NASA advanced very high resolution radiometer Pathfinder algorithm for sea surface temperature and associated matchup database. *J. Geophys. Res.*, **106**, 9179–9197.

Kubota, M., N. Iwasaka, S. Kizu, M. Konda, and K. Kutsuwada, 2002: Japanese ocean flux data sets with use of remote sensing observations (J-OFURO). *J. Oceanog.*, **58**, 213–225.

Lau, K.-M., and P. H. Chan, 1983: Short term climate variability and atmospheric teleconnection from satellite observed outgoing longwave radiation. Part II: Lagged correlations. *J. Atmos. Sci.*, **40**, 2751–2767.

Lau, K.-M., and C.-H. Sui, 1997: Mechanism of short-term sea surface temperature regulation: observations during TOGA COARE. *J. Climate*, **10**, 465–472.

Lau, K.-M., C. H. Sui, M.-D. Chou, and W. K. Tao, 1994: An inquiry into the cirrus-cloud thermostat effect for tropical sea surface temperature, *Geophys. Res. Lett.*, **21**, 1157–1160.

Lewis, M. R., M.-E. Carr, G. C. Feldman, W. Esaias, and C. McClain, 1990: Influence of penetrating solar radiation on the heat budget of the equatorial Pacific Ocean. *Nature*, **347**, 543–545.

Li, Z., and H. G. Leighton, 1993: Estimation of SW flux absorbed at the surface from TOA reflected flux. *J. Climate*, **6**, 317–330.

Lin, I.-I., C.-C. Wu, K. A. Emanuel, I.-H. Lee, C. Wu, and F. Pan, 2005: The Interaction of super typhoon Maemi (2003) with a warm ocean eddy, *Mon. Wea. Rev.*, **133**, 2635–2649.

Loschnigg, J., and P. Webster, 2000: A coupled ocean–atmosphere system of SST modulation for the Indian Ocean. *J. Climate*, **13**, 3342–3360.

Lukas, R., and E. Lindstrom, 1991: The mixed layer of the western equatorial Pacific Ocean. *J. Geophys. Res.*, **96** (Suppl.), 3343–3357.

Mather, J. H., T. P. Ackerman, W. E. Clements, F. J. Barnes, M. D. Ivey, L. D. Hatfield, and R. M. Reynolds, 1998: An atmospheric radiation and cloud station in the tropical western Pacific, *Bull. Amer. Meteorol. Soc.*, **79**, 627–642.

Niiler, P. P., and J. Stevenson, 1982: The heat budget of tropical warm pools. *J. Mar. Res.*, **40** (Suppl.), 465–480.

Palmer, T. N., and D. A. Mansfield, 1984: Response of two atmospheric general circulation models to sea-surface temperature anomalies in the tropical east and west Pacific. *Nature*, **310**, 483–485.

Paulson, C. A., and J. J. Simpson, 1977: Irradiance measurements in the upper ocean. *J. Phys. Oceanog.*, **7**, 1722–1735.

Pierrehumbert, R. T., 1995: Thermostats, radiator fins, and the local runaway greenhouse, *J. Atmos. Sci.*, **52**,1784–1806.

Picaut, J., E. Hackert, A. J. Busalacchi, R. Murtgudde, and G. S. E. Lagerloef, 2002: Mechanisms of the 1997–1998 El Niño–La Niña as inferred from space-based observations. *J. Geophys. Res.*, **107**, 3037, doi: 10.1029/2001JC000850.

Pinker, R. T., and I. Laszlo, 1992: Modeling surface solar irradiance for satellite applications on a global scale. *J. Appl. Metrorol.*, **31**, 194–211.

Pun, I.-F., I.-I. Lin, C.-R. Wu, D.-S. Ko, and W. Liu, 2007: Validation and application of altimetry-derived upper ocean thermal structure in the western North Pacific Ocean for typhoon intensity forecast, *IEEE Trans. Geosci. Rem. Sens.*, **45**, 1616–1630.

Ramanathan, V., and W. Collins, 1991: Thermodynamic regulation of ocean warming by cirrus clouds deduced from observations of the 1987 El Niño. *Nature*, **351**, 27–32.

Reynolds, R. W., and T. M. Smith, 1994: Improved global sea surface temperature analyses, *J. Climate,* **7**, 929–948.

Rossow, W. B., and R. A. Schiffer, 1999: Advances in understanding clouds from ISCCP. *Bull. Amer. Meteor. Soc.*, **64**, 779–784.

Siegel, D. A., J. C. Ohlmann, L. Washburn, R. R. Bidigare, C. T. Nosse, E. Fields, and Y. Zhou, 1995: Solar radiation, phytoplankton pigments and the radiant heating of the equatorial Pacific warm pool. *J. Geophys. Res.*, **100**, 4885–4891.

Sprintall, J., and M. Tomzak, 1992: Evidence of the barrier layer in the surface layer of the Tropics. *J. Geophys. Res.*, **97**, 7305–7316.

Suarez, M., J., and P. Schopf, 1988: A delayed action oscillator for ENSO. *J. Atmos. Sci.*, **45**, 3283–3287.

Sud, Y. C., G. K. Walker, and K.-M. Lau, 1999: Mechanisms regulating sea-surface temperatures and deep convection in the tropics, *Geophys. Res. Lett.*, **26**, 1019–1022.

Sui, C.-H., and K.-M. Lau, 1992: Multiscale phenomena in the tropical atmosphere over the western Pacific. *Mon. Wea. Rev.*, **120**, 407–430.

Sui, C.-H., X. Li, K.-M. Lau, and D. Adamac, 1997: Mutiscale air–sea interactions during TOGA COARE. *Mon. Wea. Rev.*, **125**, 448–462.

Sun, D.-Z., 2000: The heat sources and sinks of the 1986–1987 El Niño. *J. Climate*, **13**, 3533–3550.

Sun, D.-Z., and Z. Liu, 1996: Dynamic ocean–atmosphere coupling: a thermostat for the tropics. *Science*, **272**, 1148–1150.

Waliser, D. E., and N. E. Graham, 1993: Convective cloud systems and warm-pool sea surface temperatures: coupled interactions and self-regulation. *J. Geophys. Res.*, **98**, 12881–12893.

Wallace, J. M., 1992: Effect of deep convection on the regulation of tropical sea surface temperature. *Nature*, **357**, 230–231.

Wang, W., and M. J. McPhaden, 2001: Surface layer temperature balance in the equatorial Pacific during the 1997–98 El Niño and 1998–99 La Niña. *J. Climate*, **14**, 3393–3407.

Webster, P. J., and R. Lukas, 1992: TOGA COARE: the TOGA coupled ocean–atmosphere response experiment. *Bull. Amer. Meteor. Soc.*, **73**, 1377–1416.

Webster, P. J., C. A. Clayson, and J. A. Curry, 1996: Clouds, radiation, and the diurnal cycle of sea surface temperature in the tropical western Pacific. *J. Climate*, **9**, 1712–1730.

Webster, P. J., T. Palmer, M. Yanai, V. Magana, J. Shukla, R. A. Toma, and A. Yasunari, 1998: Monsoons: processes, predictability and the prospects for prediction. *J. Geophys. Res.*, **103**, (C7), 14451–14510.

Wentz, F. J., 1997: A well calibrated ocean algorithm for SSM/I. *J. Geophys. Res.*, **102**, 8703–8718.

Yu, L., and M. Rienecker, 2000: Indian Ocean warming of 1997–98. *J. Geophys. Res.*, **105**, 16923–16939.

Zhang, Y.-C., W. B. Rossow, and A. A. Lacis, 1995: Calculation of surface and top of atmosphere radiative fluxes from physical quantities based on ISCCP data sets. 1: Method and sensitivity to input data uncertainties. *J. Geophys. Res.*, **100**, 1149–1165.

Some Unsolved Problems in Atmospheric Radiative Transfer: Implication for Climate Research in the Asia–Pacific Region

K. N. Liou, Y. Gu and W. L. Lee

Department of Atmospheric and Oceanic Sciences and
Joint Institute for Regional Earth System Science and Engineering,
University of California, Los Angeles, California, USA
knliou@atmos.ucla.edu

Y. Chen

Center for Satellite Applications and Research, NESDIS,
NOAA, Camp Springs, Maryland, USA

P. Yang

Department of Atmospheric Sciences, Texas A&M University, College Station, Texas, USA

A number of unsolved problems in atmospheric radiative transfer are presented, including the light scattering and absorption by aerosols, the effect of mountains on radiation fields, and radiative transfer in the atmosphere–ocean system, with a specific application to the Asia–Pacific region. We discuss the issues of two nonspherical and inhomogeneous aerosol types, dust and black carbon, regarding climate radiative forcings. Reduction in uncertainties of their radiative forcings must begin with improvement of the knowledge and understanding of the fundamental scattering and absorption properties associated with their composition and size/shape. The effects of intensive topography, such as that in the Tibetan Plateau, on surface radiation and the radiation field above, are significant and the solution requires a three-dimensional radiative transfer program. We show that in a clear atmosphere, surface net solar flux averaged over a mesoscale domain can differ by 5–20 W/m^2 between a slope and a flat surface. Accurate calculations and parametrizations of both solar and thermal infrared radiative transfer in mountains must be developed for effective incorporation in regional and global models. The wind-driven air–sea interface is complex and affects the transfer of solar flux from the atmosphere into the ocean and the heating in the ocean mixed layer. We point out the requirement of reliable and efficient parametrizations of the ocean surface roughness associated with surface winds. The scattering and absorption properties of irregular phytoplankton and other species in the ocean must also be determined on the basis of the rigorous theoretical and experimental approaches for application to remote sensing and climate research.

1. Introduction

Radiative transfer is a subject of study in a variety of fields, including astrophysics, applied physics, optics, planetary sciences, atmospheric sciences, meteorology, and various engineering disciplines. Prior to 1950, radiative transfer was studied principally by astrophysicists, although it was also an important research area in nuclear engineering and applied physics associated with neutron transport. In his groundbreaking book, Chandrasekhar (1950) presented the subject of radiative transfer in plane-parallel (one-dimensional) atmospheres as a branch of mathematical physics and developed numerous solution methods and techniques. The field of atmospheric radiation, which evolved from the study of radiative transfer, is now concerned with the study, understanding, and quantitative analysis of the interactions of solar and terrestrial radiation with molecules, aerosols, and cloud particles in planetary atmospheres as well as the surface on the basis of the theories of radiative transfer and radiometric observations made from the ground, the air, and space (Liou, 1980, 2002). A fundamental understanding of radiative transfer processes is the key to understanding the atmospheric greenhouse effects and global warming which results from external radiative perturbations of the greenhouse gases and air pollution, and to the development of methods to infer atmospheric and surface parameters through remote sensing. In this article, we present three basic and largely unsolved radiative transfer problems in the climate and weather research with an orientation toward the Asia–Pacific region.

The contemporary scientific issues associated with anthopogenic aerosols in climate change and global warming are profound and their impacts on the environment and human health are of great concern to scientists and the lay public alike. Aerosols are globally distributed in time and space, and affect the Earth's radiation budget by the scattering and absorption of the incoming solar radiation and the emitted thermal infrared radiation in the atmosphere, referred to as the aerosol direct effect, and by modifying the microphysical and radiative properties and lifetime of clouds, referred to as the aerosol indirect effect. The radiative forcings of aerosols both direct and indirect remain the most uncertain elements in the assessment of climate and climate change (IPCC, 2001), owing to our limited knowledge of their chemical and microphysical properties that directly affect the absorption and scattering

of radiation, as well as their intricate indirect effects on cloud formation and the consequence of radiation perturbation. We have provided this review of two key aerosol types, dust and black carbon, in terms of their impact on radiative transfer in order to expand our knowledge and understanding of their fundamental scattering and absorption properties in association with satellite remote sensing and application to climate modeling and research.

In addition to aerosols, the transfer of both solar and thermal infrared radiation over mountainous regions complicates the study of radiative transfer. In particular, the temporal and spatial distribution of surface solar radiation over intensive topography results from complex interactions among the incoming direct solar beam, the atmosphere, and the surface. This distribution in turn determines the dynamics of many landscape processes, such as surface heating and moistening, evapotranspiration, photosynthesis, and snow melting. Accurate calculation of radiative transfer is crucial for many fields, including climatology, ecology, and hydrology. Significant progress has been made on land–atmosphere interactions involving vegetation; however, the effect of terrain inhomogeneity on the radiation fields of the surface and the atmosphere above has been largely neglected in weather and climate models. Many landscapes exhibit significant terrain features; for example, the Tibetan Plateau profoundly influences the general circulation of the atmosphere owing to its uplift of large scale flow patterns and its insertion of heating with a confining lower boundary at high levels.

The final issue is one of quantifying the energy transfer across the sea surface, which is essential to understanding the general circulation of the ocean. Solar radiation from the sun contributes most of the heat fluxes that penetrate the air–sea interface and are subsequently absorbed throughout the ocean mixed layer. Thus, solar radiative transfer differs from other air–sea interaction processes, such as wind stress, evaporation, precipitation, and sensible cooling, which occur only at the sea surface. Radiative transfer in the atmosphere and the ocean requires a specific treatment of the interface between air and water at which the state of the sea surface is largely controlled by surface winds. The scattering and absorption of particulates in the ocean associated with the transfer of solar radiation and the heating in the ocean mixed layer require further research from the perspectives of rigorous electromagnetic scattering theory as well as controlled laboratory experiments.

This paper is organized following the preceding three challenges of radiative transfer and we have made a specific effort to apply these radiation problems as they relate to East Asia. A summary is given in the last section.

2. Light Scattering and Absorption by Aerosols

The importance of the scattering and absorption properties of aerosols for the radiation field and the consequence of their climate impact have been briefly introduced. In this section we first illustrate the climatic radiation effect of aerosols over the East Asia region by employing a general circulation model (GCM) experiment. This is followed by a review of the composition, particle shape, and size of two types of aerosols, dust and black carbon (BC), which are particularly important in climate radiative forcing analysis.

In what follows, we present results of a climate numerical experiment using the UCLA GCM developed by Gu *et al.* (2003, 2006) to illustrate the importance of aerosol optical depth for precipitation simulation in the East Asia region. The yearly and monthly mean aerosol optical depths at the wavelength of $0.75\,\mu$m over China, determined from the data involving the daily direct solar radiation, sunshine duration, surface pressure, and vapor pressure from 1961 to 1990 (Luo *et al.*, 2001), were used to investigate the effects of the observed aerosols in China on climatic temperature and precipitation distributions. The larger aerosol optical depths are found in southern China, with a maximum value of about 0.7. In light of the available satellite data for aerosol-optical-depth climatology and for there to be compatibility and consistency with the long-term surface observations in China, we employed a background aerosol optical depth of 0.2 for the areas outside of China in the GCM simulations to focus our analysis on the effect of localized aerosols on regional climate patterns.

The effect of the aerosol optical depth in China on surface temperature and precipitation patterns was examined by comparing the two GCM experiments — the only difference is that the observed aerosol optical depths in China were used in one experiment and not the other. Since the differences between the two experiments are mainly located in the summer hemisphere associated with the position of the sun, results are presented in terms of the July means for precipitation and outgoing long-wave radiation (OLR) patterns, as shown in Fig. 1. Cooling in the mid-latitudes strengthens the meridional circulation, resulting in increased precipitation in southern China, the Arabian Sea, and the Bay of Bengal, as well as India and Myanmar. To compensate for these increases, a broad band of decreased precipitation is located to the south of the increased precipitation region, with a small decrease to the north. Increase in regional precipitation, especially in the southern part of China and over areas of Indian where abundant moisture is available in summer, appears to be related to the increased aerosol optical depth that occurred in China. Positive differences in OLR are located at higher latitudes, while decreases are found to extend from North Africa to southern China, where precipitation is enhanced, revealing increased convection in this region. In a previous study, Menon *et al.* (2002) suggested that increase in the precipitation in southern China could be related to

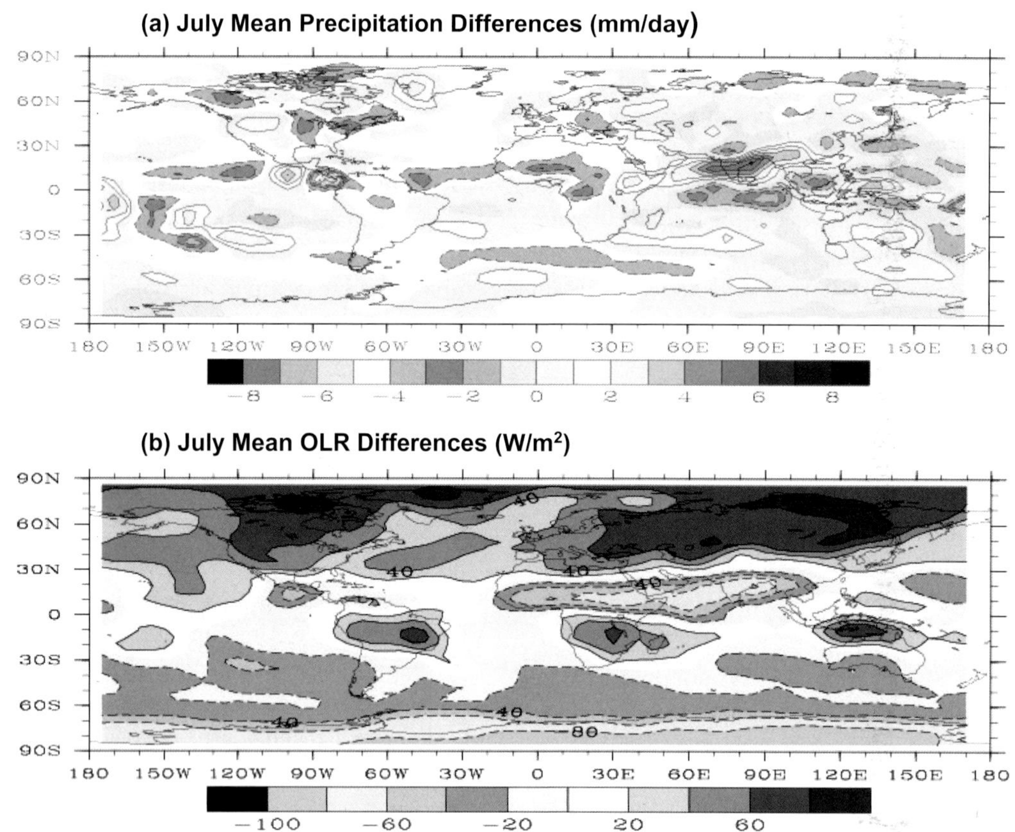

Figure 1. Effect of the aerosol optical depth in China on simulations of the July OLR and precipitation difference patterns determined from the two GCM experiments (denoted as CHIN and AERO in the figure) in which the only difference is that the observed aerosol optical depths in China were used in CHIN (after Gu *et al.*, 2006, with modification).

increase in the absorbing aerosols, a hypothesis which differs from our current finding. This difference illustrates the significance of using the aerosol optical depth (extinction) and single-scattering albedo (absorption) in GCM climate simulations.

Understanding the direct radiative forcing of aerosols requires fundamental knowledge and reliable data on their scattering and absorption properties. The single-scattering properties of aerosols are determined by their dielectric characteristics (refractive index) associated with chemical composition and morphologies in terms of external geometric shape and composition. Aerosol shapes are diverse and span from quasisphere to highly irregular geometry. In addition to complex particle geometry, the inhomogeneous composition of aerosols has been commonly observed. Dust and carbonaceous (particularly BC) particles are two major types of aerosols that profoundly affect climate.

(a) *Dust particles.* Dust can scatter and absorb solar and terrestrial radiation, leading to the positive and negative effects of aerosol direct forcing. Moreover, the solar and infrared radiative forcings of dust may be of different signs, thus further complicating the total forcing assessment. Determination of the sign and magnitude of direct radiative forcing by dust on regional and global scales remains a key unsolved problem (e.g. Sokolik *et al.*, 2001).

With respect to the chemistry and composition of dust particles, Falkovich *et al.* (2001) conducted chemical and mineralogical analysis of dust particles collected in a dust storm over Israel and found that dust particles typically occur in aggregated form with varying mineralogical composition, consisting mostly of Ca, Mg, Al, and Si. Analysis also showed that sulfur (S) and iron (Fe) adhered to the particle's surface, which was suggested to have occurred from processes within the source region. Gao *et al.* (2001) studied the characteristics of soil-based aerosols originating in China and the effects of global transport on the composition and reported that the mineralogy of the loess soil primarily consists of quartz (SiO_2), feldspars, micas, clays as well as carbonates ($CaCO_3$ in particular) and other trace minerals. Mineralogy of dust particles is a function of geographical location. Near the source region, dust assumes the composition of its parent soil; however, when dust is transported over large distances, the composition will change depending on the direction of the dust plume. For example, airborne dust passing through an urbanized region may mix with local pollutants such as black carbon or it may mix with sodium chloride crystals as it gets transported over the oceans inside the marine boundary layer. Mineral dust rarely exists as a pure mineral, but rather, owing to internal and external mixing, as a complex aggregate of several minerals. Sokolik *et al.* (1998) extensively studied the optical properties of a wide range of dust mixtures. Major minerals investigated thus far include quartz, common clays, (montmorillinite, illite, and kaolinite, to name a few), and the carbonates dolomite and hematite, the latter is responsible for the redness sometimes seen in visible observations. Large uncertainties exist as to how to best represent these minerals in an optical model for radiative flux calculations and remote sensing applications.

Characterizing the shape of a dust particle is a difficult and challenging process. Through laboratory analysis and *in situ* measurements using the scanning electron microscope, 2D particle shape can be inferred by using one of several shape parameters. One commonly used parameter is circularity, which is defined as $C = L^2/4\pi A$, where L is the particle's perimeter (μm) and A is the 2D projection area of the particle (μm^2). $C = 1$ refers to a circular particle, and a square has a C equal to 1.27. Dust partcles typically have C values ranging from 1 to as high as 4 or 5. Based on histograms from the dust samples collected in China, the bulk of C values range from 1 to 2 (Gao *et al.*, 2001). Recent individual analysis data on Saharan dust particles show that as particle size increases, as is the case during heavy dust outbreaks, so too does its circularity, indicating that dust is rarely spherical. Most dust particles are highly irregular and angular, possessing sharp corners (Koren *et al.*, 2001; Gao *et al.*, 2001). They also tend to be elongated in a direction perpendicular to their rotational axis; that is, they are oblate. Typical aspect ratios for dust particles observed during studies in China (Okada *et al.*, 2001) and the PRIDE field experiment (Reid *et al.*, 2003) are in the range of 1.4–1.9. Previous studies have made many assumptions regarding the shape of dust particles. Dust is commonly composed of quartz crystals that are typically tetrahedral in shape. Particles composed of clays have been observed to exhibit flat, platelike structures. The scattered intensity of irregular aerosols could be a factor of 1.5 larger and a factor of 2 smaller at forward scattering and backscattering angles, respectively, than those of volume-equivalent spheres (Kalashnikova and Sokolik, 2002).

Many studies have been dedicated to quantifying the size distribution of dust particles from varied global locations over the years, but data interpretation and problems with sizing techniques have large uncertainties (Reid *et al.*, 2003). d'Almeida *et al.* (1991) characterized the size spectrum assuming a lognormal distribution, which can often be bimodal. Some typical size distributions cited in contemporary works on dust research also suggested a gamma distribution with an effective radius of about $2\,\mu$m.

Satellite remote sensing of the optical depth and size of dust particles can be used to constrain GCM simulations concerning the impact of dust outbreaks on climate and climate change. As discussed above, dust particles are exclusively nonspherical. However, it has been a common practice to use the phase function computed from the Lorenz–Mie theory for "equivalence" spheres in retrieval and analysis. On the basis of the scattering results for spheroids, Mishchenko *et al.* (1995) illustrated that the spherical equivalence assumption can result in substantial errors in the aerosol optical depth retrieved from satellite observations. Dubovik *et al.* (2006) further showed that this assumption can lead to unrealistic retrievals in aerosol size distribution and the spectral dependence of the refractive index inferred from ground-based AERONET measurements. It appears that the significance of aerosol nonsphericity in remote sensing and climate study has been increasingly recognized.

In the following, we present an example of the importance of aerosol nonsphericity for satellite remote sensing and radiative forcing analysis. Figure 2 displays the phase functions at a wavelength of $0.63\,\mu$m and with a size parameter of 10 for three types of aerosol particles — a randomly oriented dustlike particle with 10 surface faces, a spheroid, and a sphere — computed with the finite-difference time domain (FDTD) approach (Yang and Liou, 2000), the T-matrix method (Mishchenko *et al.*, 1996), and the Lorenz–Mie theory, respectively. The two top panels are for oceanic particles at the two satellite sensing wavelengths of 0.63 and $0.86\,\mu$m, while the bottom panels are for mineral and soot particles using a wavelength of $0.63\,\mu$m. As shown, significant differences are evident in the backscattering directions where satellite radiometers see the atmosphere, except in the case of soot with large absorption, denoted by a significant value in the imaginary refractive index. These differences can be as large as a factor of 10 between dustlike and sphere at most of the scattering angles greater than $120°$. If the retrieval uncertainty for optical depth, τ, is 50%, defined as $[\tau(\text{dust}) - \tau(\text{sphere})]/\tau(\text{sphere})$, the aerosol solar radiative forcings at the top of the atmosphere (TOA) and the surface differ by 8 and $12\,\text{W/m}^2$, respectively, based on calculations from the Fu and Liou (1992, 1993) radiative transfer model using the standard atmospheric profiles for water vapor and other gases along with a solar constant of $1366\,\text{W/m}^2$, a surface albedo of 0.1, and a solar zenith angle of $60°$. A retrieval uncertainty factor of 2 yields radiative forcing differences of 15 and $23\,\text{W/m}^2$ at TOA and the surface, respectively. These are hypothetical large numbers for illustration purposes but nevertheless a physically reliable remote sensing of aerosol optical depth must account for the aerosol nonsphericity, particularly in the case of dust.

(b) *Carbonaceous particles.* Carbonaceous particles in the atmosphere, which are of specific interest to the climate and climate change study, strongly absorb sunlight, heat the air, and contribute to global warming (Menon *et al.*, 2002). Recent investigations indicated that the magnitude of the direct radiative forcing due to BC exceeds that from CH_4 and may be the second most important component of global warming, after CO_2, in terms of direct forcing (Jacobson, 2001). In addition to the direct climate effect, BC particles, acting as CCN, can inhibit cloud formation (Koren *et al.*, 2004). The magnitude of BC absorption depends highly on its refractive index (especially its imaginary part), shape, and size distribution. It also depends on the mixing state of BC particles.

Carbonaceous compounds make up a large but highly variable fraction of the atmospheric aerosols. Carbonaceous materials are by-products of solid, liquid, or vapor combustion,

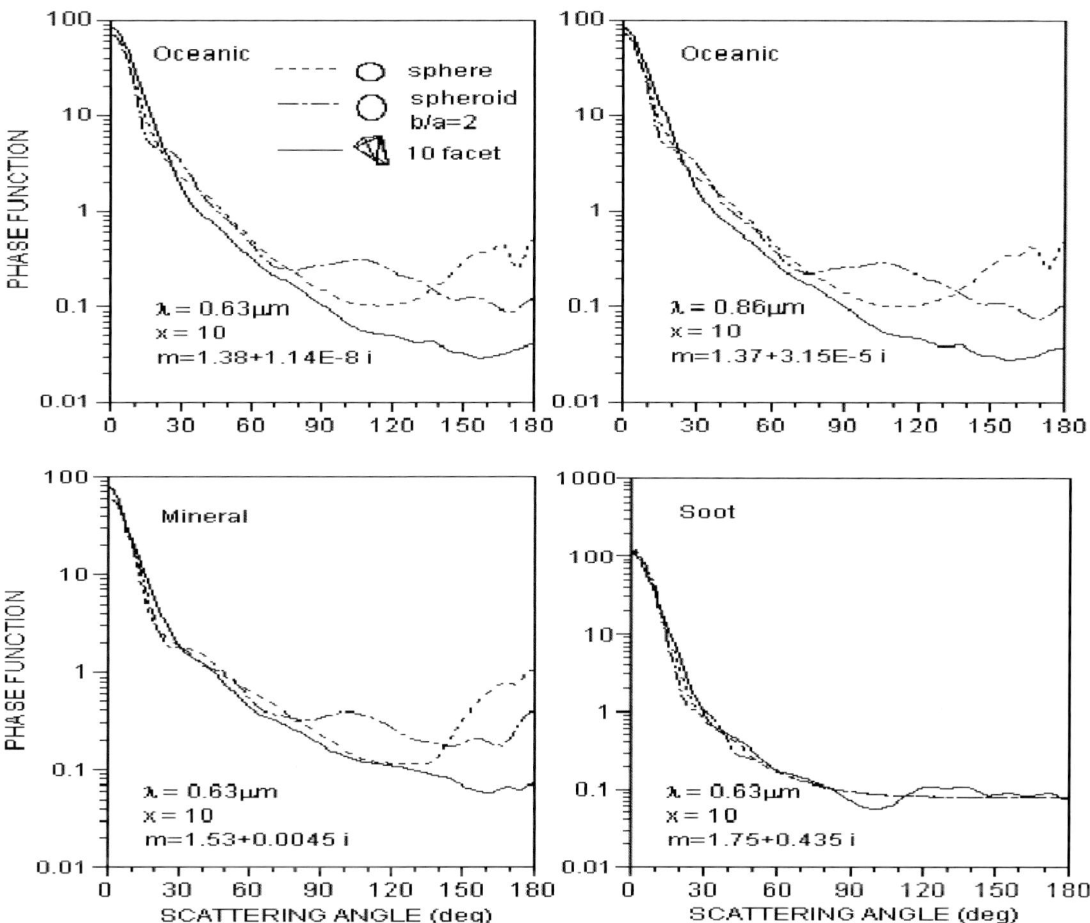

Figure 2. Phase functions at a wavelength of $0.63\,\mu$m and with a size parameter of 10 for three types of aerosol particles: a randomly oriented dustlike particle with 10 surface faces, a spheroid, and a sphere computed with the finite-difference time domain (FDTD) approach, the T-matrix method, and the Lorenz–Mie theory, respectively.

generated either directly through aggregation of molecules formed in combustion processes from coal, biomass burning, and biofuel, leading to soot particles (primary carbon), or indirectly through condensation from a supersaturated gas produced by chemical reactions and resulting in secondary organics (secondary carbon). The by-products of fossil fuel burning are injected into the atmosphere from both stationary sources, such as factories and power plants, and mobile sources, such as motor vehicles. Aerosol particles produced by fossil fuel burning are concentrated over North America, Europe, China, Japan, India, and other industrialized regions. The

primary aerosol particles released from fossil fuel burning generally fall into two categories: soot and fly ash. Soot includes both elemental (BC) and organic compounds, and is generally in the fine particle mode. Fly ash is the nonorganic by-product of coal burning. Coal contains substantial residual mineral material, including clays, shale, sulfides, carbonates, chlorides, and various trace metals. When coal is burned, these parent materials are released unreacted or thermally transformed. They take the form of spherical glassy particles in the coarse particle mode. The combination of soot and fly ash released from coal furnaces results in a bimodal size distribution of primary aerosol particles. Diesel engines are the dominant source of soot in some urban environments, but worldwide, coal burning releases about ten times as much BC into the atmosphere as diesel fuel combustion.

Primary carbon or soot aerosol consists of two subcomponents with different absorbing features: graphitic carbon with a high absorbing property and primary organics with a rather weak absorbing property (Rosen and Novakov, 1984). The number and mass concentrations of the single subcomponents are highly variable functions of space and time, as in the case of the parameters determining their size distribution. Given this importance, measurements of BC and the differentiation between BC and organic carbon still require improvement (IPCC, 2001). Recent estimates place the global emission of BC aerosols from biomass burning at 6–9 Tg/yr and from fossil fuel burning at 6–8 Tg/yr (Scholes and Andreae, 2000). The emission of BC has been especially large in China, owing to the low temperature household burning of coal and biofuels (Streets *et al.*, 2001). Giorgi *et al.* (2002) carried out a series of simulations using a coupled regional climate-chemistry/aerosol model and showed that anthropogenic fossil fuel soot exerts a positive radiative forcing of 0.5–2 W/m^2 TOA. Using a regional climate model developed in China, Wu *et al.* (2004) also reported that BC induces a positive radiative forcing at TOA and a negative forcing at the surface. Reducing uncertainties in the radiative forcing due to BC must account for its fundamental absorption and scattering properties.

BC particles can be internally and externally mixed with other nonabsorbing materials (Martins *et al.*, 1998). The same amount of BC under different types of mixing can result in fairly different absorption properties. The efficiency of absorption of a certain amount of BC is determined by its mass absorption coefficient, which depends on the size of particles and the type of mixing between BC and nonabsorbing components such as organic matter and sulfates. In the case of an external mixture where individual pure BC particles are in parallel with nonabsorbing particles, the BC mass absorption efficiency is about the same for pure BC particles. However, it is likely that BC particles that form at relatively high temperature will be coated by a nonabsorbing shell to form an internal mixture. Another type of internal mixture is a long-chain aggregate of BC particles formed at high temperature, which can also be coated with nonabsorbing materials to form an internally mixed heterogeneous structure. These opened clusters usually collapse to form closely packed sphertlike structures as a result of interactions with water vapor and clouds.

Owing to their production mechanisms, carbonaceous materials, including BC, are confined to the submicrometer size range. However, aggregation of freshly produced particles may yield particle sizes above that range. Small spherical soot particles coagulate to form chainlike aggregates. These irregularly shaped particles have been found to be fractal, and have special optical properties. The mass median diameter of BC in the atmosphere is very small, between 0.1 and 0.5 μm. In urban air, two size modes can often be detected; a smaller mode resulting from freshly formed primary aerosol and a larger mode resulting from coagulation processes. The more remote the location, the more dominant the larger size mode, reflecting the importance of coagulation during the aging of carbonaceous aerosol as it is transported from the site of combustion. BC aerosols normally have a long lifetime in the atmosphere because of their small size, usually less than 1 μm. Identifying and quantifying the purely carbonaceous material remains a difficult task. Effects of the aggregation of soot aerosols on their scattering and absorbing properties have recently been studied by Liu and Mishchenko (2005).

The state-of-the-art radiative transfer schemes that have been used in climate models have not taken into account the effects of aerosol nonsphericity and inhomogeneous structure (e.g. Gu *et al.*, 2006, and the preceding GCM example). The neglect of these effects could substantially underestimate or overestimate aerosol radiative forcing. For example, the asymmetry factor of realistic aerosols, which is directly related to the albedo effect, can be quite different from that of their spherical counterparts. It follows that a spherical "equivalence" assumption can lead to erroneous assessment of the aerosol radiative forcing. In the preceding discussion, we have demonstrated the nonspherical shape effect of dust particles on the scattering phase function and the consequence of remote sensing of their optical depth and particle size. In the case of BC particles, we have showed the complexity of inhomogeneity associated with internal inclusions, external attachments, and cluster structures. Indeed, accurate and efficient determination of the scattering and absorption properties of these two major types of nonspherical and inhomogeneous aerosols is a subject that will require in-depth research efforts.

3. The Effect of Mountains on Radiation Fields

The energy emitted by the Sun and received by the Earth's surface is determined by three sets of factors. The first includes the latitude, the solar hour angle, and the Earth's position relative to the Sun, which determine incoming solar radiation at the top of the atmosphere and can be precisely calculated (Liou, 2002). The second involves the attenuation of the solar beam by scattering and absorption caused by atmospheric gases, aerosols, and cloud particles. The last comprises terrain characteristics including elevation, slope, orientation, and surface albedo. On a sloping surface in mountainous terrain, the total solar radiation can be separated into five components according to the Sun-to-surface path (Fig. 3): (1) the direct flux contains the photons arriving on the ground directly in line from the Sun to the surface, unscattered by the atmosphere; (2) the diffuse

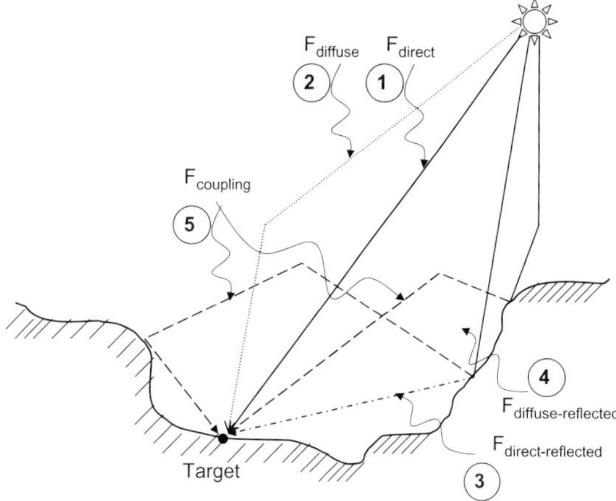

Figure 3. Schematic representation of the flux components received by the target on a sloping surface in mountainous terrain. (1) direct, (2) diffuse, (3) direct-reflected, (4) diffuse-reflected, and (5) coupling fluxes (after Chen *et al.*, 2006).

flux is composed of photons scattered by the atmosphere; (3) the direct-reflected flux is associated with photons traveling directly through the atmosphere unscattered but subsequently reflected to the target surface by neighboring slopes; (4) the diffuse-reflected flux is similar to the direct-reflected irradiance except that it is composed of photons already scattered by the atmosphere and then reflected by neighboring slopes; and (5) the coupling flux is composed of photons that have experienced multiple surface reflections and atmospheric scatterings. The direct, diffuse, and coupling components are produced when solar radiation interacts with a flat surface. However, over mountains, additional components are generated by the direct-reflected and diffuse-reflected fluxes through interaction with rough topography that are absent for the flat surface.

A variety of models of varying sophistication and complexity have been developed to compute most of these solar flux components in rugged terrain in which the direct incident beam is computed by introducing a cosine correction of the local zenith angle and considering the shadow caused by topography (e.g. Olyphant, 1986). Diffuse radiation has been modeled as being proportional to the area of the sky dome visible to the target surface (e.g. Dozier and Frew, 1990). Factors contributing to terrain-reflected flux have generally not been considered explicitly, but are approximated by assuming a first-order reflection between surrounding terrain and the target surface. The terrain-reflected radiation is particularly significant over snow surfaces (Dozier, 1980). Miesch *et al.* (1999) used a Monte Carlo approach to compute flux components at ground level and the radiance reaching the satellite-borne sensor over a very simple 2D rugged terrain with homogeneous atmospheric layers above, and concluded that a 10% error may occur when the terrain-reflected and coupling components may play a dominant role in the signal measured by a satellite sensor over poorly exposed or shadowed areas. As this brief literature survey illustrates, the effect of complex 3D mountainous terrain on the surface flux components of Fig. 3 has not been systematically and rigorously studied.

Chen *et al.* (2006) developed a 3D Monte Carlo radiative transfer model that calculates the flux components exactly given a realistic distribution of scatterers and absorbers in the atmosphere and provided the first definitive assessment of the relative importance of the flux components in a clear-sky atmosphere. The 3D Monte Carlo method for the transfer of radiation from the Sun as a point source with application to clouds has been presented in some detail by Marshak and Davis (2005). Chen *et al.* (2006) also characterized the factors controlling diurnal, seasonal, and geographical variability in the flux components and determined the significance and sources of errors in conventional radiative transfer schemes in areas of intense topography by applying the 3D radiative transfer model to a real 3D mid-latitude mountainous surface. Errors in radiative transfer schemes that have been used in GCM's with smoothed topography were assessed by comparing 3D results with the mountainous surface to identical calculations for a flat surface with the same mean elevation. It was shown that the errors are on the order of 5–20 W/m^2, depending on the time of the day, and stem from the fact that the atmosphere absorbs a different amount of solar radiation when the topography underneath it is smoothed.

The Tibetan Plateau, which covers about 2.4 million square kilometers with an average height of over 4 km, has a profound influence on the general circulation of the atmosphere, owing to its uplift of large scale flow patterns and its insertion of heating with a confining lower boundary at high levels. The wintertime dynamic effect enhances the southward transport of the Siberian high and leads to a large trough in the jet stream extending into the Pacific (e.g. Charney and Eliassen, 1949). The summertime elevated heat source has been recognized by meteorologists since early papers by T. C. Yeh (Ye, 1979) in conjunction with the formation and maintenance of the Asian summer monsoon, the onset of Meiyu, or the East Asian summer monsoon rainfall period, and the summertime heavy rainfall frequently occurring in central China. At lower levels the mechanism is similar to but more powerful than that of the summer near surface jet of the US' Great Plains. However, our knowledge of the Tibetan Plateau heating and its interactions with the complex regional dynamics and the consequent effects has remained rather limited. Although reanalysis data have been used to provide temporal and spatial distributions of differential heating, these results are severely limited by inadequacies in the formulations of diabatic processes in the models. Problems associated with the vertical profile of radiative heating rates over the mountainous region, the coupling of the surface to the lower atmosphere through boundary layer convection, and the snow albedo feedback in regional and global climate change appear to be largely unresolved and often ignored.

We applied the 3D solar radiative transfer model constructed by Chen *et al.* (2006) to a domain of $200\,\mathrm{km} \times 200\,\mathrm{km}$ centered at Lhasa over the Tibetan Plateau and carried out an analysis of the topography effects on the surface and atmospheric radiation fields. For the terrain information, we used the HYDRO1k database that was developed at the US Geological Survey's EROS Data Center to provide comprehensive and consistent global coverage of topographical datasets. The required data for radiation calculations include digital terrain elevation, slope, and orientation. Employing the ISCCP $2.5°$ by $2.5°$ resolution surface albedo map, we obtain one surface albedo value for radiation calculations. Figure 4 displays the distributions of the surface downward (upper

panels) and net (lower panels) solar fluxes at a $1\,\mathrm{km} \times 1\,\mathrm{km}$ resolution over the domain. The standard atmospheric water vapor and other trace gaseous profiles were used as input in the calcuation along with a solar zenith angle of $64°$ and two azimuthal angles correponding to early morning and late afternoon. The results displayed in these diagrams illustrate the effect of terrain and its orientation with respect to the Sun's position. We see substantial different flux values along the river valleys (two heavy solid lines). The net surface solar flux for a flat surface at an average height of abour $4\,\mathrm{km}$ is $400\,\mathrm{W/m^2}$, but it varies from about 40 to $500\,\mathrm{W/m^2}$ in the $200\,\mathrm{km} \times 200\,\mathrm{km}$ domain. Figure 5 illustrates a 3D atmospheric heating rate distribution in the domain, with the

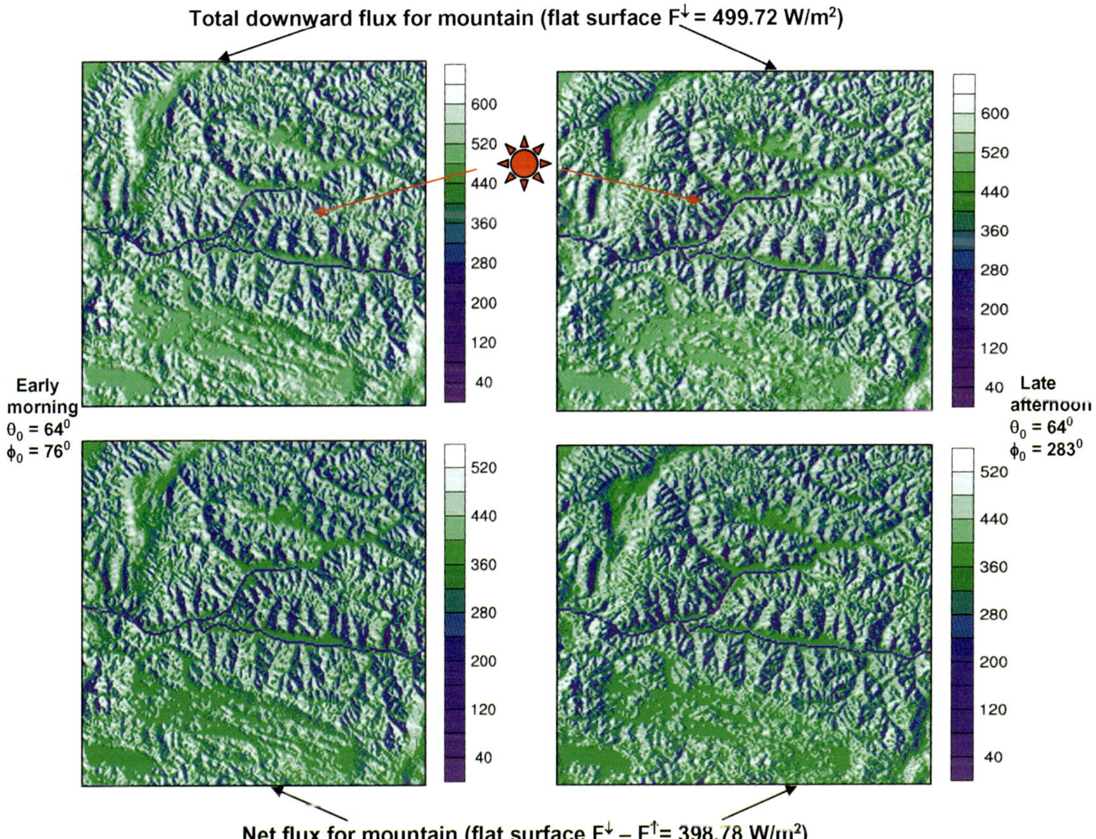

Figure 4. Distributions of the surface downward (upper panels) and net (lower panels) solar fluxes at a $1\,\mathrm{km} \times 1\,\mathrm{km}$ resolution over the domain.

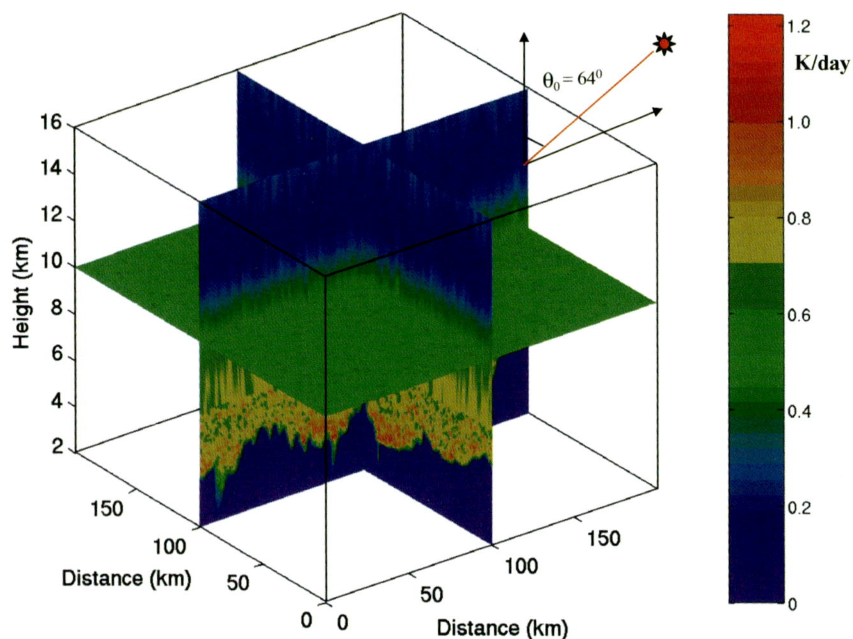

Figure 5. 3D atmospheric heating rate distribution in a 200 km × 200 km domain, with the terrain height shown at the bottom as a black shade.

terrain height shown at the bottom as a black shade. A substantial variability of the atmopheric heating rate ranging from about 0.2 to 1.2 K/day is shown. The inhomogeneous surface feature affects not only surface radiation but also the atmospheric radiation field above.

For application to regional weather and climate models, we are most interested in the flux averaged over a domain. Figure 6 illustrates the domain-averaged fluxes for flat and mountain surfaces as a function of time of day using horizontal resolutions of $50 \times 50\,\text{km}^2$, $100 \times 100\,\text{km}^2$, and $200 \times 200\,\text{km}^2$. Two surface albedos of 0.2 and 0.7, representing a snow condition, were used in the calculations. The domain-averaged fluxes were obtained by averaging the results from the $1 \times 1\,\text{km}^2$ 3D radiative transfer calculations. Net flux differences between the preceding resolutions appear to be small but, in comparison with the flat surface counterpart, a difference of 5–20 W/m^2 is shown. Figure 7 illustrates the averaged atmospheric solar heating rate over a horizatal domain of 200 km × 200 km

for the case of a flat surface with an average height of 4.7 km and two mountainous cases representing morning and afternoon conditions. Heating rate results for the two mountaineous cases are close because of the large domain average. However, significant variations from about 0 to about 0.76 K/day are shown in the regions from the lowest elevation of 2.3 km to about 6 km, resulting from the effect of mountain slopes associated with the solar zenith angles used in the calculations. These differences would have a substantial effect on the energy balance of the planetary boundary layer in GCM simulations over the Tibetan Plateau.

In summary, we have demonstrated the significance of topographical effects on surface and atmospheric radiation fields in clear atmospheres. Clouds and aerosols are anticipated to have a profound influence on the radiation budget over mountain terrain and are subject further investigation. Moreover, the thermal infrared (long wave) radiation emitted from rugged terrain with nonhomogeneous surface

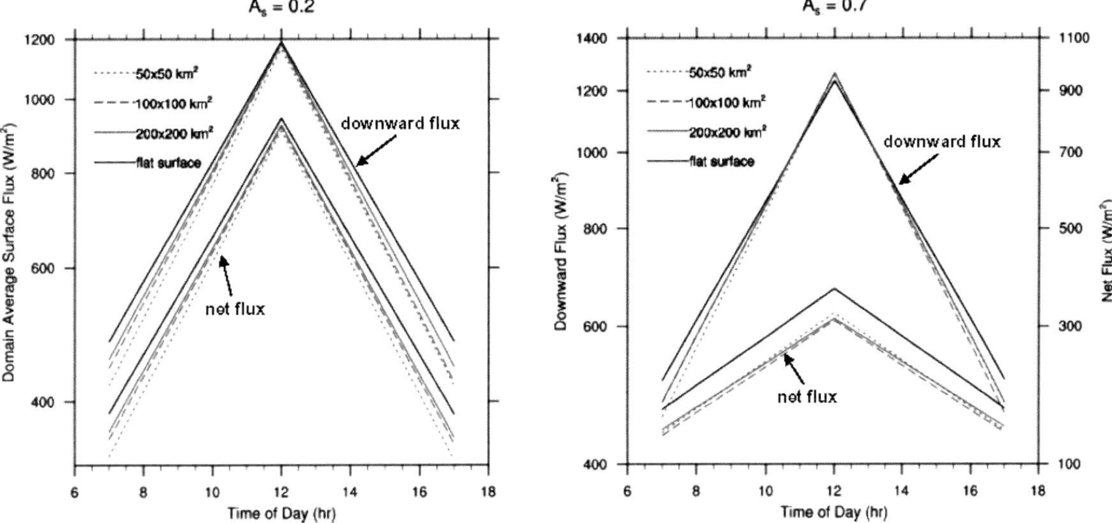

Figure 6. Domain-averaged fluxes for flat and mountain surfaces as a function of time of day using horizontal resolutions of $50 \times 50\,\text{km}^2$, $100 \times 100\,\text{km}^2$, and $200 \times 200\,\text{km}^2$. Two surface albedos of 0.2 and 0.7 (snow condition) were used in the calculations.

temperatures is another subject that requires in-depth study.

4. Radiative Transfer in the Atmosphere–Ocean System

The climatological value of solar flux penetrating the mixed layer can reach $40\,\text{W/m}^2$ in the tropical regions and produce a difference in the heating rate of a 20 m mixed layer by about $0.33°\text{C}$ a month (Ohlmann *et al.*, 1996). The vertical distribution of solar flux also influences the stability and stratification of the mixed layer and the sea surface temperature. Consequently, a quantitative understanding of the ocean solar flux profile is important to ocean model simulations. The surface albedo, which includes the surface reflectivity and the upwelling radiation from the water surface, is critical to the energy budget in the atmospheric planetary boundary layer. However, the reflectivity of the wind-blown surface is difficult to evaluate. The surface reflectance may be reduced by 50% when the solar zenith angle is $70°$ and the wind speed increases from 0 to $20\,\text{m/s}$ (Mobley, 1994).

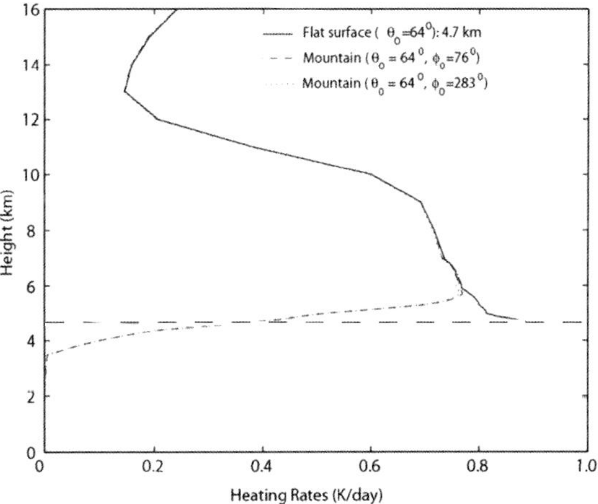

Figure 7. Averaged atmospheric solar heating rate over a horizatal domain of $200\,\text{km} \times 200\,\text{km}$ for the case of a flat surface with an average height of 4.7 km (denoted by the horizontal dashed line) and two mountainous cases representing morning and afternoon conditions.

The surface roughness also affects the upwelling radiation from the water surface and its determination requires accurate radiative transfer analysis. In addition, the oceanic pigment on radiative transfer in the ocean is also significant. A change of $0.10\,\text{mg/m}^3$ in the phytoplankton concentration in the mixed layer can result in a corresponding change of the penetrative solar flux by about $10\,\mathrm{W/m^2}$ at a level of $20\,\mathrm{m}$ depth (Siege *et al.*, 1995). Because solar radiation is the energy source for photosynthesis, it also directly affects the marine productivity.

A typical coupled radiation model deals with parametrized and simplified radiative transfer in the atmosphere and ocean separately by considering one medium as the boundary condition or the other. Computationally expensive models have also been developed for solar flux calculation (Jin *et al.*, 2002), but they are not compatible with coupled atmosphere–ocean models for climate simulation.

Lee and Liou (2007) developed a coupled atmosphere–ocean broadband solar radiative transfer model based on the analytic delta-four-stream approach originally derived by Liou (1974) and Liou *et al.* (1988). This approach is computationally efficient and at the same time can achieve acceptable accuracy for flux and heating rate calculations in the atmosphere and the oceans. Figure 8 illustrates a coupled atmosphere–ocean system with the solar beam reflected and refracted by a wind-blown sea surface. For an optically inhomogeneous system, the atmosphere and the ocean can be divided into a number of separate homogeneous layers such that each layer can be characterized by a single-scattering albedo, a scattering phase function, and an extinction coefficient. The entire radiative transfer system can be solved by matching the continuity equations for discrete intensities at the interfaces. To take into account the reflection and transmission of the wind-blown air–water interface, a Monte Carlo method (e.g. Preisendorfer and Mobley, 1985) has been employed to simulate the traveling of photons and to compute the reflectance and transmittance of direct and diffuse solar fluxes at the ocean surface using Cox-Munk's (1954) surface slope distribution. For the ocean part, existing bio-optical models, which correlate the concentration of chlorophyll and the absorption and scattering coefficients of phytoplankton and other matter, have been integrated into this coupled model.

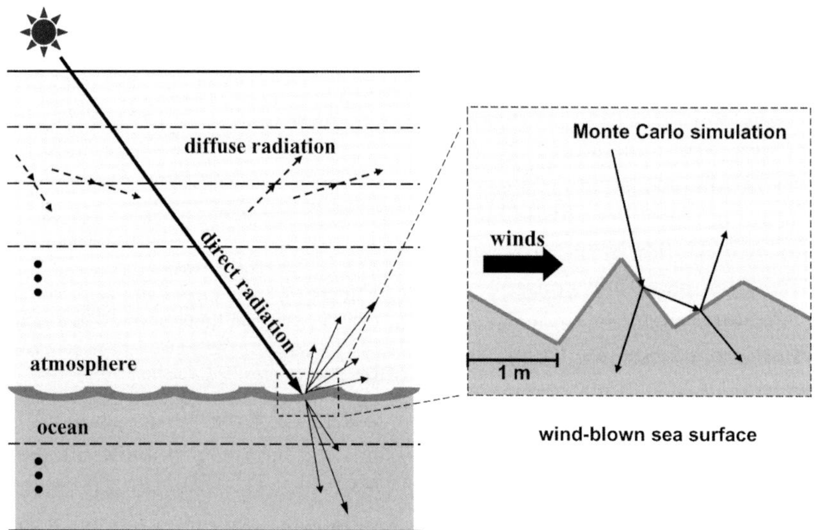

Figure 8. A coupled atmosphere–ocean system with the solar beam reflected and refracted by a wind-blown sea surface. For an optically inhomogeneous system, the atmosphere and the ocean can be divided into a number of separate homogeneous layers such that each layer can be characterized by a single-scattering albedo, a scattering phase function, and an extinction coefficient.

Lee and Liou's model was first compared to the values computed by a more exact model employing more discrete streams and illustrates that the relative accuracies of the surface albedo and total transmission in the ocean determined from the present model are generally within 5%, except in cases of the solar zenith angle being larger than 80°. The observational data collected at the CERES Ocean Validation Experiment (COVE, Jin *et al.*, 2002) site have also been used to validate this model and the results show that the relative differences of downward and upward short wave fluxes and albedo are within 10% of the observed values.

As an example, Fig. 9(a) shows the surface solar albedo as a function of the wind speed and the cosine of the solar zenith angle (SZA), μ_0, for the clear sky condition with the chlorophyll concentrations of 0 and $12\,\text{mg/m}^3$. An ocean depth of $200\,\text{m}$ with an ocean bottom albedo of 0.1 was used in the calculation and the chlorophyll concentration was considered uniformly in the whole layer. The sea surface albedo generally increases with the SZA for $\mu_0 > 0.1$, because the Fresnel reflectance increases with an increasing incident angle. However, the surface albedo decreases with the SZA for $\mu_0 < 0.1$, because when the Sun's altitude becomes lower, the intensity of direct solar radiation decreases more rapidly than that of diffuse radiation. Thus, the addition of the proportion of diffuse radiation reduces the effective SZA, leading to decrease in the surface albedo. The wind effect on the surface albedo is strong only when $\mu_0 > 0.4$. The surface albedo increases with chlorophyll concentration owing to increase in the scattering coefficient. However, this effect is not significant because increase in backscattering is canceled out by increase in the particulate absorption in sea water. Figure 9(b) shows the transmission ($z = 5\,\text{m}$) using the chlorophyll concentrations of 0 and $2\,\text{mg/m}^3$. Transmission decreases significantly with increasing chlorophyll concentration owing to increase in absorption. The transmission minimum occurs

at $\mu_0 \sim 0.25$. Owing to Rayleigh scattering, the intensity in the blue light region of direct solar radiation decreases with increase in optical depth. Sea water has a minimum absorption coefficient in the blue light region and for this reason the transmission decreases with an increasing SZA. Because there is more radiant intensity in the blue light region for diffuse radiation than the direct counterpart, the transmission increases with the SZA for the low Sun's elevation associated with more energy for diffuse radiation.

The existing bio-optical models, which correlate the concentration of chlorophyll and the absorption and scattering coefficients of phytoplankton and other matter, are highly empirical and require verification by fundamental theory and observations. For example, absorption by phytoplankton and the associated living or nonliving derivatives and absorption by the dissolved yellow matter are directly correlated with that by phytoplankton without rigorous physical fundamentals. Moreover, the Rayleigh scattering theory is not applicable to water because the distance between water molecules in the liquid phase is too short. The Einstein–Smoluchowski theory (Mobley, 1994) of scattering states that fluctuation in the molecular number density due to random molecular motions causes fluctuation in the index of refraction, which gives rise to scattering. An empirical form for the volume scattering function and the scattering coefficients between 200 and $800\,\text{nm}$ for pure sea water have been used in the field of ocean optics (Smith and Baker, 1981). In addition, the scattering coefficients for phytoplankton and other matter have been determined simply from the bio-optical model (Morel and Maritorena, 2001). The correct scattering and absorption properties for irregular phytoplankton and other species in the ocean must be determined from the theory of electromagnatic scattering along with appropriate scattering experiments in the laboratory — an exciting research area.

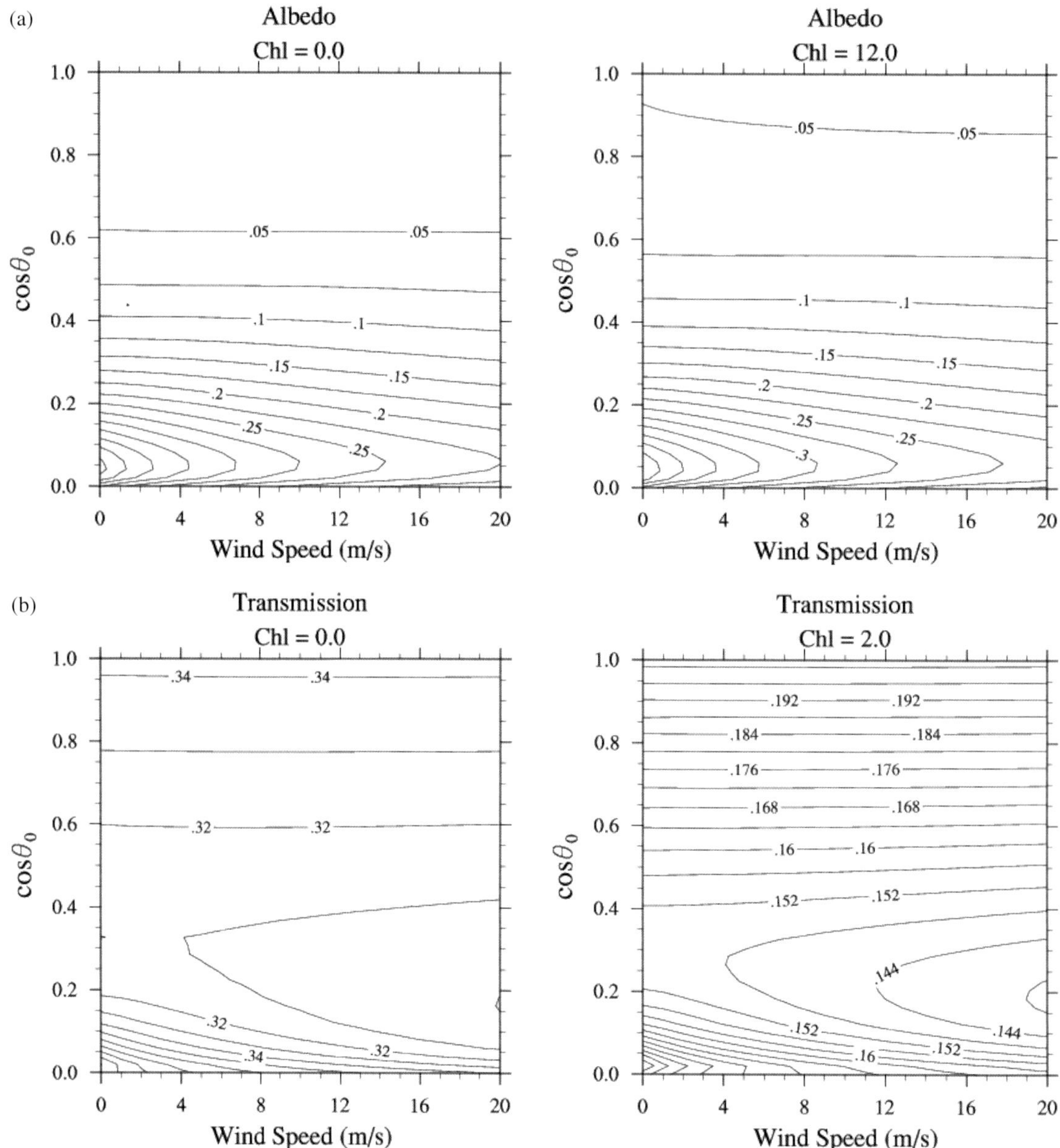

Figure 9. (a) Sea surface albedo for radiation between 0.2 and 4.0 μm as a function of the wind speed and the cosine of the solar zenith angle for the clear sky condition with chlorophyll concentrations of 0 (left panel) and 12 (right panel) mg/m^3. (b) Transmission at the ocean depth of 5 m with chlorophyll concentrations of 0 (left panel) and 2 mg/m^3 (right panel) (after Lee and Liou, 2007).

5. Summary

In this paper, we have presented three contemporary and largely unsolved problems in atmospheric radiative transfer with application to the Asian–Pacific region. Although it has been written in a review format, some new results have also been presented to illustrate important points in climate and remote sensing applications.

Aerosols are undoubtedly the most uncertain climate element in the Earth's atmosphere at the present time. Analysis and understanding of the radiative forcings owing to the direct and indirect aerosol climate effects require the basic scattering and absorption properties. The dust storms originating in China during the springtime and their global transport have a profound impact on the climate of East Asia. Moreover, China is the largest underdeveloped country in the world and has been obtaining 80% of its energy from coal combustion. The emission of BC has been especially large owing to low-temperature household coal burning. Dust and BC are nonspherical and inhomogeneous particles. Reliable and efficient determination of the scattering and absorption properties of these particles based on theoretical approach, numerical solution, and controlled laboratory experiment must be undertaken in order to reduce uncertainties in their radiative forcings by means of global and regional climate models.

The effect of terrain inhomogeneity on the radiation fields of the surface and the atmosphere above has not been accounted for in weather and climate models at this point. On the basis of a 3D radiative transfer program we illustrated that without consideration of topographical variability in typical midlatitute mountainous regions in the United States, net solar flux at the surface can be off by 5–20 W/m^2, assuming a flat surface. Many landscapes on the continents exhibit significant terrain features and in particular, the Tibetan Plateau has a profound influence on the general circulation of the atmosphere and climate in the East Asia region. The temporal and spatial distributions of surface radiation over this intensive topography determine surface dynamic processes and atmospheric radiative heating. Thus, accurate calculations and parametrizations of radiative transfer including both the solar and the thermal infrared spectra in clear and cloudy conditions must be developed for incorporation in regional and global models.

The last unsolved radiative transfer area we identified is associated with the transfer of solar radiation in the atmosphere–ocean system. The ocean covers about 70% of the Earth's surface, and the radiation interface between the atmosphere and the ocean is rather intricate. We pointed out that the vertical distribution of solar flux influences the stability and stratification of the mixed layer and the sea surface temperature. The surface roughness produced by the winds is the predominating factor controlling the penetration of solar flux into the ocean mixed layer and it must be correctly modeled for radiation calculations. Finally, we pointed out the lack of fundamental scattering and absorption information on irregular phytoplankton and other species in the ocean for heating rate calculation and remote sensing application.

Acknowledgements

The research for this work has been supported in part by NSF Grants ATM-0331550 and ATM-0437349 and DOE Grant DE-FG03-00ER62904. We thank Yoshihide Takano and Rick Hansell for their assistance during the preparation of this article.

[Received 4 January 2007; Revised 28 April 2007; Accepted 30 April 2007.]

References

Chandrasekhar, S., 1950: *Radiative Transfer*. Oxford University Press, Oxford, 393 pp.

Charney, J., and A. Elliassen, 1949: A numerical method for predicting the perturbations of middle latitude westerlies. *Tellus*, **1**(2), 38–54.

Chen, Y., A. Hall, and K. N. Liou, 2006: Application of 3D solar radiative transfer to mountains. *J. Geophys. Res.*, **111**, D21111, doi:10.1029/2006JD007163.

Cox, C., and W. Munk, 1954: Measurement of the roughness of the sea surface from photographs of the sun's glitter. *J. Opt. Soc. Amer.*, **44**, 838–850.

d'Almeida, G. A., P. Koepke, and E. P. Shettle, 1991: *Atmospheric Aerosols: Global Climatology and Radiative Characteristics*, A. Deepak Virginia, 561 pp.

Dozier, J., 1980: A clear-sky spectral solar radiation model for snow-covered mountainous terrain. *Water Resour. Res.*, **16**, 709–718.

Dozier, J., and J. Frew, 1990: Rapid calculations of terrain parameters for radiation modeling from digital elevation data. *IEEE Trans. Geosci. Remote Sens.*, **28**, 963–969.

Dubovik, O., A. Sinyuk, T. Lapyonok *et al.*, 2006: The application of spheroid models to account for aerosol particle nonsphericity in remote sensing of desert dust. *J. Geophys. Res.* **111**, D11208, doi:10.1029/2005JD006619.

Duguay, C., 1993: Radiation modeling in mountainous terrain: Review and status. *Mt Res. Dev.*, **13**, 339–357.

Falkovich, A. H., E. Ganor, Z. Levin, P. Formenti, and Y. Rudich, 2001: Chemical and mineralogical analysis of individual mineral dust particles. *J. Geophys. Res.*, **106**, 18029–18036.

Fu, Q., and K. N. Liou, 1992: On the correlated k-distribution method for radiative transfer in nonhomogeneous atmospheres. *J. Atmos. Sci.*, **49**, 2139–2156.

Fu, Q., and K. N. Liou, 1993: Parameterization of the radiative properties of cirrus clouds. *J. Atmos. Sci.*, **50**, 2008–2025.

Gao, Y., and J. R. Anderson, 2001: Characteristics of Chinese aerosols determined by individual particle analysis. *J. Geophys. Res.*, **106**, 18037–18045.

Giorgi, F., X. Bi, and Y. Qian, 2002: Direct radiative forcing and regional climatic effects of anthropogenic aerosols over East Asia: A regional coupled climate-chemistry/aerosol model study. *J. Geophys. Res.*, **107**, 4439, 10.1029/2001JD001066.

Gu, Y., J. Farrara, K. N. Liou, and C. R. Mechoso, 2003: Parameterization of cloud-radiation processes in the UCLA general circulation model. *J. Climate*, **16**, 3357–3370.

Gu, Y., K. N. Liou, Y. Xue, C. Mechoso, W. Li, and Y. Luo, 2006: Climatic effects of different aerosol types in China simulated by the UCLA atmospheric general circulation model. *J. Geophys. Res.*, **111**, D15201, doi:10.1029/2005JD006312.

IPCC, 2001: *Climate Change 2001: The Scientific Basis*, Houghton, J. T., Ding, Y., Griggs, D. J., Noguer, M., van den Linden, P. J., Dai, X., Johnson, C. A. (eds.), Intergovernmental Panel on Climate Change, Cambridge University Press, Cambridge, New York, 881 pp.

Jacobson, M. Z., 2001: Strong radiative heating owing to the mixing state of black carbon in the atmospheric aerosols. *Nature*, **409**, 695–697.

Jin, Z., T. P. Charlock, and K. Rutledge, 2002: Analysis of broadband solar radiation and albedo over the ocean surface at COVE. *J. Atmos. Oceanic Technol.*, **19**, 1585–1601.

Kalashnikova, O. V., and I. N. Sokolik, 2002: Importance of shapes and compositions of windblown dust particles for remote sensing at solar wavelengths. *Geophys. Res. Lett.*, **29**, 10.1029/2002GL014947.

Koren, I., E. Ganor, and J. H. Joseph, 2001: On the relation betweens size and shape of desert dust aerosol, *J. Geophys. Res.*, **106**, 18047–18054.

Koren, I., Y. J. Kaufman, L. A. Remer, and J. V. Martins, 2004: Measurements of the effect of Amazon smoke on inhibition of cloud formation. *Science*, **303**, 1342–1345.

Lee, W., and K. N. Liou, 2007: A coupled atmosphere–ocean radiative transfer system using the analytic four-stream approximation. *J. Atmos. Sci.*, **64**, 3681–3694.

Liou, K. N., 1974: Analytic two-stream and four-stream solutions for radiative transfer. *J. Atmos. Sci.*, **31**, 1473–1475.

Liou, K. N., 1980: *An Introduction to Atmospheric Radiation*. Academic, New York, 392 pp.

Liou, K. N., 2002: *An Introduction to Atmospheric Radiation*, 2nd edition. Academic, San Diego, 583 pp.

Liou, K. N., Q. Fu, and T. P. Ackerman, 1988: A simple formulation of the delta-four-stream approximation for radiative transfer parameterizations. *J. Atmos. Sci.*, **45**, 1940–1947.

Liu, L., and M. I. Mishchenko, 2005: Effects of aggregation on scattering and radiative properties of soot aerosols. *J. Geophys. Res.*, **110**, D11211.

Luo, Y., D. Lu, X. Zhou, W. Li, and Q. He, 2001: Characteristics of the spatial distribution and yearly variation of aerosol optical depth over China in last 30 years, *J. Geophys. Res.*, **106**, 14501–14513.

Marshak, A., and A. B. Davis (eds.), 2005: *3D Radiative Transfer in Cloudy Atmospheres*. Springer-Verlag, Berlin, 686 pp.

Martins, J. V., P. Artaxo, C. Liousse *et al.*, 1998: Effects of black carbon content, particle size, and mixing on light absorption by aerosols from biomass burning in Brazil. *J. Geophys. Res.*, **103**, 32041–32050.

Menon, S., J. Hansen, L. Nazarenko, and Y. Luo, 2002: Climate effects of black carbon aerosols in China and India. *Science*, **297**, 2250–2253.

Mishchenko, M. I., A. A. Lacis, B. E. Carlson, and L. D. Travis, 1995: Nonsphericity of dust-like tropospheric aerosols: Implications for aerosol remote sensing and climate modeling. *Geophy. Res. Lett.*, **22**, 1077–1080.

Mishchenko, M. I., L. D. Travis, and D. W. Mackowski, 1996: T-matrix computation of light scattering by nonspherical particles: A review, *J. Quant. Spectrosc. Radiat. Transfer*, **55**, 535–575.

Miesch, C., X. Briottet, Y. H. Herr, and F. Cabot, 1999: Monte Carlo approach for solving the radiative transfer equation over mountainous and heterogeneous areas. *Appl. Opt.*, **38**, 7419–7430.

Mobley, C. D., 1994: *Light and Water: Radiative Transfer in Natural Waters.* Academic, San Diego, 592 pp.

Morel, A., and S. Maritorena, 2001: Bio-optical properties of oceanic waters: A reappraisal. *J. Geophys. Res.*, **106**, 7163–7180.

Ohlmann, J. C., D. A. Siegel, and C. Gautier, 1996: Ocean mixed layer radiant heating and solar penetration: A global analysis. *J. Climate*, **9**, 2265–2280.

Okada, K., J. Heintzenberg, K. Kai, and Y. Qin, 2001: Shape of atmospheric mineral particles collected in three Chinese arid regions. *Geophys. Res. Lett.*, **28**, 3123–3126.

Olyphant, G. A., 1986: The components of incoming radiation within a mid-latitude alpine watershed during the snowmelt season. *Arct. Alp. Res.*, **18**, 163–169.

Preisendorfer, R. W., and C. D. Mobley, 1985: Unpolarized irradiance reflectances and glitter patterns of random capillary waves on lakes and seas by Monte Carlo simulation. NOAA Tech. Memo, ERL PMEL-63, Pacific Mar. Environ. Lab., Seattle, WA, 141 pp.

Reid, J. S., P. V. Hobbs, C. Liousse *et al.*, 1998: Comparisons of techniques for measuring shortwave absorption and black carbon content of aerosols from biomass burning in Brazil. *J. Geophys. Res.*, **103**, 32031–32040.

Reid, J. S., H. H. Jonsson, H. B. Maring *et al.*, 2003: Comparison of size and morphological measurements of coarse mode dust particles from Africa. *J. Geophys. Res.*, **108**, doi:10.1029/2002JD002485.

Rosen, H., and T. Novakov, 1984: Role of graphitic carbon particles in atmospheric radiation on transfer. In *Aerosols and their Climatic Effects.* H. E. Gerber and A. Deepak (eds.), Virginia, pp. 83–94.

Scholes, M., and M. O. Andreae, 2000: Biogenic and pyrogenic emissions from Africa and their impact on the global atmosphere. *Ambio*, **29**, 23–29.

Siege, D. A., J. C. Ohlmann, L. Washburn *et al.*, 1995: Solar radiation, phytoplankton pigments, and radiant heating of the equatorial Pacific warm pool. *J. Geophys. Res.*, **100**, 4885–4891.

Smith, R. C., and K. S. Baker, 1981: Optical properties of the clearest natural waters (200–800 nm). *Appl. Opt.*, **20**, 177–184.

Sokolik, I. N., O. B. Toon, and R. W. Bergstrom 1998: Modeling the radiative characteristics of airborne mineral aerosols at infrared wavelengths. *J. Geophys. Res.*, **103**, 8813–8826.

Sokolik, I. N., D. M. Winker, G. Bergametti *et al.*, 2001: Introduction to spectral section: Outstanding problems in quantifying the radiative impact of mineral dust, *J. Geophys. Res.*, **106**, 18015–18027.

Streets, D. G., S. Gupta, S. T. Waldhoff *et al.*, 2001: Black carbon emissions in China. *Atmos. Environ.*, **35**, 4281–4296.

Wu, J., W. Jiang, C. Fu, B. Su, H. Liu, and J. Tang, 2004: Simulation of the radiative effect of black carbon aerosols and the regional climate response over China. *Adv. Atmos. Sci.*, **21**, 637–649.

Yang, P., and K. N. Liou, 2000: Finite difference time domain method for light scattering by nonspherical particles. In *Light Scattering by Nonspherical Particles: Theory, Measurements, and Geophysical Applications*, M. Mishchenko *et al.* (eds.), Academic, San Diego, pp. 173–221.

Ye, D. (Yeh, T. C.), 1979: *The Meteorology of Qinghai–Xizang Plateau*, D. Ye and Y. X. Gao (eds.) Chap. 1, Science Press, Beijing (in Chinese).

Progress in Atmospheric Vortex Structures Deduced from Single Doppler Radar Observations

Wen-Chau Lee

Earth Observing Laboratory, National Center for Atmospheric Research,
Boulder, Colorado, USA*
wenchau@ucar.edu

Ben Jong-Dao Jou

Department of Atmospheric Sciences, National Taiwan University, Taipei, Taiwan

Doppler radars have played a critical role in observing atmospheric vortices including tornados, mesocyclones, and tropical cyclones. The detection of the dipole signature of a mesocyclone by pulsed Doppler weather radars in the 1960s led to an era of intense research on atmospheric vortices. Our understanding of the internal structures of atmospheric vortices was primarily derived from a limited number of airborne and ground-based dual-Doppler datasets. The advancement of single Doppler wind retrieval (SDWR) algorithms since 1990 [e.g. the velocity track display (VTD) technique] has provided an alternate avenue for deducing realistic and physically plausible two- and three-dimensional structures of atmospheric vortices from the wealth of data collected by operational and mobile Doppler radars.

This article reviews the advancement in single Doppler radar observations of atmospheric vortices in the following areas: (1) single Doppler radar signature of atmospheric vortices, (2) SDWR algorithms, in particular the VTD family of algorithms, (3) objective vortex center-finding algorithms, and (4) vortex structures and dynamics derived from the VTD algorithms. A new paradigm that improves the VTD algorithm, displaying and representing the atmospheric vortices in V_dD/R_T space, is presented. The VTD algorithm cannot retrieve the full components of the divergent wind which may be improved by either implementing physical constraints on the VTD closure assumptions or combining high temporal resolution data with a mesoscale vorticity method. For all practical perspectives, SDWR algorithms remain the primary tool for analyzing atmospheric vortices in both operational forecasts and research purposes in the foreseeable future.

---

*The National Center for Atmospheric Research is sponsored by the US National Science Foundation.

1. Introduction

Atmospheric vortices, such as tornados, meso-cyclones, and tropical cyclones (TCs), are highly correlated with severe weather and disasters threatening human lives and property. Famous examples are Typhoon Herb (1996), Hurricane Andrew (1992), Hurricane Katrina (2005), and the supertornado outbreak on 3–4 April 1974 in the United States, where hundreds of lives and billions of dollar's worth of property were lost because of damaging winds, mudslides, and floods. The spatial and temporal scales of these vortices span five orders of magnitude (from tens of meters to hundreds of kilometers, and from several minutes to ten days).

Sampling and deducing the two-dimensional and three-dimensional internal structures of these wide ranges of atmospheric vortices have been ongoing challenges for operational forecasts, scientific research, and numerical model initialization and validation (e.g. Houze *et al.*, 2006; Wang and Wu, 2004).

Atmospheric vortex structures have been observed by and inferred from *in situ* measurements, remote sensing, and photogrammetry (e.g. Davies-Jones *et al.*, 2001; Marks, 2003; Gray, 2003). Doppler radar remains the only instrument that can probe 3D internal structures of cloud and precipitation systems at a spatial resolution of 0.025–1 km and a temporal resolution of 1–30 minutes. Since Doppler radars sample only one component of the 3D motion along a radar beam, estimating 3D wind vectors requires two or more Doppler radars viewing the area of interest simultaneously, in conjunction with the mass continuity equation (e.g. Armijo, 1969). A comprehensive review of the multiple Doppler radar analysis and the subsequent dynamic retrieval techniques can be found in the literature (e.g. Ray, 1990).

Although dual Doppler analyses have provided critical insights into understanding structures of large atmospheric vortices such as TCs, dual Doppler radar analysis remains primarily an interactive and labor-intensive process with limited value for real-time applications. Only recently, encouraging real-time dual Doppler analysis of Atlantic hurricanes was performed on board the US National Oceanic and Atmospheric Administration (NOAA) WP-3D (Gamache *et al.*, 2004). This approach has potential, as the enhanced computer power enables data quality control and allows navigation corrections to be completed in real time. Until then, practical limitations prevent much wider applications to atmospheric vortices, especially in real-time settings.

To date, multiple Doppler radar datasets of atmospheric vortices are rare and remain mostly a privilege for specialized field campaigns. Suitable dual Doppler radar datasets are extremely difficult to collect. For large circulations like TCs, airborne Doppler radar datasets typically suffer from their long sampling time (>30 minutes), where stationary assumption of the circulation is questionable. In addition, the average separation of the operational Doppler radar networks in the United States is ~250 km. The useful dual Doppler lobes, if they exist, usually cover only a small portion of TCs' circulation. For small circulations like tornados and mesocyclones, dual Doppler datasets can only be obtained by two or more mobile radars deployed at very close range (e.g. <5 km), such as the 3 cm wavelength (X-band) Doppler on Wheels (DOWs, Wurman *et al.*, 1997; Wurman and Gill, 2002) and SMART-R (Biggerstaff *et al.*, 2005).

The difficulties in sampling these storms by multiple radars severely limit the monitoring and study of severe weather events associated with tornados, mesocyclones, and TCs over most of the world. As a result, 2D and 3D wind fields of atmospheric vortices can only be estimated from single Doppler radar wind retrieval (SDWR) algorithms. Although the kinematic structures of atmospheric vortices cannot be retrieved from SDWR as accurately and completely as from dual Doppler radar analysis, recent advances in SDWR algorithms have permitted the deduction of primary and secondary vortex structures in real time, expanding the analyses of atmospheric vortices far beyond the restricted dual Doppler radar domain.

Comprehensive reviews on the subject of TCs and severe weather (tornados and mesocyclones) from the radar perspective have been given by Marks (2003) and Bluestein and Wakimoto (2003). This review article focuses on the progress of SDWR algorithms, specifically their applications, in objectively identifying the vortex circulation center and deducing 2D and 3D structures of atmospheric vortices. Section 2 reviews the single Doppler radar vortex signatures and their interpretation. Section 3 reviews SDWR algorithms that can deduce structures of atmospheric vortices. Section 4 discusses objective vortex center-finding algorithms. The structures of atmospheric vortices revealed from these SDWR algorithms are reviewed in Sec. 5. Section 6 outlines how the wind field retrieved from the VTD algorithm can be used with data assimilation techniques to refine the divergent wind estimate. Section 7 discusses the generalized VTD technique. Section 8 offers outlooks.

2. Interpretation of Single Doppler Radar Vortex Signatures

The single Doppler signature of a vortex is a dipole that has been recognized since the 1960s (e.g. Donaldson, 1970). Traditionally, wind fields of atmospheric vortices have been assumed axisymmetric and modeled by a Rankine combined vortex (e.g. Rankine, 1901), characterized by two regimes: an inner core region with the velocity increasing linearly with the radius (constant vorticity), and an outer potential flow region where the velocity is inversely proportional to the radius (zero vorticity) (Fig. 1).

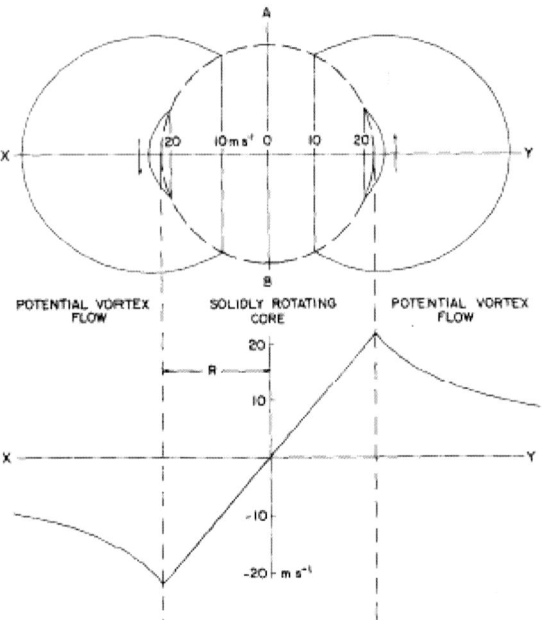

Figure 1. *Top*: Single Doppler velocity signature in a plan view of a stationary Rankine combined vortex, where the line XY is at a constant range from the radar. *Bottom*: Tangential velocity profile along the axis XY. Assuming a Doppler radar positioned below the bottom of the page, positive velocities are away from the radar, and negative toward (From Lemon *et al.*, 1978.)

Starting from a Rankine combined vortex, Brown and Wood (1991) demonstrated that Doppler velocity patterns or a ground-based radar is modulated by (1) varying the vortex intensity and the core size, (2) adding a background uniform wind field (simulating an environmental flow), (3) superimposing diverging winds onto the Rankine combined vortex, and (4) simulating double vortices and divergence sources. Figure 2 illustrates the signatures of a Rankine combined vortex and an axisymmetric radial flow as a function of the distance between the radar and the vortex, where the dipole pattern is distorted as the vortex moves closer to the radar (Wood and Brown, 1992; Lee *et al.*, 1999). The intensity and core diameter of the vortex can be estimated as the magnitude and size of the dipole. Wood and Brown (1992) showed geometrically and mathematically that (1) the dipole rotated counterclockwise when

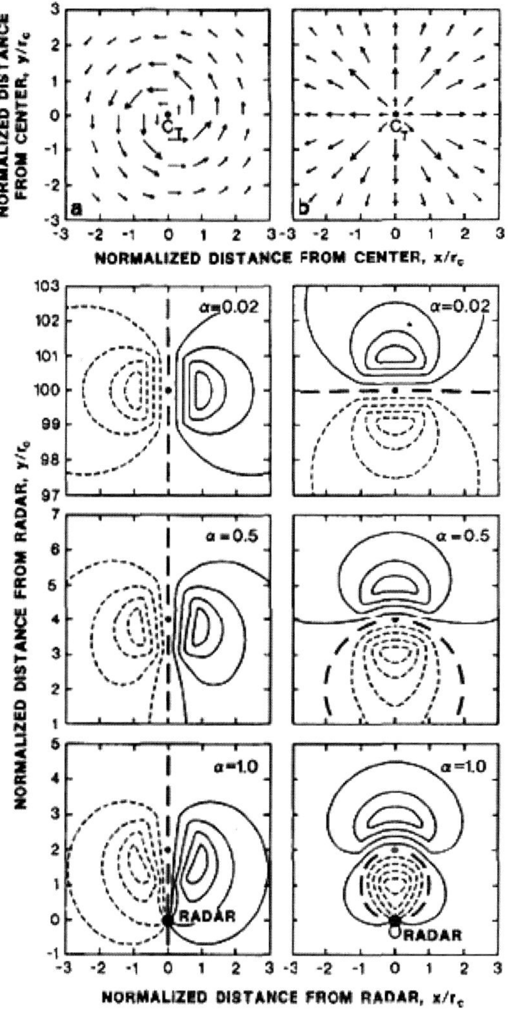

Figure 2. Horizontal (a) vortex and (b) divergence flow fields. The wind vector length is proportional to wind speed. The true circulation center is indicated by the small dot at the center of each panel. Variations of axisymmetric (a) vortex and (b) divergence Doppler velocity signatures as a function of distance north of the radar are shown beneath the respective flow fields. Solid curves represent flow away from the radar (positive Doppler velocities), short dashed curves represent flow toward the radar (negative Doppler velocities), and the thick, long-dashed curve represents flow normal to the radar viewing direction (zero Doppler velocities). Doppler velocity values are normalized relative to the peak wind speed value with contour intervals of 0.2. Radar is located at 100, 4, and 2 core radii south of the flow field (aspect ratios of 0.02, 0.5, and 1.0, respectively). Distances in the x and y directions have been normalized by the core radius. In the last two panels, radar location is indicated by the large dot. (From Brown and Wood, 1991.)

a diverging wind was superimposed onto a Rankine combined vortex, (2) the Doppler velocity patterns deformed when the aspect ratio, defined as the ratio of the vortex core diameter and the distance between the radar and the vortex center, increases (Fig. 2), and (3) the true vortex center and core diameter can be derived from the distorted Doppler velocity pattern with a correction factor.

The dipole signature has been used to identify atmospheric vortices of different scales and interpret their structures, including the mesocyclones associated with a rotating updraft core in supercell thunderstorms (e.g. Burgess, 1976), mesocyclones and tornados (e.g. Brown et al., 1978; Lemon et al., 1978), mesoscale vortices within mesoscale convective systems (e.g. Stirling and Wakimoto, 1989; Houze et al., 1989), and TCs (e.g. Baynton 1979; Wood and Marks, 1989). The Doppler velocity signatures in real atmospheric vortices in these studies [e.g. Figs. 3(a) and 3(b)] apparently deviate from that of the Rankine combined vortex illustrated in Fig. 2, including: asymmetric dipole magnitude, the zero Doppler velocity isodop which does not bisect the velocity dipole, the curved and nonparallel isodop inside the core region, and azimuthal rotation of the dipole. Attempts have been made to recognize these deviations from the Rankine combined vortex radar signatures so as to infer vortex structure (e.g. Brown and Wood, 1991; Wood and Brown, 1992; Lee et al., 1999).

Desrochers and Harris (1996) proposed a more sophisticated elliptical flow model including both the rotational and divergent winds to simulate the elongated Doppler velocity patterns of the Del City, 20 May 1997, OK mesocyclone. Lee et al., (1999) simulated Doppler velocity patterns of a Rankine vortex plus a mean flow, an axisymmetric radial flow, and tangential asymmetries from wave number 1 to 3. More importantly, the Doppler velocity pattern of an asymmetric vortex varies as a function of the radar's viewing angle

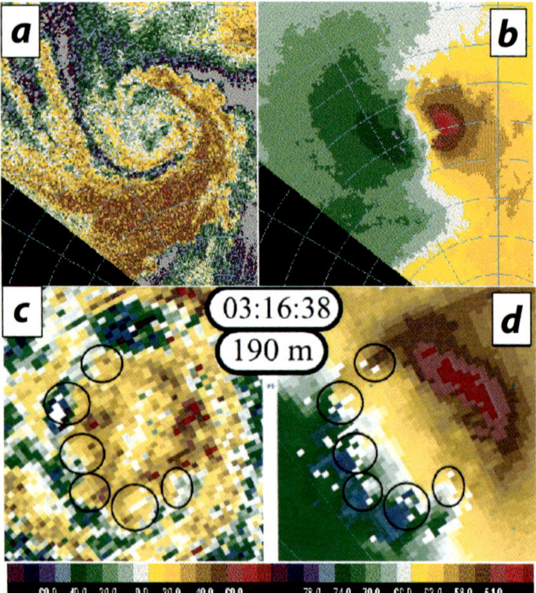

Figure 3. The upper panels show the return power (a) and Doppler velocity (b) of the Mulhall tornado observed by a DOW. Range rings are in 1 km interval. The lower panels show the enlarged corresponding tornado return power (c) and Doppler velocity (d) structure with multiple vortices (circles). (Adapted from Lee and Wurman, 2005.)

(Desrochers and Harris, 1996) and/or the phase of the asymmetric component of the vortex (Lee et al., 1999).

In summary, these Doppler velocity patterns were quite similar, and thus it is difficult to distinguish their differences visually. Although qualitative vortex structures can be inferred directly from the shape and magnitudes of the dipole signatures, deducing quantitative vortex structures is not straightforward, and will rely on more sophisticated SDWR algorithms.

3. SDWR Algorithms

Retrieval of the 2D and 3D wind fields from single Doppler observations commonly follows one of two approaches: fitting the observed radial velocities to some simplified wind model, or supplementing the radar data with some

physics (e.g. conservation, "pseudo conservation," momentum or vorticity equations) and seeking a flow field that satisfies the imposed constraints (Boccippio, 1995).

The first approach typically fits the measured radial velocities to a simple and specific (linear, nonlinear, or circular) model of the wind fields that are representative of a weather phenomenon. In the radar community, the well-known approaches are the velocity azimuth display (VAD) and volume velocity-processing (VVP) methods, where the analysis is radar-centered and assumes a linearly varying horizontal wind field in the radar sampling domain (e.g. Browning and Wexler, 1968; Caya and Zawadzki, 1992; Donaldson, 1991; Matejka and Srivastava, 1991; Waldteufel and Corbin, 1979; Koscielny et al., 1982; Boccippio, 1995). A related approach is the velocity track display (VTD) method, which fits the Doppler velocity measurements to a circular model of the wind field centered at the vortex (Lee et al., 1994; Jou et al., 1994; Roux and Marks, 1996; Lee et al., 1999; Roux et al., 2004; Liou et al., 2006).

The second approach usually constrains the single Doppler velocity field with physical equations, numerical models, or conservation of reflectivity. These SDWR algorithms typically involve multiple volumes of radar data at consecutive times. Assuming Lagrangian conservation of reflectivity, the tracking radar echoes by correlation (TREC) algorithm estimates the movement of the reflectivity field by finding the maximum cross-correlation of features in sequential radar scans or tracking echo centroids (e.g. Zawadzki, 1973; Rinehart, 1979; Symthe and Zrnic, 1983; Tuttle and Foote, 1990; Gall and Tuttle, 1999). More complicated algorithms involve a combination of the Lagrangian conservation of reflectivity and some simple assumptions about the behavior of the radial velocity field (e.g. Qiu and Xu, 1992; Xu et al., 1994; Shapiro et al., 1995; Liou, 1999a,b). The solution techniques are thus based on simple prognostic equations rather than the full set of governing equations to achieve lower computation cost and faster processing time. Recently, Lee *et al.* (2003) proposed the use of the mesoscale vorticity method (MVM) to deduce the 3D TC structures, including divergent wind and vertical velocity in the inner core region, from a frequently observed vorticity field derived from the VTD technique.

The general impression is that favorable agreement between the wind fields retrieved from these SDWR algorithms and independent observations can be achieved when the characteristics of the wind fields reasonably match the underlying geometrical and/or physical assumptions. These techniques tend not to capture the fine-scale features of the flow (i.e. higher order and/or nonlinear features), and the uncertainty of the circulation in the cross-beam (unmeasured) direction is worse than in the along-beam (measured) direction. *The wind fields retrieved from SDWR are only as good as the assumptions used in each specific situation.* In general, the algorithms in the second approach are capable of resolving more complex wind fields than the algorithms in the first approach. However, when conservation of reflectivity (a commonly used assumption in the second approach) breaks down, such as in the inner core region of atmospheric vortices, the algorithms in the first approach hold the advantage. Only those SDWR algorithms relevant to atmospheric vortices will be reviewed in this section.

3.1. *The TREC technique*

The TREC technique was first proposed by Rinehart and Garvey (1978), with modifications and improvements by Tuttle and Foote (1990). TREC works by storing two scans of low-elevation angle plan position indicator (PPI) or constant altitude PPI (CAPPI) reflectivity data measured at the same elevation angle (or at constant altitude) at two different times (Fig. 4). The time difference between two volumes is typically a few minutes for operational radars. The domain at "Time 1" is divided into equal-sized 2D arrays spaced some distance apart. Each initial array is correlated with all possible arrays of the same size at "Time 2" to find the best-matching second array with the highest correlation coefficient. The wind vector of that point is determined by the distance between the first and the second array divided by the time differences between these two scans. Such techniques are well suited for clear air returns and convective storms that can produce wind estimates,

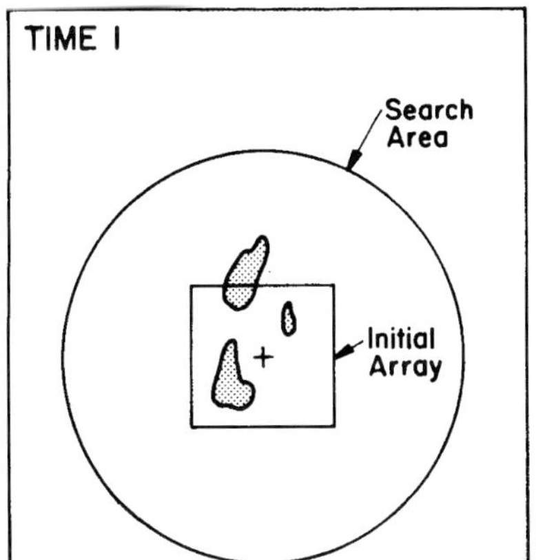

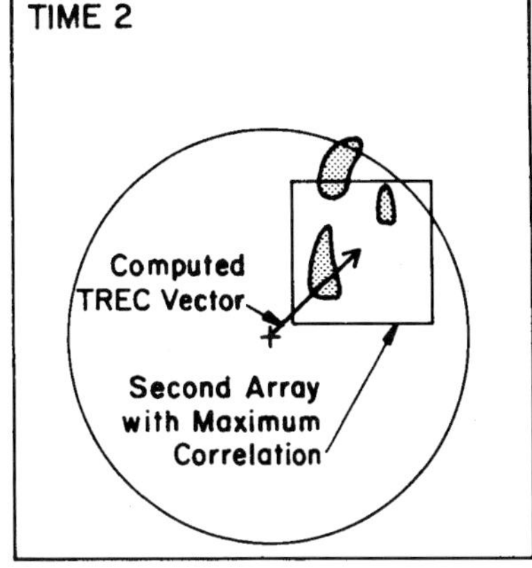

Figure 4.   Schematic showing the computation of a TREC vector to determine the motion of reflectivity echoes shaded from Time 1 to Time 2. The initial array of data at Time 1 is cross-correlated with all other second arrays of the same size at Time 2 whose center falls within the search area. The position of the second array with the maximum correlation determines the vector endpoint. (From Tuttle and Foote, 1991.)

even when the Doppler velocities are not reliable or not available (for example, beyond the Doppler range) — a significant advantage over Doppler-velocity-based SDWR algorithms. Within the Doppler range, TREC-derived winds can be combined with Doppler velocity data to improve its quality (Gall and Tuttle, 1999). When TREC-derived winds are computed at multiple altitudes, 3D kinematic structures can be obtained.

Gall and Tuttle (1999) applied TREC to three TCs — Hurricane Hugo (1989), Hurricane Erin (1995), and Typhoon Herb (1996) — and obtained reasonable cyclonic circulations (e.g. Typhoon Herb in Fig. 5). TREC tends to underestimate the wind speed in the eyewall compared with the corresponding single Doppler velocities, especially in well-organized, intense eyewalls such as Hugo and Herb. The authors attributed this discrepancy to the following factors: (1) the motion of the radar echo (precipitation features) in the eyewall may be governed by low-level convergence, and not by the wind (i.e. violation of the intrinsic assumptions of TREC); (2) in a vortex with high curvature in the wind field, it may be better for TREC to be computed in cylindrical coordinates rather than in Cartesian coordinates; (3) relatively uniform reflectivity structures in the azimuthal direction make it difficult for TREC to find unique features to track; and (4) strong radial shear may exist across the analysis box (15 km square in their examples). The first and third factors may be the most difficult to overcome in obtaining reasonable winds in the critical eyewall region. Nevertheless, TREC is effective in deducing the wind fields away from the eyewall of a TC, especially in determining the $17\,\mathrm{m\,s^{-1}}$ wind radii.

3.2. *The VTD family of techniques*

The term VTD (velocity track display), coined by Carbone and Marks (1988), refers to a display of the NOAA WP-3D tail-mounted Doppler radar velocities at the flight level during a radial hurricane penetration through its center [e.g. Hurricane Gloria (1985), Fig. 6]. When a TC is axisymmetric rotation, the dipole should be on or parallel to the flight track. Carbone and Marks (1988) noted that the peak Doppler velocities of the dipole in Fig. 6 rotated counterclockwise from the flight track. They proposed that the azimuthal rotation of the dipole resulted in a radial inflow near the eyewall. The alignment of the Doppler velocity dipole relative to the flight track is affected by the direction of the radial flow (i.e. inflow or outflow) and the ratio of the magnitude of the radial flow to the tangential flow. Hence, the axisymmetric tangential and radial winds can be estimated from the VTD.

The mathematical formulation of the VTD technique to deduce the primary circulations of a TC was given by Lee *et al.* (1994). Several variations/extensions of the technique have been developed, namely the extended

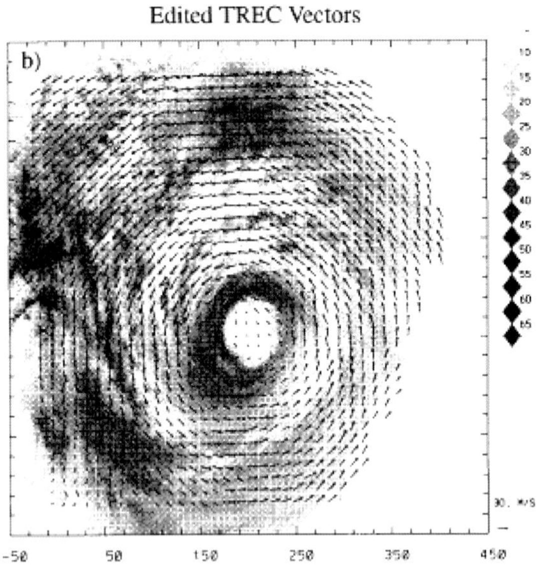

Figure 5. Edited TREC vectors for Typhoon Herb overlaid on radar reflectivity. A gray scale key for reflectivity contours is shown on the upper right and a $50\,\mathrm{m\,s^{-1}}$ reference vector on the lower right. (From Gall and Tuttle, 1999.)

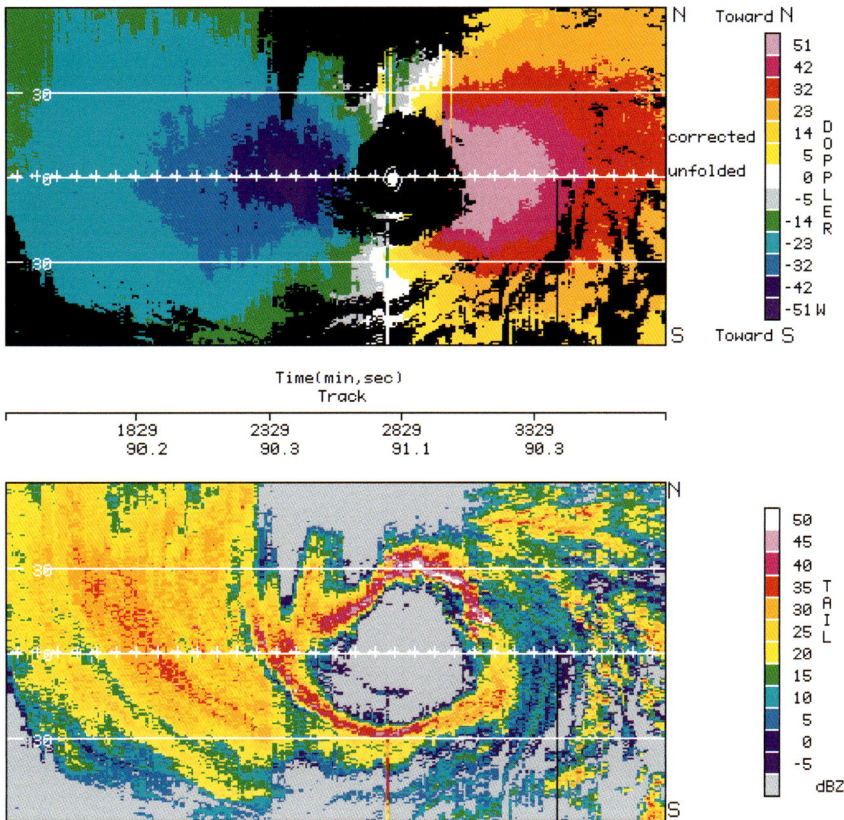

Figure 6. Flight-level (6.5 km) VTD display of Hurricane Gloria (1985) from the NOAA P3. The top panel shows the Doppler velocity while the bottom panel shows the radar reflectivity factor. The Doppler velocities on the right side of the aircraft were multiplied by −1, so the Doppler velocities do not change sign across the flight track. The aircraft moved from east to west. (From Lee *et al.*, 1994.)

VTD (EVTD) technique (Roux and Marks, 1996), the ground-based VTD (GBVTD) technique (Jou *et al.*, 1994; Lee *et al.*, 1999), the ground-based extended VTD (GB-EVTD) technique (Roux *et al.*, 2004), and the extended GBVTD (EGBVTD) technique (Liou *et al.*, 2006). This subsection provides an overview of the VTD family of techniques using the GBVTD framework.[a]

The concept of the VTD technique is illustrated in Fig. 7, where the Doppler velocity (black vectors) of an axisymmetric rotation and inflow [gray vectors in Fig. 7(a) and 7(b)] plotted

against the angle ψ form a negative sine and cosine curve, respectively. The Doppler velocity pattern of a combined axisymmetric rotation and radial inflow is a negative sine curve with a negative (clockwise) phase shift [Fig. 7(c)]. It can be shown that the sign of the phase shift corresponds to the direction of the axisymmetric radial flow at that radius, while the magnitude of the phase shift is proportional to the ratio of the axisymmetric radial and the tangential velocities. By adding an environmental mean flow to the picture, the entire curve will be shifted vertically upward or downward, depending on

---

[a]The GBVTD formulation is a more general form of the VTD formulation (Lee *et al.*, 1999; Jou *et al.*, 2008). VTD is a special case of GBVTD when a vortex is located at an infinite distance from the radar.

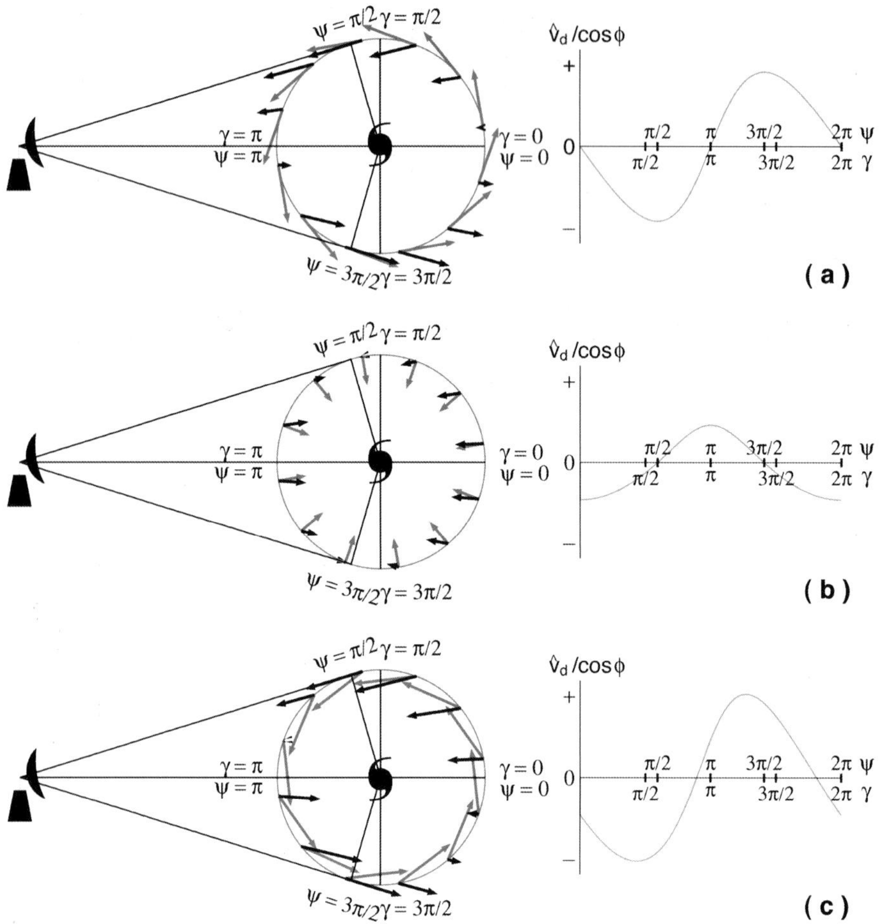

Figure 7. GBVTD concept for (a) axisymmetric rotation, (b) axisymmetric radial inflow, and (c) axisymmetric rotation and radial inflow. The left panels illustrate the circulation of axisymmetric TCs (red arrows) and their corresponding Doppler velocities (black arrows). The right panels illustrate the contributions of Doppler velocities versus azimuth angle. (From Lee *et al.*, 1999.)

the mean flow direction (i.e. the zero isodop will not bisect the dipole in an axisymmetric vortex). In fact, this simple illustration of the GBVTD concept summarizes the dipole signatures of an atmospheric vortex reviewed in Subsec. 2.2. Real atmospheric vortices also contain asymmetric structures; therefore, the resulting Doppler velocity profile is more complicated than that illustrated in Fig. 7(c). Different wave numbers can be distinguished via Fourier decomposition of the Doppler velocity profile along each radius in the vortex cylindrical coordinate. Therefore, the 3D vortex structure

can be deduced by compositing GBVTD analyses at each radius and height, providing that the circulation center at each altitude is known accurately. The issues related to vortex center finding will be discussed in Sec. 4.

Owing to insufficient independent equations in the VTD formulation, Lee *et al.* (1994) and Lee *et al.* (1999) assumed that the asymmetric radial winds are much smaller than their tangential counterparts, and acknowledged that the cross-beam mean wind component cannot be resolved, thus letting radial winds and the cross-beam mean wind alias into tangential winds to close the system. Note that if the radar is located at the center of a vortex, then only the radial wind component can be measured in that situation, and GBVTD becomes VAD. When the vortex is located at an infinite distance (i.e. the aspect ratio is small for practical purposes), the radar beams can be assumed parallel and the GBVTD formulation reduced to a simpler VTD formulation. With the knowledge of axisymmetric tangential and radial winds at each radius and altitude, several additional kinematic (convergence and vertical velocity) and dynamic (angular momentum and pressure gradient) quantities can be computed (e.g. Lee *et al.*, 2000; Lee and Wurman, 2005; Lee and Bell, 2007).

Sensitivity tests of the GBVTD technique using analytical vortices showed that accurate axisymmetric and good asymmetric vortex structures could be deduced from single Doppler observations. However, the distortion in phase and amplitude of asymmetric vortex structures worsens as the wave number increases (Lee *et al.*, 1999). Note that the analytical vortices simulated by Lee *et al.* (1999) did not contain asymmetric radial winds; thus, the biases from ignoring asymmetric radial winds could not be evaluated. Nevertheless, Lee *et al.* (1999) demonstrated that the GBVTD technique could extract plausible kinematic and wave structures embedded in the Doppler velocities of atmospheric vortices.

The EVTD technique, proposed by Roux and Marks (1996), expanded the VTD analysis to successive radial passes (legs) in order to improve the stability of the VTD analysis and extract wave number 1 radial wind information. EVTD first solves for a constant mean wind vector by performing a VTD analysis with limited coefficients on each altitude. The mean wind vector is then removed, and the residual Doppler velocities are used to solve for the coefficients at all radii simultaneously by a least-squares minimization of a cost function. This cost function minimizes three constraints: (1) the differences between mean wind relative Doppler velocities and EVTD-derived radial velocities, (2) the random variation of coefficients between the rings, and (3) the mean wind residuals resulting from the wave number 1 asymmetry. The EVTD coefficients deduced in each pass are then filtered to obtain coherent mesoscale structures over time. A similar extension of the GBVTD technique, the GB-EVTD technique, was proposed by Roux *et al.* (2004) to process tropical cyclone data collected by a ground-based Doppler radar on successive scans.

4. Atmospheric Vortex Circulation Center Deduced from Single Doppler Radar Data

Atmospheric vortex centers have different definitions based on available measurements and their associated assumptions, including the geometric center, defined as the centroid of the eyewall radar reflectivity (Griffin *et al.*, 1992); the dynamic center, defined as the minimum of the stream function, pressure, or geopotential height (e.g. Willoughby and Chelmow, 1982; Dodge *et al.*, 1999); and the vorticity (or circulation) center, defined as the point that maximizes the eyewall vorticity or circulation (e.g. Willoughby, 1992; Marks *et al.*, 1992; Lee and Marks, 2000). These centers are not necessarily collocated, and the uncertainties in estimating them, range from several kilometers to tens of kilometers in TCs (Willoughby and Chelmow, 1982; Lee and Marks, 2000; Harasti *et al.*, 2004).

Uncertainties and inconsistencies of estimated TC centers on the order of 10 km may not affect operational track forecasts, but they critically affect computing TC wind fields in a cylindrical coordinate system (Willoughby, 1992; Lee and Marks, 2000). This includes partitioning vortex wind fields from a Cartesian coordinate system to a cylindrical coordinate system (Willoughby, 1992; Marks *et al.*, 1992) and retrieving vortex wind fields from single Doppler radar data directly in cylindrical coordinates (Lee *et al.*, 1994; Roux and Marks, 1996; Lee *et al.*, 2000; Roux *et al.*, 2004; Lee and Wurman, 2005). Lee and Marks (2000) demonstrated that a vortex circulation center needs to be accurate within 1 km on a 20 km radius of maximum wind (RMW), or 5% of the RMW, to keep the nominal error of the apparent wave number 1 less than 20% of the maximum axisymmetric tangential winds (MATWs: V_{T0-max}) in the GBVTD analysis. Since the vorticity (circulation) center is directly related to the accuracy of the deduced vortex wind fields, it will be the focus of this section.

Harasti and List (2005) applied a principal component analysis (PCA) to the single Doppler velocity data arranged sequentially in a matrix according to the range and azimuth coordinates. For a Rankine vortex, the coordinates (location) of particular cusps in the curve of the first two eigenvector coefficients plotted against their indices are geometrically related to both the circulation center and the RMW. The uncertainty of the PCA method regarding asymmetric TCs is still a research topic.

The vorticity center can be computed from a given 2D horizontal vortex wind field and a simplex method (Neldar and Mead, 1965) to identify a point that maximizes the eyewall vorticity (Marks *et al.*, 1992). However, estimating TC centers using this technique is not straightforward with a single Doppler radar, because the velocity patterns do not provide a complete 2D wind field for direct vorticity computation. Lee and Marks (2000) proposed a novel approach, and invented the GBVTD-simplex algorithm, by using the simplex method to iteratively identify a center that maximized the GBVTD-derived eyewall vorticity. The search process of the simplex algorithm toward the true center of a Rankine vortex centered at (0, 60), with an RMW of 20 km and a maximum tangential wind of 50 m s^{-1}, is illustrated in Fig. 8.

The simplex method uses three operations — reflection, expansion, and contraction — based

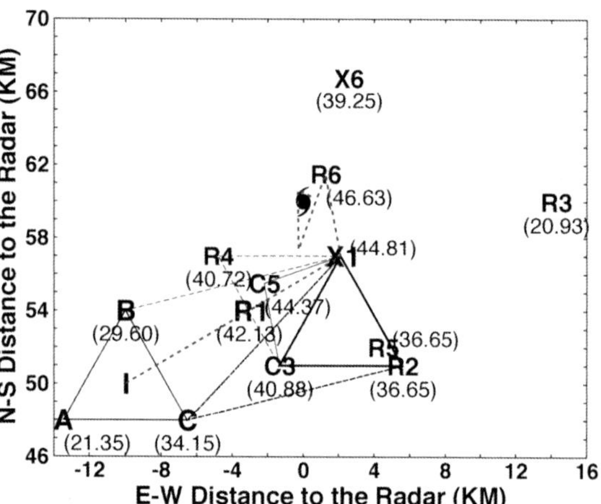

Figure 8. Illustration of the GBVTD-simplex convergence (thick dashed line) and the intermediate simplex from the initial guess at I (−10.0, 50.0). The triangle A–B–C is the initial simplex with a centroid at I. R, X, and C indicate the resulting points from reflection, expansion, and contraction operations with the number corresponding to the individual simplex. The TC center is located at (0.0, 60.0) with a hurricane symbol. Only the first six iterations are shown here. The GBVTD-derived axisymmetric tangential winds are listed in parentheses in m s^{-1}. Differently shaped lines distinguish intermediate triangles. (From Lee and Marks, 2000.)

on the value of interest to determine the subsequent search toward the maximum or minimum value. In this example, the MATWs are computed starting from vertices A, B, and C as the vortex centers of GBVTD analyses. Vertex A has the lowest MATW value, yielding a reflection point of A (R1) on segment BC. Then the MATW using R1 as a center [MATW(R1)] is computed and compared with those MATWs in A, B, and C. Subsequent operations to determine a new vertex replacing A to form the next simplex are based on whether MATW(R1) is a maximum (expansion), intermediate (R1 replaces A), or minimum (contraction). The above processes are repeated until a convergence criterion is reached. The thick, dashed line (in Fig. 8) represents the converging path of intermediate TC centers toward the true center at (0, 60).

The simplex method described here is efficient in finding a local maximum or minimum within the parameters specified by the users. In practice, it is recommended to run the simplex algorithm from various initial guesses, preferably from all quadrants of the "true" center. Without independent verifications, the grouping or scattering of the simplex results (end points) is the only statistical indication of the quality (uncertainty) of the GBVTD-simplex-derived vortex center. The mean error in the GBVTD-simplex-derived vortex centers using analytical vortices is $0.35\,\mathrm{m}$ ($<\,2\%$ of a $20\,\mathrm{km}$ RMW), which is significantly less than the $1\,\mathrm{km}$ radial grid spacing in the cylindrical coordinates. Random errors of $5\,\mathrm{m\,s^{-1}}$ added to the Doppler velocity do not change the mean error and standard deviation much. When applied to real atmospheric vortices, the standard deviations increased to $\sim\!2\,\mathrm{km}$ in Typhoon Alex ($\sim\!9\%$ of a $23\,\mathrm{km}$ RMW; Lee and Marks, 2000) and $\sim\!30\,\mathrm{m}$ in the Mulhall tornado ($\sim\!5\%$ of a $600\,\mathrm{m}$ RMW; Lee and Wurman, 2005). The accuracy of the GBVTD-simplex algorithm strongly depends on the accuracy of the GBVTD-derived MATW. Since the MATW can be biased by the unknown cross-beam mean flow and the unresolved wave number 2 radial flow, the GBVTD-simplex algorithm has experienced difficulties in deducing centers in elliptical (wave number 2) vortices such as Typhoon Herb and Nari (Michael Bell, 2005; personal communication). Further quantifying the error characteristics of the GBVTD-simplex algorithm in real atmospheric vortices requires independent measurement of the "true" vortex center from either *in situ* measurements or a second set of Doppler radar observations from a different viewing angle.

5. Atmospheric Vortex Structures

5.1. *Structure of tropical cyclones*

The first physically consistent 3D axisymmetric kinematic structure of a TC (Hurricane Gloria, 1985) deduced from NOAA WP-3D single Doppler radar data and the VTD technique was given by Lee *et al.* (1994). Figure 9 illustrates

axisymmetric structures of a TC previously only available from flight-level composite data (e.g. Jorgensen, 1984a,b), pseudo-dual Doppler analysis (e.g. Marks and Houze, 1987; Marks *et al.*, 1992; Franklin *et al.*, 1993), or numerical simulations (e.g. Rotunno and Emanuel, 1987). Key features in Fig. 9 include:

(1) An upright eyewall reflectivity ($\sim\!R = 13\,\mathrm{km}$) is accompanied by an MATW exceeding $60\,\mathrm{m\,s^{-1}}$ at an RMW of $\sim\!18\,\mathrm{km}$, while the axisymmetric tangential winds decrease rapidly both inside and outside the RMW [Fig. 9(a)]. The decrease of the MATW with height suggests a warm core structure from the thermal wind relationship.
(2) A shallow, low-level inflow ($\sim\!1\,\mathrm{km}$ deep) inside $R \sim 32\,\mathrm{km}$ peaks ($> 3\,\mathrm{m\,s^{-1}}$) as it approaches the eyewall (Fig. 9b). It converges near the eyewall [Fig. 9(c)] and turns into an updraft in the eyewall [$> 4\,\mathrm{m\,s^{-1}}$; Fig. 9(d)].
(3) Immediately outside the eyewall, a region of downdraft ($> 2\,\mathrm{m\,s^{-1}}$, between $R = 16\,\mathrm{km}$ and $40\,\mathrm{km}$) is consistent with weak reflectivity and radial inflow beneath the anvil outflow [dashed lines between $8\,\mathrm{km}$ and 13 altitude in Fig. 9(b)]. This branch of inflow beneath the anvil was seldom resolved in previous pseudo-Doppler analysis but appeared in numerically simulated TCs [Fig. 5(c) in Rotunno and Emanuel, 1987].

The VTD-derived asymmetric structure of Gloria over two successive legs suggested that the hurricane evolved from a wave number 1 [Fig. 10(a)] to a wave number 2 [Fig. 10(b)] structure within an hour. A pseudo-dual Doppler (PDD) analysis using combined data from these two legs showed a wave number 3 structure [Fig. 10(c)], suggesting temporal aliasing in the pseudo-dual Doppler analysis. *Again, the advantage of reducing the effective stationary time period from $\sim\!1$ hour (PDD) to $\sim\!20$ minutes (VTD) was clearly demonstrated.*

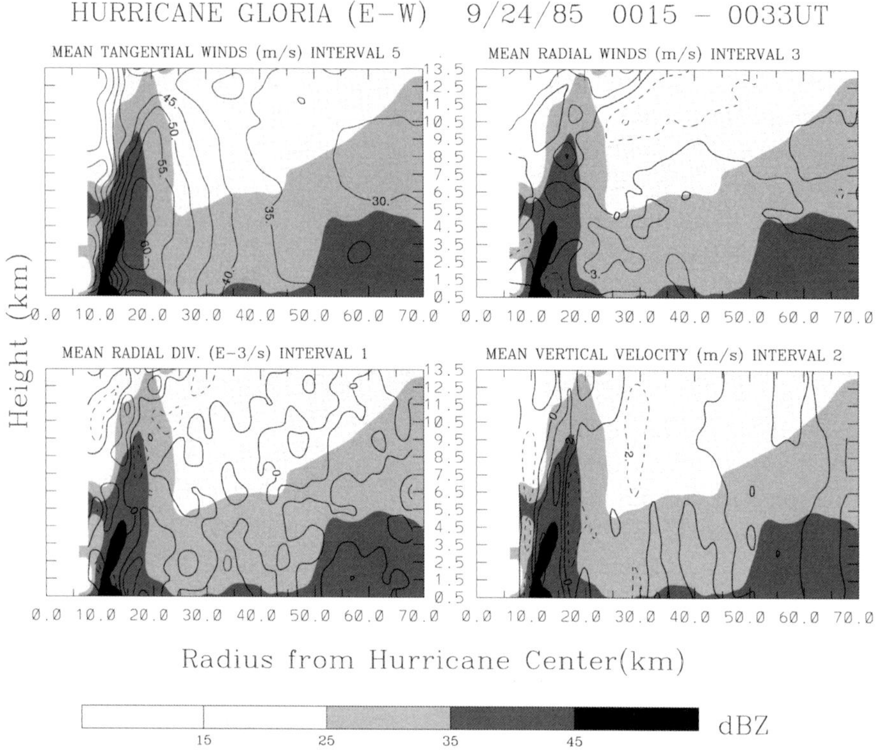

Figure 9. Radius–height cross-sections of (a) the mean tangential velocity, (b) the mean radial velocity, (c) the mean divergence, and (d) the mean vertical velocity, with a superimposed reflectivity region in shading. Positive valued contours represented by thin, solid lines; negative-valued lines are thin and dashed. A zero contour is represented by a thick, solid line. Units and the contour interval are indicated at the top of each panel. The cross-section extends from the hurricane center. (From Lee *et al.*, 1994.)

Lee *et al.* (1994) derived a hodograph of the Earth-relative horizontal wind from these two legs of data (Fig. 11) and found that Gloria's mean motion vector is ∼30° to the left of the deep-layer, density-weighted mean wind vector, consistent with those computed from the deep-layer mean wind in a much larger domain of several degrees in longitude and latitude (e.g. George and Gray, 1976; Chan and Gray, 1982; Holland, 1984; Dong and Neumann, 1986; Hanley *et al.*, 2001).

The structure and evolution of Hurricane Hugo (1989) over a 7-hour period on 17 September 1989 were presented by Roux and Marks (1996) using EVTD-derived winds from six successive eye crossings about 70 minutes apart. The axisymmetric and asymmetric structures in Hugo were qualitatively consistent with those

in Gloria, Alicia, and Norbert. More importantly, Roux and Marks (1996) demonstrated that plausible wave number 1 radial wind could be derived from the EVTD technique using multiple legs of airborne Doppler radar data. The EVTD-derived radial wind rotated from easterly at a 2 km altitude to southerly at 5 km and became southwesterly at 10 km, and compared favorably with the PDD-derived radial winds (Fig. 12).

The ability to deduce kinematic structures of landfalling TCs from coastal Doppler radar data has been demonstrated through Typhoon Alex (1987) (Lee *et al.*, 2000), the concentric eyewall (double wind maxima) in Typhoon Billis (2000) (Jou *et al.*, 2001), Hurricane Bret (1999) (Harasti *et al.*, 2003), and Hurricane Charley (2004) (Lee and Bell, 2007) using the

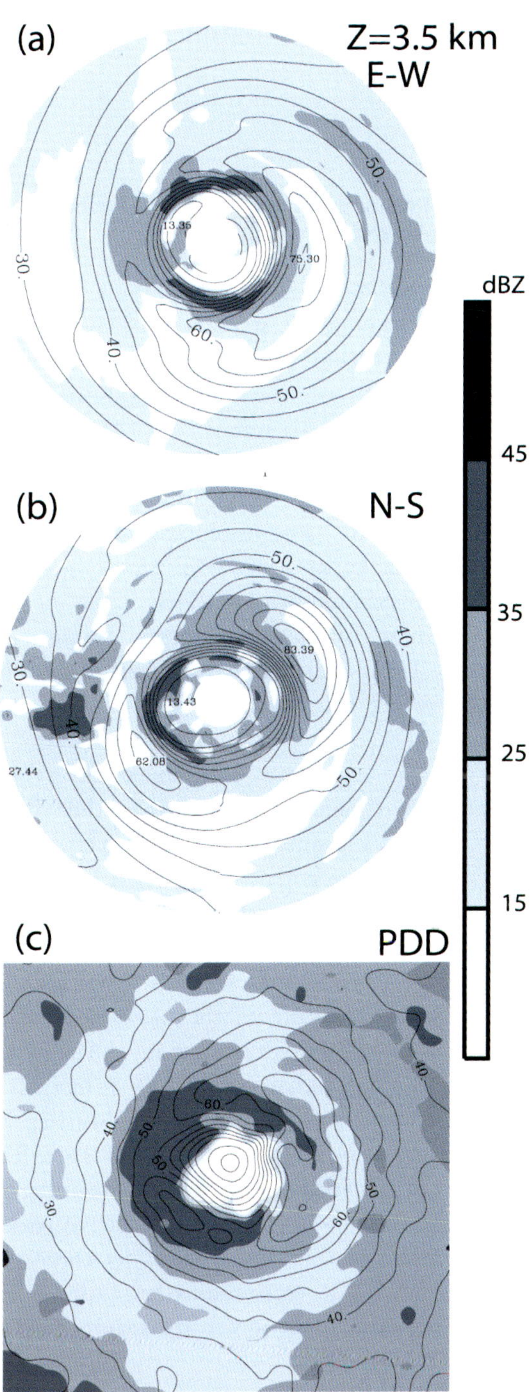

Figure 10. Total tangential winds (contours) and reflectivity (gray shades) for (a) the east–west leg, (b) the north–south leg, and (c) the PDD analysis at a 3.5 km altitude. (Adapted from Lee *et al.*, 1994.)

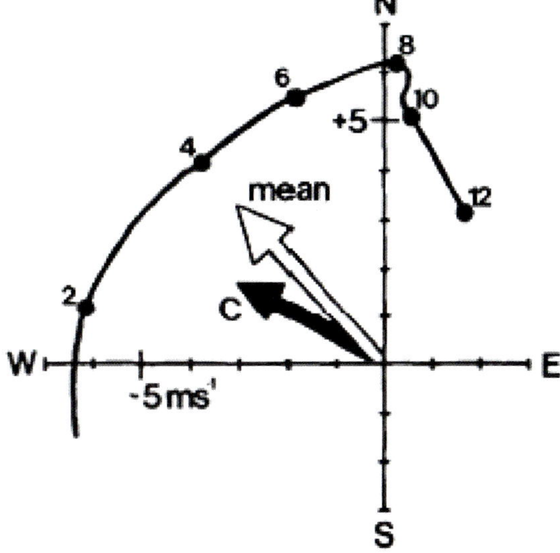

Figure 11. Hodograph of the Earth-relative horizontal winds and its deep-layer mean obtained from the VTD analysis (V_M) and the 12-hour mean motion of Hurricane Gloria (1985), V_H. (From Lee *et al.*, 1994.)

GBVTD analysis, Typhoon Nari (2001) using the EGBVTD technique (Liou *et al.*, 2006), and Cyclone Tina (2002) (Roux *et al.*, 2004) using the GB-EVTD technique. The structures of these landfalling TCs were consistent with previous studies but revealed detailed evolution of TCs (∼6–15 minute interval) compared with those deduced from airborne dual Doppler radar data (∼30–60 minute interval). Typical typhoon inner core structures are illustrated in Figs. 13(a) and 13(b). Lee *et al.* (2000) took the axisymmetric structure a step further to compute angular momentum and cyclostropic perturbation pressure in Alex (Fig. 13). The angular momentum pattern [Fig. 13(c)] is near-upright in the eyewall, similar to those deduced from dual Doppler analysis in Hurricane Norbert [Fig. 12(f) in Marks *et al.* 1992]. The retrieved perturbation surface pressure at the center [Fig. 13(d)] is 23 hPa lower than the pressure at 60 km radius. The radial perturbation pressure pattern compared favorably with the pressure pattern derived from surface pressure measurements assuming axisymmetric

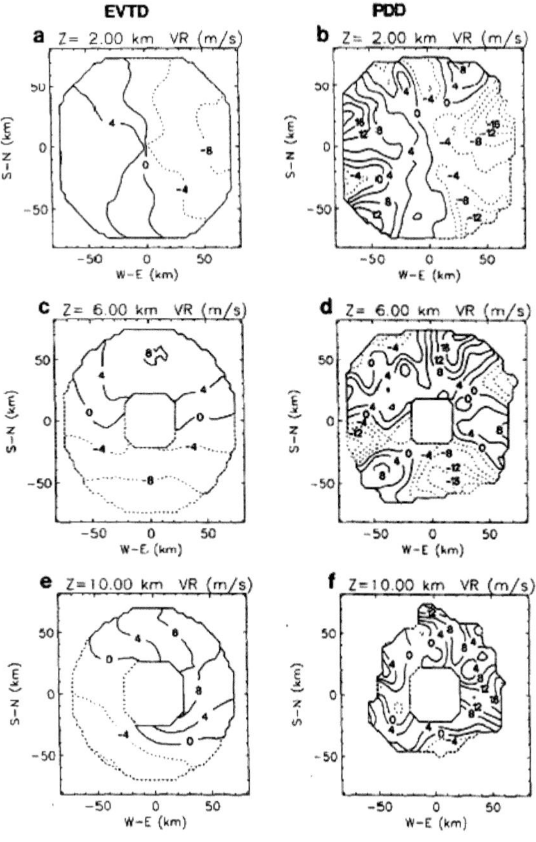

Figure 12. Horizontal cross-sections of the radial wind component at (a, b) 2 km, (c, d) 6 km, (e, f) 10 km altitudes. The left panels (a, c, e) refer to EVTD-derived values, the right panels (b, d, f) to PDD-derived values. (From Roux and Marks, 1996.)

pressure distribution from $R = 20$ km to $R = 60$ km (Fig. 14; note that there is no surface pressure measurement within 20 km of the TC center). This agreement suggests good internal consistency of the GBVTD-retrieved axisymmetric circulation and the retrieved pressure deficit, which is likely accurate to ~2–3 hPa if there is no significant aliasing from the unresolved cross-beam mean wind and wave number 2 asymmetric radial wind. Lee and Bell (2007) showed that the rapid intensification process of Hurricane Charley (2004) could be captured by GBVTD-retrieved pressure deficit in real time. The result is very encouraging, and it is feasible to estimate the central pressure of landfalling

TCs from coastal Doppler radar data. This could be a valuable tool for diagnosing the intensity of landfalling TCs.

5.2. *Structure of mesoscale vortices in MCS*

Zhao *et al.* (2007) presented the first GBVTD-derived 3D kinematic structure of a mesovortex embedded within a quasi-linear convective system (QLCS) over the northern Taiwan Strait. A hook-echo-like structure [Fig. 15(a)] developed at the southern end of a convective line segment and possessed a pronounced velocity dipole signature. The asymmetric structure of the mesovortex at 4 km [Fig. 15(a)] indicates that the maximum wind of $25\,\mathrm{m\,s^{-1}}$ was located in the rear of the mesovortex (upper left corner). The asymmetric component of the tangential wind was retrieved to wave number 2 and most of the amplitude was contained in the wave number 1 component. At a 2 km height, the kinematic structure of the vortex [Fig. 15(b)] was quite different. In the rear of the mesoscale vortex, there was a secondary maximum of tangential wind consistent with the wind distribution at 4 km. However, the most intense tangential wind of $25\,\mathrm{m\,s^{-1}}$ was located at the front quadrant of the mesoscale vortex (upper right). Through examining the sequence of the Doppler radial velocity, it is evident a strong rear-to-front flow developed during the formation period of the mesoscale vortex. The downward penetration of this rear-to-front jet enhanced the low-level cyclonic vorticity. At the leading edge of this rear-to-front jet there was a pronounced flow convergence triggering a line of convection evident in the reflectivity field at the 2 km height.

The MATW [Fig. 15(c)] at the lowest retrieved level (1 km height) is $18\,\mathrm{m\,s^{-1}}$. It is consistent with the location of the maximum mean reflectivity field. The axis of the MATW is tilted inward to the center of the vortex with

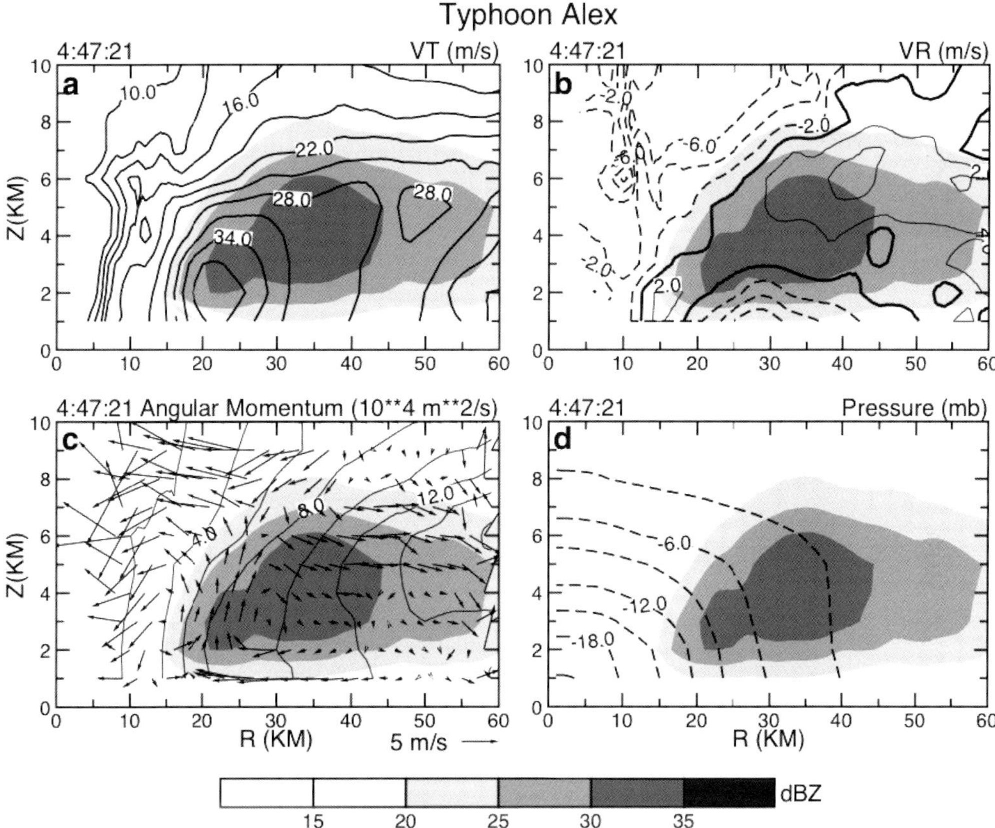

Figure 13. Axisymmetric structure of (a) tangential winds, (b) radial winds, (c) angular momentum, and (d) perturbation pressure, at 0417 LST. Solid (dashed) lines represent positive (negative) values. The thick, solid line is the zero contour. Axisymmetric reflectivity factors are in color. Vectors in (b) are composed of axisymmetric radial winds and vertical velocities. (Adapted from Lee *et al.*, 2000.)

increasing altitude. This feature stands in contrast to the inner core structure of a mature tropical cyclone in which the tilt is outward with increasing altitude. The mean radial wind [Fig. 15(d)] outside (inside) the RMW is inward (outward) with an intensity of $5\,\mathrm{m\,s}^{-1}$ $(2\,\mathrm{m\,s}^{-1})$ and convergence near the RMW. It has been demonstrated in the above presentation that the GBVTD technique is a powerful tool for retrieving the kinematic structure of mesoscale vortices.

5.3. *Structure of tornados*

The first GBVTD analysis of a tornado, near Bassett, Nebraska, was presented by

Bluestein *et al.* (2003), who illustrated coherent tornado structures over a 3.5-minute period. The W-band radar (Bluestein *et al.*, 1995) can only scan in one elevation angle; therefore, only horizontal structure can be deduced from this dataset. The Doppler velocity pattern and the corresponding GBVTD analysis of the Bassett tornado are deduced (Fig. 9 in Bluestein *et al.*, 1995). However, with a scan update time of approximately 12 seconds, unprecedented evolution of the tornado was revealed. The axisymmetric tangential and radial winds over the 3.5 minutes are portrayed in Fig. 16. The MATW is $\sim 28\,\mathrm{m\,s}^{-1}$ at R $\sim$ 170 m. The axisymmetric radial winds are positive (outflow) inside $R = 200\,\mathrm{m}$ and negative (inflow) outside

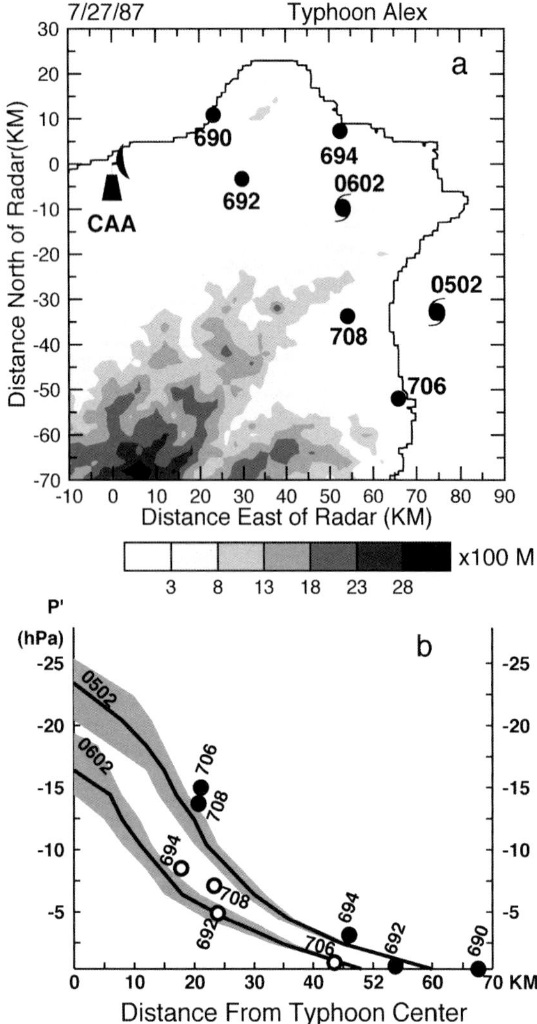

Figure 14. (a) The surface stations report hourly data in northern Taiwan. Typhoon locations at 0502 and 0602 LST are indicated by the typhoon symbols. The terrain is in gray shades. (b) The thick lines are the GBVTD-derived axisymmetric perturbation pressure (P') at the surface. Circles are the perturbation pressures computed from five surface stations. (Adapted from Lee *et al.*, 2000.)

$R = 200\,\mathrm{m}$ which suggests a two-cell circulation. The switch-over point of the inflow and outflow is outside the RMW. This may be an artifact of the additional centrifugal effects experienced by the debris (targets of the radar), not by the winds (Dowell *et al.*, 2005). The structure of another tornado, near Stockton,

Kansas, collected by the same radar, has been presented by Tanamachi *et al.* (2006). The near-stationary wave number 2 structure was attributed to the superimposition of an axisymmetric vortex onto a near-stationary deformation field (Bluestein *et al.*, 2003). Recent studies by Lee *et al.* (2006a) and Tanamachi *et al.* (2006) indicated that this near-stationary wave number 2 asymmetry might be an artifact due to the spatial aliasing of a fast-moving tornado and a relatively slow scan rate of the W-band radar in this particular situation. Research is underway to systematically examine this issue.

In contrast, DOWs are capable of scanning multiple elevation angles at a slower volume update rate ($\sim$1 minute), but the 3D structures of tornados can be retrieved. Lee and Wurman (2005) produced the first comprehensive study of tornado circulations collected by DOWs on 4 May 1999 near Mulhall, Oklahoma, USA. The axisymmetric structure of the Mulhall tornado at 0310:03 (hhmm:ss) UTC is presented in Fig. 17. The axisymmetric tangential wind [Fig. 17(a)] (hereafter, all quantities are axisymmetric unless stated otherwise) profiles resemble a miniature, intense TC (about 1/15 in length scale) and possess characteristics commonly seen in the inner core region of a mature TC (e.g. Marks *et al.*, 1992; Lee *et al.*, 1994; Roux and Marks, 1996). The depth of the inflow layer reached 1 km but the most intense inflow was clearly confined near the surface. The downdraft magnitude of $30\,\mathrm{m\,s^{-1}}$ is comparable to the estimated downdraft speed in the Dimmitt tornado with similar intensity (Wurman and Gill, 2000). The secondary (meridional) circulation [Fig. 17(b)] provided the first observational evidence that a classical two-cell circulation, commonly seen in tornado vortex chamber and numerical simulations with a swirl ratio greater than 1 (e.g. Ward, 1972; Rotunno, 1979), does exist in intense tornados. The swirl ratios of the Mulhall tornado computed from the GBVTD-derived axisymmetric circulation during the entire observational period

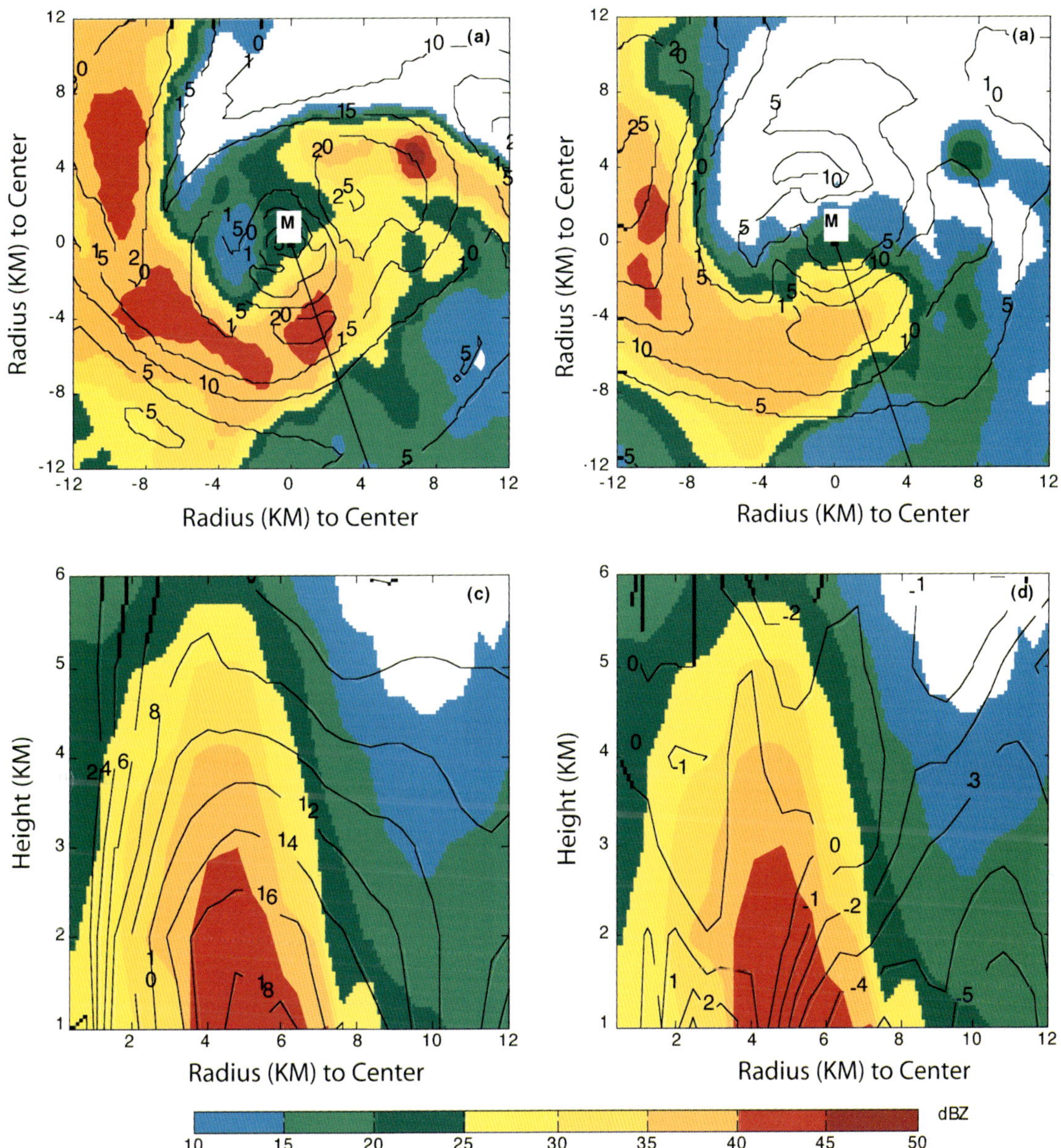

Figure 15. GBVTD-derived horizontal vortex circulations at (a) 2 km and (b) 4 km altitudes. The lower panels show (c) the axisymmetric tangential winds and (d) radial winds of the mesocyclone.

($\sim$14 minutes) are all above 2 (Table 1), consistent with the two-cell structure in the secondary circulation and the multiple vortices observed in the Mulhall tornado. A striking example is seen in Figs. 3(c) and 3(d), where six small vortices can be identified on the west side of the tornado, especially in the velocity field,

at 1316:28 (Wurman, 2002; Lee and Wurman, 2005).

The pressure deficit from advection terms [Fig. 17(c)] is consistent with the secondary circulation, a component rarely, if ever, resolved in past observational studies of tornadoes. The total pressure deficit [Fig. 17(g)] at $z = 50\,\mathrm{m}$

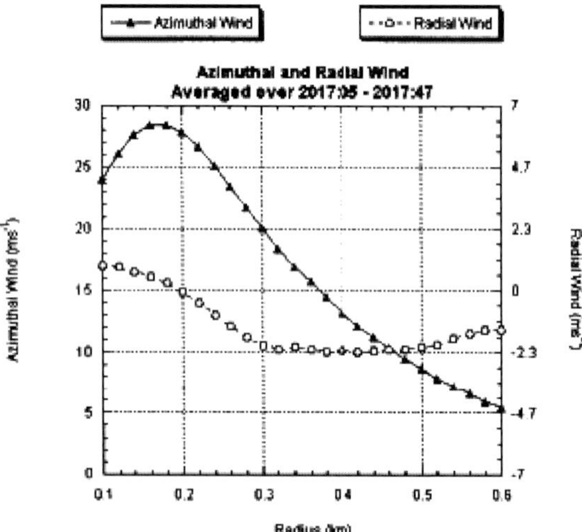

Figure 16. Averaged axisymmetric tangential and radial winds obtained by averaging across all radii. (From Bluestein *et al.*, 2003.)

is $81\,\text{hPa}$ lower than the pressure at $R = 3\,\text{km}$, which is dominated by the cyclostrophic pressure [Fig. 17(e)] with adjustment from the advection pressure [Fig. 17(c)]. These pressure deficits are comparable with rare *in situ* observations (Winn *et al.*, 1999; Lee *et al.*, 2004; Wurman and Samaras, 2004; Lee *et al.*, 2004) in strong tornadoes. This magnitude is also consistent with analytical pressure analysis from a similar wind profile of a Rankine combined vortex (see Appendix in Lee and Wurman, 2005).

The vorticity pattern in Fig. 17(f) shows an annular (or so-called "ring") vorticity profile where the peak vorticity of $0.28\,\text{s}^{-1}$ is concentrated in an annulus between 300 and $500\,\text{m}$ in radii. The radial vorticity gradient of a ring vorticity profile changes sign and satisfies the necessary condition for barotropic instability, usually accompanying mature TCs (Mallen *et al.*, 2005). The angular momentum contours [Fig. 17(h)] nearly upright inside the RMW and their value increase with the radius while the angular momentum outside the RMW increased at a slower rate. This pattern is quite similar to those resolved within a mature TC and suggests that the low-level inflow brings in higher angular

momentum and the secondary circulation maintains the vortex (e.g. Lee *et al.*, 2000; Marks *et al.*, 1992; Rotunno and Emanuel, 1987).

6. Assimilating VTD-Derived Winds into the Numerical Model

The strength of the VTD family of SDWR techniques is in providing good axisymmetric and asymmetric tangential winds but lacking asymmetric radial winds. One of the outstanding issues is whether the asymmetric divergent component can be extracted from the VTD-derived incomplete wind fields in order to initialize numerical simulations or data assimilation.

Lee *et al.* (2003) proposed an innovative concept using the mesoscale vorticity method (MVM) to retrieve inner core divergence and vertical velocity from the frequently available vertical vorticity field from MM5 simulations. The MVM derives vertical velocity from vorticity variation in space and time based on the mesoscale vorticity equation, obtained by neglecting the solenoidal term in the full equations. When the mesoscale vorticity equation is formulated in Lagrangian coordinates composed of three nonlinear ordinary differential equations, the vertical velocity can be solved numerically along each characteristic with proper boundary conditions. In their study, observing system simulation experiments were undertaken to demonstrate that the vertical velocities derived from the MVM were less susceptible to errors in horizontal winds than the kinematic method for TCs.

Lee *et al.* (2006) applied the MVM to derive the inner core vertical velocity and divergent component of the horizontal wind from single Doppler radar observations of Hurricane Danny (1997) using three consecutive volumes of GBVTD analysis. The MVM-retrieved divergent component of the wind vectors superimposed on radar reflectivity of the inner core region of Danny is illustrated in Fig. 18(a). A major convergence zone was found

Table 1. Swirl ratio (S) of the Mulhall tornado from 0310:03 to 0323:12 and the key parameters used in the calculation. See details in Lee and Wurman (2005).

UTC	R km	h km	ur m/s	vr m/s	S
0310:03	2	1.0	16	35	2
0310:57	1.5	0.5	16	40	4
0312:05	1.5	0.6	15	40	3
0310:59	1.5	0.5	10	40	6
0314:16	2.0	0.8	12	40	4
0315:13	2.5	0.6	16	30	4
0316:28	1.5	0.6	16	40	3
0317:22	1.5	0.6	8	32	5
0318:25	2.5	0.8	16	25	2
0319:18	2.8	0.7	20	20	2
0320:24	3.0	0.7	20	25	3
0321:20	3.0	1.0	18	25	2
0322:16	3.0	1.0	10	20	3
0323:12	3.0	1.0	12	20	3

at the RMW, east of the eyewall over the area of a bow-shaped echo. The updraft inside the echo drew in significant air from the surrounding area, including air from the eye, to sustain the convection. Another major convergence area associated with high reflectivity was located in the northwest of the eye. Vertical structure portrayed on an east–west cross-section through the eye is shown in Fig. 18(b). The wind vectors are composed of the density-weighted divergent E–W component and the vertical velocity. The prominent feature is the narrow band of updraft located 20 km east of the eye, where flow converges from the east and west of the line. The outward tilt of the updraft matches the reflectivity pattern well. The strength of the updraft increases with height and peaks at about a 6–7 km altitude, above which the flow diverges away from the eyewall. Also shown is a secondary circulation induced by the eyewall updraft including low-level convergence, upper-level divergence, and weak downward motion inside the eye ($x = 0$ km).

The initial success of this approach not only provides an avenue for retrieving asymmetric radial winds from the VTD analysis, but also opens the door to use the GBVTD-MVM-derived, dynamically consistent vortex structures as initial conditions for simulating atmospheric vortices. The update rate of 6 minutes will enable the use of these dynamically consistent wind fields in data assimilation. This is an area that still needs to be explored in future research.

7. The Generalized VTD (GVTD) Technique — A New Paradigm

Since the invention of weather Doppler radar, the kinematic structures of weather phenomena have been displayed exclusively in radial velocity (V_d) in a spherical coordinate system. The characteristics and limitations of atmospheric vortices displayed in V_d have been discussed in detail in this article. A new paradigm was proposed by Jou et al. (1996) to display the atmospheric vortex in V_dD/R_T space rather than in the convectional V_d space. Its potential impact on atmospheric vortex research was not realized until recently. Displaying the atmospheric vortex in V_dD/R_T space eliminates the geometric distortion of a vortex inherent in the traditional V_d space. In this framework, the dipole structure of an axisymmetric vortex is a function of the linear azimuth angle and is not distorted by a varying aspect ratio. Therefore, the center will always be the half point on the line connecting the peak inbound and outbound V_dD/R_T (Jou and Deng, 1997). Jou et al. (1996) named this center-finding algorithm the velocity distance azimuth display (VDAD) method.

The mathematic foundation of representing atmospheric vortices in V_dD/R_T space and the characteristics of the VDAD method have been documented and examined by Jou et al. (2008), who demonstrated that the GBVTD formulation when representing in V_dD/R_T space becomes exact mathematically without the need to approximate $\cos\alpha$ as required in GBVTD. This new approach is named the generalized VTD (GVTD) technique. Hence, the geometric distortion of the dipole inherent in V_d space

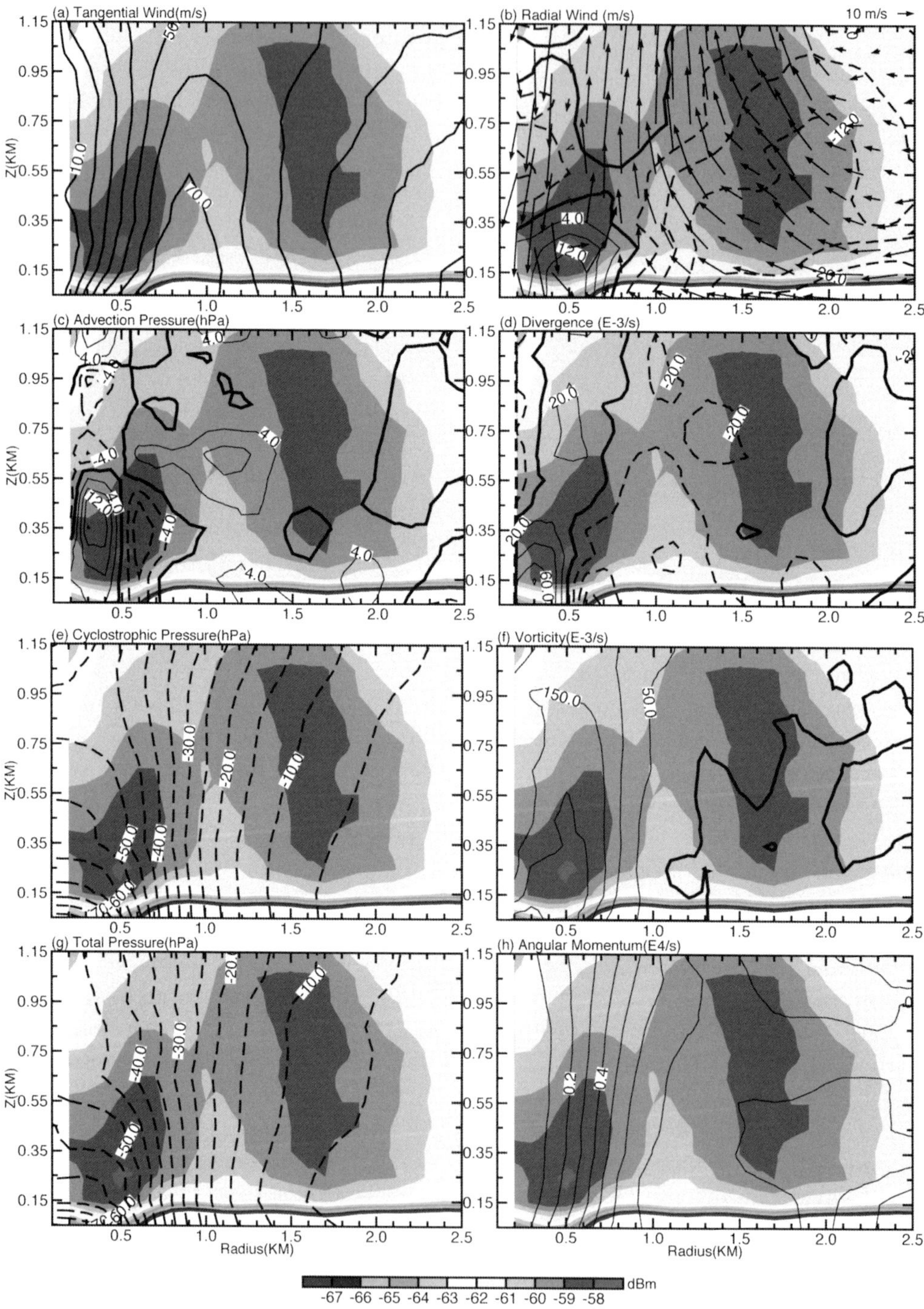

Figure 17. Axisymmetric structure (radius–height) of the Mulhall tornado at 4 May 1999 0310:03 UTC. The return power is shown in gray shades and contours represent (a) tangential wind, (b) radial wind, (c) advection pressure deficit, (d) divergence, (e) cyclostrophic pressure deficit, (f) vorticity, (g) total pressure deficit, and (h) angular momentum. The vectors in (b) illustrate the secondary circulation of the Mulhall tornado. Solid (dashed) lines represent positive (negative) values. (From Lee and Wurman, 2005.)

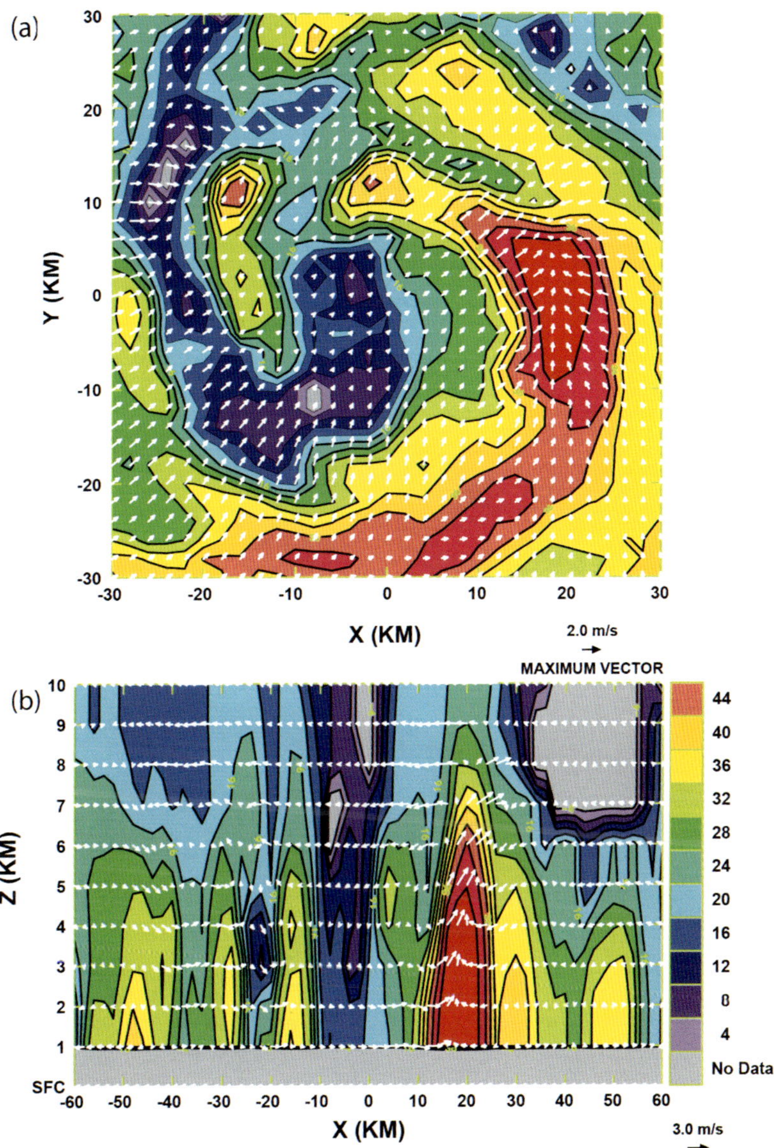

Figure 18. Radial flows superimposed on the radar echoes over the area surrounding the bow-shaped and sausage-shaped echoes. (b) Vertical structure of the secondary circulation wind vectors composed of the east–west divergent and the vertical velocities. (From Lee *et al.*, 2005.)

and the aliasing at higher wave numbers in the GBVTD-retrieved vortex circulation are effectively removed in the GVTD technique. More importantly, the constant mean wind appears as a set of parallel lines in $V_d D/R_T$ space, a much simpler signature to identify than the diverging lines in V_d space. Jou *et al.* (2008) also showed that VTD and GBVTD are special cases of GVTD. A comparison of the differences between the GBVTD- and GVTD-derived vortex structures from wave number 0 to 3 is illustrated in Fig. 19. Clear advantages of the GVTD technique can be seen in the retrieved vortex structure, especially in wave numbers 2 and 3.

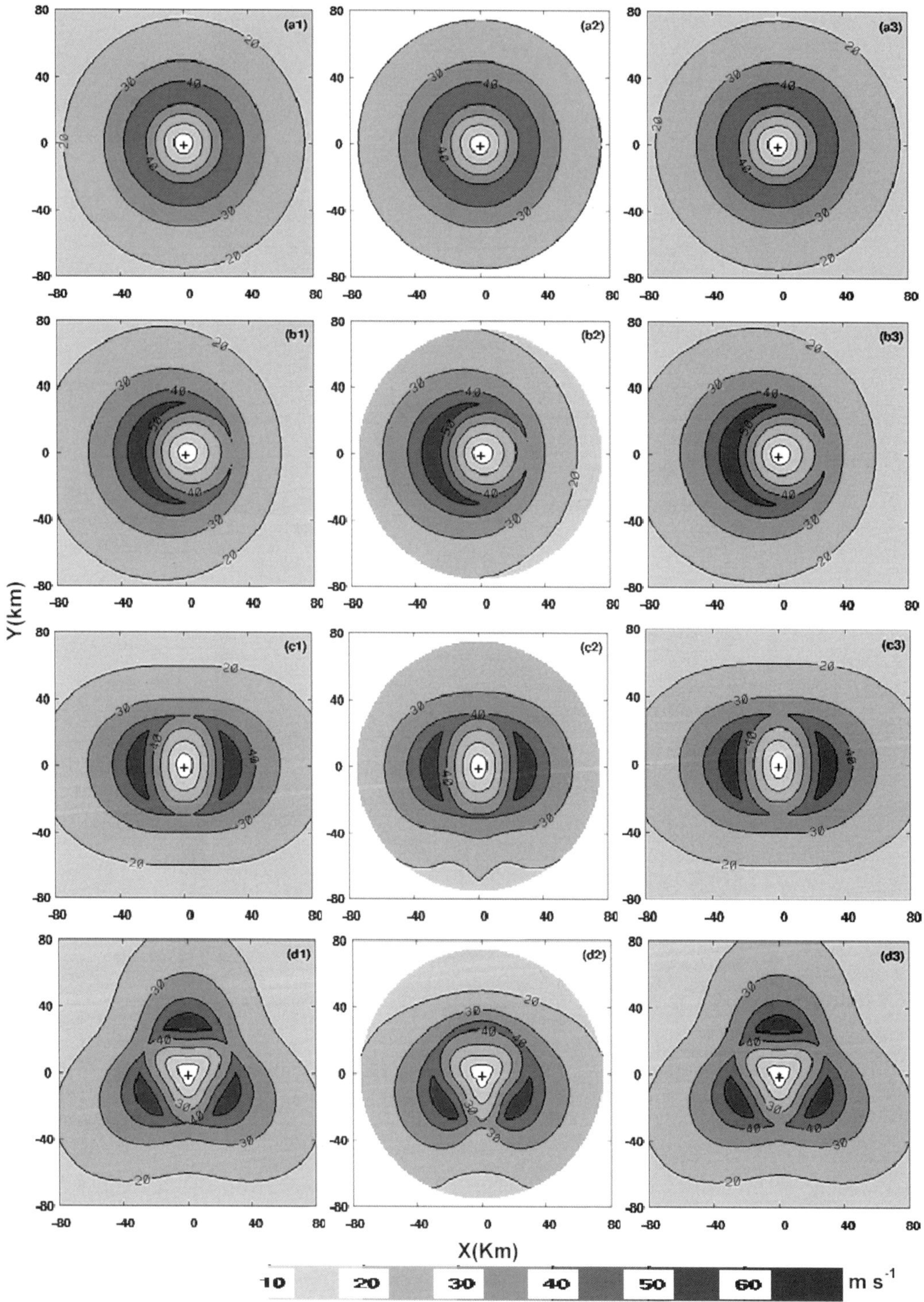

Figure 19. Comparison of GBVTD- and GVTD-retrieved vortex structure. (a1) The simulated axis-symmetry wind field, (a2) the GBVTD-retrieved wind field, and (a3) the GVTD-retrieved wind field. (b1–b3) The same as (a1–a3) but for the wave number 1 case, (c1–c3) the same as (a1–a3) but for the wave number 2 case; and (d1–d3) the same as (a1–a3) but for the wave number 3 case. (From Jou *et al.* 2008.)

GVTD also expands the analysis domain to the point where it is no longer limited by $R/R_T < 1$ as in GBVTD. The dramatic improvement occurs when $R/R_T > 1$, i.e. the radar is located inside the RMW where it does not observe the full component of the maximum tangential wind of a vortex. While the GBVTD technique cannot retrieve the full circulation in this situation, the GVTD technique can still retrieve the full tangential wind component (Fig. 20). Therefore, the GVTD technique can significantly expand the analysis domain. However, it still suffers from the VTD closure assumption, where the asymmetric tangential and radial winds cannot be separated in its current framework.

8. Outlook

SDWR algorithms remain the only practical avenue for deducing 3D atmospheric vortex structures for research and operation forecasts in the foreseeable future. While there are obvious limitations associated with the VTD family of techniques discussed in this article, it is evident that the VTD family of SDWR algorithms can provide a more complete picture of atmospheric vortices than other SDWR techniques. While this review article focuses on research results that have already appeared in refereed journals, research efforts to address these challenges are presented in this section.

8.1. *Improve mean wind estimates*

The GBVTD primary circulation can be biased by the cross-beam component of the environmental wind. This bias ultimately affects estimates of the central pressure (Lee *et al.*, 2000; Lee and Bell, 2004). Initial attempts indicated that this component of the environmental wind could be estimated by the hurricane volume velocity processing (HVVP) method (Harasti, 2003, 2004; Harasti and List, 1995; Harasti and List, 2001) and the GVTD technique. Harasti *et al.* (2005) included the mean wind derived

from HVVP and produced 3–8 hPa corrections to the GBVTD pressure values in Hurricane Charley (2004). The special signature of the mean wind in the $V_d D/R_T$ space provides a promising way to accurately estimate the mean wind (Jou *et al.*, 2008). An objective method to estimate the mean wind vector is under development. Future research will need to address the quality of the mean winds derived from these techniques and to validate the pressure retrieval results for other TCs with dropsonde data and for tornados with *in situ* measurements.

8.2. *Improving the VTD closure assumptions*

The VTD closure assumptions, based on the level of understanding of TC structures and dynamics in the early 1990s, can certainly be improved. It has been shown that vorticity waves [so-called vortex Rossby waves (VRWs); Montgomery and Kallenbach, (1997)] live and propagate along vorticity gradients near the inner core of a TC. Their associated potential vorticity mixing processes have been widely accepted as the mechanism for forming polygonal eyewalls, mesovortices, and spiral rainbands (e.g. Guinn and Schubert, 1993; Montgomery and Kallenbach, 1997; Schubert *et al.*, 1999; Kossin *et al.*, 2000). The existence of the VRW in hurricane-like vortices has been simulated in numerical models of varying complexity, ranging from a barotropic, nondivergent, three-region vortex model (e.g. Terwey and Montgomery, 2002) to full-physics, nonhydrostatic models (e.g. Wang, 2002a, b; Chen and Yau, 2001).

The VRW dynamics provides phase and amplitude relationships between asymmetric tangential wind and radial wind that may be used as closure assumptions for the VTD formulation. A simple form of the VRW is an idealized, small-amplitude, vortex edge wave traveled on the vorticity gradient of a Rankine vortex (Lamb, 1932) where a wave number 2

Wen-Chau Lee and Ben Jong-Dao Jou

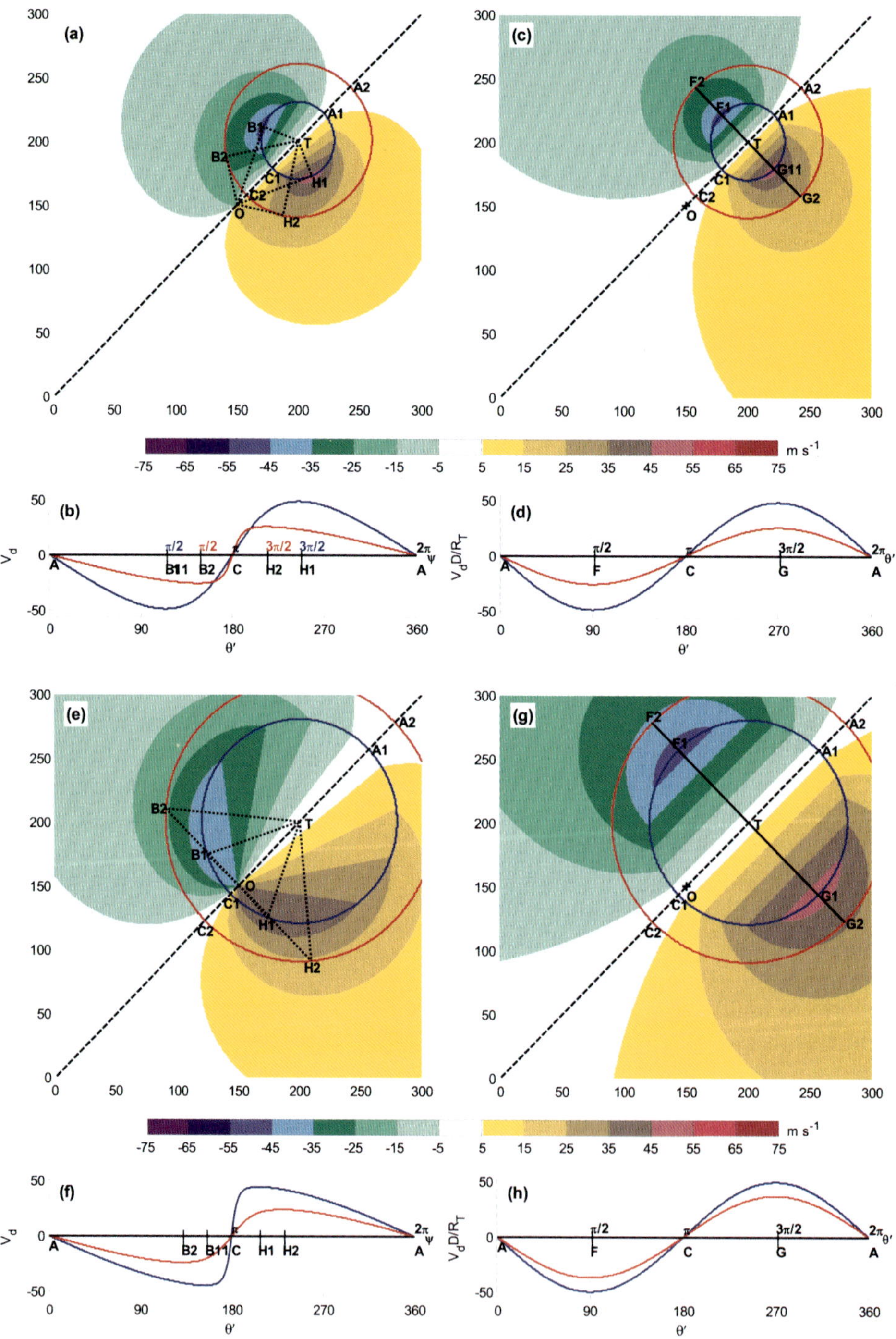

Figure 20. Comparison of V_d and $V_d D/R_T$ displays for two different radii (delineated as red and blue lines) for two vortices with different RMWs: (a) V_d display for a pure rotating vortex with $R_{\max} = 30$ km; (b) V_d profiles at $R = 30$ and 60 km; (c) same as (a) except for the $V_d D/R_T$ display; and (d) same as (b) except for the $V_d D/R_T$ profiles; (e)–(h) are the same as (a)–(d) except for $R_{\max} = 80$ km $> R_T$ and the two V_d profiles are at $R = 80$ and 110 km. The center, T, is located at $(x, y) = (200$ km, 200 km$)$ and the hypothetical Doppler radar, O, is located at $(x, y) = (150$ km, 150 km$)$ with $R_T = 50\sqrt{2}$ km. (From Jou $et~al.$ 2008.)

disturbance possesses an elliptic streamline (e.g. Lee *et al.*, 2006a). Lee *et al.* (2006a) also showed that a deformation field is essentially a wave number 2 disturbance in cylindrical coordinates. They demonstrated that the GBVTD-retrieved signatures are different between these two types of elliptical vortices. Lee *et al.* (2002) applied the simple vortex edge wave model in the GBVTD analysis of Typhoon Herb (1996) and resolved two pairs of counter rotating vortices propagating downstream, consistent with the counterclockwise rotation of the elliptical reflectivity pattern. However, the sharp vorticity gradient of the Rankine vortex created an unrealistic discontinuity near the RMW. One of the challenges is how to formulate the VRW disturbance in realistic axisymmetric, atmospheric vortices where the simple form of the equations illustrated by Lamb (1932) is no longer valid.

8.3. *Quantifying uncertainties of the GBVTD-simplex center-finding algorithms*

Efforts have been made to quantify the uncertainties of the GBVTD-simplex algorithm. Murillo *et al.* (2001) statistically evaluated the center information (including location, RMW, and MATW) over time from independently derived centers from two separate WSR-88Ds in Hurricane Danny (1997). They found that the individual tracks computed from two radars are consistent but that variations of individual centers can be large. A more consistent set of centers can be obtained by simultaneously considering time continuity in center locations RMW and MATW (Murillo *et al.*, 2002). Bell and Lee (2002, 2003) incorporated statistical information on locations, RMWs, and MATWs of past centers to objectively select the most probable center at each time, and the procedure can be automated after 6–7 time periods, when reliable statistics can be built. Future research needs to address the uncertainties in the GBVTD-simplex algorithm resulting from multiple wind maxima (e.g.

concentric or nonconcentric eyewalls) and the biases in the axisymmetric tangential winds from wave number 2 radial winds.

8.4. *Operational aspects*

Algorithms using the vortex reflectivity and dipole Doppler velocity structure to identify the vortex center and maximum wind speed have been implemented for many years (e.g. Lemon *et al.*, 1978; Wood and Brown, 1992; Wood, 1994). The EVTD-derived TC inner core structure in Hurricane Olivia (1994) was transmitted to the Tropical Prediction Center (TPC) via satellite link (Dodge *et al.*, 1995). More complicated suites of algorithms that combine these simple algorithms with several advanced techniques described in this review have drawn attention in recent years from operational centers such as the Central Weather Bureau in Taiwan (P.-L. Chang, personal communication) and the TPC (e.g. McAdie *et al.*, 2001; Harasti *et al.*, 2004). Digital radar data are available at operational centers, opening up the possibility of applying algorithms discussed in this article in operational settings in real time.

The ability to estimate the absolute central pressure and capture the rapid intensification in Hurricane Charley was attempted in Harasti *et al.* (2005) where the GBVTD-HVVP-derived pressure fields were within 2–3 hPa of the central pressure measured by the GPS dropwindsondes. A software package, Vortex Objective Radar Tracking and Circulation (VORTRAC), based on the GBVTD-HVVP framework, has been developed and transferred to the TPC and field-tested in the 2007 hurricane season (Lee *et al.*, 2006b; Harasti *et al.*, 2007). VORTRAC has been officially accepted by the TPC/NHC for operation use starting from the 2008 hurricane season.

Adequate scanning strategy selected by radar operators is essential for the success of retrieving TC wind fields using the VTD family of SDWR techniques. Radar operators have to choose the tradeoff between range folding (covering more storms) and velocity folding (detecting maximum wind). Examples shown by Lee and Bell (2004) suggested that it is more important to choose a pulse repetition frequency that gives the maximum range when sampling TCs because velocity folding is generally recoverable.

Further research aimed at integrating and capitalizing on the strengths of the individual algorithms described in this article, along with improved physical understanding of the underlying vortex dynamics, will help to better constrain the vortex solutions obtained from single Doppler radar data. These dynamically consistent wind fields can then be used for operational guidance, as initial conditions or data assimilations in numerical models, and as research tools for gaining further insight into atmospheric vortex structure.

Acknowledgements

The authors thank Mr M. Bell, Mr S. Ellis, and the two anonymous reviewers for their thoughtful comments that drastically improved this article.

[Received 26 July 2007; Revised 26 August 2007; Accepted 30 September 2007.]

References

Armijo, L., 1969: A theory for the determination of wind and precipitation velocities with Doppler radars. *J. Atmos. Sci.*, **26**, 570–573.

Baynton, H. W., 1979: The case for Doppler radars along our hurricane affected coasts. *Bull. Amer. Meteor. Soc.*, **60**, 1014–1023.

Bell, M., and W.-C. Lee, 2002: An objective method to select a consistent set of tropical cyclone circulation centers derived from the GBVTD-simplex algorithm. Preprints, 25th Conf. on Hurricane and Tropical Meteorology, (29 Apr. 1–3 May 2002, San Diego, CA, USA, Amer. Meteor. Soc.), pp. 642–643.

Bell, M., and W.-C. Lee, 2003: Improved tropical cyclone circulation centers derived from the GBVTD-simplex algorithm. Preprints, 31st Conf. on Radar Meteorology, (Seattle, Washington, USA, AMS), pp. 1012–1014.

Biggerstaff, M. I., L. J. Wicker, J. Guynes, C. Ziegler, J. M. Straka, E. N. Rasmussen, A. Doggett, IV, L. D. Carey, J. L. Schroeder, and C. Weiss, 2005: The shared mobile atmospheric research and teaching radar: a collaboration to enhance research and teaching. *Bull. Amer. Meteor. Soc.*, **86**, 1263–1274.

Bluestein, H. B., and R. M. Wakimoto, 2003: Mobile radar observations of severe convective storms. In *Radar and Atmospheric Science: A Collection of Essays in Honor of David Atlas*, eds. R. Wakimoto and R. Srivastava (Amer. Meteor. Soc.), pp. 105–136.

Bluestein, H. B., A. L. Pazmany, J. C. Galloway, and R. E. McIntosh, 1995: Studies of the substructure of severe convective storms using a mobile 3-mm-wavelength Doppler radar. *Bull. Amer. Meteor. Soc.*, **76**, 2155–2169.

Bluestein, H. B., W.-C. Lee, M. Bell, C. Weiss, and A. Pazmany, 2003: Mobile-Doppler-radar observations of a tornado in a supercell near Bassett, Nebraska on 5 June 1999. Part II: Tornado-vortex structure. *Mon. Wea. Rev.*, **131**, 2968–2984.

Boccippio, D. J., 1995: A diagnostic analysis of the VVP single-Doppler retrieval technique. *J. Atmos. Oceanic Technol.*, **12**, 230–248.

Brown, R. A., L. R. Lemon, and D. W. Burgess, 1978: Tornado detection by pulsed Doppler radar. *Mon. Wea. Rev.*, **106**, 29–38.

Brown, R. A., and V. T. Wood, 1991: On the interpretation of single-Doppler velocity patterns within severe thunderstorms. *Wea. Forecasting*, **6**, 32–48.

Browning, K. A., and R. Wexler, 1968: The determination of kinematic properties of a wind field using Doppler radar. *J. Appl. Meteor.*, **7**, 105–113.

Burgess, D. W., 1976: Single-Doppler radar vortex recognition. Part I. Mesocyclone signatures. Preprints, 17th Conf. on Radar Meteorology, (Seattle, USA, Amer. Meteor. Soc.), pp. 97–103.

Carbone, R. E., and F. D. Marks, Jr., 1988: Velocity track display (VTD): a real-time application for airborne Doppler radar data in hurricanes. Preprints, 18th Conf. on Hurricanes and Tropical Meteorology (San Diego, USA, Amer. Meteor. Soc.), pp. 11–12.

Caya, D., and I. Zawadzki, 1992: VAD analysis of nonlinear wind fields. *J. Atmos. Oceanic Technol.* **9**, 575–587.

Chan, J. C. L., and W. M. Gray, 1982: Tropical cyclone movement and surrounding flow relationships. *Mon. Wea. Rev.*, **110**, 1354–1374.

Chen, Y., and M. K. Yau, 2001: Spiral bands in a simulated hurricane. Part I: Vortex Rossby wave verification. *J. Atmos. Sci.*, **58**, 2128–2145.

Davies-Jones, R. P., R. J. Trapp, and H. B. Bluestein, 2001: Tornadoes and tornadic storms. In *Severe Convective Storms*. ed. C. A. Doswell (Amer. Meteor. Soc.), pp. 167–221.

Desrochers, P. R., and F. I. Harris, 1996: Interpretation of mesocyclone vorticity and divergence structure from single-Doppler radar. *J. Appl. Meteor.*, **35**, 2191–2209.

Dodge, P., F. D. Marks, Jr., J. Gamache, J. Griffin, and N. Griffin, 1995: EVTD Doppler radar analyses of the inner core of Hurricane Olivia (1994). Preprints, 27th Conf. on Radar Meteorology, (9–13 Oct. 1995, Vail, CO, USA, Amer. Meteor. Soc.), pp. 299–301.

Dodge, P., R. W. Burpee, and F. D. Marks Jr., 1999: The kinematic structure of a hurricane with sea level pressure less than 900 mb. *Mon. Wea. Rev.*, **127**, 987–1004.

Donaldson, R. J., Jr., 1970: Vortex signature recognition by a Doppler radar. *J. Appl. Meteor.*, **9**, 661–670.

Donaldson, R. J., Jr., 1991: A proposed technique for diagnosis by radar of hurricane structure. *J. Appl. Meteor.*, **30**, 1636–1645.

Dong, K., and C. J. Neumann, 1986: The relationship between tropical cyclone motion and environmental geostrophic flows. *Mon. Wea. Rev.*, **114**, 115–122.

Dowell, D. C., C. R. Alexander, J. Wurman, and L. J. Wicker, 2005: Centrifuging of hydrometeors and debris in tornadoes: radar-reflectivity patterns and wind-measurement errors. *Mon. Wea. Rev.*, **133**, 1501–1524.

Franklin, J. L., S. J. Lord, S. E. Feuer, and F. D. Marks, 1993: The kinematic structure of Hurricane Gloria (1985) determined from nested analyses of dropwindsonde and Doppler radar data. *Mon. Wea. Rev.*, **121**, 2433–2451.

Gall, R., and J. Tuttle, 1999: A single radar technique for estimating the wind in tropical cyclones. *Bull. Amer. Meteor. Soc.*, **80**, 653–668.

Gamache, J, F., J. S. Griffin, P. Dodge, and N. F. Griffin, 2004: Automatic Doppler analysis of three-dimensional wind fields in hurricane eyewalls. Preprints, 26th Conf. on Hurricane and Tropical Meteorology, 5D4, (3–7 May 2004, Miami, FL, USA, AMS).

George, J. E., and W. M. Gray, 1976: Tropical cyclone motion and surrounding parameter relationships. *J. Appl. Meteor.*, **15**, 1252–1264.

Gray, W. M., 2003: On the transverse circulation of the hurricane. In *Cloud Systems, Hurricanes, and the Tropical Rainfall Measuring Mission (TRMM): A Tribute to Dr Joanne Simpson*, eds. W.-K. Tao and R. Adler, pp. 149–164.

Griffin, J. S., R. W. Burpee, F. D. Marks, Jr., and J. L. Franklin, 1992: Real-time airbone analysis of aircraft data supporting operational hurricane forecasting. *Wea. Forecasting*, **7**, 480–490.

Guinn, T. A., and W. H. Schubert, 1993: Hurricane spiral bands. *J. Atmos. Sci.*, **50**, 3380–3403.

Hanley, D., J. Molinari, and D. Keyser, 2001: A composite study of the interactions between tropical cyclones and upper-tropospheric troughs. *Mon. Wea. Rev.*, **129**, 2570–2584.

Harasti, P. R., 2003: The hurricane volume velocity processing method. Preprints, 31st Conf. on Radar Meteorology, (Seattle, WA, USA, Amer. Meteor. Soc.), pp. 1008–1011.

Harasti, P. R., 2004: Separation of hurricane and environmental winds via single-Doppler radar volumetric data analysis: results from Hurricane Bret (1999). Preprints, 26th Conf. on Hurricanes and Tropical Meteorology, (Miami, FL, USA, Amer. Meteor. Soc.), pp. 124–125.

Harasti, P. R., and R. List, 1995: A single-Doppler radar analysis method for intense axisymmetric cyclones. Preprints, 27th Conf. on Radar Meteorology, (Vail, CO, USA, Amer. Meteor. Soc.), pp. 221–223.

Harasti, P. R., and R. List. 2005: Principal component analysis of doppler radar data. Part I: Geometric connections between eigenvectors and the core region of atmospheric vortices. *J. Atmos. Sci.*, **62**, 4027–4042.

Harasti, P. R., W.-C. Lee, and M. Bell, 2005: Rapid intensification of Hurricane Charley (2004) derived from WSR-88D data prior to landfall. Part II: HVVP adjustment for the environmental wind and central pressure retrieval. Preprints, 30th Int. Conf. on Radar Meteorology, (Albuquerque, New Mexico, USA, Amer. Meteor. Soc.), CD volume, paper J2J.3.

Harasti, P. R., W. C. Lee, and M. Bell, 2007: Real-time implementation of VORTRAC at the National Hurricane Center. 33rd Conf. on Radar Meteorology, (6–11 August 2007, Cairns, Australia).

Harasti, P. R., C. J. McAdie, P. P. Dodge, W.-C. Lee, J. Tuttle, S. T. Murillo, and F. D. Marks,

Jr., 2004: Real-time implementation of single-Doppler analysis methods for tropical cyclones: algorithm improvements and use with WSR-88D display data. *Wea. Forecasting*, **19**, 219–239.

Holland, G. J., 1984: Tropical cyclone motion: a comparison of theory and observation. *J. Atmos. Sci.*, **41**, 68–75.

Houze, R. A. Jr., S. A. Rutledge, M. I. Biggerstaff, and B. F. Smull, 1989: Interpretation of Doppler weather radar displays of midlatitude mesoscale convective systems. *Bull. Amer. Meteor. Soc.*, **70**, 608–619.

Houze, R. A., Jr., S. S. Chen, W.-C. Lee, R. F. Rogers, J. A. Moore, G. J. Stossmeister, J. L. Cetrone, W. Zhao, and M. M. Bell, 2006: The Hurricane Rainband and Intensity Change Experiment (RAINEX): Observations and modeling of Hurricanes Katrina, Ophelia, and Rita (2005). Submitted to *Bull. Amer. Meteor. Soc.*

Jorgensen, D. P., 1984a: Mesoscale and convective-scale characteristics of mature hurricanes. Part I: General observations by research aircraft. *J. Atmos. Sci.*, **41**, 1268–1286.

Jorgensen, D. P., 1984b: Mesoscale and convective-scale characteristics of mature hurricanes. Part II. Inner core structure of Hurricane Allen (1980). *J. Atmos. Sci.*, **41**, 1287–1311.

Jou, B. J.-D., and S.-M. Deng, 1997: Determination of typhoon center and radius of maximum wind by using Doppler radar. 22nd Conf. on Hurricanes and Tropical Meteorology, (May 1997, Fort Collins, CO, USA), pp. 104–105.

Jou, B. J.-D., B.-L. Chang, and W.-C. Lee, 1994: Analysis of typhoon circulation using ground-based Doppler radar (*in Chinese, with English abstract*). *Atmos. Sci.*, **22**, 163–187.

Jou, B. J.-D., S.-M. Deng, and B.-L. Chang, 1996: Determination of typhoon center and radius of maximum wind by using Doppler radar (*in Chinese, with English abstract*). *Atmos. Sci.*, **24**, 1–23.

Jou, B. J.-D., J.-M. Chou, P.-L. Chang, and W.-C. Lee, 2001: Landfall typhoons observed by Taiwan Doppler radar network: a case study of typhoon Billis (2000). Int. Conf. on Mesoscale Meteorology and Typhoons in East Asia, (Sept. 2001, Taipei, Taiwan), pp. 171–178.

Jou, B. J.-D., W.-C. Lee, S.-P. Liu, and Y. C. Kao, 2008: Generalized VTD (GVTD) retrieval of atmospheric vortex kinematic structure. Part I: Formulation and error analysis. *Mon. Wea. Rev.* (in press).

Koscielny, A. J., R. J. Doviak, and R. Rabin, 1982: Statistical considerations in the estimation of divergence from single-Doppler radar and applications to prestorm boundary layer observations. *J. Appl. Meteor.*, **21**, 197–210.

Kossin, J. P., W. H. Schubert, and M. T. Montgomery. 2000: Unstable interactions between a hurricane's primary eyewall and a secondary ring of enhanced vorticity. *J. Atmos. Sci.*, **57**, 3893–3917.

Lamb, H., 1932: *Hydrodynamics*, 6th edn. (Dover), 732 pp.

Lee, J. J., T. Samaras, and C. Young, 2004: Pressure measurements at the ground in an F-4 tornado. Preprints, 22nd Conf. on Severe Local Storms. (Amer. Meter. Soc.), CD volume, paper 15.3.

Lee, J. L., Y.-H. Kuo, and A. E. MacDonald, 2003: The vorticity method: extension to mesoscale vertical velocity and validation for tropical storms. *Quart. J. Roy. Meteor. Soc.*, **129**, 1029–1050.

Lee, J. L., W.-C. Lee, and A. E. MacDonald, 2006: Estimating vertical velocity and flow from Doppler radar observations of tropical cyclones. *Quart. J. Roy. Meteor. Soc.*, **132**, 125–145.

Lee, W.-C., and M. M. Bell, 2004: Kinematic structure of Hurricane Isabel (2003) near landfall from the Morehead City WSR-88D. Preprints, 26th Conf. on Hurricanes and Tropical Meteorology, (Miami, FL, USA, Amer. Meteor. Soc.), pp. 566–567.

Lee, W.-C., and M. M. Bell, 2007: Rapid intensification, eyewall contraction and breakdown of Hurricane Charley (2004) Near Landfall. *Geophys. Res. Lett.*, **34**, L02802, doi:10.1029/2006GL027889.

Lee, W.-C., and F. D. Marks, Jr., 2000: Tropical cyclone kinematic structure retrieved from single Doppler radar observations. Part II: The GBVTD-simplex center finding algorithm. *Mon. Wea. Rev.*, **128**, 1925–1936.

Lee, W.-C., and J. Wurman, 2005: Diagnosed three-dirmensional axisymmetric structure of the Mulhall tornado on 3 May 1999. *J. Atmos. Sci.*, **62**, 2373–2393.

Lee, W.-C., P. R. Harasti, and M. Bell, 2002: An improved algorithm to resolve circulations of wavenumber two tropical cyclones. Preprints, 25th Conf. on Hurricane and Tropical Meteorology, (29 Apr.-3 May 2002, San Diego, CA, USA, Amer. Meteor. Soc.), pp. 611–612.

Lee, W.-C., P. R. Harasti, and M. Bell, 2006b: VORTRAC: a utility to deduce central pressure and radius of maximum wind of landfalling tropical cyclones using WSR-88D data. Preprints, 60th Interdepartmental Hurricane Conference (20–24 Mar. 2006, Mobile, Alabama, Office of the Federal Coordinator for Meteorological Services and Supporting Research).

Lee, W.-C., F. D. Marks, Jr., and R. E. Carbone, 1994: Velocity track display (VTD): a technique to extract primary vortex circulation of a tropical cyclone in real time using single airborne Doppler radar data. *J. Atmos. Oceanic Technol.*, **11**, 337–356.

Lee, W.-C., J.-D. Jou, P.-L. Chang, and S.-M. Deng, 1999: Tropical cyclone kinematic structure retrieved from single Doppler radar observations. Part I: Interpretation of Doppler velocity patterns and the GBVTD technique. *Mon. Wea. Rev.*, **127**, 2419–2439.

Lee, W.-C., J.-D. Jou, P.-L. Chang, and F. D. Marks, 2000: Tropical cyclone kinematic structure retrieved from single-Doppler radar observations. Part III: Evolution and structure of Typhoon Alex (1987). *Mon. Wea. Rev.*, **128**, 3892–4001.

Lee, W.-C., P. R. Harasti, M. Bell, B. J.-D. Jou, and M.-H. Chang, 2006a: Doppler velocity signatures of idealized elliptical vortices. *Terr. Atmos. Oceanic Sci.*, **17**, 429–446.

Lemon, L. R., D. W. Burgess, and R. A. Brown, 1978: Tornadic storm airflow and morphology derived from single-Doppler radar measurement. *Mon. Wea. Rev.*, **106**, 48–61.

Liou, Y.-C., 1999a: A preliminary study of retrieving low level vortex circulation using a Lagrangian coordinate and single Doppler radar data, *Terr. Atmos. Oceanic Sci.*, **10**, 321–340.

Liou, Y.-C., 1999b: Single radar recovery of cross-beam wind components using a modified moving frame of reference technique. *J. Atmos. Oceanic Technol.*, **16**, 1003–1016.

Liou, Y.-C., T.-C. C. Wang, W.-C. Lee, and Y.-J. Chang, 2006: The retrieval of asymmetric tropical cyclone structures using Doppler radar observations and the method of extended GBVTD. *Mon. Wea. Rev.*, **134**, 1140–1160.

Mallen, K. J., M. T. Montgomery, and B. Wang, 2005: Re-examining tropical cyclone near-core radial structure using aircraft observations: implications for vortex resiliency. *J. Atmos. Sci.*, **62**, 408–425.

Marks, F. D., 2003: State of the science: radar view of tropical cyclones. In *Radar and Atmospheric Science: A Collection of Essays in Honor of David Atlas*, eds. R. Wakimoto and R. Srivastava (Amer. Meteor. Soc.), pp. 33–74.

Marks, F. D., and R. A. Houze, 1987: Inner core structure of Hurricane Alicia from Doppler radar observations. *J. Atmos. Sci.*, **44**, 1296–1317.

Marks, F. D., R. A. Houze, and J. F. Gamache, 1992: Dual aircraft investigation of the inner core of Hurricane Norbert. Part I: Kinematic structure. *J. Atmos. Sci.*, **49**, 919–942.

Matejka, T., and R. C. Srivastava. 1991: An improved version of the extended velocity-azimuth display analysis of single-doppler radar data. *J. Atmos. Oceanic Technol*, **8**, 453–466.

McAdie, C. J., P. R. Harasti, P. Dodge, W.-C. Lee, S. Murillo, and F. D. Marks, Jr., 2001: Real-time implementation of tropical cyclone-specific radar data processing algorithms. Preprints, 30th Int. Conf. on Radar Meteorology, (Munich, Germany, Amer. Meteor. Soc.), pp. 468–470.

Montgomery, M. T., and R. J. Kallenbach, 1997: A theory for vortex Rossby-waves and its application to spiral bands and intensity changes in hurricanes. *Quart. J. Roy. Meteor. Soc.*, **123**, 435–465.

Murillo, S. T., W.-C. Lee, and F. D. Marks, 2000: Evaluating the GBVTD-tropical center finding simplex algorithm. Preprints, 24th Conf. on Hurricane and Tropical Meteorology (29 May–2 June 2000, Fort Lauderdale, FL, USA, Amer. Meteor. Soc.), pp. 312–313.

Murillo, S. T., W.-C. Lee, F. D. Marks, and P. Dodge, 2002: Examining structural changes and circulation center of Hurricane Danny (1997) using a single-Doppler radar wind retrieval technique. Preprints, 25th Conf. on Hurricane and Tropical Meteorology (29 April–3 May 2002, San Diego, CA, USA, Amer. Meteor. Soc.), pp. 485–486.

Neldar, J. A., and R. Mead, 1965: A simplex method for function minimization. *Comp. J.*, **7**, 308–313.

Qiu, C.-J., and Q. Xu, 1992: A simple adjoint method of the wind analysis for single-Doppler data. *J. Atmos. Oceanic Technol.*, **9**, 588–598.

Rankine, W. J. M., 1901: *A Manual of Applied Mechanics.* 16th edn. (Charles Griff), pp. 574–578.

Ray, P., 1990: Convective dynamics. IN *Radar in Meteorology*, ed. D. Atlas (Amer. Metror. Soc.), pp. 348–390.

Rinehart, R., 1979. Internal storm motions from a single non-Doppler weather radar. (NCAR/TN-146+STR), 262 pp.

Rinehart, R., and E. T. Garvey, 1978: Three-dimensional storm motion detection by conventional weather radar. *Nature,* **273**, 287–289.

Rotunno, R., 1979: A study in tornado-like vortex dynamics. *J. Atmos. Sci.,* **36**, 140–155.

Rotunno, R., and K. A. Emanuel, 1987: An air-sea interaction theory for tropical cyclones. Part II: Evolutionary study using a nonhydrostatic axisymmetric numerical model. *J. Atmos. Sci.,* **44**, 542–561.

Roux, F., F. Chane-Ming, A. Lasserre-Bigorry, and O. Nuissier, 2004: Structure and evolution of intense tropical cyclone Dina near La Reunion on 22 January 2002: GB-EVTD analysis of single Doppler radar observations. *J. Atmos. Oceanic Technol.,* **21**, 1501–1518.

Roux, F., and F. D. Marks, Jr., 1996: Extended velocity track display (EVTD): an improved processing method for Doppler radar observations of tropical cyclones. *J. Atmos. Oceanic Technol.,* **13**, 875–899.

Schubert, W. H., M. T. Montgomery, R. K., Taft, T. A. Guinn, S. R. Fulton, J. P. Kossin, and J. P. Edwards, 1999: Polygonal eyewalls, asymmetric eye contraction, and potential vorticity mixing in hurricanes. *J. Atmos. Sci.,* **56**, 1197–1223.

Shapiro, A., S. Ellis, and J. Shaw, 1995: Single-Doppler retrievals with Phoenix 11 data: clear air and microburst wind retrievals in the planetary boundary layer. *J. Atmos. Sci.,* **52**, 1265–1536.

Stirling, J., and R. M. Wakimoto, 1989: Mesoscale vortices in the stratiform region of a decaying midlatitude squall line. *Mon. Wea. Rev.,* **117**, 452–458.

Symthe, G. R., and D. S. Zrnic, 1983: Correlation analysis of Doppler radar data and retrieval of the horizontal wind. *J. Climate Appl. Meteor.,* **22**, 297–311.

Tanamachi, R. L., H. B. Bluestein, W.-C. Lee, M. Bell, and A. Pazmany, 2006: Ground-based velocity track display (GBVTD) analysis of a W-band Doppler radar data in a tornado near Stockton, Kansas, on 15 May 1999. *Mon. Wea. Rev.* (accepted with revision).

Terwey, W. D., and M. T. Montgomery, 2002: Wavenumber-2 and wavenumber-*m* vortex rossby wave instabilities in a generalized three-region model. *J. Atmos. Sci.,* **59**, 2421–2427.

Tuttle, J. D., and G. B. Foote, 1990: Determination of the boundary layer airflow from a single Doppler radar. *J. Atmos. Oceanic Technol.,* **7**, 218–232.

Waldteufel, P., and H. Corbin, 1979: On the analysis of single-Doppler radar data. *J. Appl. Meteor.,* **18**, 532–542.

Wang, Y., 2002a: Vortex Rossby waves in a numerically simulated tropical cyclone. Part I: Overall structure, potential vorticity and kinetic energy budgets. *J. Atmos. Sci.,* **59**, 1213–1238.

Wang, Y., 2002b: Vortex Rossby waves in a numerically simulated tropical cyclone. Part II: The role in tropical cyclone structure and intensity changes. *J. Atmos. Sci.,* **59**, 1239–1262.

Wang, Y., and C.-C. Wu, 2004: Current understanding of tropical cyclone structure and intensity changes: a review. *Meteorol. Atmos. Phys.,* **87**, 257–278.

Ward, N. B., 1972: The exploration of certain features of tornado dynamics using a laboratory model. *J. Atmos. Sci.,* **29**, 1194–1204.

Willoughby, H. E., 1992: Linear motion of a shallow-water barotropic vortex as an initial-value problem. *J. Atmos. Sci.,* **49**, 2015–2031.

Willoughby, H. E., and M. B. Chelmow, 1982: Objective determination of hurricane tracks from aircraft observations. *Mon. Wea. Rev.,* **110**, 1298–1305.

Winn, W. P., S. J., Hunyady, and G. D. Aulich, 1999: Pressure at the ground in a large tornado. *J. Geophy. Res.,* **104: D18**, 22067–22082.

Wood, V. T., 1994: A technique for detecting a tropical cyclone center using a Doppler radar. *J. Atmos. Oceanic Technol.,* **11**, 1207–1216.

Wood, V. T., and F. D. Marks, Jr., 1989: Hurricane Gloria: simulated land-based Doppler velocities reconstructed from airborne Doppler radar measurements. Preprints, 18th Conf. on Hurricane and Tropical Meteorology (San Diego, USA, Amer. Meteor. Soc.), pp. 115–116.

Wood, V. T., and R. A. Brown, 1992: Effects of radar proximity on single-Doppler velocity of axisymmetric rotation and divergence. *Mon. Wea. Rev.,* **120**, 2798–2807.

Wurman, J., 2002: The multiple vortex structure of a tornado. *Wea. Forecasting,* **17**, 473–505.

Wurman, J., and S. Gill, 2000: Finescale radar observations of the Dimmitt, Texas (2 June 1995), tornado. *Mon. Wea. Rev.,* **128**, 2135–2164.

Wurman, J., and T. Samaras, 2004: Comparison of *in-situ* and DOW Doppler winds in a tornado and RHI vertical slices through 4 tornadoes during 1996–2004. Preprints, *22nd Conf. on Severe Local Storms* (Amer. Meteor. Soc.), CD volume, paper 15.4.

Wurman, J., J. M. Straka, E. N. Rasmussen, M. Randall, and A. Zahrai, 1997: Design and deployment of a portable, pencil-beam, pulsed, 3-cm Doppler radar. *J. Atmos. Oceanic Technol.*, **14**, 1502–1512.

Xu, Q., C. Qiu, and J. Yu, 1994: Adjoint method retrieving of horizontal wind from single-Doppler reflectivity measured during Phoenix II. *J. Atmos. Oceanic Technol.*, **11**, 275–288.

Xu, Q., and C. J. Qiu, 1994: Simple adjoint methods for single-Doppler wind analysis with a strong constraint of mass conservation. *J. Atmos. Oceanic Tech.*, **11**, 289–298.

Yamauchi, H., O. Suzuki, and Akaeda, 2002: Asymmetry in wind field of Typhoon 0115 analyzed by triple Doppler radar observations. Preprints, Int. Conf. on Mesoscale Convective Systems and Heavy Rainfall/Snowfall in East Asia (Tokyo, Japan, Japan Science and Technology Corporation and Chinese Academy of Meteorological Sciences), pp. 245–250.

Zawadzki, I. I., 1973: Statistical properties of precipitation patterns. *J. Appi. Meteor.*, **12**, 459–472.

Zhao, K., B. J.-D. Jou, Y.-J. Pan, and W.-Z. Ge, 2007: Structural characteristics of a meso-vortex in Taiwan Strait from single Doppler radar analysis. *Acta Meteor Sinica* (in Chinese), in press.

Typhoon–Ocean Interactions Inferred by Multisensor Observations

I-I Lin and Chun-Chieh Wu

Department of Atmospheric Sciences, National Taiwan University, Taipei, Taiwan
iilin@webmail.as.ntu.edu.tw

The western North Pacific Ocean and the surrounding seas are among the world's oceans where tropical cyclones, highest both in number and in intensity, are found. There has long been interest in studying the typhoon–ocean interaction processes in this vast oceanic region. However, observations are rare and it has been difficult to study these complex, dynamic, and interdisciplinary processes. With the advancement of satellite remote sensing, especially microwave remote sensing with cloud-penetrating capabilities, it has finally become possible to catch a glimpse of some of these processes in the western North Pacific. In this article, we review a number of recent papers using these new satellite observations to study (1) the interaction between typhoons and warm ocean eddies, (2) enhancement of ocean primary production induced by typhoons, and (3) posttyphoon air–sea interaction.

1. Introduction

The interaction between tropical cyclones and the ocean is complex, dynamic, and inter-disciplinary. Tropical cyclones form on the ocean, which is the energy source that fuels a cyclone's intensification (Emanuel, 1986, 1988, 1991, 1995; Holland, 1997; Black *et al.*, 2007). As cyclones intensify, they impact back on the ocean and cause cold water from the deeper ocean to be entrained and upwelled to the upper ocean layer (Chang and Anthes, 1979; Price, 1981; Stramma *et al.*, 1986; Shay *et al.*, 1992; Dickey *et al.*, 1998; Jacob *et al.*, 2000). This self-induced ocean cooling in turn plays a critical negative feedback role in the cyclone's intensi-fication (Gallacher *et al.*, 1989; Emanuel, 1999; Schade and Emanuel, 1999, Bender and Ginis, 2000; Cione and Uhlhorn, 2003; Emanuel *et al.*, 2004; Lin *et al.*, 2005; Zhu and Zhang, 2006; Wu *et al.*, 2007). After the cyclone's departure, cold wakes, typically more evident to the right of the cyclone tracks (Change and Anthes, 1978;

Price, 1981; Cornillon *et al.*, 1987; Bender and Ginis, 2000; Monaldo *et al.*, 1997; Wentz *et al.*, 2000; Lin *et al.*, 2003b), are left behind. These cold wakes can exist in the ocean for days to weeks and generate continual feedback with the atmosphere (Emanuel, 2001; Lin *et al.*, 2003a).

Besides the above physical interactions, there can be biogeochemical interactions between cyclones and the ocean because the intense cyclone wind mixes and transports not only the cold water from the deeper ocean but also nutrients in the deeper layer (Lin *et al.*, 2003b; Babin *et al.*, 2004; Davis and Yan, 2004; Siswanto *et al.*, 2007). As most of the tro-pical and subtropical oceans are oligotrophic, i.e. nutrient-poor (Eppley, 1989; Behrenfeld and Falkowski, 1997; McGillicuddy *et al.*, 1998; McGillicuddy *et al.*, 2001; Uz *et al.*, 2001), the nutrients brought to the upper ocean are cri-tical for phytoplankton, inducing phytoplankton blooms (Lin *et al.*, 2003b; Babin *et al.*, 2004; Siswanto *et al.*, 2007). Also, since phytoplankton is the base of the marine food chain, these blooms may produce a chain reaction to affect the fishery yield. This phenomenon has not been lost on experienced fishermen who harvest in the wake long after a tropical cyclone's passage. Besides being the base of the ocean food chain, these phytoplankton blooms can have even more profound impact on the earthy system. Like all other plants, phytoplankton uses carbon dioxide, sunlight, and nutrients to photosynthesize; additionally, half of the world's oxygen is produced by phytoplankton. Therefore, phytoplankton production also affects the uptake of carbon dioxide, an important greenhouse gas and a major cause of natural and man-made climate changes (Eppley and Peterson, 1979; Eppley, 1989; Bates *et al.*, 1998; McGillicuddy *et al.*, 1998; Lin *et al.*, 2003b; Uz *et al.*, 2001).

Pre-existing ocean features are known to cause even more complex air–sea interactions in the tropical cyclone and ocean system (Qiu, 1999; Shay *et al.*, 2000; Goni and Trinanes 2003; Emanuel *et al.*, 2004; Lin *et al.*, 2005; Scharroo *et al.*, 2005; Oey *et al.*, 2007; Wu *et al.*, 2007). It has been reported that some major category 5[a] storms, e.g. Hurricane Opal of 1995 (Shay *et al.*, 2000), Katrina of 2005 (Scharroo *et al.*, 2005), supertyphoon Maemi of 2003 (Lin *et al.*, 2005), and Dianmu of 2004 (Pun *et al.*, 2007), rapidly intensified when encoun tering warm ocean features.

The western North Pacific basin and the surrounding seas are where most tropical cyclone (i.e. typhoon) activities are located. But due to the lack of *in situ* and airborne observations, it has been difficult to credibly analyze the complex typhoon–ocean interaction processes in this basin. Available opportune *in situ* observations in this region are too sparse to lend meaningful spatial correlations. Visible and IR satellite remote sensing images (e.g. Advanced Very High Resolution Radiometer, AVHRR) are frequently contaminated by clouds (Wentz *et al.*,

2000), especially near typhoons, limiting the usefulness of these images for our purposes. Therefore, observations of typhoon–ocean interactions in the northwestern Pacific have been extremely difficult. With recent advances in microwave remote sensing (Fu *et al.*, 1994; Fu and Cazenave, 2001; Liu *et al.*, 1998; Wentz *et al.*, 2000), however, some inroads have been made on exploring the processes that occur at the air–sea interface as typhoons translate over the ocean.

The major advantage of using microwave remote sensing is its ability to penetrate clouds and its independence from sunlight. Therefore, observations can be made both during the day and the night without bias toward fairweather at the air–sea interface. For the 50th anniversary of the Department of Atmospheric Sciences of the National Taiwan University, we demonstrate in this review the use of three types of microwave remote sensors and one optical remote sensor to study a number of little-studied or infrequently observed typhoon–ocean interaction processes in the western North Pacific based on a series of recent papers (Lin *et al.*, 2003a; Lin *et al.*, 2003b; Lin *et al.*, 2005). The three types of microwave data are (1) the ocean surface wind vector from the QuikSCAT active microwave scatterometer (Liu *et al.*, 1998), (2) the sea surface temperature (SST) data from the TRMM (Tropical Rainfall Measuring Mission) passive microwave imager and the Advanced Microwave Scanning Radiometer (AMSR-E) (Wentz *et al.*, 2000), and (3) the sea surface height anomaly (SSHA) data from the TOPEX-Poseidon and JASON-1 active microwave altimeters (Fu *et al.*, 1994; Fu and Cazenave, 2001). The optical remote sensor used is the ocean color data from the NASA Sea-viewing Wide Field-of-view (SeaWiFS) sensor (O'Reilly *et al.*, 1998). In this work, the QuikSCAT, TOPEX Poseidon, and JASON-1

---

[a]Saffir–Simpson Tropical Cyclone Scale, based on the 1 min maximum sustained winds: Category 1: 34–43 ms^{-1}; Category 2: 44–50 ms^{-1}; Category 3: 51–59 ms^{-1}; Category 4: 59–71 ms^{-1}; and Category 5: > 71 ms^{-1}.

data are from the daily level 2 product of the NASA/Jet Propulsion Laboratory. The SeaWiFS data are the daily level 2 chlorophyll-a data of the NASA Goddard Space Flight Center while the TRMM/SST data are the daily product of the Remote Sensing Systems (Wentz *et al.*, 2000). All products have been validated with *in situ* observations and readers are referred to the original references (Fu *et al.*, 1994; Liu *et al.*, 1998; O'Reilly *et al.*, 1998; Wentz *et al.*, 2000) for their respective accuracies.

In Sec. 2, we will discuss the interaction between supertyphoon Maemi (2003) and a warm ocean eddy (Lin *et al.*, 2005), and show that warm ocean eddies play a critical role in Maemi's intensification to category 5. In Sec. 3, we will show the drastic biological response induced by typhoon Kai-Tak (2000) in the South China Sea (Lin *et al.*, 2003b). In Sec. 4, we present observations in previously unobserved posttyphoon air–sea interaction processes that the cold wakes, left behind by typhoons, can have evident feedback to the atmosphere by reducing the ocean surface wind speed (Lin *et al.*, 2003a). In Sec. 5, a summary is presented.

2. The Interaction between Supertyphoon Maemi (2003) and a Warm Ocean Eddy (Based on Lin *et al.*, 2005)

Since the observation of a noticeable number of intense category 4 or 5 Atlantic/Gulf of Mexico hurricanes [e.g. Opal (1995), Mitch (1998), Bret (1999)] rapidly intensified to category 5 when encountering warm mesoscale ocean eddies (Shay *et al.*, 2000; Goni and Trinanes, 2003; Emanuel *et al.*, 2004), there has been much interest in studying the interaction between tropical cyclones and ocean features (e.g. eddies and currents) (Shay *et al.*, 2000; Hong *et al.*, 2000; Goni and Trinanes, 2003; Emanuel *et al.*, 2004; Lin *et al.*, 2005; Scharroo *et al.*, 2005; Oey *et al.*, 2006; Pun *et al.*, 2007; Wu *et al.*, 2007). In particular, it is suggested that the warm eddy in the Gulf of Mexico may have induced the intensification of the devastating hurricane Katrina of 2005 to category 5 before its landfall (Scharroo *et al.*, 2005). The western North Pacific Ocean is among the world's oceans in which the greatest number of intense category 4 and 5 cyclones are found.[b] It is interesting to find out whether these intense typhoons are associated with ocean features. In this section, we review the first event observed in the western North Pacific of such an encounter.

In September 2003, typhoon Maemi passed directly over a prominent warm ocean eddy in the western North Pacific, as observed by the satellite SSHA data from the TOPEX/Poseidon and JASON-1 altimeters (Fu *et al.*, 1994). This warm ocean eddy is around $700\,\mathrm{km} \times 500\,\mathrm{km}$ in size and is characterized by its large positive SSHA[c] of 10–45 cm (Fig. 1). Joint analysis with the best track data from the Joint Typhoon Warning Center (JTWC) shows that during the 36 h of the Maemi eddy encounter, Maemi's intensity (in 1 min sustained wind) shot up from the modest category 1 ($41\,\mathrm{ms^{-1}}$) to its peak in category 5 ($77\,\mathrm{ms^{-1}}$). As can be observed in Fig. 1, Maemi entered the eddy region at 1800 UTC, 8 September 2003, when its intensity was at category 1 (green bullet). It then rapidly intensified to category 2, (blue bullet) in 6 h and jumped to category 4 (yellow bullet) in the next 6 h. At 0000 UTC, 10 September 2003 (i.e. 36 h from category 1), Maemi reached its peak at category 5 ($77\,\mathrm{ms^{-1}}$; black bullet) and became the most intense tropical cyclone globally in 2003. As Maemi left the eddy region, its intensity declined.

---

[b]From the best-track data of the Joint Typhoon Warning Center during 1960–2005.
[c]From the existing literature, warm ocean features with SSHA > 8 cm are considered prominent. (Qiu, 1999; Shay *et al.*, 2000).

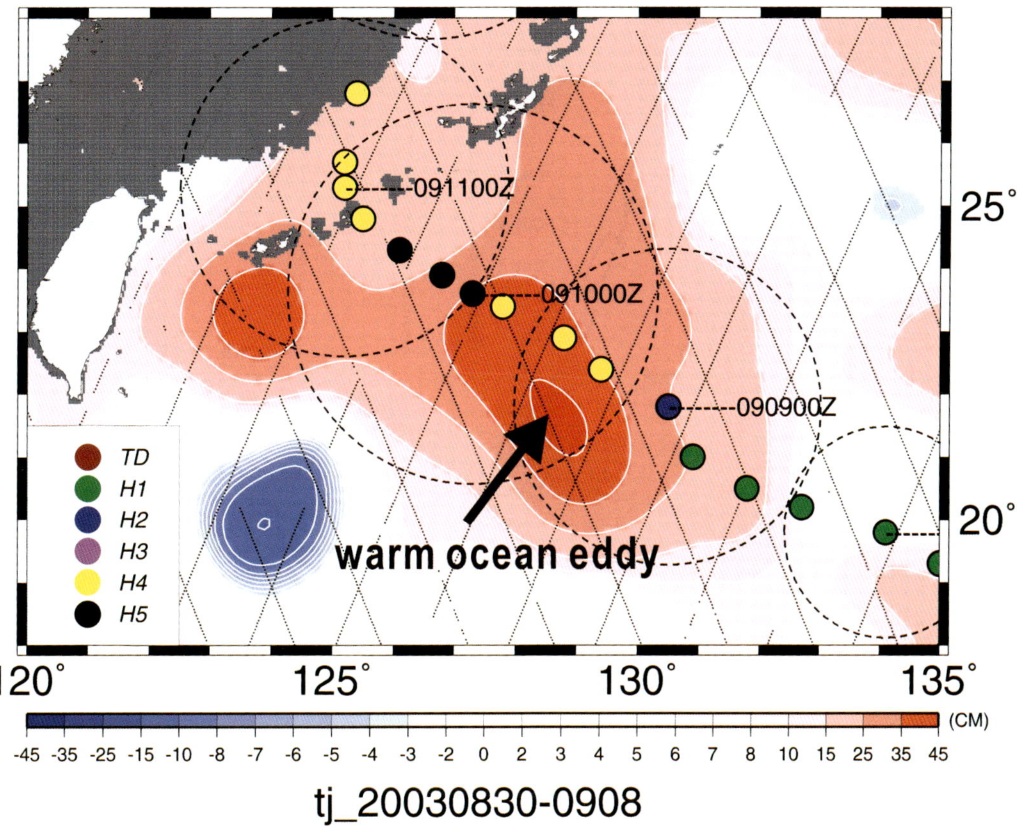

Figure 1. The pre-Maemi SSHA map (based on the 30 August–8 September 2003 cycle of TOPEX/Poseidon and Jason-1 measurements) showing the warm ocean eddy (characterized by a positive SSHA of 10–45 cm). Maemi's track, intensity (on the Saffir–Simpson scale, illustrated as color bullets) and radius of maximum wind are also shown. The storm position is denoted every 6 h. (After Lin *et al.*, 2005.)

Using a simple coupled typhoon–ocean model, i.e. the CHIPS (Coupled Hurricane Intensity Prediction System) model (Emanuel, 1999; Emanuel *et al.*, 2004), numerical experiments were conducted to assess the influence of the warm eddy on the intensity evolution of Maemi. Numerical experiments are run with and without the input of the eddy information derived from the satellite SSHA field. The run without the eddy input (denoted as CTRL) uses monthly climatological ocean mixed layer depth (Levitus, 1982). The one with the eddy input (denoted as EDDY) uses one cycle (ten days) of the observed pretyphoon satellite SSHA measurement as input to an algorithm developed by

Shay *et al.* (2000) to estimate a new mixed layer depth (Emanuel *et al.*, 2004) (Fig. 2). The best-track intensity data from the JTWC are also shown in Fig. 2(a). All the runs were initialized and tuned according to the best-track data for the first 24 h.

As in Fig. 2(a), it is evident that the intensity evolutions, including the eddy-adjusted mixed layer depth, are much closer to the best-track intensity than the run without the eddy (i.e. CTRL) in both the magnitude and timing of the peak intensity. The peak intensity is $68\,\mathrm{ms}^{-1}$ (i.e. category 4) for the CTRL run, but the observed peak is $77\,\mathrm{ms}^{-1}$, occurring at 0300 UTC, 10 September [Fig. 2(a)]. Maemi

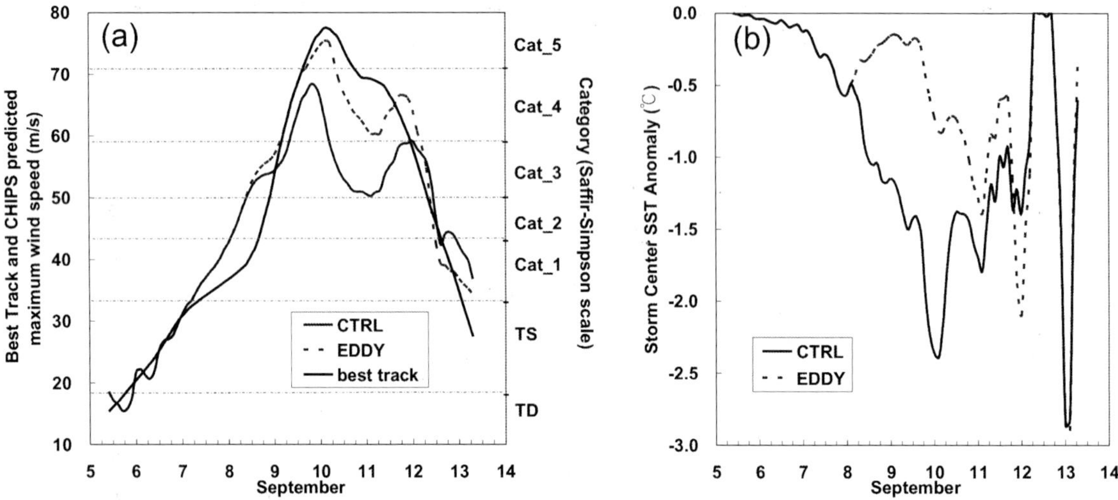

Figure 2. Results of the CHIPS runs showing the two primary experiments, i.e. CTRL (a controlled run using standard CHIPS input) and EDDY (a run incorporating the warm ocean eddy information in SSHA). JTWC's best-track intensity is shown in black. (a) The intensity (ms^{-1}) evolution. (b) The typhoon-induced sea surface temperature anomaly at the storm center. (After Lin *et al.*, 2005.)

reaches its maximum intensity in CTRL at 1800 UTC, 9 September, 9 h earlier than the OBS. When the warm eddy is included to initialize a deeper mixed layer in CHIPS, an improvement in the intensity hindcast is evident. The peak intensity for the EDDY run reaches the category 5 scale of $75\,\mathrm{ms}^{-1}$, well matched by the observed best-track intensity peak. The timing for peak intensity is also correctly captured [Fig. 2(a)].

Further analysis of the CHIPS results finds that the reason why the EDDY run can match well with the observed intensity is that the upper ocean thermal structure in the warm eddy is correctly represented by the satellite SSHA in the EDDY run. As can be seen in Fig. 3, in the eddy region, warm water of $\geq 26^\circ\mathrm{C}$ extends well downward to about 120–130 m. In contrast, outside the eddy region the subsurface warm layer is much shallower, so that the warm water of $\geq 26^\circ\mathrm{C}$ extends only to 40 m or so (i.e. the background). Also, it is noted that the difference between the eddy and the background is not in the SST, but in the subsurface thermal structure. As can be seen in Fig. 3, both background and eddy profiles show similar SST

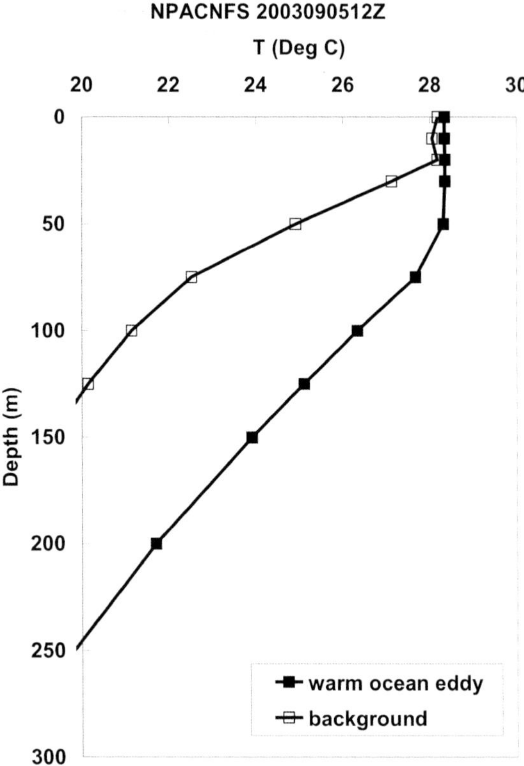

Figure 3. The ocean's depth–temperature profiles (based on the US Naval Research Lab's NPACNFS ocean model; Ko *et al.*, 2003) in the warm ocean eddy and the reference background region. (After Lin *et al.*, 2005.)

values of around 28.2°C. In other words, what makes the difference is the thickness of the subsurface warm layer, but not the SST.

Figure 2(b) shows that with the very deep warm mixed layer in the warm eddy, typhoon-induced ocean surface cooling is significantly suppressed. The typhoon-induced ocean cooling is only ≤ 0.5°C [dashed curve in Fig. 2(b)] throughout the intensification period before the storm reaches peak intensity at 0300 UTC, 10 September. As a consequence, the storm is able to intensify without being hampered by the cooler sea surface temperature induced by the typhoon, as the negative feedback mechanism (Emanuel, 1999, Bender and Ginis, 2000; Emanuel *et al.*, 2004) is reduced in the presence of the warm eddy. In contrast, in the CTRL run (solid curve), the self-induced ocean cooling is much stronger, with the SST anomaly $\sim$1.5–2.5°C [Fig. 2(b)], during the intensification period (0000 UTC, 5 September, to the peak at 0300 UTC, 10 September) without the warm eddy and its deep warm mixed layer. The increased cooling of the sea surface in the CTRL experiment contributes to the reduction of the maximum wind speed by $\sim$10 ms^{-1} of the simulated storm, as is evident in Fig. 2(a) (solid curve).

In the current literature, three other cyclone–warm ocean feature interaction cases in the Atlantic (i.e. hurricanes Opal, Mitch, and Bret) have been studied using atmosphere–ocean coupled models (Hong *et al.*, 2000; Emanuel *et al.*, 2004). Opal (1995) was simulated by Hong *et al.* (2000) using the U.S. Naval Research Laboratory's Coupled Ocean–Atmosphere Mesoscale Prediction System (COAMPS) as the atmospheric component and the Geophysical Fluid Dynamics Laboratory's Modular Ocean Model, version 2 (MOM2) as the ocean component. The other two cases, Mitch (1998) and Bret (1999), were run using the CHIPS model (Emanuel *et al.*, 2004). Table 1 compares the results of these three Atlantic cases with the Maemi case.

In Table 1, it is clear that in all cases without the warm ocean eddy information, the intensity is too low as compared to the best-track intensity. With the addition of the eddy information in the numerical experiments, the predicted intensity is evidently improved. In the case of Opal, the peak of the without-eddy simulation reaches only category 4 (i.e. 932-hPa). When the eddy is included, the peak intensity reaches 916 hPa, in agreement with the best-track peak at category 5 (Hong *et al.*, 2000). Consistent results are also found in Mitch, Bret, and Maemi: without the inclusion of the ocean feature in the simulation, the intensity is about 26–31% lower (typically one category lower) as compared to the best-track peak (Table 1). With the inclusion of the eddy information in the simulation, the peak intensity can be correctly simulated (Table 1).

3. The Ocean's Biogeochemical Response to the Typhoon (Based on Lin *et al.*, 2003b)

When passing over land, tropical cyclones can affect human lives and activities. Over the ocean, they can also affect other forms of life, i.e. ocean primary production. As introduced in Sec. 1, ocean primary production plays a significant role in the Earth's ecological and environmental system, especially because it affects the uptake of the important greenhouse gas carbon dioxide (Eppley and Peterson, 1979; Eppley, 1989; Behrenfeld and Falkowski, 1997; Bates *et al.*, 1998; McGillicuddy *et al.*, 1998; Uz *et al.*, 2001; Babin *et al.*, 2004). Primary production takes place mainly in the euphotic zone of the ocean, i.e. the top 50–150 m of the water column where there is abundant light for photosynthesis and when nutrients are available. Marine nutrients, however, are mostly located in the deeper ocean. Therefore, the vertical entrainment due to mixing and the induced upwelling in the ocean, caused by tropical cyclone winds, are crucial mechanisms

Table 1. Comparison of the intensification parameters based on coupled model results for Maemi (2003), Opal (1995), Bret (1999), and Mitch (1998).

	Maemi (this work)	Opal (Shay *et al.*, 2000; Hong *et al.*, 2000)	Bret (Goni and Trinanes, 2003; Emanuel *et al.*, 2004)	Mitch (Goni and Trinanes, 2003; Emanuel *et al.*, 2004)
Location	NWPO	Gulf of Mexico	Gulf of Mexico	Gulf of Mexico
RI Period	2003/09/08/06Z–2003/09/10/06Z	1995/10/03/12Z–1995/10/04/12Z	1999/08/21/06Z–1999/08/22/06Z	1998/10/24/00Z–1998/10/27/00Z
RI Duration (hours)	48 h	24 h	24 h	72 h
Obs. Peak (best track)	77 m/s (cat-5)	916 hPa (cat-5)	64 m/s (cat-4)	79 m/s (cat-5)
Obs. intensity at the beginning of RI	38 m/s (cat-1)	969 hPa (cat-2)	38 m/s (cat-1)	27 m/s (TS)
Intensity increase during RI	77-38 = 39 m/s	916-969 = −53 hPa	64-38 = 26 m/s	79-27 = 52 m/s
Coupled model used	CHIPS	COAMPS-MOM2	CHIPS	CHIPS
Model Estimated Peak – from with warm eddy (feature) run	75 m/s (cat-5)	917 hPa (cat-5)	64 m/s (cat-4)	78 m/s (cat-5)
Peak underestimation & underestimation percentage in RI from with warm eddy (feature) run	77-75 = 2 m/s $2/39 = 5\%$	916-9171 hPa = −1 hPa $(-1)/(-53) = 2\%$	64-64 = 0 $0/26 = 0\%$	79-78 = 1 m/s $1/52 = 2\%$
Model Estimated Peak – from without eddy/feature run	67 m/s (cat-4)	932 hPa (cat-4)	56 m/s (cat-3)	65 m/s (cat-4)
Peak underestimation & underestimation percentage in RI from without warm eddy (feature) run	77-67 = 10 m/s $10/39 = 26\%$	916-9321 hPa = −16 hPa $(-16)/(-53) = 30\%$	64-56 = 8 m/s $8/26 = 31\%$	79-65 = 14 m/s $14/52 = 27\%$

in the Earth system by which the deep, cold, nutrient-rich water can be brought up from the deeper layer to the light-replete euphotic zone. The nutrients can then fuel photosynthetic activities and cause enhancement in primary production (i.e. phytoplankton bloom) (Eppley, 1989; Marra *et al.*, 1990; Dickey *et al.*, 1998; Lin *et al.*, 2003b). Such processes are difficult to monitor and measure, and a quantitative determination of the change of the marine primary production induced by tropical cyclones is elusive.

During the three days, from 5 to 8 July 2000, that typhoon Kai-Tak translated over the South China Sea, it triggered a huge phytoplankton bloom with an average of a 30-times-over increase in surface chlorophyll-a (Chl-a) concentration, as observed by the SeaWiFS (O'Reilly *et al.*, 1998) sensor. In this section, major findings of this event based on Lin *et al.*, 2003b are introduced. This is one the first quantitative events documenting such a typhoon-induced biogeochemical response in the western North Pacific and adjacent seas.

Kai-Tak was a moderate, category 2 typhoon on the Saffir–Simpson hurricane scale. It lingered at a nearly stationary speed of 0–$1.4\,\mathrm{ms}^{-1}$ on the northern South China Sea (from 5 to 8 July 2000) before it proceeded speedily ($\sim 6.1\,\mathrm{ms}^{-1}$) northward [Fig. 4(b)]. The biological response to the passing of Kai-Tak

was depicted by changes in the surface distribution of Chl-a. The pretyphoon condition was illustrated in the SeaWiFS composite from 27 June to 4 July 2000 [Fig. 4(a)], which showed the typical summer surface Chl-a concentrations of predominantly $\leq 0.1\,\mathrm{mgm}^{-3}$. After Kai-Tak's passage (5–8 July), the first available cloud-free SeaWiFS image composite (12–15 July) illustrated an evident enhancement of biological activity, as revealed by the Chl-a concentration [Fig. 4(b)]. The bloom patch (117.5–120°E, 19.3–20.7°N), predominantly of Chl-a concentrations of around $10\,\mathrm{mgm}^{-3}$, coincided with Kai-Tak's trajectory and its radius of intense wind ($\geq 14\,\mathrm{ms}^{-1}$). At certain locations (e.g. 118.4°E, 20°N), the Chl-a concentrations reached as high as $30\,\mathrm{mgm}^{-3}$, 300-fold the pretyphoon condition as depicted in the Chl-a distribution (on the log scale) along tr1 [Fig. 4(c)]. The pretyphoon [from Fig. 4(a)] and the three-year (1998, 1999, and 2001) monthly average of the July Chl-a concentrations along tr1 are also depicted for comparison in Fig. 4(c).

Another drastic response, which can be observed in Fig. 5, is the drop in the SST. Before Kai-Tak's arrival, the SCS was characterized by a warm SST predominantly above 30°C [Fig. 5(a)]. Immediately after Kai-Tak's departure, on 9 July, a cold SST (21.5–24°C) pool (118–120°E, 19–20.5°N) of a size comparable to Kai-Tak's 150 km radius of intense

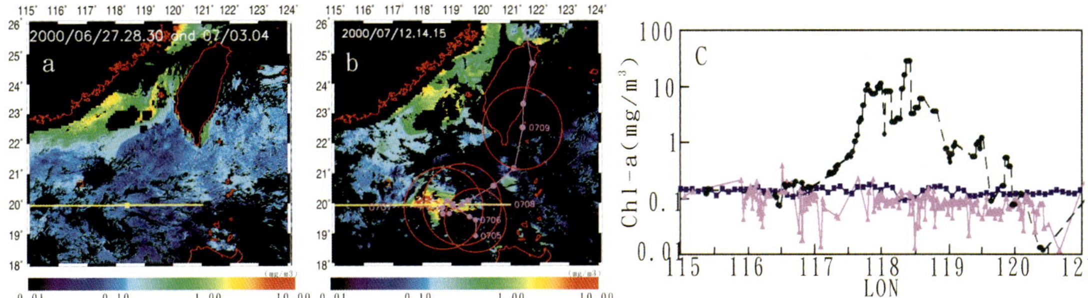

Figure 4.　SeaWiFS surface Chl-a image composite on (a) 27 June–4 July 2000 (before Kai-Tak) and (b) 12, 14, 15 July 2000 (after Kai-Tak). The circle denotes Kai-Tak's radius of intense wind (defined as $\geq 14\,\mathrm{ms}^{-1}$ in this work). The location of the transect tr1 crossing the longitude is also depicted. (c) Comparison of the surface Chl-a distribution along tr1: pink — before [from Fig. 4(a)]; green — after [from Fig. 4(b)]; blue — the three-year (1998, 1999, 2001) average of surface Chl-a concentration for the month of July.

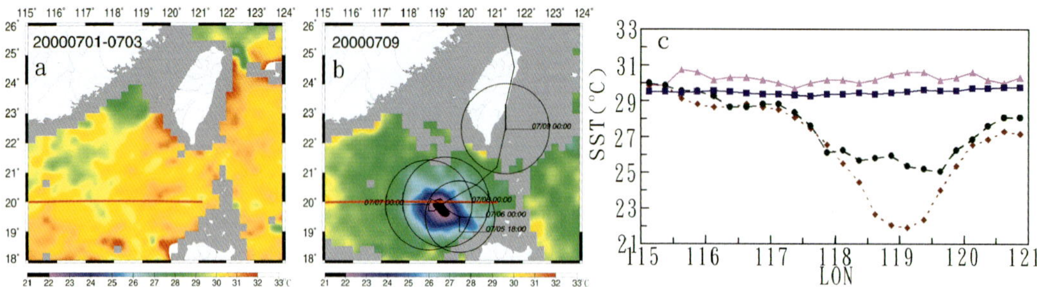

Figure 5. Same as Fig. 4 but for the TRMM TMI/SST image on (a) 1–3 July 2000 (before Kai-Tak) and (b) 9 July 2000 (after Kai-Tak). (c) Comparison of the SST distribution along tr1: pink — before [from Fig. 5(a)]; brown — 1-day after [from Fig. 5(b)]; green — 4–6 days after (image not shown); blue — the three-year (1998, 1999, 2001) average of SST for the month of July.

wind ($\geq 14\,\mathrm{ms}^{-1}$), colocated with the typhoon's track, was observed [Fig. 5(b)]. The minimum SST of 21.5°C was found at the center (118.9°E, 19.9°N) of the cold pool. In comparison with the pretyphoon condition (30.7°C), the SST dropped by as much as 9°C. The distributions of SSTs along the cross-section tr1 (depicted in Fig. 5) over the cold pool before and after the passing of the typhoon, and the three-year (1998, 1999, and 2001) mean for July are shown in Fig. 5(c). Since the 1960s, there have been a number of observational and modeling studies of typhoon-induced upper ocean cooling responses (Chang and Anthes, 1978; Price, 1981; Stramma *et al.*, 1986; Cornilon *et al.*, 1987; Monaldo *et al.*, 1997; Dickey *et al.*, 1998; Bender and Ginis, 2000; Wentz *et al.*, 2000) with reported SST reduction generally spanning 0.5–6°C. The 9°C cooling inferred by the TRMM microwave imager here is one of the strongest-ever observed. In the original paper (Lin *et al.*, 2003b), entrainment mixing and upwelling velocity are estimated to show that due to the shallow thermocline in the South China Sea during summer, it is possible for a near-stationary typhoon to induce such a drastic cooling response.

Using the observed Chl-a and SST data as input to a marine primary production model (Behrenfeld and Falkowski, 1997), it is possible to estimate the contribution of Kai-Tak to marine primary production. The changes in SST, surface Chl-a, and depth-integrated primary production (IPP) with time at the center of the phytoplankton bloom (Lin *et al.*, 2003b) are shown in Fig. 6. The temperature depression and the phytoplankton bloom, as indicated by the elevation of the Chl-a concentration, could be tracked for about one month. The pretyphoon IPP was $300\,\mathrm{mg\,C\,m^{-2}d^{-1}}$, similar to the annual mean (Liu *et al.*, 2002) IPP of $350\,\mathrm{mg\,C\,m^{-2}d^{-1}}$. After the passage of the typhoon, IPP increased by almost an order of magnitude, to $2800\,\mathrm{mg\,C}$ $\mathrm{m^{-2}d^{-1}}$. By integrating IPP over the bloom patch through each time interval, the carbon fixation resulting from this single event (12 July – 16 August) was about 0.8 Mt (1 Mt = 10^{12} g) of carbon. Taking the 200 m bathymetry as the lower boundary of the oligotrophic waters, the area of the oligotrophic South China Sea is $2.76 \times 10^6\,\mathrm{km}^2$, or about 80% of the total area of the South China Sea. If the f ratio in the South China Sea is similar to those in other oligotrophic waters, 0.06–0.14 (Eppley, 1989), typhoon Kai-Tak would have accounted for 2–4% of the annual marine primary production in the oligotrophic South China Sea.[d]

---

[d]The oligotrophic part of the South China Sea in this work is defined as the basin (i.e., open ocean part) of the South China Sea where the bathymetry is typically > 200 m.

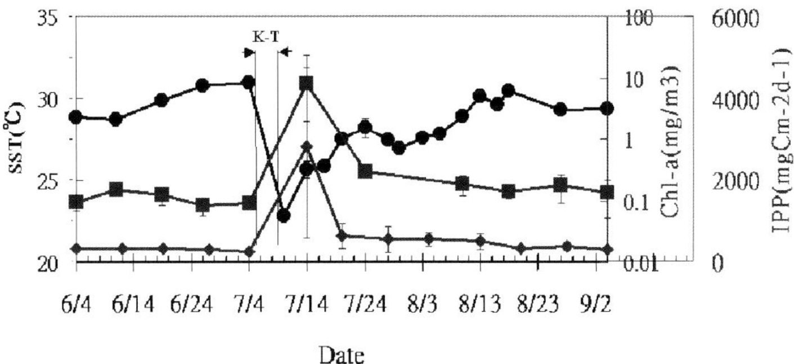

Figure 6. Changes in SST (circles), surface Chl-a (squares; log scale), and IPP (diamonds) of the bloom patch before and after Kai-Tak's passage (5–8 July).

Due to the lack of definitive observations, the contribution of tropical cyclones to primary production has long been treated as negligible. Our result, based on the synergy of three recently available satellite data sets, proves, on the contrary, that tropical cyclones induce significant contributions to the overall primary production in tropical seas.

4. Posttyphoon Air–Sea Interaction (Based on Lin *et al.*, 2003a)

As discussed before, tropical cyclones can cool the ocean surface and mixed layer by induced entrainment and upwelling as deep, cold water is brought to the upper ocean layer. After a typhoon's departure, a cold wake is left behind. As in Sec. 3, the cold patches in the wake may be as much as 9°C cooler than the surrounding warm ocean (Lin *et al.*, 2003b). Therefore they top represent a sizable perturbation of the SST in an otherwise relatively uniform warm ocean environment. This presents a unique natural laboratory for investigating the nature of ocean–atmosphere coupling.

Following on from Sec. 3, the case of typhoon Kai-Tak is chosen to study the posttyphoon air–sea interaction. In this work, colocated and near-coincident TRMM SST and QuikSCAT wind vectors are intercompared. It can be observed in Fig. 7(a) that prior to Kai-Tak, the northern South China Sea was under typical summer conditions, with the SST in the range of 30.5–33.0°C. The corresponding wind field [Fig. 8(a)] is characterized by a higher wind speed (9–11 ms^{-1}) at the region north of 19°N, while

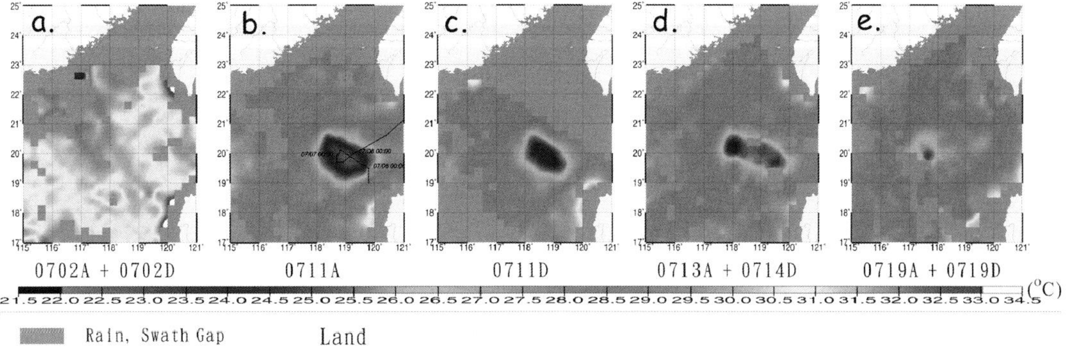

Figure 7. Sequence of representative TMI/SST images showing the evolution of typhoon Kai-Tak's cold SST patch: (a) before the typhoon on 2 July 2000; (b) after the typhoon on 11 July at 0100 UTC; (c) at 0900 UTC on 11 July; (d) composite of 13 and 14 July passes; (e) on 19 July. The trajectory of Kai-Tak is depicted in Fig. 7(b).

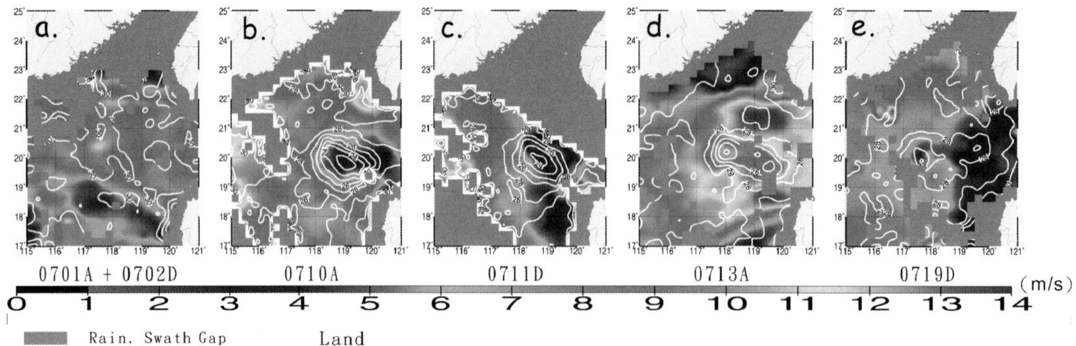

Figure 8. Same as Fig. 7, but for the matching QuikSCAT wind speed image: (a) before the typhoon on 1 and 2 July; (b) after the typhoon on 10 July; (c) on 11 July; (d) on 13 July; (e) on 19 July.

south of 19°N the wind speed is lower (3–6 ms^{-1}). No evident association between SST and wind is observed. Between 5 and 8 July, Kai-Tak passed over the South China Sea [Fig. 7(b)]. Though the maximum SST cooling occurred on 9 July [as seen in Sec. 3 and Fig. 4(b)], the wind fields were then still under the influence of the typhoon. Posttyphoon response is studied from 11 July onward. From Fig. 7(b), one sees that the cold SST patch (118–120°E, 19–21°N) has the dimension of around 150–200 km and that the minimum SST of 22°C is found at the center (118.9°E, 19.9°N) of the oval cold patch and increases outward to 26°C toward the edge. The surrounding SST is around 29–30°C. The corresponding wind field [Fig. 8(b)] shows a distinctive minimum spatially coincident with the cold patch. The wind speed inside the cold patch is between 2.5 and 6 ms^{-1}, while the surrounding wind speed is between 8 and 11 ms^{-1}.

Observing Figs. 7(c)/8(c), 7(d)/8(d), and 7(e)/8(e), 9 similar correlation between the cold SST patch and reduction in wind speed can be found till 19 July, persisting for eight days. For example, on 13 July, the induced cold SST pattern had weakened and elongated [Fig. 7(d)], but the corresponding wind speed [Fig. 8(d)] evolved into similar shapes, illustrating the close correlation between SST and wind, with a relatively high (≥ 12 ms^{-1}) wind north and south of the cold patch. This again shows that the wind speed over the cold SST patch remains relatively low, in contrast to the high wind in the adjacent areas to the north and south of it. On 19 July, the cold patch had greatly diminished into a small, circular feature at around 117.5°E, 20°N [Fig. 7(e)], and disappeared shortly thereafter. This is consistent with the mechanism proposed by Wallace *et al.* (1989), that cool SST is correlated with low surface wind because of a decrease in atmospheric boundary layer stability. Over colder waters, the marine boundary layer is stable, vertical mixing is suppressed, vertical wind shear increases, and the surface wind speed is reduced.

In the existing literature, a number of studies report a similar correlation between SST and surface wind speed on a much longer time ($\sim$20–40 days) and spatial scale ($\sim$1000–2000 km), i.e. in the case of tropical instability waves and the eastern Pacific Ocean cold tongue (Xie *et al.*, 1998; Wentz *et al.*, 2000; Chelton *et al.*, 2001). Our results support the Wallace *et al.* (1989) hypothesis of the SST–wind coupling in a different situation, namely in the cold SST wakes of typhoons. What we show in our examples discussed above are the small space and short time scales on which this mechanism can act. In the typhoon-induced cold wake situation, the coupling manifests itself within a day and on the spatial scale of 100–400 km. In the original paper (Lin *et al.*, 2003a), the relationship between wind speed and SST anomalies is further investigated. Also, a similar correlation between wind and SST in the cold wake of supertyphoon Bilis (2000) is found. Interested readers are referred to Lin *et al.* (2003a) for further details.

5. Summary

In this review, we have introduced applications of new satellite observations of previously rarely observed typhoon–ocean interaction phenomena in the western North Pacific and adjacent seas, as presented by Lin *et al.* (2003a,b; 2005). Examples for three different phenomena have been given to illustrate the interaction between the typhoon and the warm ocean eddy, enhancement of ocean primary production induced by the typhoon, and posttyphoon air–sea interaction. These new observations show that:

(1) The presence of warm ocean eddies plays a critical role in supertyphoon Maemi's intensification. In the presence of a warm ocean eddy with deep warm ocean water, typhoon self-induced ocean cooling is much reduced. As a result, Maemi (2003) was able to reach category 5, due to the minimal negative feedback. Numerical experiments by the CHIPS (Emanuel, 1999; Emanuel *et al.*, 2004) coupled typhoon–ocean model have found that without the presence of a warm ocean eddy, Maemi's intensity could only reach category 4.

(2) Typhoons can induce drastic biological responses in the ocean, and hence may play a significant role in the marine primary production and carbon fixation. As observed in the case of typhoon Kai-Tak (2000), during its short, three-day meandering in the South China Sea, this moderate, category 2 typhoon caused an average of a 30-fold increase in the ocean surface chlorophyll-a concentration. The estimated carbon fixation resulting from this event alone is 0.8 Mt, or 2–4% of the South China

Sea's annual new production. Each year, about 14 cyclones pass over the South China Sea, suggesting that the long-neglected contribution of typhoons to the South China Sea's annual new production may be as much as 20–30%.

(3) There is an interesting air–sea coupling long after the typhoon's passage. The cold wake of the typhoon existed for more than a week, as observed in the case of typhoon Kai-Tak (2000). Intercomparison of coincident/colocated QuikSCAT ocean surface wind vectors finds clear and consistent weakening of the surface wind speed over the cold SST wake. This suggests that the boundary layer stability is increased because of the cold ocean surface, and the surface wind speed is reduced due to diminished vertical mixing (Wallace *et al.*, 1989; Xie *et al.*, 1998; Chelton *et al.*, 2001). In particular, our result suggests that this mechanism can act on a relatively small spatial scale ($\sim$100 km) and short ($\sim$1 day) time scale, in contrast to the previously reported much longer time scale ($\sim$20–40 days) and large spatial scale of $\sim$1000–2000 km (Xie *et al.*, 1998; Wentz *et al.*, 2000; Chelton *et al.*, 2001).

As the above works are based on studies of an individual case, ongoing efforts are being made to conduct systematic analysis in studying the typhoon–ocean interaction problems in the western North Pacific.

[Received 9 April 2007; Revised 30 August 2007; Accepted 15 September 2007.]

References

Babin, S. M., J. A. Carton, T. D. Dickey, and J. D. Wiggert, 2004: Satellite evidence of hurricane-induced phytoplankton blooms in an oceanic desert. *J. Geophys. Res.*, **109**, C03043, doi:10.1029/2003JC001938.

Bates, N. R., A. H. Knap, and A. F. Michaels, 1998: Contribution of hurricanes to local and global estimates of air–sea exchange of CO_2. *Nature*, **395**, 58–61.

Behrenfeld, M. J., and P. G. Falkowski, 1997: Photosynthetic rates derived from satellite based chlorophyll concentration. *Limnol. and Oceanog.*, **42**, 1–20.

Bender, M. A., and I. Ginis, 2000: Real-case simulations of hurricane–ocean interaction using a high-resolution coupled model: effects on hurricane intensity. *Mon. Wea. Rev.*, **128**, 917–946.

Black, P. G., E. A. D'Asaro, W. M. Drennan, J. R. French, P. P. Niller, T. B. Sanford, E. J. Terrill, E. J. Walsh, and J. A. Zhang, 2007: Air–sea exchange in hurricanes, synthesis of observations from the Coupled Boundary Layer Air–Sea Transfer Experiment. *Bull. Amer. Meteor. Soc.*, **88**, 357–374.

Chang, S. W., and R. A. Anthes, 1978: Numerical simulations of the ocean's nonlinear, baroclinic response to translating hurricanes. *J. Phys. Oceanog.*, **8**, 468–480.

Chang, S. W., and R. A. Anthes, 1979: The mutual response of the tropical cyclone and the ocean. *J. Phys. Oceanog.*, **9**, 128–135.

Chelton, D. B., S. K. Esbensen, M. G. Schlax, N. Thum, M. H. Freilich, F. J. Wentz, C. Gentemann, M. J. McPhaden, and P. S. Schopfet, 2001: Observations of coupling between surface wind stress and sea surface temperature in the Eastern Tropical Pacific. *J. Climate*, **14**, 1470–1498.

Cione, J. J., and E. W. Uhlhorn, 2003: Sea surface temperature variability in hurricanes: implications with respect to intensity change. *Mon. Wea. Rev.*, **131**, 1783–1796.

Cornillon, P., L. Stramma, and J. F. Price, 1987: Satellite measurements of sea surface cooling during hurricane Gloria. *Nature*, **326**, 373–375.

Davis, A., and X. H. Yan, 2004: Hurricane forcing on chlorophyll-a concentration off the northeast coast of the U.S., *Geophys. Res. Lett.*, **31**, L17304, doi: 10.1029/2004GL020668.

Dickey, T., D. Frye, J. McNeil, D. Manov, N. Nelson, D. Sigurdson, H. Jannasch, D. Siegel, A. Michaels, and R. Johnson, 1998: Upper-ocean temperature response to Hurricane Felix as measured by the Bermuda Testbed Mooring. *Mon. Wea. Rev.*, **126**, 1195–1201.

Emanuel, K. A., 1986: An air–sea interaction theory for tropical cyclones Part I. *J. Atmos. Sci.*, **42**, 1062–1071.

Emanuel, K. A., 1988: The maximum intensity of hurricanes. *J. Atmos. Sci.*, **45**, 1143–1155.

Emanuel, K. A., 1991: The theory of hurricanes. *Annual Rev. Fluid Mech.*, **23**, 179–196.

Emanuel, K. A., 1995: Sensitivity of tropical cyclones to surface exchange coefficients and a revised steady-state model incorporating eye dynamics. *J. Atmos. Sci.*, **52**, 3969–3976.

Emanuel, K. A, 1999: Thermodynamic control of hurricane intensity. *Nature*, **401**, 665–669.

Emanuel, K. A., 2001: The contribution of tropical cyclones to the oceans' meridional heat transport. *J. Geophys. Res.*, **106**, 14771–14781.

Emanuel, K. A., C. DesAutels, C. Holloway, and R. Korty, 2004: Environmental control of tropical cyclone intensity. *J. Atmos. Sci.*, **61**, 843–858.

Eppley, R. W., and B. J. Peterson, 1979: Particulate organic matter flux and planktonic new production in the deep ocean. *Nature*, **282**, 677–680.

Eppley, R. W., 1989: New production: history, methods, problems. In *Productivity of the Ocean: Present and Past*, eds. W. H. Berger, V. S. Smetacek, and G. Wefer (Wiley, Chichester), pp. 85–97.

Fu, L. L., E. J. Christensen, C. A. Yamarone, M. Lefebvre, Y. Menard, M. Dorrer, and P. Escudier, 1994: TOPEX/POSEIDON mission overview. *J. Geophys. Res.*, **99**, 24369–24381.

Fu, L. L., and A. Cazenave, 2001: *Satellite Altimetry and Earth Sciences: A Handbook of Techniques and Applications, International Geophysics Series*, Vol. 69 (Academic San Diego).

Gallacher, P. C., R. Rotunno, and K. A. Emanuel, 1989: Tropical cyclogenesis in a coupled ocean-atmosphere model. Preprints, *18th Conf. on Hurricanes and Tropical Meteorology* (Amer. Meteor. Soc., San Diego), pp. 121–122.

Goni, G. J., and J. A. Trinanes, 2003: Ocean thermal structure monitoring could aid in the intensity forecast of tropical cyclones. *Eos, Trans. Amer. Geophys. Union*, **84**, 573–580.

Holland, G. J., 1997: The maximum potential intensity of tropical cyclones. *J. Atmos. Sci.*, **54**, 2519–2541.

Hong, X., S. W. Chang, S. Raman, L. K. Shay, and R. Hodur, 2000: The interaction between Hurricane Opal (1995) and a warm core eddy in the Gulf of Mexico. *Mon. Wea. Rev.*, **128**, 1347–1365.

Jacob, S. D., L. K. Shay, A. J. Mariano, and P. G. Black, 2000: The 3-D oceanic mixed layer response to hurricane Gilbert. *J. Phys. Oceanog.*, **30**, 1407–1429.

Ko, D. S., R. H. Preller, G. A. Jacobs, T. Y. Tang, and S. F. Lin, 2003: Transport reversals at Taiwan Strait during October and November 1999. *J. Geophys. Res.*, **108**, C11, 3370.

Levitus, S., 1982: *Climatological Atlas of the World Ocean.* NOAA Prof. Paper 13, 173 pp., and 17 microfiche.

Lin, I. I., W. T. Liu, C. C. Wu, J. C. H. Chiang, and C. H. Sui, 2003: Satellite observation of modulation of surface layer winds by typhoon-induced upper ocean cooling. *Geophys. Res. Lett.*, **30**, 1131, doi:10.1029/2002GL015674.

Lin, I. I., W. T. Liu, C. C. Wu, G. T. F. Wong, C. Hu, Z. Chen, W. D. Liang, Y. Yang, and K. K. Liu, 2003: New evidence for enhanced ocean primary production triggered by tropical cyclone. *Geophys. Res. Lett.*, **30**, 1718, doi:10.1029/2003GL017141.

Lin, I. I., C. C. Wu, K. A. Emanuel, I. H. Lee, C. R. Wu, and I. F. Pun, 2005: The interaction of supertyphoon maemi (2003) with a warm ocean eddy. *Mon. Wea. Rev.*, **133**, 2635–2649.

Liu, K. K., S. Y. Chao, P. T. Shaw, G. C. Gong, C. C. Chen, and T. Y. Tang, 2002: Monsoon-forced chlorophyll distribution and primary production in the South China Sea. *Deep-Sea Res., Part I*, **49**, 1387–1412.

Liu, W. T., W. Tang, and P. Polito, 1998: NASA scatterometer provides global ocean–surface wind fields with more structures than numerical weather prediction. *Geophys. Res. Lett.*, **25**, 761–764.

Marra, J., R. R. Bidigare, and T. D. Dickey, 1990: Nutrients and mixing, chlorophyll and phytoplankton growth. *Deep-Sea Res.*, **37**, 127–143.

McGillicuddy, D. J., A. R. Robinson, D. A. Siegel, H. W. Jannasch, R. Johnson, T. D. Dickey, J. McNeil, A. F. Michaels, and A. H. Knap, 1998: Influence of mesoscale eddies on new production in the Sargasso Sea. *Nature*, **394**, 263–266.

McGillicuddy, D. J., V. K. Kosnyrev, J. P. Ryan, and J. A. Yoder, 2001: Covariation of mesoscale ocean color and sea-surface temperature patterns in the Sargasso Sea, *Deep Sea Res., Part II*, **48**, 1823–1836.

Monaldo, F. M., T. D. Sikora, S. M. Babin, and R. E. Sterner, 1997: Satellite imagery of sea surface temperature cooling in the wake of hurricane Edouard (1996). *Mon. Weather Rev.*, **125**, 2716–2721.

Oey, L. Y., T. Ezer, D. P. Wang, X. Q. Yin, and S. J. Fan, 2007: Hurricane-induced motions and interaction with ocean currents. *Cont. Shelf Res.*, **27**, 1249–1263.

O'Reilly, J. E., S. Maritorena, B. G. Mitchell, D. A. Siegel, K. L. Carder, S. A. Garver, M. Kahru, and C. McClain, 1998: Ocean color chlorophyll algorithms for SeaWiFS. *J. Geophys. Res.*, **103**, 24937–24953.

Price, J. F., 1981: Upper ocean response to a hurricane. *J. Phys. Oceanog.*, **11**, 153–175.

Pun, I. F., I. I. Lin, C. R. Wu, D. S. Ko, and W. T. Liu, 2007: Validation and application of altimetry-derived upper ocean thermal structure in the western North Pacific Ocean for typhoon intensity forecast. *IEEE Trans. Geosci. Remote Sens.* **45**(6), 1616-1630.

Qiu, B., 1999: Seasonal eddy field modulation of the North Pacific Subtropical Countercurrent: TOPEX/Poseidon observations and theory. *J. Phys. Oceanog.*, **29**, 1670–1685.

Schade, L. R., and K. A. Emanuel, 1999: The ocean's effect on the intensity of tropical cyclones: results from a simple coupled atmosphere–ocean model. *J. Atmos. Sci.*, **56**, 642–651.

Scharroo, R., W. H. F. Smith, and J. L. Lillibridge, 2005: Satellite altimetry and the intensification of hurricane Katrina. *EOS, Trans. Amer. Geophys. Union*, **86**, 366–367.

Shay, L. K., P. G. Black, A. J. Mariano, J. D. Hawkins, and R. L. Elsberry, 1992: Upper ocean response to hurricane Gilbert. *J. Geophys. Res.*, **12**, 20277–20248.

Shay, L. K., G. J. Goni, and P. G. Black, 2000: Effects of a warm oceanic feature on Hurricane Opal. *Mon. Wea. Rev.*, **128**, 1366–1383.

Siswanto, E., J. Ishizaka, K. Yokouchi, K. Tanaka, and C. K. Tan, 2007: Estimation of interannual and interdecadal variations of typhoon-induced primary production: a case study for the outer shelf of the East China Sea. *Geophys. Res. Lett.*, **34**, L03604, doi:10.1029/2006GL028368.

Stramma, L., P. Cornillon, and J. F. Price, 1986: Satellite observations of sea surface cooling by hurricanes, *J. Geophys. Res.*, **91**, 5031-5035.

Uz, B. M., J. A. Yoder, and V. Osychny, 2001: Pumping of nutrients to ocean surface waters by the action of propagating planetary waves. *Nature*, **409**, 597–600.

Wallace, J. M., T. P. Mitchell, and C. Deser, 1989: The influence of sea surface temperature on surface wind in the eastern equatorial Pacific: Seasonal and interannual variability. *J. Climate*, **2**, 1492–1499.

Wentz, F. J., C. Gentemann, D. Smith, and D. Chelton, 2000: Satellite measurements of sea surface temperature through clouds. *Science*, **288**, 5467, 847–850.

Wu, C. C., C. Y. Lee, and I. I. Lin, 2007: The effect of the ocean eddy on tropical cyclone intensity. *J. Atmos. Sci.* (accepted).

Xie, S. P., M. Ishiwatari, H. Hashizume, and K. Takeuchi, 1998: Coupled ocean–atmospheric waves on the equatorial front. *J. Geophys. Res. Lett.*, **25**, 3863–3966.

Zhu, T., and D. L. Zhang, 2006: The impact of the storm-induced SST cooling on hurricane intensity. *Advances in Atmospheric Sciences*, **23**, 1, 14–22.

Outbreaks of Asian Dust Storms: An Overview from Satellite and Surface Perspectives

Si-Chee Tsay*

Laboratory for Atmospheres, Goddard Space Flight Center, Greenbelt, Maryland, USA
si-chee.tsay@nasa.gov

Taiwan, located downwind of dust storm outbreaks from China, in the sink region of biomass-burning aerosols from Southeast Asia, and at the outflow of urban-industrial pollutants from the Pearl and Yangtze River Delta, is exposed to a seasonal milieu of natural and anthropogenic aerosols in the atmosphere. In the springtime, outbreaks of Asian dust storms occur frequently in the arid and semiarid areas of northwestern China — about 1.6×10^6 square kilometers, including the Gobi and Taklimakan deserts — with continuous expansion of spatial coverage. These airborne dust particles, originating in desert areas far from polluted regions, interact with anthropogenic sulfate and soot aerosols emitted from Chinese megacities during their transport over the mainland. Adding the intricate effects of clouds and marine aerosols, dust particles reaching the marine environment can have drastically different properties than those from their sources.

Together with anthropogenic pollutants, airborne dust particles may alter regional hydrological cycles by aerosol direct/indirect radiative forcing, influence fisheries by causing nutrient deposition anomalies, and increase adverse health effects on humans by trace metal enrichment. In addition to their local-to-regional impact, these dust aerosols can be transported swiftly across the Pacific Ocean to reach North America in less than a week, resulting in an even larger scale effect. Asian dust aerosols can be distinctly detected by their colored appearance on modern Earth-observing satellites [e.g. MODerate-resolution Imaging Spectroradiometer, (MODIS), Total Ozone Mapping Spectrometer (TOMS), Sea-viewing Wide Field-of-view Sensor (SeaWiFS), Geostationary Meteorological Satellites (GMSs)], and their evolution monitored by satellites and surface networks [e.g. AErosol RObotic NETwork (AERONET), Micro-Pulse Lidar Network (MPLNET)]. However, these essential observations are incomplete due to the unique properties of the data constituted unilaterally on either the spatial (snapshot global coverage) or temporal (long-term point sites) dimension. Comprehensive modeling is required to bridge these spatial and temporal observations, and to serve as an integrator for our understanding of the effects of dust's physical, optical, and radiative properties on various forcing, response, and feedback processes occurring in the Earth–atmosphere system.

Recently, many field experiments (e.g. international ACE-Asia and regional follow-on campaigns) have been conducted to shed light on characterizing the compelling variability of dust aerosols on spatial and temporal scales, especially near the source and downwind regions. As a result of synergizing satellite, aircraft, and surface observations, our understanding of the distributions and properties of airborne dust aerosols has advanced significantly. It is our goal/hope to continue combining observational and theoretical studies to investigate in depth the changes of regional climate, hydrological budget, tropospheric chemistry, wind erosion, and dust properties in Asia. These regional changes (e.g. aerosol loading, cloud amount, precipitation rate) constitute a vital part of global change, and our success or failure in developing reliable predictions of, as well as adequate responses to, the changes will determine the prospective course for sustainable civilization. Consequently, the lessons learned will help strengthen our ability to issue early warnings of Asian dust storms and minimize further desertification in the future.

---

*With contributions from Gin-Rong Liu, Nai-Yung Hsu, Wen-Yih Sun, Neng-Huei Lin, Qiang Ji, Guang-Yu Shi, Myeong-Jae Jeong, Tang-Huang Lin, Chi-Ming Peng, Sheng-Hsiang Wang, and Jr-Shiuan Yang.

1. Background

Having entered the new millennium, there is no doubt that globalization is the wave of the future in societal and economic activities, including the byproducts associated with these activities. An excellent example is the recent intensified outbreaks of Asian dust storms and air pollution that have brought even broader public attention to their wide-ranging impacts, from the macro scale of aerosol radiative forcing on weather and climate to the micro scale of pathogens/minerals on the spread of human/agricultural diseases. Thus, aerosol emissions in one area can cause damage to other regions through transborder and even transcontinental transport. The stunning image of Living Earth (the Earthrise, first captured by NASA's Apollo-8 mission on 22 December 1968) rising above the Moon's horizon truly indicates the dynamical and restless nature of our home planet. Starting from the last century, however, the growth of the world population, the development of modern technology, the demand for consumable energy, and drastic land-cover/land-use changes, among other factors, have revealed an exponential trend that presents a major environmental stress that tips Earth away from sustainable balance. This urgent call for the saving of our planet has been echoed overwhelmingly by the recent documentary film *An Inconvenient Truth*, presented by Al Gore, former US vice-president and 2007 Nobel Peace laureate.

Taiwan, a role model for the economic miracle of developing countries during the mid-20th century, now faces many critical challenges regarding environmental protection. For instance, along with a multitude of other effects, the outbreaks of Asian dust storms seasonally alter the chemistry in the atmosphere and the surrounding seas, pollute the breathing air (impacting human health), and degrade the visibility (affecting traffic safety). In this overview article, we first give an introduction to the outbreak of Asian dust storms with historical and statistical prospects, as well as the environmental conditions associated with the dust source, transport, and sink. The societal impact and scientific significance of dust aerosols are presented in Sec. 2. Section 3 discusses the properties of dust aerosols based on available remote sensing measurements, *in situ* observations, and theoretical modeling to gain a better understanding of what role dust aerosols play in the Earth–atmosphere system. Finally, concluding remarks are given on what lessons have been learned and what further action strategies may be taken.

1.1. *Dust storm outbreaks*

Utilizing advanced instrumentation from space, large-scale dust storm outbreaks are commonly observed on the Earth and other planets in our solar system. For example, on 26 June 2001, NASA's Hubble Space Telescope spotted an enormous dust storm outbreak on Mars that quickly enveloped the whole planet and subsequently raised the temperature of the glacial Martian atmosphere by about 30°C. Although the sum of Earth's continents is roughly equal to the total Martian surface area, dust storm outbreaks on Earth are generally smaller because it is not a global desert like Mars. Nearly one-third of Earth's lands are deserts or arid regions, which receive annual precipitation of either less than $25\,\text{cm}$ ($\sim$10 inches) or less than half of the evaporation. Based on this definition, the polar regions ($\sim 2.8 \times 10^7\,\text{km}^2$) comprise Earth's largest desert, followed by the Saharan ($\sim 9.1 \times 10^6\,\text{km}^2$) and Arabian ($\sim 2.6 \times 10^6\,\text{km}^2$) deserts in subtropical Africa/Mideast, and then China's Gobi–Taklimakan deserts ($\sim 1.6 \times 10^6\,\text{km}^2$) in midlatitude Asia.

A dust storm (or sandstorm in some contexts) outbreak has many names across the desert regions, such as "*simoom*" for the African Sahara, "*haboob*" for deserts in the Arabian Peninsula, "*shachenbao*" for the Chinese Gobi–Taklimakan deserts, and "*kosa*" for downwind

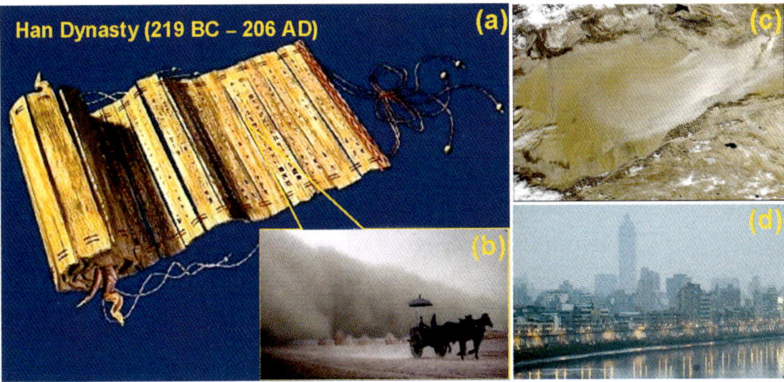

Figure 1. (a) Official document recording on bamboo strips in ancient China; (b) reconstructed image of a fierce dust storm originating at the Taklimakan desert that destroyed a stagecoach during the Chinese Han Dynasty, reported by a surviving officer (*Xinhua News*, May 2002); (c) a typical Taklimakan dust storm appearing in pale beige and sweeping toward the northeast, captured by MODIS/Aqua (image courtesy: NASA Earth Observatory on 3 December 2005); and (d) a few days after violent dust storm outbreaks from the Gobi–Taklimakan deserts, inhalable dust aerosols blanket the downwind regions of the Korean peninsula, Japan, and Taiwan (photo courtesy: *Taipei Times* on 20 March 2006 in Taiwan, with the Taipei-101 Tower, currently the world's tallest building, in the background).

Korea–Japan regions. The prime mechanism causing these dust storm outbreaks differs noticeably: those from the Sahara are mainly driven by instability induced by strong solar heating at the subtropical surface (e.g. Karyampudi *et al.*, 1999), with year-round frequency except during the African monsoon season; those forming at the Arabian deserts arise chiefly from the collapse of thunderstorms (also known as a downburst) during subtropical summer (e.g. Tindale and Pease, 1999); and those originating at the Gobi–Taklimakan deserts are largely associated with cold air outbreaks causing the Mongolian cyclonic depression and frontal activities in the spring (e.g. Qian *et al.*, 2002).

As much as one-third to one-half of global dust emission, estimated to be about 800 Tg (Zhang *et al.*, 1997), is introduced annually into Earth's atmosphere from various deserts in China. Asian dust storm outbreaks are believed to have persisted for hundreds of thousands of years over the vast territory of north and northwest China, but it was not until recent decades that many studies for compelling evidence for recognizing the importance of these eolian dust particles for forming the Chinese Loess Plateau (e.g. Derbyshire *et al.*, 1998),

and biogeochemical cycling in the North Pacific Ocean (e.g. Husar *et al.*, 1997), to as far as in the Greenland ice-sheets (e.g. Svensson *et al.*, 2000) through long-range transport. Recently, in the vicinity of Dunhuang, China — the gateway to the ancient Silk Road — Chinese archaeologists unearthed a "bamboo document" from the early Han Dynasty [see Fig. 1(a)], which described officially and explicitly the earliest dust storm event in Chinese written records near the source regions.

The Taklimakan ("place of no return" in Turkish) desert, a predominant land feature in the Tarim Basin, is enclosed by three major mountain ranges in western China: the Tien-Shan to the north, the Pamirs to the west, and the Kunlun-Shan to the south, with a narrow opening of saline marshy depression (the lowest area in the region is 150 m below sea level) in the east [see Fig. 1(c) for the topography]. Snow and glacier melt-waters from the surrounding mountain ranges supply all the rivers in the Tarim Basin, but these waters never find the sea. Added to this extraordinary geography, the location of the Taklimakan desert far from oceans further removes any rainfall from Asian monsoons. Having an annual precipitation rate of less than 10 mm and plenty of dry river/lake sediments, Taklimakan ($\sim 0.34 \times 10^6\,\mathrm{km}^2$) constitutes the second-largest shifting-sand desert on Earth (e.g. Sun and Liu, 2006). As a result, it is hardly surprising that the Tarim Basin experiences more dust storms than any other place on Earth, with up to 100–174 events annually (e.g. Washington *et al.*, 2003), as depicted in Fig. 1(c) for a typical dust storm outbreak. Another region of frequent dust storm outbreaks in China is Asia's largest desert area, the great Gobi ("very large and dry" in Mongolian, or simply "big desert" in Manchu), which is bounded by the Altai-Shan and Mongolia grasslands/steppes to the north, the Tibetan Plateau to the southwest, and the North China Plain to the southeast. Unlike the unique geography of the Taklimakan desert, the great Gobi ($\sim 1.29 \times 10^6\,\mathrm{km}^2$), by and large consisting of gravel and bare rock, nonetheless has numerous distinct ecological and geographic regions due to variations in local climate and topography. Outbreaks of dust storms in north and northwest China not only cover massive areas in source regions, but also transport airborne dust particles downwind to the Korean peninsula, Japan, Taiwan [see Fig. 1(d)], and further beyond (e.g. VanCuren and Cahill, 2002).

The spatial and temporal distributions of dust storm outbreaks in China have been the subject of numerous studies (e.g. Wang *et al.*, 2004, and references therein). Generally, meteorological observers report a *dust storm* event when the horizontal surface visibility is reduced to 1 km or below, a *blowing dust* event for surface visibility in the range of 1–10 km, and a *hazy sky* for visibility less than 10 km with aeolian dust particles suspended homogeneously in the air. As an example, Fig. 2(a) depicts 30-year (1951–1980), monthly-mean dust events from surface observations at Dunhuang, China, located between the Taklimakan and great Gobi deserts. It is clear that the peak season of dust events over Dunhuang is during the boreal spring (March–May) and

the lowest during the autumn (September–November). Analyzing modern satellite observations, such as the TOMS Aerosol Index (e.g. Hsu *et al.*, 1996), of a $2^\circ \times 2^\circ$ region over Dunhuang during 1979–2000, the statistical features vary slightly, as illustrated in Fig. 2(b). The major discrepancy between these two temporal–spatial datasets is the seasonal trend of minimal dust events. Although satellite observations provide sizeable spatial coverage compared to that of surface *in situ*, they often experience an insensitivity of phenomena in the planetary boundary layer. Since the peak season of dust events over Dunhuang is evidently prolonged from the analyses of two completely different datasets, it is likely that during the winter season ground-based observations are more sensitive to aerosol loading in the atmosphere near the surface than those of satellite measurements.

Wang *et al.* (2004) presented a comprehensive overview of modern dust storm outbreaks in China and concluded that the most active geographic regions are (1) the Taklimakan desert, (2) the Hexi Corridor to the west Inner Mongolia Plateau, and (3) the central Inner Mongolia Plateau. Although the compiled results reveal seasonal changes of dust storm occurrence in different regions, the peak season is indicated during the boreal spring for all. In fact, the temporal distribution averaged over all regions in China closely mimics that of Dunhuang [see Fig. 2(a)]. However, with regard to the long-term trend of dust storm outbreaks in China, it is relatively divisive among studies reviewed by Wang *et al.* (2004). Overall, the highest frequency of dust storms occurred in the 1960s and 1970s; but they concluded that there were no significant statistical correlations between the frequency of dust storms and wind energy, or annual precipitation and evaporation. Besides the natural variability of weather/climate systems and other controlling factors, Wang *et al.* (2004) considered human activities, such as breaking up naturally wind-resistant surfaces and wiping out protective

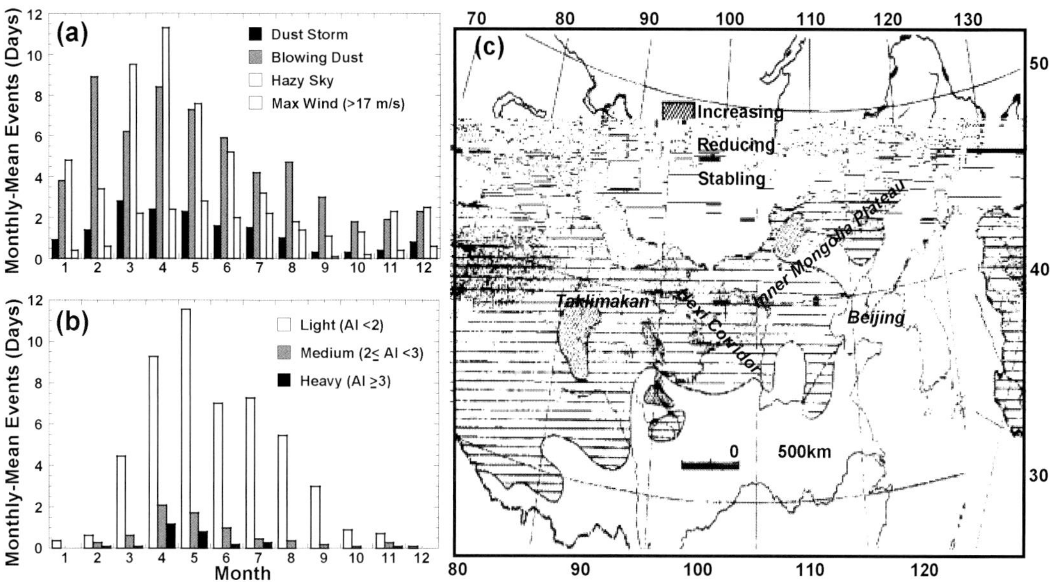

Figure 2. (a) Surface observations of dust events at Dunhuang (40° 2′ N, 94° 7′ E), China, during 1951–1980, based on the criteria of horizontal surface visibility; (b) satellite analyses of dust events in the vicinity of Dunhuang (39–41°N, 93–95°E) during 1979–2000, anchored in deviations of spectral reflectance (i.e. Aerosol Index; Hsu *et al.*, 1996) for indicating aerosol abundance in the column atmosphere; and (c) the annual trend (at 95% significance level) of dust storm distributions (Wang *et al.*, 2004) in China during 1954–2000, utilizing records of 681 key stations operated by the China Meteorological Administration.

vegetation from lands, to have played the most imperative role in the long-term trend of dust storm outbreaks in China over the last 50 years. As illustrated in Fig. 2(c), the four regions, which had experienced increases in the frequency of dust storm outbreaks, were largely caused by regional land desertification due to human activities.

1.2. *Source/sink and transport pathway*

Airborne dust particles can be found everywhere; the dust that falls in our backyard (the *sink*) may have originated in arid regions (the *source*) somewhere on Earth but traveled thousands of miles (the *transport*). Based on the analyses of 40 years of meteorological data (1960–1999 at 174 stations), Sun *et al.* (2001) concluded that springtime dust storms originating in China are highly associated with the activities of frontal systems and the Mongolian cyclonic depression. With some additions, their results on the routes/frequencies of cold air and dust storm outbreaks are summarized in Fig. 3.

From these 40-year statistics, the springtime dust storms in China were also revealed to have a peak occurrence in April, which was about three times those having taken place in either March or May. Overall, the frequencies of cold air outbreaks during springtime, originating from the west, north, and northwest, range from relatively comparable (Fig. 3, upper-left graph; Sun *et al.*, 2001) to nearly doubled to quadrupled (respectively, 13.8%, 27.8%, and 58.6%; Gao *et al.*, 2006), which all resulted in triggering dust storms from the great Gobi deserts along the Hexi Corridor and Inner Mongolia Plateau. However, only the western routes of cold air outbreaks frequently cause dust storms from the Taklimakan desert. Dust particles lifted from the great Gobi deserts (sum of frequencies > 80%; Fig. 3, upper-right graph) are commonly elevated up to 3 km, which is favorable for

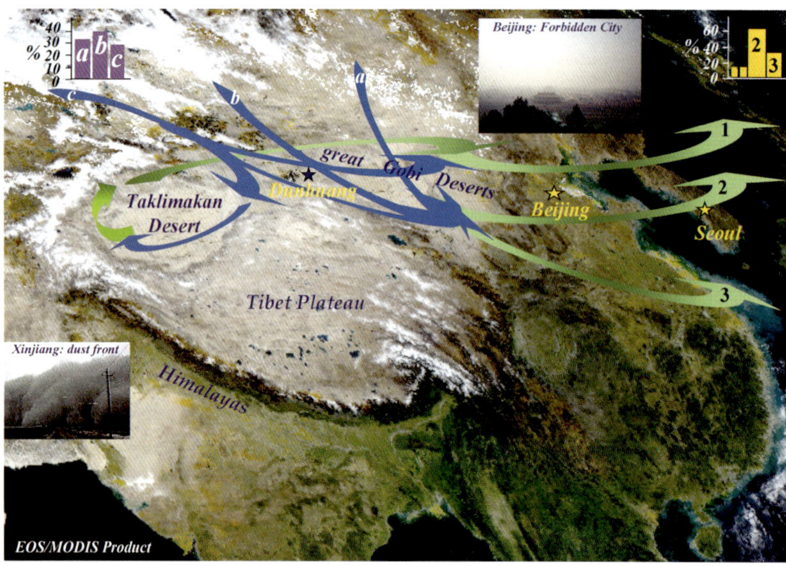

Figure 3. Statistical analyses of the source/sink regions and transport pathways for dust storm outbreaks in China (after Sun *et al.*, 2001), overlying an EOS/MODIS product of a clear-sky, true-color composite image. Examples of dust aerosols near the source (lower-left picture, showing a passage of a dust front in Xinjiang province, western China) and over the sink (upper-right picture depicting dust fallouts over the Forbidden City in Beijing, China) regions are presented, as well as the routes/frequencies of cold air outbreaks (blue arrows/bars) and patterns/frequencies of transport pathways (brown arrows/bars).

regional transport. Consequently, their impacts are limited to the regional scale, from the proximal Loess plateau and metropolitan Beijing to downwind areas of the Korean peninsula, Japan, Taiwan, and the nearby Pacific Ocean; whereas those dust aerosols originating mainly from the Taklimakan desert and rarely from the great Gobi deserts (total <20%) are frequently entrained to an elevation higher than 5 km and can be transported over long distances (e.g. ∼5,000 km) by the prevailing westerly jet streams.

Atmospheric circulations over the Taklimakan desert are extremely complex due to the influence of bounding terrain [see Fig. 1(c)] which induces topographic channeling of the winds. The dominant moving direction of Taklimakan sand dunes, running up against the Tibetan plateau, is observed as either easterly or northeasterly, which clearly indicates the existence of prevailing low-level, easterly

airflows. Thus, without a strong vertical lifting, the dust-laden atmosphere is poorly ventilated and the dust aerosols remain trapped in the enclosed basin. Recent lidar profiling of aerosol vertical distributions, such as the Geoscience Laser Altimeter System (GLAS) aboard the Ice, Cloud, and land Elevation Satellite (ICESat), is very useful for interpreting such cases. Figure 4(a) shows a Terra/MODIS red–green– blue composite image, depicting an outbreak of a Taklimakan dust storm. This dust storm began to be visible from space a few days before this image was taken. Approximately 6 hours prior to the closest satellite overpass, ICESat/GLAS captured this dust episode [Fig. 4(b)] when crossing the Taklimakan desert, indicated as a black line on the Terra/MODIS image (Fig. 4a). The dust layer was located about 3 km above ground, forming an arch across the desert but not lifted high enough to escape out of the basin. In addition, the dust layer in the southern

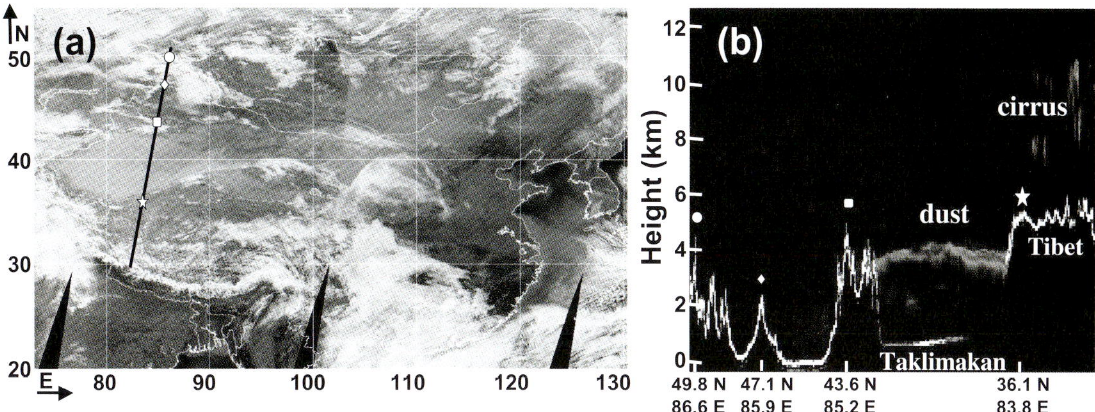

Figure 4. (a) Terra/MODIS red–green–blue composite image acquired on 14 March 2006, and (b) ICESat/GLAS vertical profile at 532 nm wavelength of a dust storm in the Taklimakan desert, with its corresponding geolocation marked as a black line on (a). Symbols on the transect line denote various mountain peaks in the Altai-Shan (circle and diamond), Tien-Shan (square), and Kunlun-Shan (star) ranges.

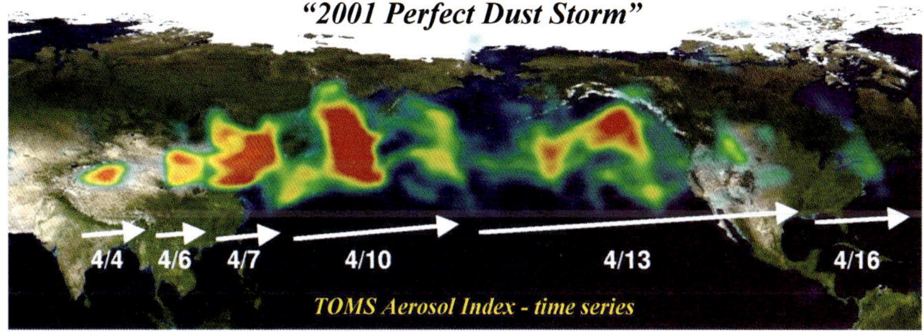

Figure 5. Composite of TOMS Aerosol Index, a surrogate for aerosol loading, on 4–16 April 2001, depicting the long-range transport of Asian dust particles with a pathway across the Pacific Ocean and North America to the Atlantic Ocean. The arrows/dates indicate the size and location of the dust aerosols observed, with yellow–green–red colors in increasing order of density for dust loadings.

portion of the arch against the Tibetan plateau was thick enough to prevent complete penetration of the laser beam, impairing estimations of the column dust loading and ground level detection. This also portrays the difficulties involved in monitoring, assessing, and analyzing complicated cases such as the Asian dust storm outbreaks when utilizing satellite sensors alone.

Nevertheless, satellite measurements provide dramatic results for the discovery that airborne dust particles can travel remarkably vast distances, as illustrated in Fig. 5, and induce various environmental impacts along their transport pathway. In the spring, Asian dust particles together with other anthropogenic and natural aerosols, once generated over the source regions, can be moved out of the boundary layer into the free troposphere. Sometimes riding with the westerly jet streams, they can travel thousands of miles across the Pacific into the United States, Canada, and beyond. As the infamous "2001 perfect dust storm" demonstrated, it took just less than a week to mobilize these dust plumes across the Pacific Ocean.

Part of these dust aerosols have been observed to subside in the Arctic, with some aloft in layers up to 10 km over Texas, and others linger over the NASA/Goddard Space Flight Center in Maryland (see NASA press release in April 2001).

2. Consequence

The processes of generating, transporting, and dissipating airborne dust particles are global phenomena — African dust regularly reaches the Alps; Asian dust seasonally crosses the Pacific into North America, and ultimately the Atlantic into Europe. One of the vital biogeochemical roles which dust storms play in Earth's ecosystem is the routine mobilization of mineral dust, as a source of iron (e.g. Meskhidze *et al.*, 2005), from deserts into oceans for fertilizing the growth of phytoplankton — the basis of the oceanic food chain. Similarly, these dust-laden airs supply crucial nutrients for the soil of tropical rain forests, the so-called *womb of life*, which hosts 50–90% of the species on Earth. Historically, dust storm outbreaks are mainly natural events, but recent decades of increases in surface disturbances associated with anthropogenic activities worldwide may have altered considerably the net amount of airborne dust distributions. Unrestrained land use (e.g. deforestation, overgrazing, urbanization) coupled with long periods of drought can lead to massive deterioration of land cover, which can in turn increase the frequency of dust storm outbreaks.

More than 60% of the world's population resides in Asia, and China alone contains more than 20% ($\sim$1.3 billion) of the world's population. China relies primarily on agricultural products harvested from about 8–9% of Earth's arable and permanent cropland. In comparison, the United States, with a similar size of cropland, supports only a population of $\sim$0.3 billion. In addition, as one of the major sources of protein, China accounts for 33% of the global fish and seafood consumption (e.g. Pauly *et al.*, 2003), while domestic fisheries from the lakes, rivers, and coastal seas/oceans account only for 15% of the world's fish catch. As a result of meeting this demand, the production of aquaculture freshwater fish has leapt sharply, at the cost of reducing China's cropland. The balance of China's ecosystem is particularly delicate; thus, a fatal perturbation (e.g. drastic change of land cover and land use, depleted fisheries) can upset the balance and have an enormous impact on the global society due to the extremely large population size.

Synthesizing available data and records, Liu and Diamond (2005) presented and discussed in detail China's changing environment and socioeconomic challenges in the context of global interdependence. On the issues of natural disasters, it was noted that from AD 300 to 1949 northwestern China was exposed to major dust storms once every 31 years on average. Additionally, based on the statistics studied by the Chinese Academy of Sciences, the average frequency increased to about once per year from the 1950s to the 1990s. Since 2000, the average number of dust storms for the same region has escalated to 5–6 per year, including 8 fallouts in Beijing in 2006. In the drier northwest, overgrazing and overplowing severely degraded the vast natural grasslands in China, especially in Qinghai province and the Inner Mongolia Autonomous Region, two of the major source regions of dust storms. Liu and Diamond (2005) further concluded that 90% of China's grasslands have been degraded and declining at a rate of $\sim$1.5 $\times$ 10^4 km^2 per year since the early 1980s. Moreover, the current status of desertification in China is very serious. The total area of desertification covers about 2.6 $\times$ 10^6 km^2 ($\sim$27% of China's territory), of which 1.6 $\times$ 10^6 km^2 of desertified land is caused by wind erosion. Studley (1999) estimated that the spreading rate of desertification is $\sim$2,460 km^2 per year. Similar figures were acknowledged by the Chinese authorities (i.e. *Xinhua News*), and appeared in a press conference held on 26 May 2001.

The consequence of desertification and topsoil erosion in China, in terms of dust storm outbreaks, has had a worldwide impact and appeared frequently in the springtime headline news of various media. A particular example is the catastrophic losses of human lives and property, crops, and livestock in the 5 May 1993 dust storm (Yang, 2001). The ways that airborne dust particles affect humankind's sustainable developments are wide-ranging and can be classified broadly under societal impact and scientific significance.

2.1. *Societal impact*

With massive amounts of dust lifted from China's desert regions and injected into the atmosphere, these dust storms often affect daily activities in dramatic ways: pushing grit through windows and doors, forcing people to stay indoors, causing breathing problems, reducing visibility and delaying flights, and by and large creating chaos. Essentially, their long-lasting consequences are:

- *Busting crop yields.* Airborne dust particles not only decrease the crop yields through the reduction of solar radiation reaching the plants for photosynthesis (Chameides *et al.*, 1999), but also contribute to the spread of agricultural diseases, such as bacteria and fungi of plant pathogens that are primarily devastating to rice and wheat — the staple food grains in Asia. Given the projections of limited and declining cultivable croplands and a steadily rising population with increasing per capita food consumption, this absolutely threatens food security in China in the coming decades (Brown, 2002).
- *Posing health risks.* Toxic effects on people near the source/sink and along the pathway of dust storms are evident, since dust contains aluminum, zinc, iron, and other trace metals, which irritate the eyes and respiratory system. Two to three days after dust storm events, the escalation in hospital admissions for asthma (Yang *et al.*, 2005a) and cardiovascular disease (Yang *et al.*, 2005b) is prominent and statistically significant. Furthermore, these adverse effects of air pollution including inhalable dust aerosols on human health are blamed for increases in bronchitis, meningitis, and even premature mortality (e.g., Xu *et al.*, 1998) in the Asia-Pacific region. According to the *World Bank's 2000 Annual Review*, the health costs in major Asian cities can reach as much as 15–18% of urban income.
- *Causing economic hazards.* During its life cycle — generated at the source region, transported along the pathway, and dissipated over the sink area — a dust storm with an average strength could have numerous direct economic impacts (e.g. injuring people and destroying properties, damaging crops and forests, suffocating livestock, reducing semiconductor yields), as well as influence the economy in many indirect ways (e.g. increasing health costs, depleting fisheries due to a triggered red tide, disrupting transportation and communication, shortening the functional life of hydroelectric dams and river channels). The combined economic loss was estimated by the United Nations Environment Program to be in the range of $6.5–9.1 billion each year — roughly equivalent to 0.6–0.9% of China's gross domestic product in 2001.

2.2. *Scientific significance*

Since the early 1970s there has been a series of successful launches of Landsat satellites capable of global observations, so it is no wonder that dust science at the planetary scale is on the rise. This momentum has been fueled by the dawn of the Earth Observing System (EOS) era in the late 1990s — currently, there are 16 (and counting), active polar-orbiting satellites that provide unprecedented views of the size, scale, distribution, and movement of the dust storms from space. The unique vantage provided by satellites in detecting and tracking the progress of airborne dust has helped to shed new light on how humankind has affected the development and magnitude of these storms. A synergy of multisatellite observations suggested an increase in both the intensity and the frequency of Asian dust storm outbreaks that was paralleled by the spatial and temporal scales of manmade development, and subsequent change in the qualities of the land surface, occurring in northwestern China.

From a macro-scale perspective, satellite measurements of dust characteristics offer vital information for determining how dust particles affect the weather and climate by redistributing solar energy within Earth's atmosphere and by changing the thermal contrast between land and ocean. When interacting with sunlight, airborne dust particles not only absorb solar radiation but also reflect it back to space. Generally, this results in a net warming in the dust-laden atmosphere and a net cooling on Earth's surface. However, the magnitude of its cooling over the ocean surface differs from that over the land. Exactly how dust particles modify Earth's radiation budget depends on their color, size, shape, and chemical composition. Furthermore, airborne dust particles may have very complex interactions with atmospheric waters and aerosols from a micro-scale perspective. In some instances, atmospheric aerosols including dust particles serve as nuclei for condensing raindrops, eventually leading to enhancement/acceleration of precipitation processes (e.g. Shepherd and Burian, 2003), but in other cases those aerosols just stifle precipitation (e.g. Rosenfeld, 2000). It is hardly surprising to find that the interactions of dust particles, among other types of aerosols, with the key factors of meteorology (water vapor supply), dynamics (diffusion, collision, and coalescence), and microphysics (water and ice nucleation), can be very complicated (e.g. Li and Yuan, 2006).

To understand the profound, complex, and far-reaching dust impacts on the Earth–atmosphere system, it is of paramount importance to assimilate the spatial and temporal variability of airborne dust properties. Given the daunting diversity of airborne dust properties, a combined observational and theoretical approach is required to better understand and quantify the effects of dust. In doing so, many questions arise:

- Could these increasing airborne dust particles, together with anthropogenic pollutants, drastically alter the regional cloud distributions and hydrological cycles through aerosol direct and indirect effects (*variability and forcing*)?
- Would essential fisheries in this region be critically impacted by dust storms through the influence of the nutrient deposition pattern and extent, in terms of primary productivity of plankton (*response and consequence*)?
- How could we strengthen our ability, through a better understanding of dust properties and interactions with regional meteorology, to issue early warnings of dust storms and of adverse health effects on humans (*prediction*)?
- To what extent could we assess the effectiveness of increasing vegetation/trees (e.g. a reforestation project on the outskirts of Beijing) in preventing further desertification (*feedback and prediction*)?

3. Dust Properties

Solar radiation is the sole large-scale source of diabatic heating that drives the weather and climate system on planet Earth. The emission of terrestrial radiation back to space keeps the planet in balance to make it habitable for all forms of life. Aerosols play an important role in modifying the distributions of solar and terrestrial radiation (IPCC, 2001). Four major types of aerosols — dust particles, biomass-burning smoke, air pollutants, and sea salts — commonly occur in the atmosphere. How light is scattered, absorbed, and emitted by aerosols depends critically on their physical and chemical properties, including refractive index, species, mixture, hygroscopicity, size distribution, shape, and orientation. A thorough understanding of aerosol properties, as well as their temporal and spatial distribution, is imperative for comprehending how Earth's atmosphere maintains its current state of equilibrium and how anthropogenic activities can potentially ruin that balance. Information obtained from coordinated ground-based (temporal scale) and spaceborne (spatial scale) measurements will allow scientists to study in detail the properties of dust particles from sources to sinks and along transport pathways.

3.1. *Ground-based observation and analysis*

Spaceborne remote sensing observations are often plagued by contamination of surface signatures. Thus, ground-based *in-situ* and remote sensing measurements, where signals come directly from the atmospheric constituents, the Sun, and/or the Earth–atmosphere interactions,

provide additional information for comparisons that confirm quantitatively the usefulness of the integrated surface, aircraft, and satellite datasets. Under the auspices of the International Global Atmospheric Chemistry Program, a most comprehensive field campaign in East Asia, the Aerosol Characterization Experiment–Asia (ACE-Asia; Huebert *et al.*, 2003) was conducted in the spring of 2001 and beyond to study the Asian dust and pollutant aerosols. During the ACE-Asia intensive observation period (IOP), many surface sites, together with three aircraft, two research ships, and numerous EOS satellite overpasses, made simultaneous measurements of aerosol chemical, physical, optical, and radiative properties under a variety of environmental conditions. Subsequent to the ACE-Asia IOP, a few surface sites (e.g. the Gosan site in S. Korea, the Mt. Bamboo site in Taiwan) continued to operate so as to acquire *in situ* and column-integrated aerosol properties to assess their spatial and temporal variability.

Near the source regions of dust storm outbreaks, as depicted in Fig. 6(a), a subset of the NASA SMART(Surface-sensing Measurements for Atmospheric Radiative Transfer)–COMMIT (Chemical, Optical and Microphysical Measurements of *In situ* Troposphere) facility was deployed at the Dunhuang site, located

Figure 6. (a) An early-stage SMART-COMMIT instrumental setup at Dunhuang, China, during the ACE-Asia campaign on a dusty day; (b) a newly established LABS instrumental setup on the building rooftop; and (c) distant view of the LABS facility located at an altitude of ∼3 km near the top of Lulin mountain, Taiwan.

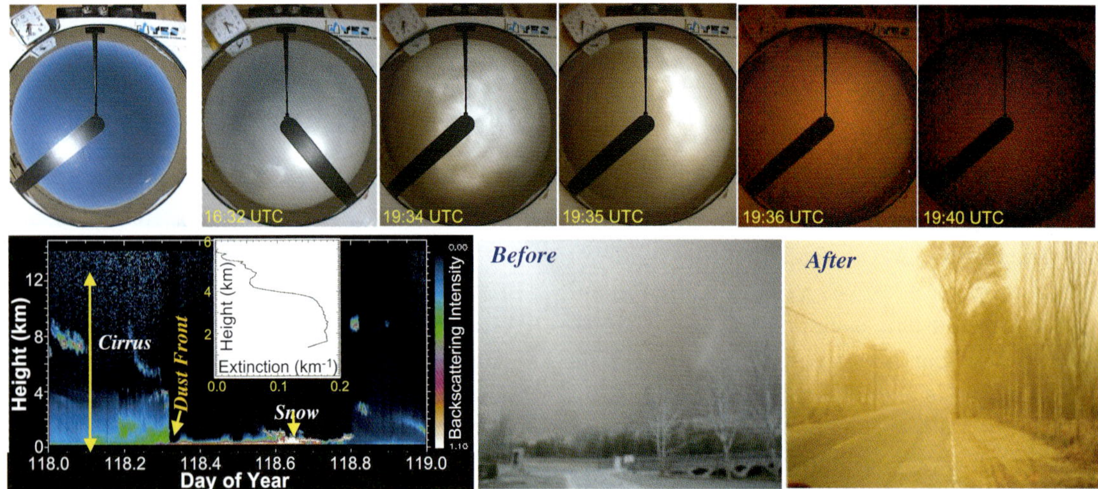

Figure 7. On 28 April 2001, a fierce dust storm passed right over the Dunhuang site. A total-sky imager captured a series of sky conditions (upper panel) for a nearly clear sky (for comparison), a few hours before the storm (16:32 UTC), and the other four images quick succession as the storm passed over. The vertical distribution of dust properties (e.g. aerosol extinction profile) is measured by a micropulse lidar (lower-left panel) during transport. Two photos contrast the changes of airborne dust particles, before (lower-center, 1:30 p.m. local time) and 1 hour after the storm passed (lower-right).

between the Taklimakan and Gobi deserts. SMART-COMMIT (see http://smart-commit.gsfc.nasa.gov for more details) is a mobile laboratory, consisting of many commercially available and in-house-developed remote sensing instruments, as well as a variety of *in situ* probes, for measuring aerosol and precursor properties and meteorological parameters. Likewise, a suite of similar instrumentation, with additional sensors particularly for cloud water chemistry, was deployed for ACE-Asia and ADSE (Asian Dust Storm Experiment, 2001–2004) at the sink areas of dust aerosols in northern Taiwan, known as the Mt. Bamboo site, at an ∼1.1 km altitude. Furthermore, recognizing the vital environmental impacts of aerosol long-range transport, the Lulin Atmospheric Background Station (LABS) was established and has been operational since 13 April 2006, as shown in Figs. 6(b) and 6(c). Situated near the mountaintop (∼3 km height) in central Taiwan and equipped with advanced/refined instrumentation from the Mt. Bamboo site, measurements from the LABS (see http://labs.org.tw for more

details) are observing key parameters representing the free troposphere after the ACE-Asia and ADSE projects.

Among many research topics, Fig. 7 illustrates, for the first time, how dramatically a fierce storm generated a dense blanket of dust, and was observed by a suite of sophisticated ground-based instruments and spaceborne sensors (not shown). On 28 April 2001, a total-sky imager captured the entire passage of this fast-approaching dust storm at ∼3 p.m. local time right over the ACE-Asia Dunhuang site. Daily vertical distributions of dust aerosols (Fig. 7, lower-left panel) were documented by a micropulse lidar, form which an aerosol extinction profile can be retrieved quantitatively. Dense dust particles completely obstructed the lidar backscattering signals immediately after the dust front passage (∼5 minutes), followed by the snowfall. Also shown in Fig. 7 is a photo (lower-center) taken at 1:30 p.m. local time for the approaching dust storm, while another photo (lower-right) demonstrates the dense blanket of dust in the air approximately 1 hour after the storm passed. The yellowish color clearly indicates different chemical compositions (e.g. iron content) in the dust particles, as compared to those of brownish Saharan dust. From satellite observations, this dust storm began to form on 27 April 2001, intensified on 28 April, moved eastward on 29–30 April, and dissipated on 1 May.

During transport, airborne dust particles can interact with anthropogenic sulfate and soot aerosols from heavily polluted urban areas. Added to the complex effects of clouds and natural marine aerosols, dust particles reaching the downwind and sink areas can have drastically different properties than those from the sources. In ACE-Asia, micrographs of airborne dust particles clearly reveal changes of their compositions during transport. Dust particles sampled at the Dunhuang site (source regions) contain predominately silicates, with additional clay minerals, carbonates, feldspars, and gypsum present, while they become dirtier by the time passing over polluted areas. Aircraft observations of dust micrographs by J. Anderson (Arizona State University) show that many different forms of soot balls and nonsoot carbonaceous particles aggregate with the mineral dust particles.

When these dust particles are transported downwind to the sink areas associated with frontal activities, the dry/wet removal processes involving interactions with clouds, biomass-burning aerosols, and local pollutants further complicate atmospheric composition and tropospheric chemistry. During ADSE (e.g. Lin and Peng, 1999), cloud waters were collected hourly from the Mt. Bamboo site at an altitude of $\sim 1.1\,$km, which was frequently immersed in liquid water clouds due to either frontal passage or topographic lifting. Considering the location and altitude of the Mt. Bamboo site and the prevailing northeast-monsoon winds, cloud waters sampled upwind should not be contaminated by megacity pollutants emitted from Taipei. Subsequently, the collected cloud waters were measured to obtain their acidity (in terms of pH values), and conductivity, and were analyzed for ion concentrations of Cl^-, NO_3^-, SO_4^{2-}, NH_4^+, Na^+, K^+, Mg^{2+}, and Ca^{2+} using ion chromatography (e.g. Lin *et al.*, 1999) to determine cloud chemical composition.

Figure 8 shows an assessment of dust and other anthropogenic aerosols influencing cloud chemistry by means of colored cloud waters (unusually brown-to-black), compared to a normally transparent color from natural clouds. Furthermore, the pH value of one sample

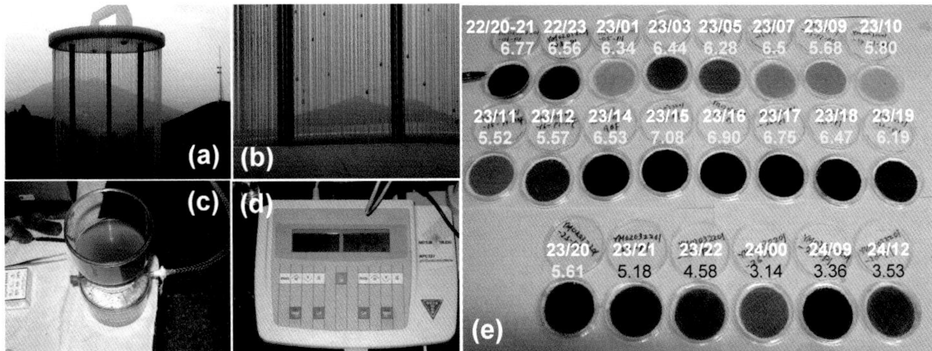

Figure 8. On 22–23 March 2002, surface measurements indicated variations of cloud water chemistry due to a dust storm event over the Mt. Bamboo site, Taiwan. (a) Cloud water collector; (b) close-up view of contaminated cloud water strings; (c) hourly collected cloud water; (d) conductivity and pH measurements; and (e) hourly filter papers (date/local-time in white) revealing changes of color appearance and pH values (gray-to-black numbers).

reached as high as 7.89, alkalized typically by the calcium ion (Ca^{2+}), a major composition of dust particles. By contrast, the pH values of cloud waters were generally 3.0–4.5 at the Mt. Bamboo site. The conductivity of this sample was also measured to as high as 962 $\mu\Omega^{-1}cm^{-1}$, ~10 times more than that of rainwater collected in Taipei, indicating exceedingly contaminated conditions. The time series filter papers of cloud water [Fig. 8(e)] clearly depict the varying degrees of aerosol contaminations: prevailing north–northeast winds carried dust-dominant aerosols (pH values $\geq \sim 6$), but abruptly changed (at 21–22 local time on 23 March) to south–southwest winds with biomass-burning and megacity aerosols (pH values $\leq$ ~5) respectively from Southeast Asia and locally. Essentially, the brown and black colors represent respectively the dust and biomass-burning aerosols. The former tends to alkalize the cloud waters, causing lower acidity in comparison with the CO_2-equilibrated pH value of 5.6. However, biomass-burning and megacity aerosols generally consist of black and organic carbons and associated acidic substances (e.g. $SO_4^=$), resulting in increasing acidity of the cloud waters.

From satellite observations combined with a five-day backward trajectory analysis, this dust storm outbreak started in Inner Mongolia on 18 March 2002, moved eastward on 19–20 March, and blanketed the entire Korean peninsula (the highest measured PM_{10} concentration exceeded $3,000\,\mu g\,m^{-3}$ and thus the instrumental detection limit) on 21–22 March. This dust outflow, associated with a frontal passage, reached northern Taiwan in the late evening of 22 March and lasted until midday of 24 March. Meanwhile, widespread biomass-burning activities in Southeast Asia had also been observed since 19 March. During the period of this event, ADSE micropulse lidar of NASA/MPLNET detected two layers of strong backscattering signals (i.e., lower-level airborne dust particles advected from the north–northeast and

higher-level, ~2–3 km, biomass-burning aerosols from the south–southwest). However, the frontal system played an important role in the mixing of dust/biomass-burning aerosols with clouds. The complex interactions among aerosols, clouds, and atmospheric constituents are an important topic to be studied of atmospheric composition and tropospheric chemistry in ACE-Asia.

Radiative forcing is an area of keen scientific interest, because it is a key parameter in understanding the perturbations that drive the weather and climate system. To quantify the energetics of the surface–atmosphere system, accurate surface measurements of broadband shortwave (0.3–2.8 μm) and longwave (4.0–50 μm) irradiance by flux radiometers are required. By combing irradiance measurements from satellites at the top of the atmosphere and those from ground-based radiometers with aerosol optical thickness, the radiation budget of the surface–atmosphere system can be determined over an extensive area (e.g. Hansell $et\ al.$, 2003). However, due to the temperature gradients between the filter dome and detector in flux radiometers, surface measurements must be corrected to account for the thermal dome effect (e.g. Ji and Tsay, 2000), which may range from 5 to 20 watts m^{-2} in magnitude, depending on the state and condition of the atmosphere. Applying the synergy of the surface/satellite multisensor approach, Hsu $et\ al.$ (2000) demonstrated that the presence of Saharan dust results in a net cooling over ocean and a net warming over land. However, care should be taken in screening out cloud contamination when one extends this approach to studying the radiative forcing of Asian dust, which is generally associated with cloudy, moisture-laden weather fronts. On the other hand, with radiance sensors acquiring spectral or narrowband measurements in the visible, shortwave-infrared, longwave-infrared, and microwave regions and lidar backscattering intensity at the surface, accurate knowledge of the atmospheric aerosols and constituents can be extracted, such as aerosol optical thickness and corresponding vertical profile, columnar size distribution, column water vapor/liquid amount, and ozone abundance. These retrieved parameters can be used to initialize the atmospheric aerosol profile in the forward calculation of radiation models, and to evaluate the results of numerical modeling studies.

3.2. Satellite monitoring and retrieval

Among all atmospheric properties to be monitored and retrieved from space, tropospheric aerosols are especially important, since they have a relatively short lifetime with large temporal and spatial variations. Atmospheric aerosols affect various aspects of solar and terrestrial radiative transfer in spectral (λ), spatial (x, y, z), angular (θ, ϕ), and temporal (t) domains; in turn, the operational satellites use one or more of these four aspects for monitoring, assessing, and retrieving the aerosol properties and effects.

The longest record of aerosol observations from satellites can be dated back to late 1978 with TOMS, and continues to the present time of Aura/OMI (Ozone Monitoring Instrument), in the context of an aerosol index by inverting spectral measurements. The TOMS aerosol index (Hsu *et al.*, 1996) is determined from a pair of ultraviolet spectra that respond negligibly to ozone absorption but strongly to Rayleigh scattering (e.g. 340, 360, and 380 nm). Because the spectral reflectivity of cloud and surface varies, weakly in the ultraviolet wavelengths, the TOMS aerosol index can be used unambiguously to differentiate aerosols from clouds, and to detect absorbing aerosols over arid and semiarid surface. Unlike the thermal-contrast method using infrared spectra, the detection of mineral aerosols by TOMS is not susceptible to water vapor absorption and surface temperature variation. As illustrated in Fig. 5, the TOMS aerosol index can be used to monitor the evolution of dust storms after

dust particles are lifted from the source regions. Also, the frequency statistics of the aerosol index can be utilized for obtaining the source information attributed to multiple types of aerosols, such as airborne dust particles, biomass-burning smoke, and air pollution. However, since the aerosol index represents a measure of how atmospheric molecules intervene with absorbing particulates and the concentration of molecules strongly depends on atmospheric pressure (e.g. Penndorf, 1957), the resulting signals are quite sensitive to the height of particulates residing in the atmosphere. Thus, at the current stage the aerosol index is considered to be an extremely valuable qualitative product.

Although the TOMS aerosol index provides much information about absorbing aerosols, the once-per-day observation limits its timely applications, such as issuing near-real-time warnings of dust storm outbreaks. Utilizing the spectral and temporal aspects of a geostationary satellite, Liu and Lin (2004) successfully developed an automatic operational system for Asian dust storm detection and monitoring, based on measurements of GMS-5 S-VISSR (Stretched-Visible and Infrared Spin Scan Radiometer) from the Japan Meteorological Agency. The efficacy of GMS-5 S-VISSR observations for timely dust storm monitoring is made possible by the broad spectral channels (visible — $0.73\,\mu$m; water vapor — $6.75\,\mu$m; infrared split-window — 11–$12\,\mu$m) and high temporal resolutions (hourly).

Figure 9(a) depicts the scatter plot for GMS-5 S-VISSR digital counts of visible versus infrared channels, which clearly identify distinct atmospheric and environmental aspects such as dust, cloud, land, and ocean surface. With wide spatial coverage and high temporal resolution of the geostationary satellite, the dust-free background data can be established before a dust storm occurs. Thus, the source regions of the dust storm and dust-affected areas can be determined through the variances between observed and background data. However, comparable

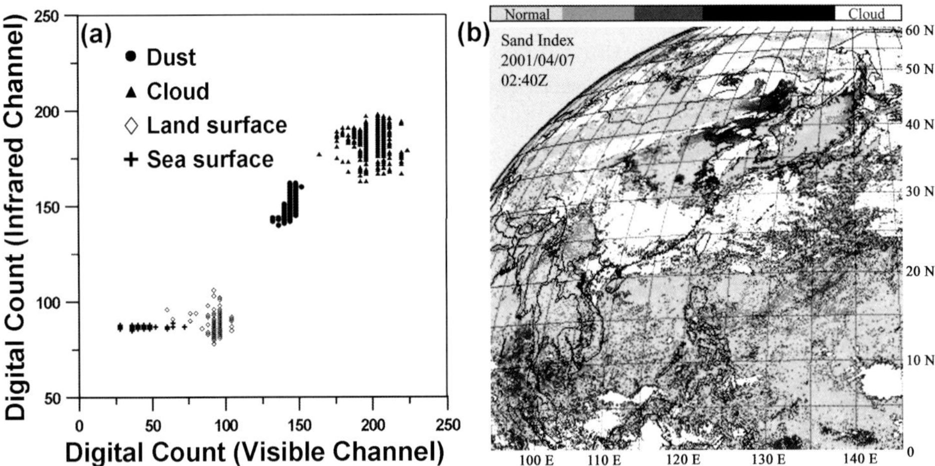

Figure 9. GMS-5 S-VISSR measurements at 02:40Z 7 April 2001: for (a) a scatter plot of digital counts between visible and infrared channels (note the higher the counts the brighter for visible or the colder for infrared scene) and (b) dust storm detection illustrated by the Sand Index gray scale, in which light gray stands for dust-free and white for cloudy areas, while an increasing order of the gray scale represents the abundance of the atmospheric dust loading.

albedo values between dust particles and cloudy neighborhoods, especially around low cloud-cover regions, may lead to major misdetections. To overcome this difficulty, the split-window technique (brightness temperature difference of 11–12 μm) is employed to provide water vapor information, since the main difference between dusty and cloudy scenes is the water vapor content. An example for detecting and monitoring the formation of a dust storm and delineating dust-affected regions is presented in Fig. 9(b). These results clearly demonstrate the superiority by using both infrared channels of GMS-5 to mitigate misdetections of dust–cloud neighboring regions.

To assist policymakers in making accurate and timely decisions, this automatic detection/monitoring system is readily embedded in the GMS-5 satellite receiving system for expediting the provision of information about dust storm outbreaks, evolution, and affected areas. There are still two caveats: the low dynamical range of the S-VISSR visible channel leaves little sensitivity for differentiating dust-contaminated clouds from clean clouds, and the usage of the visible channel prevents its nighttime applications.

Aerosol optical thickness ($0 \leq \mathrm{AOT} < \infty$), an optical measure of aerosols loading, is the most quantitative and fundamental property in describing the consequence of aerosols interacting with spectral light. The earliest reliable retrievals of AOT from space, as shown in Fig. 10(a), have been using two spectral radiance measurements of NOAA/AVHRR (Advanced Very High Resolution Radiometer) since 1983, but have only been available over global oceans (Geogdzhayev *et al.*, 2002). Due to limited spectral channels, the other parameter retrieved from AVHRR is the Ångström exponent (α), which indicates the size groups of dominant aerosols — the larger the α, the smaller the size (e.g. $\alpha = 4$, for molecules; $\alpha \approx \pm 0$, for dust particles). Starting from the EOS era in the late 1990s, measurements from advanced satellite sensors with wide spatial coverage and multispectral channels, as well as innovative algorithms for analyses, have created a new horizon for retrieving aerosol properties. An excellent example is the MODIS sensors aboard

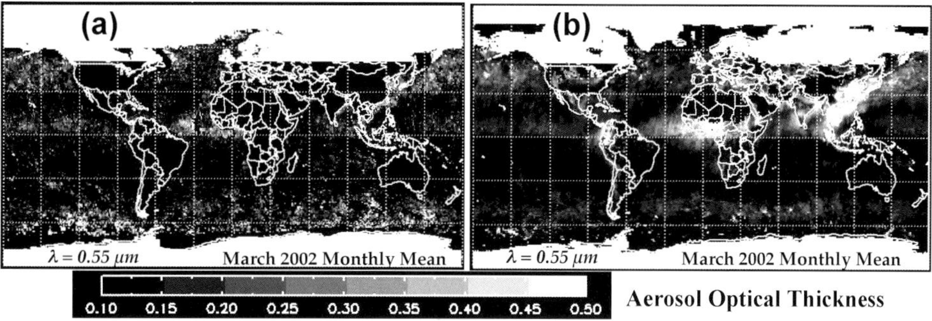

Figure 10. Examples of (a) the monthly mean of aerosol optical thickness at $0.55\,\mu$m wavelengths retrieved over global oceans using NOAA/AVHRR measurements (Geogdzhayev *et al.*, 2002); (b) same as in (a), but retrieved over global vegetated land and open oceans (no retrievals over bright-reflecting surfaces, denoted in black over land) using EOS/MODIS measurements.

both the NASA/EOS Terra and Aqua satellites, which are successfully making near-global measurements daily permitting the retrieval of spectral AOT and aerosol size parameters over both land and ocean (e.g. Remer *et al.*, 2005). Figure 10(b) depicts one of the current MODIS AOT products over global vegetated land and open oceans.

New information obtained from the advanced sensors now allows scientists to better understand the optical and microphysical properties of mineral dust which help to improve the mapping and prediction of dust storm outbreaks. In particular, measurements made by different satellite sensors passing overhead at different times have been instrumental in studying the creation and evolution of dust plumes over time. However, as illustrated in Fig. 10(b), aerosol properties in the vicinity of major desert regions, where dust storms frequently originate, are still largely missing due to large uncertainties in surface emissivity and/or reflectivity, as well as uncertainties in vertical profiles of aerosol and water vapor that are required in the conventional retrieval algorithms. Although Terra/MISR (Multi-angle Imaging SpectroRadiometer) utilizes the spectral and angular aspects to overcome this difficulty of retrieving aerosol properties over bright-reflecting surface (Diner *et al.*, 2005), its relatively narrow spatial

coverage significantly impacts the resulting statistics of airborne dust properties near the source regions (Hsu *et al.*, 2006).

Aerosol retrievals over bright-reflecting surfaces (e.g. airborne dust particles near desert regions) have been a challenging problem ever since the applications of satellite remote sensing. Essentially, the radiance received by a satellite sensor (or apparent radiance) comprises the direct scattered/emitted radiance by the scene and the path radiance that represents the contribution of scattered/emitted radiance directly and/or diffusely by the atmosphere and surface. Over the solar spectral wavelengths, the presence of aerosols in the atmosphere would practically brighten the scene over dark surfaces and darken the scene over bright-reflecting surfaces. The principle of the newly developed Deep Blue algorithm (Hsu *et al.*, 2004) for retrieving aerosols over bright-reflecting surfaces takes advantage of the darker properties of such a surface at the blue spectra, as depicted in Fig. 11. For the first time, the Deep Blue retrievals, with optimal spectral wavelengths from the past, current, and future MODIS-like sensors (e.g. GLI, SeaWiFS, MODIS, VIIRS), can provide comprehensive aerosol properties that permit scientists to quantitatively track the evolution of dust and fine-mode anthropogenic aerosols from source to sink regions, with only

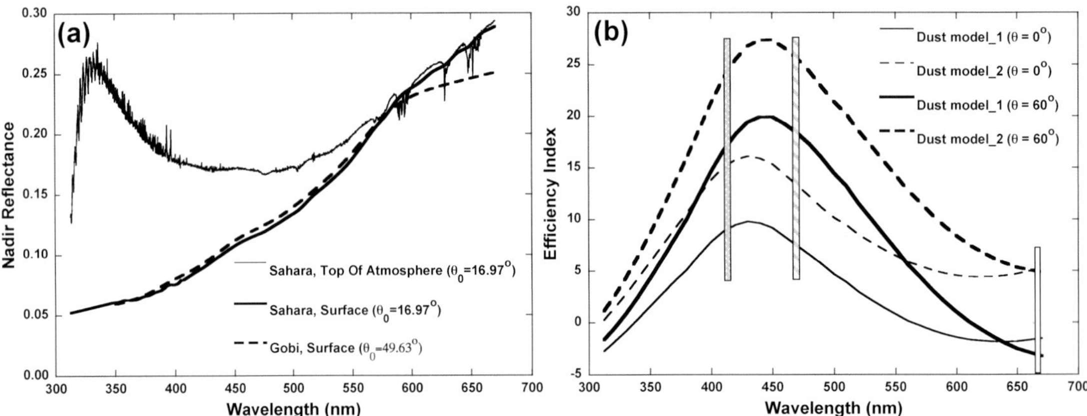

Figure 11. (a) Nadir spectral reflectance acquired over the Sahara desert (top of atmosphere — thin line; without molecular scattering — thick line) and over the surface of Gobi deserts (thick, dashed line), and (b) aerosol efficiency index, indicating the percentage change of the apparent spectral reflectance in comparing a Rayleigh (aerosol-free) atmosphere to a Rayleigh atmosphere containing aerosols (the higher the values, the more sensitive the spectra in detecting the presence of aerosols). Two dust models (D2 for regular and D1 for extremely absorbing dust aerosols) and two operational satellite-viewing geometries (0° for nominal and 60° for limb) are used to cope with the sensitivity ranges of dust effects. The optimal and currently available Deep Blue channels for aerosol retrievals are indicated as vertical bars (e.g. 412, 470, and 670 nm).

moderate sensitivity to uncertainties in aerosol plume height (Hsu *et al.*, 2006).

An example of MODIS Deep Blue AOT retrievals on 6 April 2001 is given in Fig. 12(c), which clearly reveals extensive dust plumes occurring over Mongolia, the Taklimaken desert, and Inner Mongolia. The corresponding Terra/MODIS red–green–blue image (10:30 a.m. overpass) and current MODIS-operational AOT products are also shown in Figs. 12(a) and 12(b), respectively. Because of the bright surface properties over large parts of the land area in East Asia, the current MODIS-operational algorithm cannot retrieve aerosol information over the entire dust source region. By comparing the Deep Blue aerosol product with the current MODIS-operational product, it is evident that there is a consequential enhancement in the spatial coverage where aerosol information can be retrieved from MODIS Deep Blue. Since Asian deserts play a significant role in contributing to the overall global budget of dust loading, it is imperative to use the Deep Blue products to monitor the dust storms as soon as the dust particles are lifted into the

atmosphere, and to follow their evolution along the transport pathways.

Furthermore, additional dust properties can be obtained consistently by applying the Deep Blue algorithm to other MODIS-like sensors, such as SeaWiFS, which has a different overpass time (noon) and viewing geometry. As depicted in Figs. 12(d)–12(f), the spatial distribution of SeaWiFS-retrieved AOT, following that of MODIS in Fig. 12(c), clearly confirms the progress of dust storms, being transported toward the very populated East China. Also shown in Fig. 12(f) is the retrieved dust scattering probability (or single scattering albedo, $0 \leq \omega_0 \leq 1$), with significant differences observed over different dust plumes. Apparently, the northern dust plumes over the Mongolian Gobi appear to be more absorbing ($\omega_0 \sim 0.87$) at 412 nm than the southern plumes spread over central Inner Mongolia and the Taklimakan ($\omega_0 \sim 0.95$). The corresponding spectral signatures of satellite-measured reflectance are dramatically different between these two dust plumes throughout the entire spectral range from the visible to the near-IR in both MODIS and SeaWiFS measurements.

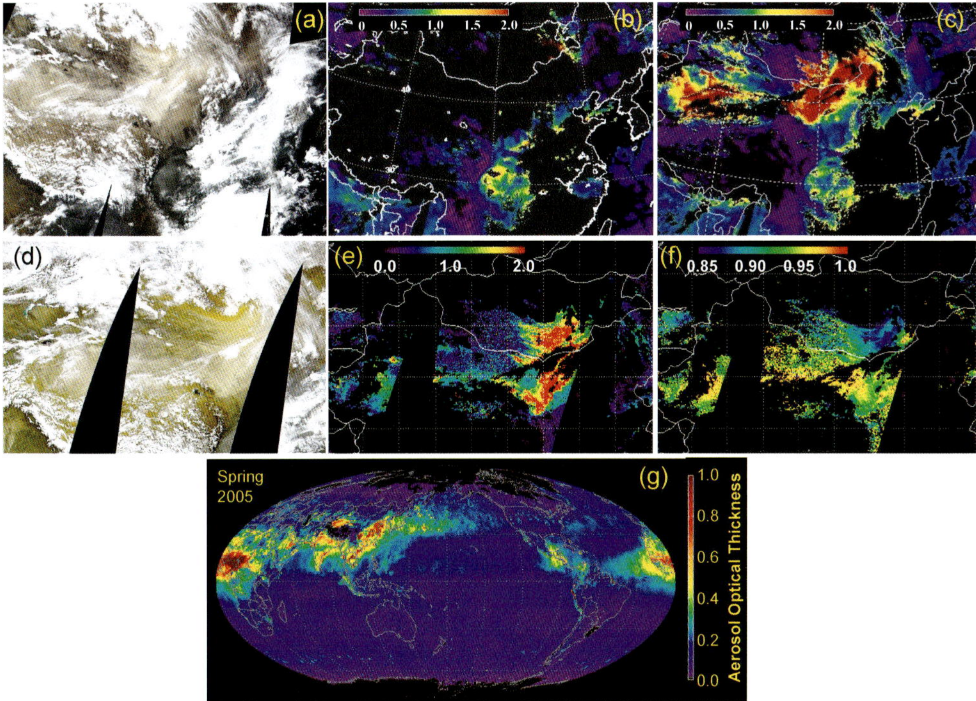

Figure 12. On 6 April 2001, showing (a) the MODIS red–green–blue image, with Rayleigh scattering removed; (b) the current MODIS aerosol product of optical thickness; (c) the new MODIS Deep Blue product of aerosol optical thickness; (d) the SeaWiFS red–green–blue image, with Rayleigh scattering removed; (e) Deep Blue retrievals of SeaWiFS aerosol optical thickness; (f) SeaWiFS scattering probability (or single scattering albedo) at 412 nm; and (g) global-seasonal mean aerosol optical thickness at 0.55 μm for the spring of 2005, as a composite of Deep Blue AOT retrieved over bright-reflecting areas and current MODIS AOT products over vegetated land and open oceans.

Therefore, this disparity in spectral characteristics of the two plumes may well indicate that the different dust compositions (e.g. iron content; Alfaro *et al.*, 2004) are related to their source regions and are not the result of plume height differences. The differential solar heating between the atmosphere and the surface, in the presence of absorbing dust aerosols by such a magnitude of sensitivity in ω_0, may be influential in triggering an aerosol-monsoon large-scale circulation and water-cycle feedbacks (Lau *et al.*, 2008). To close the gaps in MODIS aerosol products, the Deep Blue algorithm enables aerosol retrievals, particularly over, but not limited to, bright-reflecting surfaces. This is evident from the smooth transition near the boundary where surface types change, as depicted in Fig. 12(g). The products of

the aerosol optical thickness (AOT), Ångström exponent (α), and scattering probability (ω_0) from Deep Blue are now available through the Goddard Distributed Active Archive Center (http://ladsweb.nascom.nasa.gov).

3.3. *Numerical modeling and interpretation*

To bring closure, remote sensing and *in situ* observations are incomplete due to the unique properties of their data constituted in either the spatial (satellite global maps) or the temporal (surface point sites) dimension only. Thus, comprehensive modeling is required to bridge these temporal and spatial observations, and to serve as the integrator of our understanding of many physical processes in the Earth–atmosphere system. Many regional-to-global scale models have been developed to investigate dust production, spatial distribution, and long-range transport. The handling of dust aerosols in the numerical models can be categorized into *off-line* (e.g. Ginoux *et al.*, 2001; Tegen *et al.*, 2002) and *on-line* (e.g. Woodward, 2001; Perlwitz *et al.*, 2001). The off-line models are generally common and easy to apply for a better understanding of specific processes, but the impact of aerosols on weather and climate can truly be considered only in the on-line models.

While there has been advancement in characterizing the importance of mineral dust in global-scale processes, there has been less progress in identifying the large-scale dust sources, the environmental processes shaping dust generation in these source regions, and the meteorological parameters influencing the subsequent transport (e.g. Prospero *et al.*, 2002). Because of the short lifetime of mineral dust in the atmosphere (1–2 weeks), which does not allow complete mixing at the global scale, an estimate of dust emissions and spatial distributions requires a relatively high resolution in time and space. To accurately simulate the spatial and temporal variations of dust aerosols during the dust storm event, and to quantify the impacts of these dust aerosols at the regional scale by taking account of the radiative effects, on-line coupled regional-dust models are required to perform on-line dust generation and transport, as well as real-time radiative feedback to the atmospheric modeling.

Based on the comprehensive 3D Purdue Regional Model (PRM; Sun, 1993), an on-line coupled PRM dust model has been developed in an attempt to shed light on the understanding of Asian dust storm outbreaks and their regional impacts. The PRM is equipped with prognostic equations for wind, equivalent ice potential temperature, surface pressure, turbulent kinetic energy, and all phases of water, including vapor, cloud water, ice, snow, rain, and supercooled water (Sun and Chern, 2000). Other features include: radiation parametrization (Chou and Suarez, 1994) and cumulus parametrizations (Sun and Haines, 1996). The use of a local reference is adopted in the model to calculate the pressure gradient so as to reduce the error near steep topography, which is vital in this region of interest. Furthermore, the land-vegetation package (Bosilovich and Sun, 1998) includes Richards' equation and the diffusion equation to predict the moisture and temperature within the soil. It also considers the partial coverage of vegetation, biophysical resistance to the release of water, and evapotranspiration of liquid water (either intercepted precipitation or dew formation) from the vegetated surface. Among many other successful simulations of severe events in North America and Asia (see Sun, 2002 for more details), the robustness of the PRM can be demonstrated in Fig. 13, where the model reproduces the development of the cold front, precipitation, and daily evolution of weather in Asia after two-weeks-or-longer continuous

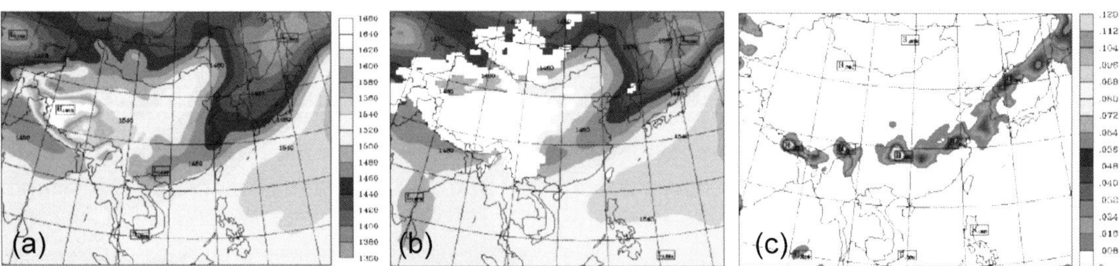

Figure 13. Asian cold air outbreak for comparison of 850mb geopotential at 00Z 24 April 1998 by (a) ECMWF TOGA analysis with (b) those from PRM simulation, as well as (c) corresponding PRM-simulated instantaneous precipitation.

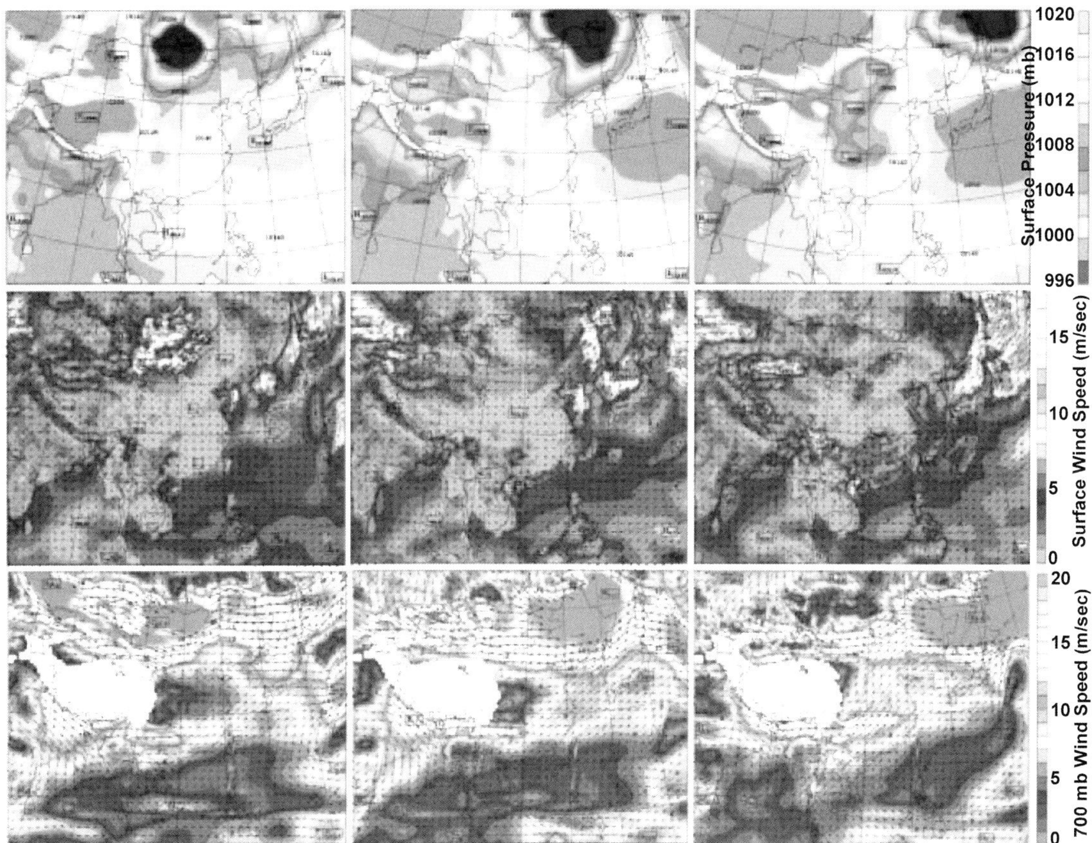

Figure 14.   On-line PRM dust model simulations at 00Z from 19 to 21 April 1998 (three-day results from left to right, respectively) for mean-sea-level pressure (top panels), surface wind field (middle panels), and 700 mb wind field (bottom panel).

integration without nudging (Sun and Hsu, 1988).

For simulating the lifting and transport processes, an on-line dust emission module (Ginoux *et al.*, 2001) with prescribed dust source functions, an on-line transport scheme within the PRM, wet and dry removal processes, and dust aerosol optical properties calculations are fully implemented in the PRM dust model using the meteorological parameters driven by the PRM. In the aerosol module, the PRM dust has also included seven different sizes of aerosols (with effective radii of 0.15, 0.25, 0.4, 0.8, 1.5, 2.5, and 4 μm) to simulate the evolution of dust particles along the transport pathway in East Asia. Samples of the simulation results for the spring

of 1998 over East Asia from the PRM dust model are displayed in Figs. 14 and 15. During this time period, numerous intensive large-scale dust outbreaks were observed from space. Figure 14 reveals the meteorological conditions of the dust event during 19–21 April 1998, including the mean-sea-level surface pressure field, surface wind field and 700 mb wind field. On 19 April, a low-pressure system was passing over eastern Mongolia while a strong, high-pressure system was located over western Mongolia as part of a cold weather system shown in Fig. 14 (top panels). This strong pressure gradient produced high winds over the Gobi desert. The modeled surface winds near Gobi were over $13\,\mathrm{m\,s^{-1}}$ and the 700 mb winds over $27\,\mathrm{m\,s^{-1}}$, as

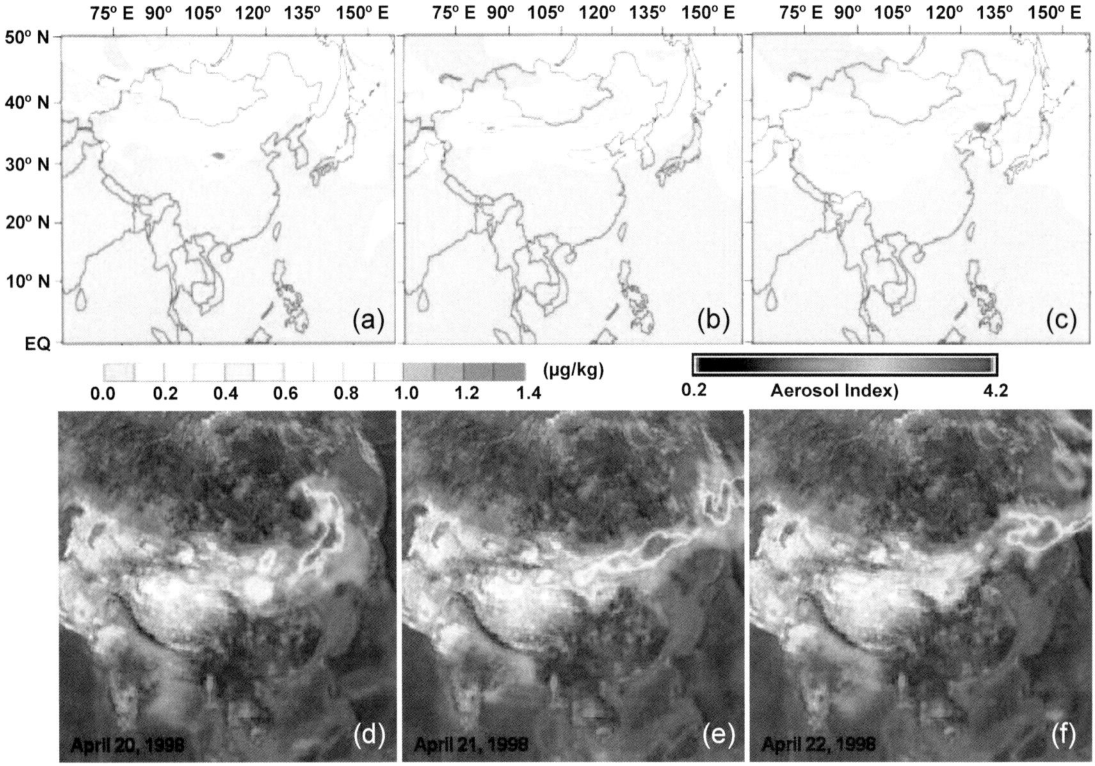

Figure 15. On 20–22 April 1998, comparisons of simulated dust mixing ratios (in units of μg kg^{-1}; top panels) at the 0.498 — pressure level ($\sim$5 km) against TOMS aerosol index maps (bottom panels; http://toms.gsfc.nasa.gov).

depicted respectively in the middle and bottom panels of Fig. 14. As a result, loose dust particles were moved from the surface into the atmospheric boundary layer or above. Once the dust particles were uplifted, the even stronger winds at the higher altitudes could transport them along the airflow. Such an alignment of strong wind fields at the surface and upper-level atmosphere created a favorable environment for generating dust storms during the springtime of 1998.

The resulting extensive dust plumes simulated by the PRM dust model reflect the presence of strong wind fields associated with the passage of cold fronts during 20–22 April 1998, as shown in Fig. 15. For comparison purposes, the TOMS aerosol index maps for the corresponding days are also depicted in Fig. 15 (bottom panels), against the distributions of

PRM dust simulated dust mixing ratios (μg kg^{-1}; top panels) at the 0.498 σ pressure level ($\sim$550 mb). On 20 April, high winds in the great Gobi deserts caused a huge amount of dust particles to be uplifted from the ground and to reach high altitudes. On 21 April, violent dust storms started to be generated in the Taklimakan desert, joining the dust clouds from the Gobi and getting transported further east. On 22 April, very large dust clouds at high altitudes were formed and transported with the airflows to East Asia. The modeled transport and distribution patterns are generally consistent with the satellite observations. Based on the PRM dust model results, excessive dust mass concentrations were simulated at the elevated altitude ($\sim$5 km), peaking at $\sim$2.5 μg kg^{-1}. This confirms that airborne dust particles may frequently be uplifted to high altitudes and then transported

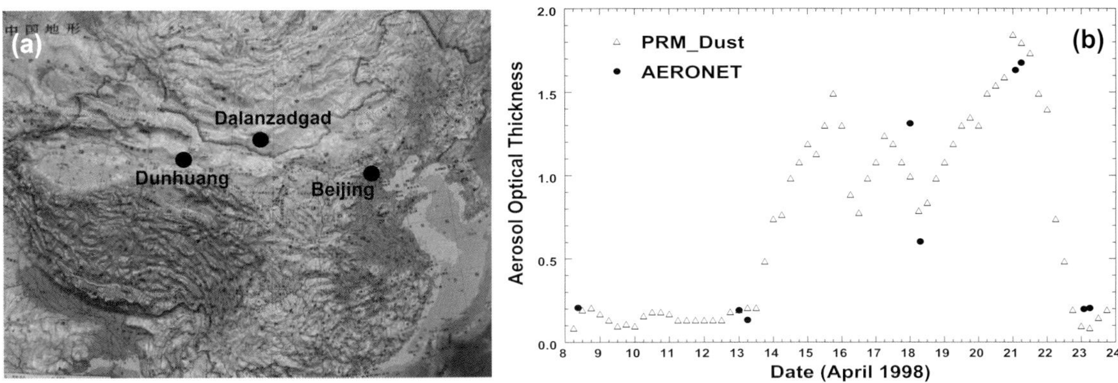

Figure 16. (a) Sun photometer measurements from an AERONET site at Dalanzadgad, Mongolia (43.58°N, 104.42°E) selected for comparisons, and (b) time series of column aerosol optical thickness at 500 nm from the PRM dust model (6-hour resolutions) and nearest AERONET (15-minute cloud-free) observations.

by the westerly jet streams across the Pacific Ocean, even reaching the United States.

Incorporating a mass-optical conversion scheme, the regional dust AOT modeled by the PRM dust radiation module can be compared directly with that from ground-based Sun photometer measurements. The NASA AERONET project (AErosol RObotic NETwork; http://aeronet.gsfc.nasa.gov), which imposes standardization of instrumentation and calibration/processing, provides globally distributed observations of spectral AOT and other inversion products in geographically diverse aerosol regimes. Among all regional AERONET sites, measurements from Dalanzadgad, Mongolia are the most suitable for comparison with the PRM dust results, as shown in Fig. 16, due to the location and time span. Apparently, the timing and intensity of the dust events which occurred during 14–22 April 1998 and were observed by AERONET are well captured by the PRM-dust model simulations.

The regional/global models, such as the PRM dust one, can serve as a critical tool for *providing* 3D pictures of aerosol distributions and aiding in interpreting satellite/*in-situ* observations of aerosol properties. They also give invaluable continuous temporal information to bridge the limited satellite snapshot measurements, making it possible to characterize the role of each source region in attributing the overall dust effects on regional and global scales. The amount of dust in the atmosphere can be affected by the changes in climate, such as wind strength, precipitation, and regional moisture balance, as well as the changes in dust sources caused by either anthropogenic or climatically induced changes in vegetation cover. The atmospheric dust can in turn shape the climate through changes of the radiation budget. Thus, full coupling between dust cycle models and atmospheric models is required to provide a comprehensive assessment of the importance of interactions and feedbacks that have been postulated involving mineral dust.

4. Concluding Remarks

All substances are poisons; there is none which is not a poison. The right dose differentiates a poison and a remedy.
— P. A. Paracelsus (1493–1541), one of the fathers of modern medicine

Thus, both *increasing* and *decreasing* concentrations of doses result in harmful biological effects; so do airborne dust particles, damaging our living Earth. Abnormal Asian dust storm outbreaks create significant impacts worldwide (Liu and Diamond, 2005), which are particularly severe in those regions in China and East Asia under their transport pathways.

4.1. *Lessons learned*

Besides the natural climatic variability in atmospheric water cycles and wind fields, anthropogenic activities, in disturbing the land surface and stripping it of its vegetation, play a key role in mobilizing dust. Apart from the steady growth of its population, China's livestock has been increasing rapidly since its 1978 economic reforms. According to the statistics in the *World Bank's 2000 Annual Review*, China has more than 400 million livestock, four times that of the United States. As China's northwestern provinces are already suffering from overplowing and overgrazing, expanding livestock populations are stripping off ever more vegetation, particularly in the semiarid or marginal land regions. The extra demand on the land's carrying capacity results in fast-deteriorating grassland, intensifying wind erosion, and further desertification.

In addition to exceeding the sustainability of the land, decades-long "North drought and South flood" climate patterns in China have further depleted the already-limited water resources in the arid and semiarid regions. When water tables fall, lakes vanish and rivers run dry early, before the arrival of the arid season. Analyzing about 30 years of Landsat measurements, scientists have reached the conclusion that literally thousands of lakes in North China have disappeared (e.g. Brown *et al.*, 2002). Consequently, using airborne dust particles $6.3\,\mu$m in diameter as a size qualifier, scientists at the Cold and Arid Regions Environment and Engineering Research Institute, Chinese Academy of Sciences (G. Yang, *China Youth Daily* press release on 9 June 2002), have examined samples from 19 different surface types in areas that dust storms frequently originate. Their findings indicate that samples from natural deserts and desert margins contain, on average, less than 10% of dust particles smaller than $6.3\,\mu$m. However, the percentage increases sharply to more than 30% from barren farm fields and up to 50–60% from deteriorated grasslands and dried-out lake basins and riverbeds. These provide hard evidence that recent increases in Asian dust storm outbreaks are attributable largely to regional intensified anthropogenic movements, but without sufficient management of land and water resources. To mitigate the negative consequences of dust storm outbreaks, identifying and controlling those factors owing to human activities would be the first step. The success or failure in developing a better understanding and reliable predictions of, as well as adequate responses to, this environmental change will determine the prospective course for a sustainable civilization, regionally at the very least.

4.2. *Action strategy*

Successful rehabilitation and management are rooted in a solid understanding of the ecohydrological processes that cause desertification in the ecosystems of the regions concerned. Turning China's environmental trend from deterioration to improvement requires a huge endeavor, but the options are clear. Reversing this degradation entails stabilizing the human population, reducing livestock inhabitants, and, most critically, *balancing* afforestation and reforestation to help create a virtuous cycle within the ecosystem. The term "afforestation and reforestation" is defined as restoring/converting highly erodible cropland to its original (or even additional) grassland or woodland and planting shrub/tree shelterbelts across the windswept areas of productive cropland. In the last two decades, afforestation and reforestation plans have commenced in southern South America as a promising for mitigating desertification and reducing concentrations of carbon dioxide, one of the atmospheric greenhouse gases capable of triggering various global warming scenarios.

Figure 17. (a) Vast grasslands severely degraded near the boundary of Inner Mongolia, then the dynasties of Yuan, Ming and Qing for more than two years' imperial hunting ground (e.g. Chengde Mulanweichang); (b) an example of a successful reforestation project on the outskirts of Inner Mongolia ~400 km northwest of Beijing; and (c) a close-up view of very rich topsoil, 1–2 cm beneath the regional deteriorated grassland in (a) and (b).

One of China's most ambitious coping strategies for desertification was announced (*Xinhua News*, press conference on 26 May 2001 by Mr Li Hung, Deputy Director-General of the China Meteorological Administration) and has since been underway, with a scope of RMB100-billion funding for the 10-year program to plant 170,000 square miles of trees, a coverage equivalent to the size of California. As illustrated in Fig. 17, the reforestation on the outskirts of Inner Mongolia, about 400 km northwest of Beijing, has proved very successful — perhaps due partly to the remaining rich topsoil beneath and partly to the early stage of sandification. However, countermeasures to combat further desertification in oases or pastures of sandy land (e.g. in the vicinity of the Dunhuang site; ACE-Asia 2001) require more aggressive practices, such as biological and engineering shelterbelts. Straws, brushwood or branches are first engineered to act as the sand barriers; then, the biological enclosures are built to fence in arid and semiarid areas, prohibiting human activities and animal use, and gradually restoring natural vegetation. Supported by China and international funds, as well as subsidies from the World Bank, the sand enclosure program in northwestern China started in 1978 and is expected to be completed in 2050, and has been assessed as high-efficiency and low-cost.

In general, afforestation and reforestation leads to a reduction in albedo and upsurges of the leaf area index, surface roughness length, and root-depth distributions (e.g. Sellers, 1992). Alterations to these parameters can modify the near-surface energy fluxes; in turn, the modulations of near-surface temperature, sensible and latent heat transport (e.g. Xue and Shukla, 1996; Bosilovich and Sun, 1998). It is believed that vegetation can prevent desertification and reduce effectively on the dust storm outbreaks, because vegetation coverage reduces the exposition of bare soil and increases the surface roughness, which suppresses wind speed near the surface. However, the planetary boundary layer is most sensitive to the amount of the soil water content, and while vegetation coverage is not the most sensitive parameter at the surface, its influence on the surface energy and hydrological balance is crucial. Afforestation and reforestation can affect hydrology and biogeochemistry (Jackson *et al.*, 2005), but the degree of the criticalness of its influence on precipitation is largely related to the geographical location and extent, regional meteorology, and biophysical parameters of the land-use/land-cover change.

Exploring precursors and transport of a dust storm and determining its regional-to-global impact require a synergy of satellite, aircraft, and ground-based measurements, together with physics-based modeling. Comprehensive observations and information on dust aerosols are increasingly available over source/sink regions and on transport pathways. Regional models with detailed physics and surface parametrization, such as the PRM dust model, are capable of predicting soil temperature and wetness, as well as interactions among vegetation, soil moisture, and atmospheric forcing, which is particularly important in forecasting dust-lifting events and also the development of the planetary boundary layer. The transport, vertical dispersion, and deposition of dust particles depend highly on the evolution of the weather system and the terrain blocking. Moreover, regional models with fine resolution can perform well in predicting the movement of cold fronts, precipitation systems, and blocking flows, and therefore provide a better prediction of dust aerosols interacting with meteorological forcing.

However, uncertainty exists in all regional models. To evaluate the effect of land-cover change on regional climate and dust storm outbreaks, it would be necessary to examine the model differences between extreme situations (e.g. varying degrees of desertification from Mongolia to the coastline of Bo-Hai) and those with modified land surface (e.g. varying degrees of success in afforestation and reforestation in the semiarid region of Inner Mongolia). It is the utmost importance to understand whether the extension of dry soil will reduce the local precipitation, making the land drier and hotter to cause more dust storms, or whether the extension of dry land may create a stronger land–sea contrast in the warm seasons to bring in more moist air and precipitation from the ocean, and subsequently the dry land becomes moist gradually. If precipitation increases, then it is expected that vegetation can grow and desertification can be reversed. Consequently, the occurrence of dust storm will reduce. Otherwise, vegetation eventually will die and the region will return to being an arid or semiarid one. Thus, more dust storms will prevail.

The research on complex relations among natural climatic variability, anthropogenic activities, and dust storm outbreaks should be carried out further. Consequently, the lessons learned will help strengthen our ability to issue early warnings of Asian dust storm outbreaks and minimize further desertification in the future. It was concluded by Brown *et al.* (2002). "…If China cannot quickly arrest the trends of deterioration, the growth of the dust bowl could acquire an irreversible momentum. What is at stake is not just China's soil, but also its future.…"

Acknowledgements

The lead author is grateful for the continuous support of this research on Asian dust storm outbreaks, as part of many federal grants from the NASA/Radiation Sciences Program (managed by Dr H. Maring), NASA/Modeling, Analysis and Prediction Program (by Dr D. Anderson), ONR/Marine Meteorology Program (by Dr R. Ferek), and US/EPA–China/East Asia Program (by Mr D. Thompson) under the NASA-US/EPA interagency agreement with Taiwan/EPA–Department of Environmental Monitoring and Information (by Ms H.-C. Hsiao, Director-General).

[Received 23 July 2007; Revised 8 December 2007; Accepted 16 December 2007.]

References

Alfaro, S. C., S. Lafon, J. L. Rajot, P. Formenti, A. Gaudichet, and M. Maillé, 2004: Iron oxides and light absorption by pure desert dust: an experimental study. *J. Geophys. Res.*, **109**, D08208, doi:10.1029/2003JD004374.

Bosilovich, M. B., and W. Y. Sun, 1998: Simulation of soil moisture and temperature. *J. Atmos. Sci.*, **55**, 1170–1183.

Brown, L. R., J. Larsen, and B. Fischlowitz-Roberts, 2002: *The Earth Policy Reader* (W. W. Norton, New York), 282 pp.

Chameides, W. L., H. Yu, S. C. Liu, M. Bergin, X. Zhou, L. Mearns, G. Wang, C. S. Kiang, R. D. Saylor, C. Luo, Y. Huang, A. Steiner, and F. Giorgi, 1999: Case study of the effects of atmospheric aerosols and regional haze on agriculture: an opportunity to enhance crop yields in China through emission controls? *Proc. Nat. Acad. Sci.*, **96**, 13626–13633.

Chou, M.-D., and M. J. Suarez, 1994: An efficient thermal infrared radiation parameterization for use in general circulation models. *NASA Tech. Memo. 104606*, **3**, 85 pp.

Derbyshire, E., X. Meng, and R. A. Kemp, 1998: Provenance, transport, and characteristics of modern aeolian dust in western Gansu province, China, and interpretation of the quaternary loess record. *J. Arid Environ.*, **39**, 497–516.

Diner, D. J., J. V. Martonchik, R. A. Kahn, B. Pinty, N. Gobron, D. L. Nelson, and B. N. Holben, 2005: Using angular and spectral shape similarity constraints to improve MISR aerosol and surface retrievals over land. *Rem. Sens. Environ.*, **94**, 155–171.

Gao, T., Y. Xu, Y. Bo, and X. Yu, 2006: Synoptic characteristics of dust storms observed in Inner Mongolia and their influence on the downwind area (the Beijing–Tianjin region). *Meteorol. Appl.*, **13**, 393–403, doi: 10.1017/S1350482706002404.

Geogdzhayev, I. V., M. I. Mishchenko, W. B. Rossow, B. Cairns, and A. A. Lacis, 2002: Global two-channel AVHRR retrievals of aerosol properties over the ocean for the period of NOAA-9 observations and preliminary retrievals using NOAA-7 and NOAA-11 data. *J. Atmos. Sci.*, **59**, 262–278.

Ginoux, P., M. Chin, I. Tegen, J. M. Prospero, B. Holben, O. Dubovik, and S.-J. Lin, 2001: Sources and distributions of dust aerosols simulated with the GOCART model. *J. Geophys. Res.*, **106**, 20255–20273.

Hansell, R., S. C. Tsay, Q. Ji, K. N. Liou, and S. Ou, 2003: Surface aerosol radiative forcing derived from collocated ground-based radiometric observations during PRIDE, SAFARI, and ACE-Asia. *Appl. Opt.*, **42**, 5533–5544.

Hsu, N. C., J. R. Herman, P. K. Bhartia, C. J. Seftor, O. Torres, A. M. Thompson, J. F. Gleason, T. F. Eck, and B. N. Holben, 1996: Detection of biomass burning smoke from TOMS measurements. *Geophys. Res. Lett.*, **23**, 745–748.

Hsu, N. C., J. R. Herman, and C. Weaver, 2000: Determination of radiative forcing of Saharan dust using combined TOMS and ERBE data. *J. Geophys. Res.*, **105**, 20649–20661.

Hsu, N. C., S. C. Tsay, M. D. King, and J. R. Herman, 2004: Aerosol properties over bright-reflecting source regions. *IEEE Trans. Geosci. Remote Sens.*, **42**, 557–569.

Hsu, N. C., S. C. Tsay, M. D. King, and J. R. Herman, 2006: Deep-Blue retrievals of Asian aerosol properties during ACE-Asia. *IEEE Trans. Geosci. Remote Sens.*, **44**, 3180–3195.

Huebert, B. J., T. Bates, P. B. Russell, G. Shi, Y. J. Kim, K. Kawamura, G. Carmichael, and T. Nakajima, 2003: An overview of ACE-Asia: strategies for quantifying the relationships between Asian aerosols and their climatic impacts. *J. Geophys. Res.*, **108**, doi:10.1029/2003JD003550.

Husar, R. B., J. M. Prospero, and L. L. Stowe, 1997: Characterization of tropospheric aerosols over the oceans with the NOAA advanced very high-resolution radiometer optical thickness operational product. *J. Geophys. Res.*, **102**, 16889–16909.

IPCC (Intergovernmental Panel on Climate Change), 2001: *Climate Change 2001: The Scientific Basis* (Cambridge University Press), 881 pp.

Ji, Q., and S. C. Tsay, 2000: On the dome effect of Eppley pyrgeometers and pyranometers. *Geophys. Res. Lett.*, **27**, 971–974.

Karyampudi, V. M., S. P. Palm, J. A. Reagen, H. Fang, W. B. Grant, R. M. Hoff, C. Moulin, H. F. Pierce, O. Torres, E. V. Browell, and S. H. Melfi, 1999: Validation of the Saharan dust plume conceptual model using lidar, Meteosat, and ECMWF data. *Bull. Amer. Meteor. Soc.*, **80**, 1045–1075.

Lau, W. K. M., V. Ramanathan, G.-X. Wu, Z. Li, S. C. Tsay, N. C. Hsu, R. Sikka, B. Holben, D. Lu, G. Tartari, M. Chin, P. Koudelova, H. Chen, Y. Ma, J. Huang, K. Taniguchi, and R. Zhang, 2008: Aerosol-hydrologic cycle interaction: a new challenge for monsoon climate research. *Bull. Amer. Meteor. Soc.*, March, 1–15.

Li, Z., and T. Yuan, 2006: Exploring aerosol-cloud-climate interaction mechanisms using the new generation of earth observation system data. In *Current Problems in Atmospheric Radiation*, eds. H. Fischer and B.-J. Song. (Deepak), pp. 1–4.

Lin, N.-H., and C.-M. Peng, 1999: A semi-quantitative scheme for estimating the contribution of below-cloud scavenging by raindrops based on observations of cloud chemistry at a mountain site in Taipei, Taiwan. *Terr. Atmos. Ocean. Sci.*, **10**, 693–704.

Lin, N.-H., H.-M. Lee, and M.-B. Chang, 1999: Evaluation of the characteristics of acid precipitation in Taipei, Taiwan using cluster analysis. *Water, Air Soil Pollut.*, **113**, 241–260.

Liu, G.-R., and T.-H. Lin, 2004: Application of geostationary satellite observations for monitoring dust storms of Asia. *Terrestrial, Atmos. Oceanic Sci.*, **15**, 825–837.

Liu, J., and J. Diamond, 2005: China's environment in a globalizing world: How China and the rest of the world affect each other. *Nature*, **435**, 1179–1186.

Meskhidze, N., W. L. Chameides, and A. Nenes, 2005: Dust and pollution: a recipe for enhanced ocean fertilization? *J. Geophys. Res.*, **110**, D03301, doi:10.1029/2004 JD005082.

Pauly, D., J. Alder, E. Bennett, V. Christensen, P. Tyedmers, and R. Watson, 2003: The future for fisheries. *Science*, **302**, 1359–1361.

Penndorf, R., 1957: Tables of the refractive index for standard air and the Rayleigh scattering coefficient for the spectral region between 0.2 and 20.0 μm and their application to atmospheric optics. *J. Opt. Soc. Amer.*, **47**, 176–182.

Perlwitz, J., I. Tegen, and R. L. Miller, 2001: Interactive soil dust aerosol model in the GISS GCM. 1: Sensitivity of the soil dust cycle to radiative properties of soil dust aerosols. *J. Geophys. Res.*, **106**, 18167–18192.

Prospero, J. M., P. Ginoux, O. Torres, S. E. Nicholson, and T. E. Gill, 2002: Environmental characterization of global sources of atmospheric soil dust identified with the Nimbus 7 Total Ozone Mapping Spectrometer (TOMS) absorbing aerosol product. *Rev. Geophys.*, **10**(1), 1002, doi:10.1029/2002RG000095.

Qian, W., L. Quan, and S. Shi, 2002: Variations of the dust storm in China and its climatic control. *J. Climate*, **15**, 1216–1229.

Remer, L. A., Y. J. Kaufman, D. Tanré, S. Mattoo, D. A. Chu, J. V. Martins, R.-R. Li, C. Ichoku, R. C. Levy, R. G. Kleidman, T. F. Eck, E. Vermote, and B. N. Holben, 2005: The MODIS aerosol algorithm, products, and validation. *J. Atmos. Sci.*, **62**, 947–973.

Rosenfeld, D., 2000: Suppression of rain and snow by urban and industrial air pollution. *Science*, **287**, 1793–1796.

Sellers, P. J., 1992: Biophysical models of land surface processes. In *Climate System Modeling*, ed. K. E. Trenberth (Cambridge University Press), pp. 451–490.

Shepherd, J. M., and S. J. Burian, 2003: Detection of urban-induced rainfall anomalies in a major coastal city. *Earth Interactions*, **7**, 1–14.

Studley, J. 1999: Environmental degradation in China. *China Review*, **12**, 28–33.

Sun, J., M. Zhang, and T. Liu, 2001: Spatial and temporal characteristics of dust storms in China and its surrounding regions, 1960–1999: relations to source area and climate. *J. Geophys. Res.*, **106**, 10325–10333.

Sun, J., and T. Liu, 2006: The age of the Taklimakan desert. *Science*, **312**, 1621.

Sun, W. Y., 1993: Numerical experiments for advection equation. *J. Comput. Phys.*, **108**, 264–271.

Sun, W. Y., and P. A. Haines, 1996: Semi-prognostic tests of a new mesoscale cumulus parameterization scheme. *Tellus*, **48A**, 272–289.

Sun, W. Y., and W. R. Hsu, 1988: Numerical study of cold air outbreak over the warm ocean. *J. Atmos. Sci.*, **45**, 1205–1227.

Sun, W. Y., and J. D. Chern, 2000: Regional climate simulation of 1998 summer flood in East Asia. In *Proc. Atmosphere Second International Ocean and Atmosphere Conference, COAA 2000.* (10–12 July; Central Weather Bureau, Taipei, Taiwan) pp. 40–45.

Sun, W. Y., 2002: Numerical modeling in the atmosphere. *J. Korean Meteorol. Soc.*, **38**, 237–251.

Svensson, A., P. E. Biscaye, and F. E. Grousset, 2000: Characterization of late glacial continental dust in the Greenland Ice Core Project ice core. *J. Geophys. Res.*, **105**, 4637–4656.

Tegen, I., S. P. Harrison, K. Kohfeld, I. C. Prentice, M. Coe, and M. Heimann, 2002: Impact of vegetation and preferential source areas on global dust aerosol: results from a model study. *J. Geophys. Res.*, **107**(D21), AAC 14–1–AAC 14–27.

Tindale, N. W., and P. P. Pease, 1999: Aerosols over the Arabian Sea: Atmospheric transport pathways and concentration of dust and sea salt. *Deep-Sea Research II*, **46**, 1577–1595.

Van Curen, R., and T. Cahill, 2002: Asian aerosols in North America: frequency and concentration of fine dust. *J. Geophys. Res.*, **107**, doi:10.1029/2002JD002204.

Wang, X., Z. Dong, J. Zhang, and L. Liu, 2004: Modern dust storms in China: an overview. *J. Arid Environ.*, **58**, 559–574.

Washington, R., M. Todd, N. J. Middleton, and A. S. Goudie, 2003: Dust-storm source areas determined by the Total Ozone Monitoring Spectrometer and surface observations.

Ann. Assoc. Amer. Geog., **93**, doi:10.1111/1467-8306.9302003.

Woodward, S., 2001: Modeling the atmospheric life cycle and radiative impact of mineral dust in the Hadley Centre climate model. *J. Geophys. Res.*, **106**(D16), 18155–18166.

Xu, X., L. Wang, and T. Niu, 1998: Air pollution and its health effects in Beijing. *Ecosyst. Health*, **4**, 199–209.

Xue, Y., and J. Shukla, 1996: The influence of land surface properties on Sahel climate. Part II: Afforestation. *J. Climate*, **9**, 3260–3275.

Yang, C.-Y., S.-S. Tsai, C.-C. Chang, and S.-C. Ho, 2005a: Effects of Asian dust storm events on daily admissions for asthma in Taipei, Taiwan. *Inhal. Toxicol.*, **17**, 817–821.

Yang, C.-Y., Y.-S. Chen, H.-F. Chiu, and W.B. Goggins, 2005b: Effects of Asian dust storm events on daily stroke admissions in Taipei, Taiwan. *Environ. Res.*, **99**, 79–84.

Yang, Y., 2001: Dust-sandstorms: inevitable consequences of desertification — a case study of desertification disasters in the Hexi Corridor, NW China. In: Yang, Squires, and Lu (eds.). *Global Alarm: Dust and Sandstorms from the World's Drylands* (United Nations, New York), pp. 228.

Zhang, X. Y., R. Arimoto, and Z. S. An, 1997: Dust emission from Chinese desert sources linked to variations in atmospheric circulation. *J. Geophys. Res.*, **102**, 28041–28047.

An Overview of Satellite-Measured Cloud Layer Structure and Cloud Optical and Microphysical Properties

Fu-Lung Chang

Langley Research Center, NASA, Hampton, Virginia, USA
Fu-Lung.Chang-1@nasa.gov

Satellite observations are the only means of obtaining a continuous survey of cloud properties on global scales. Many climate and weather forecasting models have been evaluated using satellite-derived products. For clouds, the evaluation has been limited mainly to cloud amount, in part because conventional satellite remote sensing techniques cannot provide detailed information on cloud vertical structure, and in part because conventional satellite retrieval algorithms assume a single cloud layer in their retrievals. Increasing attention is being paid to the vertical structure of cloud fields. However, observations of cloud layer data on global scales are scarce and unreliable for model evaluation. The single-layer assumption does not emcompass the overlapping of cloud layers or the inhomogeneity of cloud vertical structure. In nature, overlapped upper-level cirrus and low-level stratus clouds occur frequently. As such, comparisons of cloud vertical structures derived from the satellite measurements and models remain a challenge. This article discusses some issues related to the single-layer assumption used in current satellite cloud remote sensing techniques. The assumption represents one major deficiency in satellite observations of cloud vertical structure. Some differences caused by different satellite retrieval algorithms and some improvements on enhanced satellite retrieval algorithms are discussed based on the research work by the author. Credible global cloud and radiation measurements are needed in order to evaluate and improve the performance of global climate and weather forecast models. The recent launch of the CloudSat radar and Cloud-Aerosol Lidar and Infrared Pathfinder Satellite Observations (CALIPSO) mission in a formation flight of the "A-Train" satellite constellation is expected to provide large-scale spatial and temporal observations on the vertical profile of cloud properties.

1. Introduction

Clouds play a dominant role in the Earth's climate and its changes. They strongly affect the balances of radiant energy and the water cycle. Accurate representation of clouds in climate models is one of the central issues in improving global climate modeling. Curently, satellite remote sensing is the only means of observing cloud and other climate variables on a global scale. Satellite observations have provided a wealth of information pertaining to cloud horizontal distribution and column-integrated optical properties. However, there is a dearth of information concerning the climatology of cloud vertical structure. Neglecting cloud vertical distribution can have a large impact on both the Earth's shortwave and longwave radiative fluxes (Wielicki *et al.*, 1995, 1996; Chou *et al.*, 1998, 1999). General circulation models (GCMs) simulate cloud fraction and cloud water/ice content at various levels. Using different assumptions of cloud overlap schemes in GCMs can lead to considerable variations in the cloud radiative forcing and modeled climate.

Prior to the Moderate-Resolution Imaging Spectroradiometer (MODIS) instruments on board the NASA Terra and Aqua Earth Observing System (EOS) satellites, our knowledge of clouds had been gained primarily by means of the International Satellite Cloud Climatology Project (ISCCP) data products (Rossow *et al.*, 1991; Rossow and Schiffer, 1999) using a series of polar-orbiting and geostationary satellite measurements made by international space instruments like the NOAA Advance Very High Resolution Radiometer (AVHRR) and the Geostationary Operational Environmental Satellite (GOES), the Japanese Geostationary Meteorological Satellite (GMS), and the European Geostationary Meteorological Satellite (METEOSAT). Due to the limited spectral channels provided by these satellite instruments, the retrievals of cloud properties have assumed single-layer clouds for retrieving either their bulk (e.g. cloud-column optical depth) or cloud-top (e.g. temperature, droplet effective radius) properties. There is little information available concerning the observations of cloud vertical structure.

The dearth of information on cloud vertical structure results in part from a lack of sensitivity to vertical structure in the current passive satellite sensors, as well as the inadequacy of the inversion algorithms employed in current satellite retrieval methods (Chang and Li, 2005a). For instance, cloud-top height data in the existing satellite products give either an effective cloud emission height, as in the ISCCP cloud data products, or the height of the uppermost cloud layer, as in the MOD06 of the MODIS cloud data products (King *et al.*, 2003; Platnick *et al.*, 2003). These cloud-top heights cannot reveal the vertical structure of clouds because their retrieval methods assume only a single-layer cloud. As such, comparison of the cloud vertical distributions derived from

satellites and climate models has faced major difficulties.

Knowledge of the vertical distribution of clouds is in demand in all aspects of model simulations. For instance, in testing cloud overlap schemes in GCMs, it is necessary to understand not only the horizontal distribution of clouds but also their vertical distribution. In testing the results of cloud-resolving models (CRMs), it is necessary to have observations of cloud altitude distribution at high spatial resolution. Also, in testing chemical transport models that include the conversions of SO_2 into sulfate particles within cloud layers, accurate cloud layering information is required.

In 2006, the National Aeronautics and Space Administration (NASA) of the United States launched two active satellite sensors: the CloudSat radar (Stephens *et al.*, 2002) and Cloud-Aerosol Lidar and Infrared Pathfinder Satellite Observations (CALIPSO; Winker *et al.*, 2002). The combined CloudSat and CALIPSO mission is a satellite experiment primarily designed to provide measurements of the vertical structure of clouds from space. The launches of CloudSat and CALIPSO join the Afternoon 'A-Train' Satellite Constellation in a tight formation flight. The A-Train formation flight consists of six satellites flying in close proximity. Given that both sensors provide only a nadir-pointing view of the Earth along the satellite tracks, a period of observation is required to accumulate enough cloud samples to develop a credible climatology of cloud layering data on global scales. The important measurements from CloudSat and CALIPSO are the vertical profiles of cloud liquid and ice water contents and related cloud physical and radiative properties, which are unavailable on the global scale but needed for the quantitative evaluation of global models.

2. Cloud Problems in the Climate System

Clouds are the ever-changing features in the Earth's atmosphere as seen from space. They not only act as sources and sinks of the global water cycle, but also dominate the solar energy flows into the climate system while at the same time redistributing the diabatic heating in the system (Webster and Stephens, 1984; Stephens *et al.*, 2002). Current modeling activities in accurately capturing the change of clouds due to climate warming have been slow. Little progress has been made in studying cloud feedbacks, as discussed in a critical review by Stephens (2005). As clouds can redistribute the energy and water in the climate system, even a small disturbance within them can have a large impact on the Earth's climate.

State-of-the-art GCMs still cannot convincingly simulate cloud processes (Rossow and Schiffer, 1991; Mitchell, 1993a,b). Comparisons of GCMs from around the world (Cess *et al.*, 1990; Arking, 1991) have documented how sensitive the GCM simulations are to the different assumptions made about clouds. The differences lead to very different top-of-atmosphere (TOA) radiative forcing and climate responses. Most GCMs show a tendency for cloud cover to decrease and cloud height to increase with warming, especially for low and middle-level clouds. Radiative–convective models built on observations of increasing cloud liquid water with temperature suggest a negative cloud optical depth feedback (Somerville and Remer, 1984; Betts and Harshvardhan, 1987). On the other hand, satellite cloud retrievals suggest a tendency for low-cloud optical depth to decrease with warming (Tselioudis *et al.*, 1992; Chang and Coakley, 2007). This is contrary to the model tendencies. GCMs can produce either a positive or negative cloud optical feedback, depending on their cloud parametrizations. Current models cannot convincingly determine whether cloudiness will increase or decrease as the climate warms. The shortcomings are physical reliable cloud observations that adequately constrain the treatment of cloud processes in GCMs. Such observations are crucial for determining cloud forcing, cloud variability, and cloud change as the climate warms.

3. Current Assessment of the Satellite Measured Clouds

The most widely used satellite cloud products for current assessment of GCM clouds are from the ISCCP of the World Climate Research Programme. The ISCCP provides a long history (since 1983) of global observations of cloud horizontal coverage (cloud amount), cloud top height (temperature), and cloud optical depth. Rossow and Schiffer (1999) reported that the annual-mean global average cloud amount was 63% based on the C-series of the ISCCP data, rising to 68% with the new D-series of the ISCCP products. Stowe *et al.* (2002) and Jacobowitz *et al.* (2003) compiled a 10-year (1985–1994) cloud climatology from the CLAVR (Clouds from AVHRR) dataset of the NOAA AVHRR Pathfinder Atmosphere Project (PATMOS), and found a rather constant, 48%–52% monthly-mean global cloud amount. The different cloud amounts between ISCCP and CLAVR are attributed to the different cloud detection techniques used by the two algorithms. Stowe *et al.* (2002) suggest that CLAVR tends to be more conservative in preserving the overcast radiances, so a large portion of variable cloudy pixels were classified as mixed pixels, which were assigned a cloud cover of 50%. On the other hand, ISCCP tends to treat the mixed pixels as completely overcast with 100% cloud coverage. As the majority of the mixed pixels were not fully covered by clouds, the ISCCP-derived cloud amounts tend to be overestimated (Chang and Coakley, 1993; Coakley *et al.*, 2005).

3.1. *Cloud top height and optical depth*

The ISCCP cloud top height is retrieved using the infrared (IR) window measurement at the $11\,\mu$m spectral channel, which is a common channel available from all weather satellite instruments, including AVHRR, GOES, VIRS, GMS, and METEOSAT. There are pros and cons concerning the use of the $11\,\mu$m radiance as a measure of cloud top height (Menzel *et al.*, 1992; Jin *et al.*, 1996; Chang and Li, 2005a). A major concern with the $11\,\mu$m channel is that the $11\,\mu$m brightness temperature is not a good measure of cloud top location for semitransparent clouds such as cirrus (Liou, 1986).

The MODIS instrument provides 36 channels of high-resolution imagery data (King *et al.*, 2003; Platnick *et al.*, 2003). The MODIS-measured radiances facilitate the use of the CO_2-slicing method to determine cloud layer pressures (Chahine, 1974; Smith and Platt, 1978; Wylie and Menzel, 1989). The CO_2-slicing method is based on multiple sounding channels in the $15\,\mu$m CO_2-absorbing band. The method is suitable for retrieving cloud layer pressures for mid-level to upper-level clouds. One advantage of the CO_2-slicing method is its ability to determine the altitudes for

semitransparent cirrus clouds. The disadvantage of this method is that its application is limited to mid-level and high-level clouds, but not the low-level clouds because of their low signal-to-noise ratios (Wielicki and Coakley, 1981). Thus, the MODIS operational cloud product (MOD06) uses the CO_2-slicing method to retrieve cloud top pressure above $700\,$hPa, but uses the $11\,\mu$m channel to determine the altitude for low clouds below $700\,$hPa (Menzel *et al.*, 2002).

A new method recently developed by Chang and Li (2005a) combines the MODIS multispectral CO_2-slicing method with conventional visible and IR window methods to determine the cloud top heights for single-layer and overlapped clouds. Applying the new method to near-global (polar areas excluded) Terra/MODIS data in 2001, they reported a bimodal distribution of cloud top height, with clouds lying predominantly in a high-cloud regime (200–350 hPa) and a low-cloud regime (650–800 hPa). The bimodal distribution shows a minimum in cloudiness between 500 and 600 hPa.

To illustrate the differences of the three methods, Fig. 1 shows comparisons of the frequency of cloud top pressure and cloud optical depth retrieved by the new method of Chang and Li (2005a) [Fig. 1(a)], the MODIS

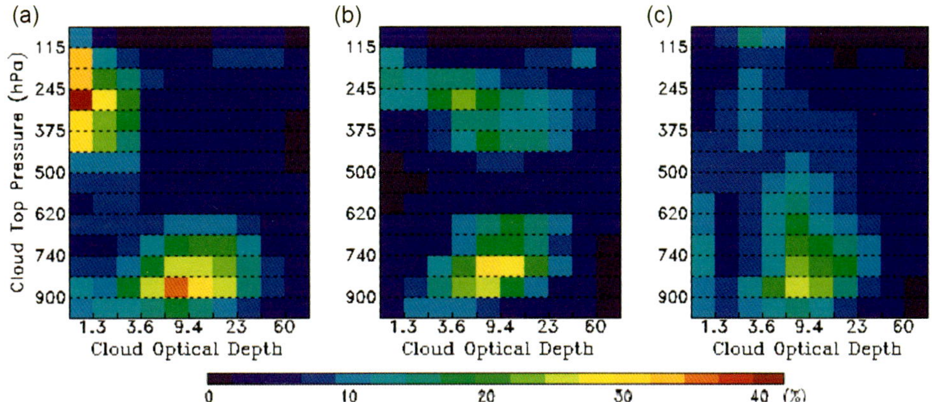

Figure 1. Comparisons of cloud top pressure and cloud optical depth derived by applying three different retrieval algorithms: (a) Chang and Li (2005a), (b) MODIS (MOD06), and (c) simulation of the ISCCP algorithm to one day (6 March 2003) of the MODIS radiance data.

operational method (MOD06 Collection 004) [Fig. 1(b)], and a simulation of the ISCCP method [Fig. 1(c)]. The figure shows results obtained by applying three different methods to the same MODIS radiance data for 6 March 2003, from 60°S to 60°N. The three methods resulted in significantly different cloud layer structures. As discussed by Chang and Li (2005a), their new method deals with both single-layer and overlapped clouds whereas the MODIS operational product ignores the under-lying low clouds when overlapped by upper-level cirrus. As suggested by Chang and Li (2005b), the MODIS operational product (MOD06) [Fig. 1(b)] underestimates the occur-rence of low-level clouds, especially in regions with abundant cirrus clouds, like the tropics and midlatitude storm tracks. This is because when high-level clouds are retrieved by the MODIS CO_2-slicing method, the MOD06 product ignores any potential low clouds. On the other hand, the simulated ISCCP method based on the $11\,\mu m$ channel places high-level cirrus clouds at lower altitudes. For cirrus

overlapping low clouds, the ISCCP method retrieves a biased single-layer mid-level cloud. The biased mid location depends on the opacity of the cirrus and underlying low-level cloud.

3.2. *Cirrus overlapping lower clouds*

Cirrus clouds often overlap lower-level clouds. Surface observations from ships rarely report cirrus clouds existing alone (Warren *et al.*, 1985). When cirrus clouds are present, most satellite retrieval methods ignore the lower clouds beneath the cirrus by assuming only the presence of cirrus clouds. As a result, the radiances from the cirrus-overlapped lower clouds can influence satellite-measured signals through cirrus. For instance, the underlying low-cloud optical depth can be misinterpreted as being associated with the upper cirrus.

Figure 2 shows a case study for a cloud field containing cirrus-and-stratus overlapped systems that were observed on 2 April 2001, 1715(UTC), when Terra MODIS passed

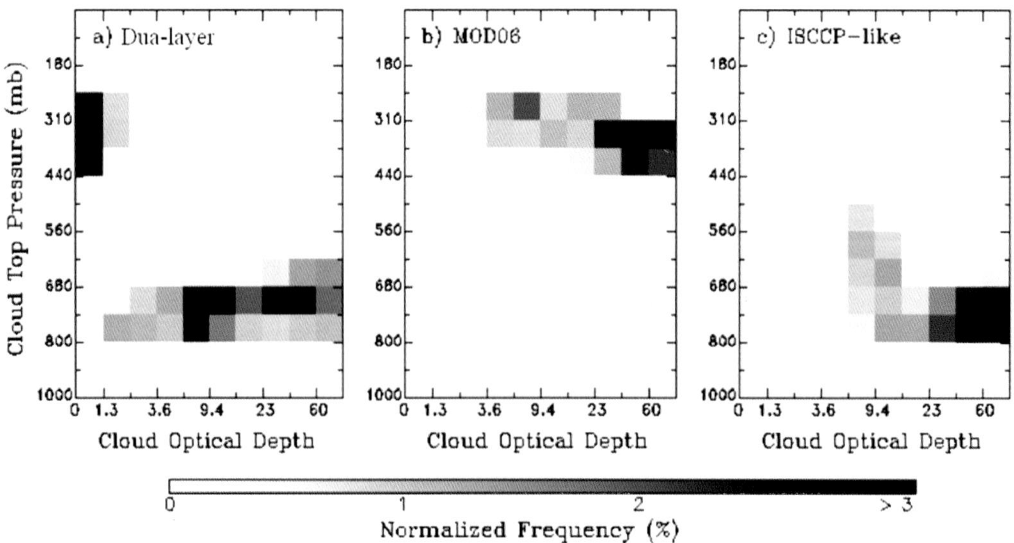

Figure 2. Comparisons of the cloud top pressures and cloud optical depths derived from (a) the method of Chang and Li (2005a), (b) MODIS operational products (MOD06), and (c) a simulation of the ISCCP algorithm for a cirrus layer over stratus clouds observed for a $(500\,\mathrm{km})^2$ area centered at the ARM Central Facility site in northern Okhlahoma.

over the U.S. Department of Energy (DOE) Atmospheric Radiation Measurement (ARM) Program's central facility ground site in northern Oklahoma (Ackerman and Stokes, 2003). The figure shows comparisons of the joint histogram of cloud top pressure and cloud optical depth retrieved using the three methods. As confirmed by the ground-based radar and lidar measurements, the method of Chang and Li (2005a) [Fig. 2(a)] retrieved an upper-level cirrus layer with small optical depths and a low-level stratus layer with larger optical depths. The MODIS product (MOD06 Collection 004) [Fig. 2(b)] also retrieves the upper-level cirrus layer, but no low-level cloud. It also misinterprets the overlapped system as an optically thick layer of upper-level clouds. The simulated ISCCP method [Fig. 2(c)] misinterprets the overlapped system as a mid-to-low-level cloud system which does not separate the upper and lower clouds. The single-layer assumption caused the disappearance of the underlying low-level clouds in the MODIS product and the mixture of the upper-level cirrus with the lower-level stratus cloud into a single midlevel cloud in the simulated ISCCP retrievals. The discrepancies among the three cloud layer pressures as shown in Fig. 2 can have a significant impact on the evaluation of the vertical distribution of clouds generated by models. The differing resultant optical depths for upper-level clouds can also have a significant impact on cloud forcing at TOA and the heating profile within the atmosphere.

3.3. *The bimodal distribution of cloud top altitudes*

Chang and Li (2005b) suggest a distinct bimodal distribution of cloud top altitudes when examining the Terra MODIS data (excluding height-latitude regions) for January, April, July, and October 2001. Their algorithm discriminates high clouds, defined by a cloud top pressure <500 hPb, into three categories: (1) High1 for single-layer cirrus, (2) High2 for overlapped cirrus with underlying low cloud, and (3) High3 for optically thick high clouds that cannot be determined for the overlapping situation. Table 1 gives the different categories of single-layer and overlapped high clouds, along with two categories of low clouds defined by a cloud top pressure $\geq$500 hPa, where Low1 is for the single-layer low cloud not masked by any upper cloud and Low2 is for the overlapped low clouds co-occurring with High2.

Figure 3 shows the frequency of occurrence of cloud top pressures obtained for the single-layer and overlapped categories that were obtained over ocean (upper subpanels) and over land (bottom subpanels) for each of the four months in 2001. It is noted that because of the challenges in dealing with broken clouds that do not fill a partly cloudy imager pixels, the frequency distributions of cloud top pressures as shown in Fig. 3 are obtained only for each of the 5 km regions that are overcast. The 5 km overcast regions are determined using the MODIS Collection 004 cloud mask products (MOD35; Ackerman *et al.*, 1998). The monthly-mean overcast cloud

Table 1. Classification of single-layer, overlap, and high thick clouds and the associated cloud-top pressure (P_c) and 11 μm cloud emissivity (ε_c).

Cloud type	Cloud properties
High1	Single-layer cirrus ($P_c < 500$ hPa and $\varepsilon_c < 0.85$)
High2	Overlapped cirrus ($P_c < 500$ hPa and $\varepsilon_c < 0.85$) with Low2
High3	High thick cloud ($P_c < 500$ hPa and $\varepsilon_c \geq 0.85$)
Low1	Single-layer low cloud ($P_c \geq 500$ hPa)
Low2	Overlapped low cloud ($P_c \geq 500$ hPa) with High2

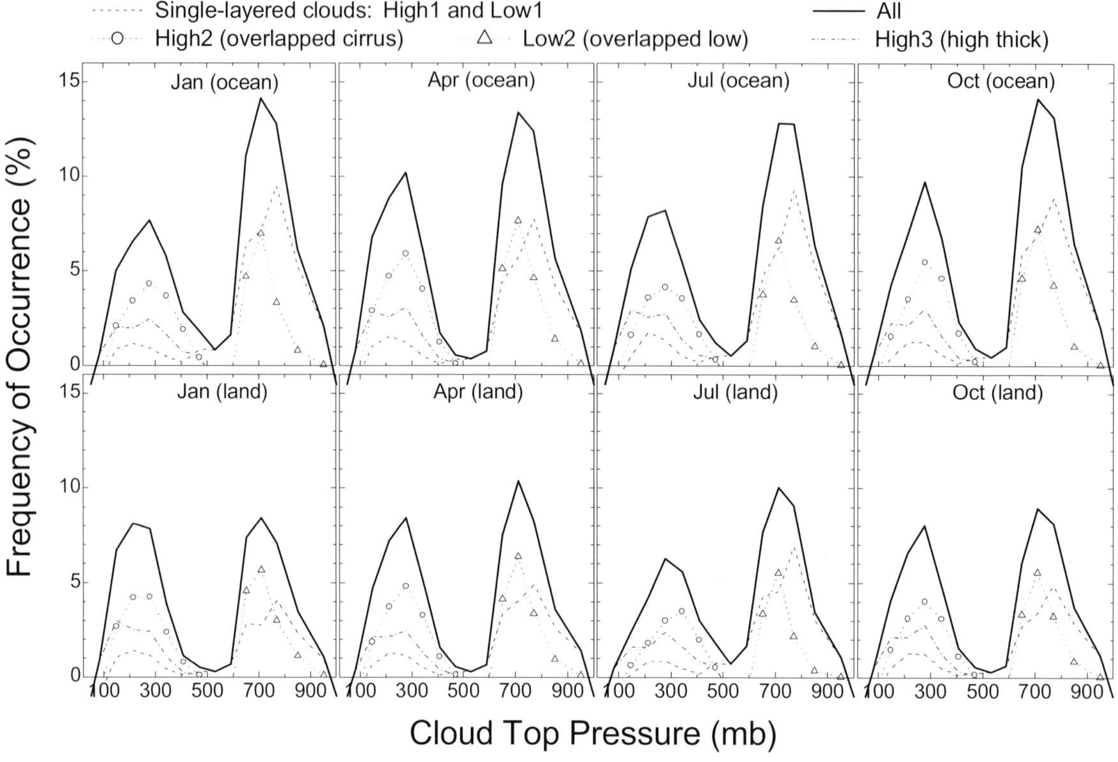

Figure 3. Frequency distributions of cloud top pressure obtained for ocean (top row) and land (bottom row) between 60°S and 60°N from Terra MODIS data.

fraction obtained at the 5 km scale is approximately 47%. By dividing the cloud top pressures into high (<440 hPa), mid (440–680 hPa), and low (>680 hPa) levels, the overcast cloud fractions calculated for the high, mid, and low pressure levels are about 24%, 9%, and 29%, respectively. Since MODIS operational products made the single-layer assumption, the corresponding high, mid, and low overcast cloud amounts are about 24%, 6%, and 17%, which show fewer mid clouds and much fewer low clouds. The simulated ISCCP retrievals have about 15% in high, 15% in mid, and 17% in low for the corresponding dataset, which shows much fewer high and low clouds but much more mid clouds in comparisons with the Chang and Li method. The ISCCP operational products have been used for comparisons with GCM results and the ISCCP shows substantially more mid-cloud amounts than the GCMs (Webb *et al.*, 2001; Zhang *et al.*, 2005). The exceptional ISCCP mid-cloud amounts may result from misidentifications of semitransparent upper-level clouds that appeared warmer in the satellite 11 μm images and lower in altitude (Webb *et al.*, 2001; Chang and Li, 2005a).

In a recent assessment of GCMs by the State University of New York at Stony Brook and the DOE ARM Cloud Parameterization and Modeling Working Group, Zhang *et al.* (2005) compared the high, mid, and low cloud climatologies generated by ten popular GCMs (NCAR, GFDL, GISS, GSFC, UKMO, etc.) along with two satellite products from the ISCCP and the Cloud and Earth's Radiant Energy System (CERES) program. The CERES cloud algorithm uses the 11 μm approach like the ISCCP to determine cloud top altitude, but it differs from the ISCCP in that it uses four instead of two channels to retrieve cloud phase, optical depth, particle size, and temperatures (Minnis *et al.*, 1995; Minnis *et al.*, 1998). Also different is that CERES products are based on the pixel radiance data from MODIS.

While large discrepancies exist among the GCMs, all GCMs produce less mid clouds but more high clouds than both the ISCCP and CERES. They show that GCMs generally produce more high-cloud amounts and less mid cloud amounts than the ISCCP and CERES. The results are consistent with those found in the MODIS and analysis by Chang and Li (2005b). The GISS (NASA Goddard Institute for Space Study) model produced the least high-cloud amount ($<15\%$) while the GFDL (Geophysical Fluid Dynamics Laboratory) model produced the greatest ($>35\%$). The GISS and GFDL were the two greatest in producing mid-level clouds, whereas all other models had significantly less mid clouds ($<10\%$). As for low clouds, the GISS also had the greatest low-cloud amount and was closest to the ISCCP and CERES, whereas other models had much less low-cloud amounts. Nonetheless, all GCMs showed that their low-cloud amounts were two to three times larger than their mid-cloud amounts. Chang and Li (2005b) suggest that more than 30% of the low-level clouds were obscured by high-level clouds, and thus the low-level cloud amounts from satellite operational products like the ISCCP and MODIS were underestimated due to the single-layer assumption.

3.4. *Cloud optical depth feedback*

Cloud optical depth feedback is another major uncertainty in climate studies (Stephens, 2005). Tselioudis *et al.* (1992) and Tselioudis and Rossow (1994) used the ISCCP cloud data and reported a negative cloud optical depth (τ) and temperature (T) relationship. As such, they suggest a positive cloud optical feedback to climate warming. Concerns with

the ISCCP-derived results are associated with the frequently observed partly cloudy pixels in nature (Wielicki and Parker, 1992; Chang and Coakley, 1993). The partly cloudy pixels can cause negative biases in τ retrievals due to broken clouds (Harshvardhan *et al.*, 2004; Coakley *et al.*, 2005; Chang and Coakley, 2007). Whether cloud optical depth would increase or decrease in response to climate warming is still an open question. A negative τ–T relationship implies that clouds would become thinner as climate warms. Thinner clouds would allow more solar radiation to enter and warm the climate system, leading to a positive feedback. Based on their studies with the GISS GCM simulations, Tselioudis *et al.* (1998) and Yao and Del Genio (1999) also show that low-level clouds exert an overall positive cloud optical depth feedback.

On the contrary, early arguments on the cloud optical depth feedback were related to the variation of cloud liquid water content (LWC) with temperature. For example, Somerville and Remer (1984) suggested that cloud optical depth would increase with temperature, if all other cloud variables remain constant. Their suggestion is based on aircraft *in situ* measurements made over the former Soviet Union (Feigelson, 1978), where the measured cloud LWC data showed an increase rate of $d\ln(\text{LWC})/dT = +0.04$ to $+0.05$ for T between $-25°C$ and $+5°C$. The observed rate of increase for $d\ln(\text{LWC})/dT$ is supported by Betts and Harshvardhan (1987) based on the moist adiabatic calculations. Although the rate of increase is calculated to be slightly smaller at $+0.02$ to $+0.03$, Betts and Harshvardhan (1987) also suggested that cloud optical depth will increase with climate warming.

To examine effects of partly cloudy pixels in the ISCCP cloud retrievals, Chang and Coakley (2007) explore the differences in cloud properties between overcast and partly cloudy pixels for low-level, single-layer cloud systems using the AVHRR data obtained between 55°S and 55°N over the Pacific in March 1989. For single-layer systems, they separate overcast pixels from partly cloudy pixels through the use of the spatial coherence method (Coakley and Bretherton, 1982). Chang and Coakley (2007) also follow the ISCCP cloud detection method by applying the visible–IR threshold technique to the AVHRR data to simulate the ISCCP-like cloudy pixels that include both overcast and partly cloudy pixels. They compare the mean properties derived from the overcast pixels and threshold cloudy pixels for the Pacific region and show that the threshold cloudy pixels had substantially smaller ($\sim 50\%$) cloud optical depths and larger ($\sim 3^\circ$C) cloud top temperatures than the overcast pixels.

Having found the differences between overcast and threshold cloudy pixels, Chang and Coakley (2007) reexamine the τ–T relationships through the use of the overcast pixels and threshold cloudy pixels. Despite the differences, correlations for the overcast and threshold cloudy pixels show similar negative τ–T relationships and their corresponding values of $d\ln\tau/dT$ derived for the midlatitudes and subtropical regions agree in magnitude with those derived using the ISCCP data as reported by Tselioudis *et al.* (1992). Chang and Coakley (2007) suggest that the cloud thickness thinning is the main reason why similar negative τ–T relationships were found in both overcast and threshold cloudy pixels.

Chang and Coakley (2007) further examine the variable effects of cloud liquid water content (L), droplet effective radius (r_e), cloud layer temperature (T_c), and cloud geometrical thickness (D) based on the fundamental relationship between τ, r_e and cloud liquid water path (LWP) given by (Hansen and Travis, 1974)

$$\tau \cong \frac{3\,\mathrm{LWP}}{2\,r_e} = \frac{3\,LD}{2\,r_e}. \qquad (1)$$

By taking logarithmic derivatives of Eq. (1), the sensitivities of τ, L, D, and r_e to the 740 hPa atmospheric temperature (T) are given by

$$\frac{d\ln\tau}{dT} = \frac{d\ln L}{dT} + \frac{d\ln D}{dT} - \frac{d\ln r_e}{dT}, \qquad (2)$$

where

$$\frac{d\ln L}{dT} = \frac{d\ln L}{dT_c} \times \frac{dT_c}{dT}. \qquad (3)$$

Following the study by Chang and Coakley (2007), which focused on midlatitudes and subtropics, here the values of $d\ln\tau/dT$ [Fig. 4(a)], $d\ln r_e/dT$ [Fig. 4(b)], and dT_c/dT [Fig. 4(c)] are presented for all latitudes from 55°S to 55°N.

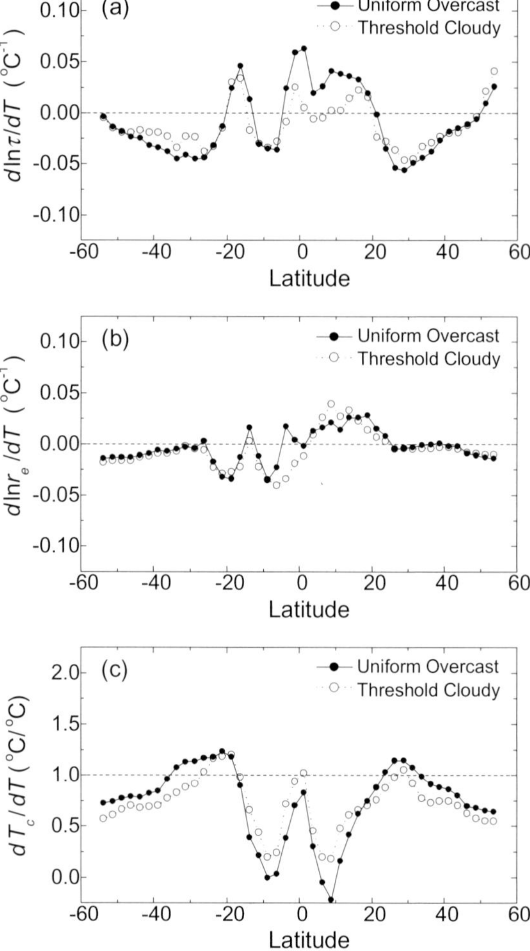

Figure 4. Latitudinal variations of $d\ln\tau/dT$ (a), $d\ln r_e/dT$ (b), and dT_c/dT (c) obtained for low-level, single-layer clouds over the Pacific Ocean between 55°S and 55°N. The results for overcast and threshold cloudy pixels are compared.

The sensitivities to the atmospheric temperature at 740 hPa in Fig. 4 are also compared between the overcast (solid circles) and threshold cloudy pixels (open circles). It is seen from the figure that the negative τ–T relationships are mostly found in midlatitudes and appear to be relatively unaffected by the choice of overcast and partly cloudy pixels, even though partly cloudy pixels have significantly smaller τ than the overcast pixels. There are positive τ–T relationships found in the tropics, and the differences between overcast and threshold cloudy pixels are also found to be larger in the tropics. The positive or negative τ–T relationship may be determined by the different properties of clouds in different regimes, which requires more detailed investigations.

To examine the sensitivity of L, two approaches are used here: (1) the moist adiabatic liquid water content as documented by the theoretical study of Betts and Harshvandhan (1987) for low clouds (>680 hPa), and (2) the subadiabatic liquid water content as documented by the observational study of Somerville and Remer (1984). Figure 5 shows the best-fit regression for the values of $d\ln L/dT_c$ derived for the moist adiabatic (open points; data taken from Betts and Harshvandhan, 1987) and for the subadiabatic (solid points; data taken from Somerville and Remer, 1984). The regression for the moist adiabatic is given by

$$\frac{d\ln L}{dT_c} = 0.0436 - 1.39 \times 10^{-3}T_c + 1.63 \times 10^{-5}T_c^2, \tag{4}$$

and for the subadiabatic it is given by

$$\frac{d\ln L}{dT_c} = 0.0267 - 1.05 \times 10^{-3}T_c + 2.36 \times 10^{-5}T_c^2. \tag{5}$$

It is noted that the positive values of $d\ln L/dT_c$ used here required further investigation, but it is unfortunate that there are few complete observational studies for determining the temperature-dependent cloud LWC. There is also no complete observational data for quantifying the temperature dependence of the cloud droplet effective radius and cloud geometrical thickness. Based on Eqs. (2) and (3), the value of $d\ln D/dT$ can be estimated by

$$\frac{d\ln D}{dT} = \frac{d\ln \tau}{dT} + \frac{d\ln r_e}{dT} - \frac{d\ln L}{dT}. \tag{6}$$

With the values of $d\ln \tau/dT$ being negative, the values of $d\ln L/dT$ being positive, and the values of $d\ln r_e/dT$ being near zero, the values of $d\ln D/dT$ are expected to be negative.

Figure 6 shows the estimated values of $d\ln D/dT$ for the data shown in Fig. 4, where calculations made with the adiabatic value are plotted in open squares, and calculations made with the subadiabatic value are plotted in solid squares. The range of these calculated values for $d\ln D/dT$ reaches between $-0.05°C^{-1}$ and $-0.10°C^{-1}$ in midlatitudes and subtropics, but shows some positive values at about $0.05°C^{-1}$ in the tropics. The negative value implies that cloud geometrical thickness would have decreased by about 0.05–0.10 km per kilometer cloud thickness if the cloud LWC varied at the

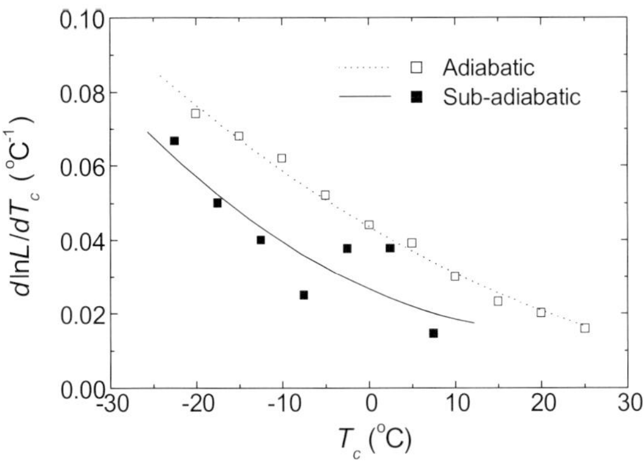

Figure 5. $d\ln L/dT_c$ and T_c adapted from the adiabatic cloud liquid water content of Betts and Harshvandhan (1987) (open squares) and the subadiabatic cloud liquid water content of Somerville and Remer (1984) (solid squares). The lines represent the polynomial regression fits to the individual dataset.

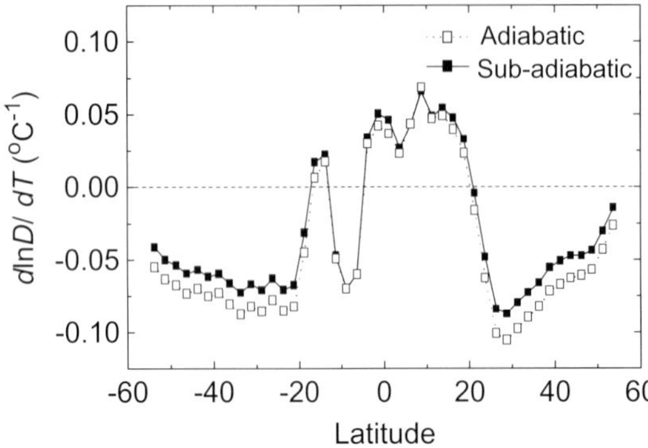

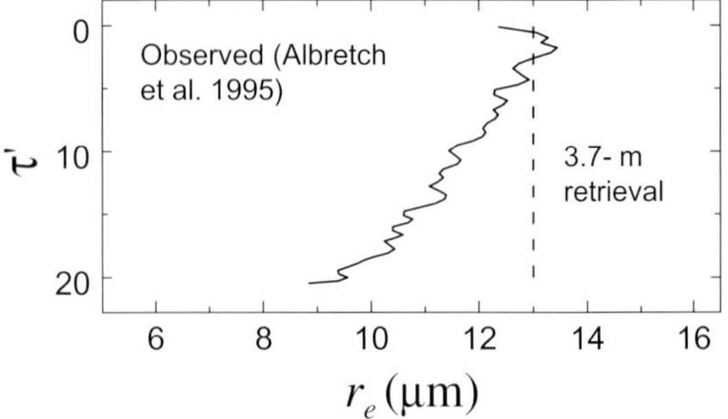

Figure 7. Vertical profiles of the cloud droplet effective radius from observation (solid line) and a 3.7 μm retrieval algorithm (dashed line).

Figure 6. Latitudinal variation of $d\ln D/dT$ obtained for low-level, single-layer clouds over the Pacific Ocean between 55°S and 55°N. Results are obtained for the overcast pixels based on the adiabatic and subadiabatic values of the liquid water content.

subadiabatic or adiabatic rate. Similarly, the positive value in the tropics implies that cloud geometrical thickness would have increased by about 0.05 km per kilometer cloud thickness. However, there is no information available for the cloud geometrical thicknesses. Also, there is no information concerning the absolute values of the cloud LWC and their rate of change for the clouds under the investigation.

3.5. *Cloud droplet effective radius and liquid water path*

Current operational satellite retrieval algorithms employ the near-IR radiances from a single channel at either 3.7 μm (e.g. AVHRR algorithm) or 2.1 μm (e.g. MODIS algorithm) for the retrieval of the cloud droplet effective radius (r_e). The retrieved r_e is used together with the retrieved cloud optical depth, τ, to compute the cloud LWP, given by (Hansen and Travis, 1974)

$$\text{LWP} \cong \frac{4\rho_w}{3Q_e}\tau \times r_e, \qquad (7)$$

where ρ_w is the density of water and $Q_e \approx 2$ is the extinction efficiency. The computation assumes that r_e is invariant with height and the assumption can incur significant bias errors in computing LWP (Chen *et al.*, 2007) because

r_e usually varies with height. As illustrated in Fig. 7, retrieval at a single near-IR channel like 3.7 μm may only represent a shallow layer near the cloud top.

Chang and Li (2002, 2003) have developed a new satellite retrieval method to account for the vertical variation of the r_e profile. It employs a combination of multiple near-IR spectral measurements at the 3.7 μm, 2.1 μm, and 1.6 μm channels, available from MODIS, to retrieve a linear r_e profile. The linear r_e profile is assumed to be a linear function of the in-cloud optical depth (τ'), which is given by

$$r_e(\tau') = r_{e\_top} + (r_{e\_base} - r_{e\_top})\frac{\tau'}{\tau_{\text{total}}}, \qquad (8)$$

where τ_{total} is the total cloud optical depth and r_{e_top} and r_{e_base} represent the r_e at the cloud top ($\tau' = 0$) and the cloud bottom ($\tau' = \tau_{\text{total}}$), respectively. Its retrieval principle lies in the fact that photons at the three near-IR wavelengths are sensitive to different depths within a cloud. This method is applicable to single-layer water clouds and is most effective for clouds with near-linear variation of r_e from cloud top to cloud base (Miles *et al.*, 2000). However, the estimation of r_e for the cloud base would have a large bias if the r_e profile is nonlinear or if the cloud is thick, because the r_e signal from the cloud base becomes weak (Chang and Li, 2002).

Based on the linear r_e profile, Chang and Li (2003) show that the calculation of cloud

LWP can be improved when compared with the calculation based on a constant r_e profile. The improved cloud LWP calculation is given by

$$\text{LWP} = \int \text{LWP}(\tau')d\tau' = \int \frac{2}{3}\rho_w r_e(\tau')d\tau'. \tag{9}$$

Chen *et al.* (2007) applied the linear-r_e retrieval method to the Aqua-MODIS-observed water clouds ($T_c > 273\,\text{K}$) over tropical and subtropical oceans between 40°S and 40°N. The Aqua satellite also carries an Advanced Microwave Scanning Radiometer (AMSR-E) instrument for measuring cloud LWP and precipitation information. Passive microwave radiometers have been used to measure cloud LWP (Curry *et al.*, 1990; Greenwald *et al.*, 1993; Wentz, 1997; Grody *et al.*, 2001). Since the microwave cloud LWP retrieval is based on the radiances emitted by cloud droplets, it is applicable for observations during day and night. However, the microwave retrieval has difficulty over land due to the large variability of land surface emissivity at microwave frequencies. Another disadvantage of the microwave measurements is their large footprints, which have a much coarser spatial resolution ($\sim 15\,\text{km}$) than cloud imagers like MODIS ($\sim 1\,\text{km}$). The MODIS-retrieved r_e profile along with the τ retrieval can be useful for measuring cloud LWP for both land and ocean areas.

Chen *et al.* (2007) compared one day (1 January 2003) of coincident Aqua MODIS and AMSR-E overcast regions and reported the effects of the r_e vertical variation on the MODIS-derived cloud LWP. Their study shows that while the overall mean values are similar between the MODIS and AMSR-E, the instantaneous cloud LWP retrievals differ markedly, which may be caused by precipitation and/or uncertainties in both the MODIS and AMSR-E retrieval algorithms. In comparisons of the AMSR-E retrievals and MODIS standard products (MOD06) that assumed a constant r_e as given by Eq. (7), the MODIS cloud LWP tended to be underestimated when the r_e profile increased from cloud top to cloud base, and overestimated when the r_e profile decreased from cloud top to cloud base. But in comparisons with the MODIS-improved cloud LWP retrievals that accounted for the r_e vertical variation, the mean biases and root-mean-square errors between the AMSR-E- and MODIS-derived LWPs were reduced.

Chen *et al.* (2007) further suggest that the manner in which r_e varies with height has the potential for discriminating precipitative and nonprecipitative warm water clouds. For precipitating clouds, the r_e at the cloud top is smaller than the r_e at the cloud base; whereas for nonprecipitating clouds, the r_e is smaller toward the cloud base. Figure 8 illustrates the differences of the MODIS-retrieved r_e at the cloud top minus the r_e at the cloud base for the overcast warm regions ($\sim 13\,\text{km}$) based on the AMSR-E pixel scale. The data were obtained from Aqua MODIS on 2 July 2003 and were divided into precipitating [Fig. 8(a)] and nonprecipitating [Fig. 8(b)] clouds using the operational AMSU-E rainfall products (Wentz and Meissner, 1999). For AMSR-E-flagged precipitating clouds, r_e at the cloud top are generally larger than r_e at the cloud base; whereas for AMSR-E-flagged nonprecipitating clouds, it is generally the opposite. There is the possibility that precipitating clouds are not detected by the operational AMSR-E algorithm, as many r_e at the cloud base are significantly larger than r_e at the cloud top [*cf.* Fig. 8(b)].

The conventional satellite remote sensing techniques for detecting precipitation rely on the measurements of cold cloud temperatures and/or large ice particles for convective clouds (Arkin and Ardanuy, 1989). They mainly detect deep convection and heavy precipitation clouds, but have overlooked many warm precipitating clouds within shallow stratocumulus cloud systems. As such, the total precipitation on global scales may be underestimated. Liu *et al.* (1995) utilized a microwave emission technique to detect precipitating warm clouds, and found that clouds with $T_c > 273\,\text{K}$ contributed to 14%

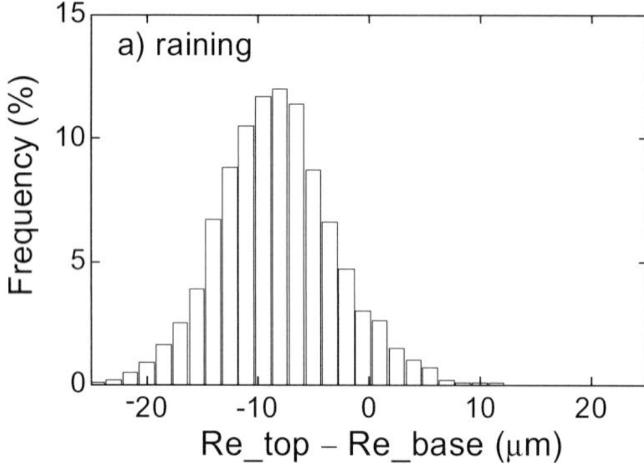

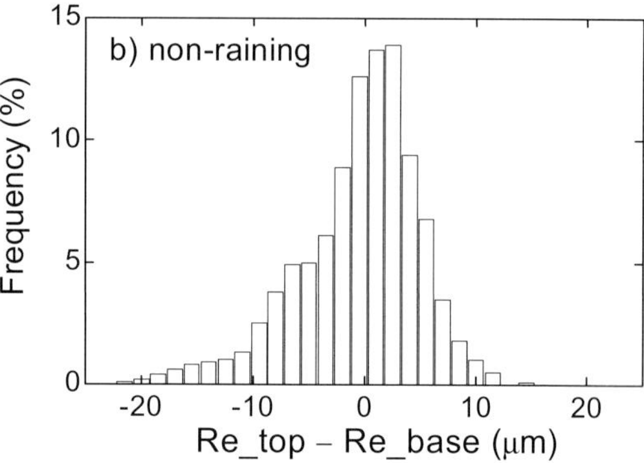

Figure 8. Frequency distributions of the MODIS-retrieved r_e for the cloud top minus r_e for the cloud base for the overcast warm ($> 273\,\mathrm{K}$) clouds. The raining (a) and nonraining (b) categories are determined based on the AMSR-E products.

of the total rainfall in the western equatorial Pacific Ocean. They suggest that this percentage may be underestimated due to certain inabilities of the microwave measurements. The amount of precipitation from warm clouds that reaches the ground on global scales is a key question in understanding the hydrological cycle.

4. Concluding Remarks

Clouds remain a major source of uncertainty in climate simulation and weather forecast models. Improvements in cloud modeling have been slow, in part because the conventional satellite observations cannot provide detailed information on cloud vertical structure for model evaluation and cloud vertical structure represents a critical gap in determining how clouds interact with the local and large-scale environments. Measuring the vertical profile of cloud properties is fundamentally important, and requires the combination of both active and passive instruments.

The active satellite remote sensing of CloudSat and CALIPSO has promised to provide the Earth science community with the first data product of satellite-measured profiles of cloud physical and optical properties. While the radar can measure the vertical profile of cloud hydrometer particle information, the lidar can detect thin cirrus layers that may not be detectable by the radar. Many of the

scientific questions discussed above are expected to be addressed with the data acquired from a combination of the two active radar and lidar instruments. For example, concerning the cloud vertical distributions: What are the frequencies of occurrence of high-level, mid-level, and low-level clouds? What are the frequencies of occurrence of multilayer clouds? Are they composed of two layers or more layers of clouds? What are the percentages of single-layer and overlapped cirrus clouds? How different or similar are the cloud vertical structures observed by CloudSat and CALIPSO in comparisons with the MODIS-retrieved distributions of cloud top pressure? What are the vertical structures of cloud hydrometer phase and particle size, and can we use the droplet effective radius to enhance the detection of warm precipitating clouds? The CloudSat observations may help answer these questions.

While the two active satellite instruments observe the global cloud vertical structure, their observations are limited to a nadir-viewing direction that does not convey the horizontal observations of clouds. Fortunately, CloudSat and CALIPSO will be flying in constellation with the Aqua A-Train platform. The cloud vertical structure observed by CloudSat and CALIPSO can be combined with cloud and radiation quantities measured by the Aqua MODIS, AMSR-E, and CERES instruments. The MODIS provides a wealth of information on the horizontal distributions of cloud top height and cloud optical and microphysical properties. The AMSR-E provides the column-integrated total water path needed to constrain the CloudSat algorithm for retrieving the vertical profile of cloud optical and microphysical properties (Austin and Stephens, 2001).

The CERES incorporates the MODIS-retrieved cloud properties and provides enhanced top-of-atmosphere (TOA) radiative flux measurements (Wielicki *et al.*, 1996; Loeb *et al.*, 1999, 2000). Development of improved angular distribution models requires satellite measurements from different cloud scene-type identifications in obtaining accurate estimates of the radiative fluxes or albedos of the Earth–atmosphere system (Loeb *et al.*, 1999, 2000; Chang *et al.*, 2000a,b). Areas affected by optically thick clouds can have a net cloud radiative cooling of $-100\,\mathrm{W/m^2}$ and the net global cloud radiative forcing is on the order of $-15\,\mathrm{W/m^2}$ at the tropopause (Wielicki *et al.*, 1995). The interaction of radiation with cloud properties, particularly cloud optical depth, has a significant impact on the climate, which is several times the impact of a CO_2 doubling in the atmosphere. A small change in cloud optical properties can modify the amounts of radiative fluxes that enter and leave the Earth–atmosphere system and alter the climate.

To date, climate simulations agree that clouds have a strong influence on climate change, but we are far from knowing the magnitude, and even the sign, of this influence. It is indeed a daunting task in climate simulations on cloud feedback. However, progress toward an improved understanding of the cloud–climate feedback is beginning to emerge through the coordination of many world-class research programs in advancing the observations and models; for example, the efforts under the international Global Energy and Water Cycle Experiment (GEWEX) Cloud System Study (GCSS) Program and the Intergovernmental Panel on Climate Change (IPCC).

Acknowledgements

The author thanks Dr Jim Coakley, Jr of the Oregon State University and another, anonymous reviewer for their valuable comments and suggestions for improving the quality of this article.

[Received 31 December 2005; Revised 13 September 2007; Accepted 16 September 2007.]

References

Ackerman, S. A., K. I. Strabala, W. P. Menzel, R. A. Frey, C. C. Moeller, and L. E. Gumley, 1998: Discriminating clear-sky from clouds with MODIS. *J. Geophys. Res.*, **103**, 32141–32158.

Ackerman, T. P., and G. Stokes, 2003: The atmospheric radiation measurement program. *Phys. Today*, **56**, 38–45.

Arkin, P. A., and P. Ardanuy, 1989: Estimating climatic-scale precipitation from space: a review. *J. Climate*, **2**, 1229–1238.

Arking, A., 1991: The radiative effects of clouds and their impact on climate. *Bull. Amer. Meteor. Soc.*, **72**, 795–813.

Austin, R. T., and G. L. Stephens, 2001: Retrieval of stratus cloud microphysical parameters using millimeter-wave radar and visible optical depth in preparation for CloudSat. I. Algorithm formulation, *J. Geophys. Res.*, **106**, 28233–28242.

Betts, A. K., and Harshvandhan, 1987: Thermodynamic constraint on the liquid water feedback in climate models. *J. Geophys. Res.*, **92**, 8483–8485.

Cess, R. D., and coauthors, 1990: Interpretation of climate feedback processes in 19 atmospheric general circulation models. *J. Geophys. Res.*, **95**, 16601–16615.

Chahine, M. T., 1974: Remote sounding of cloudy atmospheres. I. The single cloud layer. *J. Atmos. Sci.*, **31**, 233–243.

Chang, F.-L., and J. A. Coakley Jr, 1993: Estimating errors in fractional cloud cover obtained with infrared threshold methods. *J. Geophys. Res.*, **98**, 8825–8839.

Chang, F.-L., Z. Li, and A. P. Trishchenko, 2000a: The dependence of TOA reflectance anisotropy on cloud properties inferred from ScaRaB satellite data. *J. Appl. Meteorol.*, **39**, 2480–2493.

Chang, F.-L., Z. Li, and S. A. Ackerman, 2000b: Examining the relationship between cloud and radiation quantities derived from satellite observations and model calculations. *J. Climate*, **13**, 3842–3859.

Chang, F.-L., and Z. Li, 2002: Estimating the vertical variation of cloud droplet effective radius using multispectral near-infrared satellite measurements. *J. Geophys. Res.*, **107**, AAC 7 1–12.

Chang, F.-L., and Z. Li, 2003: Retrieving vertical profiles of water-cloud droplet effective radius: algorithm modification and preliminary application. *J. Geophys. Res.*, **108**, AAC 3, 1–11.

Chang, F.-L., and Z. Li, 2005a: A new method for detection of cirrus overlapping water clouds and determination of their optical properties. *J. Atmos. Sci.*, **62**, 3993–4009.

Chang, F.-L., and Z. Li, 2005b: A near-global climatology of single-layer and overlapped clouds and their optical properties retrieved from Terra/MODIS data using a new algorithm. *J. Climate*, **18**, 4752–4771.

Chang, F.-L., and J. A. Coakley Jr, 2007: Relationships between marine stratus cloud optical depth and temperature: inferences from AVHRR observations, *J. Climate*, **20**, 2022–2036.

Chen, R., F.-L. Chang, Z. Li, R. Ferraro, and F. Weng, 2007: Impact of the vertical variation of cloud droplet size on the estimation of cloud liquid water path and rain detection. *J. Atmos. Sci.*, **64**, 3843–3853.

Chou, M.-D., M. J. Suarez, C.-H. Ho, M.-H. Yan, and K.-T. Lee, 1998: Parameterizations for cloud overlapping and shortwave single-scattering properties for use in general circulation and cloud ensemble models. *J. Climate*, **11**, 202–214.

Chou, M.-D., K.-T. Lee, S.-C. Tsay, and Q. Fu, 1999: Parameterization for cloud longwave scattering for use in atmospheric models. *J. Climate*, **12**, 159–169.

Coakley, Jr J. A., and F. P. Bretherton, 1982: Cloud cover from high-resolution scanner data: detecting and allowing for partially filled fields of view. *J. Geophys. Res.*, **87**, 4917–4932.

Coakley, Jr J. A., M. A. Friedman, and W. R. Tahnk, 2005: Retrieval of cloud properties for partly cloudy imager pixels. *J. Atmos. Ocean. Tech.*, **22**, 3–17.

Curry, J. A., C. D. Ardeel, and L. Tian, 1990: Liquid water content and precipitation characteristics of stratiform clouds as inferred from satellite microwave measurements. *J. Geophys. Res.*, **95**, 16659–16671.

Feigelson, E. M., 1978: Preliminary radiation model of a cloudy atmosphere. 1. Structure of clouds and solar radiation. *Beitr. Phys. Atmos.*, **51**, 203–229.

Greenwald, T. J., G. L. Stephens, T. H. Vonder Haar, and D. L. Jackson, 1993: A physical retrieval of cloud liquid water over the global oceans using Special Sensor Microwave/Imager (SSM/I) observations. *J. Geophys. Res.*, **98**, 18471–18488.

Grody, N. C., J. Zhao, R. Ferraro, F. Weng, and R. Boers, 2001: Determination of precipitable water and cloud liquid water over oceans from the NOAA-15 advanced microwave sounding unit. *J. Geophys. Res.*, **106**, 2943–2954.

Hansen, J. E., and L. D. Travis, 1974: Light scattering in planetary atmospheres. *Space Sci. Rev.*, **16**, 527–610.

Harshvardhan, B. A. Wielicki, and K. M. Ginger, 1994: The interpretation of remotely sensed cloud properties from a model parameterization perspective. *J. Climate*, **7**, 1987–1998.

Jin, Y., W. B. Rossow, and D. P. Wylie, 1996: Comparison of the climatologies of high-level clouds from HIRS and ISCCP. *J. Climate*, **9**, 2850–2879.

Jacobowitz, H., L. L. Stowe, G. Ohring, A. Heidinger, K. Knapp, and N. R. Nalli, 2003: The Advanced Very High Resolution Radiometer Pathfinder Atmosphere (PATMOS) climate dataset: a resource for climate research. *Bull. Amer. Meteor. Soc.*, **84**, 785–793.

King, M. D., and coauthors, 2003: Cloud and aerosol properties, precipitable water, and profiles of temperature and humidity from MODIS. *IEEE Trans. Geosci. Remote Sens.*, **41**, 442–458.

Liou, K.-N., 1986: Influence of cirrus clouds on weather and climate processes: a global perspective. *Mon. Wea. Rev.*, **114**, 1167–1199.

Liu, G., J. A. Curry, and R.-S. Sheu, 1995: Classification of clouds over the western equatorial Pacific Ocean using combined infrared and microwave satellite data. *J. Geophys. Res.*, **100**, 13811–13826.

Loeb, N. G., P. O'R. Hinton, and R. N. Green, 1999: Top-of-atosphere albedo estimation from angular distribution models: a comparison between two approaches. *J. Geophys. Res.*, **104**, 31, 255–260.

Loeb, N. G., F. Parol, J.-C. Buriez, and C. Vanbauce, 2000: Top-of-atmosphere albedo estimation from angular distribution models using scene identification from satellite cloud property retrievals. *J. Climate*, **13**, 1269–1285.

Menzel, W. P., D. P. Wylie, and K. I. Strabala, 1992: Seasonal and diurnal changes in cirrus clouds as seen in four years of observations with the VAS. *J. Appl. Meteor.*, **31**, 370–385.

Menzel, W. P., B. A. Baum, K. I. Strabala, and R. A. Frey, 2002: Cloud top properties and cloud phase — Algorithm Theoretical Basis Document: ATBD-MOD-04. 61 pp. Available at http://modis-atmos.gsfc.nasa.gov/docs/atbd_mod04.pdf.

Miles, N. L., J. Verlinde, and E. E. Clothiaux, 2000: Cloud droplet size distributions in low-level stratiform clouds, *J. Atmos. Sci.*, **57**, 295–311.

Minnis, P., and coauthors, 1995: Cloud Optical Property Retrieval (Subsystem 4.3). Clouds and the Earths Radiant Energy System (CERES) Algorithm Theoretical Basis Document, III: Cloud Analyses and Radiance Inversions (Subsystem 4), NASA RP 1376, Vol. 3, ed. CERES Science Team, pp. 135–176.

Minnis, P., D. P. Garber, D. F. Young, R. F. Arduini, and Y. Takano, 1998: Parameterization of reflectance and effective emittance for satellite remote sensing of cloud properties. *J. Atmos. Sci.*, **55**, 3313–3339.

Mitchell, J. F. B., 1993a: Simulation of climate. *Renewable Energy*, **3**, 421–432.

Mitchell, J. F. B., 1993b: Simulation of climate change. *Renewable Energy*, **3**, 433–445.

Platnick, S., and coauthors, 2003: The MODIS cloud products: algorithms and examples from Terra, *IEEE Trans. Geosci. Remote Sens.*, **41**, 459–473.

Rossow, W. B., and R. A. Schiffer, 1991: ISCCP cloud data products, *Bull. Amer. Meteor. Soc.*, **72**, 2–20.

Rossow, W. B., L. C. Garder, P. J. Lu, and A. Walker, 1991: International Satellite Cloud Climatology Project (ISCCP) documentation of cloud data. *WMO/TD No. 266* (World Meteorological Organization, Geneva), 76 pp.

Rossow, W. B., and R. A. Schiffer, 1999: Advances in understanding clouds from ISCCP, *Bull. Amer. Meteor. Soc.*, **80**, 2261–2287.

Smith, W. L., and C. M. R. Platt, 1978: Comparison of satellite-deduced cloud heights with indications from radiosonde and ground-based laser measurements. *J. Appl. Meteorol.*, **17**, 1796–1802.

Somerville, R. C. J., and L. A. Remer, 1984: Cloud optical thickness feedbacks in the CO_2 climate problem. *J. Geophys. Res.*, **89**, 9668–9673.

Stephens, G. L., and coauthors, 2002: The Cloudsat mission and the A-train: a new dimension of space-based observations of clouds and precipitation. *Bull. Amer. Meteor. Soc.*, **83**, 1771–1790.

Stephens, G. L., 2005: Cloud feedbacks in the climate system: a critical review. *J. Climate*, **18**, 237–273.

Stowe, L. L., H. Jacobowitz, G. Ohring, K. R. Knapp, and N. R. Nalli, 2002: The Advanced Very High Resolution Radiometer (AVHRR) Pathfinder Atmosphere (PATMOS) climate dataset: initial analyses and evaluations. *J. Climate*, **15**, 1243–1260.

Tselioudis, G., W. B. Rossow, and D. Rind, 1992: Global patterns of cloud optical thickness variation with temperature. *J. Climate*, **5**, 1484–1495.

Tselioudis, G., and W. B. Rossow, 1994: Global, multiyear variations of optical thickness with temperature in low and cirrus clouds. *Geophys. Res. Lett.*, **21**, 2211–2214.

Tselioudis, G., A. D. Del Genio, W. Kovari Jr, and M.-S. Yao, 1998: Temperature dependence of low cloud optical thickness in the GISS GCM: contributing mechanisms and climate implications. *J. Climate*, **11**, 3268–3281.

Warren, S. G., C. J. Hahn, and J. London, 1985: Simultaneous occurrence of different cloud types, *J. Climate Appl. Meteor.*, **24**, 658–667.

Webb, M., C. Senior, S. Bony, and J.-J. Morcrette, 2001: Combining ERBE and ISCCP data to assess clouds in the Hadley Centre, ECMWF and LMD atmospheric climate models, *Climate Dynamics*, **17**, 905–922.

Webster, P. J., and G. L. Stephens, 1984: Cloud–radiation interaction and the climate problem. *The Global Climate*, ed. J. Houghton (Cambridge University Press, New York), pp. 63–78.

Wentz, F. J., 1997: A well-calibrated ocean algorithm for SSM/I. *J. Geophys. Res.*, **102**, 8703–8718.

Wentz, F. J., and T. Meissner, 1999: AMSR ocean algorithm, version 2. RSS Tech. Rep. 121599A, 66 pp. (Available from Remote Sensing Systems, 438 First Street, Suite 200, Santa Rosa, CA 95401, USA.)

Wielicki, B. A., and J. A. Coakley Jr, 1981: Cloud retrieval using infrared sounder data: error analysis. *J. Appl. Meteorol.*, **20**, 157–169.

Wielicki, B. A., and L. Parker, 1992: On the determination of cloud cover from satellite sensors: the effect of sensor spatial resolution. *J. Geophys. Res.*, **97**, 12799–12823.

Wielicki, B. A., R. D. Cess, M. D. King, D. A. Randall, and E. F. Harrison, 1995: Mission to Planet Earth: role of clouds and radiation in climate. *Bull. Amer. Meteorol. Soc.*, **76**, 2125–2153.

Wielicki, B. A., B. R. Barkstrom, E. F. Harrison, R. B. Lee, III, G. L. Smith, and J. E. Cooper, 1996: Cloud and the Earth radiant energy system (CERES): an earth observing system experiment. *Bull. Amer. Meteorol. Soc.*, **77**, 853–868.

Winker, D. M., J. Pelon, and M. P. McCormick, 2002: The CALIPSO mission: spaceborne lidar for observation of aerosols and clouds. SPIE Asia–Pacific Symposium on Remote Sensing of the Atmosphere, Environment and Space (23–27 Oct. 2002; Hangzhou, China).

Wylie, D. P., and W. P. Menzel, 1989: Two years of cloud cover statistics using VAS. *J. Climate*, **2**, 380–392.

Yao, M.-S., and A. D. Del Genio, 1999: Effects of cloud parameterization on the simulation of climate changes in the GISS GCM. *J. Climate*, **12**, 761–779.

Zhang, M. H., and coauthors, 2005: Comparing clouds and their seasonal variations in 10 atmospheric general circulation models with satellite measurements, *J. Geophys. Res.*, **110**, D15S02, doi:10.1029/2004JD005021.

Recent Advances in Research on Micro- to Storm-Scale Ice Microphysical Processes in Clouds

Pao K. Wang, Hsinmu Lin, Hui-Chun Liu, Mihai Chiruta and Robert E. Schlesinger

Department of Atmospheric and Oceanic Sciences,
University of Wisconsin, Madison, Wisconsin, USA
pao@windy.aos.wisc.edu

In this paper, we summarize our recent research results from ice processes in the atmosphere from storm- to microscale. First, we tested the sensitivity of ice processes in a simulated US High Plains supercell storm using a 3-dimensional cloud model by specifying three different ice physics schemes; namely, the control run, the all-liquid run with normal latent heats, and the all-liquid run with latent heat of sublimation. We showed that the absence of ice processes would result in a substantially shorter lifespan of the storm. Furthermore, we showed that this impact is due to the microphysical properties of ice rather than the thermodynamics of the cloud due to latent heat release.

Secondly, we tested the sensitivity of ice crystal habits on the development of thin cirrus clouds using a 2-dimensional cirrus model with detailed microphysics. Four different ice crystal habits were studied: plates, columns, rosettes and spheres. The results show that the cirrus development is indeed greatly influenced by the habit of ice crystals in the cloud. The largest differences exist between cirrus consisting of rosettes and that consisting of spheres. The largest impact is in the long-wave heating rates where the peak heating rates between these two cases differ by more than 6 times. Other cloud properties are also significantly influenced by the different habits.

Finally, calculations of the ice crystal capacitance for three ice habits: rosettes, solid columns and hollow columns were performed using finite element techniques. The results show that the homomorphic solid and hollow columns of same dimensions have nearly the same capacitance, implying that the mass growth rates of the two are nearly the same but the hollow column will have greater linear growth rate due to its hollowness. The results for rosettes show that the capacitance of rosettes is a nonlinear function of the number of lobes, and hence previous assumptions that their capacitances can be approximated by spheres or prolate/oblate spheroids of the same diameter may result in substantial errors. The computed capacitances can be used in the calculations of crystal growth rates in ice clouds.

1. Introduction

Ice processes in clouds have great impacts on our atmosphere, but their significance was not appreciated by the atmospheric science community until late into the 20th century. The early studies of ice processes were limited to the basic physical phenomena, such as the nucleation phenomenon, the Bergeron–Findeisen process of light rain initiation, the formation of hail, and a few minor aspects. One of the main reasons is most likely that ice processes occur mostly in the upper troposphere, and few airborne platforms at the time could reach such altitudes to perform *in situ* observations. Remote sensing techniques such as radar and lidar have their own difficulties in observing ice processes.

In the 1980s, however, it was increasingly recognized that the Earth–atmosphere system must be viewed as a whole, and that the components of this system act upon one another. A casual look at any zonal mean cross-section of the atmospheric temperature profile would easily reveal that the total volume of the global tropospheric air that is colder than $0°C$ is greater than the warmer one. This implies that the spatial probability of ice particle formation in atmospheric clouds is greater than that of liquid drop formation, although one needs to consider further the probability of supercooling to ascertain this point. In any case, atmospheric ice layers represent such a prominent part of the troposphere that it is only logical to think that it would have great influences on the behavior of our atmosphere. Thus, Ramanathan *et al.* (1983) pointed out that high cirrus clouds, which were traditionally considered as thin and tenuous, and hence were of minor importance to the atmosphere, could have great impact on the global climate process because of their strong interaction with solar and terrestrial radiations (see, for example, Liou 2002).

Ice processes exert their influences via the constituent ice particles in clouds. These ice particles may interact strongly with long and short wave radiations according to their size and shape, and impact the heat budget of the atmosphere. To understand such impacts accurately, we need to know the habit of these particles and their growth modes and rates. Alternatively, these particles may influence the development of the cloud by changing their dynamic environment via their drag on the cloudy air and evaporation-driven downdrafts. Of course, the influences mentioned above are not independent of each other, but intertwined in a complex way. Careful studies are necessary to unravel these influences and the roles ice particles play in them.

This article will summarize some recent studies (from 2000 to the present) performed by our research group on the micro- to storm-scale ice processes in clouds. We will first describe studies at the cloud scale (including both convective and cirrus clouds), followed by microscale studies.

2. The Role of Ice Processes in the Lifespan of US High Plains Thunderstorms

The majority of thunderstorm studies performed by meteorologists in recent decades are concerned with the dynamical and thermodynamic processes, and rarely about microphysical processes. Tao *et al.* (1991) performed numerical simulations of a subtropical squall line over the Taiwan Strait using a two-dimensional cloud model. In that study, they also performed a sensitivity study of ice microphysics and found that the storm starts to dissipate earlier without ice processes. Johnson *et al.* (1993) examined the effects of cloud microphysics on the overall dynamical behavior of thunderstorms using a three-dimensional nonhydrostatic cloud-resolving model equipped with explicit cloud microphysics (including both liquid and ice processes) to perform a sensitivity study of the effect of the ice process. They performed model simulations with and without the ice process. By comparing the two results, they found that the simulated storm with full ice physics lasts longer than the one without, and the different behavior can be attributed to the density of ice particles being lower than that of liquid hydrometeors. Barth and Parsons (1996) studied the microphysical processes associated with intense frontal rainbands using a 2D cloud model, and found that the melting and sublimation of ice hydrometeors increase the intensity of the downdraft and cold pool, which strengthens the squall line convection. Similar conclusions were obtained by Phillips *et al.* (2007), who also used a 2D cloud model with a refined melting scheme to simulate the dynamics and precipitation production in maritime and continental storm clouds. Liu *et al.* (1997) also performed a study on two tropical squall lines using 2D cloud model simulation, and found that the addition of ice microphysics did produce more intense storms. Aside from these simulation studies, Heymsfield *et al.* (2005) performed a study of the homogeneous ice nucleation in subtropical and tropical convection based on observational data. The importance of many ice processes was deduced from the study.

In the above studies, Johnson *et al.* (1993) were the only team to use a 3D model, which is important for studying the severe convective storms, as the lack of the third dimension may impose severe restrictions on the dynamics and hence influence the interpretation of the results. However, that study did not definitely resolve whether the observed effects are completely due to microphysics or thermodynamics. The discussions to be summarized in this section are based on a follow-up study that removed the ambiguity.

2.1. *The cloud model and the CCOPE supercell*

The tool utilized for the present study is the Wisconsin Dynamical/Microphysical Model (WISCDYMM). The WISCDYMM predicts the three wind components — turbulent kinetic energy, potential temperature, pressure deviation — and mixing ratios for water vapor, cloud water, cloud ice, rain, graupel/hail and snow. It adapts the quasi-compressible, nonhydrostatic primitive equation system of Anderson *et al.* (1985), rearranging the mass continuity equation to predict the pressure deviation much as in the fully compressible 3D cloud model of Klemp and Wilhelmson (1978), but allows time steps approximately three times larger by assigning acoustic waves a reduced pseudo-sound speed roughly twice the maximum anticipated wind speed. As in Klemp and Wilhelmson, subgrid transports are parametrized via 1.5-order "K-theory" to predict turbulent kinetic energy, from which a time- and space-dependent eddy coefficient is diagnosed for momentum, and set 35% larger for the heat and moisture predictands (Straka, 1989).

As elaborated by Straka (1989), a version of the WISCDYMM called the HPM (Hail Parametrization Model) features a bulk microphysics parametrization that entails water vapor and five hydrometeor types: cloud water, cloud ice, rain, graupel/hail and snow, with 37 individual transfer rates (source/sink terms). Adapted largely from Lin *et al.* (1983) and Cotton *et al.* (1982, 1986), this package treats all hydrometeors as spheres except for cloud ice, which is treated as small hexagonal plates. Cloud water and cloud ice are assumed monodisperse, with zero fallspeed relative to the air. All three precipitation classes have inverse-exponential size distributions, with temperature-dependent intercepts for snow and graupel/hail, while the intraspectral variation of particle fallspeed versus diameter for each is assumed to satisfy a power law.

If necessary, the WISCDYMM can also be run in the HCM (Hail Category Model) mode, in which the evolution of hailstones can be tracked by studying the growth of hail sizes in a number (e.g. 25) of size bins. The HCM has been tested successfully by Straka (1989).

The WISCDYMM is also programmed to activate one or more of the following three iterative microphysical adjustments (Straka, 1989) where and if needed:

A saturation adjustment is performed to: (a) condense cloud water (or depose cloud ice) to eliminate supersaturation, releasing latent heat of evaporation (or sublimation), or (b) evaporate cloud water (or sublimate cloud ice) in subsaturated air until either saturation is reached or the cloud water (or cloud ice) is exhausted, absorbing latent heat instead. Cloud water is adjusted first and cloud ice second, incrementing the water vapor and temperature to suit. In-cloud saturation mixing ratios are weighted between their values with respect to water and ice in proportion to the relative amounts of cloud water and cloud ice respectively. Where no cloud is present, saturation is taken w.r.t. water or ice if the temperature is above or below 0°C respectively. More than three iterations are rarely needed. Prior to the saturation adjustment, any cloud ice at temperatures above 0°C is melted, and any cloud water at temperatures below −4°C is frozen, respectively absorbing or releasing latent heat of fusion.

After partial update of the moisture fields by advection and turbulent mixing, the decrement in each hydrometeor field due to the net sink (the sum of the individual microphysical sink terms) is compared to its available supply, defined as its partially updated mixing ratio plus the increment due to its net source (the sum of the individual microphysical source terms). If the net sink of a hydrometeor class exceeds its available supply, it is prorated downward along with each of its components so as to not exceed 25% of the available supply. The procedure is iterative because prorating down the sinks of one class also reduces the corresponding source terms for one or more other classes, but more than two iterations are rarely needed.

The storm chosen for the simulation in this study is a supercell that passed through the center of the Cooperative Convective Precipitation Experiment (CCOPE) observational network in southeastern Montana on 2 August 1981. The storm and its environment were intensively observed for more than 5 h by a combination of 7 Doppler radars, 7 research aircraft, 6 rawinsonde stations and 123 surface recording stations as it moved east–southeastward across the CCOPE network. Again, the observational history of this storm has been reported previously (Miller *et al.*, 1988; Wade, 1982), and readers are referred to these sources for further details. The storm has also been successfully simulated using the WISCDYMM, and the general dynamical and microphysical behaviors were reported by Johnson *et al.* (1993, 1995). The current study uses the simulated CCOPE storm that was initialized by the same conditions as in Johnson *et al.* (1993, 1995) and with the same resolution, i.e. $1 \times 1 \times 0.5\,\mathrm{km}^3$.

Figure 1 shows the sounding used as the initial conditions for starting the simulation.

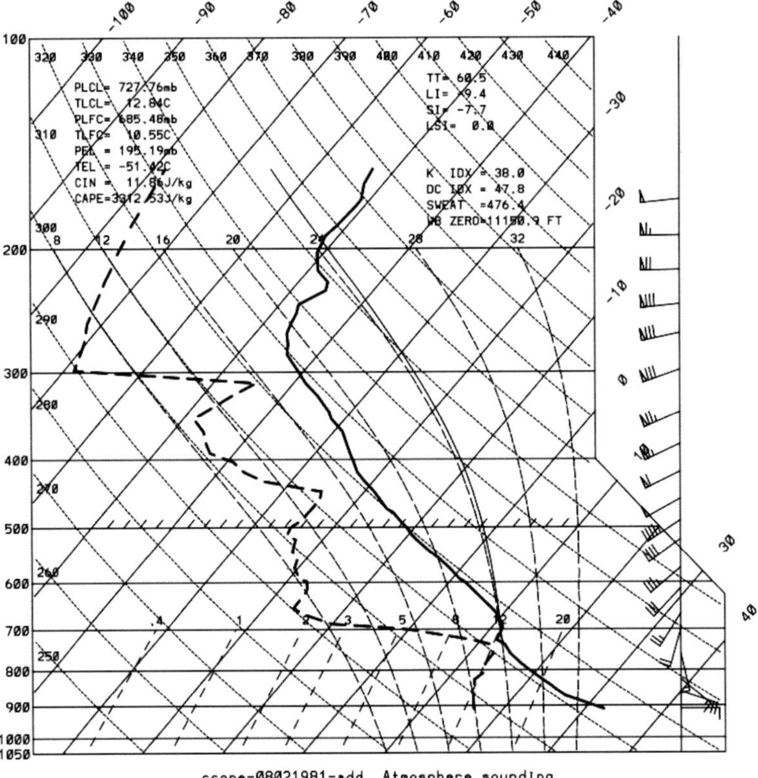

Figure 1. The 1746 MDT Knowlton, Montana sounding on 2 August 1981. The solid curve is for temperature and dashed curve for dew point. The portion of dew point curve above 300 hPa, which was missing in the original sounding, is constructed using an average August 1999 HALOE water vapor profile over 40–60N.

This is the same as that used by Johnson *et al.* (1993, 1995) and is a typical case for producing US High Plains supercells. In the present study, we performed three different runs:

(a) Full Physics Run (FPR): This is a simulation with the complete set of cloud microphysics, including both liquid and ice physics.

(b) Normal Liquid Run (NLR): This is a simulation in which the ice physics is suppressed. The only condensed phase is liquid and the latent heat released during the phase change is the latent heat of evaporation.

(c) Enhanced Liquid Run (ELR): As in the case of the NLR, this is a simulation in which the ice physics is suppressed and the only condensed phase is liquid. However, the latent heat released during the phase change is the latent heat of sublimation. This is to say that we artificially enhance the latent heat released in order to test the sensitivity of it on the lifespan of the supercell under consideration.

2.2. *Results and discussions*

The results of both the FPR and the NLR were reported by Johnson *et al.* (1993), and the main conclusion was that the suppression of ice processes in this supercell would shorten its lifespan dramatically. The reason for this phenomenon was thought to be the density difference between ice and liquid hydrometeors in the cloud. In the case of the FPR, the most abundant hydrometeors in the upper tropospheric level are the snowflakes. The lighter density of the flakes allows them to distribute in a much wider volume, and the eventual descending branch of the storm circulation occurs at much larger distances downstream from the ascending inflow of the unstable air. Thus, the descending and ascending branches of the storm circulation do not interfere destructively, but instead enhance each other, producing a long-lasting supercell. In contrast, the NLR results show that the

higher density of liquid hydrometeors, consisting largely of raindrops, makes them fall very close to the main ascending branch of the storm circulation, effectively cutting off the inflow of the unstable air necessary for the life of the storm. As a result, the NLR storm starts to dissipate only after 1.5 h into the simulation.

While the above explanation points to the density difference of hydrometeors as the main reason for the quick dissipation of the pure liquid NLR storm, there remains the possibility that the smaller latent heat release in this storm as compared to that in the FPR storm could also contribute to its shorter lifespan. This is the motivation for performing the ELR storm simulation, as mentioned previously. By artificially enhancing the latent heat release during condensation, we hope to eliminate the above-mentioned ambiguity.

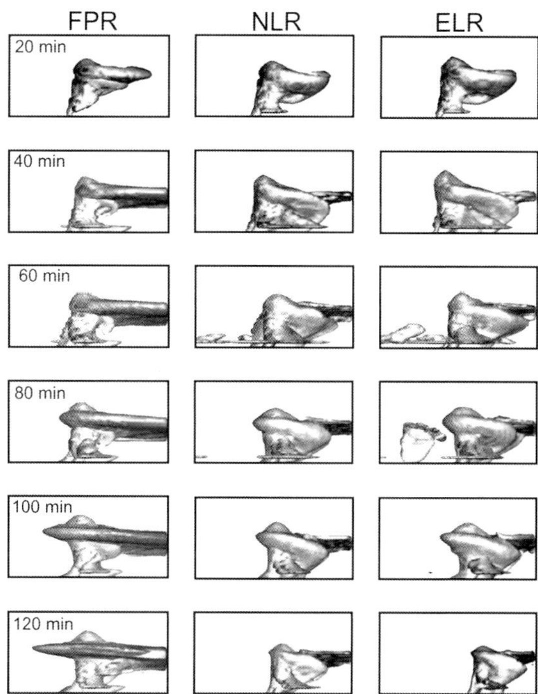

Figure 2. The rendered contour surfaces of hydrometeor mixing ratio $0.1\,\mathrm{g\,m^{-3}}$ of the three simulated storms as viewed from south for $t = 20$ to $120\,\mathrm{min}$. The vertical axis range is 0–$20\,\mathrm{km}$ and the horizontal range is 0–$55\,\mathrm{km}$.

Figure 2 shows a comparison of the development history of the three simulated storms from $t = 20$ to $120\,\text{min}$. It demonstrates that the ELR storm is the one that dissipates the earliest, and hence the enhanced latent heat obviously does not help prolong the lifespan of this storm; rather, it does the opposite. This establishes which the long life of the FPR storm is indeed due to the microphysical property of ice particles (their lighter densities) which influences the internal dynamics of the storm. It appears that the additional heat energy prompts the formation of a large amount of liquid hydrometeors faster than the NLR case, and hence cuts off the inflow of unstable air sooner, resulting in the quicker dissipation of the ELR storm.

Figure 3 shows the same development history but viewed from the top. As is consistent with the descriptions above, the FPR storm evolves into a steady state supercell and hence maintains an extensive anvil. Both NLR and ELR storms dissipate after more than $90\,\text{min}$; the anvils break up and become disorganized. We are conducting a more thorough analysis of all three simulated storms and will report the results in a future paper.

Our conclusions here appear to be consistent with earlier findings using 2D model simulations on squall line systems, as mentioned at the beginning of Sec. 2, although the two situations are not identical. Thus the role of ice processes in the overall lifespan of thunderstorms seems to be valid for these different storm types.

3. Impacts of Ice Crystal Habit on Cirrus Development

Thin cirrus clouds are highly important to climate studies. In spite of their tenuous appearance, cirrus clouds have a pronounced influence on climate because of their effect on the radiation (Ramanathan *et al.*, 1983). They are usually located high in the troposphere where the temperatures are cold. By virtue of their cold temperatures, they will interact strongly with the upwelling and downwelling infrared radiation in the atmosphere, as demanded by the Kirckhoff law. In addition, satellite observations indicate that cirrus clouds cover extensive areas of the Earth (e.g. Wylie and Menzel, 1989). These two factors together imply that cirrus clouds can significantly influence the radiative budget of the Earth–atmosphere system. The radiative effects of cirrus clouds can be highly variable because of the variability in thin radiative and microphysical properties. Either cooling or warming can occur, depending on the cloud radiative properties, cloud height, and its thermal contrast with the surface (e.g. Manabe *et al.*, 1965).

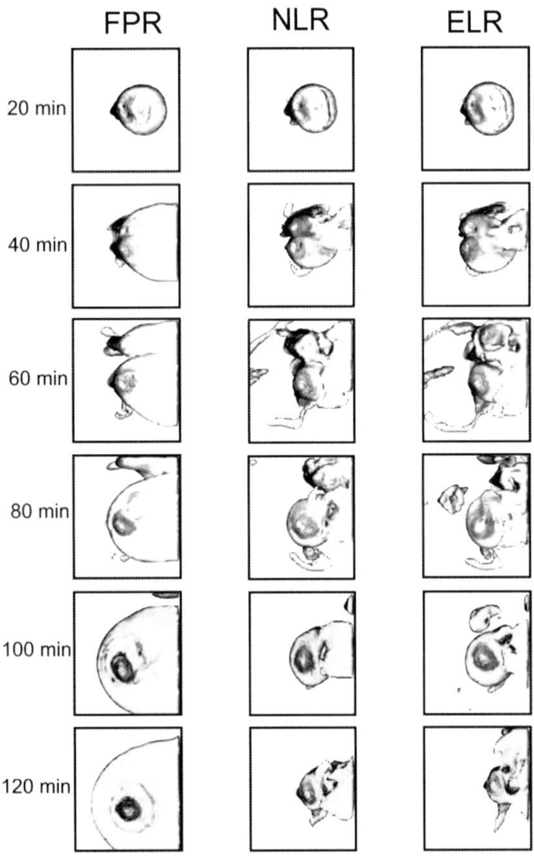

Figure 3. The rendered contour surfaces of hydrometeor mixing ratio $0.1\,\text{gm}^{-3}$ of the three simulated storms as viewed from top for $t = 20$ to $120\,\text{min}$. The horizontal ranges in both x and y-directions are 0–55 km.

Randall *et al.* (1989) performed simulation studies using a general circulation model, and showed that upper tropospheric clouds have dramatic impacts on large-scale circulation in the tropics, with the attendant effects on precipitation and water vapor amounts. Ramaswamy and Ramanathan (1989) also performed GCM studies, and suggested that the discrepancies between the previous simulations and observed upper tropospheric temperature structure in the tropics and subtropics can be explained by the radiative heating effects of cirrus cloud systems. These studies point out that cirrus clouds are likely to have great impacts on the radiation, and hence the intensity of the general circulation.

At present, the cloud forcing in GCMs is a major uncertainty factor. Cess *et al.* (1989) compared the outputs of 14 GCMs simulating an equivalent climate change scenario, and found that the results of global temperature change in response to an imposed sea surface temperature change were relatively uniform when clear sky conditions were assumed for radiative computations. The results were very different when the radiative effects of clouds were included. They also found that the effect of cloud feedback was comparable in magnitude to that due to imposed forcing, i.e. the change in sea surface temperature, but the sign could be positive or negative, depending on the model chosen. Needless to say, this does not help to build confidence in the model predictions, and there is an urgent need to reduce the uncertainty in the cloud radiative forcing in GCM and climate models.

The radiative properties of cirrus clouds depend on their microphysical characteristics: ice crystal size, concentration, habit, spatial distribution, etc. The uncertainty about the radiative properties comes from our inadequate understanding of the cirrus microphysical behavior and associated subgrid (unresolved) dynamics in GCMs. One way to improve our understanding of cirrus microphysics is to perform model studies — provided, of course, that the model is adequately realistic.

In the following, we will briefly summarize our findings that are relevant to the impact of ice microphysics on cirrus clouds. The focus will be on the impacts of ice habit on the cirrus development. We have developed a cirrus model for this study, and performed various sensitivity studies using the model. The details of the physics and mathematics of the model can be found in Liu *et al.* (2003a,b). In the following subsections, the model will only be briefly described, and the main results relevant to the present article will be summarized.

3.1. *The model physics*

The basic idea of the cirrus model used for the present study was derived from an earlier work by Starr and Cox (1985), but the details differ significantly. In this subsection, the main elements of the model will be discussed. The three physical processes that have been identified to be essential for the development of cirrus clouds are the *dynamical, microphysical and radiative processes*. Cirrus clouds often form during the lifting of moist air associated with large-scale motions. Small ice crystals are formed in the updraft. If the upward motion persists long enough to cause further cooling of the layer, ice crystals will grow to sizes with substantial fall velocities, and precipitation will occur. As ice crystals grow larger, the radiative effect becomes more significant. The resulting radiative heating profile will change the temperature lapse rate, and thus the dynamics in the cloud will be different. A change in the cloud dynamics will affect the microphysical processes to alter the size distribution of ice crystals. This, in turn, will further modify the radiative heating profiles within the cloud. It is clear that these processes are interactive. None of the three processes should be ignored in developing a cirrus model.

The dynamics model is a modified version of the dynamics framework used in the WISCDYMM, as described in the previous section. The main modification is the use of the six-order Crowley scheme (see Tremback *et al.*, 1987) to calculate the advection term for turbulent kinetic energy, water vapor and potential temperature. The numerical method used to calculate the advection of hydrometeors is the total variation diminishing (TVD) scheme, as described by Yee (1987).

The microphysical module focuses on the detailed ice microphysics. A double moment scheme is used to predict the evolution of the size distribution of ice crystals at each grid point. Both the mixing ratio and the number concentration of ice crystals are prognostic variables. The distribution mean diameter is then diagnosed from the mixing ratio and the number concentration. This is more realistic than predicting the mixing ratio only, as in most models with bulk microphysics. The growth rate of an ice crystal is explicitly calculated in this model. In the growth equation of ice crystals, both capacitance and ventilation coefficients are a function of ice crystal shape. Ventilation coefficients for different shapes of ice crystals that are commonly observed in cirrus are computed. The important microphysical process, homogeneous freezing nucleation, is included in our model since it has been recognized as a very effective source producing ice crystals in cirrus.

A radiation module is also implemented, as it is important to the development of cirrus clouds. Because we are examining optically thin cirrus clouds here, the mean free path for a photon colliding with a particle in cirrus is much larger than that in a typical stratocumulus; the radiative heating is distributed through the entire cirrus cloud body instead of being distributed like two Dirac functions with opposite signs at the cloud top and bottom as in a typical stratocumulus cloud (Ackerman *et al.*, 1988). Moreover, the volume absorption coefficient and the volume extinction coefficient are very sensitive to the ice crystal size distribution. As the cloud evolves, the change in the ice crystal size distribution causes changes in the radiative heating rates not only in the interior, but also below and above the cloud deck. It is therefore important to correctly represent the ice crystal optical properties. For this purpose, a modified anomalous diffraction theory (MADT) proposed by Mitchell (1996) is employed. Its parametrization is based on the physics of how incident rays interact with a particle. By using anomalous diffraction theory, analytical expressions are developed describing the absorption and extinction coefficients and the single scattering albedo as functions of the size distribution parameter, ice crystal shape, wavelength and refractive index. Therefore, the optical properties calculated are not based on an effective radius that has little physical meaning. Another advantage of the MADT is that the scattering properties in the thermal infrared spectral range can be explicitly calculated, so that the scattering is not ignored. The radiative fluxes are calculated using a two-stream model. More details of the cloud microphysics and radiative modules are given in Liu *et al.* (2003a,b).

3.2. *Model setup and initializations*

The model domain illustrated in Fig. 4 represents a cross-section of cirrus advected by a nonsheared mean wind. The initial supersaturated layer is about 1 km thick. Temperature in the supersaturated layer is randomly perturbed between $-0.02°C$ and $0.02°C$ to initialize the convective disturbance. The background vertical motion is set to $3\,cm\,s^{-1}$, a typical value for large scale lifting, which is uniformly imposed throughout the model domain at all times.

The model domain is 20 km wide and 6 km deep, with respective spatial resolutions of 200 m and 100 m. When a test was conducted with the grid mesh reduced to 100 m in the horizontal and 50 m in the vertical, the results did

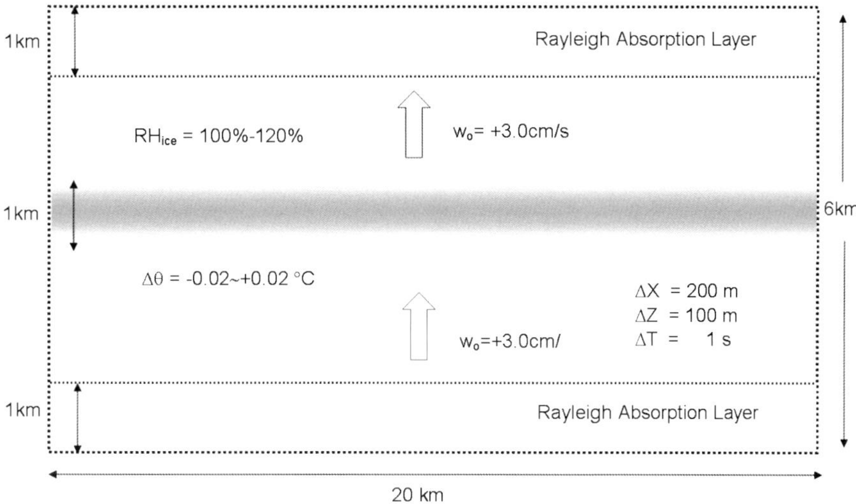

Figure 4. Schematic of the cirrus model domain (from Liu *et al.*, 2003b).

not show any obvious differences from those obtained using the $200 \times 100\,\mathrm{m}^2$ grid cell configuration. There are four different time steps used in this model. In the dynamic module, a time step of 1.0 s is imposed for the large time step, while 0.1 s is used for the small time step. In order to reduce the computational time spent on radiative transfer, the radiation time step is usually set larger than the dynamic time step (Lin, 1997). So, the radiative transfer module is evaluated at 30 s. This indicates that the radiative heating rates remain constant during this 30 s period. The cloud microphysics module was evaluated every 0.5 s.

Liu *et al.* (2003b) have summarized the many simulations performed, including four kinds of atmospheric background profiles: warm unstable, warm stable, cold unstable and cold stable. The warm cirrus profiles are based on the US Standard Spring Atmosphere at 45°N. The cold cirrus profiles are based on the US Standard Summer Atmosphere at 30°N. Surface temperature is 15°C for warm cirrus, and 31.4°C for cold cirrus. The background tropospheric temperature lapse rate is 6.5°C km^{-1} in the former standard profile, while there is more structure in the latter. The tropopause occurs at

10.5 km (-56.5°C) and 15.5 km (-76°C) respectively. The background tropospheric relative humidity is set to 40% with respect to a plane water surface. Temperature lapse rates for the warm unstable case are ice pseudoadiabatic from 8.0 to 8.5 km, and 1°C km^{-1} greater than ice-pseudoadiabatic from 8.5 to 9.0 km. Similarly, temperature lapse rates for the cold unstable case are ice pseudoadiabatic from 13 to 13.5 km, and 1°C km^{-1} greater than ice-pseudoadiabatic from 13.5 to 14 km. For both statically stable cases, the lapse rates in these layers are set to 8°C km^{-1}. Relative humidity with respect to ice is 100% at the base of the ice-neutral layer, and increases linearly with height to 120% at the base of the conditionally unstable layer, above which it is constant.

Some care must be taken regarding solar geometry, which is defined via a specification of latitude, date and initial local solar time (LST). We choose March 21 (vernal equinox) for the warm cirrus case, and June 21 (summer solstice) for the cold cirrus case, running the simulations from 1300 to 1600 LST. Since the simulations are run for 3 h, the corresponding solar zenith angle varies from 47.8° to 69.9° for the warm cirrus cases at 45°N, and from 14.9° to 53.4° for

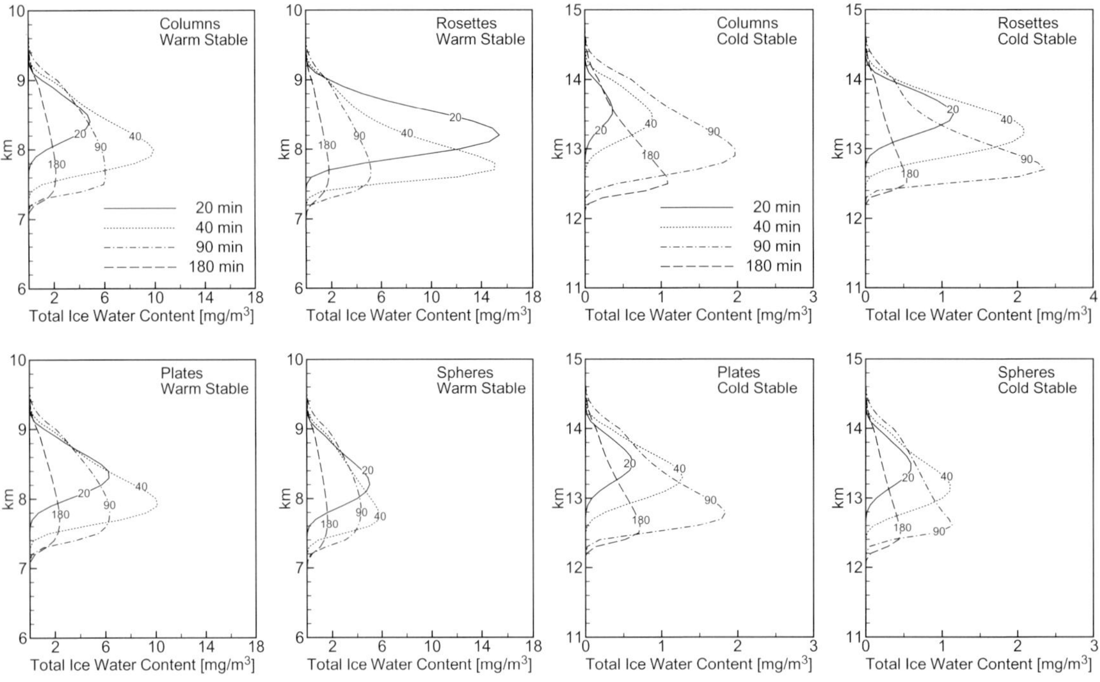

Figure 5. Profiles of horizontally averaged ice water content for four different ice crystal types. The warm stable case is shown in the left two columns, and the cold stable case is shown in the right two columns (adopted from Liu *et al.*, 2003b).

the cold cirrus cases at 30°N. In addition, we assume that the surface albedo is 0.2, approximating a climatological average. The input parameters for the simulation sets are summarized in Table 3 of Liu *et al.* (2003b).

3.3. *Results and discussions on the effect of ice crystal habit on the development of cirrus*

Four types of ice crystal are considered here: columns, plates, bullet rosettes and spheres. The first three types are commonly observed in cirrus clouds, while the last one is a simplified approximation to ice particles. To focus on the effect of habit, we will consider only the development of cirrus in cold and warm stable atmospheres, but with the aggregation process turned off because of its insignificance in the stable environment. The results for the cases of cirrus in warm and cold unstable atmospheres are similar, but with

different magnitudes. They will not be summarized here.

The sensitivity of cirrus development to ice crystal habit can be seen in Fig. 5 for warm and cold cirrus. For the warm stable cases, the peak amplitudes of the ice water content all occur around 40 min, but their values are strongly habit-dependent. The maximum ice water content for cirrus consisting of rosettes is more than twice as large as for spheres. These maxima are similar for columns and plates, and are larger than for spheres but smaller than for rosettes. The results for the cold stable case are similar except that the ice water content reaches its maximum value around 90 min.

The corresponding profiles of mean crystal size for the warm and cold stable cases are shown in Fig. 6. Rosettes are largest, followed by plates and columns, while spheres are smallest. The difference in mean size between rosettes and spheres is as large as fourfold.

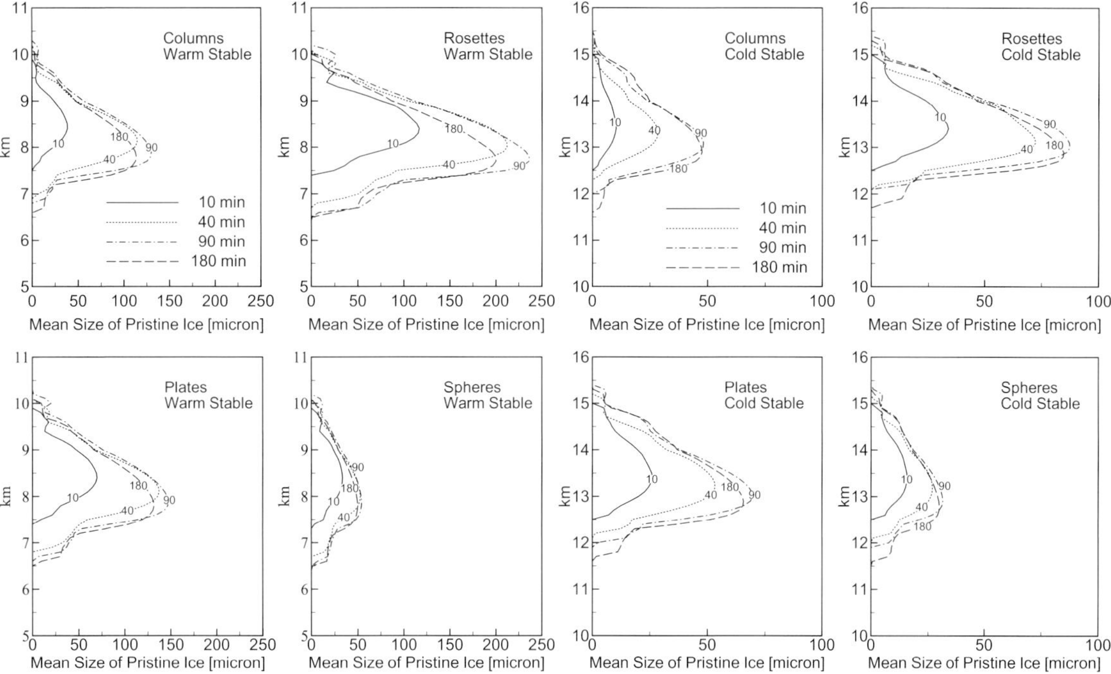

Figure 6. Profiles of horizontally averaged mean ice crystal size for four different habits. The warm stable case is shown in the left two columns, and the cold stable case is shown in the right two columns (adopted from Liu *et al.*, 2003b).

The corresponding radiative heating rates for the warm and cold stable cases are shown in Fig. 7. Although the ice water content profiles for columns and plates are similar (Fig. 5), the corresponding solar and IR heating rates are slightly larger for plates than for columns. It can be seen that the IR heating rate is more sensitive to ice crystal habit than the solar radiative heating rate is.

The differences among simulations with different ice habits may be explained on the basis of ice crystal capacitance. The growth rate of ice crystals is calculated based on the concept of electrostatic analogy as explained by Pruppacher and Klett (1997). In this analogy, the central quantity determining the ice growth rate is the capacitance, which is parametrized as a function of the maximum crystal diameter and habit in this model. A plot of the capacitance as a function of mass and habit is shown in Fig. 8. For ice particles with the

same mass, bullet rosettes have the largest capacitance, hence the greatest diffusional growth rate, followed by plates and columns, whereas spheres have the smallest. We have seen that the ice water content is generally largest for rosettes, with columns ranked second, plates third, and spheres last (Fig. 4). Although the capacitance of columns is larger than that of plates, the resulting ice water contents are almost the same. This may be because the differential radiative heating induced by plates is greater than that for columns, resulting in slightly stronger upward vertical motion for the plates, and thus producing more ice water content.

The above discussions have made it clear that assuming spherical particles in cirrus modeling would have resulted in seriously underestimating the radiative impact. To make an accurate assessment of the radiative impact of cirrus clouds, it is important to have good

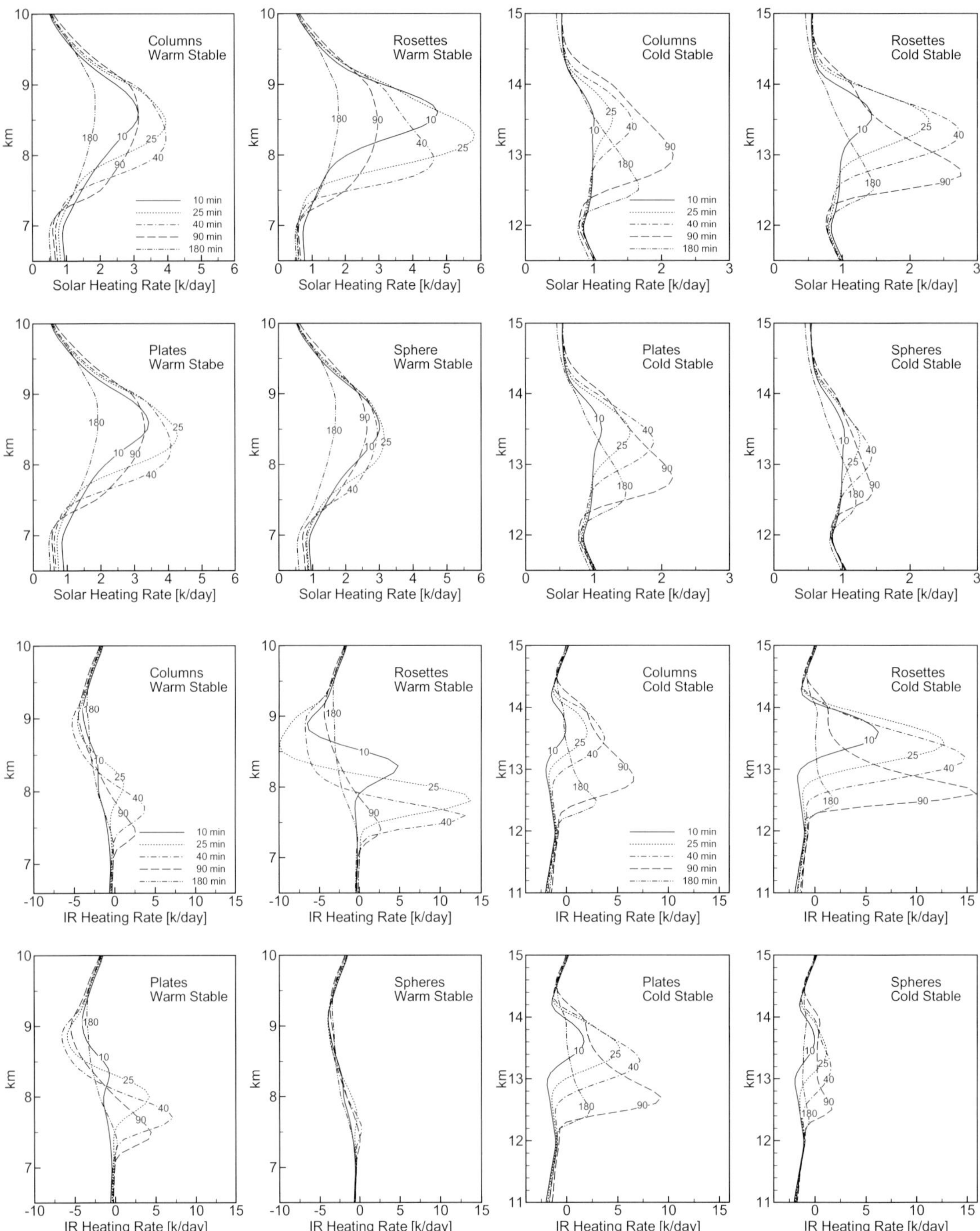

Figure 7. Profiles of horizontally averaged solar heating rates and IR heating rates for four different ice crystal types. The warm stable case is plotted in left two panels, and the cold stable case is shown in right two panels (adopted from Liu *et al.*, 2003b).

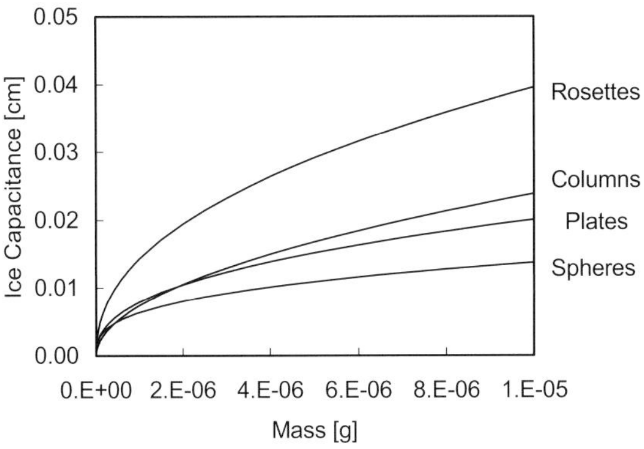

Figure 8. Ice crystal capacitance as a function of ice crystal mass and habit (adopted from Liu *et al.*, 2003b).

observational data on not only the size but also the shape distributions of ice crystals in them.

The domain-averaged ice water content and number concentration are shown in Fig. 9 for both the warm and cold stable cases. The effect of ice crystal habit is evident here, too. During the growing stage, rosettes tend to consume more water vapor and thus suppress the homogeneous nucleation process (the homogeneous freezing of haze solution droplets, as parametrized in Part I of the article). Consequently, the ice number concentration for rosettes is less than that for the other ice habits. However, the

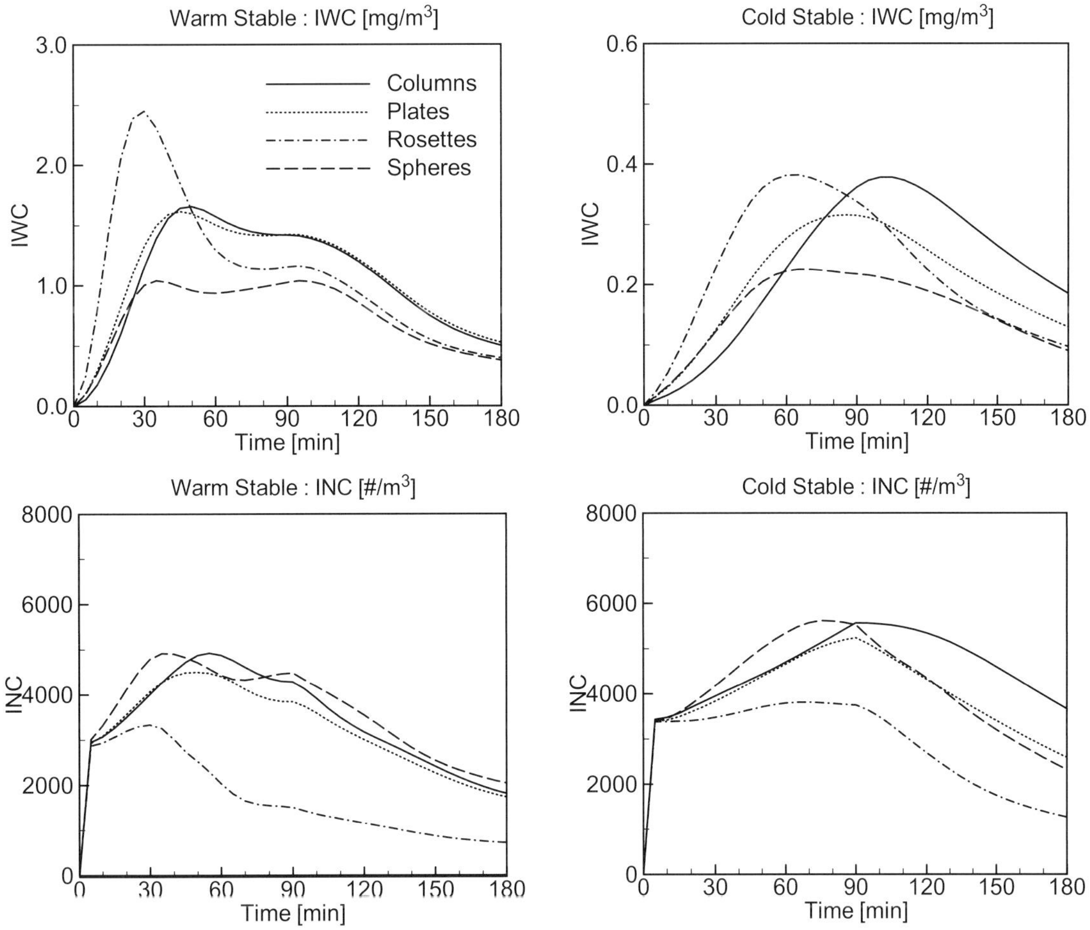

Figure 9. Domain averaged ice water content and number concentration for warm stable and cold stable case (adopted from Liu *et al.*, 2003b).

resulting ice water content still exceeds those for the other ice habits because their large growth rates produce correspondingly larger crystals. This is confirmed by the profiles of mean size for rosettes in Fig. 6, where the warm case at 10 min already shows a mean size in excess of $100\,\mu$m for rosettes, while the other ice types still average less than $70\,\mu$m.

3.4. *Summary and conclusions*

In summary, it has been found that the cloud development is very sensitive to ice crystal habit. This is mainly because the different ice crystal habits have distinctly different capacitances and optical properties, and all these factors contribute differently to ice crystal growth rates and hence to cloud development. The most important factor is the capacitance — among ice particles of a given mass, rosettes have the greatest capacitance and hence the largest growth rate, with successively smaller capacitances for columns, hexagonal plates and spherical particles. Bullet rosettes generally grow 2–3 times faster than other crystal types. Ice crystal habit influences the homogeneous nucleation process. Bullet rosettes consume large amounts of the water vapor available in the cloud layer, due to their largest growth rate and thus severely suppress the initial nucleation process. As a result, cirrus clouds consisting of bullet rosettes have significantly fewer ice crystals nucleated than those with other ice crystal types. The IR heating rate is also more sensitive to ice crystal habit than the solar heating rate is. The role of aggregation in the development of cirrus has also been examined; it tends to reduce the optical depth of the cloud and is thus likely to reduce the radiative destabilization of the cloud layer.

4. Capacitance of Ice Crystals

The results in the previous section indicate that the growth rate of ice crystals has great impacts on the behavior of cloud radiative and dynamical properties. Since in most cloud scale models, as in our cirrus model, the ice growth rate is calculated using the electrostatic analogy, the capacitance of the ice crystal is the central quantity in this analogy. Clearly, it is necessary to obtain the values of capacitances of ice crystals considered in the cloud scale model if their growth rates are to be determined accurately.

The capacitance can be determined either experimentally or theoretically. Experimentally, one can construct metallic models of ice crystals; charge them with electricity and then measure the capacitance of the model directly. Some earlier measurements were done this way (e.g. McDonald, 1963; Podzimek, 1966). Aside from the fact that such techniques are usually quite involved, it is difficult to make models that cover a wide variety of shapes and sizes of ice crystals in clouds. In the following, we report on the theoretical calculation results of ice capacitances performed recently by Chiruta and Wang (2003, 2005).

Chiruta and Wang (2003) gave an outline of this technique. To calculate the capacitance of an ice crystal (assumed to be an electrical conductor), we need to determine the electric potential distribution around an ice crystal. Obviously the potential has to satisfy the Laplace equation (no space charge needs to be considered). This is a differential equation, and boundary conditions are needed to uniquely determine the solutions. The surface potential is assumed to be known (and is usually taken as 1 in the nondimensional formulation), whereas the potential at infinity is 0. Since numerical techniques are used in most cases (because of the complicated crystal shape), the "infinity" simply means the outer boundary of the computational domain. In this way, the potential distribution around an ice crystal is determined. Integrating the potentials over a certain equipotential surface will give the total charge inside this surface (which, in reality, resides on the crystal surface). Knowing the total charge and the potential difference, we can determine the capacitance.

The central difficulty in this technique is how to specify the inner boundary surface which coincides with the crystal surface. For simpler shapes such as hexagonal columns and plates, it is feasible to map the boundary surface points directly onto the computational grids, as was done by Chiruta and Wang (2005). For more complicated shapes such as rosettes, it is convenient to use mathematical formulas to generate the surface points when such formulas are available. This was done by Chiruta and Wang (2003), who used the mathematical formula described by Wang (1999). In both papers, Chiruta and Wang (2003, 2005), the numerical technique used to solve the Laplace equation was the finite element technique.

In this way, Chiruta and Wang computed the capacitances of seven rosette ice crystals and nine hexagonal columns (both solid and

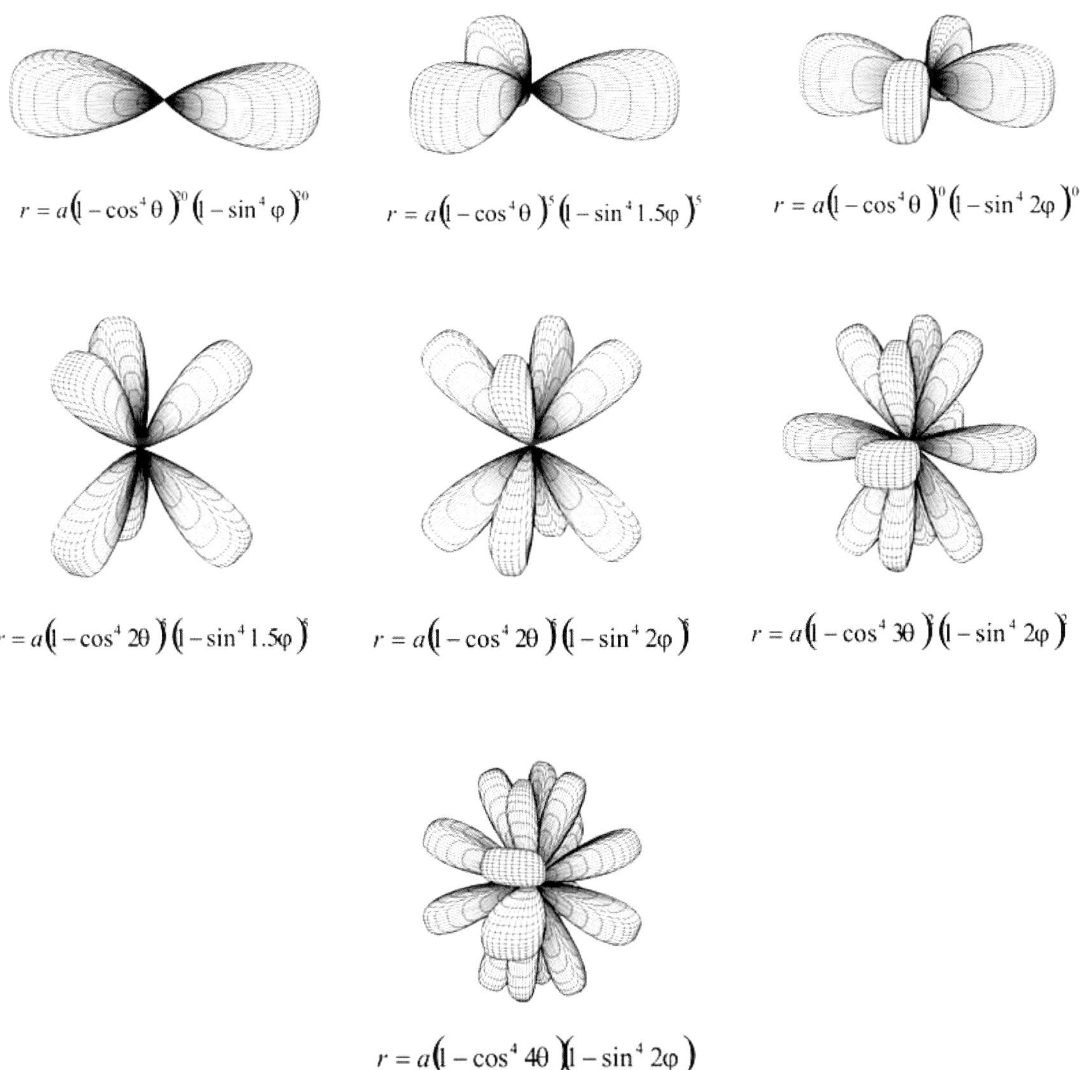

Figure 10. The seven bullet rosette ice crystals considered in Chiruta and Wang (2003). Each individual crystal is represented by the generating formula in spherical coordinates.

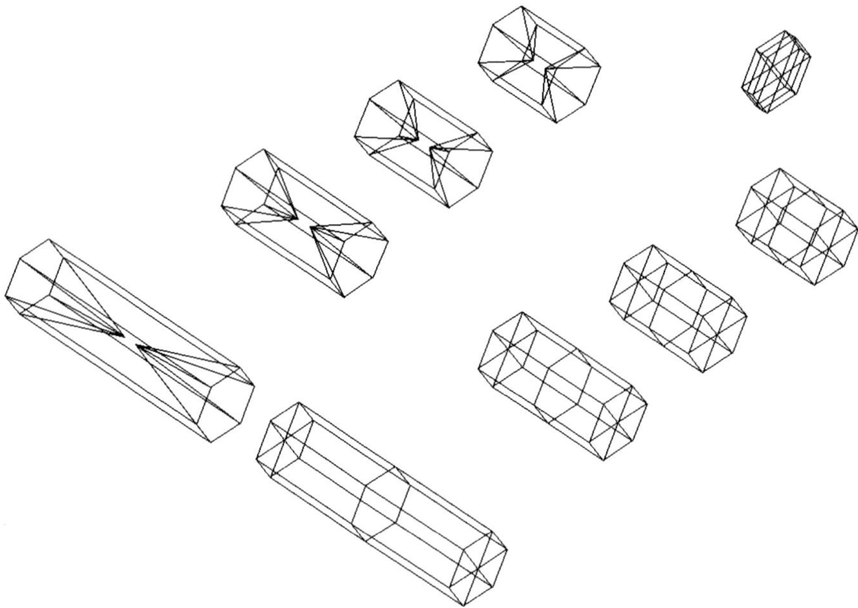

Figure 11. The nine simulated columnar ice crystals considered in Chiruta and Wang (2005). The hollow columns (left) and solid columns (right) have the same external dimensions. The small hexagonal disk (upper right) is a disk whose thickness is the distance between the tips of the two opposing cavities. Its capacitance is also calculated to serve as a reference (adopted from Chiruta and Wang, 2005).

hollow). Figures 10 and 11 show the seven bullet rosettes and nine solid and hollow columns that they treated in their papers. Geometrical properties of these ice crystals have been given in the above-cited papers.

Figures 12 and 13 show the calculated capacitances of the rosettes and columns, respectively. Figure 12 demonstrates that the capacitance of rosettes cannot be approximated by spheres, oblate or prolate spheroids. Rather, it is a nonlinear function of the number of lobes. The more lobes a rosette has, the closer is its capacitance to that of a sphere of the same radius. This seems to be logical, as one can imagine that a rosette with an infinite number of lobes would be effectively the same as a sphere. For rosettes with few lobes, the capacitances are smaller. This implies that their growth rates are smaller correspondingly, and may result in significant impact on the radiative properties of the cloud over an extended period.

Figure 13 demonstrates that the capacitance of a hollow hexagonal column is the same as that of a solid column of the same dimension and aspect ratio. This implies that their mass growth rates are the same. But since the hollow column has a smaller mass than the solid column, this implies that the hollow column will grow faster linearly (assuming that the main direction of growth is along the c axis, which is likely the case). The different linear dimensions of ice crystals would also result in different magnitudes of radiative interaction and hence different radiative properties of the cloud.

Naturally, the remarks made in the previous two paragraphs represent only the effect of the ice growth rate alone. To really assess the overall impacts, one needs to incorporate these results into a cloud scale model so that the interactions between the dynamics, radiation and cloud microphysics can be considered together. Our group is working on this research currently.

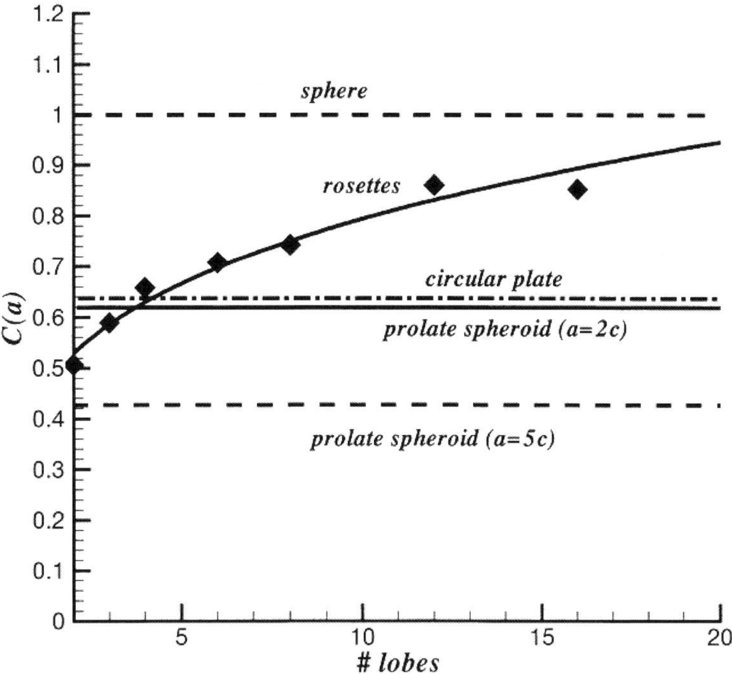

Figure 12. Computed capacitance (diamonds) of rosettes as a function of number of lobes. Thick solid curve represents power fit by Eq. (15). Also shown are the capacitances of a sphere (thin solid line), a circular plate (dashed), and two prolate spheroids with semi-axis ratio $a/b = 2$ and 5, respectively. The rosettes, the sphere and circular plate all have a radius a. The unit of capacitance is a (the radius of the crystal) (adopted from Chiruta and Wang, 2003).

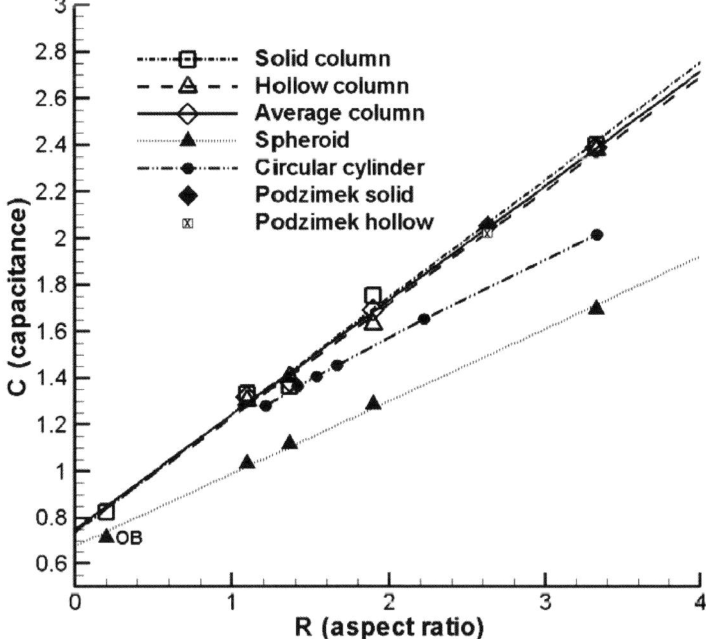

Figure 13. The capacitances of hexagonal ice columns, prolate and oblate (the one indicated by OB) spheroids, and circular cylinders as a function of the aspect ratio R. Experimental results are taken from Podzimek (1966). The unit of capacitance is a (the radius of the crystal) (adopted from Chiruta and Wang, 2005).

Acknowledgements

We are grateful for the support of NSF Grants ATM-0234744 and ATM-0244505. The lead author would also like to acknowledge the support of the National Science Council, Taiwan, for a summer visit to the National Taiwan University in 2005.

[Received 28 December 2005; Revised 22 August 2007; Accepted 27 September 2007.]

References

Ackerman, T. P., N. K. Liou, F. P. J. Valero, and L. Pfister, 1988: Heating rates in tropical anvils. *J. Atmos. Sci.*, **45**, 1606–1623.

Barth, M. C., and D. B. Parson, 1996: Microphysical processes associated with intense frontal rainbands and the effect of evaporation and melting on frontal dynamics. *J. Atmos. Sci.*, **53**, 1569–1586.

Cess, R. D., G. L. Potter, J. P. Blanchet, G. J. Boer, S. J. Ghan, J. T. Khiel, H. Letreut, Z. X. Li, X. Z. Liang, J. F. B. Mitchell, J. J. Morcrette, D. A. Randall, M. R. Riches, E. Roeckner, U. Schlese, A. Slingo, K. E. Taylor, W. M. Washington, R. T. Weatherald, and I. Yagai, 1989: Interpretation of cloud-climate feedback as produced by 14 atmospheric general-circulation models. *Science*, **245**, 513–516.

Chiruta, M., and P. K. Wang, 2003: On the capacitance of bullet rosette crystals. *J. Atmos. Sci.*, **60**, 836–846.

Chiruta, M., and P. K. Wang, 2005: The capacitance of solid and hollow hexagonal ice columns. *Geophys. Res. Lett.*, **32**, L05803, doi:10.1029/2004GL021771, 2005.

Heymsfield, A. J., *et al.*, 2005: Homogeneous ice nucleation in subtropical and tropical convection and its influence on cirrus anvil microphysics. *J. Atmos. Sci.*, **62**, 41–64.

Johnson, D. E., P. K. Wang, and J. M. Straka, 1993: Numerical simulation of the 2 August 1981 CCOPE supercell storm with and without ice microphysics. *J. Appl. Meteor.*, **32**, 745–759.

Johnson, D. E., P. K. Wang, and J. M. Straka, 1995: A study of microphysical processes in the 2 August 1981 CCOPE supercell storm. *Atmos. Res.*, **33**, 93–123.

Lin, R. F., 1997: A numerical study of the evolution of nocturnal cirrus by a two-dimensional model with explicit microphysics. PhD dissertation, (Department of Meteorology, Pennsylvania State University), 198 pp.

Lindzen, R. S., M. D. Chou, and A. Y. Hou, 2001: Does the earth have an adaptive infrared iris? *Bull. Amer. Meteor. Soc.*, **82**, 417–432.

Liou, K. N., 2002: *An Introduction to Atmospheric Radiation.* (Academic), 853 pp.

Liu, C., M. W. Moncrieff, and E. J. Zipser, 1997: Dynamical influence on microphysics in tropical squall lines: A numerical study. *Mon. Wea. Rev.*, **125**, 2190–2210.

Liu, H. C., P. K. Wang, and R. E. Schlesinger, 2003a: A numerical study of cirrus clouds. Part I: Model description. *J. Atmos. Sci.*, **60**, 1075–1084.

Liu, H. C., P. K. Wang, and R. E. Schlesinger, 2003b: A numerical study of cirrus clouds. Part II: Effects of ambient temperature and stability on cirrus evolution. *J. Atmos. Sci.*, **60**, 1097–1119.

Manabe, S., J. Smagorinsky, and R. F. Strickler, 1965: Simulated climatology of a general circulation model with a hydrologic cycle. *Mon. Wea. Rev.*, **93**, 769–798.

McDonald, J. E., 1963: Use of the electrostatic analogy in studies of ice crystal growth. *Z. Angew. Math. Phys.*, **14**, 610–620.

Miller, L. J., Tuttle, J. D., and Knight, C. A., 1988. Airflow and hail growth in a severe northern High Plain supercell. *J. Atmos. Sci.*, **45**, 736–762.

Mitchell, D. L., 1996: Use of mass- and area-dimensional power laws for determining precipitation particle terminal velocities. *J. Atmos. Sci.*, **53**, 1710–1723.

Phillips, A. T. J., A. Pokrovsky, and A. Khain, 2007: The influence of time-dependent melting on the dynamics and precipitation production in maritime and continental storm clouds. *J. Atmos. Sci.*, **64**, 338–359.

Podzimek, J. 1966: Experimental determination of the "capacity" of ice crystals. *Studia Geophys. et Geodet.*, **10**, 235–238.

Pruppacher, H. R., and J. D. Klett, 1997: *Microphysics of Clouds and Precipitation.* (Kluwer), 954 pp.

Ramanathan, V., E. J. Pitcher, R. C. Malone, and M. L. Blackmon, 1983: The response of a spectral general circulation model to refinements in radiative processes, *J. Atmos. Sci.*, **40**, 605.

Ramaswamy, V., and V. Ramanathan,1989: Solar absorption of cirrus clouds and the maintenance of the ropical upper troposphere thermal structure. *J. Atmos. Sci.*, **46**, 2293–2310.

Randall, D. A., Harshvadan, D. A. Dazlich, and T. G. Gorsetti, 1989: Interactions among radiation, convection, and large-scale dynamics in a general circulation model. *J. Atmos. Sci.*, **46**, 1943–1970.

Starr, D. O'C., and S. K. Cox, 1985: Cirrus clouds. I, A cirrus cloud model. *J. Atmos. Sci.*, **42**, 2663–2681.

Tao, W. K., J. Simpson, and S. T. Soong, 1991: Numerical simulation of a subtropical squall line over the Taiwan Strait. *Mon. Wea. Rev.*, **119**, 2699–2723.

Tremback, C. J. Powell, W. R. Cotton, and R. A. Pielke, 1987: The forward-in-time upstream advection scheme: Extension to higher order. *Mon. Wea. Rev.*, **115**, 540–555.

Wade, C. G., 1982: A preliminary study of an intense thunderstorm which moved across the CCOPE research network in southeastern Montana. *Proc.*

Ninth Conf. on Weather Forecasting and Analysis (Seattle, WA, Am. Meteor. Soc, Boston, MA, USA, 1982), pp. 388–395.

Wang, P. K., 1999: Three-dimensional representations of hexagonal ice crystals and hail particles of elliptical cross-sections. *J. Atmos. Sci.*, **56**, 1089–1093.

Wang, P. K., 2003: Moisture plumes above thunderstorm anvils and their contributions to cross tropopause transport of water vapor in midlatitudes. *J. Geophys. Res.*, **108(D6)**, 4194, doi: 10.1029/2003JD002581, 2003.

Wylie, D. P., and P. Menzel, 1989: Two years of cloud cover statistics using VAS. *J. Clim.*, **2**, 380–392.

Yee, H. C., 1987: Construction of explicit and implicit symmetric refinements in TVD schemes and their applications. *J. Comput. Phys.*, **68**, 151–179.

A Review of Ozone Formation in Megacities of East Asia and Its Potential Impact on the Ozone Trends

Shaw Chen Liu

*Research Center for Environmental Changes, Academia Sinica and
Department of Atmospheric Sciences, National Taiwan University, Taipei, Taiwan
shawliu@rcec.sinica.edu.tw*

In this work, we review the formation of high levels of ozone in megacities of East Asia and their potential impact on the increasing trends of tropospheric ozone concentration. A series of intensive observation experiments were conducted in Kaohsiung, Taipei and Taichung by scientists of the Research Center for Environmental Changes (RCEC) of Academia Sinica, in collaboration with colleagues from the National Taiwan University and the National Central University, to study the ozone formation and the strategy for its control in 2003–2005. In addition, RCEC scientists participated in three large-scale international experiments led by Peking University in the Pearl River Delta (PRD) and in Beijing in 2004 and 2006. A one-dimensional model and an observation-based method (OBM) were used to analyze data from these experiments to examine the photochemical processes of ozone formation and the relationship of ozone to its precursors. We found that the O_3 production rate was NMHCs-limited, and that controlling NMHCs was more efficient than controlling NO_x in reducing ozone levels in all the megacities studied.

The increasing trends of ozone formation at the background stations in Taiwan, Hong Kong and Japan from the 1980s to 2005 strongly suggest that photochemical production of ozone in Asia has been increasing due to anthropogenic emissions of ozone precursors. In addition, we believe that the increasing trend of ozone concentration in Mauna Loa (4.1% per decade between 1973 and 2004) could be a good indicator/measure and be useful for inverse-modeling the trend of the background ozone level of the entire Asia and even the Northern Hemisphere.

1. Introduction

Tropospheric ozone is an important gas in many aspects. Near the surface, ozone is a major air pollutant because at high concentrations it is hazardous to human health and can damage vegetation. In addition, tropospheric ozone plays a critical role in atmospheric chemistry because it controls the oxidation capacity of the atmosphere, both directly and indirectly through the generation of hydroxyl radicals (OH). It is well known that OH is responsible for the oxidation of the majority of reduced gases in the atmosphere. For example, CO, H_2 and CH_4 and most nonmethane hydrocarbons (NMHCs) are all oxidized by OH radicals. Ozone is also an important greenhouse gas because it absorbs infrared radiation near $9.6\,\mu$m, in the window of carbon dioxide and water vapor. According to the IPCC Report (2001), changes in tropospheric ozone since the industrial revolution (i.e. 1750–2000) have made it the third-most-important greenhouse gas, behind carbon dioxide and methane.

Tropospheric ozone was thought to come mostly from the stratosphere and to be deposited at the surface (Regener, 1949) until the early 1970s, when Crutzen (1973) and Chameides and Walker (1973) proposed that photochemical production of ozone can be important. In the presence of OH, CO (or NMHC) and NO_x, O_3 can be produced via the following simplified catalytic reaction scheme:

$$OH + CO(NMHC) + O_2 \rightarrow HO_2 + CO_2 + RO_2$$
$$HO_2(RO_2) + NO \rightarrow NO_2 + OH(RO)$$
$$NO_2 + h\upsilon \rightarrow NO + O$$
$$O + O_2 + M \rightarrow O_3 + M$$
$$Net : CO(NMHC) + 2O_2 + h\upsilon$$
$$\rightarrow CO_2 + O_3 + (RO)$$

NMHC, CO and NO_x are called precursors of O_3. The OH radicals needed to initiate the ozone production reactions are generated through ozone's photolysis by solar ultraviolet radiation at a wavelength of less than 320 nm:

$$O_3 + h\upsilon(< 320 \, nm) \rightarrow O(1D) + O_2$$

The electronically excited atomic oxygen O(1D) is highly reactive. It can react with water vapor and form the hydroxyl radicals:

$$O(1D) + H_2O \rightarrow 2OH$$

The key step in the simplified ozone production scheme is the $HO_2(RO_2) + NO$ reaction, which splits the O_2 bond. The $HO_2(RO_2) + NO$ reaction is the so-called rate-limiting process of O_3 production in most of the troposphere, because NO_x has a shorter lifetime than the average lifetime of hydrocarbons and CO. The lifetimes of CH_4 and CO are so long that there is plenty of CO and CH_4 in the background troposphere to produce O_3. As a result, NO_x is the rate-limiting precursor of O_3 (i.e. it controls the production rate of O_3) in most parts of the troposphere. In fact, NMHC becomes the rate-limiting precursor only in some urban and suburban areas where concentrations of NO_x are high and the ratios of NMHC to NO_x are low

(e.g. Seinfeld and Pandis, 1998; Blanchard et al., 1999).

There is a significant diffused source of NO_x from lightning, such that the concentration of NO_x in the background troposphere is often greater than the nominal breakeven level of NO_x of about 10 pptv, at which the gross photochemical O_3 production equals the gross photochemical O_3 loss. In addition to the lightning source, there is a small but significant stratospheric source of NO_x (and NO_y) in the upper troposphere, making the gross O_3 production frequently comparable to the gross O_3 loss in the middle and the upper troposphere. As a result, the "apparent lifetime" of ozone, which is defined as $O_3/(\text{gross production} - \text{gross loss})$, is usually very long. It can last for many months in the middle and the upper troposphere. Once O_3 is produced or transported into the middle and the upper troposphere, it can be transported like an inert species over a long distance.

The gross loss of ozone can be calculated from the following reactions:

$$OH + CO + O_2 \rightarrow HO_2 + CO_2$$
$$H + O_2 + M \rightarrow HO_2 + M$$
$$HO_2 + O_3 \rightarrow OH + 2O_2$$
$$Net: CO + O_3 \rightarrow CO_2 + 2O_2$$

The estimated value of the gross loss of ozone for the Northern Hemisphere is about $50 \times 10^{28} \, S^{-1}$. This is about four times the downward flux of ozone from the stratosphere, and is thus a key basis for arguing that photochemical processes play a controlling role in the tropospheric budget of ozone (Fehsenfeld and Liu, 1993).

Changes of ozone ranging from the local to the regional and even the global scale have been reported extensively (e.g. Lee et al., 1998; Oltmans et al., 1998; Jacob et al., 1999; Akimoto, 2003; Parrish et al., 2004). The various scales of changes occur because of the relatively long photochemical lifetime of ozone in the atmosphere which ranges from a few days in the boundary layer to a few months in the upper troposphere. For example, Jacob et al.

(1999) suggested that tripling Asian anthropogenic emissions from 1985 to 2010 would lead to an increase in the monthly mean O_3 concentration by 2–6 ppbv in the western US and by 1–3 ppbv in the eastern US. Akimoto *et al.* (2003) discussed that intercontinental transport of O_3 and hemispheric ozone pollution jeopardized agriculture and ecosystems worldwide and had a strong effect on climate. Both studies emphasize that international initiatives to mitigate global air pollution from both developed and developing countries are necessary. Observational evidence of a regional increase of O_3 over East Asia is estimated to be 7–10 ppb from the 1970s to the 1980s (Lee *et al.*, 1998). Also, the increase in Asian continental outflows is suggested to account for the increase of O_3 in the northern Atlantic Ocean region (Parrish *et al.*, 2004).

Rapid economic development in Asia in the last few decades has brought about an unprecedented level of prosperity. On the other hand, the environmental price of prosperity is also overwhelmingly high. As a result Asian cities are among the most-polluted metropolitan areas in the world. High levels of surface O_3 in major metropolitan areas are of particular concern because of their impact on a large segment of the population. For example, in the Beijing–Tianjin area, the Pearl River Delta (PRD) and the Yangtze Delta, pollution episodes with O_3 concentration of over 100 ppb have been frequently observed (Shao *et al.*, 2006; Wang *et al.*, 2003; Zhang *et al.*, 1998). In addition, satellite images of high levels of anthropogenic aerosols, CO, NO_2 and HCHO are strikingly visible over large areas in Asia (Kunhikrishnan *et al.*, 2004). Because NO_2, CO and HCHO are the precursors of ozone and both CO and HCHO are surrogates of NMHCs, it is not surprising to see a large regional scale of high concentrations of ozone detected from the satellite (Ziemke *et al.*, 2006). These regional scale high O_3 concentrations eventually link up with the previously isolated high levels of O_3 in other megacities of East Asia, such as Tokyo, Osaka and Taipei (Wakamatsu *et al.*, 1996; Chou *et al.*, 2006), and form the East Asian ozone plume, which, in combination with the ozone in South Asia, can have a potential impact on the entire Northern Hemisphere. Therefore it is extremely important to understand the ozone formation processes and determine the rate-limiting precursor in the megacities of East Asia.

In this work, we review the formation of high levels of ozone in major metropolitan areas or megacities of East Asia and explore their possible impacts on the long term increasing trends of tropospheric ozone concentration in East Asia. The review is based on recent research efforts of scientists at the Research Center for Environmental Changes (RCEC) of Academia Sinica, in collaboration with colleagues from the National Taiwan University (NTU) and the National Central University (NCU), as well as a large group of international scientists coordinated by Peking University in the last few years. A series of intensive observation experiments were conducted in Taiwan centered in Kaohsiung, Taipei and Taichung by scientists of the RCEC, NTU and NCU to study the ozone formation and its strategy for control in 2003–2005. In addition, RCEC scientists participated in three large scale international experiments led by Peking University in the Pearl River Delta and in Beijing in 2005 and 2006. In particular, the experiment in Beijing in August 2006 was designed to address air pollution problems in preparation for the 2008 Olympics. We will use field experiments conducted in southern Taiwan, a one-dimensional model and an observation-based method (OBM) as an example to examine the relationship between O_3 and its precursors (Shiu *et al.*, 2007). The OBM is then applied to similar field experiments conducted in the Pearl River Delta and in Beijing (Chang *et al.*, 2008a,b). The results are important for understanding the budget of ozone and developing an effective ozone control strategy for megacities in East Asia as well as for the whole region.

On the long term trends of tropospheric ozone concentration, we will first examine the trends of ozone formation observed in Taiwan and compare them to trends observed in other areas of East Asia. We will show that long range transport of ozone from the Asian continent plays a key role in the ozone distribution over East Asia. Furthermore, we have noticed that the increasing trend of ozone formation observed in Mauna Loa, Hawaii, from 1973 to 2004 is consistent with the trends in East Asia. Given the long lifetime of ozone in the free troposphere, the trend of ozone formation observed in Mauna Loa may have an important implication not only for East Asia but also for the entire Northern Hemisphere.

2. Ozone Formation in Megacities

Southern Taiwan at the Kaohsiung urban center is a multitown "megacity" spanning over $30\,km \times 50\,km$, with more than 6 million inhabitants. Oxidants such as ozone and PAN have become a serious environmental problem. In addition to large vehicle emissions, there are significant emission sources from four industrial parks nearby. During autumn and winter when high pressure conditions prevail, severe deteriorations of air quality with O_3 over 120 ppbv frequently occur in this region.

2.1. *Field measurements*

Two intensive field experiments were conducted by the RCEC in the Kaohsiung–Pingtung (KaoPing) area in southern Taiwan during 26–29 September 2003 and 23–28 October 2003 (hereafter referred to as the RCEC 2003 Experiment). NMHCs were collected daily with stainless canisters side by side with the air quality instruments at 12 Taiwan EPA (TEPA) air quality monitoring stations in two periods, namely 9–10 a.m. and 11 a.m. to 12 noon, during the 10-day field campaign. The measurements between 11 a.m. and 12 a.m. were intended to examine the NMHCs' composition just before

the peak of ozone, which usually occurs between 12 noon and 1 p.m. Measurements between 9 a.m. and 10 a.m. were designed to study the composition of NMHCs near the start of the daily photochemical cycle to get a measure of their ozone production potential for that day. The EPA monitoring stations have routine hourly measurements of O_3, CO, NO_x, SO_2, PM10 and meteorological parameters. Detailed sampling strategy and geographical information can be found in Chang *et al.* (2005).

Ozone is measured using an ultraviolet photometric instrument with a precision of 1 ppbv. The oxides of nitrogen (NO and NO_2) are measured using a chemiluminescence instrument. The manufacturer's suggested value of precision is 0.4 ppbv. However, our independent evaluation of the accuracy showed that the signal-to-noise level was about 1 at 3 ppbv for NO, significantly worse than the manufacturer's suggested value (Chou *et al.*, 2006). Similar accuracy was found for NO_2. This can cause serious problems for NO measurements, as NO values frequently dip below 5 ppbv during photochemically active hours. It is not a serious problem for NO_2, because its values are usually greater than 10 ppbv. On the other hand, it is well known that the catalytic converter converting NO_2 to NO suffers interference from non-NO_2 compounds such as PAN and even HNO_3. This problem will be elaborated later. A total of 56 C_2–C_{10} NMHCs were analyzed by a GC/MS system. It is estimated that species not resolved by the GC/MS system are about 30–60% of the 56 NMHCs in terms of reactivity toward OH (Chang *et al.*, 2003). A detailed description regarding the accuracy of the GC/MS instrument can be found in Chang *et al.* (2003).

Meteorological parameters observed by the Central Weather Bureau station in Kaohsiung showed that the weather was fair during the two field experiments and there was a typical land–sea breeze with westerly to northwesterly winds in the daytime, and easterly to northeasterly winds at night. The diurnal variation of wind speed was consistent with the influence of the land–sea breeze. The daytime maximum temperature during the first period of the experiment (26–29 September 2003) was around 32°C, and 28°C during the second period (23–28 October 2003). The atmospheric moisture contents for the second period were lower than those for the first period.

Hourly time series of concentrations of O_3, NO, NO_2, CO and PM10 for three EPA stations are shown in Fig. 1. Two urban stations, Lernwu and Zhaoin, are located in the Kaohsiung city, while Chaochou is a suburban station in Pingtung County. The day-to-day variations of O_3 for different stations show a similar pattern, but the primary pollutants like CO, NO and NO_2 distributions change greatly from station to station. Concentrations of NO, NO_2 and CO for Chaochou are usually smaller than those for Lernwu and Zhaoin because of a low level of industrial activities and a smaller population density.

2.2. *One-dimensional model*

A one-dimensional model is developed to help with the analysis and interpretation of measurements. It is an updated version of the 1D photochemical model developed by Trainer *et al.* (1987, 1991). Photochemical mechanisms and reaction rate constants are updated following those in Wang *et al.* (2000). The vertical

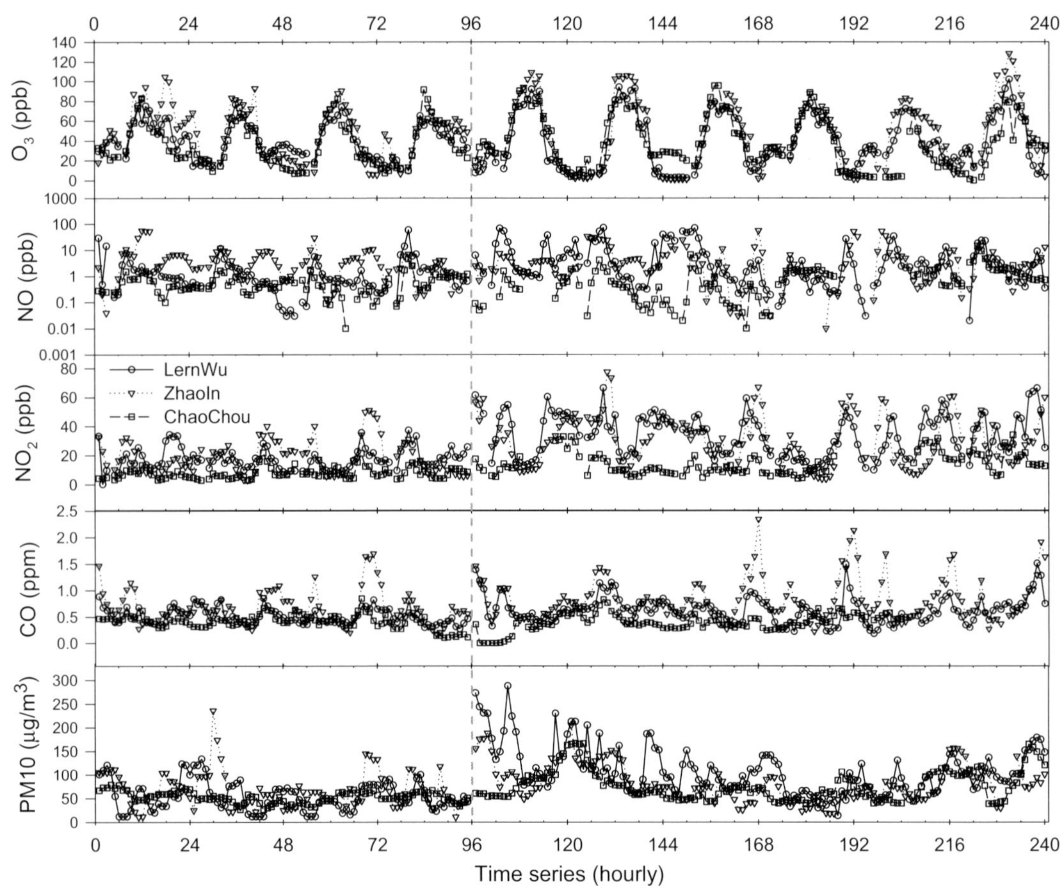

Figure 1. Hourly time series of O_3 NO, NO_2, CO and PM1O for the two experiments are shown for three stations: Lernwu, Zhaoin and Chaochou.

resolution of the model decreases from $2\,m$ near the surface to $1\,km$ at a $12\,km$ altitude. Species with a photochemical lifetime shorter than 10 min are calculated by neglecting the transport. The time step for model integration for transport and chemistry is 30 s. Initial vertical distributions of trace species are prescribed by assuming constant mixing ratios equal to observed values in the mixed layer. The initial values of short-lived species in the mixed layer are not important to the model results, because emissions and photochemistry quickly redefine the concentrations after a few days of model simulations. Above the mixed layer, mixing ratios of trace gases are assumed to be negligible except for CO and O_3, which are assumed to be their observed seasonal average values. The value of the eddy diffusion coefficient (K_z) in the surface layer is calculated from the similarity theory (Trainer *et al.*, 1987). The value of K_z above the surface layer is specified to be $60\,m\,s^{-2}$. Dry deposition and heterogeneous removal processes are included in the same way as in Trainer *et al.* (1987). The surface resistance is set at $2\,s/cm$ for O_3 and peroxides, $0.1\,s/cm$ for HNO_3 and $3.3\,s/cm$ for carbonyls and organic nitrates. Lifetimes of heterogeneous losses of soluble gases HNO_3, H_2O_2, ROOH, CH_3OOH, CH_2O, HCOOH, CH_3COOH, SO_2, SO_4 and CRESOL are assumed to be 5 days (Logan *et al.*, 1981). Results of the layer of $16\,m$ height, at which most surface measurements were made, are used to make a comparison with measurements at surface stations and the OBM results.

The emission rates of O_3 precursors are based on the Taiwan Emission Data System (TEDS 5.0, 2000) published by the TEPA (http://www.aqmc.org.tw). Area and line sources in the KaoPing area are averaged to obtain the average emission rates. The emission rates of NO_x, CO and SO_2 are calculated to be $3.4 \cdot 10^{16}$, $5.0 \cdot 10^{16}$ and $3.0 \cdot 10^{15}$ molecules $m^{-2}\,s^{-1}$, respectively. The emission rates of ethane, propane, butane, ethane, propene

and toluene are assumed to be $3.4 \cdot 10^{15}$, $2.5 \cdot 10^{15}$, $2.7 \cdot 10^{15}$, $5.9 \cdot 10^{15}$, $8.4 \cdot 10^{15}$ and $2.6 \cdot 10^{15}$ molecules $m^{-2}\,s^{-1}$, respectively. The isoprene emission rate is fixed at $3.6 \cdot 10^{15}$ molecules $m^{-2}\,s^{-1}$. The solar zenith angle is set to be that of September 30, at $22.5°N$. A few weeks' difference in the zenith angle is negligible for photodissociation rates. The photodissociation rate constants are generated by the Tropospheric Ultra-Violet (TUV) radiation model (TUV 4.2; Madronich and Flocke, 1998). The climatological values of total ozone observed by the Brewer ozone meter at Chengkung station ($121.3°E$, $23.1°N$, close to the latitude of northern Kaohsiung County) and an aerosol optical depth of 0.4 at $550\,nm$ are used. The model calculation is carried out for 3 days to reach a quasi-steady state condition. Atmospheric concentrations at 11 a.m.–12 noon of the day 4 simulation are used to make a comparision with the OBM results.

2.3. *Observation-based method/model*

Following the approach developed by Liu *et al.* (1987) and Trainer *et al.* (2000), the net production rate of O_3, $P(O_x)$, can be expressed as

$$P(O_x) = \frac{d[O_x]}{d[NO_x]}\frac{d[NO_x]}{dt} \approx \frac{\Delta[O_x]}{\Delta[NO_x]}\frac{\Delta[NO_x]}{\Delta t}$$

$$= \varepsilon\frac{\Delta[NO_x]}{\Delta t}. \tag{1}$$

The net O_x production derived from observation of $\Delta[O_x]$ is equal to the gross O_x production minus the gross photochemical loss, deposition and dispersion of O_x.

$$\varepsilon = \frac{\Delta[O_x]}{\Delta[NO_x]} \tag{2}$$

is the ozone production efficiency. It is the number of O_3 molecules produced per molecule of NO_x consumed photochemically and a key factor in understanding the photochemical formation of O_3 (Liu *et al.*, 1987; Trainer *et al.*, 2000). Typical values of ε over the US are in the range of 1–$20\,mol/mol$ (Jacob, 1999).

The most straightforward way to evaluate the net production rate of O_3 is to use a model. However, modeling results usually contain large uncertainties that exist in the emission inventories, meteorology parameters, photochemical processes, etc. which are essential components of all models. In the following, we use the data from field measurements to derive the ozone production efficiency and $P(O_x)$. This is an observation based method/model (OBM) which can reduce the uncertainties by using observations to bypass some of the processes, including emission inventories, photochemical processes and meteorology parameters.

In the daytime, the major removal process of NO_x is the oxidation of NO_2 by OH that produces HNO_3. HNO_3 is readily scavenged by gas-to-particle conversions, washout or dry deposition. A minor sink of NO_x is the formation of peroxyacetylnitrate (PAN). However, PAN can undergo rapid thermal decomposition and regenerate NO_x at the prevailing temperature during the experiments. Nighttime reactions of NO_3 and N_2O_5 are significant but neglected because the region studied in this work has substantial fresh emissions in the morning and we are mainly dealing with the production of O_3 a few hours after the emissions of O_3 precursors. Therefore we consider the reaction of NO_2 with OH as the only removal process for NO_x and assume that the removal of NO_x is pseudo-first-order, as shown below. In this case, following a Lagrangian trajectory, we have

$$[NO_x] = [NO_x]_0 \exp\left(-k \int_0^t [OH]dt\right), \quad (3)$$

where k is the reaction rate constant for NO_x with OH. The reaction rate constant for NO_2 with OH is $1.04 \cdot 10^{-11}\,\mathrm{cm^3\,s^{-1}}$ at 25°C and one atmosphere pressure (Sander $et\ al.$, 2002). Since NO_2 is part of the NO_x, the value of k should be scaled down by the ratio NO_2/NO_x. The average of the NO_2/NO_x ratio is about 0.6, and thus k for NO_x is prescribed at $6.0 \cdot 10^{-12}\,\mathrm{cm^3\,s^{-1}}$.

The ratio of two hydrocarbons which have the same emission sources but different reactivities with OH can be used as a measure of photochemical oxidation by OH (Calvert, 1976; Singh $et\ al.$, 1977; Robert $et\ al.$, 1984; Parrish $et\ al.$, 1992; Mckeen and Liu, 1993; Kramp $et\ al.$, 1997). We choose a pair of aromatic hydrocarbons, ethylbenzene and m,p-xylene, for this purpose (Chang $et\ al.$, 2007). Again following the Lagrangian trajectory, we have

$$\frac{E}{X} = \left(\frac{E_0}{X_0}\right) \exp\left(-\int_0^t (k_e - k_x)[OH]dt\right), \quad (4)$$

where E and X represent concentrations of ethylbenzene and m,p-xylene at time t, respectively; E_0 and X_0 are their corresponding initial concentrations; k_x and k_e are their reaction rate constants with OH, and are equal to $2.17 \cdot 10^{-11}\,\mathrm{cm^3\,s^{-1}}$ and $7.0 \cdot 10^{-12}\,\mathrm{cm^3\,s^{-1}}$, respectively (Atkinson, 1990). The value of the ratio E_0/X_0 is about 0.4 for the emission sources in southern Taiwan. This value is usually observed in the early morning, around 7–8 a.m. In fact, even around 9–10 a.m., when we collect the first daily VOC samples, the E/X ratios are fairly close to 0.4 because the photochemical activities are relatively weak before 9–10 a.m. Using the observed value of E/X at 11 a.m. –12 noon, $\int_0^t [OH]dt$ between early morning and 11 a.m.–12 noon can be calculated from Eq. (5). Then, using observed values of NO_x at 11 a.m.–12 noon, initial values of NO_x in the early morning (along the backward Lagrangian trajectory) can be calculated from Eq. (4), and the photochemically consumed values can be obtained from the difference between initial values and the values at 11 a.m.–12 noon. Similarly, NMHCs consumed by OH and initial values can be calculated from Eq. (5) (Chang $et\ al.$, 2007).

$$[NMHCs] = [NMHCs]_0 \exp\left(-k_i \int_0^t [OH]dt\right),$$

$$(5)$$

where k_i is the reaction rate constant of the individual NMHC reaction with OH. It is important to note that any NMHCs derived from Eq. (5) along the backward Lagrangian trajectory, e.g. initial NMHCs and consumed NMHCs, are a sum of OH reactivity-weighted NMHCs, not a straight sum of NMHCs.

2.4. *Ozone production rate*

Values of the ozone production rate, $P(O_x)$, derived from Eq. (1) are shown as a function of NO_x concentration in Fig. 2. OBM-based $P(O_x)$ in the KaoPing area can range from a few to 10 ppbv/h when the NO_x concentration is less than 10 ppbv. When the NO_x concentration is increased to 30 ppbv, $P(O_x)$ can be as large

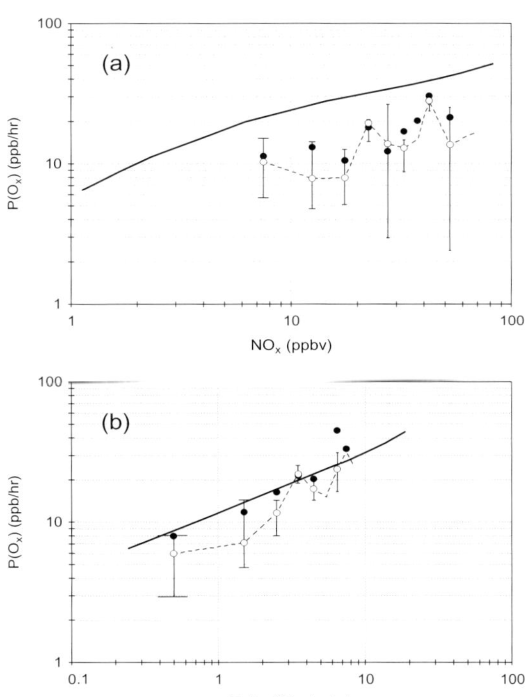

Figure 2. The median values and 75th percentile of the total oxidant production rate derived from Jacob (1999) (dotted line and bars) and from the OBM (dashed line and bars) are plotted as a function of the NO_x concentration. Also shown is the ozone production rate calculated by the ID model (solid line). (b) Same as (a) but plotte against ΔNO_x.

as 30 ppbv/h. Because the data for OBM are from 12 different sites in the KaoPing area that include upwind, urban center and downwind sites, the scattered distribution of $P(O_x)$ vs NO_x is expected. $P(O_x)$ can also be calculated from the Jacob equation. Values of $P(O_x)$ from this method agree very well with those of the OBM approach. There are some outliers estimated by the OBM with low values less than 2 ppbv/h. As discussed earlier, they are likely the result of fresh emissions.

Also shown in Fig. 2 is the 1D-model-simulated $P(O_x)$ versus NO_x. The median values of $P(O_x)$ obtained by the OBM and the Jacob equation are about 60% and 40% smaller than model-simulated values, respectively.

We have compared our values of $P(O_x)$ from the OBM with those derived by Frost *et al.* (1998). Unfortunately, our values are at much higher concentrations of NO_x than those of Frost *et al.* (1998), as theirs are aircraft measurements while ours are from surface stations in urban and suburban areas in the morning between 9–10 a.m. and 11 a.m.–12 noon. It is well known that the top of the mixed layer usually grows in the morning and does not reach the top of the boundary layer of the previous day ($\sim$1.5 km) until around noon. As a result, surface stations tend to be significantly influenced by fresh emissions in the morning. This is evident from the observed decreasing concentrations of CO, NO_x and NMHCs in the morning at nearly all stations. Part of the decrease of NO_x and NMHCs is due to photochemical consumption; the decreases of CO is due to the dilution of fresh emissions when the mixed layer grows in thickness.

$P(O_x)$ simulated by our 1D model is very similar to that of Frost *et al.* (1998) below a few ppbv of NO_x. Above a few ppbv of NO_x, $P(O_x)$ in the model of Frost *et al.* (1998) levels off and starts to decrease, while $P(O_x)$ in our model continues to increase with NO_x (Fig. 2). One reason for the difference is that their NMHC mix has significantly higher isoprene which does not

446 Shaw Chen Liu

increase with NO_x. Another reason is probably that we have a more reactive mix of NMHCs. In this context, we note that Kleinman *et al.* (2002a) compared $P(O_x)$ vs NO_x in five urban areas in the US and found that the peak $P(O_x)$ occurred at a NO_x concentration of about a few to 20 ppbv for those cities. At highly polluted sites, $P(O_x)$ was larger than 30 ppbv/h and decreasing values of $P(O_x)$ with increasing NO_x were not observed, consistent with results of our 1D model and the OBM.

From the discussion on the ozone production efficiency, it is clear that $P(O_x)$ should be related to the consumed NO_x rather than NO_x itself. In Fig. 2, $P(O_x)$ is plotted against ΔNO_x, the consumed NO_x between 9–10 a.m. and 11 a.m.–12 noon. As expected, the correlation of $P(O_x)$ with ΔNO_x improves significantly because ΔNO_x is a better measure of the photochemical activity. Values of $P(O_x)$ from the OBM agree well with those from the 1D model at ΔNO_x greater than 3 ppbv. Below 3 ppbv of ΔNO_x, values of $P(O_x)$ from the OBM are about 50% lower. Considering the uncertainties involved in all three methods, the discrepancies are surprisingly small. In fact, the high level of agreement between the 1D model and Jacob's approach is likely to be fortuitous, as the latter is the gross photochemical production of O_x, while the former is the net O_x photochemical production. The OBM values are expected to be less than those of the other two approaches, because the OBM also includes the loss of O_x due to processes such as heterogeneous loss, dispersion and surface deposition of O_3 and NO_2. The effect of these processes is expected to be more significant at low production rates of O_x, consistent with the feature in Fig. 2 that $P(O_x)$ from the OBM starts to become lower than for the other two approaches at about 5 ppbv of ΔNO_x.

The agreement of $P(O_x)$ from Jacob (1999) with the 1D model is even better. Given the independent nature of the three methods in evaluating $P(O_x)$, the significant level of agreement among them provides strong support for the validity of all three methods.

2.5. *Ozone–precursors relationship*

A widely used approach to formulating the ozone control strategy is the so-called empirical kinetic modeling approach (EKMA) or ozone isopleth diagrams (e.g. Milford *et al.*, 1989; Gery and Crouse, 1990; Altshuler *et al.*, 1995; Seinfeld and Pandis, 1998). The EKMA diagram usually depicts daily peak O_3 mixing ratios as a function of initial concentrations or emission rates of O_3 precursors. An example is given in Fig. 3, which shows peak O_3 concentrations calculated for various reduction scenarios in NO_x and NMHC emissions in southern Taiwan (Chang, G.-H., private communication, 2004). One can easily see that reduction of NMHC emission while holding the NO_x emission constant (i.e. sliding to the left along the top horizontal edge) is quite effective in lowering the ozone level. On the other hand, reducing NO_x emission while holding NMHC emission constant (i.e. sliding down along the right vertical edge) will raise the ozone level.

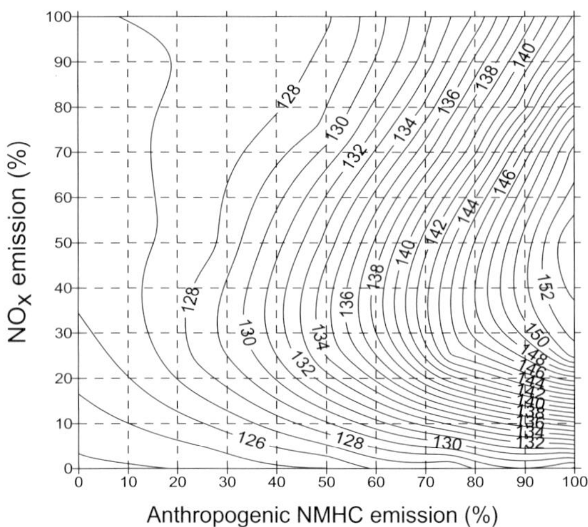

Figure 3. Peak O_3 concentrations calculated for various reductions in NO_x and NMHC emissions in southern Taiwan. 100% represents current emissions.

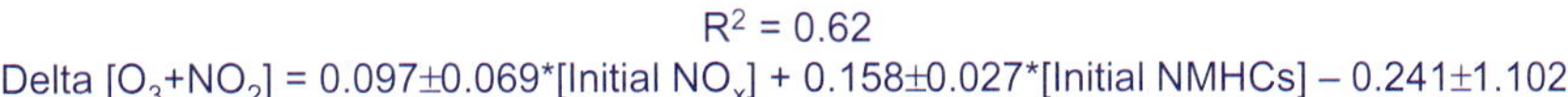

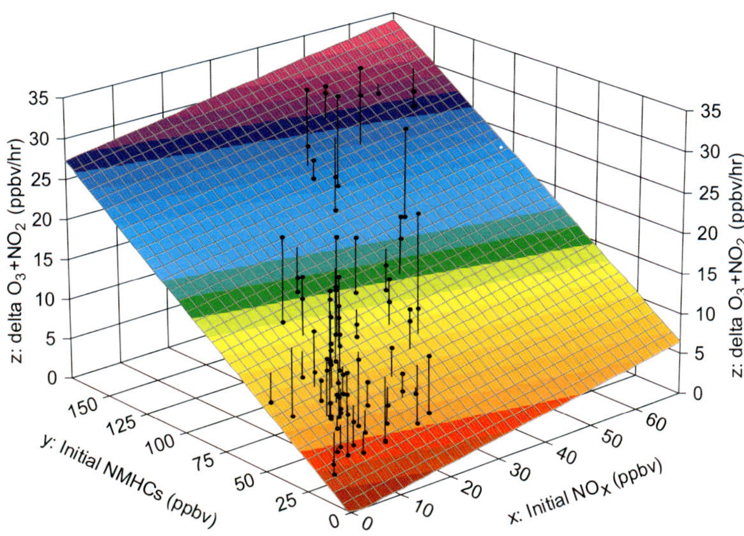

Figure 4. The total oxidant production rate, i.e. $\Delta[O_3 + NO_2]$ between 10 a.m. and 12 noon (black dots), is plotted in a three-dimensional space as a function of initial concentrations of NO_x and NMHCs. The shaded plane is a linear surface fit to the black dots. The equation listed represents the surface as a function of initial NO_x and NMHCs. R^2 is the correlation coefficient of the fit. Lines between the black dots and the plane denote the deviation from the plane.

Results from the last section can be applied to formulating ozone pollution control strategies in a similar way to EKMA. Figure 4 shows $\Delta[O_3 + NO_2]$, or the O_x production rate, as a function of initial NO_x and initial NMHC concentrations. $\Delta[O_3 + NO_2]$ is calculated from $[O_3 + NO_2]$ at 11 a.m.–12 noon minus the values at 10–11 a.m. and 9–10 a.m. As with the earlier calculation, the initial concentration of NO_x at each site is derived from the consumed NO_x plus the concentration of NO_x observed at 11 a.m.–12 noon. Initial NMHCs are calculated similarly. Consumed CO, CH_4 and isoprene are also taken into account. In addition effects of NMHCs not resolved in the GC are included by multiplying the 56 NMHCs by a factor of 1.5.

Figure 4 is similar to the concept of EKMA. In fact, it is equivalent to a three-dimensional OBM ozone production rate isopleth diagram as a function of initial concentrations of ozone precursors. Initial concentrations are used because they are directly proportional to the emissions of ozone precursors.

In order to obtain a simple relationship between the O_x production rate and the precursors of O_3, we have made a flat surface (linear) fit to data points of RCEC 2003. Multivariate statistical regression is used to evaluate whether the correlations versus initial NO_x and NMHCs have statistical significance as separate entities and uncertainty ranges associated with the slopes versus initial NO_x and NMHCs. The correlation coefficient (R^2) for the surface fit is 0.62. The slopes of $\Delta[O_3 + NO_2]$ versus initial NO_x and NMHCs are 0.097 ± 0.069 and 0.158 ± 0.027, respectively. The statistical result of the F test shows that the regression surface is statistically significant and the value of the slope is significant against initial NMHCs but not against initial NO_x. This lack of significance can weaken the conclusions in the following interpretation of Fig. 4.

The slightly greater slope against NMHCs suggests that NMHCs are slightly more effective than NO_x in controlling the production rate of the total oxidant. Moreover, because NO_2 is a part of the total oxidant O_x, reducing NO_x emissions will reduce the titration of O_3 and thus increase O_3. For example, a 20 ppbv reduction in the initial concentration of NO would reduce the titration of O_3 by 20 ppbv, which could be converted to about 5 ppbv/h of the ozone production rate over the morning hours ($\sim 4\,h$). In comparison, the reduction of the O_x production rate for the 20 ppbv reduction in the initial concentration of NO is about $0.11 \cdot 20 = 2.2$ ppbv/h. This means that the production rate of O_3 (not O_x) will increase by 2.8 ppbv/h, i.e. the O_3 concentration will increase when the NO_x initial value is decreased.

The effect of titration on O_3 in KaoPing has been substantiated by an independent study (Chou *et al.*, 2006) on the trends of urban O_3 levels at three major metropolitan centers in Taiwan in 1993–2003. The study showed that urban O_3 levels in Taiwan increased significantly between 1993 and 2003 due to reduced titration by NO when NO_x emissions were reduced. Figure 5 illustrates this point for the KaoPing area. One can see that nearly 90% of the increase of O_3 in 1993–2003 can be accounted for by the reduction in NO_2, i.e. by the reduction of titration. We also use the 1D model to generate a surface similar to the diagram in Fig. 4 (not shown). The slope of O_x against initial NMHCs is more than a-factor-of-3 greater than the slope against initial NO_x, consistent with the OBM result. Therefore the balance of evidence suggests that the O_3 production rate is likely to be NMHCs-sensitive and that controlling NMHCs is more efficient than controlling NO_x in reducing urban ozone production.

The effectiveness of our OBM can be demonstrated by comparing Fig. 4 with Fig. 6, in which we replace the initial concentrations of precursors with the concentrations of precursors observed at the daily maximum values of $O_3 + NO_2$. In addition, the vertical coordinate is replaced with the daily maximum values of $O_3 + NO_2$ instead of $\Delta[O_3 + NO_2]$. One can see that no meaningful relationship can be derived. The change to $\Delta[O_3 + NO_2]$ does not affect the value of the correlation coefficient. The difference between Fig. 6 and Fig. 4 clearly shows that the daily maximum concentrations of the total oxidant are photochemically controlled by the initial early morning abundances of precursors which are derived by the OBM along Lagrangian trajectories, but are affected little by the concurrent concentrations of precursors.

2.6. *Applying the OBM to Beijing and PRD experiments*

A plot corresponding to Fig. 4 for observations at the station at Peking University during the summer 2006 campaign is shown in Fig. 7 (Chang *et al.*, 2008a). It can be seen that the O_x production at the Peking University station is even more NMHCs-sensitive as the slope of delta O_x against initial NO_x is negative, meaning that O_x production will decrease with NO_x. The increased sensitivity to NMHCs at Peking University compared with southern Taiwan is probably due to the 50%-smaller ratio

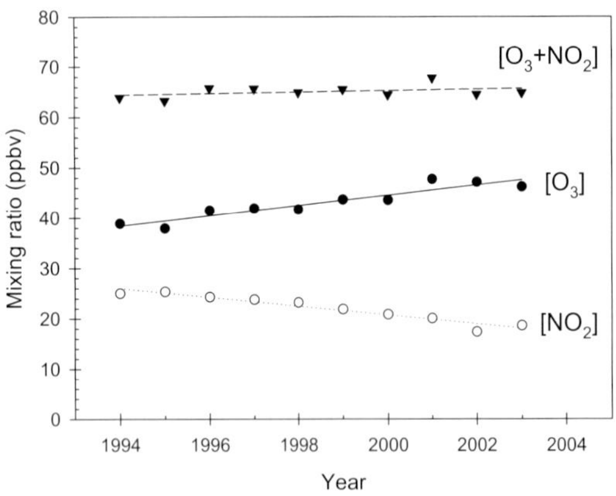

Figure 5. Linear trends of $O_3 NO_2$ and $O_3 + NO_2$ in the KaoPing area from 1994 to 2003. The increase in O_3 is nearly equal to the decreased titration by NO.

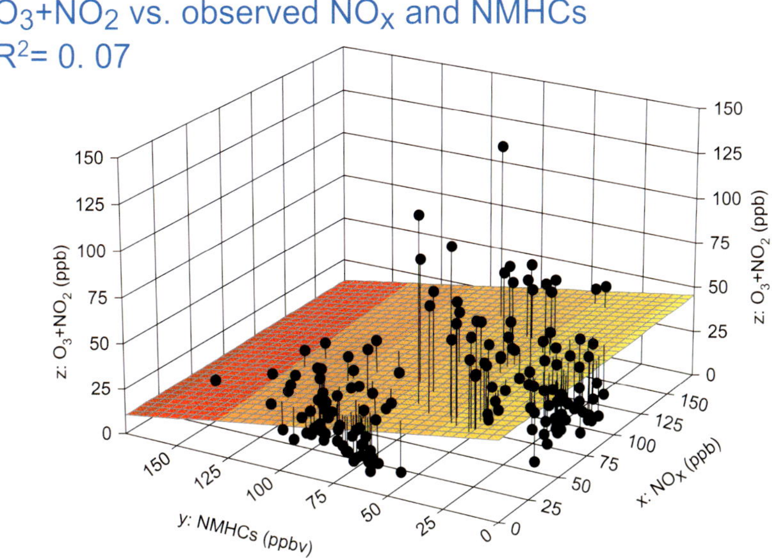

Figure 6. Same as Fig. 4 except that the horizontal coordinates are changed to O_3 precursors observed at the daily maximum values of $O_3 + NO_2$, which are plotted as the vertical coordinate.

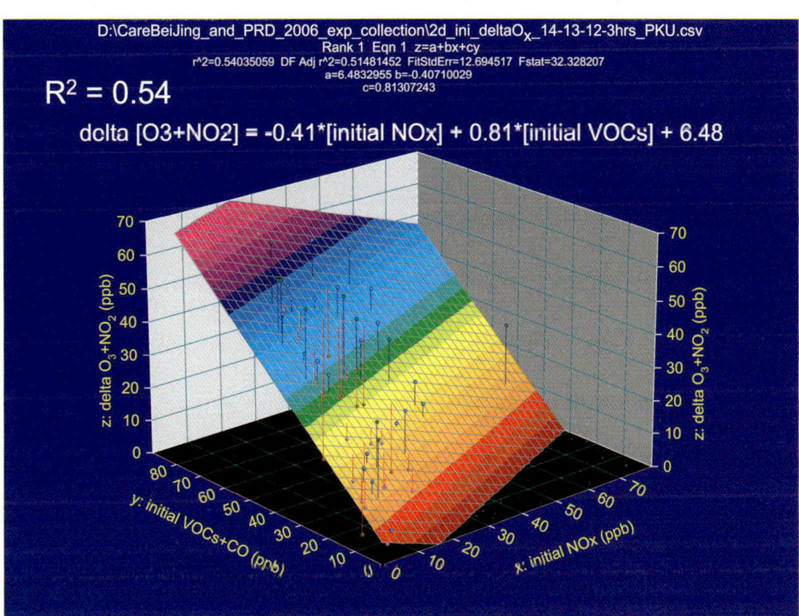

Figure 7. Same as Fig. 4 except for the Peking University station during the summer 2006 campaign.

of NMHC/NO$_x$ at the former station, which makes photochemical activity more limited by the availability of NMHCs than NO$_x$. Therefore controlling the emissions of NMHC and CO would be the most effective strategy for the control of ozone levels during the 2008 summer Olympics, while controlling NO$_x$ would be counterproductive.

Similar efforts for other stations, including two stations in PRD, Guangzhou and Backgarden, have yielded results that are very close to those of the Peking University station. However, the correlation coefficients are significantly worse than for the KaoPing and Peking University cases. This is shown for the case at the Backgarden station in Fig. 8 (Chang *et al.*, 2008b). The flat plane fit to the points has slopes nearly the same as those at the Peking University station, indicating similar sensitivity to NMHCs. However, it is not statistically meaningful, because the correlation coefficient (R^2) is only 0.24. Fortunately, when the horizontal coordinates are changed to photochemically consumed NMHC and NO$_x$ (Fig. 9), the correlation coefficient (R^2) increases to 0.65, supporting the sensitivity to NMHCs. The increase in the correlation coefficient is expected because the photochemically consumed precursors are more closely related to the photochemical formation of O$_3$ than the initial concentrations of precursors. This is another piece of evidence showing that the OBM is an effective method for examining the relationship between ozone and its precursors.

The Backgarden station is in a rural area about 65 km north of Guangzhou, yet concentrations of O$_3$ precursors in the morning are around 30 ppbv, only about a factor of 2 less than those of the Guangzhou station. In fact, the Backgarden station has nearly the same concentrations of NMHCs and NO$_x$ as Peking University, which is only about 10 km from the center of Beijing. So it is not surprising that the Backgarden station is still in the NMHCs-sensitive

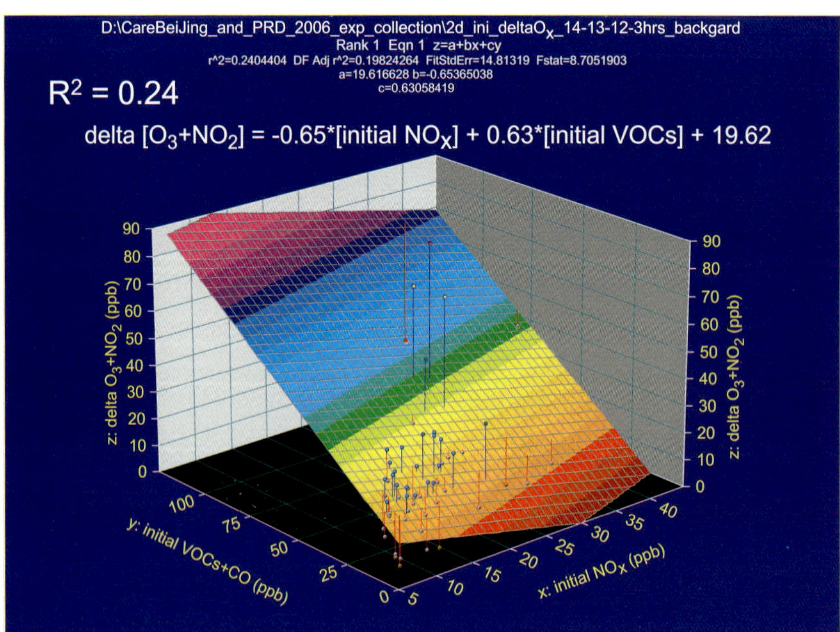

Figure 8. Same as Fig. 7 except for the Backgarden station in PRD.

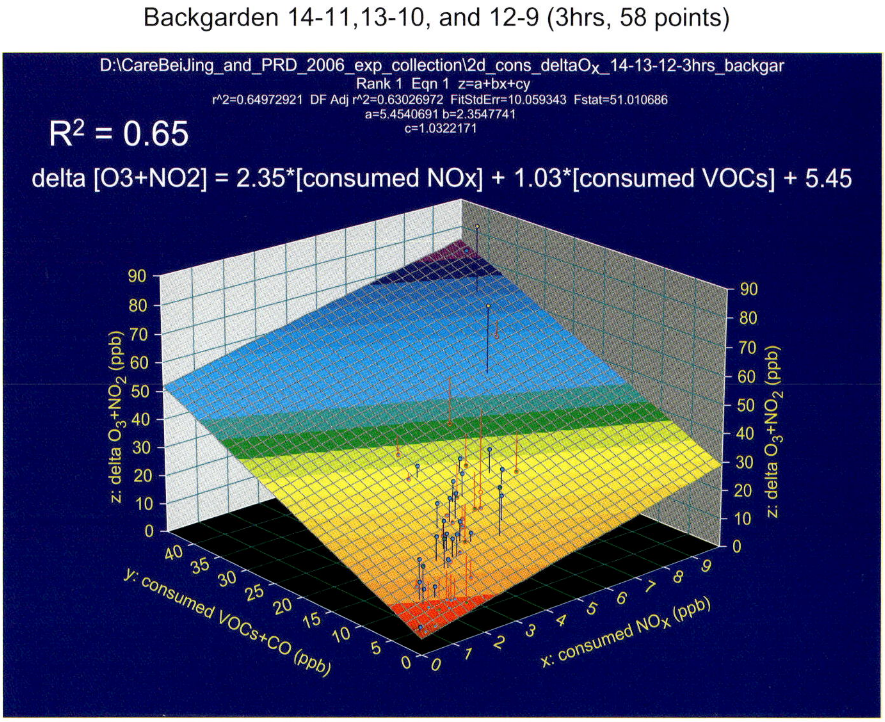

Backgarden 14-11,13-10, and 12-9 (3hrs, 58 points)

D:\CareBeiJing_and_PRD_2006_exp_collection\2d_cons_deltaOx_14-13-12-3hrs_backgar
Rank 1 Eqn 1 z=a+bx+cy
r^2=0.64972921 DF Adj r^2=0.63026972 FitStdErr=10.059343 Fstat=51.010686
a=5.4540691 b=2.3547741
c=1.0322171

$R^2 = 0.65$

delta [O3+NO2] = 2.35*[consumed NOx] + 1.03*[consumed VOCs] + 5.45

Figure 9. Same as Fig. 8 except that horizontal coordinates are changed to photochemically consumed precursors of ozone.

regime. In order to find the regime of NO_x sensitivity, one may need to be at least 200 km away from Guangzhou, i.e. the PRD megacity plume which is in the NMHCs-sensitive regime may be as large as $400 \times 400\,km^2$. In this context, given the widespread emissions of ozone precursors in the vast plain in northern China, it may be difficult to find any rural area in northern China where the ozone formation is limited by NO_x. This has a profound implication for the ozone control strategy for China in general and for the 2008 Beijing Olympics in particular.

Large amounts of ozone and ozone precursors are transported from megacities to rural and background atmospheres where the ozone production is known to be limited by NO_x. Therefore, reduction of NO_x emissions must be considered with regard to the synoptic-to-global-scale ozone problem.

3. Ozone Trends

3.1. Ozone trends in Taiwan

Taiwan is located in the southern part of East Asia, where the background ozone is influenced by long range transport of air pollutants from the Asian continent, which has undergone rapid economic development since the 1970s. Taiwan itself has experienced extraordinarily rapid economic development in the last few decades. The local photochemical production due to large emissions of ozone precursors plays a controlling role in the ozone distribution. In recent years the high levels of surface ozone have increasingly become a major air pollution problem. Since 2001, ozone has replaced particulate matter (PM10) as the major air pollutant responsible for most violations of the ambient air quality standards. In fact, efforts to control air pollutant emissions have significantly lowered the ambient

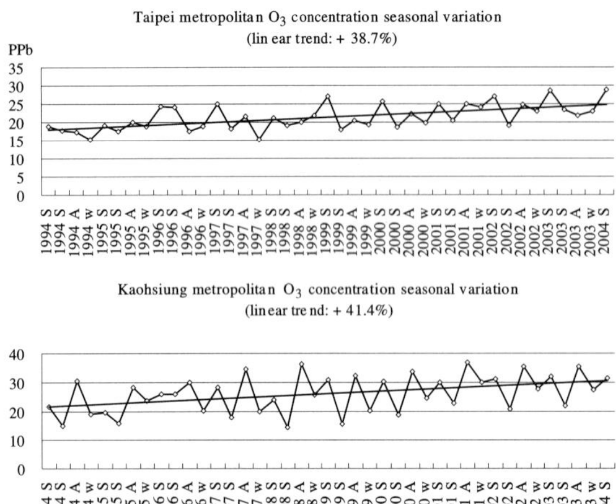

Figure 10. Increasing trends of ozone concentration in Taipei and Kaohsiung metropolitan areas from 1994 to 2004.

Table 1. Trends of O_3 and $O_3 + NO_2$ for three large metropolitan areas (Taipei, Taichung, Kaohsiung), one small city (Ilan) and three background stations (Yangming, Wanli, Lanyu).

Station	O_3 trend (%/decade)	$O_3 + NO_2$ trend (%/decade)
Taipie	21.6	−1.7
Taichung	22.5	5.4
Kaohsiung	20.2	2.1
Ilan	20.4	7.4
Yangming	13.2	13
Wanil	14.0	10.2
Lanyu	14.1	13.1

levels of nearly all air pollutants in Taiwan during the past decade. The only exception to this improving trend is the increase in ambient ozone concentration. For example, average ozone mixing ratios in the two largest metropolitan areas, Taipei and Kaohsiung, have increased by 39% and 41%, respectively for the period of 1994–2004 (Fig. 10).

With the reduction of emissions of ozone precursors, one would expect the photochemical production to decrease. What factors have contributed to the large increases in Taipei and Kaohsiung? Chou *et al.* (2006) showed that the reduction in the titration effect of NO due to decreasing emissions of NO played an important role in the increasing trends of ozone concentration. Another likely factor contributing to the trends is an increase in the long range transport of ozone. The impact of long range transport of ozone can be seen in Table 1, which shows greater increases in O_3 but smaller increases in $O_3 + NO_2$ at urban stations compared to background stations. Since the lifetime of NO_2 is less than a day, its concentration is controlled by local processes. The small trends for O_3+NO_2 at urban stations are consistent with the reduction

in the titration effect of NO due to its decreasing emissions, as suggested by Chou *et al.* (2006). We note that the increasing trends at three background stations are nearly equal to one another (at 14% per decade for O_3) for three different locations (Wanli is along the northern coast, Lanyu is on an island off the southeastern coast, and Yangming is a mountain site about 1 km above sea level). Moreover, the trends exist in all four seasons (not shown). The consistent trends for O_3 and $O_3 + NO_2$ at the three background stations suggest that the trends are large-scale and most likely the result of long range transport of ozone from the Asian continent. This is supported by similar trends found in Hong Kong (Fig. 11) and Okinawa (Lee *et al.*, 1998). The increases in Hong Kong ($\sim$40% per decade) and Okinawa ($\sim$25% per decade) are significantly greater than the 14% per decade increase observed in Taiwan. The greater increase in Hong Kong may be the result of the rapid development of the Pearl River Delta region. However, local effect should not be a factor in the ozone trend in Okinawa. Its increase must be because of the influence of long range transport from continental Asia. The same argument can be applied to the background stations in Taiwan. However, at this moment, we do not have a good explanation for the difference between Okinawa and the background stations in Taiwan.

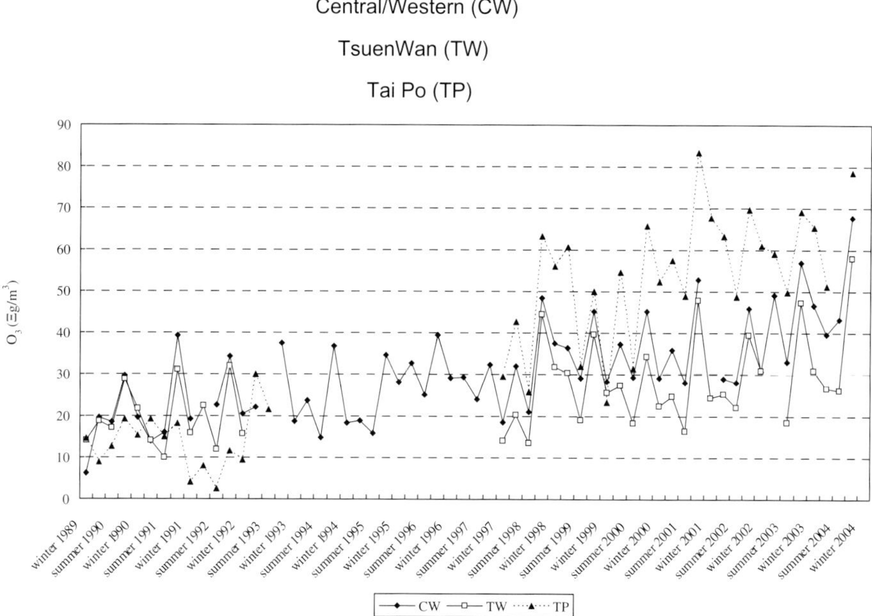

Figure 11.   Trends of ozone concentration observed at three stations in Hong Kong. The Tai Po station is a rural station, and the other two are urban stations.

3.2. *Ozone trends in East Asia and Mauna Loa, Hawaii*

The above discussion shows that the ozone trend at a background station is a valuable measure of the regional budget of ozone. Because of the large change in the photochemical lifetime of ozone from a few days in the boundary layer to a few months in the upper troposphere, the representativeness of a station depends critically on the location and altitude of the station. The ground station in Okinawa and the background stations in Taiwan should reflect mainly the ozone budget of East Asia. However, the influence of Southeast Asia, South Asia and further upwind regions could be significant too. In this regard, we notice that the increase of the ozone mixing ratio in the boundary layer measured by ozonesondes over Japan from the 1970s to the 1980s was about 30%, consistent with similar measurements in Hohenpeissenberg and Payern in Europe (Akimoto, 2003). The consistency suggests that the increase in ozone is extremely large-scale, probably over the entire midlatitudes in the Northern Hemisphere. The

cause of the increase is most likely the increasing emissions of ozone precursors from industrial developments in Europe, the US and Asia. This increase must have had a significant impact (on the order of a few ppbv) on the concentration of ozone over Taiwan. Moreover, the impact should still exist today because the emissions of ozone precursors have been increasing since the 1980s (Street *et al.*, 2001). Unfortunately, the impact on the concentration of ozone over Taiwan could not be verified because there was no reliable measurement of ozone in Taiwan before 1993, when the EPA established the current air quality monitoring stations.

The large scale impact of Northern Hemispheric industrial developments on the ozone concentration is probably best-recorded at the Mauna Loa, Hawaii observatory, which is located 3.4 km above the sea level and 20°N. The observatory observes free tropospheric air during the night when down-slope flow prevails. It usually is downwind of Asia, especially in winter and spring. Nevertheless, given the long lifetime of ozone in the free troposphere

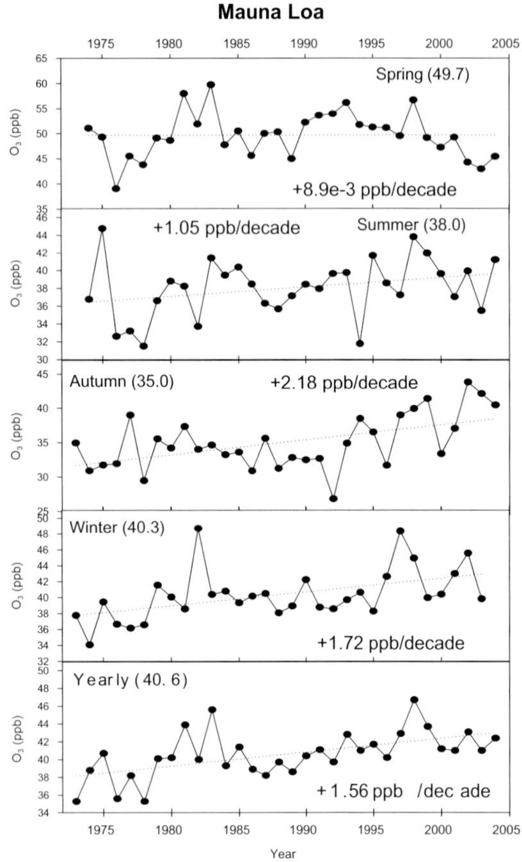

Figure 12. Trends of ozone concentration observed at Mauna Loa, Hawaii (3.4 km above the sea level and 20°N) between 1973 and 2004.

(more than a month), ozone trends observed at the observatory should reflect the integrated impact of Northern Hemispheric industrial developments. Figure 12 shows the ozone trends in different seasons and the annual mean from 1973 to 2004. The increasing trends are obvious except in the spring. The lack of increase in the spring is probably due to the relatively strong influence of stratospheric intrusion, which brings down stratospheric ozone to the troposphere. The persistent increases in ozone at Mauna Loa are consistent with the upward emission trends of NO_x in the US, Europe and East Asia (Street *et al.*, 2003). The fact that NO_x emissions in East Asia have increased the most may also contribute to the persistent ozone trends because East Asia is directly upwind of Mauna Loa. In

addition there are substantial increases in NO_x emissions in South Asia and Southeast Asia, which may also have a significant impact on the ozone trends at Mauna Loa.

What does the persistent increase in ozone at Mauna Loa imply for Taiwan? Taiwan is slightly north of Mauna Loa but much closer to the Asian continent. So we expect a substantially greater influence of the emissions from East Asia for Taiwan, i.e. a substantially greater increase in ozone. This is borne out by the fact that the annual average trend for background stations in Taiwan (~14% increase per decade between 1993 and 2004) is about a-factor-of-3.5 greater than the trend observed at Mauna Loa (~4.1% increase per decade between 1973 and 2004). If this ratio of 3.5 could be extended back to 1973 for Taiwan, it would imply a 43% increase for the background concentrations of ozone over Taiwan from 1973 to 2004.

In summary, the increasing trends of ozone concentration at background stations in Taiwan, Hong Kong and Japan from the 1980s to 2005 strongly suggest that photochemical production of ozone in Asia has been increasing due to increasing anthropogenic emissions of ozone precursors. In addition, we believe that the increasing trend of ozone concentration (4.1% increase per decade between 1973 and 2004) at Mauna Loa (3.4 km) is strong evidence supporting the theory that the background ozone level over Asia and most of the Northern Hemisphere has been rising in the same period. An observation-constrained three-dimensional photochemical transport model should be able to use the measurements at Mauna Loa to inverse-model the trends of background ozone concentration in Asia. More importantly, one can even inverse-model the global trend of ozone concentration between 1973 and 2004.

4. Concluding Remarks

In this work we have reviewed the formation of high levels of ozone in large metropolitan areas or megacities of East Asia and explored their possible impacts on the long term increasing trends of tropospheric ozone concentration in East Asia. The review is based on recent research efforts of scientists at the RCEC of Academia Sinica, in collaboration with colleagues from NTU and NCU as well as a large group of international scientists coordinated by Peking University in the last few years.

A series of intensive observation experiments were conducted in Kaohsiung, Taipei and Taichung by scientists of the RCEC, NTU and NCU to study the ozone formation and the strategy for its control in 2003–2005. In addition, RCEC scientists participated in three large scale international experiments led by Peking University in the Pearl River Delta and in Beijing in 2004 and 2006. Data from these measurements have been analyzed and interpreted by using a three-dimensional observation-based method to investigate the ozone production and its relationship with ozone precursors. It has been shown that the ozone production efficiency is consistent with values found in other regions. Although there is considerable uncertainty, the balance of evidence suggests that the O_3 production rate is likely to be NMHCs-sensitive and that controlling NMHCs is more efficient than controlling NO_x in lowering ozone levels both in southern Taiwan and in Beijing. This conclusion may have important implications for the ozone control strategy for the 2008 Olympics in Beijing.

Large amounts of ozone and ozone precursors are transported from megacities to rural and background atmospheres where the ozone production is known to be limited by NO_x. Therefore, reduction of NO_x emissions should also be considered with regard to the synoptic-to-global-scale ozone problem.

The increasing trends of ozone concentration at the background stations in Taiwan, Hong Kong and Japan from the 1980s to 2005 suggest that photochemical production of ozone in Asia

has been increasing due to anthropogenic emissions of ozone precursors. In addition, we believe that the increasing trend of ozone concentration (4.1% increase per decade between 1973 and 2004) at Mauna Loa (3.4 km) is a strong piece of evidence supporting that the background ozone level over Asia and most of the Northern Hemisphere has been rising in the same period. An observation-constrained three-dimensional photochemical transport model should be able to use the measurements at Mauna Loa to inverse-model the trends of background ozone concentration in Asia. More importantly, one could inverse-model the global trend of ozone concentration between 1973 and 2004.

[Received 6 September 2007; Revised 10 February 2008; Accepted 13 February 2008.]

References

Akimoto, H., 2003: Global air quality and pollution. *Science*, **302**, 1716–1719.

Altshuler, S. L., T. D. Arcado, and D. R. Lawson, 1995: Weekday vs weekend ambient ozone concentrations: discussion and hypotheses with focus on northern California. *J. Air Waste Mgmt. Assoc.*, **45**(12), 967–972.

Atkinson, R., 1990: Gas-phase tropospheric chemistry of organic compounds: a review. *Atmos. Environ.*, **24**A, 1–42.

Blanchard, C. L., F. W. Lurmann, P. M. Roth, H. E. Jeffries, and M. Korc, 1999. The use of ambient data to corroborate analyses of ozone control strategies. *Atmos. Environ.*, **33**(3), 369–381.

Calvert, J. G., 1976. Hydrocarbon involvement in photochemical smog formation in Los Angeles atmosphere. *Environ. Sci. Technol.*, **10**, 256–262.

Chameides, W. L., and J. C. G. Walker, 1973: A photochemical theory of tropospheric ozone. *J. Geophys. Res.*, **78**: 8751–8760.

Chang, C.-C., S.-J. Lo, J.-G. Lo,, and J.-L. Wang, 2003: Analysis of methyl tert-butyl ether (MTBE) in the atmosphere and implications as an exclusive indicator of automobile exhaust. *Atmos. Environ.*, **37**, 4747–4755.

Chang, C.-C., S. C. Liu, C.-J. Shiu, C.-J. Lai, and J.-L. Wang, 2007: A concept for estimating consumption of ozone precursors and assessing correlation between oxidant and precursors. *Atmos. Environ.* (submitted).

Chang, C.-C., T.-Y. Chen, C.-Y. Lin, C.-S. Yuan, and S. C. Liu, 2005: Effects of reactive hydrocarbons on ozone formation in southern Taiwan. *Atmos. Environ.*, **39**, 2867–2878.

Chang *et al.*, 2008a: Photochemistry of ozone formation in Beijing. In preparation.

Chang *et al.*, 2008b: Photochemistry of ozone formation in Pearl River Delta. In preparation.

Chou, C. C.-K., S. C. Liu, C.-Y. Lin, C.-J. Shiu, and K.-H. Chang, 2006: Trends of ozone and its precursors in Taiwan: implications for ozone control strategies. *Atmos. Environ.*, **40**, 3898–3908.

Crutzen, P. J., 1973: A discussion of the chemistry of some minor constituents in the stratosphere and troposphere. *Pure Appl. Geophys.*, **106**, 1385–1399.

Fehsenfeld, F. C., and S. C. Liu, 1993: Tropospheric ozone: distribution and sources. In *Global Atmospheric Chemical Change*, C. N. Hewitt, and W. T. Sturges (eds.), Elsevier, London, New York, pp. 169–231.

Frost, G. J., *et al.*, 1998: Photochemical ozone production in the rural southeastern United States during the 1990 Rural Oxidants in the Southern Environment (ROSE) program. *J. Geophys. Res.*, **103**, 22491–22508.

Gery, M., and R. Crouse, 1990. User's Guide for Executing OZIPR, Order No. 9D2196NASA, US Environmental Protection Agency, Research Triangle Park, NC, 27711.

IPCC, 2001: Climate change 2001: the scientific basis. Contribution of Working Group I to the Third Assessment Report of the Intergovernmental Panel on Climate Change. Cambridge University Press, 881 pp.

Jacob, D. J., 1999: *Introduction to Atmospheric Chemistry*, Princeton University Press, 266 pp.

Jacob, D. J., J. A. Logan, and P. P. Murti, 1999: Effect of rising Asian emissions on surface ozone in the United States. *Geophys. Res. Lett.* **26**, 2175–2178.

Kleinman, L. I., P. H. Daum, D. Imre, Y.-N. Lee, L. J. Nunnermacker, S. R. Springston, J. Weinstein-Lloyd, and J. Rudolph, 2002a: Ozone production rate and hydrocarbon reactivity in 5 urban areas: a case of high ozone concentration in Houston. *Geophys. Res. Lett.*, **29**(10), 1467, 10.1029/2001GL014569.

Kunhikrishnan, T., M. G. Lawrence, R. von Kuhlmann, A. Richter, A. Ladstatter-Weisenmayer, and J. P. Burrows, 2004: Analysis of tropospheric NO_x over Asia using the model of atmospheric transport and chemistry (MATCH-MPIC) and GOME-satellite observations. *Atmos. Environ.*, **38**, 581–596.

Kramp, F., and A. Volz-Thomas, 1997: On the budget of OH radicals and ozone in an urban plume from the decay of C5–C8 hydrocarbons and NO_x. *J. Atmos. Chem.*, **28**, 263–282.

Lee, S. H., H. Akimoto, H. Nakane, S. Kurnosenko, and Y. Kinjo, 1998: Lower tropospheric ozone trend observed in 1989–1997 at Okinawa, Japan. *Geophys. Res. Lett.*, **25**(10), 1637–1640.

Liu, S. C., M. Trainer, F. C. Fehsenfeld, D. D. Parrish, E. J. Williams, D. W. Fahey, G. Hubler, and P. C. Murphy, 1987: Ozone production in the rural troposphere and the implications for regional and global ozone distributions. *J. Geophys. Res.*, **92**(D4), 4191–4207.

Logan, J. A., M. J. Prather, S. C. Wofsy, and M. B. Mcelroy, 1981: a Tropospheric chemistry: global perspective. *J. Geophys. Res.*, (D86), 7210–7254.

Madronich, S., and S. Flocke, 1998: The role of solar radiation in atmospheric chemistry. In *Handbook of Environmental Chemistry*, P. Boule (ed.), Springer-Verlag, Heidelberg, pp. 1–26.

McKeen, S. A., and S. C. Liu, 1993: Hydrocarbon ratios and photochemical history of air masses. *Geophys. Res. Lett.*, **20**, 2363–2366.

Milford, J. B., A. G. Russell, and G. J. McRae, 1989: A new approach to photochemical pollution control: implications of spatial patterns in pollutant responses to reductions in nitrogen oxides and reactive organic gas emissions. *Environ. Sci. Technol.*, **23**, 1290–1301.

Oltmans, S. J., A. S. Lefohn, H. E. Scheel, J. M. Harris, H. Levy II, I. E. Galbally, E.-G. Brunke, C. P., Meyer, J. A. Lathrop, B. J. Johnson, D. S. Shadwick, E. Cuevas, F. J. Schmidlin, D. W. Tarasick, H. Claude, J. B. Kerr, O. Uchino, and V. Mohnen, 1998: Trends of ozone in the troposphere. *Geophys. Res. Lett.*, **25**(2), 139–142.

Parrish, D. D., C. J. Hahn, E. J. Williams, R. B. Norton, F. C. Fehsenfeld, H. B. Singh, J. D. Shetter, B. W. Gandrud, and B. A. Ridley, 1992: Indications of photochemical histories of Pacific air masses from measurements of atmospheric trace species at Point Arena. *California J. Geophys. Res.*, **97**, 15833–15901.

Parrish, D. D., *et al.*, 2004: Changes in the photochemical environment of the temperate North Pacifictroposphere in response to increased Asian emissions. *J. Geophys. Res.*, **109**, D23S18, doi: 10.1029/2004JD004978.

Regener, E., 1949: Ozonschicht und atmospharische turbulenz. *Ber. Deut. Wetterdienstes US-Zone*, **11**, 45–57.

Robert, J. M., F. C. Fehsenfeld, S. C. Liu, M. J. Bollinger, C. Hahn, D. L. Albritton, and R. E. Sievers, 1984: Measurements of aromatic hydrocarbon ratios and NO_x concentrations in the rural troposphere: observation of air mass photochemical aging and NO_x removal. *Atmos. Environ.*, **18**(11), 2421–2432.

Seinfeld, J. H., and S. N. Pandis, 1998: *Atmospheric Chemistry and Physics from Air Pollution to Climate Change*, John Wiley & Sons, 1326.

Shao, M., X. Tang, *et al.*, 2006: City clusters in China: air and surface water pollution. *Front. Ecol. Environ.*, **4**, 353–361.

Shiu, C.-J., S. C. Liu, C.-C. Chang, J.-P. Chen, Charles C. K. Chou, C.-Y. Lin, and C.-Y. Young, 2007: Photochemical production of ozone and control strategy for Southern Taiwan. *Atmos. Environ.*, in press.

Singh, H. B., 1977: Atmospheric halocarbons: evidence in favor of reduced average hydroxyl radical concentration in the troposphere. *Geophys. Res. Lett.*, **4**, 101–104.

Streets, D. G., *et al.*, 2003: An inventory of gaseous and primary aerosol emissions in Asia in the year 2000. *J. Geophys. Res.*, **108**(D21), 8809, doi: 10.1029/2002JD003093.

Trainer, M., *et al.*, 1991: Observations and modeling of the reactive nitrogen photochemistry at a rural site. *J. Geophys. Res.*, **96**, 3045–3063.

Trainer, M., E. Y. Hsie, S. A. Mckeen, R.Tallamaju, D. D. Parrish, F. C. Fehsenfeld, and S. C. Liu, 1987: Impact of natural hydrocarbons on hydroxyl and peroxy radicals at a remote site. *J. Geophys. Res.*, **92**(D10), 11879–11894.

Trainer, M., B. A. Ridley, M. P. Buhr, G. Kok, J. Walega, G. Hübler, D. D. Parrish, and F. C. Fehsenfeld, 1995: Regional ozone and urban plumes in the southeastern United States: Birmingham, a case study. *J. Geophys. Res.*, **100**(D9), 18823–18834.

Wakamatsu, S., I. Toshimasa, and U. Itsushi, 1996: Recent trends in precursor concentrations and oxidant distributions in the Tokyo and Osaka areas. *Atmos. Environ.*, **30**, 715–721.

Wang, Y.-H., S. C. Liu, H.-B. Yu, S. T. Sandholm, T. Y. Chen, and D. R. Blake, 2000: Influence of convection and biomass burning outflow on tropospheric chemistry over the tropical Pacific. *J. Geophys. Res.*, **105**(D7), 9321–9333.

Wang, T., C. N. Poon, *et al.*, 2003: Characterizing the temporal variability and emission patterns of pollution plumes in the Pearl River Delta of China. *Atmos. Environ.*, **37**: 3539–3550.

Zhang, Y., K. Shao, *et al.*, 1998: Study on photochemical smog in the cities of China. *Acta Sci. Nat. Univ. Pekin.*, **34**(2–3): 392–400.

Ziemke, J. R., S. Chandra, B. N. Duncan, L. Froidevaux, P. K. Bhartia, P. F. Levelt, and J. W. Waters, 2006: Tropospheric ozone determined from Aura OMI and MLS: evaluation of measurements and comparison with the Global Modeling Initiative's Chemical Transport Model. *J. Geophys. Res.*, **111**(D19303), 1–18.

Impact of FORMOSAT-3/COSMIC Data on Typhoon and Mei-yu Prediction

Ying-Hwa Kuo, Hui Liu and Yong-Run Guo

National Center for Atmospheric Research, Boulder, Colorado, USA
kuo@ucar.edu

Chuen-Teyr Terng and Yun-Tien Lin

Central Weather Bureau, Taipei, Taiwan

The Formosa Satellite Mission #3/Constellation Observing System for Meteorology, Ionosphere and Climate (FORMOSAT-3/COSMIC), hereafter referred to as COSMIC, is a Taiwan–US mission launched in April 2006. COSMIC consists of six small satellites that employ the Global Positioning System (GPS) radio occultation (RO) technique to sound the neutral atmosphere and ionosphere with uniform global coverage. As of January 2008, COSMIC provides approximately 2,000 RO soundings per day to support the research and operational communities. For East Asian countries, the COSMIC soundings are particularly valuable for the study of typhoons and Mei-yu convective systems, as they provide observations over the data-sparse Western Pacific Ocean and the South China Sea.

In this study, we assimilate COSMIC GPSRO soundings and examine their impact on the prediction of Typhoon Shanshan (2006). The assimilation is first carried out using the WRF-Var (3D-Var) system. We find that in order for COSMIC GPSRO soundings to have an impact, it is critical to perform continuous assimilation through cycling. With one-hour cycling over a one-day period, COSMIC GPSRO soundings significantly improve the track forecast. However, the assimilation of only seven COSMIC GPSRO soundings in a cold-start experiment produces virtually no impact. The continuous cycling assimilation is able to incorporate 110 GPSRO soundings, and has a profound impact. We also find that the assimilation of typhoon bogus soundings improves the typhoon intensity and track forecast, particularly during the first two days.

To assess the impact of data-assimilation systems, we compare the performance of the WRF 3D-Var system with the WRF/DART ensemble filter system for the assimilation of COSMIC GPSRO soundings. The results show that the WRF/DART ensemble filter system can assimilate the GPSRO data more effectively than the WRF 3D-Var method. In particular, the WRF/DART ensemble filter system is able to produce a storm with more coherent typhoon structure after one day of continuous assimilation, while a much weaker and less coherent storm is produced by WRF 3D-Var.

In addition to Typhoon Shanshan (2006), we assimilate GPSRO soundings from COSMIC during the two-week period of 1–14 June 2007, associated with a Mei-yu system, using the WRF/DART ensemble filter data-assimilation system. We find that the assimilation of COSMIC data significantly strengthens the Western Pacific Subtropical High, and consequently improves the prediction of Mei-yu precipitation over southern China and Taiwan.

1. Introduction

The radio occultation (RO) technique, which makes use of radio signals transmitted by Global Position System (GPS) satellites, has emerged as a powerful and relatively inexpensive approach to sound the global atmosphere with high precision, accuracy and vertical resolution in all weathers and over both land and ocean. This was first demonstrated by the proof-of-concept GPS/MET (GPS Meteorology) experiment in 1995–1997 (Ware *et al.*, 1996), and substantiated by the CHAMP (CHAllenging Minisatellite Payload; Wickert *et al.*, 2001) and SAC-C (Satellite de Aplicaciones Cientificas-C; Hajj *et al.*, 2004) missions. In April 2006, the Formosa Satellite Mission #3/Constellation Observing System for Meteorology, Ionosphere and Climate (FORMOSAT-3/COSMIC, hereafter referred to as COSMIC) was successfully launched into initial orbits of 512 km from the Vandenberg Air Force Base (Kuo *et al.*, 1999; Rocken *et al.*, 2000; Cheng *et al.*, 2006). COSMIC started collecting GPSRO soundings eight days after the launch, and three months later started providing data to the international science communities. By the end of 2007, COSMIC satellites had already been deployed to their operational orbits at 800 km elevation, evenly spaced with 30° separation. This allows COSMIC soundings to be distributed uniformly around the globe. As of 5 March 2008, the COSMIC Data Analysis and Archive Center (CDAAC) has processed 941,933 neutral atmospheric profiles and 1,227,682 ionospheric profiles, and delivered them to the operational and research communities. Even though COSMIC was launched less than two years ago, it has already contributed significantly to global weather prediction, climate monitoring, and ionospheric research (Anthes *et al.*, 2008).

In comparison with previous GPSRO missions, COSMIC offers three major advantages. First, COSMIC uses the advanced open-loop (OL) tracking technique (Sokolovskiy, 2001). All previous GPSRO missions (with the exception of SAC-C, which was used to test the OL tracking) used phase-lock-loop (PLL) tracking. With the use of PLL tracking, only a small fraction of GPSRO soundings penetrate to below 1 km. Those that do penetrate to the lower troposphere are often affected by tracking errors, especially over the tropical atmospheric boundary layer (ABL). With OL tracking, 70% of COSMIC soundings penetrate to the lower tropical troposphere, while those over the higher latitudes, more than 90% of COSMIC soundings, penetrate to below 1 km (Sokolovskiy *et al.*, 2006a; Anthes *et al.*, 2008). The deep penetration of GPSRO soundings allows us to monitor the variation of the height of the ABL (Sokolovskiy *et al.*, 2006b), and atmospheric river events over the Eastern Pacific Ocean (Neiman *et al.*, 2008).

COSMIC represents the world's first constellation designed to provide GPSRO soundings with uniform global distribution. The six satellites were launched with one single rocket. The differential precession technique is used to deploy the satellites to their final orbits at 800 km using the on-board propulsion system, with 30° separation (Yen *et al.*, 2008). The evenly spaced orbital plans allow the COSMIC GPSRO soundings to be distributed uniformly around the globe in local solar time. This is very important for observing the diurnal cycle in the neutral atmosphere and ionosphere (where it is especially strong) and for preventing aliasing of the diurnal cycle in climate signals (Zeng *et al.*, 2008). Moreover, with each of the six satellites capable of performing both rising and setting occultation, COSMIC provides an-order-of-magnitude more soundings than a single satellite mission, such as CHAMP.

The third advantage of COSMIC is the availability of GPSRO data in near real time to support operational applications. With the support of two high-latitude ground stations, each COSMIC satellite can download its data once every orbit (100 min). After 5 min of data transfer and another 15 min of data processing, these data are available to support global operational numerical weather prediction through the Global Telecommunication System (GTS; Rocken *et al.*, 2000). After a few months of testing, the ECMWF (European Centre for Medium Range Forecasts) began the operational assimilation of COSMIC GPSRO data on 12 December 2006, the NCEP (National Centers for Environmental Prediction) on 1 May 2007, the UKMO (United Kingdom Meteorological Office) on 15 May 2007, and Meteo France on 1 September 2007. All these operational centers have reported positive impact with the operational assimilation of COSMIC GPSRO data (Cucurull and Derber, 2008; Healy, 2008; Poli *et al.*, 2008).

Taiwan is located at subtropical latitudes, and is affected by severe weather in every season. In particular, the heavy rainfall events during late spring and early summer, a period known as the Mei-yu (Kuo and Chen, 1990), and the typhoons in the summer (Wu and Kuo, 1999) are major weather-related disasters that can cause significant loss of lives and property. With the availability of the GPSRO observations from COSMIC, this provides an opportunity to improve the forecasting of these severe weather events, and to improve our understanding of the regional climate in the vicinity of Taiwan (Wu *et al.*, 2000). In this article, we examine the impact of COSMIC GPSRO soundings on the prediction of Typhoon Shanshan (2006) and a Mei-yu heavy rainfall event that took place in June 2007.

In Sec. 2, we describe the Typhoon Shanshan (2006) case and the assimilation of COSMIC GPSRO soundings using the WRF-Var (3D-Var) system. We examine how the details of assimilation procedures could influence the impact of GPSRO soundings on typhoon prediction. In Sec. 3, we compare the performance of WRF-3D-Var and the WRF/DART (Data Assimilation Research Testbed) ensemble filter data-assimilation system in the assimilation of GPSRO soundings and their impact on Typhoon Shanshan prediction. In Sec. 4, we examine the impact of COSMIC GPSRO soundings on the prediction of a Mei-yu system, and its associated heavy rainfall over Taiwan. A summary and closing remarks are given in the final section.

2. WRF 3D-Var Assimilation of COSMIC GPSRO Data and Its Impact on the Prediction of Typhoon Shanshan (2006)

2.1. *Typhoon Shanshan case*

Typhoon Shanshan formed on 9 September 2006, about 500 km north–northeast of Yap, near 14°N, 139°E (Fig. 1). The storm moved northwestward and went through rapid development, becoming a Category 4 storm by 12 September 2006. It then moved westward and northward, skirting to the east of Taiwan on 15 and 16 September. It reached its peak intensity near 0000 UTC 16 September, with a central pressure of 919 mb, and a peak wind speed of about $60 \, \mathrm{m \, s^{-1}}$. Typhoon Shanshan made landfall on the island of Kyushu on 17 September. It was downgraded from a Category 4 storm to a Severe Tropical Storm by 0000 UTC 18 September, just before it crossed the island of Hokkaido. Later, it was transformed into an extratropical cyclone after it interacted with a midlatitude system. Figure 1 shows the best track and storm central pressure as a function of time. For this study, we will focus on the period from 0000 UTC 13 September through 17 September. In particular, we will assimilate the COSMIC GPSRO soundings over a one-day period, from 0000 UTC 13 to 0000 UTUC 14 September, and assess their impact on the prediction of Typhoon Shanshan.

2.2. *Assimilation of COSMIC data using WRF 3D-Var*

Through collaboration between NCAR and the Central Weather Bureau (CWB), a WRF

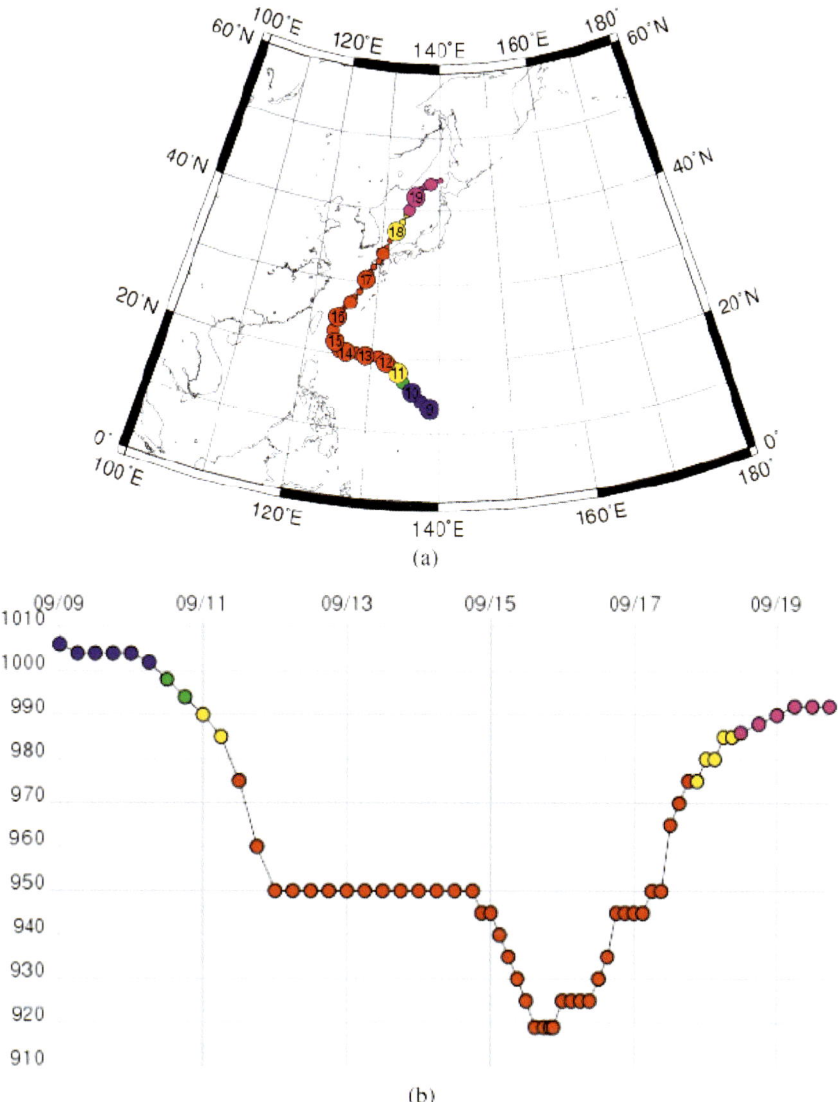

Figure 1. (a) Best track and (b) central pressure time series of Typhoon Shanshan (courtesy of Japan Meteorological Agency). The color of the points represents the storm intensity: blue — tropical depression; green — tropical storm; yellow — severe tropical storm; red — typhoon; magenta — extratropical cyclone.

(weather research and forecasting model) and WRF 3D-Var (three-dimensional variational data assimilation) system (Barker *et al.*, 2004) has been established for operation, starting from July 2007. The operational WRF model consists of three nested domains, with grid sizes of 45, 15 and 5 km, respectively. In this study, we perform experiments only on the 45 km domain (without nesting). Additional information on the WRF-Var system can be found at http://www.mmm.ucar.edu/wrf/WG4/wrfvar/wrfvar-tutorial.htm

Figures 2(a) and 2(b) show the distribution of COSMIC GPSRO soundings, both in time and in space, over the one-day period of 0000 UTC 13 to 0000 UTC 14 September 2006. During this 24 h period, there are very few GPSRO soundings in the vicinity of Typhoon

Ying-Hwa Kuo *et al.*

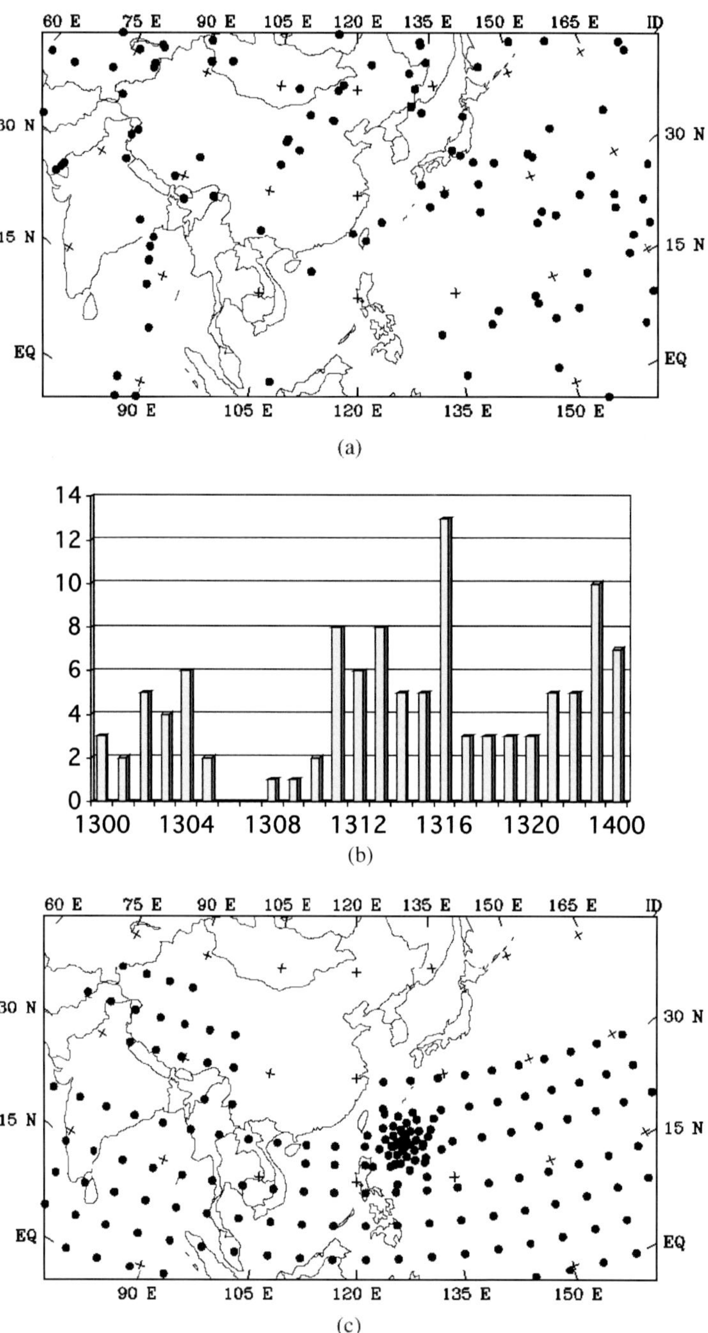

Figure 2. (a) The spatial and (b) temporal distribution of COSMIC GPSRO soundings, and (c) the loci of CWB global and typhoon bogus soundings.

Table 1. List of WRF 3D-Var experiments.

Name	Cycling	Bogus data	COSMIC GPSRO data	Remarks
NODA	No	No	No	No assimilation
COLDNBNG	No	No	No	Cold start, no bogus, no GPSRO
COLDNB	No	No	Yes	Cold start, no bogus
COLDALL	No	Yes	Yes	Cold start, all data
CYCLNBNG	1 h cycling	No	No	Cycling, no bogus, no GPSRO
CYCLNB	1 h cycling	No	Yes	Cycling, no bogus
CYCLALL	1 h cycling	Yes	Yes	Cycling, all data

Shanshan. Moreover, the temporal distribution of the soundings data is not homogeneous. It varies from zero at 0600 UTC 13 September to 13 soundings at 1600 UTC 13 September. A total of seven experiments are conducted (see Table 1).

The schematic diagram of all data-assimilation experiments is shown in Fig. 3, and the experimental domain is shown in Fig. 2(a) with the size of 222×128 in a 45 km grid distance. There are a total of 45 full vertical model levels with the η values of 1.0, 0.995, 0.988, 0.98, 0.97, 0.96, 0.945, 0.93, 0.91, 0.89, 0.87, 0.85, 0.82, 0.79, 0.76, 0.73, 0.69, 0.65, 0.61, 0.57, 0.53, 0.49, 0.45, 0.41, 0.37, 0.34, 0.31, 0.28, 0.26, 0.24, 0.22, 0.2, 0.18, 0.16, 0.14, 0.12, 0.10, 0.082, 0.066, 0.052, 0.04, 0.03, 0.02, 0.01, 0.0. The model top is 30 hPa, and the time step is 180 s. The moist physics include the new Kain–Fritsch cumulus parametrization scheme

and the WSM 5-class explicit scheme. The YSU PBL scheme and the Monin–Obukhov scheme are used as the boundary and surface layers' physics, and the Noah land-surface model is selected for the low-boundary condition over the land. The long-wave and short-wave radiation schemes, which are called every 30 min during the model integration, are the RRTM scheme and the Goddard scheme, respectively. The first experiment (NODA) is a no-data-assimilation experiment. The initial condition for this experiment is obtained from the NCEP AVN global analysis at 0000 UTC 13 September. The purpose of the NODA experiment is to serve as a benchmark for illustrating the impact of data assimilation.

In the data-assimilation experiments with WRF-Var (see the section on WRF-Var at the link http://www.mmm.ucar.edu/wrf/users/tutorial/tutorial_presentation.htm), the

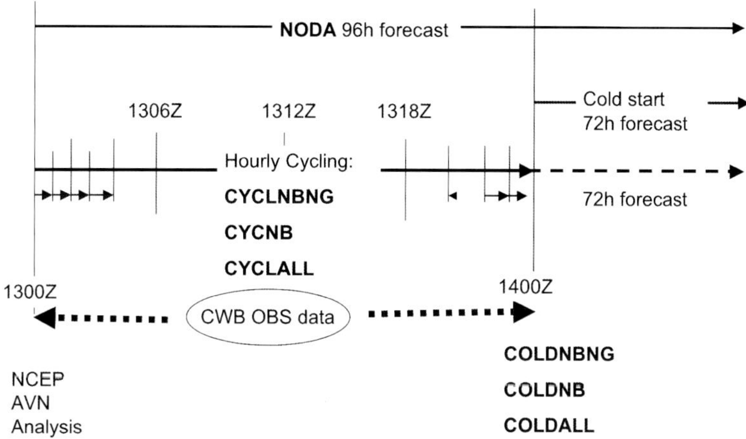

Figure 3. Schematic diagram illustrating the design of data-assimilation experiments.

background error statistics (BES) are obtained by interpolation from a 41-level CV Option-5 BES file derived based on the three months' (August, September, and October 2006) forecast data over the same domain with the NMC method. The observations are obtained from CWB (referred to as "CWB-obs"), and include the upper-air soundings, PILOT, surface data (SYNOP, SHIPS, METAR, BUOY), aircraft data (AIREP), and the satellite-derived or -retrieved data (satellite wind, QuikScat wind, SATEM, etc.). In addition to CWB-obs, the bogus data are also provided by CWB [Fig. 2(c)].

Two sets of data-assimilation experiments are conducted. The first set is cold-start experiments. For this set, data assimilation is performed at only one time, which is 0000 UTC 14 September. COLDNBNG is an experiment that does not assimilate bogus data or COSMIC GPSRO data. For CWB operations, two types of bogus data are used. The first is called "global bogus." To minimize "systematic drifting" of regional analysis, CWB extracts a set of bogus soundings from their global analysis, and treats them as synthetic sounding data. These soundings are located at regular intervals. Based on past experience, the assimilation of these global bogus soundings has been effective in maintaining consistence between regional and global analysis and in preventing regional analysis from drifting away from global analysis. When a typhoon is found within the analysis domain, CWB also creates a set of typhoon bogus soundings. These consist of wind soundings in the vicinity of the typhoon. For the Typhoon Shanshan case, a total of 134 global bogus soundings and 40 typhoon bogus soundings are available at 6 h intervals from 0000 UTC 13 to 0000 UTC 14 September 2006 [see Fig. 2(c)]. For the COLDNBNG (meaning cold-start, no-bogus, no-GPSRO soundings) experiment, only CWB-obs data are assimilated. Neither the bogus nor the GPSRO soundings are assimilated.

For the COLDNB experiment, global and typhoon bogus data are not assimilated. However, a total of 7 COSMIC GPSRO soundings (available within a 1 h interval centered at 0000 UTC 14 September) are assimilated, in addition to CWB-obs data. In COLDALL, the 134 global and 40 typhoon bogus soundings are assimilated, in addition to the 7 COSMIC GPSRO soundings and CWB-obs data.

The second set of data-assimilation experiments involves cycling data assimilation experiments. In these experiments, continuous assimilation is performed from 0000 UTC 13 to 0000 UTC 14 September at 1 h intervals. Basically, the 1 h forecast is used as the first guess for the next analysis cycle. Then data that fall within $+/-$ 30 min of the particular hour are assimilated. This procedure is repeated over the 24 h period (see Fig. 3). Obviously, the cycling experiments will be able to assimilate a lot more data than the cold-start experiments. For CYCLNBNG, neither bogus nor COSMIC GPSRO soundings are assimilated. However, more CWB-obs data are assimilated, in comparison with COLDNBNG. For CYCLNB and CYCLALL, they are similar to COLDNB and COLDALL, with the exception of continuous cycling. CYCLNB assimilates a total of 110 COSMIC GPSRO soundings, and CYCLALL assimilates an additional 870 (174×5) bogus soundings (which are available at 6 h intervals) in addition to the GPSRO soundings. With the assimilation of a much larger number of soundings, we would expect the cycling data-assimilation experiments to have a larger impact on the forecast.

Figure 4 shows the track and central pressure of Typhoon Shanshan in no-data assimilation, and cold-start series of WRF 3D-Var experiments. The best track and the observed central pressure are also shown in the figure. The no-data-assimilation (NODA) experiment has the worst performance, as it is essentially a 24–96 h forecast using the NCEP AVN analysis at 0000 UTC 13 September 2006 as the

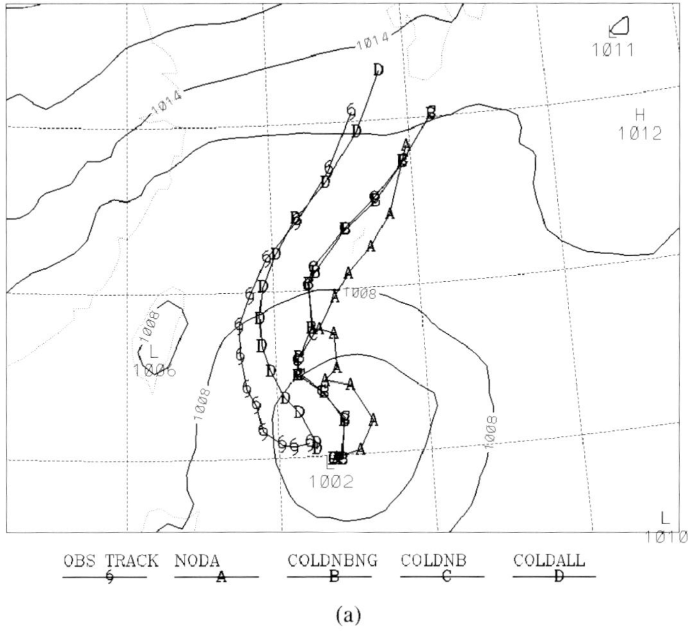

(a)

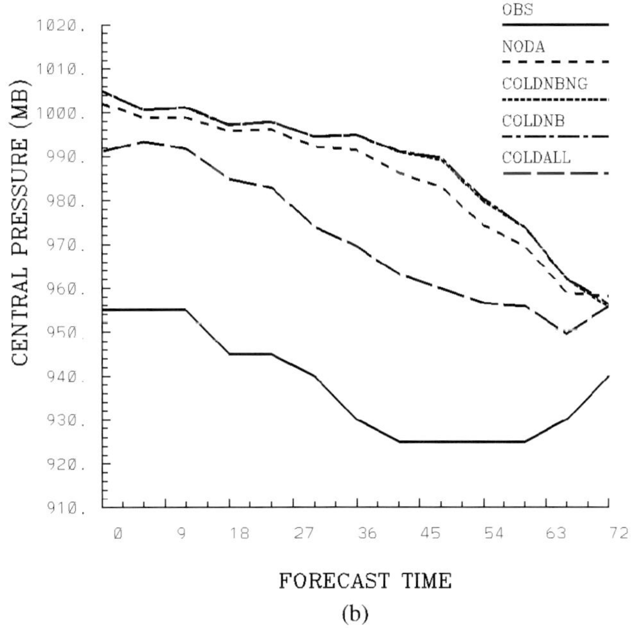

(b)

Figure 4. (a) Track (the letters A, B, C, D, etc. indicate the 6 h storm positions, and the contours are SLP from NODA at 0000 UTC 14 September 2006) and (b) central pressure of cold-start WRF 3D-Var experiments.

initial condition. The averaged track forecast error over the period of 0300 UTC 14 to 0000 UTC 17 September is 273 km. With the assimilation of CWB-obs data, the COLDNBNG experiment had a noticeable improvement over the NODA experiment. The three-day averaged track forecast error is reduced from 273 km to 233 km. The assimilation of seven COSMIC

GPSRO soundings at 0000 UTUC 14 September had only very minor impact on the track and intensity forecasts. The results of COLDNB are almost identical to those of COLDNBNG. However, the assimilation of 134 global bogus soundings and 40 typhoon bogus soundings at 0000 UTC 14 September in the COLDALL experiment has a major impact on the forecast. The three-day averaged track forecast error is reduced to 140 km, almost half of those in the COLDNB experiment. Figure 4 also clearly shows that the track in COLDALL is much closer to that of the best track and the intensity closer to the observation. These results suggest that, for the cold-start experiment, the assimilation of bogus data (particularly the typhoon bogus data) has a positive and profound influence on typhoon track forecasting.

The results of cycling data-assimilation experiments are shown in Fig. 5. In comparison with the cold-start experiments, the continuous assimilation improves the results considerably. For example, the three-day averaged track errors for CYCLNBNG and CYCLNB are 144 and 111 km, while they are 233 and 232 km for the COLDNBNG and COLDNB experiments. Figure 5 also shows that the tracks of Typhoon Shanshan in CYCLNBNG and CYCLNB are much closer to the best track compared with their counterparts in the cold-start experiments. It is interesting to note that the assimilation of 110 COSMIC GPSRO soundings has produced a noticeable impact. The three-day averaged track error is reduced by 33 km (from 144 km to 111 km) with WRF 3D-Var. This is a 23% improvement. Also, the assimilation of COSMIC GPSRO soundings produces a storm with stronger intensity (by about 10 mb), which is closer to the observation. These results suggest that in order for the COSMIC GPSRO soundings to have an impact, it is essential that a continuous assimilation approach is used.

The incorporation of bogus data in the cycling experiments produces improvements for the first two days. However, for day 3, the CYCLALL experiment performs worse than without the assimilation of bogus data. Figure 5 shows that the storm track is biased westward after one day. The storm then moves much slower than the observation, ending with a larger track error on day 3. The exact reasons for this poorer performance with the assimilation of the bogus data are not known. However, we suspect that this might be related to the details of the bogusing procedure and the background error statistics used in WRF 3D-Var, etc. For the CWB typhoon bogusing procedure, 40 wind soundings are extracted from a symmetric Rankine vortex plus the large-scale environmental flow. These soundings are assimilated into the system without further data quality control. For a real typhoon, the storm would often develop asymmetric structures. The assimilation of the bogus soundings would destroy these real asymmetric structures, and the model storm must recreate them. A continuous assimilation of bogus data implies that these counteracting procedures are repeated throughout the assimilation windows. One cannot make a general statement that the assimilation of a bogus vortex does not improve the forecast. However, it is clear that improvement of the typhoon bogusing procedure is needed to improve WRF 3D-Var performance, particularly in cycling experiments.

2.3. *Comparison of WRF 3D-Var and WRF/DART ensemble assimilation*

Over the past several years, a Data Assimilation Research Testbed (DART), a community data-assimilation facility for geosciences, has been developed at NCAR. DART includes a wide variety of ensemble filter assimilation algorithms (Anderson, 2003) which can be applied to a wide range of geosciences problems, including those of the atmosphere, oceans, atmospheric chemistry, and ionosphere. Details on DART can be found at http://www.image.ucar.edu/DAReS.

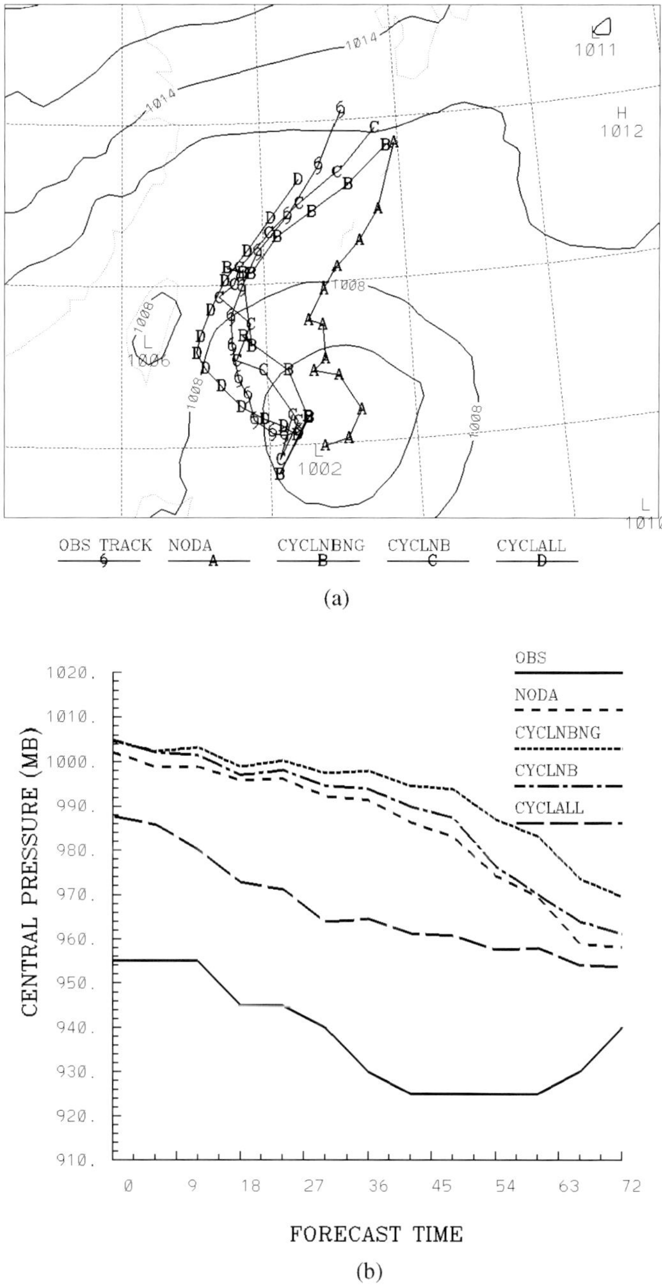

(a)

(b)

Figure 5.   Same as Fig. 4 but for 3D-VAR cycling run experiments.

DART makes it easy to implement deterministic ensemble filter data-assimilation approaches with various types of numerical models. A number of models have already been implemented with DART, including the WRF model and the CCSM-3 Atmospheric Model (CAM-3). Both CAM-3/DART and WRF/DART have been used for the assimilation of GPSRO data. For example, using the CAM-3/DART system, Liu *et al.* (2007) demonstrated the importance of forecast error multivariate correlations, between specific humidity, temperature, and surface pressure, for the assimilations of GPSRO data. Liu *et al.* (2008) used the WRF/DART system to compare the performance of a nonlocal observation operator (Sokolovskiy *et al.*, 2005) and of a local refractivity observation operator in the assimilation of GPSRO soundings. Since both the WRF 3D-Var and the WRF/DART ensemble filter data-assimilation system are available for the assimilation of GPSRO soundings, it would be desirable to compare the performance of these two data-assimilation systems for the same Typhoon Shanshan case. For such a comparison, we try to make the two systems as compatible as possible. For example, they use the same local refractivity observation operator, the same observational data sets, and the same model domain and grid configurations. For this article, WRF/DART assimilated only the key observation types, including the radiosonde, satellite wind, QuikScat wind from CWB, and GPSRO data from COSMIC/CDAAC. The WRF/DART system is also set up for continuous 1 h cycling, in ways similar to that of WRF 3D-Var cycling experiments. For the WRF/DART system, we do not assimilate the global and typhoon bogus data.

Figure 6 compares the tracks and central pressures of Typhoon Shanshan in the WRF/DART and WRF 3D-Var experiments. The track map shows that the WRF/DART experiments (DARTNBNG and DARTNB) follow the best track closely. This is also reflected in Table 2. For a three-day average, DARTNBNG and DARTNB have track errors of 75 and 62 km, respectively. They represent almost a 50% improvement over the WRF 3D-Var experiments (e.g. CYCLNBNG and CYCLNB), which have 144 and 111 km, respectively. The central pressure time series also indicates that the WRF/DART experiments produce stronger typhoons, although they differ only by less than 5 mb initially (at 0000 UTC 14 September). By 36 h, they differ by more than 20 mb, with

WRF/DART producing typhoons with more realistic intensity. It is interesting to note that in terms of track errors, WRF/DART without the assimilation of global and typhoon bogus data performs better than WRF 3D-Var which assimilates everything, including global and typhoon bogus data, after one day [as shown in Table 2]. Figure 6(c) also indicates that the assimilation of COSMIC GPSRO soundings with the WRF/DART system improves the track forecasts. The three-day averaged track error is reduced from 75 km to 62 km (Table 2), a 17% improvement.

The more realistic simulation of Typhoon Shanshan by WRF/DART experiments is illustrated in Fig. 7. Here we show vertically integrated cloud water (which serves as surrogate cloud fields) for CYCLNB, CYCLALL, and DARTNB experiments, together with the observed IR satellite images. Even with a horizontal resolution of 45 km, the DARTNB experiment clearly shows an eye for Typhoon Shanshan and the eyewall clouds. In contrast, the corresponding WRF 3D-Var data-assimilation experiment, CYCLNB, does not show an eye. The CYCLALL experiment gives the hint of an eye, with less cloud water in the center of the storm. However, its position is biased considerably westward when compared with that of the observed storm.

The more realistic structure of Typhoon Shanshan after one day of data assimilation with WRF/DART is illustrated in Fig. 8, which shows the potential temperature and tangential winds along a north–south cross section (along 125.8°E) that cuts across the center of the storm. The WRF/DART experiment without the assimilation of GPSRO produces a vortex, with a radius of maximum wind of about 250 km. The maximum tangential wind exceeds $15\,\mathrm{m\,s^{-1}}$. With the assimilation of GPSRO data from COSMIC, the tangential winds are increased, and the radius of maximum winds is decreased slightly. In contrast, WRF 3D-Var produces a much weaker

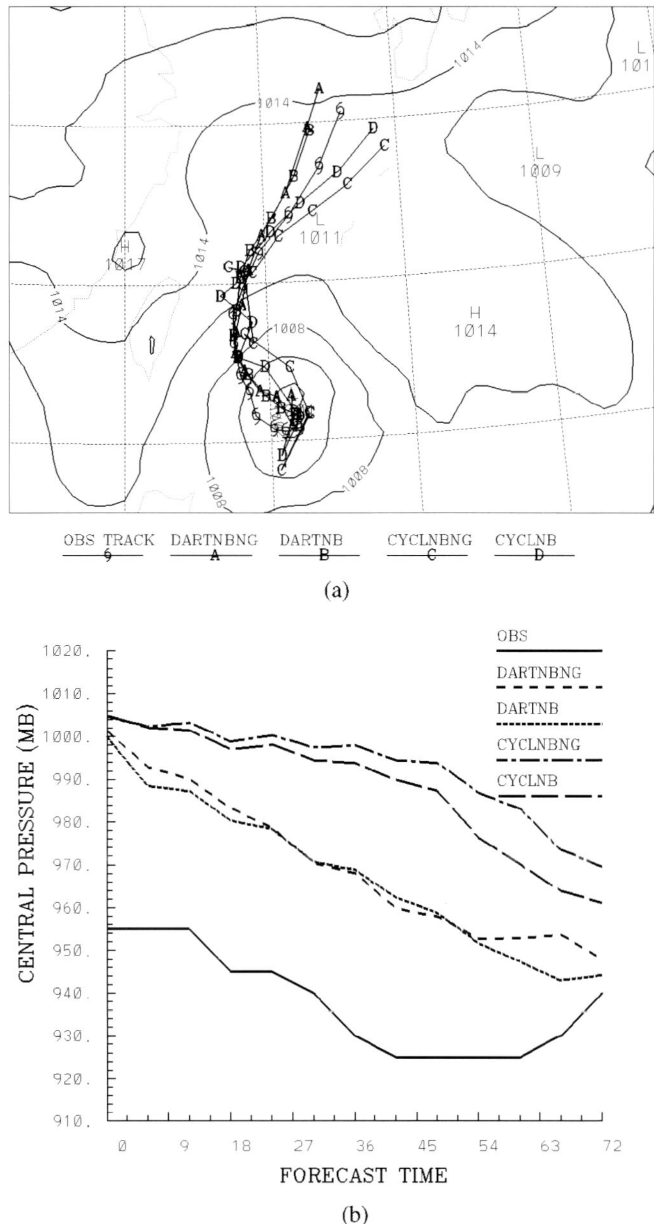

Figure 6. (a) Track of cycling data-assimilation experiments, including DARTNBNG, DARTNB, CYCLNBNG, CYCLNB, and the best track; the SLP field is obtained from the DARTNBNG at 0000 UTUC 14 September 2006; (b) the corresponding central pressure time series; and (c) track forecast errors for the four cycling data-assimilation experiments.

storm, with a radius of maximum wind of about 350 km and the maximum tangential wind just a little over $10\,\mathrm{m\,s^{-1}}$. The assimilation of GPSRO soundings does not seem to produce an appreciable improvement to the storm intensity and structure.

The corresponding plots of vorticity for DARTNB and CYCLNB are shown in Fig. 9.

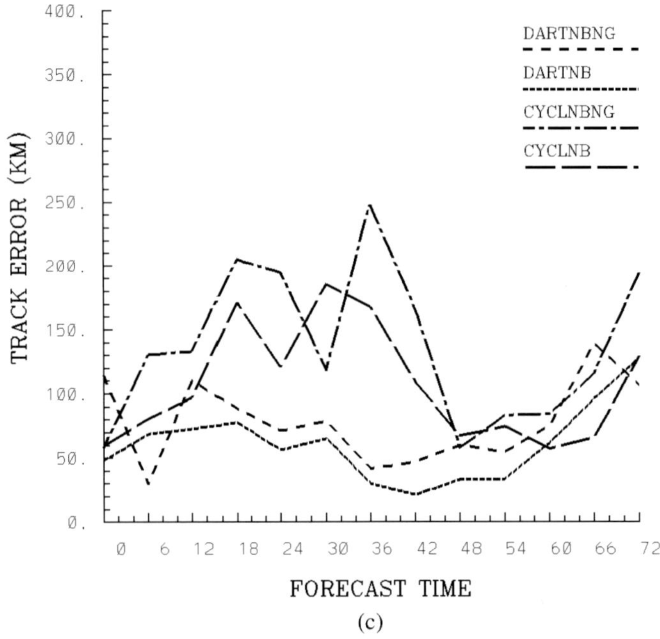

(c)

Figure 6. (*Continued*)

Table 2. Track forecast errors averaged over different forecast periods for cold-start and cycling WRF 3D-Var experiments, and cycling WRF/DART experiments.

Experiment	3–24 h forecast	27–48 h forecast	51–72 h forecast	3–72 h forecast
NODA	276	294	248	273
COLDNBNG	197	237	264	233
COLDNB	199	233	265	232
COLDALL	111	150	160	140
CYCLNBNG	166	147	119	144
CYCLNB	117	132	82	111
CYCLALL	61	124	211	132
DARTNBNG	75	57	94	75
DARTNB	69	38	80	62

The vorticity plots verify our assessment. With the assimilation of GPSRO soundings, DARTNB produced a storm with a maximum vorticity of $24 \times 10^{-5}\,\mathrm{s}^{-1}$. The corresponding WRF 3D-Var experiment has a value of $9 \times 10^{-5}\,\mathrm{s}^{-1}$, less than 50% of that of the WRF/DART experiment. Moreover, the assimilation of COSMIC GPSRO soundings produces an increase of $3 \times 10^{-5}\,\mathrm{s}^{-1}$ in WRF/DART, which extends from 900 mb to 600 mb (Fig. 10). It is clear that the assimilation of COSMIC GPSRO soundings produces a stronger and more robust typhoon vortex. On the other hand, the impact of COSMIC GPSRO assimilation with WRF 3D-Var is much weaker, and visible only in the lowest 1 km.

Similar results are found for the moisture, temperature, and height fields. The WRF/DART system produces a stronger typhoon and a well-defined warm core. Also, the assimilation of GPSRO soundings from COSMIC increases the temperature at the

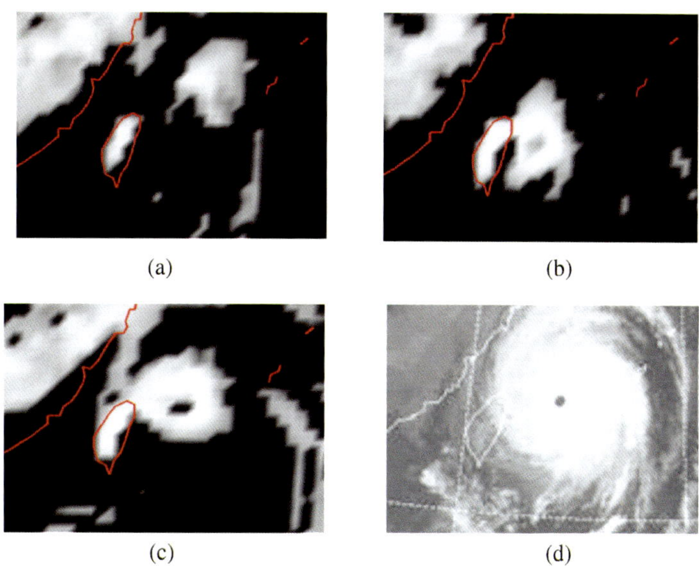

Figure 7.   Integrated cloud water forecasts from (a) CYCLNB, (b) CYCLALL, and (c) DARTNB experiments valid at 0000 UTC 16 September 2006. The verifying observed IR image at 0000 UTC 16 September 2006 is shown in (d).

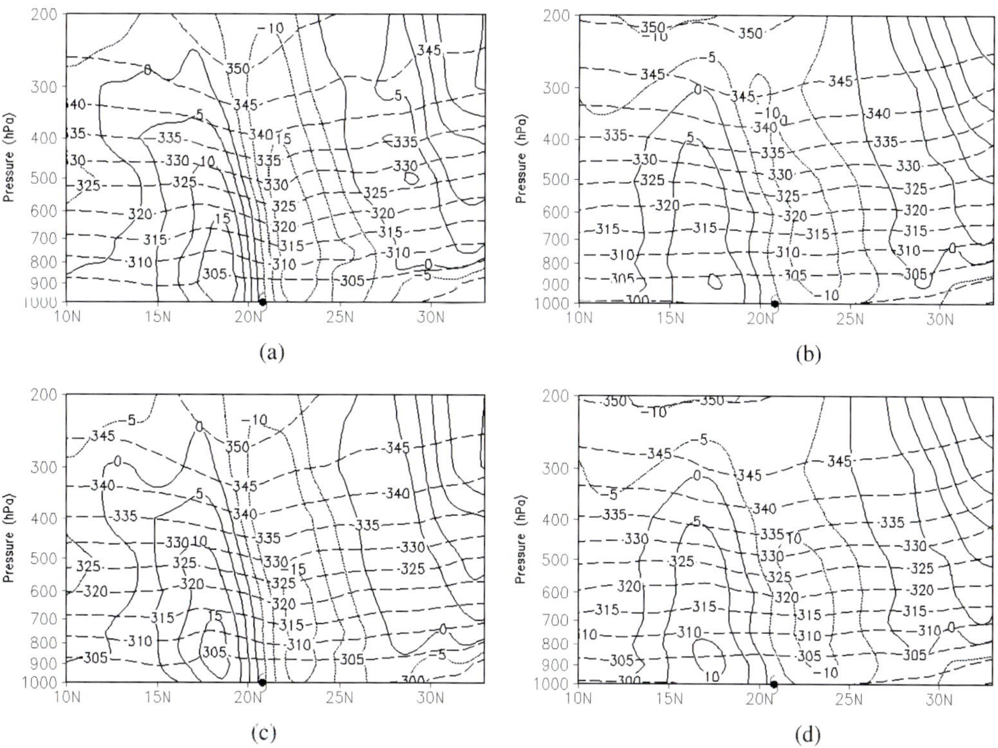

Figure 8.   Zonal wind (solid fine) and potential temperature (dashed line) analysis along 125.8° E from (a) DARTNB, (b) CYCLNB (3D-Var), (c) DARTNBNG, and (d) CYCLNBNG (3D-Var) at 0000 UTC 14 September 2006. Units: zonal wind (m/s), temperature (K). The location of the typhoon is marked with the typhoon symbol.

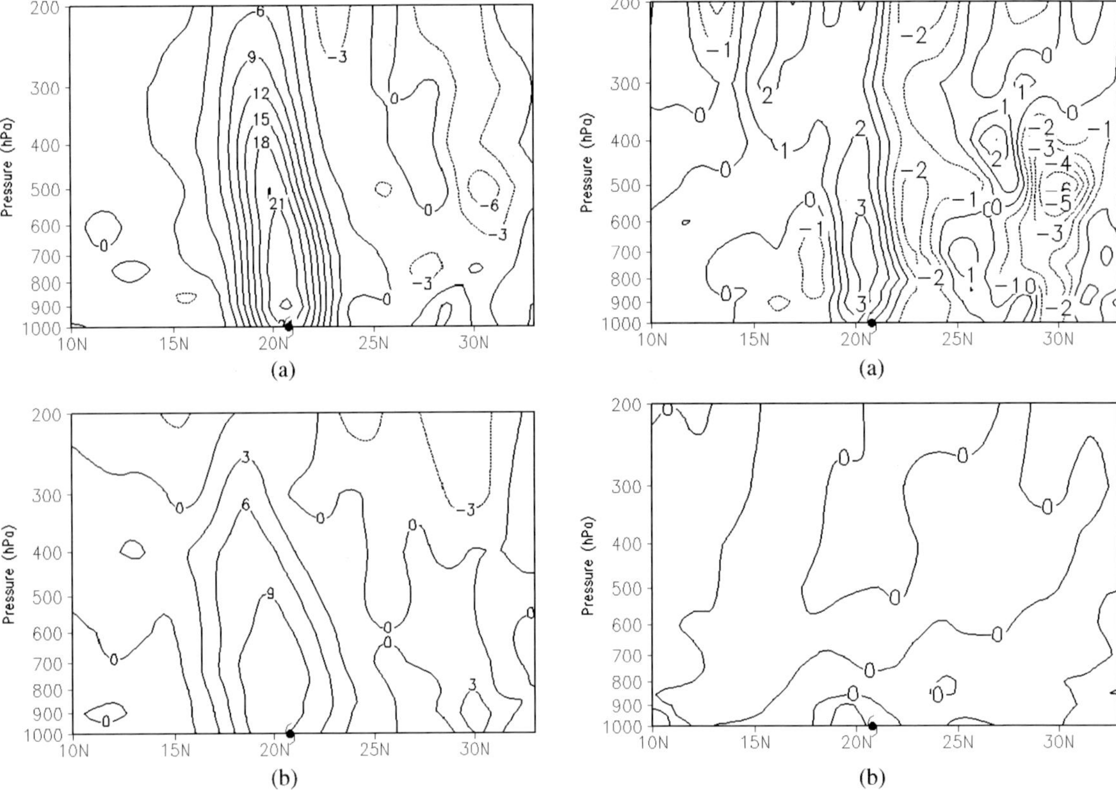

Figure 9. Relative vorticity analysis along 125.8°E from (a) DARTNB and (b) CYCLNB (3D-Var) at 0000 UTC 14 September 2006. Units: $1.0 \times 10^{-5}\,\mathrm{s}^{-1}$. The location of the typhoon is marked with the typhoon symbol.

Figure 10. Relative vorticity difference analysis from (a) DARTNB–DARTNBNG and (b) CYCLNB-CYCLNBNG (3D-Var) at 0000 UTC 14 September 2006. Units: $1,0 \times 10^{-5}\,\mathrm{s}^{-1}$. The location of the typhoon is marked with the typhoon symbol.

center of the typhoon by about 0.5°C with WRF/DART, while the impact is not apparent in WRF 3D-Var (not shown). The WRF/DART assimilation of GPSRO soundings produces profound changes of water vapor over the western Pacific, with amounts varying from 1.5 to $2\,\mathrm{g\,kg}^{-1}$ [Fig. 11(a)]. The corresponding changes in WRF 3D-Var are much more modest, with amounts on the order of $0.5\,\mathrm{g\,kg}^{-1}$ or less [Fig. 11(b)].

The track of typhoons over the western Pacific is strongly affected by the subtropical high. Naturally, we would be interested in seeing how the analysis of the Western Pacific Subtropical High is influenced by the data-assimilation systems. Figure 12 shows the 500 mb

geopotential height for both the DARTNB experiment and the CYCLNB (WRF 3D-Var) experiment. We see a much stronger high-pressure system in the vicinity of the typhoon. For example, the height field to the northeast of Typhoon Shanshan has a geopotential height contour of 5920 m in the WRF/DART experiment [Fig. 12(a)], while that in the WRF 3D-Var experiment is only 5900 m [Fig. 12(b)]. Moreover, the assimilation of GPSRO soundings produces a 500 mb potential height difference of 30 m with WRF/DART [Fig. 12(c)], while the impact of COSMIC GPSRO data assimilation is barely visible with WRF 3D-Var [Fig. 12(d)].

One may ask: Why should WRF/DART perform better than WRF 3D-Var? This is a

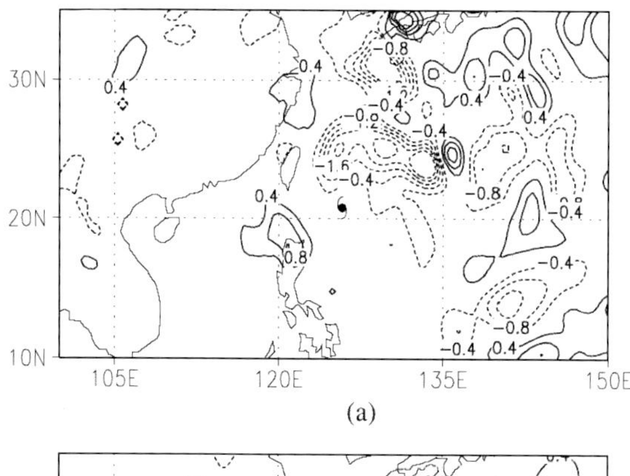

(a)

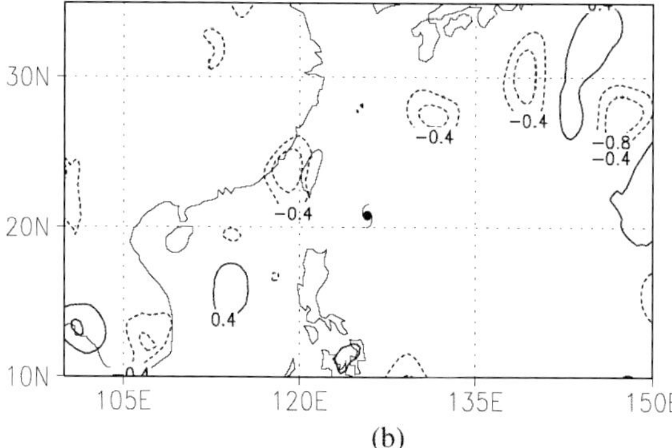

(b)

Figure 11. Water vapor difference fields from (a) DARTNB–DARTNBNG and (b) CYCLNB–CYCLNBNG (3D-Var) at 0000 UTC 14 September 2006. Units: $1.0 \times 10^{-3}\,\mathrm{g\,kg^{-1}}$. The location of the typhoon is marked with the typhoon symbol.

very important question and will require considerably more analysis before we can fully answer it. However, we make the following observations. First of all, the ensemble data-assimilation system (i.e. WRF/DART) uses flow-dependent background error covariances, while the background-error covariances used in WRF 3D-Var are not flow dependent. Second, the WRF/DART system takes into account the forecast multivariate error correlations between specific humidity and temperature, as well as surface pressures, while they are not taken into consideration in WRF 3D-Var. Of course, one advantage of WRF 3D-Var is the significantly reduced computational cost. For one-day assimilation with 1 h cycling, the WRF/DART system with 32 ensemble members takes 5.5 h of wall-clock time on a machine with 32 IBM Power

5 processors. The corresponding cost for WRF 3D-Var is about 0.4 h. So, the WRF/DART system is more than one order of magnitude more expensive than WRF 3D-Var. Therefore, the improved analysis comes with an increased computational cost.

3. Impact of COSMIC GPSRO Soundings on Mei-yu Prediction

The Western Pacific Subtropical High (WPSH) has a profound influence on weather systems over East Asia, in particular the East Asia monsoons and the tropical cyclones. In late spring and early summer, Taiwan and southern China are significantly influenced by a quasi-stationary Mei-yu front. Mesoscale convective systems embedded within the Mei-yu front travel eastward along the front, and can produce heavy precipitation. The location and intensity of the Mei-yu front and the formation and development of mesoscale convective systems are strongly affected by the intensity and position of the WPSH, as well as the southwesterly monsoon flows that originate from the Indian Ocean and the South China Sea. Because of the lack of observations over the Pacific Ocean, the South China Sea, and the Indian Ocean, weather analysis, particularly moisture analysis, is often subject to significant uncertainty. Because of this lack of observations over the Pacific Ocean, the intensity and location of the WPSH are often not accurately analyzed by global models. With the availability of GPSRO soundings, COSMIC provides an opportunity to improve the analysis of the WPSH, the southwesterly monsoons, and the Mei-yu front.

In this study, we assimilate GPSRO soundings from COSMIC over a two-week period, from 1 to 14 June 2007, using the WRF/DART ensemble filter data-assimilation system. For this study, we use a WRF/DART system at a horizontal resolution of 36 km, with 35 vertical levels. The number of ensemble members is 32. Figure 13 shows the distribution of COSMIC

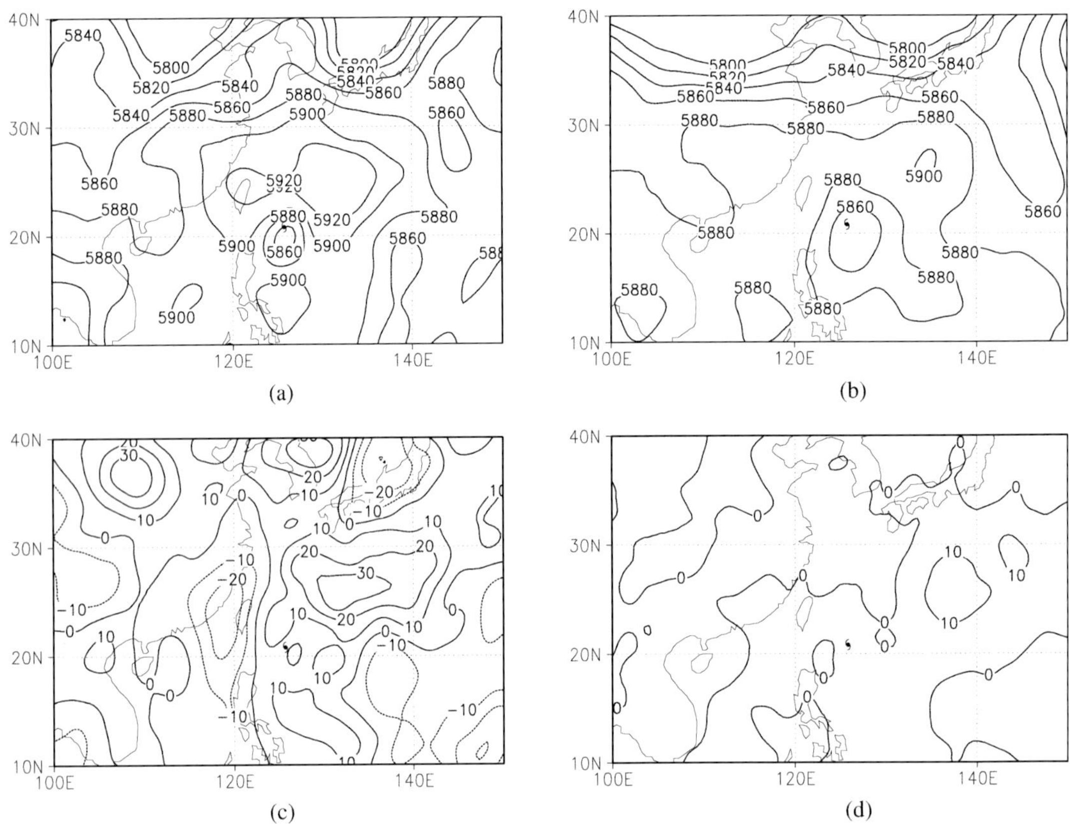

Figure 12. 500 hPa height analysis from (a) DARTNB and (b) CYCLNB at 0000 UTC 14 September 2006; difference fields of (c) DARTNB–DATNBNG and (d) CYCLNB–CYCLNBNG at the same time. Unit: m. The location of the typhoon is marked with the typhoon symbol.

GPS RO soundings during this period and the experimental domain. There are a total of 1,567 GPSRO soundings uniformly distributed over the model domain. In addition to GPSRO soundings, we assimilate upper-air soundings, surface reports, satellite winds, and the cloud-free AIRS-retrieved temperature data. Two experiments are performed. The first is a NoGPS experiment, which assimilates the conventional operational data from NCEP, which include radiosondes, satellite cloud motion winds, and QuikScat surface winds. The AIRS-standard-retrieved temperature profiles (at 50 km resolution) from NASA/JPL are also used. The other experiment is the GPS experiment, which assimilates COSMIC GPSRO soundings in addition to all the aforementioned data. The

assimilation experiments are done with 3 h cycling for the entire two-week period.

Figure 14 shows the 850 mb wind fields averaged over the two-week period of 1–14 June 2007 for the NoGPS and GPS experiments. At first glance, they look almost identical, aside from some subtle differences in the flow fields over the western Pacific to the east of Taiwan. The 850 mb wind fields show that the WPSH is extended to about 110°E over the South China Sea. During this period, Taiwan and southern China are under the influence of two confluent flows: one is the southwesterly monsoon flow originating from the Bay of Bengal, and the other is the southerly returning flow associated with the WPSH. The difference fields between the NoGPS and GPS experiments show

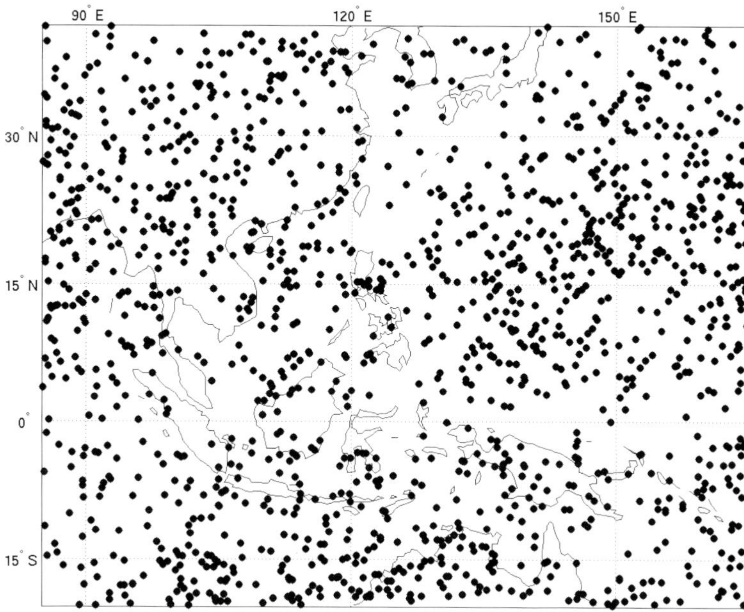

Figure 13.   Distribution of 1,567 COSMIC GPSRO soundings from 1 to 14 June 2007 over the experiment domain.

an anticyclonic gyre located at about 145°E and 25°N. This gyre is of a scale of about 2,500 km, and has a northeast and southwest orientation. The difference fields suggest that the assimilation of COSMIC GPSRO soundings has enhanced the WPSH over the western Pacific. Note that there is little difference over the northern and eastern lateral boundaries of the model domain. This is because identical lateral boundary conditions, obtained from the NCEP AVN global analysis, are used for the GPS and NoGPS experiments. It is possible that if we use a much larger model domain that covers the entire Pacific Ocean, the COSMIC GPSRO soundings will have an even bigger impact on the entire Pacific subtropical high.

The intensity of the WPSH will have a profound influence on the moisture fluxes, which could have a significant impact on clouds and precipitation. Figure 15 shows the mean moisture fluxes at 850 mb averaged over the two-week period of 1–14 June 2007. The 850 mb moisture flux indicates that a significant amount of moisture originates from the

Bay of Bengal, climbs over the Indochina peninsula, and converges with the returning moisture flow associated with the WPSH. Taiwan and southern China are under the strong influence of the southwesterly moist flow, after these two air streams converge. The impact of COSMIC GPSRO simulation can be visualized by examining the differences in moisture flux between the NoGPS and GPS experiments. Again, an anticyclonic gyre is clearly visible. Over the western part of this gyre, moisture is being transported toward Taiwan. This suggests that the assimilation of COSMIC GPSRO soundings produces an improved analysis, with more moisture being transported to the Taiwan area.

During the period of 6–9 June, Taiwan was under the influence of a Mei-yu front. Mesoscale convective systems propagated from west to east, and produced a significant amount of precipitation. Heavy precipitation took place at first on the west coast of Taiwan on 6 June 2007 [Fig. 16(a)]. It then migrated over northwestern Taiwan, with 24 h accumulated rainfall exceeding 150 mm on 7 June

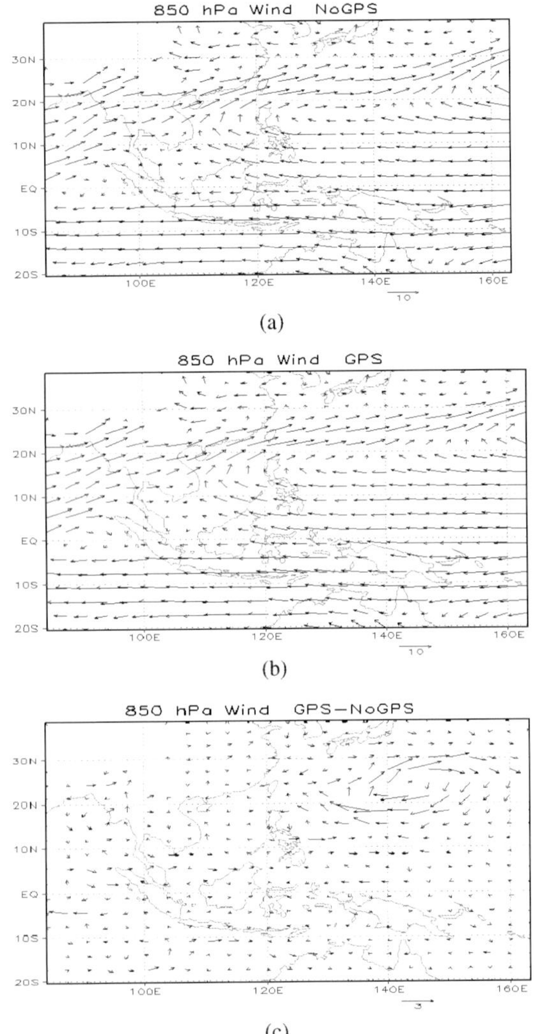

Figure 14. 850 mb wind fields for (a) the NoGPS experiment and (b) the GPS experiment, and (c) their differences.

[Fig. 16(b)]. This continued into the day of 8 June [Fig. 16(c)], with significant precipitation over the Taichung area, as well as northern Taiwan and southern Taiwan immediately to the west of the Central Mountain Range. The maximum 24 h accumulated rainfall ending at 0000 UTC 9 June exceeded 300 mm over the Taichung area [Fig. 16(c)]. By 0000 UTC 10 June, most of the precipitation, with weaker amounts, had fallen over the Central Mountain Range

[Fig. 16(d)]. An interesting question is: Would the assimilation of COSMIC GPSRO soundings help improve rainfall forecasts?

To answer this question, we show in Fig. 17 the 850 mb moisture flux at 0000 UTC 8 June 2007 for the GPS run, and the differences between the GPS and NoGPS experiments. The basic flow pattern and moisture flux pattern are essentially the same as those of the two-week averages. The difference field of the 850 mb moisture flux shows an interesting structure. The anticyclonic gyre is already established at this time, although the center of the gyre is located further to the west from its two-week mean position. More interestingly, significant eastward moisture fluxes are found to the west of Taiwan and southern China. There is also enhanced moisture flux convergence over southern China and the Taiwan Strait. This should contribute to increased precipitation over this region.

Figure 18 shows the 850 mb moisture analysis for the NoGPS and GPS experiments, and their differences (GPS – NoGPS) at 0000 UTC 8 June. The moisture content is much larger and robust in the GPS experiment. For example, the $16\,\mathrm{g\,kg^{-1}}$ contour is found on the east coast of China near Fujiang province in the GPS experiment, while a weaker amount is found in the NoGPS experiment at the same location. The difference in moisture over the southeastern China coast exceeds $1.0\,\mathrm{g\,kg^{-1}}$ on the east coast of China, which is directly related to the precipitation event [Fig. 18(c)]. Figures 17 and 18 suggest that the assimilation of COSMIC GPSRO soundings over a one-week period (1–8 June 2007) has produced noticeable changes in moisture distribution and moisture fluxes associated with the Mei-yu system. One should expect that such changes would have an influence on precipitant forecasts.

Indeed, this is the case. Figure 19 shows the 24 h precipitation forecast from the WRF model at 12 km, which is initialized with the WRF/DART 36 km analysis at 0000 UTC 8 June from the NoGPS and GPS experiments.

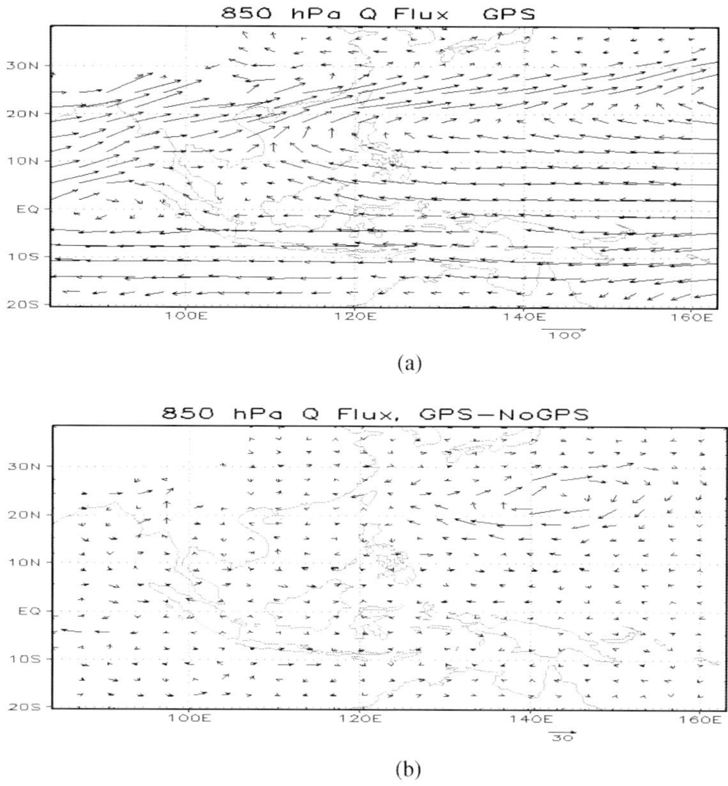

Figure 15. (a) 850 mb moisture flux in the GPS experiment, and (b) differences in the 850 mb moisture flux between the GPS and NoGPS experiments averaged over the two-week period from 1 to 14 June 2007.

The location and intensity of precipitation over southern China and the Taiwan Strait are very different. In particular, the WRF model initialized with the WRF/DART analysis that assimilates GPSRO soundings produces more intense precipitation over the Taiwan Strait. The difference field shows that more than 30 mm additional precipitation falls over western and southern Taiwan. In comparison with the available precipitation analysis over Taiwan (Fig. 16), we find that the more intense precipitation as a result of COSMIC GPSRO data assimilation compares more favorably with the observed rainfall. The 24 h accumulated rainfall over mainland China ending at 0000 UTC 9 June (Fig. 20) also compares more favorably with the GPS experiment, which gives precipitation further to the south than the NoGPS experiment.

4. Summary and Conclusions

The successful launch of the FORMOSAT-3/COSMIC mission marked the beginning of a new era in GPS atmospheric remote sensing. By providing more than 2,000 GPS radio occultation soundings per day uniformly distributed around the globe in near real time, COSMIC provides much needed data over data-sparse regions of the world, including the tropical oceans and the polar regions. For weather forecasting over Taiwan and East Asia, COSMIC provides valuable data over the western Pacific and the South China Sea. The assimilation of COSMIC GPSRO soundings can contribute to improved forecasting of typhoons and heavy precipitation associated with the Mei-yu front. In this article, we have examined the impact of COSMIC GPSRO soundings on the prediction

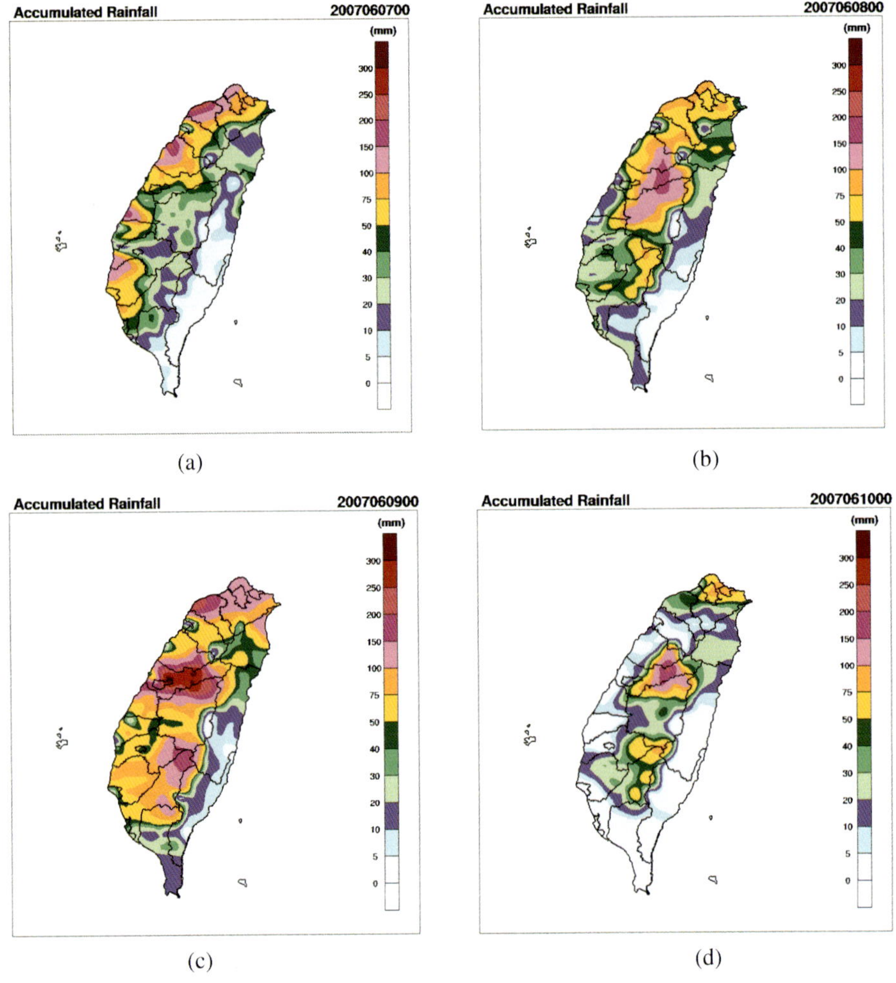

Figure 16. 24 h accumulated precipitation over Taiwan ending at 0000 UTC on 7, 8, 9, and 10 June 2007.

of Typhoon Shanshan (2006) and the heavy precipitation event associated with a Mei-yu front in early June 2007. Our study has led to the following conclusions:

(1) It is essential to perform continuous assimilation through cycling in order for COSMIC GPSRO soundings to have a significant impact on typhoon track prediction. One needs to realize that even with 2,000 GPSRO soundings per day, the data density is still relatively low. Over the CWB 45 km domain, there are approximately 100 GPSRO soundings over a 24 h period, or 25 GPSRO soundings over a 6 h period. Continuous assimilation with a relatively narrow assimilation window (1 h) allows more COSMIC soundings to be assimilated at the time close to observations. Cold-start-type data assimilation is usually not effective, since only a limited number of soundings are used.

(2) The assimilation of COSMIC GPSRO soundings using the WRF/DART ensemble filter method is found to be more effective than the WRF 3D-Var assimilation method for the Typhoon Shanshan case. The WRF/DART system produces a much

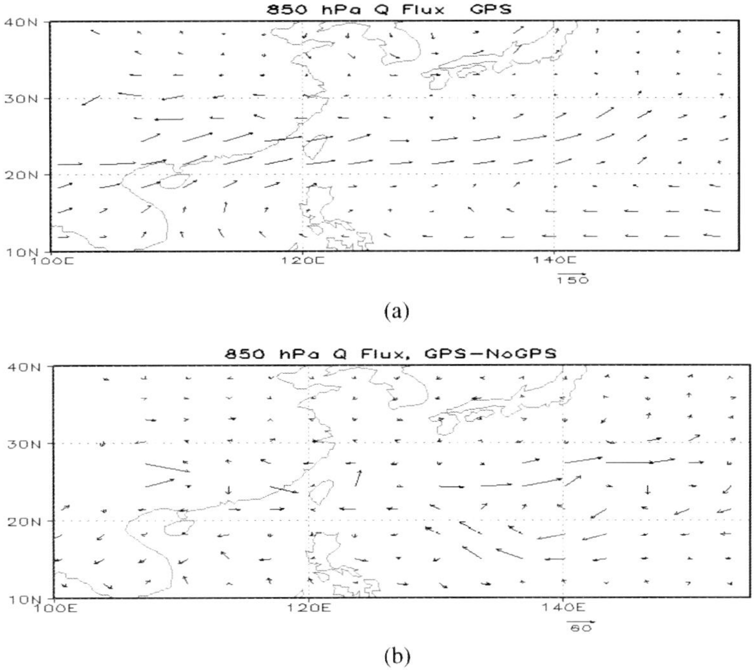

Figure 17. (a) 850 mb moisture flux in the GPS experiment, and (b) differences in the 850 mb moisture flux between the GPS and NoGPS experiments, ending at 0000 UTC 8 June 2007.

stronger typhoon than the WRF 3D-Var system with or without the assimilation of COSMIC GPSRO data. Moreover, the assimilation of COSMIC GPSRO data with the WRF/DART system produces much more profound changes than the WRF 3D-Var system. In other words, the WRF/DART ensemble system can extract more information from the same GPSRO data than the WRF 3D-Var system, and subsequently has a larger analysis increment. The superior performance of the WRF/DART ensemble system is attributed to the fact that WRF/DART uses flow-dependent background error covariances, and takes into consideration the forecast multivariate error correlations among moisture, temperature, and surface pressure. On the other hand, WRF 3D-Var uses background error covariances derived from historical forecasts, which do not contain information directly related to the case at hand.

Of course, we should also recognize that the improved performance of WRF/DART ensemble filter assimilation is obtained at an increased computational cost. For the Typhoon Shanshan case, the WRF/DART system requires more than an order of magnitude more computing resources than the WRF 3D-Var system.

(3) The assimilation of typhoon bogus soundings is found to be quite important for improved typhoon track forecasting, particularly for cold-start experiments. Because the NCEP AVN global analysis often underestimates the intensity of the storm, forecasts without the assimilation of typhoon bogus data usually give large track errors. However, for cycling experiments, we found that the assimilation of bogus soundings does not always give better results, particularly for longer-range forecasting. We conclude that improved typhoon bogusing procedures should be developed. Better yet,

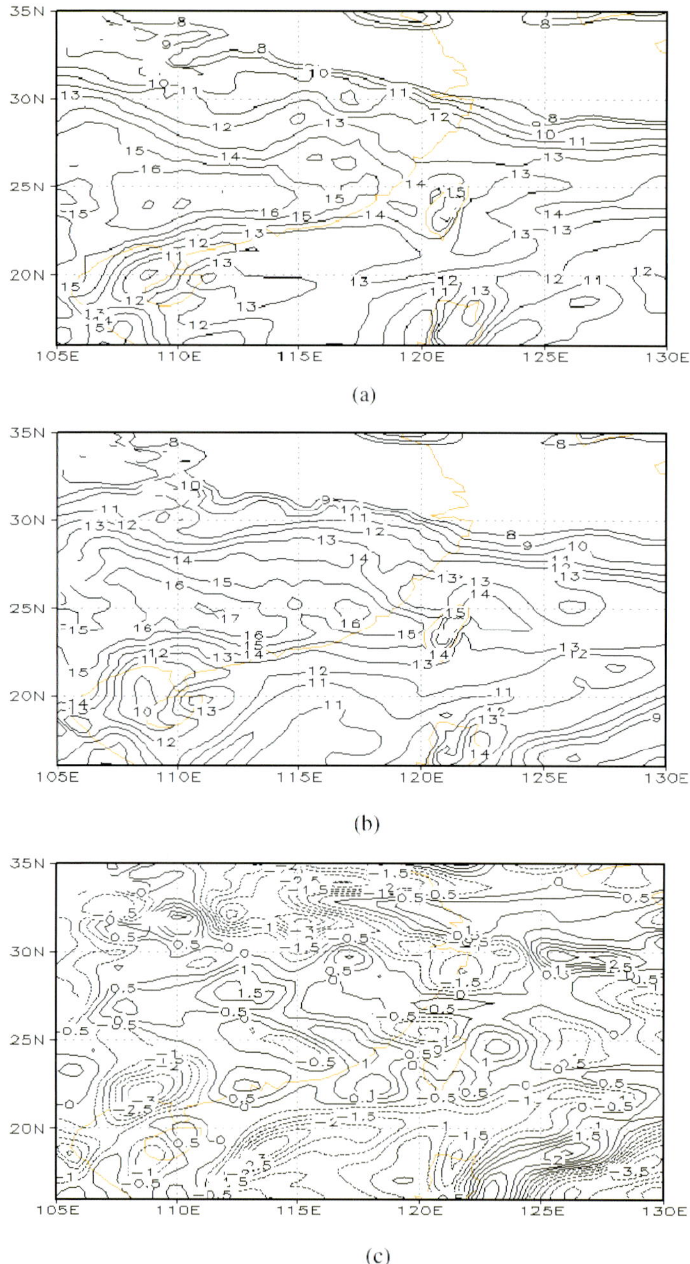

Figure 18. 850 mb specific humidity for (a) the NoGPS run, (b) the GPS run, and (c) the difference between the GPS and the NoGPS run (in $g\,kg^{-1}$) at 0000 UTC 8 June 2007.

we should achieve improved forecasting by assimilating more real observations and avoiding the use of typhoon bogus soundings, which is the current practice at the ECMWF.

(4) The assimilation of COSMIC GPSRO soundings with the WRF/DART system over a two-week period has made a profound impact on the analysis of the Western Pacific Subtropical High. In particular, the

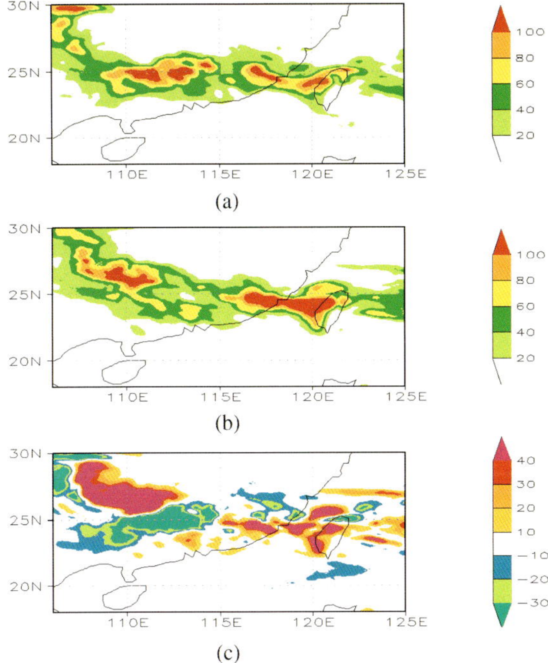

Figure 19. 24 h accumulated rainfall ending at 0000 UTC 9 June for (a) the NoGPS experiment, (b) the GPS experiment, and (c) their differences.

COSMIC GPSRO assimilation strengthens the WPSH, increases the moisture content of the Mei-yu front, and increases moisture flux convergence along the Mei-yu front. These changes made a significant positive impact on short-range precipitation forecasts over both Taiwan and southern China.

Although these results are very encouraging, they need to be verified with additional case studies. Moreover, the assimilation of GPSRO soundings should be tested over an extended period before it can be used operationally. These efforts are currently being carried out at Taiwan's Central Weather Bureau. Given the importance of the COSMIC GPSRO soundings for the typhoon and Mei-yu prediction, it is important that we continue the operation of the COSMIC mission through its life. We should also begin planning for a follow-on mission to replace COSMIC, which is expected to last through 2011 with a five-year mission life.

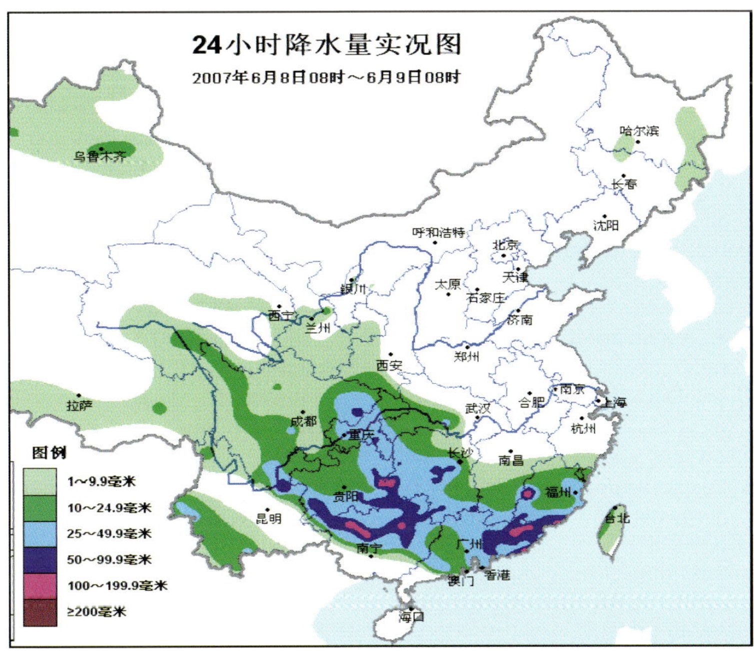

Figure 20. Observed rainfall over China ending at 0000 UTC 9 June 2007.

Acknowledgements

We thank Jeff Anderson for fruitful discussions on the results of WRF/DART assimilations. The research results presented in this article are supported by the Central Weather Bureau, the National Space Organization, the National Science Foundation under awards #ATM-0410018 and INT-0129369, and the National Aeronautics and Space Administration under grant NNX08AI23G.

[Received 10 February 2008; Revised 12 March 2008; Accepted 14 March 2008.]

References

Anderson, J. L., 2003: A local least squares framework for ensemble filtering. *Mon. Wea. Rev.*, **131**, 634–642.

Anthes, R. A., and coauthors, 2008: The COSMIC/FORMOSAT-3 mission: early results. *Bull. Amer. Meteor. Soc.*, **89**, 313–333.

Barker, D. M., W. Huang, Y.-R. Guo, A. J. Bourgeois, and Q. N. Xiao, 2004: A three-dimensional variational data assimilation system for MM5: implementation and initial results. *Mon. Wea. Rev.*, **132**, 897–914.

Cheng, C.-Z., Y.-H. Kuo, R.A. Anthes, and L. Wu, 2006: Satellite constellation monitors global and space weather. *EOS*, **87**(17), 166–167.

Cucurull, L., and J. C. Derber, 2008: Operational implementation of COSMIC observations into the NCEP's global data assimilation system. *Wea. Forecasting* (in press).

Cucurull, L., J. C. Derber, R. Treadon, and R. J. Purser, 2008: Assimilation of Global Positioning System radio occultation observations into NCEP's Global Data Assimilation System. *Mon Wea. Rev.* (in press).

Hajj, G., and coauthors, 2004: CHAMP and SAC-C atmospheric occultation results and intercomparisons. *J. Geophys. Res.*, **109**, D06109, doi:10.1029/2003JD003909.

Healy, S., 2008: Forecast impact experiment with a constellation of GPS radio occultation receivers. *Geophys. Res. Lett.* (in press).

Kuo, Y.-H., B. Chao, and L. Lee, 1999: A constellation of microsatellites promises tohelp in a range of geoscience research. *EOS Trans.* American Geophysical Union, **80**(40), 467–471.

Kuo, Y.-H., and G. T.-J. Chen, 1990: Taiwan Area Mesoscale Experiment: an overview. *Bull. Amer. Meteor. Soc.*, **71**, 488–505.

Liu, H., J. L. Anderson, Y.-H. Kuo, and K. Raeder, 2007: Importance of forecast error multivariate correlations in idealized assimilation of GPS radio occultation data with the ensemble adjustment filter. *Mon. Wea. Rev.*, **135**, 173–185, 10.1175/MWR3270.1

Liu, H., J. L. Anderson, Y.-H. Kuo, C. Snyder, and A. Caya, 2008: Evaluation of a non-local observation operator in assimilation of CHAMP radio occultation refractivity with WRF. *Mon. Wea. Rev.*, **136**, 242–256.

Neiman, P. J., F. M. Ralph, G. A. Wick, Y.-H. Kuo, W.-K. Wee, Z. Ma, G. H. Taylor, and M. D. Dettinger, 2008: Diagnosis of an intense atmospheric river impacting the Pacific Northwest: Storm summary and offshore vertical structure observed with cosmic satellite retrievals, *Mon. Wea. Rev.* (in press).

Poli, P., P. Moll, D. Puech, F. Rabier, and S. B. Healy, 2008: Quality control, error analysis, and impact assessment of FORMOSAT-3/COSMIC in numerical weather prediction. *Terr. Atmos. Oceanic Sci.* (in press).

Rocken, C., Y.-H. Kuo, W. Schreiner, D. Hunt, S. Sokolovskiy, and C. McCormick, 2000: COSMIC system description. *Terr. Atmos. Ocean.*, **11**, 21–52.

Sokolovskiy, S., 2001: Tracking tropospheric radio occultation signals from low Earth orbit. *Radio Sci.*, **36**(3), 483–498.

Sokolovskiy, S., Y.-H. Kuo, and W. Wang, 2005: Assessing the accuracy of a linearized observation operator for assimilation of radio occultation data: case simulations with high-resolution weather model. *Mon. Wea. Rev.*, **133**, 2200–2212.

Sokolovskiy S., C. Rocken, D. Hunt, W. Schreiner, J. Johnson, D. Masters, and S. Esterhuizen, 2006a: GPS profiling of the lower troposphere from space: inversion and demodulation of the open-loop radio occultation signals. *Geophys. Res. Lett.*, **33**, L14816, doi: 10.1029/2006GL026112.

Sokolovskiy S., Y.-H. Kuo, C. Rocken, W. S. Schreiner, D. Hunt, and R. A. Anthes, 2006b: Monitoring the atmospheric boundary layer by GPS radio occultation signals recorded in the open-loop mode. *Geophys. Res. Lett.*, **33**, L12813, doi: 10.1029/2006GL025955.

Ware, R., and coauthors, 1996: GPS sounding of the atmosphere from low Earth orbit: preliminary results. *Bull. Amer. Meteor. Soc.*, **77**, 19–40.

Wickert, J., and coauthors, 2001: Atmosphere sounding by GPS radio occultation: first results from CHAMP. *Geophys. Res. Lett.*, **28**(17), 3263–3266.

Wu, C.-C., and Y.-H. Kuo, 1999: Typhoons affecting Taiwan: current understanding and future challenges. *Bull. Ameri. Meteor. Soc.*, **80**, 67–80.

Wu, C.-C., H.-C. Kuo, H.-H. Hsu, and B. J.-D. Jou, 2000: Weather and climate research in Taiwan: potential applications of GPS/MET data. *Terr. Atmos. Oceanic Sci.*, **11**, 211–234.

Yen, N., and coauthors, 2008: FORMOSAT-3/ COSMIC spacecraft constellation system, mission results and prospect for follow-on mission. *Terr. Atmos. Oceanic Sci.* (in press).

Zeng, Z., W. Randel, S. Sokolovskiy, C. Deser, Y.-H. Kuo, M. Hagan, J. Du, and W. Ward, 2008: Detection of migrating diurnal tide in the tropical UTLS using the CHAMP radio occultation data. *JGR-Atmosphere J. Geophys. Res.* **113**, D03102, doi: 10.1029/2007JD008725.

Author Index

Wang, Pao K., 419
Wong, Teo Suan, 172
Wu, Ching-Chi, 200
Wu, Chun-Chieh, 49, 136, 358

Yang, Jr-Shiuan, Kate, 200
Yang, Ming-Jen, 66

Yang, P., 307
Yeh, Kaosan, 200
Yildirim, Ahmet, 200
Yu, Jin-Yi, 3
Yu, Tsann-Wang, 182
Yu, Yi-Chiang, 200